C.W.F. Everitt • M.C.E. Huber • R. Kallenbach •
G. Schäfer • B.F. Schutz • R.A. Treumann
Editors

# Probing The Nature of Gravity

## Confronting Theory and Experiments in Space

Previously published in *Space Science Reviews* Volume 148, Issues 1–4, 2009
Fischbach et al., Time-Dependent Nuclear Decay Parameters: New Evidence for New Forces? was previously published in *Space Science Reviews* Volume 145, Issues 3–4, 2009

C.W.F. Everitt
Stanford University
Stanford, CA, USA

M.C.E. Huber
Paul-Scherrer Institute (PSI)
Villigen, Switzerland

R. Kallenbach
Max-Planck Institute for Solar System Research
Katlenburg-Lindau, Germany

G. Schäfer
Friedrich-Schiller University Jena
Jena, Germany

B.F. Schutz
Albert Einstein Institute
Potsdam, Germany

R.A. Treumann
Munich University
Munich, Germany

ISBN 978-1-4419-1361-6
Springer New York Dordrecht Heidelberg London

Library of Congress Control Number: 2010922455

©Springer Science+Business Media, LLC 2010
All rights reserved. This work may not be translated or copied in whole or in part without the written permission of the publisher (Springer Science+Business Media, LLC, 233 Spring Street, New York, NY 10013, USA), except for brief excerpts in connection with reviews or scholarly analysis. Use in connection with any form of information storage and retrieval, electronic adaptation, computer software, or by similar or dissimilar methodology now known or hereafter developed is forbidden.
The use in this publication of trade names, trademarks, service marks, and similar terms, even if they are not identified as such, is not to be taken as an expression of opinion as to whether or not they are subject to proprietary rights.

*Cover illustration*: The spacecraft of ‘Gravity Probe B’ (launched in 2004 to measure gravitomagnetism, upper left) and of ‘Microscope’ (a mission to be launched within the next three years to test the equivalence principle to an accuracy of $10^{-15}$, lower right). The Earth in the background is partly covered by a representation of quantum spacetime at the level of loop quantum gravity. Cover image provided courtesy of Sascha Rieger/Milde Marketing/Albert Einstein Institute.

©Albert Einstein Institute (AEI)

Printed on acid-free paper

Springer is part of Springer Science+Business Media (www.springer.com)

# Contents

Space Sci Rev (2009) 148: 1–2
DOI 10.1007/s11214-009-9626-2

# Introduction

**Martin C.E. Huber · Rudolf A. Treumann**

Received: 21 December 2009 / Accepted: 21 December 2009 / Published online: 12 January 2010
© Springer Science+Business Media B.V. 2010

In the week of October 6–10, 2008, the International Space Science Institute (ISSI) welcomed a new community: About forty physicists and astronomers came to Bern to debate about "Probing the Nature of Gravity—Confronting Theory and Experiment in Space". Following the tradition of ISSI Workshops, data obtained in space were the focus of the event. For the first time, physics experiments carried out in space provide data adding to the knowledge of fundamental physics that we have up till now got mainly through observations and from laboratory experiments. Presentation and discussion of all these data was embedded into the theoretical framework of gravity and its recent advances. As could have been expected, an important part of the discussions on this led to a look into the future of gravitational theory and experiment.

Two introductory talks on the status of experimental tests of General Relativity and on the transition to Quantum Gravity set the stage. Gravitomagnetism was then addressed, first by a theoretical exposé on this particular consequence of General Relativity and subsequently by measurement and observation—both as it arises from the rotation of the Earth and as it manifests itself in massive astrophysical objects. Experimentally, Gravity Probe B (GP-B), with its direct measurement of the Lense-Thirring effect, set a new standard of experimental elegance and refinement, but the gravitomagnetic results from analysis of observations of the LAGEOS system of satellites remind us that tiny effects crucial to fundamental physics may well be visible in high-precision observations of astronomical bodies.

Next, the Workshop turned to tests of basic laws and principles by exploring possible violations of Newton's inverse-square laws and of the equivalence principle. Ground-based laboratory investigations were included here as well. Cosmological observations that could

M.C.E. Huber
Paul-Scherrer-Institute (PSI), 5232 Villigen, Switzerland
e-mail: mceh@bluewin.ch

R.A. Treumann (✉)
Department Geophysics and Environmental Sciences, Munich University, Theresienstrasse 41, 80333 Munich, Germany
e-mail: rudolf.treumann@geophysik.uni-muenchen.de

set constraints on gravitational theory rounded off the choice of subjects. All along, presentations of the theoretical foundations substantiated the analysis of results.

The last day of the Workshop was devoted to examining the crucial question that is persistently asked by space officials, who are sceptical about pursuing fundamental physics in space, namely: “To what extent are high accuracy measurements in space possible?” The measurement uncertainties of GP-B were scrutinised thoroughly and the specialists involved reported in detail on the way, in which the data sent back by GP-B were being analysed. The Workshop then continued with assessments of the accuracy of results that can be provided by future fundamental physics missions, and came to a close with a summary lecture.

According to the custom of ISSI workshops, the workshop participants were asked to supply a written version of their talks. These papers are reproduced here after having gone through peer review. For two additional talks, by Nicolas Gisin on ‘Entanglement and Decoherence’ and by Alan Watson, who spoke about ‘The Problem of Creating Particles with Energies above $10^{21}$ eV’, it was agreed to give references to papers that have been published elsewhere—D. Salart, A. Baas, C. Branciard, N. Gisin, H. Zbinden, Testing the speed of ‘spooky action at a distance’, Nature **454**, 861–864 (2008); A. A. Watson, Highlights from the Pierre Auger Observatory—the Birth of the Hybrid Era, *Proceed. 30th International Cosmic Ray Conference, Mérida, México, 2007, Vol. 6: Invited and Rapporteur Papers*, ed. by R. Caballero et al. (Universidad Nacional Autónoma de México, 2009) arXiv:0801.2321v1 (astro-ph)—, because these topics are only indirectly related to gravity. On the other hand, we include the written version of the pre-dinner speech on ‘Wolfgang Pauli and Modern Physics’ by Norbert Straumann, where he shows how Pauli—referred to by colleagues as the conscience of the community of theoretical physicists—inspired and anticipated the development of physics not only through formal publications but in many cases through detailed letters to his colleagues.

In the preparatory stage of the Workshop, the Science Committee of ISSI has asked the Convenors to concentrate on gravity. Important discussions about the Workshop programme also took place with the ISSI Directorate: the one-day session of examining uncertainty estimates is one of the results. We are greatly indebted to Rüdeger Reinhard, who was instrumental as Convenor in the early stages of the Workshop, and helped shape the programme. We also wish to thank Brigitte Schutte-Fasler, Andrea Fischer, Saliba F. Saliba, Katja Schüpbach, Silvia Wenger and the other members of the ISSI staff for actively and cheerfully supporting us whenever we needed to use the infrastructure of ISSI.

Space Sci Rev (2009) 148: 3–13
DOI 10.1007/s11214-009-9541-6

# The Confrontation Between General Relativity and Experiment

**C.M. Will**

Received: 13 November 2008 / Accepted: 20 May 2009 / Published online: 4 June 2009
© Springer Science+Business Media B.V. 2009

**Abstract** We review the experimental evidence for Einstein's general relativity. A variety of high precision null experiments confirm the Einstein Equivalence Principle, which underlies the concept that gravitation is synonymous with spacetime geometry, and must be described by a metric theory. Solar system experiments that test the weak-field, post-Newtonian limit of metric theories strongly favor general relativity. Binary pulsars test gravitational-wave damping and aspects of strong-field general relativity. During the coming decades, tests of general relativity in new regimes may be possible. Laser interferometric gravitational-wave observatories on Earth and in space may provide new tests via precise measurements of the properties of gravitational waves. Future efforts using X-ray, infrared, gamma-ray and gravitational-wave astronomy may one day test general relativity in the strong-field regime near black holes and neutron stars.

**Keywords** General relativity · Gravitational experiments

## 1 Introduction

During the late 1960s, it was frequently said that "the field of general relativity is a theorist's paradise and an experimentalist's purgatory". The field was not without experiments, of course: Irwin Shapiro, then at MIT, had just measured the relativistic retardation of radar waves passing the Sun (an effect that now bears his name), Robert Dicke of Princeton was claiming that the Sun was flattened by rotation in an amount whose effects on Mercury's perihelion advance would put general relativity afoul of experiment, and Joseph Weber of the University of Maryland was busy building gravitational wave antennas out of massive

C.M. Will (✉)
Department of Physics, McDonnell Center for the Space Sciences, Washington University, St. Louis, MO, USA
e-mail: cmw@wuphys.wustl.edu

C.M. Will
Gravitation et Cosmologie, Institut d'Astrophysique de Paris, 98 bis Bd. Arago, 75014 Paris, France

aluminum cylinders. Nevertheless the field was dominated by theory and by theorists. The field *circa* 1970 seemed to reflect Einstein's own attitudes: although he was not ignorant of experiment, and indeed had a keen insight into the workings of the physical world, he felt that the bottom line was the *theory*. As he once famously said, if experiment were to contradict the theory, he would have "felt sorry for the dear Lord".

Since that time the field has been completely transformed, and today experiment is a central, and in some ways dominant component of gravitational physics. The breadth of current experiments, ranging from tests of classic general relativistic effects, to searches for short-range violations of the inverse-square law, to a space experiment to measure the relativistic precession of gyroscopes, attest to the ongoing vigor of experimental gravitation.

The great progress in testing general relativity during the latter part of the 20th century featured three main themes:

- The use of advanced technology. This included the high-precision technology associated with atomic clocks, laser and radar ranging, cryogenics, and delicate laboratory sensors, as well as access to space.
- The development of general theoretical frameworks. These frameworks allowed one to think beyond the narrow confines of general relativity itself, to analyse broad classes of theories, to propose new experimental tests and to interpret the tests in an unbiased manner.
- The synergy between theory and experiment. To illustrate this, one needs only to note that the LIGO-Virgo Scientific Collaboration, engaged in one of the most important general relativity investigations—the detection of gravitational radiation—consists of over 700 scientists. This is big science, reminiscent of high-energy physics, not general relativity!

Today, because of its elegance and simplicity, and because of its empirical success, general relativity has become the foundation for our understanding of the gravitational interaction. Yet modern developments in particle theory suggest that it is probably not the entire story, and that modifications of the basic theory may be required at some level. String theory generally predicts a proliferation of additional fields that could result in alterations of general relativity similar to that of the Brans-Dicke theory of the 1960s. In the presence of extra dimensions, the gravity that we feel on our four-dimensional "brane" of a higher dimensional world could be somewhat different from a pure four-dimensional general relativity. Some of these ideas have motivated the possibility that fundamental constants may actually be dynamical variables, and hence may vary in time or in space. However, any theoretical speculation along these lines *must* abide by the best current empirical bounds. Still, most of the current tests involve the weak-field, slow-motion limit of gravitational theory.

Putting general relativity to the test during the 21st century is likely to involve three main themes

- Tests of strong-field gravity. These are tests of the nature of gravity near black holes and neutron stars, far from the weak-field regime of the solar system.
- Tests using gravitational waves. The detection of gravitational waves, hopefully during the next decade, will initiate a new form of astronomy but it will also provide new tests of general relativity in the highly dynamical regime.
- Tests of gravity at extreme scales. The detected acceleration of the universe, the observed large-scale effects of dark matter, and the possibility of extra dimensions with effects on small scales, have revealed how little precision information is known about gravity on the largest and smallest scales.

In this paper we will review selected highlights of testing general relativity during the 20th century and will discuss the potential for new tests in the 21st century. We begin in

Sec. 2 with the "Einstein equivalence principle", which underlies the idea that gravity and curved spacetime are synonymous, and describe its empirical support. Section 3 describes solar system tests of gravity in terms of experimental bounds on a set of "parametrized post-Newtonian" (PPN) parameters. In Sect. 4 we discuss tests of general relativity using binary pulsar systems. Section 5 describes tests of gravitational theory that could be carried out using future observations of gravitational radiation, and Sect. 6 describes the possibility of performing strong-field tests of general relativity. Tests of gravity at cosmological and sub-millimeter scales are significant topics in their own right and are beyond the scope of this paper. Concluding remarks are made in Sect. 7. For further discussion of topics in this paper, and for references to the primary literature, the reader is referred to *Theory and Experiment in Gravitational Physics* (Will 1993) and to the "living" review articles by Will (2006a), Stairs (2008), Psaltis (2008) and Mattingly (2005).

## 2 The Einstein Equivalence Principle and Metric Theories of Gravity

The Einstein equivalence principle (EEP) is a powerful and far-reaching principle, which states that (i) test bodies fall with the same acceleration independently of their internal structure or composition (Weak Equivalence Principle, or WEP); (ii) the outcome of any local non-gravitational experiment is independent of the velocity of the freely-falling reference frame in which it is performed (Local Lorentz Invariance, or LLI); and (iii) the outcome of any local non-gravitational experiment is independent of where and when in the universe it is performed (Local Position Invariance, or LPI).

The Einstein equivalence principle is central to gravitational theory, for it is possible to argue convincingly that if EEP is valid, then gravitation must be described by "metric theories of gravity", which state that (i) spacetime is endowed with a symmetric metric, (ii) the trajectories of freely falling bodies are geodesics of that metric, and (iii) in local freely falling reference frames, the non-gravitational laws of physics are those written in the language of special relativity.

General relativity is a metric theory of gravity, but so are many others, including the Brans-Dicke theory.

To illustrate the high precisions achieved in testing EEP, we shall review tests of the weak equivalence principle, where one compares the acceleration of two laboratory-sized bodies of different composition in an external gravitational field. A measurement or limit on the fractional difference in acceleration between two bodies yields a quantity $\eta \equiv 2|a_1 - a_2|/|a_1 + a_2|$, called the "Eötvös ratio", named in honor of Baron von Eötvös, the Hungarian physicist whose experiments carried out with torsion balances at the end of the 19th century were the first high-precision tests of WEP (see Fig. 1). Later classic experiments by Dicke and Braginsky in the 1960s and 1970s improved the bounds by several orders of magnitude. Additional experiments were carried out during the 1980s as part of a search for a putative "fifth force", that was motivated in part by a reanalysis of Eötvös' original data (the range of bounds achieved during that period is shown schematically in the region labeled "fifth force" in Fig. 1).

The best limit on $\eta$ currently comes from the "Eöt–Wash" experiments carried out at the University of Washington, which used a sophisticated torsion balance tray to compare the accelerations of bodies of different composition toward the Earth, the Sun and the galaxy. Another strong bound comes from Lunar laser ranging (LLR), which checks the equality of free fall of the Earth and Moon toward the Sun. The results from laboratory and LLR experiments are:

$$\eta_{\text{Eöt-Wash}} < 3 \times 10^{-13}, \qquad \eta_{\text{LLR}} < 3 \times 10^{-13}. \tag{1}$$

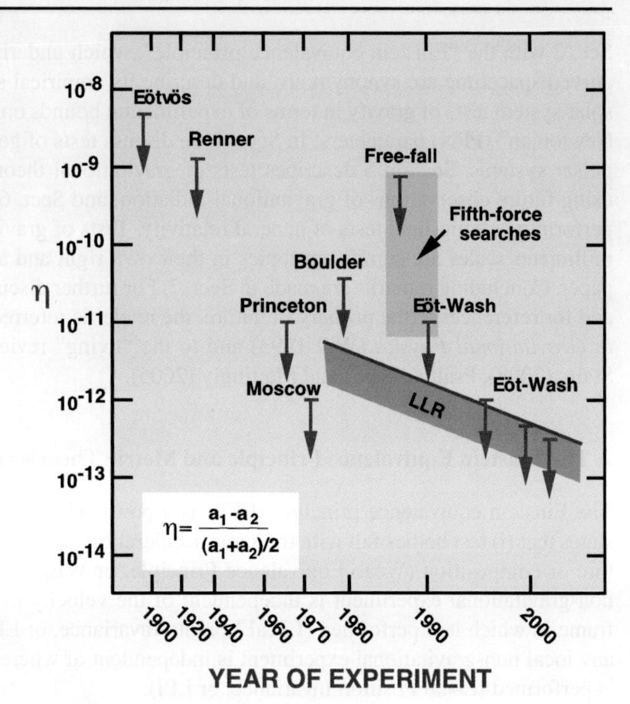


**Fig. 1** Selected tests of the Weak Equivalence Principle, showing bounds on the fractional difference in acceleration of different materials or bodies. *Blue line* and shading shows evolving bounds on WEP for the Earth and the Moon from Lunar laser ranging (LLR)

In fact, by using laboratory materials whose composition mimics that of the Earth and Moon, the Eöt–Wash experiments permit one to infer an unambiguous bound from Lunar laser ranging on the universality of acceleration of gravitational binding energy at the level of $9 \times 10^{-4}$ (test of the Nordtvedt effect—see Sect. 3 and Table 1).

The Apache Point Observatory for Lunar Laser-ranging Operation (APOLLO) project, a joint effort by researchers from the Universities of Washington, Seattle, and California, San Diego, plans to use enhanced laser and telescope technology, together with a good, high-altitude site in New Mexico, to improve the Lunar laser-ranging bound by as much as an order of magnitude.

High-precision WEP experiments, can test superstring inspired models of scalar-tensor gravity, or theories with varying fundamental constants in which weak violations of WEP can occur via non-metric couplings. The project MICROSCOPE, designed to test WEP to a part in $10^{15}$ has been approved by the French space agency CNES for a 2012 launch. A proposed NASA-ESA Satellite Test of the Equivalence Principle (STEP) seeks to reach the level of $\eta < 10^{-18}$.

Very stringent constraints on Local Lorentz Invariance have been placed, notably by experiments that exploited laser-cooled trapped atoms to search for variations in the relative frequencies of different types of atoms as the Earth rotates around our velocity vector relative to the mean rest frame of the universe (as determined by the cosmic background radiation). For a review, see Will (2006b). Local Position Invariance has also been tested by gravitational redshift experiments and by tests of variations with cosmic time of fundamental constants. For a review of such tests, see Uzan (2003).

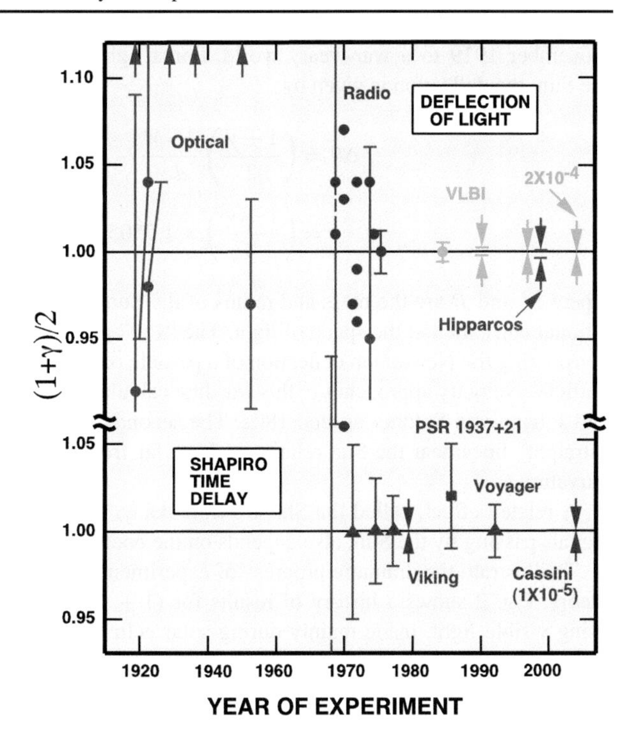


**Fig. 2** Measurements of the coefficient $(1+\gamma)/2$ from observations of the deflection of light and of the Shapiro delay in propagation of radio signals near the Sun. The general relativity prediction is unity

## 3 Solar-System Tests of Weak-Field Gravity

It was once customary to discuss experimental tests of general relativity in terms of the "three classical tests", the gravitational redshift, which is really a test of the EEP, not of general relativity itself; the perihelion advance of Mercury, the first success of the theory; and the deflection of light, whose measurement in 1919 made Einstein a celebrity. However, the proliferation of additional tests as well as of well-motivated alternative metric theories of gravity, made it desirable to develop a more general theoretical framework for analyzing both experiments and theories.

This "parametrized post-Newtonian (PPN) framework" dates back to Eddington in 1922, but was fully developed by Nordtvedt and Will in the period 1968–72. When we confine attention to metric theories of gravity, and further focus on the slow-motion, weak-field limit appropriate to the solar system and similar systems, it turns out that, in a broad class of metric theories, only the values of a set of numerical coefficients in the expression for the spacetime metric vary from theory to theory. The framework contains ten PPN parameters: $\gamma$, related to the amount of spatial curvature generated by mass; $\beta$, related to the degree of non-linearity in the gravitational field; $\xi$, $\alpha_1$, $\alpha_2$, and $\alpha_3$, which determine whether the theory violates local position invariance or local Lorentz invariance in *gravitational* experiments (violations of the Strong Equivalence Principle); and $\zeta_1$, $\zeta_2$, $\zeta_3$ and $\zeta_4$, which describe whether the theory has appropriate momentum conservation laws. In general relativity, $\gamma = 1$, $\beta = 1$, and the remaining parameters all vanish. For a complete exposition of the PPN framework see Will (1993).

To illustrate the use of these PPN parameters in experimental tests, we cite the deflection of light by the Sun, an experiment that made Einstein an international celebrity when the sensational news of the Eddington-Crommelin eclipse measurements was relayed in

November 1919 to a war-weary world. For a light ray which passes a distance $d$ from the Sun, the deflection is given by

$$\begin{aligned}\Delta\theta &= \left(\frac{1+\gamma}{2}\right)\frac{4GM}{dc^2}\\ &= \left(\frac{1+\gamma}{2}\right)\times 1.7505\left(\frac{R}{d}\right)\text{ arcsec,}\end{aligned} \tag{2}$$

where $M$ and $R$ are the mass and radius of the Sun and $G$ and $c$ are the Newtonian gravitational constant and the speed of light. The "1/2" part of the coefficient can be derived by considering the Newtonian deflection of a particle passing by the Sun, in the limit where the particle's velocity approaches $c$; this was first calculated independently by Henry Cavendish and Johann von Soldner around 1802. The second "$\gamma/2$" part comes from the bending of "straight" lines near the Sun relative to lines far from the Sun, as a consequence of space curvature.

A related effect, called the Shapiro time delay, an excess delay in travel time for light signals passing by the Sun, also depends on the coefficient $(1+\gamma)/2$.

To illustrate the dramatic progress of experimental gravity since the dawn of Einstein's theory, Fig. 2 shows a history of results for $(1+\gamma)/2$. "Optical" denotes measurements using visible light, made mainly during solar eclipses, beginning with the 1919 measurements of Eddington and his colleagues. Arrows denote values well off the chart from one of the 1919 eclipse expeditions and from others through 1947. "Radio" denotes interferometric measurements of radio-wave deflection, and "VLBI" denotes Very Long Baseline Radio Interferometry, culminating in a global analysis of VLBI data on 541 quasars and compact radio galaxies distributed over the entire sky, which verified GR at the 0.02 percent level. "Hipparcos" denotes the European optical astrometry satellite. Shapiro time delay measurements began in the late 1960s, by bouncing radar signals off Venus and Mercury; the most recent test used tracking data from the *Cassini* spacecraft on its way to Saturn, yielding a result at the 0.001 percent level.

Other experimental bounds on the PPN parameters, all consistent with general relativity, came from measurements of the perihelion-shift of Mercury, bounds on the "Nordtvedt effect" (a possible violation of the weak equivalence principle for self-gravitating bodies) via Lunar laser ranging, and geophysical and astronomical observations. The current bounds are summarized in Table 1.

The perihelion advance of Mercury, the first of Einstein's successes, is now known to agree with observation to a few parts in $10^3$. Although there was controversy during the 1960s about this test because of Dicke's claims of an excess solar oblateness, which would result in an unacceptably large Newtonian contribution to the perihelion advance, it is now known from helioseismology that the oblateness is of the order of a few parts in $10^7$, as expected from standard solar models, and too small to affect Mercury's orbit, within the experimental error.

Scalar-tensor theories of gravity are characterized by a coupling function $\omega(\phi)$ whose size is inversely related to the "strength" of the scalar field relative to the metric. In the solar system, the parameter $|\gamma-1|$, for example is equal to $1/(2+\omega(\phi_0))$, where $\phi_0$ is the value of the scalar field today outside the solar system. Solar-system experiments (primarily the Cassini results) constrain $\omega(\phi_0) > 40000$.

Proposals are being developed for advanced space missions which will have tests of PPN parameters as key components, including GAIA, a high-precision astrometric telescope (successor to Hipparcos), which could measure light-deflection and $\gamma$ to the $10^{-6}$ level, the

**Table 1** Current limits on the PPN parameters

| Parameter | Effect | Limit | Remarks |
|---|---|---|---|
| $\gamma - 1$ | (i) time delay | $2.3 \times 10^{-5}$ | Cassini tracking |
| | (ii) light deflection | $4 \times 10^{-4}$ | VLBI |
| $\beta - 1$ | (i) perihelion shift | $3 \times 10^{-3}$ | assumes $J_2 = 10^{-7}$ from helioseismology |
| | (ii) Nordtvedt effect | $2.3 \times 10^{-4}$ | $\eta = 4\beta - \gamma - 3$ assumed |
| $\xi$ | Earth tides | $10^{-3}$ | gravimeter data |
| $\alpha_1$ | orbital polarization | $10^{-4}$ | Lunar laser ranging PSR J2317+1439 |
| $\alpha_2$ | solar spin precession | $4 \times 10^{-7}$ | alignment of Sun and ecliptic |
| $\alpha_3$ | pulsar acceleration | $2 \times 10^{-20}$ | pulsar $\dot{P}$ statistics |
| $\eta$[a] | Nordtvedt effect | $9 \times 10^{-4}$ | Lunar laser ranging |
| $\zeta_1$ | – | $2 \times 10^{-2}$ | combined PPN bounds |
| $\zeta_2$ | binary motion | $4 \times 10^{-5}$ | $\ddot{P}_p$ for PSR 1913+16 |
| $\zeta_3$ | Newton's 3rd law | $10^{-8}$ | Lunar acceleration |
| $\zeta_4$ | – | – | not independent |

[a]Here $\eta = 4\beta - \gamma - 3 - 10\xi/3 - \alpha_1 - 2\alpha_2/3 - 2\zeta_1/3 - \zeta_2/3$

Laser Astrometric Test of Relativity (LATOR), a mission involving laser ranging to a pair of satellites on the far side of the Sun, which could measure $\gamma$ to a part in $10^8$, and could possibly detect second-order effects in light propagation, and BepiColombo, a Mercury orbiter mission planned for the 2012 time frame, which could, among other measurements, determine $J_2$ of the Sun to $10^{-8}$, and improve bounds on a time variation of the gravitational constant. The ground-based laser ranging program APOLLO could improve the bound on the Nordtvedt parameter $\eta$ to the $3 \times 10^{-5}$ level.

The NASA mission called Gravity Probe-B completed the flight portion of its mission to measure the Lense-Thirring and geodetic precessions of gyroscopes in Earth orbit in the fall of 2005. Launched on April 20, 2004 for a 16-month mission, it consisted of four spherical fused quartz rotors coated with a thin layer of superconducting niobium, spinning at 70–100 Hz, in a spacecraft containing a telescope continuously pointed toward a distant guide star (IM Pegasi). Superconducting current loops encircling each rotor measured the change in direction of the rotors by detecting the change in magnetic flux through the loop generated by the London magnetic moment of the spinning superconducting film. The spacecraft orbited the Earth in a polar orbit at 650 km altitude. The proper motion of the radio-emitting guide star relative to distant quasars was measured before, during and after the mission using VLBI. The primary science goal of GPB was a one-percent measurement of the 41 milliarcsecond per year frame dragging or Lense-Thirring effect caused by the rotation of the Earth; its secondary goal was to measure to six parts in $10^5$ the larger 6.6 arcsecond per year geodetic precession caused by space curvature. Final results of the experiment were expected toward the end of 2009.

A complementary test of the Lense-Thirring precession, albeit with lower accuracy than the GPB goal, involved measuring the precession of the orbital planes of two Earth-orbiting laser-ranged satellites called LAGEOS, using up-to-date models of the gravitational field of the Earth in order to subtract the dominant Newtonian precession with sufficient accuracy to yield a measurement of the relativistic effect. The resulting measurement of frame dragging is at the 10 percent level.

**Table 2** Parameters of selected binary pulsars

| Parameter | Symbol | Value[a] in PSR1913+16 | Value[a] in J0737-3039 |
|---|---|---|---|
| *Keplerian parameters* | | | |
| Eccentricity | $e$ | 0.6171338(4) | 0.087778(2) |
| Orbital Period | $P_b$ (day) | 0.322997462727(5) | 0.1022515628(2) |
| *Post-Keplerian parameters* | | | |
| Periastron advance | $\langle\dot{\omega}\rangle$ ($^{\circ}\mathrm{yr}^{-1}$) | 4.226595(5) | 16.900(2) |
| Redshift/time dilation | $\gamma'$ (ms) | 4.2919(8) | 0.39(2) |
| Orbital period derivative | $\dot{P}_b$ ($10^{-12}$) | −2.4184(9) | −1.20(8) |
| Shapiro delay ($\sin i$) | $s$ | | 0.9995(4) |

[a]Numbers in parentheses denote errors in last digit

## 4 Binary Pulsars

The binary pulsar PSR 1913+16, discovered in 1974 by Joseph Taylor and Russell Hulse, provided important new tests of general relativity, specifically of gravitational radiation and of strong-field gravity. Through precise timing of the pulsar "clock", the important orbital parameters of the system could be measured with exquisite precision. These include non-relativistic "Keplerian" parameters, such as the eccentricity $e$, and the orbital period (at a chosen epoch) $P_b$, as well as a set of relativistic "post-Keplerian" parameters (see Table 2). The first PK parameter, $\langle\dot{\omega}\rangle$, is the mean rate of advance of periastron, the analogue of Mercury's perihelion shift. The second, denoted $\gamma'$ is the effect of special relativistic time-dilation and the gravitational redshift on the observed phase or arrival time of pulses, resulting from the pulsar's orbital motion and the gravitational potential of its companion. The third, $\dot{P}_b$, is the rate of decrease of the orbital period; this is taken to be the result of gravitational radiation damping (apart from a small correction due to galactic differential rotation). Two other parameters, $s$ and $r$, are related to the Shapiro time delay of the pulsar signal if the orbital inclination is such that the signal passes in the vicinity of the companion; $s$ is a direct measure of the orbital inclination $\sin i$. According to general relativity, the first three post-Keplerian effects depend only on $e$ and $P_b$, which are known, and on the two stellar masses which are unknown. By combining the observations of PSR 1913+16 with the general relativity predictions, one obtains both a measurement of the two masses, and a test of the theory, since the system is overdetermined. The results are

$$\begin{aligned} m_1 &= 1.4414 \pm 0.0002\, M_{\odot}, \qquad m_2 = 1.3867 \pm 0.0002\, M_{\odot}, \\ \dot{P}_b^{\mathrm{GR}} / \dot{P}_b^{\mathrm{OBS}} &= 1.0013 \pm 0.0021. \end{aligned} \tag{3}$$

The accuracy in measuring the relativistic damping of the orbital period is now limited by uncertainties in our knowledge of the relative acceleration between the solar system and the binary system as a result of galactic differential rotation.

The results also test the strong-field aspects of metric gravitation in the following way: the neutron stars that comprise the system have very strong internal gravity, contributing as much as several tenths of the rest mass of the bodies (compared to the orbital energy, which is only $10^{-6}$ of the mass of the system). Yet in general relativity, the internal structure is "effaced" as a consequence of the Strong Equivalence Principle (SEP), a stronger version of EEP that includes *gravitationally* bound bodies and local *gravitational* experiments. As a result, the orbital motion and gravitational radiation emission depend *only* on the masses $m_1$ and $m_2$, and not on their internal structure, apart from standard tidal and spin-coupling effects. By contrast, in alternative metric theories, SEP is not valid in general, and internal-structure effects can lead to significantly different behavior, such as the emission of dipole gravitational radiation. Unfortunately, in the case of scalar-tensor theories of gravity, because the neutron stars are so similar in PSR 1913+16 (and in other double-neutron star binary pulsar systems), dipole radiation is suppressed by symmetry; the best bound on the coupling parameter $\omega(\phi_0)$ from PSR 1913+16 is in the hundreds. By contrast, the close agreement of the data with the predictions of general relativity constitutes a kind of "null" test of the effacement of strong-field effects in that theory.

However, the recent discovery of the relativistic neutron star/white dwarf binary pulsar J1141-6545, with a 0.19 day orbital period, may ultimately lead to a very strong bound on dipole radiation, and thence on scalar-tensor gravity. The remarkable "double pulsar" J0737-3039 is a binary system with two detected pulsars, in a 0.10 day orbit seen almost edge on, with eccentricity $e = 0.09$, and a periastron advance of 17° per year. A variety of novel tests of relativity, neutron star structure, and pulsar magnetospheric physics will be possible in this system. For a review of binary pulsar tests, see Stairs (2008) and Damour (2009).

## 5 Gravitational-Wave Tests of Gravitation Theory

The detection of gravitational radiation by ground-based laser interferometers, such as LIGO in the USA, or VIRGO and GEO in Europe, by the proposed space-based laser interferometer LISA, will usher in a new era of gravitational-wave astronomy (for a review, see Hough and Rowan 2000). Furthermore, it will yield new and interesting tests of general relativity in its radiative regime.

One important test is of the speed of gravitational waves. According to general relativity, in the limit in which the wavelength of gravitational waves is small compared to the radius of curvature of the background spacetime, the waves propagate along null geodesics of the background spacetime, i.e. they have the same speed, $c$, as light. In other theories, the speed could differ from $c$ because of coupling of gravitation to "background" gravitational fields. For example, in some theories with a flat background metric $\boldsymbol{\eta}$, gravitational waves follow null geodesics of $\boldsymbol{\eta}$, while light follows null geodesics of $\boldsymbol{g}$. In brane-world scenarios, the apparent speed of gravitational waves could differ from that of light if the former can propagate off the brane into the higher dimensional "bulk". Another way in which the speed of gravitational waves could differ from $c$ is if gravitation were propagated by a massive field (a massive graviton), in which case $v_g$ would depend on the wavelength $\lambda$ of the gravitational waves according to $v_g/c \approx 1 - \lambda^2/2\lambda_g^2$, where $\lambda_g = h/m_g c$ is the graviton Compton wavelength ($\lambda_g \gg \lambda$ is assumed).

The most obvious way to test for a massive graviton is to compare the arrival times of a gravitational wave and an electromagnetic wave from the same event, e.g. a supernova. For a source at a distance of hundreds of megaparsecs, and for a relative arrival time between the two signals of order seconds, the resulting bound would be of order $|1 - v_g/c| < 5 \times 10^{-17}$. A bound such as this cannot be obtained from solar system tests at current levels of precision.

However, there is a situation in which a bound on the graviton mass can be set using gravitational radiation alone. That is the case of the inspiralling compact binary, the final stage of evolution of systems like the binary pulsar, in which the loss of energy to gravitational waves has brought the binary to an inexorable spiral toward a final merger. Because the frequency of the gravitational radiation sweeps from low frequency at the initial moment of observation to higher frequency at the final moment, the speed of the gravitational waves emitted will vary, from lower speeds initially to higher speeds (closer to $c$) at the end. This will cause a distortion of the observed phasing of the waves as a function of wavelength. Furthermore, through the technique of matched filtering, whereby a theoretical model of the wave and its phase evolution is cross-correlated against the detected signal, the parameters of the compact binary can be measured accurately, along with the parameter, the graviton Compton wavelength, that governs the distortion.

Advanced ground-based detectors of the LIGO-VIRGO type could place a bound $\lambda_g$ on the order of several times $10^{12}$ km, and the proposed space-based LISA antenna could place a bound on the order of $10^{16}$ km. These potential bounds can be compared with the solid bound $\lambda_g > 2.8 \times 10^{12}$ km, derived from solar system dynamics, which limit the presence of a Yukawa modification of Newtonian gravity of the form $V(r) = (GM/r)\exp(-r/\lambda_g)$, and with the model-dependent bound $\lambda_g > 6 \times 10^{19}$ km from consideration of galactic and cluster dynamics.

## 6 Tests of Gravity in the Strong-Field Regime

One of the main difficulties of testing GR in the strong-field regime is the possibility of contamination by uncertain or complex physics. In the solar system, weak-field gravitational effects could in most cases be measured cleanly and separately from non-gravitational effects. The remarkable cleanliness of the binary pulsar permitted precise measurements of gravitational phenomena in a strong-field context. Unfortunately, nature is rarely so kind. Still, under suitable conditions, qualitative and even quantitative strong-field tests of GR could be carried out.

One example is the exploration of the spacetime near black holes and neutron stars via accreting matter. Studies of certain kinds of accretion known as advection-dominated accretion flow (ADAF) in low-luminosity binary X-ray sources may yield the signature of the black hole event horizon. The spectrum of frequencies of quasi-periodic oscillations (QPO) from galactic black hole binaries may permit measurement of the spins of the black holes. Aspects of strong-field gravity and frame-dragging may be revealed in spectral shapes of iron fluorescence lines from the inner regions of accretion disks around black holes and neutron stars. Measurements of the detailed shape of the infrared image of the accretion flow around the massive black hole in the center of our galaxy could also provide tests of the spacetime black hole metric. Because of uncertainties in the detailed models, the results to date of studies like these are suggestive at best, but the combination of higher-resolution observations and better modelling could lead to striking tests of strong-field predictions of GR. For a review of such tests see Psaltis (2008).

The best tests of GR in the strong field limit may come from gravitational-wave observations. The ground-based interferometers may detect the gravitational waves from the final inspiral and merger of a pair of stellar-mass black holes. Comparison of the observed waveform with the predictions of GR from a combination of analytic and numerical techniques may provide a test of the theory in the most dynamical, strong-field limit. The proposed space antenna LISA may observe as many as 100 mergers of massive black holes per year, with large signal-to-noise ratio. Such observations could provide precise measurements of black hole masses and spins, and could test the "no hair" theorems of black holes by detecting the spectrum of quasi-normal "ringdown" modes emitted by the final black hole. Observations by LISA of the hundreds of thousands of gravitational-wave cycles emitted when a small black hole inspirals onto a massive black hole could test whether the geometry of a black hole actually corresponds to the "hair-free" Kerr metric.

## 7 Conclusions

Einstein's general theory of relativity altered the course of science. It was a triumph of the imagination and of theory, but in the early years following its formulation, experiment played a secondary role. In the final four decades of the 20th century, we witnessed a second triumph for Einstein, in the systematic, high-precision experimental verification of general relativity. It has passed every test with flying colors.

But the work is not done. During the 21st century, we may look forward to the possibility of tests of strong-field gravity in the vicinity of black holes and neutron stars. Gamma-ray, X-ray and gravitational-wave astronomy will play a critical role in probing this largely unexplored aspect of general relativity.

General relativity is now the "standard model" of gravity. But as in particle physics, there may be a world beyond the standard model, beyond Einstein. Quantum gravity, strings and branes may lead to testable effects beyond standard general relativity. Experimentalists and observers can be counted on to continue a vigorous search for such effects using laboratory experiments, particle accelerators, gravitational-wave detectors, space telescopes and cosmological observations, well into the 21st century.

**Acknowledgements** This work was supported in part by the US National Science Foundation, Grant No. PHY 06-52448 and by the National Aeronautics and Space Administration, Grant No. NNG-06GI60G. We are grateful for the hospitality of the Institut d'Astrophysique de Paris, where this paper was prepared. This review is based in large measure on a paper prepared for the proceedings of the conference "Beyond Einstein: Perspectives on Geometry, Gravitation and Cosmology in the Twentieth Century", held in Mainz, Germany, September 22-26, 2008.

## References

T. Damour, in *Physics of Relativistic Objects in Compact Binaries: from Birth to Coalescence: Astrophysics and Space Science Library*, vol. 359, ed. by M. Colpi, P. Casella, V. Gorini, U. Moschella, A. Possenti, (Springer, New York, 2009, in press)

J. Hough, S. Rowan, Living Rev. Relativ. **3**, 3 (2000) [On-line article] Cited on 1 November 2008, www.livingreviews.org/lrr-2000-3

D. Mattingly, Living Rev. Relativ. **8**, 5 (2005) [On-line article] Cited on 1 November 2008, www.livingreviews.org/lrr-2005-5

D. Psaltis, Living Rev. Relativ. **11**, 9 (2008) [On-line article] Cited on 1 November 2008, www.livingreviews.org/lrr-2008-9

I.H. Stairs, Living Rev. Relativ. **6**, 5 (2008) [On-line article] Cited on 1 November 2008, www.livingreviews.org/lrr-2003-5

J.-P. Uzan, Rev. Mod. Phys. **75**, 403 (2003)

C.M. Will, *Theory and Experiment in Gravitational Physics* (Cambridge University Press, Cambridge, 1993)

C.M. Will, Living Rev. Relativ. **9**, 3 (2006a) [Online article]: Cited on 1 November 2008, www.livingreviews.org/lrr-2006-3

C.M. Will, in *Einstein 1905–2005: Poincaré Seminar 2005*, ed. by T. Damour, O. Darrigol, B. Duplantier, V. Rivasseau (Birkhäuser, Switzerland, 2006b), p. 33

Space Sci Rev (2009) 148: 15–23
DOI 10.1007/s11214-009-9575-9

# From Classical Theory to Quantum Gravity

**Bernard F. Schutz**

Received: 6 August 2009 / Accepted: 3 September 2009 / Published online: 6 January 2010
© The Author(s) 2009. This article is published with open access at Springerlink.com

**Abstract** While it has been generally accepted for decades that general relativity and quantum theory are inconsistent, and that therefore general relativity must yield to a new theory of gravity compatible with quantum principles, the way to the new theory is still very unclear. A major obstacle is the lack of any experimental indicators as to where general relativity might break down. I speculate on the different ways in which a quantum theory including gravity might emerge from present-day theory and use these speculations to examine a number of promising lines for new experiments.

**Keywords** Fundamental physics · General relativity · Tests of gravitation · Quantum gravity

## 1 Introduction

Ever since the early days of quantum theory it has been clear that general relativity and quantum theory are incompatible, and that some new theory of gravity needs to be found that will be compatible with quantum principles. However, I am aware of no proof of this, no explicit demonstration that it is impossible to marry general relativity with, say, electroweak theory, that using them together leads to a mathematical contradiction. In fact, both general relativity and our best quantum theories (take electroweak theory, for example) have internal problems: singularities evolve in general relativity from regular initial data, and there are solutions that have naked singularities that lead to acausal evolution (the Big Bang being an example); and infinite energies in standard quantum field theory can only be removed by the device of renormalization, which looks very worrying when one expects gravity to be created by all forms of energy. Moreover, quantum measurement theory as currently understood requires a distinction between the "system" and the "observer", a distinction that will be difficult to maintain if one wants a quantum formulation of the entire universe.

B.F. Schutz (✉)
Albert Einstein Institute, Potsdam, Germany
e-mail: bernard.schutz@aei.mpg.de

Perhaps partly because it is unclear exactly where the inconsistency between these two pillars of physics lies, there is no general agreement among theoretical physicists as to how to find the desired quantum theory of gravity. Many theorists expect that we will only have truly consistent and causal theories of gravity and matter when we find a unified quantum description of them all. String theory is today the most popular framework within which these hopes are being pursued. Others, however, look for a quantum description of spacetime itself, on the grounds that such a theory might look very different from our current paradigm about how to form a quantum theory of a matter field, and therefore one can only do the job of integrating matter fields into the picture later, once one knows what the picture looks like.

In both approaches it is generally expected that the transition from classical gravity to quantum gravity occurs at the Planck scale, variously expressed as $\ell_{\rm Pl} = (G\hbar/c^3)^{1/2} = 1.6 \times 10^{-33}$ cm, $t_{\rm Pl} = \ell_{\rm Pl}/c = 5 \times 10^{-44}$ s, $m_{\rm Pl} = c^2\ell_{\rm Pl}/G = 2 \times 10^{-5}$ g, and $E_{\rm Pl} = m_{\rm Pl}c^2 = 1.2 \times 10^{28}$ eV. This is basically a dimensional argument, so it is not difficult to devise scenarios where the transition occurs at lower energies and longer length scales. We shall meet some alternatives below. Wherever the transition might occur, presumably at smaller length scales spacetime loses its continuum character and is replaced by a probabilistic tangle of fibers, strings, loops, foam, or something even more exotic.

Although there are many ideas about how this might work, it has been remarkably difficult to find a theory or theoretical framework which remains self-consistent when developed fully. String theory is popular because it is essentially the only candidate in the spirit of a field theory that seems to be finite when all fields are included; to remain finite it needs ten dimensions. The matter fields are confined to our 4-dimensional spacetime embedded in this larger manifold, and only gravity seems to extend into the other dimensions. This picture offers rich freedom to suggest new phenomenology, depending on assumptions about the nature of the extra dimensions and the physics of the strings themselves. Particularly popular lately have been brane-world scenarios. Many string-related scenarios offer experimental access to quantum gravitational effects at scales much larger than $\ell_{\rm Pl}$.

The main alternative to string theory is loop quantum gravity, which follows the idea of quantizing gravity essentially by itself. It lives in four dimensions and has no other natural length scales, so essentially all quantum effects in loop quantum gravity occur at the Planck scale. Loop quantum gravity therefore has so far not been such a fertile source of new tests of quantum gravity. However, if it replaces continuum spacetime with a fundamentally different structure, then Lorentz invariance should be a casualty, and it is certainly at least possible that this is visible over larger distances. Moreover, the recent demonstration that, in a restricted version of loop quantum gravity, it is possible to construct a cosmological model that goes through the Big Bang without a singularity (Bojowald 2008), is exciting and offers the possibility that observations of the cosmic microwave background or of some earlier epoch might show characteristics of this process.

Experimental tests may in fact be the only way forward in the quantum-gravity program; at the very least new experimental results would be huge progress. The problem is daunting if one believes that new results will only be found on the Planck scale. The highest energies accessible to us at present are from ultra-high-energy cosmic rays, the highest of whose energies approach $10^{22}$ eV. (See the contribution from Watson in this volume.) But it is possible that at low energies there are some residual effects of quantization that are still visible, or that some of the extra dimensions in string theory are larger than the Planck length, which would make new experimental tests more accessible. Another type of theory, emergent gravity, offers the same possibility. I will discuss all these possibilities below.

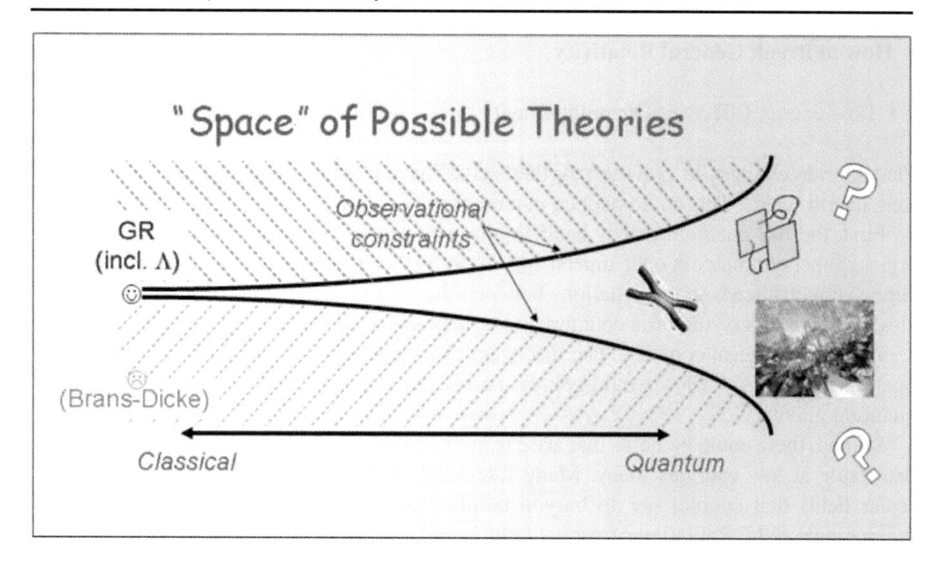


**Fig. 1** Experimental constraints at low energies point very strongly to general relativity with a cosmological constant, excluding for example Brans-Dicke Theory for small values of the coupling parameter. At high energies, as one approaches the quantum regime, the allowed space of theories widens and is constrained more by mathematical consistency than by experiment

The central problem for the experimental search for quantum gravity is that, so far, there are *no* observed deviations of general relativity from experiment. Unlike the situation at the beginning of quantum theory, where there was a host of unexplained spectroscopic evidence and some very important particle-scattering experiments, today we have nothing to try to explain or fit. General relativity with a cosmological constant ($\Lambda$GR) is all we need at present. (See the contribution from Clifford Will in this volume.) The situation is illustrated in Fig. 1: constraints on gravitation theories at low energies are very strong, but as we approach the quantum scale these same constraints provide less and less guidance.

There is, of course, some room for worry. The cosmological constant is very small compared to the "natural" or non-renormalized vacuum energy density that quantum field theory would suggest, so its non-zero value might be a strong clue to the integration of quantum theory and gravity; however, no-one yet has worked out how to use that clue successfully. Alternatively, the dark energy might not be a cosmological constant; it might evolve with time in a way that depends on the energy density or expansion rate of the universe. That would also be a clue, at least to a new cosmological field that might be a product of the marrying of gravity and quantum theory; here the problem is that theorists have proposed a wealth of such "quintessence" fields, and there is not enough precision in the cosmological observations so far to pin down which, if any, is right. One could also take the evidence for dark matter as evidence instead for modified gravity. Since we have not yet identified the dark matter particle, perhaps some other theory that modifies gravity at large length scales (Milgrom 2008; Bekenstein 2004) is better than general relativity, and then its extra fields would be explained as low-energy effects of a quantum gravity theory. All of these alternatives will be discussed in this volume.

The rest of this paper is organized into two main parts. The first reviews the different approaches to extending general relativity into a quantum theory and what observable differences they might produce. The second part then suggests what kinds of experiments might provide insight into the path beyond general relativity.

## 2 How to Break General Relativity

### 2.1 Low-energy Effects of Quantum Gravity

Planck-scale changes in gravity may themselves have residual effects at low energies and long length scales. This could happen in a number of ways.

First, the full quantum theory itself might have unusual features that persist at low energies. Supersymmetry is one: superstring theory is the preferred variant at present because supersymmetry leads to cancellations between the zero-point energies of the normal particles and their supersymmetric counterparts (sparticles), leading naturally to a zero or at least a small cosmological constant. The discovery of a sparticle when the LHC reaches its full operating energy at CERN would greatly narrow down the space of possible approaches to quantum gravity.

Second, there could be fields that arise from the full theory but which are still present and detectable at low energies today. Many low-energy effective theories introduce massless scalar fields that couple, say, to baryon number and therefore behave very like a scalar gravitational field. Radiation of such a field would carry energy away from a gravitational wave source, altering its long-term evolution and waveform in potentially detectable ways, while the radiation itself could be distinguished from standard tensor gravitational waves by its distinctive polarization pattern. It might happen that the full theory has an effective Lagrangian that has a Chern-Simons-type term, breaking parity invariance. This would have distinctly observable effects in gravitational waves from binary systems: see Alexander and Yunes (2008).

It is also possible that, if the full theory includes grand unification at very high energies, then spontaneous symmetry breaking during the expansion of the universe could have produced, besides standard matter, a completely separate matter sector that has its own internal interactions (analogous to our strong, weak, and electromagnetic interactions) but whose only coupling to standard matter is through gravity. Such "shadow matter" would be detectable by gravitational means but not otherwise.

Third, there could be fossils of high-energy systems that still live on. For example, Damour and Vilenkin (2005) have suggested that fundamental strings, which are the building blocks of particles in string theory, could be turned into macroscopic cosmic strings by inflation, in the same way that inflation turns quantum fluctuations into the real density fluctuations that are measured in the cosmic microwave background. These cosmic strings would emit gravitational waves with a distinctive signature, and their detection would be a big step toward verifying the string theory framework.

Low-energy effects can also be searched for by performing highly sensitive null experiments, such as looking for temporal variations in the fine-structure constant, for violations of the local isotropy of space, or for violations of local Lorentz invariance. While we are unlikely to measure Planck-level corrections to $g_{00}$ any time soon, an extra field introduced by quantum effects might be visible where general relativity predicts a zero.

A special kind of null experiment is the nulling of fringes in an interferometer. Precisely here there has recently been a suggestion from Hogan (2008) that quantum spacetime effects could produce a fundamental noise with a spectral noise density $S_h \sim t_{\mathrm{Pl}}$. It turns out that modern gravitational wave interferometers operate near this level, and in fact that the GEO600 interferometer has for years had an unexplained noise of this size. Hogan derives his effect from a holographic approach to spacetime information theory; as such it does not fit into standard quantum gravity schemes. If this noise could be proved to exist, it would revolutionize the search for quantum gravity.

Finally, it is well to recall that in at least one area we are only about 6 orders of magnitude away from the Planck scale. Ultra-high-energy cosmic rays (UHCRs) have been measured at energies approaching $10^{22}$ eV. If these consist of single particles, then their individual energies are as large as $10^{-6} E_{\mathrm{Pl}}$. If there is a violation of Lorentz invariance at Planck energies, which seems very possible, then ultra-precise observations of these particles (hard to do!) might reveal it. Of course, when they interact with atmospheric hadrons the center-of-mass energies are much smaller, of order $10^{15}$ eV. This is still above the energy of the LHC, so supersymmetry and other effects might be manifest in the collisions of UHCRs with atmospheric nuclei, but the energies are probably too far from Planck for direct effects of quantum gravity to be visible.

## 2.2 Models Involving Extra Dimensions

Our universe is a 4-dimensional subspace of the 10- or 11-dimensional space in which string theory is finite. In the most conservative picture, the extra dimensions are all curled up, as in Kaluza-Klein models, on a Planck length-scale. But larger dimensions are possible. A lot of attention has been paid recently to brane-world models, where we live on a 4-brane and there could be other branes embedded in this larger space. It is possible that one or more dimensions is larger, and either wraps up over a longer length scale or terminates on another brane some distance away. In such a picture, the conventional three interactions of particle physics would be confined to our brane, but gravity would extend into the other dimensions (called the bulk) as well. To be consistent with current limits on the Newtonian inverse-square-law, the size of these "large" dimensions would need to be smaller than a millimeter or so. Nevertheless, this would be 32 orders of magnitude longer than $\ell_{\mathrm{Pl}}$, and would potentially bring new effects into physics at energies as low as $10^{-4}$ eV.

The most direct test of large extra dimensions is to measure the gravitational inverse-square law at sub-millimeter lengths, a big challenge. (See the contribution of Gundlach to this volume.) The gravitational acceleration produced by a point mass in $n$ dimensions falls off as $r^{-(n-1)}$ (for $n > 1$), so if below some length-scale gravity could sense the extra dimensions, then it would get strong faster at smaller distances than one expects in three dimensions.

Other tests are possible, however. There have been plenty of novel predictions and speculations. A number of scenarios within the brane picture are reviewed by Maartens (2004). For example, branes could collide, even giving rise to the Big Bang itself, or creating gravitational waves. Nearby branes could influence our own brane gravitationally, so that we might detect gravitational waves radiated by a binary system in a parallel universe!

## 2.3 Alternative Approaches

Although most work on quantum gravity today is being done within the string theory and loop quantum gravity frameworks, there are alternatives, and each makes distinct predictions.

There is a program to derive continuum spacetime as a limit of discrete causal sets due to Sorkin (2007). This approach may make predictions about a variety of testable violations, such as Lorentz invariance and the time-behavior of the dark energy.

Penrose (2002) has argued that measurement theory will be affected by gravity, and in particular that the distinction between the observer and the system will come about in a natural way by decoherence of the wave-function induced by gravity. If this could be shown experimentally, it would have a very strong influence on how quantum gravity progressed.

General relativity might not be the fundamental field we think it is. Instead it could be an *emergent phenomenon*, analogous to the way that superfluidity emerges from detailed molecular interactions: see Volovik (2003). In this case, the Planck scale is an illusion created by the weakness of gravity, whereas the fundamental physics is happening at other energy scales, perhaps much lower than $E_{\mathrm{Pl}}$. I am not aware of concrete predictions from this point of view, but certainly any anomalous experimental result showing a violation of general relativity could point in this direction.

### 2.4 Dark Energy

The cosmological constant $\Lambda$ represents a field whose energy density $\rho$ is invariant under Lorentz transformation. In order to have this property, it must necessarily have negative pressure $p = -\rho$. It arises naturally in quantum field theories, since any zero-point energy density of a relativistic field must be Lorentz invariant. But typically these are large and have to be renormalized away. Supersymmetry does this automatically, perhaps too efficiently, since it might be hard to get any non-zero $\Lambda$ in the end.

Theorists have therefore invented other physical fields that have almost but not quite invariant energy densities. Generally these are called quintessence theories. They can have an equation of state (relation of $p$ and $\rho$) that changes with time. Precision cosmological observations are necessary in order to distinguish these from a cosmological constant and from one another.

Emergent theories also lead to a cosmological constant, but not to a particularly small one.

## 3 Where to Look for Broken General Relativity

After this survey of the theoretical perspectives on quantum gravity, a number of experimental directions seem natural.

Cosmological observations with high precision are very important, and should be done out to redshifts up to 2 or beyond. If it turns out that the dark energy really is a cosmological constant, then that kicks the ball back to conventional theories. But if there is some kind of evolution in the dark energy, that could be a real clue. Relevant space missions include JDEM and LISA. Ground-based observing programs can also make decisive contributions.

Observations of a cosmological background of gravitational waves would also give strong clues: the waves might arise from inflation, but stronger backgrounds could come from symmetry breaking as the universe expands, from brane-world effects, from shadow matter, and so on. Here one looks to the gravitational wave detectors LIGO (Abbott 2009), VIRGO (Acernese et al. 2008), and LISA, and even more to the next-generation detector now being called the Einstein Telescope (ET).

There may be clues in gravitational wave propagation effects. If the graviton has some kind of effective mass, this could lead to dispersion and anomalous time-delays within signals, measurable by LISA, LIGO, VIRGO, or ET. A Chern-Simons term, mentioned earlier, could make the two components of circular polarization propagate at different speeds, leading to observable effects in observations of signals from binary systems detected by LISA (Alexander and Yunes 2008).

Experiments that look for violations of local Lorentz invariance and other principles of local physics are potentially very informative; they are null experiments and so even when they produce no observable violation they produce interesting constraints on the terms in a theory that lead to violations. The equivalence principle is expected to be violated at some level; astonishingly accurate laboratory experiments (Schlamminger et al. 2008) will soon be superseded by space missions like MICROSCOPE (Touboul et al. 2006), GG (Nobili et al. 2009), and possibly STEP (Overduin et al. 2009) that could push the limits to parts in $10^{18}$. Temporal or spatial variations in the fundamental constants would give evidence of how spontaneous symmetry breaking happens, or of unexpected couplings between fields, or of incomplete homogenization by inflation. Temporal variations can be explored in local experiments, and there are several interesting proposed space missions. Spatial variations presumably happen only over cosmological distances, and to detect them requires very precise spectroscopic observations of distant sources like quasars. Space could be anisotropic even locally, which would be evidence of new tensor fields. And, perhaps most interestingly, Lorentz invariance could fail. If this happens on macroscopic scales, it might indicate preferred frames or new long-range forces; if it happens in tiny domains then this might indicate the length-scale of the transition to quantum gravity.

Testing general relativity itself has been an important activity ever since the first observations of the deflection of light by the Sun. Solar-system tests in the PPN framework have so far not revealed any difficulties with general relativity, but they are only just beginning to probe second post-Newtonian effects. Radio observations of pulsars in binaries, and particularly of the double pulsar system PSRJ 0737-3039, test the radiative part of general relativity, are becoming accurate enough also to test higher-order post-Newtonian effects on the orbital motions, and are also able to probe wider aspects of gravity, such as preferred-frame effects Lorimer (2008), Stairs (2003). Observations of binary inspiral using gravitational wave detectors will someday, at the latest when LISA flies, provide strong constraints on post-Newtonian gravity. Such observations can also test other aspects of general relativity, such as whether waves have just two transverse tensorial polarization states, whether there is frequency dispersion of gravitational waves, and whether right- and left-handed circularly polarized waves travel at the same speed. LISA and ET will also make observations of binary black hole mergers with sufficient precision to place good constraints on the black hole uniqueness theorem, the Hawking area theorem, and the cosmic censorship conjecture.

As we have seen, testing Newtonian gravity at short range could reveal the length scale where quantum effects set in, if large-size extra dimensions play a role (Gundlach et al. 2009, this issue). Newtonian gravity over long ranges could also show up quantum effects. Some or all of the anomalous gravitational effects attributed to dark matter could in principle be an indication of the violation of general relativity over long distances.

Finally, experimenting on quantum physics over large distance scales might possibly reveal anomalies. If gravity plays a role in decoherence then this could be testable in space. There are interesting mission proposals for experiments on entangled states over very large distances that could shed light on how standard quantum measurement theory should be extended to quantum gravity. This area has had less theoretical attention than quantization itself has (Penrose 2002).

## 4 Conclusions

While it is plausible that pointers to quantum gravity exist at low energies, there is no guarantee, and so one should be cautious before deciding to invest in a delicate and possibly expensive experiment. We have no clues so far to the strength of violations of general relativity: theories have too many adjustable parameters. It could happen that general relativity is "clean" all the way down to the Planck scale!

That would, however, be very surprising in light of the complexity of physics at energies accessible to us. If some form of grand unification holds, then it would be remarkable if we already see all the physics there is, and nothing new emerges right up to the unification energy that could have effects at our energies. So the motivation to find new effects is strong, and the difficulty is to know which direction to look.

Fortunately, some of the ways of investigating these problems are by-products of experiments and observation programs that have other motivations. Cosmological observations of the dark energy and gravitational wave observations fit into these categories. So do precise spectroscopic observations of quasars. Any of these could turn up the first evidence and point the way to more dedicated experiments in fundamental physics.

Alongside these, there is strong motivation to do laboratory experiments to look for violations of our standard expectations, such as the equivalence principle or the inverse-square force. These are constant sources of technological innovation. And as null experiments, even a null result is a useful constraint on theory.

In the last decade a number of space missions have been proposed to test general relativity and look for quantum effects. GP-B (Silbergleit et al. 2009, this issue) and the LAGEOS (Ciufolini 2007) observations measure a non-zero effect, the gravitomagnetic part of general relativity. MICROSCOPE will be the first null experiment for gravity in space. The European Space Agency's call for mission proposals in the Horizons 2000+ program produced a set of very interesting proposals using cold atom technology to do null experiments and to study quantum measurement theory. While none were selected for study in the first round, funding is being made available for technology development; there is a strong motivation to do so, not just because of the insight these experiments will give into quantum gravity, but because of the potential further applications of these new technologies.

Ultimately, despite the uncertainty about what the right direction might be for an experiment, the motivation for doing the experiments is that finding a quantum theory of gravitation is the single most important outstanding problem in our understanding of fundamental physics today. And the search for this quantum theory is starved of experimental data to guide it.

Every experiment is a step in the right direction.

**Open Access** This article is distributed under the terms of the Creative Commons Attribution Noncommercial License which permits any noncommercial use, distribution, and reproduction in any medium, provided the original author(s) and source are credited.

## References

B.P. Abbott (The LIGO Scientific Collaboration), LIGO: the laser interferometer gravitational-wave observatory. Rep. Progr. Phys. **72**(7), 076901 (2009)

F. Acernese, M. Alshourbagy, P. Amico, F. Antonucci, S. Aoudia, P. Astone, S. Avino, L. Baggio, F. Barone, L. Barsotti, M. Barsuglia, T.S. Bauer, S. Bigotta, S. Birindelli, M. Bizouard, C. Boccara, F. Bondu, L. Bosi, S. Braccini, C. Bradaschia, A. Brillet, V. Brisson, D. Buskulic, G. Cagnoli, E. Calloni, E. Campagna, F. Cavalier, R. Cavalieri, G. Cella, E. Cesarini, E. C-Mottin, A.-C. Clapson, F. Cleva, E. Coccia, C. Corda, A. Corsi, F. Cottone, J.-P. Coulon, E. Cuoco, S. D'Antonio, A. Dari, V. Dattilo, M. Davier, R. DeRosa, M. DelPrete, L. Di Fiore, A. Di Lieto, A. Di Virgilio, B. Dujardin, M. Evans, V. Fafone, I. Ferrante, F. Fidecaro, I. Fiori, R. Flaminio, J.-D. Fournier, S. Frasca, F. Frasconi, L. Gammaitoni, F. Garufi, E. Genin, A. Gennai, A. Giazotto, L. Giordano, V. Granata, C. Greverie, D. Grosjean, G. Guidi, S. Hamdani, S. Hebri, H. Heitmann, P. Hello, D. Huet, S. Kreckelbergh, P. La Penna, M. Laval, N. Leroy, N. Letendre, B. Lopez, M. Lorenzini, V. Loriette, G. Losurdo, J.-M. Mackowski, E. Majorana, M. Mantovani, F. Marchesoni, F. Marion, J. Marque, F. Martelli, A. Masserot, F. Menzinger, L. Milano, Y. Minenkov, C. Moins, J. Moreau, N. Morgado, S. Mosca, B. Mours, C.N. Man, I. Neri, F. Nocera,

G. Pagliaroli, C. Palomba, F. Paoletti, S. Pardi, A. Pasqualetti, R. Passaquieti, D. Passuello, F. Piergiovanni, L. Pinard, R. Poggiani, M. Punturo, P. Puppo, P. Rapagnani, T. Regimbau, A. Remillieux, F. Ricci, I. Ricciardi, A. Rocchi, L. Rolland, R. Romano, P. Ruggi, G. Russo, S. Solimeno, A. Spallicci, M. Tarallo, R. Terenzi, A. Toncelli, M. Tonelli, E. Tournefier, F. Travasso, C. Tremola, G. Vajente, J.F.J. van der Brand, S. van der Putten, D. Verkindt, F. Vetrano, A. Vicerè, J.-Y. Vinet, H. Vocca, M. Yvert, VIRGO: a large interferometer for gravitational wave detection started its first scientific run. J. Phys. Conf. Ser. **120**(3), 032007 (2008)

S. Alexander, N. Yunes, Chern-Simons modified general relativity. Phys. Rev. D **77**, 124040 (2008)

J.D. Bekenstein, Relativistic gravitation theory for the modified Newtonian dynamics paradigm. Phys. Rev. D **70**(8), 083509 (2004)

M. Bojowald, Loop quantum cosmology. Living Rev. Relativ. **11**(4) (2008)

I. Ciufolini, Dragging of inertial frames. Living Rev. Relativ. **449**, 41–47 (2007)

T. Damour, A. Vilenkin, Gravitational radiation from cosmic (super)strings: bursts, stochastic background, and observational windows. Phys. Rev. D **71**, 063510 (2005)

J.H. Gundlach, S. Schlamminger, T. Wagner, Laboratory tests of the equivalence principle at the University of Washington. Space Sci. Rev. (2009). doi:10.1007/s11214-009-9609-3

C.J. Hogan, Indeterminacy of holographic quantum geometry. Phys. Rev. D **78**(8), 087501 (2008)

D.R. Lorimer, Binary and millisecond pulsars. Living Rev. Relativ. **11**(8) (2008)

R. Maartens, Brane-world gravity. Living Rev. Relativ. **7**, 7 (2004)

M. Milgrom, The MOND paradigm. arXiv:0801.3133v2 (2008)

A.M. Nobili, G.L. Comandi, S. Doravari, D. Bramanti, R. Kumar, F. Maccarrone, E. Polacco, S.G. Turyshev, M. Shao, J. Lipa, H. Dittus, C. Laemmerzhal, A. Peters, J. Mueller, C.S. Unnikrishnan, I.W. Roxburgh, A. Brillet, C. Marchal, J. Luo, J. van der Ha, V. Milyukov, V. Iafolla, D. Lucchesi, P. Tortora, P. de Bernardis, F. Palmonari, S. Focardi, D. Zanello, S. Monaco, G. Mengali, L. Anselmo, L. Iorio, Z. Knezevic, "Galileo Galilei" (GG) a small satellite to test the equivalence principle of Galileo, Newton and Einstein. Exp. Astron. **23**, 689–710 (2009)

J. Overduin, F. Everitt, J. Mester, P. Worden, The science case for STEP. Adv. Space Res. **43**, 1532–1537 (2009)

R. Penrose, Gravitational collapse of the wavefunction: an experimentally testable proposal, in *The Ninth Marcel Grossmann Meeting*, ed. by V.G. Gurzadyan, R.T. Jantzen, R. Ruffini (2002), pp. 3–6

S. Schlamminger, K.-Y. Choi, T.A. Wagner, J.H. Gundlach, E.G. Adelberger, Test of the equivalence principle using a rotating torsion balance. Phys. Rev. Lett. **100**(4), 041101 (2008)

A. Silbergleit, J. Conklin, D. Debra, M. Dolphin, G. Keiser, J. Kozaczuk, D. Santiago, M. Salomon, P. Worden, Polhode motion, trapped flux, and the GP-B science data analysis. Space Sci. Rev. (2009). doi:10.1007/s11214-009-9548-z

R.D. Sorkin, Is the cosmological "constant" a nonlocal quantum residue of discreteness of the causal set type? in *Particles, Strings, and Cosmology-PASCOS 2007*, ed. by A. Rajantie, C. Contaldi, P. Dauncey, H. Stoica. American Institute of Physics Conference Series, vol. 957 (AIP, New York, 2007), pp. 142–153

I.H. Stairs, Testing general relativity with pulsar timing. Living Rev. Relativ. **6**(5) (2003)

P. Touboul, R. Chhun, M. Rodrigues, B. Foulon, E. Guiu, MICROSCOPE, equivalence principle test with a micro-satellite, in *36th COSPAR Scientific Assembly*. COSPAR, Plenary Meeting, vol. 36 (2006), p. 3585

G.E. Volovik, *The Universe in a Helium Droplet* (Oxford University Press, Oxford, 2003)

Space Sci Rev (2009) 148: 25–36
DOI 10.1007/s11214-009-9486-9

# Wolfgang Pauli and Modern Physics

**N. Straumann**

Received: 13 November 2008 / Accepted: 31 December 2009 / Published online: 23 January 2009
© Springer Science+Business Media B.V. 2009

**Abstract** In this written version of a pre-dinner-speech at the workshop "The Nature of Gravity" at ISSI I illustrate Pauli's science primarily with material that has not formally been published by him, but was communicated in detailed letters to eminent colleagues and friends.

**Keywords** Relativity · Quantum theory · Gauge theory · Kaluza-Klein theory

## 1 Introduction

Wolfgang Pauli was one of the most influential figures in twentieth-century science. In the foreword of the memorial volume to Pauli, edited by Markus Fierz and Victor Weisskopf (Fierz and Weisskopf 1960)—two former assistants of Pauli—Niels Bohr wrote about Pauli: "At the same time as the anecdotes around his personality grew into a veritable legend, he more and more became the very conscience of the community of theoretical physicists." There are few fields of physics on which Pauli's ideas have not left a significant imprint. From Pauli's enormous correspondence, edited by Karl von Meyenn (von Meyenn 1979–2005), and his studies in historical, epistemological and psychological questions, it becomes obvious that his searching mind embraced all aspects of human endeavor.

I knew Pauli only as a student. Beside attending his main courses, I saw him in action in the joint Theoretical Physics Seminar of ETH and the University, and in our general Physics Colloquium. Although it was a bit too early for me, I also visited some specialized lectures. In addition, I vividly remember a few public talks, like the famous one "On the earlier and more recent History of the Neutrino" that was given by Pauli immediately after the discovery of parity violation (Pauli 1994). Therefore, I can only talk about Pauli's science. Pauli was obviously a difficult and complex personality. Markus Fierz, who knew and understood him particularly well, once said in a talk: "Whoever knew him also felt that in this man the opposites of heavenly light and archaic darkness were having a tremendous impact."

N. Straumann (✉)
Institute for Theoretical Physics, University of Zurich, Zurich, Switzerland
e-mail: norbert.straumann@gmail.com

And in a letter Markus Fierz wrote to me that a true biography would have to be written by a physicist with poetic gifts. In certain circles there is now a lot of interest in Pauli's special relationship with the psychiatrist Carl Gustav Jung. Only few physicist colleagues knew about this and corresponded with him on Jungian ideas and psychology in general. Pauli attached great importance to the analysis of his dreams and wrote them down in great number. Hundreds of pages of Pauli's notes are not yet published. There are scholars who are convinced that Pauli's thoughts about psychology are important. Others regard all this as mystical mumbo-jumbo. At any rate, Jung successfully helped Pauli to overcome his life crisis after his first marriage had broken up in 1930.

It would be hopeless and pointless to give an overview of Pauli's most important scientific contributions, especially since Charles Enz, the last of a prestigious chain of Pauli assistants, has published a very complete scientific biography (Enz 2002). Much of what Pauli has achieved has become an integral part of physics.[1] You are all aware that he was one of the founders of quantum electrodynamics and quantum field theory in general, and he is, of course, the father of the neutrino. After some biographical remarks, and a sketch of his early work, I will select some important material that appeared only in letters.[2]

## 2 A Brief Biography

Let me begin with a few biographical remarks. Pauli was born in 1900, the year of Planck's great discovery. During the high school years Wolfgang developed into an infant prodigy familiar with the mathematics and physics of his day.

Pauli's scientific career started when he went to Munich in autumn 1918 to study theoretical physics with Arnold Sommerfeld, who had created a "nursery of theoretical physics". Just before he left Vienna on 22 September he had submitted his first published paper, devoted to the energy components of the gravitational field in general relativity (Pauli 1919a). As a 19-year-old student he then wrote two papers (Pauli 1919b, 1919c) about the recent brilliant unification attempt of Hermann Weyl (which can be considered in many ways as the origin of modern gauge theories). In one of them he computed the perihelion motion of Mercury and the light deflection for a field action which was then preferred by Weyl. From these first papers it becomes obvious that Pauli mastered the new field completely. Hermann Weyl was astonished. Already on 10 May, 1919, he wrote to Pauli from Zürich: "I am extremely pleased to be able to welcome you as a collaborator. However, it is almost inconceivable to me how you could possibly have succeeded at so young an age to get hold of all the means of knowledge and to acquire the liberty of thought that is needed to assimilate the theory of relativity."

Sommerfeld immediately recognized the extraordinary talent of Pauli and asked him to write a chapter on relativity in *Encyklopädie der mathematischen Wissenschaften*. Pauli was in his third term when he began to write this article. Within less than one year he finished this demanding job, beside his other studies at the university. With this article (Pauli

[1] For a review, see Straumann (2000).

[2] Pauli's Scientific Correspondence, admirably edited by Karl von Meyenn, is a source of wonderful insights. (Hopefully, the letters written in German will one day be translated into English.) In an obituary for Pauli his former assistant Rudolf Peierls wrote in 1960: "It would be impossible to list all the ideas, constructive or critical, by which he influenced the work of pupils and colleagues in innumerable letters. (...) In these letters, as well as in conversations, he would often discuss conjectures and intuitive judgements, which went far beyond anything he would regard as worthy of publication."

1921, 1981) of 237 pages and almost 400 digested references Pauli established himself as a scientist of rare depth and surpassing synthetic and critical abilities. Einstein's reaction was very positive: "One wonders what to admire most, the psychological understanding for the development of ideas, the sureness of mathematical deduction, the profound physical insight, the capacity for lucid, systematic presentation, the knowledge of the literature, the complete treatment of the subject matter or the sureness of critical appraisal."

Pauli studied at the University of Munich for six semesters. At the time when his Encyclopedia article appeared, he obtained his doctorate with a dissertation on the hydrogen molecule ion $H_2^+$ in the old Bohr-Sommerfeld theory. In it the limitations of the old quantum theory showed up. About the faculties of the young Pauli Lise Meitner wrote to Pauli's widow Franca on 22 June 1959: "I often thought of and also have told it that in the fall of 1921 I have met Sommerfeld in Lund, and that he told me he had such a gifted student that the latter could not learn anything any more from him, but because of the university laws valid in Germany he had to sit through (*absitzen*) 6 semesters in order to make his doctorate. Therefore he, Sommerfeld had set his student on an encyclopedia article (...)." (From Enz 2002, p. 25.)

In the winter semester of 1921/22 Pauli was Max Born's assistant in Göttingen. During this time the two collaborated on the systematic application of astronomical perturbation theory to atomic physics. Already on 29 November, 1921, Born wrote to Einstein: *"Little Pauli is very stimulating: I will never have again such a good assistant."* Well, Pauli's successor was Werner Heisenberg.

## 3 Discovery of the Exclusion Principle

Pauli's next stages were in Hamburg and Copenhagen. His work during these crucial years culminated with the proposal of his exclusion principle in December 1924. This was Pauli's most important contribution to physics, for which he received a belated Nobel Prize in 1945.

The discovery story begins in fall 1922 in Copenhagen when Pauli began to concentrate his efforts on the problem of the anomalous Zeeman effect. He later recalled: 'A colleague who met me strolling rather aimlessly in the beautiful streets of Copenhagen said to me in a friendly manner, "You look very unhappy"; whereupon I answered fiercely, "How can one look happy when he is thinking about the anomalous Zeeman effect?"'.

In a Princeton address in 1946 (Pauli 1946), Pauli tells us how he felt about the anomalous Zeeman effect in his early days:

> The anomalous type of splitting was on the one hand especially fruitful because it exhibited beautiful and simple laws, but on the other hand it was hardly understandable, since very general assumptions concerning the electron, using classical theory as well as quantum theory, always led to a simple triplet. A closer investigation of this problem left me with the feeling that it was even more unapproachable (...). I could not find a satisfactory solution at that time, but succeeded, however, in generalizing Landé's analysis for the simpler case (in many respects) of very strong magnetic fields. This early work was of decisive importance for the finding of the exclusion principle.

This is not the place to even only sketch how Pauli arrived at his exclusion principle.[3] At the time—before the advent of the new quantum mechanics—it was not at all on the horizon,

[3]For a detailed description, see, e.g., Straumann (2007), and references therein.

because of two basic difficulties: (1) There were no general rules to translate a classical mechanical model into a coherent quantum theory, and (2) the spin degree of freedom was unknown. It is very impressive indeed how Pauli arrived at his principle on the basis of the fragile Bohr-Sommerfeld theory and the known spectroscopic material.

Initially Pauli was not sure to what extent his exclusion principle would hold good. In a letter to Bohr of 12 December 1924 Pauli writes "The conception, from which I start, is certainly nonsense. (...) However, I believe that what I am doing here is no greater nonsense than the hitherto existing interpretation of the complex structure. My nonsense is conjugate to the hitherto customary one." The exclusion principle was not immediately accepted, although it explained many facts of atomic physics. A few days after the letter to Bohr, Heisenberg wrote to Pauli on a postcard: "Today I have read your new work, and it is certain that I am the one who *rejoices most* about it, not only because you push the swindle to an unimagined, giddy height (by introducing *individual* electrons with 4 degrees of freedom) and thereby have broken all hitherto existing records of which you have insulted me. (...)."

At the end of his final paper (Pauli 1925) on the way to the exclusion principle, Pauli expresses the hope that a deeper understanding of quantum mechanics might enable us to derive the exclusion principle from more fundamental hypothesis. To some extent this hope was fulfilled in the framework of quantum field theory. Pauli's much later paper from 1940 (Pauli 1940) on the spin-statistics connection ends with:

> In conclusion we wish to state, that according to our opinion the connection between spin and statistics is one of the most important applications of the special theory of relativity.

For the letters of Pauli on the exclusion principle, and the reactions of his influential colleagues, I refer to Vol. I of the *Pauli Correspondence*, edited by Karl von Meyenn (von Meyenn 1979–2005). Some passages are translated into English in the scientific biography by Charles Enz (2002).

### 3.1 Some Side Remarks

Let me end this brief account with some remarks about Pauli's fruitful Hamburg time. I begin with recollections of Otto Stern from a recorded interview with Res Jost—one of my most important teachers and a later close colleague—in Zurich that took place on December 2, 1961. During his Hamburg time, Pauli collaborated closely with Stern, who in 1922 had become professor for physical chemistry at the University of Hamburg. Stern said about Pauli: "But, of course, it was very nice with Pauli for, although he was thus highly learned, one could all the same really discuss physics with him. And ... you know, he was not allowed to enter our laboratory, because of the Pauli effect. Don't you know the famous Pauli effect? Jost: I know it all right, but I didn't know that this led to such consequences. Stern: Yet, now, as I said, we always went eating together, he always fetched me. But he did not enter, instead he only knocked, and I then came to the door and said I'm coming. Oh yes, we were very superstitious at the time. Jost: Did something ever happen? Stern: Alas, many things did happen. The number of Pauli effects, the guaranteed (verbürgten) Pauli effects, is enormously large." (Translation from Enz 2002, p. 149.)

In his obituary for Stern Rabi wrote: "Some of Pauli's great theoretical contributions came from Stern's suggestions, or rather questions; for example, the theory of magnetism of free electrons in metals." From Charly Enz and Armin Thellung—Pauli's last two assistants—I have learned that Pauli has also discussed the question of zero point energies extensively with Stern during his Hamburg time, before the advent of the new quantum mechanics. The following remarks may be of some interest since they are related to things discussed at our conference here.

As background I recall that Planck had introduced the zero-point energy with somewhat strange arguments in 1911. The physical role of the zero-point energy was much discussed in the early years of quantum theory. There was, for instance, a paper by Einstein and Stern in 1913 (Einstein 1987, Vol. 4, Doc. 11; see also the Editorial Note, p. 270 ff) that aroused widespread interest. In this two arguments in favor of the zero-point energy were given. The first had to do with the specific heat of rotating (diatomic) molecules. The authors developed an approximate theory of the energy of rotating molecules and came to the conclusion that the resulting specific heat agreed much better with recent experimental results by Arnold Eucken, if they included the zero-point energy. The second argument was based on a new derivation of Planck's radiation formula. In both arguments Einstein and Stern made a number of problematic assumptions, and in fall 1913 Einstein retracted their results. At the second Solvay Congress in late October 1913 Einstein said that he no longer believed in the zero-point energy, and in a letter to Ehrenfest (Einstein 1987, Vol. 5, Doc. 481) he wrote that the zero-point energy was "dead as a doornail".

In Hamburg Stern had calculated, but never published, the vapor pressure difference between the isotopes 20 and 22 of Neon (using Debye theory for the solid phase). He came to the conclusion that without zero-point energy this difference would be large enough for easy separation of the isotopes, which is not the case in reality. These considerations penetrated into Pauli's lectures on statistical mechanics (Pauli 1973) (which I attended). The theme was taken up in an article by Enz and Thellung (1960). This was originally written as a birthday gift for Pauli, but because of Pauli's early death, appeared in a memorial volume of *Helv. Phys. Acta*.

From Pauli's discussions with Enz and Thellung we know that Pauli estimated the influence of the zero-point energy of the radiation field—cut off at the classical electron radius—on the radius of the universe, and came to the conclusion that it "could not even reach to the moon".

When, as a student, I heard about this, I checked Pauli's unpublished[4] remark by doing the following little calculation (which Pauli must have done):

In units with $\hbar = c = 1$ the vacuum energy density of the radiation field is

$$\langle \rho \rangle_{vac} = \frac{8\pi}{(2\pi)^3} \int_0^{\omega_{\max}} \frac{\omega}{2} \omega^2 d\omega = \frac{1}{8\pi^2} \omega_{\max}^4,$$

with

$$\omega_{\max} = \frac{2\pi}{\lambda_{\max}} = \frac{2\pi m_e}{\alpha}.$$

The corresponding radius of the Einstein universe in Equation (2) would then be ($M_{pl} \equiv 1/\sqrt{G}$)

$$a = \frac{\alpha^2}{(2\pi)^{\frac{2}{3}}} \frac{M_{pl}}{m_e} \frac{1}{m_e} \sim 31 \text{ km}.$$

This is indeed less than the distance to the moon. (It would be more consistent to use the curvature radius of the static de Sitter solution; the result is the same, up to the factor $\sqrt{3/2}$.)

---

[4] A trace of this is in Pauli's Handbuch article (Pauli 1933) on wave mechanics in the section where he discusses the meaning of the zero-point energy of the quantized radiation field. The last sentence of a brief paragraph ends with "(...) and, as is evident from experience, neither does it produce any gravitational field".

Our present estimates of the vacuum energy, that possibly is responsible for an accelerated expansion of the universe, are not much better.

### 3.2 Exclusion Principle and the New Quantum Mechanics

On August 26, 1926, Dirac's paper containing the Fermi-Dirac distribution was communicated by R. Fowler to the Royal Society. This work was the basis of Fowler's *theory of white dwarfs*. I find it remarkable that the quantum statistics of identical spin-1/2 particles found its first application in astrophysics. Pauli's exclusion principle was independently applied to *statistical thermodynamics* by Fermi.[5] In the same year 1926, Pauli simplified Fermi's calculations, introducing the grand canonical ensemble into quantum statistics. As an application he studied the behavior of a gas in a magnetic field (paramagnetism).

Heisenberg and Dirac were the first who interpreted the exclusion principle in the context of Schrödinger's wave mechanics for systems of more than one particle. In these papers it was not yet clear how the spin had to be described in wave mechanics. (Heisenberg speaks of spin coordinates, but he does not say clearly what he means by this.) The definite formulation was soon provided by Pauli in a beautiful paper (Pauli 1927), in which he introduced his famous *spin matrices* and two-component spinor wave functions.

At this point the foundations of non-relativistic quantum mechanics had been completed in definite form. For a lively discussion of the role of the exclusion principle in physics and chemistry from this foundational period, I refer to Ehrenfest's opening laudation (Ehrenfest 1931) when Pauli received the Lorentz medal in 1931. This concluded with the words: "You must admit, Pauli, that if you would only partially repeal your prohibitions, you could relieve many of our practical worries, for example the traffic problem on our streets." According to Ehrenfest's assistant Casimir who was in the audience, Ehrenfest improvised something like this: "and you might also considerably reduce the expenditure for a beautiful, new, formal black suit" (quoted in Enz 2002, p. 258).

These remarks indicate the role of the exclusion principle for the stability of matter in bulk. A lot of insight and results on this central issue, both for ordinary matter (like stones) and self-gravitating bodies, have been obtained in more recent times, beginning with the work of Dyson and Lenard in 1967 (Dyson and Lenard 1967; Dyson and Lenard 1968). For further information, I highly recommend the review articles in Lieb's Selecta (Lieb 1991). (For a brief description, see Straumann 2007.)

## 4 Pauli's Discovery of the Relation between Matrix Mechanics and Wave Mechanics (Letter to P. Jordan)

On April 12, 1926 Pauli wrote a very remarkable letter to P. Jordan (von Meyenn 1979–2005, Vol. I, letter 131), just after the first communication of Schrödinger had appeared.

B.L. van der Waerden devoted his talk at the "Dirac conference" in Trieste in 1972 (van der Waerden 1973) almost entirely to this letter. In this Pauli established the connection between wave and matrix mechanics in a "logically irreproachable way, independent of Schrödinger. He never published the contents of this letter, but signed a carbon copy (which

[5] According to Max Born, Pascual Jordan was actually the first who discovered what came to be known as the Fermi-Dirac statistics. Unfortunately, Born, who was editor of the *Zeitschrift für Physik*, put Jordan's paper into his suitcase when he went for half a year to America in December of 1925, and forgot about it. For further details on this, I refer to the interesting article by E.L. Schucking (1999).

is quite unusual) and he kept the letter in a plastic cover until his death" (van der Waerden's words).

I would like to go through this letter, which is also remarkable in other respects. At the same time it gives an impression of the enormous influence Pauli had through his extensive correspondence. Pauli's letters are an integral part of his work and thinking. It is also a wonderful experience to read at least some of them. The letter begins with

> Dear Jordan,
> Many thanks for your last letter and for looking through the proof sheets. Today I want to write neither about my Handbuch-Article nor about multiple quanta; I will rather tell you the results of some considerations of mine connected with Schrödinger's paper 'Quantisierung als Eigenwertproblem' which just appeared in the *Annalen der Physik*. I feel that this paper is to be counted among the most important recent publications. Please read it carefully and with devotion.
> Of course I have at once asked myself how his results are connected with those of the Göttingen Mechanics. I think I have now completely clarified this connection. I have found that the energy values from Schrödinger's approach are always the same as those of the Göttingen Mechanics, and that from Schrödinger's functions $\psi$, which describe the eigenvibrations, one can in a quite simple and general way construct matrices satisfying the equations of the Göttingen Mechanics. Thus at the same time a rather deep connection between the Göttingen Mechanics and the Einstein-de Boglie Radiation Field is established.
> To make this connection as clear as possible, I shall first expose Schrödinger's approach, styled a little differently.

Pauli does not start with Schrödinger's stationary equation of his 'First Communication', whose justification I find, by the way, rather obscure.[6] Pauli first derived what we now call the Klein-Gordon equation.[7] He starts from the relativistically invariant Einstein-de Broglie relations $\mathbf{p} = \hbar\mathbf{k}$, $E = \hbar\omega$, and inserts these into the relativistic mechanical equation

$$E - V = E_{kin} = \sqrt{c^2\mathbf{p}^2 + (mc^2)^2} \Rightarrow \mathbf{p}^2 = \hbar^2\mathbf{k}^2 = \frac{1}{c^2}\left[(E-V)^2 - (mc^2)^2\right].$$

If this is inserted into the stationary wave equation

$$(\Delta + k^2)\psi = 0$$

one obtains the stationary Klein-Gordon equation

$$\left[\Delta + \left(\frac{E-V}{\hbar c}\right)^2 - \left(\frac{mc}{\hbar}\right)^2\right]\psi = 0.$$

(Actually, Pauli first arrives at a time-dependent equation, which is, however, different from the Klein-Gordon equation, except in the free case.) Then Pauli considers the non-relativistic limit and writes: "This equation is given in Schrödinger's paper, and he also shows how it can be derived from a Variational Principle". After a remark about the analogy with the difference between Geometrical Optics and Wave Optics, he says:

[6] A profound justification, based on the mechanical-optical analogy, was given in the Second Communication.

[7] From Schrödinger's research notes we know that he studied this equation before he had the Schrödinger equation, but abandoned it because it gave the wrong fine structure for hydrogen.

"Next comes my own contribution, namely the connection with the Göttingen Mechanics." With the complete set of eigenfunctions of the stationary Schrödinger equation he associates to operators of the Schrödinger theory matrices, and verifies that these "satisfy the equations of the Göttingen Mechanics". Since we are all familiar with this, no further comments are necessary.

The following remark is noteworthy. Pauli treats the wave function without saying as a complex field, while Schrödinger maintained that "$\psi$ is surely fundamentally a real function …". Only in the fourth communication he arrived at the conviction that it is actually complex, but states in the last of his major papers on wave mechanics "that a certain difficulty no doubt still lies in the use of a *complex* wave function." (For more on this, see Straumann 2001.)

During the short time after the First Communication of Schrödinger had appeared, Pauli did more:

> I have calculated the oscillator and rotator according to Schrödinger. Further the Hönl-Kronig-formulae for the intensity of Zeeman components are easy consequences of the properties of the spherical harmonics. Perturbation theory can be carried over completely into the new theory, and the same thing holds for the transformation to principal axes, which in general is necessary if degenerations (multiple eigenvalues) are cancelled by external fields of force. At the moment I am occupying myself with the calculation of transition probabilities in hydrogen from the eigenfunctions calculated by Schrödinger. For the Balmer lines finite rational expressions seem to come out. For the continuous spectrum the situation is more complicated: the exact mathematical formulation is not yet quite clear to me.

On this Pauli was again a bit too late to submit a paper. The one by Schrödinger was submitted on March 18, who wrote a month later to Sommerfeld: "With Pauli I have exchanged a few long letters. He is indeed a phenomenal fellow. How fast he has obtained everything! In a tenth of the time I have been using." [Mit Pauli habe ich ein paar lange Briefe gewechselt. Er ist schon ein phänomenaler Kerl. Wie der wieder alles schnell heraussen gehabt hat! In einem Zehntel der Zeit, die ich dazu gebraucht hab.]

The letter of Pauli to Jordan ends with: "Cordial greetings for you and the other people in Göttingen (especially to Born, in case he is back from America; please show him this letter)."

*Supplementary Remarks.* From a letter of Pauli to Schrödinger late in 1926 it is clear that he independently discovered the gauge invariance. In this letter Pauli begins by saying that at first sight the relativistic wave equation does not only contain the field strengths, but also the absolute values of the 4-potential. However, he adds: "Thanks God this is only apparent", and he gives the formulae for what we call gauge invariance of the relativistic Kein-Gordon equation. Again, Pauli did not publish the content of this letter, because he learned from Schrödinger's answer about a paper Schrödinger had just submitted to the 'Annalen' (two days before Pauli had written his letter). However, Schrödinger says in his paper nothing about gauge invariance.

## 5 On Pauli's Invention of non-Abelian Kaluza-Klein Theory in 1953

There are documents which show that Wolfgang Pauli constructed in 1953 the first consistent generalization of the five-dimensional theory of Kaluza, Klein, Fock and others to a higher dimensional internal space. Because he saw no way to give masses to the gauge bosons, he refrained from publishing his results formally. This is still a largely unknown chapter of the early history of non-Abelian gauge and Kaluza-Klein theories.

Pauli described his detailed attempt of a non-Abelian generalization of Kaluza-Klein theories extensively in some letters to A. Pais, which have been published in Vol. IV, Part II of Pauli's collected letters (Pauli 1999), and also in two seminars in Zürich on November 16 and 23, 1953. The latter have later been written up in Italian by Pauli's pupil P. Gulmanelli (Gulmanelli 1954). An English translation of these notes by P. Minkowski is now available on his home page. By specialization (independence of spinor fields on internal space) Pauli got all important formulae of Yang and Mills, as he later (Feb. 1954) pointed out in a letter to Yang (Pauli 1954), after a talk of Yang in Princeton. Pauli did not publish his study, because he was convinced that "one will always obtain vector mesons with rest mass zero" (Pauli to Pais, 6 Dec., 1953).

## 5.1 The Pauli Letters to Pais

At the Lorentz-Kammerlingh Onnes conference in Leiden (22–27 June 1953) A. Pais talked about an attempt of describing nuclear forces based on isospin symmetry and baryon number conservation. In this contribution he introduced fields, which do not only depend on the spacetime coordinates $x$, but also on the coordinates $\omega$ of an internal isospin space. The isospin group acted, however, globally, i.e., in a spacetime-independent manner.

During the discussion following the talk by Pais, Pauli said:

> I would like to ask in this connection whether the transformation group with constant phases can be amplified in a way analogous to the gauge group for electromagnetic potentials in such a way that the meson-nucleon interaction is connected with the amplified group (...).

Stimulated by this discussion, Pauli worked on the problem, and wrote on July 25, 1953 a long technical letter to Pais (Pauli 1953a), with the motto: "Ad *usum Delfini* only". This letter begins with a personal part in which Pauli says that "the whole note for you is of course written in order to drive you further into the real virgin-country". The note has the interesting title:

> Written down July 22–25 1953, in order to see how it looks. Meson-Nucleon Interaction and Differential Geometry.

In this manuscript, Pauli generalizes the original Kaluza-Klein theory to a six-dimensional space and arrives through dimensional reduction at the essentials of an $SU(2)$ gauge theory. The extra-dimensions form a two-sphere $S^2$ with space–time dependent metrics on which the $SU(2)$ operates in a space–time-dependent manner. Pauli emphasizes that this transformation group "seems to me therefore the *natural generalization of the gauge-group* in case of a two-dimensional spherical surface". He then develops in 'local language' the geometry of what we now call a fibre bundle with a homogeneous space as typical fiber (in this case $SU(2)/U(1)$). Since it is somewhat difficult to understand exactly what Pauli did, we give some details, using more familiar formulations and notations (O'Raifeartaigh and Straumann 2000).

Pauli considers the six-dimensional total space $M \times S^2$, where $S^2$ is the two-sphere on which $SO(3)$ acts in the canonical manner. He distinguishes among the diffeomorphisms (coordinate transformations) those which leave the space–time manifold $M$ pointwise fixed and induce space–time-dependent rotations on $S^2$:

$$(x, y) \rightarrow (x, R(x) \cdot y). \qquad (1)$$

Then Pauli postulates a metric on $M \times S^2$ that is supposed to satisfy three assumptions. These led him to what is now called the non-Abelian Kaluza-Klein ansatz: The metric $\hat{g}$ on the total space is constructed from a space–time metric $g$, the standard metric $\gamma$ on $S^2$, and a Lie-algebra-valued 1-form,

$$A = A^a T_a, \quad A^a = A^a_\mu dx^\mu, \tag{2}$$

on $M$ ($T_a$, $a = 1, 2, 3$, are the standard generators of the Lie algebra of $SO(3)$) as follows: If $K^i_a \partial/\partial y^i$ are the three Killing fields on $S^2$, then

$$\hat{g} = g - \gamma_{ij}[dy^i + K^i_a(y)A^a] \otimes [dy^j + K^j_a(y)A^a]. \tag{3}$$

In particular, the non-diagonal metric components are

$$\hat{g}_{\mu i} = A^a_\mu(x)\gamma_{ij}K^j_a. \tag{4}$$

Pauli does not say that the coefficients of $A^a_\mu$ in (4) are the components of the three independent Killing fields. This is, however, his result, which he formulates in terms of homogeneous coordinates for $S^2$. He determines the transformation behavior of $A^a_\mu$ under the group (1) and finds in matrix notation what he calls "the generalization of the gauge group":

$$A_\mu \to R^{-1} A_\mu R + R^{-1}\partial_\mu R. \tag{5}$$

With the help of $A_\mu$, he defines a covariant derivative, which is used to derive "*field strengths*" by applying a generalized curl to $A_\mu$. This is exactly the field strength that was later introduced by Yang and Mills. To our knowledge, apart from Klein's 1938 paper, it appears here for the first time. Pauli says that "this is the *true* physical field, the analog of the *field strength*" and he formulates what he considers to be his "main result":

> The vanishing of the field strength is necessary and sufficient for the $A^a_\mu(x)$ in the whole space to be transformable to zero.

It is somewhat astonishing that Pauli did not work out the Ricci scalar for $\hat{g}$ as for the Kaluza-Klein theory. One reason may be connected with his remark on the Kaluza-Klein theory in Note 23 of his relativity article (Pauli 1981) concerning the five-dimensional curvature scalar (p. 230):

> There is, however, no justification for the particular choice of the five-dimensional curvature scalar $P$ as integrand of the action integral, from the standpoint of the restricted group of the cylindrical metric (gauge group). The open problem of finding such a justification seems to point to an amplification of the transformation group.

In a second letter (Pauli 1953b), Pauli also studies the dimensionally reduced Dirac equation and arrives at a mass operator that is closely related to the Dirac operator in internal space $(S^2, \gamma)$. The eigenvalues of the latter operator had been determined by him long before (Pauli 1939). Pauli concludes with the statement: "So this leads to some rather unphysical *shadow particles*."

Pauli's main concern was that the gauge bosons had to be massless, as in quantum electrodynamics. He emphasized this mass problem repeatedly, most explicitly in the second letter (Pauli 1953b) to Pais on December 6, 1953, after he had made some new calculations and had given the two seminar lectures in Zurich already mentioned. He adds to the Lagrangian what we now call the Yang-Mills term for the field strengths and says that "one will always obtain vector mesons with rest-mass zero (and the rest-mass if at all finite, will always remain zero by all interactions with nucleons permitting the gauge group).” To this Pauli adds: “One could try to find other meson fields”, and he mentions, in particular, the scalar fields which appear in the dimensional reduction of the higher-dimensional metric. In view of the Higgs mechanism this is an interesting remark.

Pauli learned about the related work of Yang and Mills in late February, 1954, during a stay in Princeton, when Yang was invited by Oppenheimer to return to Princeton and give a seminar on his joint work with Mills. About this seminar Yang reports (Yang 1983): “Soon after my seminar began, when I had written down on the blackboard $(\partial_\mu - i\epsilon B_\mu)\Psi$, Pauli asked: *What is the mass of this field* $B_\mu$*?*, I said we did not know. Then I resumed my presentation, but soon Pauli asked the same question again. I said something to the effect that was a very complicated problem, we had worked on it and had come to no conclusion. I still remember his repartee: ‘That is no sufficient excuse.’ I was so taken aback that I decided, after a few moments’ hesitation to sit down. There was general embarrassment. Finally Oppenheimer said, ‘we should let Frank proceed.’ Then I resumed and Pauli did not ask any more questions during the seminar.” (For more on this encounter, see Yang 1983.)

In a letter to Yang (Pauli 1954) shortly after Yang’s Princeton seminar, Pauli repeats: “But I was and still am disgusted and discouraged of the vector field corresponding to particles with zero rest-mass (I do not take your excuses for it with ’complications’ seriously) and the difficulty with the group due to the distinction of the electromagnetic field remains.” Formally, Pauli had, however, all important equations, as he shows in detail, and he concludes the letter with the sentence: “On the other hand you see, that your equations can easily be generalized to include the $\omega$-space” (the internal space). As already mentioned, the technical details have been written up by Pauli’s pupil P. Gulmanelli (Gulmanelli 1954) and have recently been translated by P. Minkowski from Italian to English.

I hope that my scattered remarks have at least indicated that Wolfgang Pauli was a great man of uncompromising scientific honesty, to whom his own words (Pauli 1984) on Einstein apply equally well: “His life anticipating the future will forever remind us of the ideal—under threat in our time—of spiritual, contemplative man, his thoughts calmly and unswervingly bent on the great problems of the structure of the cosmos.”

**Acknowledgements** I wish to thank the ISSI Institute and the local organizers, Martin Huber and Rudolf Treumann, for the opportunity to attend such an interesting workshop. I also thank Engelbert Schücking for comments.

## References

F.J. Dyson, A. Lenard, Stability of matter, I and II. J. Math. Phys. **8**, 423–434 (1967)
F.J. Dyson, A. Lenard, Stability of matter, I and II. J. Math. Phys. **9**, 698–711 (1968)
P. Ehrenfest, Ansprache zur Verleihung der Lorentzmedaille an Professor Wolfgang Pauli am 31. Oktober 1931, Versl. Akad. Amsterdam 40, 121–126 (1931)
A. Einstein, *The Collected Papers of Albert Einstein*, Vols. 1–10 (Princeton University Press, 1987). See also: http://www.einstein.caltech.edu/
Ch.P. Enz, *No Time to Be Brief: A Scientific Biography of Wolfgang Pauli* (Oxford University Press, New York, 2002)
C.P. Enz, A. Thellung, Helv. Phys. Acta **33**, 839 (1960)
M. Fierz, V. Weisskopf (eds.), *Theoretical Physics in the Twentieth Century: A Memorial Volume to Wolfgang Pauli* (Interscience, New York, 1960)
P. Gulmanelli, *Su una Teoria dello Spin Isotropico, Pubblicazioni della Sezione di Milano dell’istituto Nazionale di Fisica Nucleare* (Casa Editrice Pleion, Milano, 1954)
E.H. Lieb, *The Stability of Matter: From Atoms to Stars, Selecta of Elliott H. Lieb* (Springer, Berlin, 1991)
L. O’Raifeartaigh, N. Straumann, Rev. Mod. Phys. **72**, 1–23 (2000)

W. Pauli Jr, Physikalische Zeitschrift **20**, 25 (1919a)
W. Pauli Jr, Physikalische Zeitschrift **20**, 457 (1919b)
W. Pauli Jr, Verhandlungen der Deutschen Physikalischen Gesellschaft **21**, 742 (1919c)
W. Pauli, in *Encyklopädie der mathematischen Wissenschaften*, vol. 19 (Teubner, Leipzig, 1921), pp. 539–775 new German Edition: Edited and annotated by D. Giulini, including Pauli's own supplementary notes from 1956 (Springer, Berlin, 2000)
W. Pauli, Über den Zusammenhang des Abschlusses der Elektronengruppen im Atom mit der Komplexstruktur der Spektren. Z. Phys. **31**, 765–783 (1925)
W. Pauli, Zur Quantenmechanik des magnetischen Elektrons. Z. Physik **43**, 601–623 (1927)
W. Pauli, *Die allgemeinen Prinzipien der Wellenmechanik, Handbuch der Physik*, vol. XXIV (1933). New edition by N. Straumann (Springer, 1990); see Appendix III, p. 202
W. Pauli, Helv. Phys. Acta **12**, 147 (1939)
W. Pauli, On the connection between spin and statistics. Phys. Rev. **58**, 716 (1940)
W. Pauli, Remarks on the history of the exclusion principle. Science **103**, 213–215 (1946)
Pauli to Pais, Letter [1614] in (Pauli 1999)
Pauli to Pais, Letter [1682] in (Pauli 1999)
Pauli to Yang, Letter [1727] in (Pauli 1999)
W. Pauli, in *Pauli Lectures on Physics*, ed. by C.P. Enz, vol. 4 (MIT, Cambridge, 1973) especially Sect. 20
W. Pauli, *Theory of Relativity* (Dover, New York, 1981)
W. Pauli, Translation from the article Impressionen über Albert Einstein, in Aufsätze und Vorträge über Physik und Erkenntnistheorie (Vieweg, Braunschweig, 1961), p. 81. Reprinted as Physik und Erkenntnistheorie (Vieweg, Braunschweig, 1984)
W. Pauli, in *Writings on Physics and Philosophy*, ed. by Ch.P. Enz, K. von Meyenn (Springer, Berlin, 1994), translated by R. Schlapp
W. Pauli, in (Meyenn 1979–2005), Vol. IV, Part II
E.L. Schucking, Jordan, Pauli, Politics, Brecht, and a variable gravitational constant, Physics Today, October 1999, pp. 26–31
N. Straumann, On Wolfgang Pauli's most important contributions to physics. Opening talk at the symposium: Wolfgang Pauli and Modern Physics, in honor of the 100th anniversary of Wolfgang Pauli's birthday, ETH (Zurich), May 4–6 2000. arXiv:physics/0010003
N. Straumann, Schrödinger's Entdeckung der Wellenmechanik. arXiv:quant-ph/0110097
N. Straumann, The role of the exclusion principle for atoms to stars: a historical account. Int. Rev. Phys. **1**, 184–196 (2007). arXiv:quant-ph/0403199
B.L. van der Waerden, in *The Physicist's Conception of Nature*, ed. by J. Mehra (Reidel, Dortrecht, 1973) p. 315
K. von Meyenn (ed.), *Wolfgang Pauli: Scientific Correspondence with Bohr, Einstein, Heisenberg*, a.O., Vol. I-IV, volume 2,6,11,14,15,17,18 of Sources in the History of Mathematics and Physical Sciences (Springer, Heidelberg and New York, 1979–2005)
C.N. Yang, *Selected papers 1945–1980 with Commentary* (Freeman, San Francisco, 1983), p. 525

Space Sci Rev (2009) 148: 37–52
DOI 10.1007/s11214-009-9537-2

# Gravitomagnetism in Physics and Astrophysics

**Gerhard Schäfer**

Received: 8 May 2009 / Accepted: 12 May 2009 / Published online: 28 May 2009
© Springer Science+Business Media B.V. 2009

**Abstract** Based on general relativity, the article reviews gravitomagnetism in physics and astrophysics. Emphasis is put on observational effects. Accelerated reference frames in flat spacetime are discussed to illuminate the gravitomagnetic field. Compact insight into the dynamics of gravitationally interacting non-spinning and spinning objects is achieved by employing the Hamilton formalism.

**Keywords** General Relativity · Gravitomagnetism · Spinning objects · Hamiltonian formulation

## 1 Introduction

Gravitomagnetism is a phenomenon of relativistic gravity where mass-currents generate gravitational fields. Compared with the Newtonian gravitational field those fields are weak because the velocities of macroscopic bodies are small compared with the speed of light $c$. Though Newtonian gravity contains no gravitomagnetic field, the Coriolis field emerging e.g. in rotating reference frames like on the surface of the Earth resembles the gravitomagnetic field. More precisely, according to Einstein's equivalence principle, which states the general local equivalence between inertial and gravitational forces, Coriolis fields can not be discriminated from gravitomagnetic fields, locally (e.g. Ciufolini and Wheeler 1995).

The most important sources for gravitomagnetic fields are rotating bodies, i.e. spinning bodies, and bodies orbiting about companions. Therefore, in our first section spinning objects will be treated in flat spacetime (Minkowski space) to uncover fundamental aspects of those objects in relativity. The gravitational interaction between the spinning and orbiting objects will then be introduced through general relativity.

The presentation in the article will be based mainly on the Hamiltonian canonical formalism. The great advantage of this setting is the well-known fact that all dynamical evolutions, orbital and rotational, are generated by one single scalar function, the Hamiltonian.

G. Schäfer (✉)
Theoretisch-Physikalisches Institut, Friedrich-Schiller-Universität Jena, Max-Wien-Platz 1, 07743 Jena, Germany
e-mail: G.Schaefer@tpi.uni-jena.de

For isolated systems, the total Hamiltonian is conserved in time and takes part in the global Poincaré algebra.

For differently based presentations see, e.g. Mashhoon (2001), Lämmerzahl and Neugebauer (2001); for some more experimental aspects, Ruggiero and Tartaglia (2002).

## 2 Spinning Objects in Flat Spacetime

In flat spacetime physical phenomena are controlled by the Poincaré group, also called inhomogeneous Lorentz group, which is the symmetry group of Minkowski space (e.g. Schwinger 1998). Even in curved spacetime, in case of isolated systems, the Poincaré group is valid in the flat space-asymptotic regime (e.g. Regge and Teitelboim 1974). The ten generators of the group fulfill the famous Poincaré algebra

$$\{P_i, H\} = \{J_i, H\} = 0, \tag{1}$$

$$\{J_i, P_j\} = \varepsilon_{ijk} P_k, \qquad \{J_i, J_j\} = \varepsilon_{ijk} J_k, \tag{2}$$

$$\{J_i, G_j\} = \varepsilon_{ijk} G_k, \tag{3}$$

$$\{G_i, H\} = P_i, \tag{4}$$

$$\{G_i, P_j\} = \frac{1}{c^2} H \delta_{ij}, \tag{5}$$

$$\{G_i, G_j\} = -\frac{1}{c^2} \varepsilon_{ijk} J_k, \tag{6}$$

where $P_i$, $J_i$, $H$, $G_i$ are the total linear momentum, angular momentum, Hamiltonian, and the center-of-mass vector, respectively. Latin indices are running from 1 to 3, $\epsilon_{ijk} = (i - j)(j - k)(k - i)/2$, and $\{,\}$ denotes the Poisson brackets. The conserved boost vector reads

$$K_i = G_i - t P_i, \qquad dK_i/dt = \partial K_i/\partial t + \{K_i, H\} = -P_i + \{G_i, H\} = 0. \tag{7}$$

'Point' masses with spin have the expressions

$$\mathbf{P} = \sum_a \mathbf{p}_a, \qquad \mathbf{J} = \sum_a (\mathbf{x}_a \times \mathbf{p}_a + \mathbf{S}_a), \tag{8}$$

$$\mathbf{G} = \sum_a \left( p_a^0 \mathbf{x}_a + \frac{1}{(p_a^0 + m_a) c^2} \mathbf{p}_a \times \mathbf{S}_a \right), \tag{9}$$

$$H = c^2 \sum_a p_a^0, \quad \text{with } p_a^0 = \sqrt{m_a^2 + \mathbf{p}_a^2 / c^2}, \tag{10}$$

where $m_a$, $\mathbf{p}_a = (p_{ai})$, $\mathbf{x}_a = (x_a^i)$, and $\mathbf{S}_a = (S_{ai})$ respectively denote the mass, linear momentum, position, and spin of the $a$th particle.

The Poincaré algebra is fulfilled provided the standard commutation relations hold,

$$\begin{aligned} &\{x_a^i, p_{bj}\} = \delta_{ab}\delta_{ij}, \qquad \{x_a^i, x_b^j\} = 0, \qquad \{p_{ai}, p_{bj}\} = 0, \\ &\{S_{ai}, S_{bj}\} = \delta_{ab}\epsilon_{ijk} S_{ak}, \qquad \{x_a^i, S_{bj}\} = 0, \qquad \{p_{ai}, S_{bj}\} = 0. \end{aligned} \tag{11}$$

The definition of a canonical center-of-mass vector $\mathbf{R}$ is by far not an easy task (e.g. Schwinger 1998). Not even for spinless particles the usual definition $\mathbf{X} = \sum_a p_a^0 \mathbf{x}_a c^2/H$ (e.g. Landau and Lifshitz 1985) results in a canonical center-of-mass position vector (e.g. Hanson and Regge 1974).

For a single particle with mass $m$, (canonical) position $\mathbf{x}$, linear momentum $\mathbf{p}$, and spin $\mathbf{S}$, the vector $\hat{\mathbf{x}} = \mathbf{x} + (p^0(p^0+m)c^2)^{-1}\mathbf{p} \times \mathbf{S}$ can be called the center-of-mass (energy) vector, with $p^0 = (m^2 + \mathbf{p}^2/c^2)^{1/2}$, whereas $\mathbf{x}$ is known as the center-of-spin vector (e.g. Fleming 1965). Also a center-of-inertia vector different from both other vectors can be defined. It is the Lorentz boosted center of energy of the rest frame. The definition of the center-of-energy (energy) vector is made in the moving frame and the center-of-spin vector is defined through locality condition, i.e. vanishing Poisson brackets of its components. The reason for the existence of various center vectors is the necessarily finite extension of a spinning object. Its radius of extension orthogonal to its spin direction has to be at least $|\mathbf{S}|/mc$ (e.g. Fleming 1965; Møller 1969: Misner et al. 1973). The various centers imply different definitions for the spin. The canonical spin vector is related with the center-of-spin, the covariant one with the center-of-inertia, and the spin defined with the aid of the center-of-mass (energy) vector is neither canonical nor covariant. The various spin vectors can also be characterized by specific spin supplementary conditions. Evidently, spinning objects are not simple, not even in Minkowski space (e.g. Hanson and Regge 1974).

## 3 Curvilinear Coordinates in Flat Spacetime

The introduction of curvilinear coordinates in the Minkowski space changes the well-known line-element in quasi-cartesian coordinates, $ds^2 = \eta_{\mu\nu}dx^\mu dx^\nu$ with $\eta_{\mu\nu} = \mathrm{diag}(-1, 1, 1, 1)$, into the form

$$ds^2 = g_{\mu\nu}dx^\mu dx^\nu = g_{00}c^2dt^2 + 2g_{0i}cdtdx^i + g_{ij}dx^i dx^j, \tag{12}$$

where $x^0 = ct$. $g_{\mu\nu} = g_{\nu\mu}$ denotes the metric tensor. More details to this section can be found in e.g. the review by Schäfer (2005). A congruence of observers at rest in these coordinates has the four-velocity field $u^\mu$, with $u^\mu u^\nu g_{\mu\nu} = -1$ and $u^\mu = (1/\sqrt{-g_{00}}, 0, 0, 0)$. We may think therefore in terms of observers, or reference frames, if we talk about coordinate systems.

The following two orthogonal decompositions of the line element are most useful,

$$ds^2 = (-g_{00})[-(cdt - A_i dx^i)^2 + (G_{ij} + A_i A_j)dx^i dx^j] \tag{13}$$

and

$$ds^2 = -N^2c^2dt^2 + g_{ij}(dx^i + N^i cdt)(dx^j + N^j cdt), \tag{14}$$

where

$$A_i = -\frac{g_{0i}}{g_{00}}, \qquad G_{ij} = -\frac{g_{ij}}{g_{00}}, \tag{15}$$

$$N^2 = -\frac{1}{g^{00}} = -g_{00} + g_{ij}N^i N^j, \quad N^i = -\frac{g^{0i}}{g^{00}}. \tag{16}$$

$N$ and $N^i$, $(N^i) = \mathbf{N}$, are called lapse and shift functions, respectively. The first decomposition is adapted to an observer congruence at rest. $cdt = A_i dx^i$ just gives the coordinate time difference between two simultaneous (with respect to the observers at rest in the coordinate system) events of coordinate vector separation $dx^i$. The second decomposition is adapted to the $t$ = const slices. $dx^i = -N^i cdt$ describes the velocity $(dx^i/dt)$ of an observer at rest in the $t =$ const slices.

The symmetry of the metric coefficients $g_{\mu\nu}$ implies a "square root", a tetrad (or vierbein) field $e^\mu_\alpha$, $\alpha = (0), (1), (2), (3)$,

$$g^{\mu\nu}(x) = e^\mu_\alpha(x) e^\nu_\beta(x) \eta^{\alpha\beta}, \qquad g_{\mu\nu}(x) e^\mu_\alpha(x) e^\nu_\beta(x) = \eta_{\alpha\beta} \tag{17}$$

with anholonomic coordinates $de^\alpha = e^\alpha_\mu dx^\mu$. $de^a$ is an inexact differential. Below, several other differentials will be inexact too. Evidently, $ds^2 = \eta_{\alpha\beta} de^\alpha de^\beta$. The tetrad field easily allows the formulation of local Lorentz transformations, $e'^\mu_\alpha(x) = L^\beta{}_\alpha(x) e^\mu_\beta(x)$, with $L^\alpha{}_\gamma(x) \eta_{\alpha\beta} L^\beta{}_\delta(x) = \eta_{\gamma\delta}$. In the so-called time gauge, $N = e^{(0)}_0$ and $N^i = -e^{(0)}_0 e^i_{(0)}$ hold, connecting lapse and shift with specific components of the tetrad field. The introduction of tetrad fields is crucial for the implementation of spin, classical or quantum, into gravity, see Sect. 4.2 below.

### 3.1 The Coriolis Field

In a rigidly rotating reference frame in flat spacetime the metric reads (e.g. Møller 1969),

$$ds^2 = -(c^2 - \Omega^2 \varrho^2) dt^2 + 2\Omega \varrho^2 dt d\phi + (d\varrho^2 + \varrho^2 d\phi^2 + dz^2), \tag{18}$$

where the rotation is about the $z$-axis (cylindrical coordinates are used). Obviously, $g_{0\phi} = \frac{\Omega \varrho^2}{c}$ or, in vector notation, $\mathbf{g} = \frac{1}{c} \mathbf{\Omega} \times \mathbf{r}$; $\Omega^2 \varrho^2 = (\mathbf{\Omega} \times \mathbf{r})^2$. The lapse and shift functions read, respectively,

$$N = 1, \qquad \mathbf{N} = -\frac{\mathbf{\Omega} \times \mathbf{r}}{c}. \tag{19}$$

The Newtonian centrifugal and Coriolis potentials respectively have the well-known forms, $-(\mathbf{\Omega} \times \mathbf{r})^2/2$ and $\mathbf{v} \cdot (\mathbf{\Omega} \times \mathbf{r})$.

A particle at rest in Minkowski space has velocity $\mathbf{v}$ in the rotating frame given by $\mathbf{v} = -\mathbf{N}c$. In the rotating frame, the Hamiltonian of a freely moving particle with spin $\mathbf{S}$ reads

$$H = \frac{\mathbf{p}^2}{2m} - \mathbf{\Omega} \cdot \mathbf{J}, \tag{20}$$

where $\mathbf{J} = \mathbf{r} \times \mathbf{p} + \mathbf{S}$ is its total angular momentum. From the equations of motion for spin follows

$$\frac{d\mathbf{S}}{dt} = \{\mathbf{S}, H\} = -\mathbf{\Omega} \times \mathbf{S} \tag{21}$$

(e.g. Landau and Lifschitz 1981, 1985). The precession equation is different from the *Thomas precession* of a spinning particle fixed to the rotating frame. That precession frequency reads

$$\mathbf{\Omega}_{\rm Th} = -\frac{1}{2c^2} (\mathbf{\Omega} \times \mathbf{r}) \times \mathbf{a}, \tag{22}$$

where $\mathbf{a}$ denotes the particle's acceleration (e.g. Misner et al. 1973).

## 4 Einstein Gravity

All equations with general metric coefficients $g_{\mu\nu}$ given in Sect. 3 remain valid in the Einstein gravity, i.e. general relativity. The only difference are the more general metric functions $g_{\mu\nu}$. If only generated from $\eta_{\mu\nu}$ through four coordinate transformations, the Riemann tensor (or curvature tensor), built up from them, stays zero. If the Riemann tensor is different from zero, Einstein gravity is in action. In flat spacetime, the metric coefficients $g_{\mu\nu}$ are just the potentials of the inertial forces, in curved spacetime they also include the gravitational potentials and are much richer in structure. Gravitomagnetism and Coriolis effects rely on the functions $A_i$ or, $N^i$.

### 4.1 Particle Dynamics in Curved Spacetime

As already mentioned in the Introduction, the dynamics of physical objects we present in Hamiltonian form. Introducing the action $W$, for one single object, the action one-form reads

$$dW = -H\,dt + p_i\,dx^i = p_\mu\,dx^\mu, \quad \text{with } -cp_0 = H. \tag{23}$$

$H$ denotes the Hamiltonian if $H = H(p_i, x^i, t)$ is given.

The condition for $dW$ of being an exact differential, and thus $W$ being a function of time $t$ and space variables $x^i$ only, can be written as $\delta dW = d\delta W$, where $\delta$ is a second variation independent from $d$ with $\delta dx^\mu = d\delta x^\mu$ (in the calculus of exterior differential forms, $(\delta d - d\delta)W = 0$ gets replaced with $ddW = 0$; without loss of generality, exact and closed differential forms are considered to coincide). This results in the Hamilton equations of motion in standard form (e.g. Hagihara 1970),

$$\frac{dp_i}{dt} = -\frac{\partial H}{\partial x^i} = \{p_i, H\}, \qquad \frac{dx^i}{dt} = \frac{\partial H}{\partial p_i} = \{x^i, H\}, \qquad \frac{dH}{dt} = \frac{\partial H}{\partial t}. \tag{24}$$

Then also $p_\mu = \partial_\mu W$ ($\partial_\mu \equiv \partial/\partial x^\mu$) is valid.

The coupling to gravity is supplied by the constraint equation

$$g^{\mu\nu} p_\mu p_\nu = -m^2c^2, \quad \text{where } g^{\mu\lambda} g_{\lambda\nu} = \delta^\mu_\nu, \tag{25}$$

which can also be called *equation of state*. Hereof $H = H(p_i, x^i, t)$ follows in the form

$$H = Nc\sqrt{m^2c^2 + \gamma^{ij} p_i p_j} - N^i c p_i, \qquad \gamma^{il} g_{lj} = \delta^i_j \; (= \delta_{ij}) \tag{26}$$

or, expanded in powers of $1/c$,

$$H = mc^2 + \frac{1}{2m}(\mathbf{p} - m\mathbf{N}c)^2 + m\Phi + \cdots, \qquad N = 1 + \frac{\Phi}{c^2} + \cdots. \tag{27}$$

In the weak-field slow-motion approximation, particles couple to gravity in exactly the same way as they couple to the electromagnetic field. The metric functions $N, N^i, \gamma^{ij}$ appearing in (26) have to be calculated with the aid of the Einstein field equations.

The exact coupling of a charged particle with electric charge $e$ to the *electromagnetic field* is quite different from the gravitational coupling. Here the potential one-form $dA = A_\mu\,dx^\mu$, which is an inexact differential (only gauge transformations are exact differentials), plays a central rôle. The coupling reads

$$dW = dW_k + \frac{e}{c}dA = \left(\pi_\mu + \frac{e}{c}A_\mu\right)dx^\mu, \quad \text{or} \quad \pi_\mu + \frac{e}{c}A_\mu \equiv p_\mu = \partial_\mu W, \tag{28}$$

where $dW_k$ is the action one-form of the kinetic momentum $\pi_\mu$. Its constant four-dimensional length $\pi_\mu \pi^\mu = -m^2c^2$ results in

$$g^{\mu\nu}\left(p_\mu - \frac{e}{c}A_\mu\right)\left(p_\nu - \frac{e}{c}A_\nu\right) = -m^2c^2 \tag{29}$$

and the Hamiltonian turns into

$$H = Nc\sqrt{m^2c^2 + \gamma^{ij}\left(p_i - \frac{e}{c}A_i\right)\left(p_j - \frac{e}{c}A_j\right)} - N^i cp_i - eA_0. \tag{30}$$

In the coupling to gravity an important, still implicit aspect should be pointed out. Under the condition of fulfillment of the Hamilton equations of motion the variation of the action one-form $\delta W$ along the particle trajectory is zero, i.e.

$$0 = d\delta W = dp_\mu \delta x^\mu + p_\mu \, d\delta x^\mu, \tag{31}$$

indicating a *conservation law*. In view of the symmetry property $d\delta x^\mu = \delta dx^\mu$ we may write

$$d\delta x^\mu = -\Gamma^\mu_{\rho\sigma} dx^\rho \delta x^\sigma \tag{32}$$

with $\Gamma^\mu_{\rho\sigma} = \Gamma^\mu_{\sigma\rho}$, later, the Christoffel symbols of second kind. Hereof the equation of *parallel transport* for $p_\mu$ along the motion increment $dx^\mu$ follows,

$$dp_\lambda - \Gamma^\mu_{\rho\lambda} p_\mu dx^\rho = 0, \quad \text{or} \quad \frac{dp_\lambda}{dt} - \Gamma^\mu_{\rho\lambda} p_\mu \frac{dx^\rho}{dt} = 0. \tag{33}$$

The comparison with the Hamilton equations of motion yields

$$\Gamma^\mu_{\rho\sigma} = \frac{1}{2} g^{\mu\lambda}(\partial_\rho g_{\lambda\sigma} + \partial_\sigma g_{\lambda\rho} - \partial_\lambda g_{\rho\sigma}). \tag{34}$$

The fundamental relation between the Christoffel symbols and the metric coefficients also results from the relations

$$\delta dx^\mu + \Gamma^\mu_{\rho\sigma} dx^\rho \delta x^\sigma = 0, \qquad d\delta x^\mu + \Gamma^\mu_{\rho\sigma} \delta x^\rho dx^\sigma = 0, \tag{35}$$

indicating the commutativity of the parallel transport of $dx^\mu$ along $\delta x^\mu$ and of $\delta x^\mu$ along $dx^\mu$, through the conditions of equal length of the vectors $dx^\mu$ and $dx^\mu + \delta dx^\mu$ on the one side and $\delta x^\mu$ and $\delta x^\mu + d\delta x^\mu$ on the other. Closing up infinitesimal *parallelograms* are typical for spaces without torsion (e.g. DeWitt 1965). Sufficient for vanishing torsion is the symmetry of the Christoffel symbols.

### 4.2 Classical Spin and Gravity

If particles carry spin, the action one-form turns into

$$dW = -H\,dt + p_i\,dx^i + \frac{1}{2}S_{ab}\,d\theta_{ab}, \tag{36}$$

with the inexact differential $d\theta_{ab} = \lambda_{ca} d\lambda_{cb} = -d\theta_{ba}$ and $H = H(p_i, x^i, S_{ab}, t)$; $a, b, c =$ (1), (2), (3). The matrix $\lambda_{ab}$ represents the angle degrees of freedom of the spin, $\lambda_{ac}\lambda_{bc} =$

$\delta_{ab} = \lambda_{ca}\lambda_{cb}$. The antisymmetric spin tensor has constant three-dimensional length, $S_{ab}S_{ab} = 2s^2$. The action one-form, (36), is given already in reduced form, i.e. only the canonical degrees of freedom are involved. The original one-form, with *linear-in-spin* coupling, reads

$$dW = \left(\pi_\mu + \frac{1}{2}\tilde{S}_{\alpha\beta}\omega_\mu^{\alpha\beta}\right)d\tilde{x}^\mu + \frac{1}{2}\tilde{S}_{\alpha\beta}\, d\tilde{\theta}^{\alpha\beta}, \tag{37}$$

where $d\tilde{\theta}^{\alpha\beta} = \tilde{\lambda}_\gamma{}^\alpha d\tilde{\lambda}^{\gamma\beta}$. $\tilde{S}_{\alpha\beta}$ denotes a four-dimensional covariant antisymmetric spin tensor with constant four-dimensional length $\tilde{S}_{\alpha\beta}\tilde{S}^{\alpha\beta} = 2s^2$ and spin-supplementary constraint $\pi_\mu \tilde{S}^{\mu\nu} = 0$; again $\pi_\mu \pi^\mu = -m^2c^2$. $\tilde{x}^\mu$ and $\tilde{\lambda}^{\beta\alpha}$ respectively are the spin-related position and angle-type variables (cf. Sect. 2), and the Ricci rotation coefficients $\omega_\mu^{\alpha\beta}$ are given by

$$\omega_\mu^{\alpha\beta} = e^{\alpha\rho}e^\beta_\sigma \Gamma^\sigma_{\mu\rho} - e^{\alpha\nu}\partial_\mu e^\beta_\nu. \tag{38}$$

Our action ansatz with constraints is analogous to a scheme investigated and developed by Frenkel (1926) for the coupling of the magnetic moment, or spin, to the electromagnetic field apart from his Routh function setting (Lagrangian for center-of-mass motion, Hamiltonian for spin); see also Yee and Bander (1993).

Introducing the (precession) frequency $\Omega_{ab}dt = d\theta_{ab}$ the equations of motion for the spin read

$$\Omega_{ab} = \frac{\partial H}{\partial S_{ab}} \tag{39}$$

and

$$\frac{dS_{ab}}{dt} = S_{ac}\Omega_{cb} - \Omega_{ac}S_{cb} = \{S_{ab}, H\}, \tag{40}$$

where use has been made of (cf. Hanson and Regge 1974)

$$\delta d\theta_{ab} - d\delta\theta_{ab} = \delta\theta_{ac}d\theta_{cb} - d\theta_{ac}\delta\theta_{cb}. \tag{41}$$

The transition to the spin and the spin-precession frequency vectors $S_a$ and $\Omega_a$, respectively, is achieved by $2S_a = \epsilon_{abc}S_{bc}$ and $2\Omega_a = \epsilon_{abc}\Omega_{bc}$. Later, $S_a \equiv S_{(i)} = S_i$ and $(S_i) = \mathbf{S}$ will be used because of $e_{ik}e_{kj} = g_{ij}$, with $e_{ij} = e_{ji}$ (symmetric square root of the three-metric), and $e_{ij} = e_{(i)j}$ (Steinhoff et al. 2008a).

### 4.3 Clock Effect and Gravitomagnetism

The influence of gravitomagnetism on the motion of particles is most easily seen with the representation via the particle's proper time $\tau$ or, action $W = -mc^2\tau$. Integration along the particle trajectory gives ($d\tau^2 = -ds^2/c^2$),

$$\tau(t) - \tau(t_0) = \int_{t_0}^t \sqrt{N^2 - g_{ij}\left(N^i + \frac{v^i}{c}\right)\left(N^j + \frac{v^j}{c}\right)}\, dt, \quad v^i = \frac{dx^i}{dt}. \tag{42}$$

This difference in proper time is known as clock effect. In the weak-field slow-motion approximation the clock effect reads

$$\tau(t) - \tau(t_0) = \int_{t_0}^t N dt - \frac{1}{c}\int_{x(t_0)}^{x(t)} \delta_{ij}N^i dx^j. \tag{43}$$

This equation shows the leading order influence of the gravitomagnetic field on the particle's proper time. The transition to quasi-classical quantum aspects is supplied by the matter-wave phase $W/\hbar = -mc^2\tau/\hbar$; see also the next section.

### 4.4 Phase Shift and Gravitomagnetism

This section investigates the impact of the gravitational field onto the phase of a light ray. Let us denote the phase by $\psi$, then the angular frequency $\omega$ reads $\frac{\partial\psi}{\partial t} = -\omega$, and the change of the phase between two spacetime points $(t, x^i)$ and $(t_0, x_0^i)$ is given by

$$\psi(t, x^i) - \psi(t_0, x_0^i) = -\int_{t_0}^{t} \omega dt + \int_{x_0}^{x} \frac{\omega}{c}\Big(\sqrt{(G_{ij} + A_i A_j) dx^i dx^j} + A_i dx^i\Big) \tag{44}$$

provided the eikonal equation

$$g^{\mu\nu}\frac{\partial\psi}{\partial x^\mu}\frac{\partial\psi}{\partial x^\nu} = 0 \tag{45}$$

holds. Along the path of the light ray the equations $\psi = \text{const}$ and $ds^2 = 0$ are valid, i.e.

$$\int_{t_0}^{t} \omega dt = \int_{t_0}^{t} \frac{\omega}{c}\Big(\sqrt{(G_{ij} + A_i A_j) v^i v^j} + A_i v^i\Big) dt, \quad v^i = \frac{dx^i}{dt}. \tag{46}$$

Hereof with $\omega = \text{const}$ (stationary field) the usual *Sagnac effect* of flat spacetime follows from the term linear in $A_i$ (only this term depends on the sign of the velocity). Generalized to curved spacetime, the phase shift reads, subtracting from each other the phases connected with two different but closing curves,

$$\Delta\psi = -\omega\Delta t, \quad \text{with } \Delta t = \frac{1}{c}\oint A_i dx^i, \tag{47}$$

where the integration is taken over the closed curve in question. To leading order $A_i = -N^i$ holds. The comparison with the previous section, (43), clearly shows the Sagnac effect for matter waves. Though ring lasers are high precision tools for measuring the Sagnac effect (e.g. Stedman 1997), its gravitomagnetic aspect is still not accessible to them (e.g. Klügel et al. 2007). The HYPER mission aims at measuring the gravitomagnetic field in space using atomic interferometers (Jentsch et al. 2004).

## 5 The Gravitomagnetic Field of Rotating Objects

The exterior gravitational field of a rotating body to first post-Newtonian approximation is given by (e.g. Misner et al. 1973),

$$ds^2 = -\left(c^2 - \frac{2GM}{r} + \frac{2G^2M^2}{c^2r^2}\right)dt^2 - \frac{4GS_z\varrho^2}{c^2r^3}dt d\phi + \left(1 + \frac{2GM}{c^2r}\right)(d\varrho^2 + \varrho^2 d\phi^2 + dz^2). \tag{48}$$

$M$ is the mass of the gravitating body and $\mathbf{S} = (0, 0, S_z)$ its proper angular momentum (spin). Obviously, $g_{0\phi} = -\frac{2GS_z\varrho^2}{c^3r^3}$ or, in vector notation, $\mathbf{g} = -\frac{2G}{c^3r^3}\mathbf{S}\times\mathbf{r}$. The shift function reads

$$\mathbf{N}c = \frac{2G}{c^2r^3}\mathbf{S} \times \mathbf{r} \tag{49}$$

and a particle moving with speed $\mathbf{v} = -\mathbf{N}c$ stays at rest with respect to the $t$ = const slices, i.e. it follows the drag of the gravitomagnetic field ("frame dragging").

The gravitomagnetic field strength $\mathbf{H}$ is given by

$$\mathbf{H} = \nabla \times \mathbf{N}c = \frac{2G}{c^2r^3}\left(\mathbf{S} - 3\frac{(\mathbf{r}\cdot\mathbf{S})\mathbf{r}}{r^2}\right). \tag{50}$$

In the space outside the matter, the field strength allows the additional representation

$$\mathbf{H} = \nabla\chi, \quad \chi = \frac{2G}{c^2r^3}\mathbf{r}\cdot\mathbf{S}. \tag{51}$$

As orthogonal trajectories of the $\chi$ = const lines, the gravitomagnetic field lines are easily obtained, reading

$$\frac{r_{\max} - r}{r_{\max}} = \frac{(\mathbf{r}\cdot\mathbf{S})^2}{r^2S^2}, \tag{52}$$

where $r_{\max}$ is the maximum value of $r$ for a given field line (Schäfer 2004).

The precession frequency of the spin of a freely moving particle results in the form

$$\boldsymbol{\Omega}_{\mathrm{S}} = -\frac{1}{2}\mathbf{H}. \tag{53}$$

## 6 Precession Effects in Gravitating Many-Body Systems

The Hamiltonian of the leading-order spin-orbit coupling in a binary system, in the center-of-mass frame, takes the form (e.g. Barker and O'Connell 1979),

$$H_{\mathrm{SO}} = \frac{2G}{c^2R^3}(\mathbf{S}\cdot\mathbf{L}) + \frac{3GM_1M_2}{2c^2R^3}(\mathbf{b}\cdot\mathbf{L}), \tag{54}$$

where $\mathbf{S} = \mathbf{S}_1 + \mathbf{S}_2$, $\mathbf{b} = \frac{\mathbf{S}_1}{M_1^2} + \frac{\mathbf{S}_2}{M_2^2}$, $\mathbf{L} = \mathbf{R}\times\mathbf{P}$, $\mathbf{P} \equiv \mathbf{p}_1 = -\mathbf{p}_2$. $M_1$ and $M_2$ denote the masses of the bodies and $R = |\mathbf{R}|$ their relative euclidean separation.

The leading-order spin(1)-spin(2) coupling Hamiltonian reads,

$$H_{\mathrm{S}_1\mathrm{S}_2} = \frac{G}{c^2R^3}\left(\frac{3(\mathbf{S}_1\cdot\mathbf{R})(\mathbf{S}_2\cdot\mathbf{R})}{R^2} - (\mathbf{S}_1\cdot\mathbf{S}_2)\right). \tag{55}$$

Binary spinning black holes, to the first post-Newtonian order in their relative interaction, are described by the Hamiltonian (e.g. Damour 2001),

$$H_{\mathrm{SS}} = \frac{GM_1M_2}{2c^2R^3}\left(\frac{3(\mathbf{a}\cdot\mathbf{R})(\mathbf{a}\cdot\mathbf{R})}{R^2} - (\mathbf{a}\cdot\mathbf{a})\right), \quad \mathbf{a} = \frac{\mathbf{S}_1}{M_1} + \frac{\mathbf{S}_2}{M_2}. \tag{56}$$

The coefficients in the spin-squared terms are typical for black holes. The corresponding coefficients for neutron stars are matter equation-of-state dependent.

6.1 Lense-Thirring Effect

The Lense-Thirring effect dates back to 1918 (Lense and Thirring 1918). It describes the precession of the orbital plane of a satellite originating from the gravitomagnetic field of the central object. Introducing the Runge-Lenz (-Laplace-Lagrange) vector (e.g. Damour and Schäfer 1988).

$$\mathbf{A} = \mathbf{P} \times \mathbf{L} - \frac{GM\mu^2}{R}\mathbf{R} \tag{57}$$

which points along the semi-major axis of the orbital Kepler ellipse, the precession of the orbital plane can be easily described. After averaging over one orbital period, the precession equation for orbital angular momentum and Runge-Lenz vector read,

$$\left\langle \left( \frac{d\mathbf{L}}{dt} \right)^{\mathrm{SO}} \right\rangle_t = \boldsymbol{\Omega}_{\mathrm{SO}} \times \mathbf{L}, \tag{58}$$

$$\left\langle \left( \frac{d\mathbf{A}}{dt} \right)^{\mathrm{SO}} \right\rangle_t = \boldsymbol{\Omega}_{\mathrm{SO}} \times \mathbf{A}, \tag{59}$$

where

$$\boldsymbol{\Omega}_{\mathrm{SO}} = \frac{2G}{c^2} \left\langle \frac{1}{R^3} \right\rangle_t \left( \mathbf{S}_{\mathrm{eff}} - 3\frac{(\mathbf{L} \cdot \mathbf{S}_{\mathrm{eff}})\mathbf{L}}{L^2} \right), \quad \mathbf{S}_{\mathrm{eff}} \equiv \mathbf{S} + \frac{3}{4} M_1 M_2 \mathbf{b}. \tag{60}$$

For the LAGEOS satellites 31 mas/yr are predicted; for measurements, see this volume.

6.2 Schiff Effect

The Schiff effect dates back to 1959/60 (Pugh 1959; Schiff 1960a, 1960b). It is called Schiff effect because of Leonard Schiff's merits in experimental gravity, particularly the satellite experiment GP-B. The Schiff effect, also called Lense-Thirring effect for spin or *frame dragging* effect, means the precession of a freely moving spinning object in the gravitomagnetic field of a central object. The spin precession equation reads

$$\left( \frac{d\mathbf{S}_1}{dt} \right)^{\mathrm{S_1S_2}} = \boldsymbol{\Omega}_{\mathrm{S}_2} \times \mathbf{S}_1, \quad \boldsymbol{\Omega}_{\mathrm{S}_2} = \frac{G}{c^2 R^3} \left( 3\frac{(\mathbf{R} \cdot \mathbf{S}_2)\mathbf{R}}{R^2} - \mathbf{S}_2 \right). \tag{61}$$

For GP-B 42 mas/yr are predicted; for recent measurement, see this volume.

6.3 de Sitter Effect

The de Sitter effect dates back to 1916 (de Sitter 1916). It is also called Fokker effect or *geodetic precession*. It describes the precession of the orbital plane of a binary system when orbiting a central nonrotating object. The precession equation reads

$$\left( \frac{d\mathbf{S}_1}{dt} \right)^{\mathrm{SO}} = \boldsymbol{\Omega}^{\mathrm{s}}_{\mathrm{SO}} \times \mathbf{S}_1, \quad \boldsymbol{\Omega}^{\mathrm{s}}_{\mathrm{SO}} = \frac{2G}{c^2 R^3} \left( 1 + \frac{3M_2}{4M_1} \right) \mathbf{L}, \tag{62}$$

where $\mathbf{L} = \frac{M_1 M_2}{M_1 + M_2} \mathbf{R} \times \mathbf{V}$. For the Earth-Moon system 19 mas/yr were predicted and for GP-B 6,600 mas/yr (for recent measurement, see this volume). Observations have shown the correctness of de Sitter's prediction for the Earth-Moon system with relative precision of 0.02 (Bertotti et al. 1987; Shapiro et al. 1988).

## 7 Effects in the Field of Rotating Black Holes

Black holes are the most compact objects in nature, far more compact than elementary particles (e.g., the Schwarzschild radius $2GM/c^2$ of an electron amounts to $10^{-55}$ cm). Additionally, for rotating black holes (Kerr black holes) the gravitomagnetic field may become extremely strong, so strong that frame dragging can not be resisted. It is the *ergosphere* wherein the gravitomagnetic field plays this unusual rôle. The ergosphere (static limit) starts at radial coordinate $r = M + \sqrt{M^2 - S^2\cos^2\theta/M^2}$ and terminates at the event horizon $r = M + \sqrt{M^2 - S^2/M^2}$. At the poles ($\theta = 0$), the ergosphere and the event horizon touch each other.

For rotating black holes, in the equatorial plane, *marginally stable circular orbits* for massive particles are located in the following regimes (here, $G = c = 1$) (e.g. Shapiro and Teukolsky 1983),

$$M \le r_{\text{mp}} \le 6M(\text{direct}), \qquad 6M \le r_{\text{mp}} \le 9M(\text{retrograde}), \tag{63}$$

and for photons in the regimes

$$M \le r_{\text{ph}} \le 3M(\text{direct}), \qquad 3M \le r_{\text{ph}} \le 4M(\text{retrograde}), \tag{64}$$

where $6M$ and $3M$ are the limiting values for the nonrotating Schwarzschild black hole. The orbital angle velocity equation for circular motion of massive particles reads

$$\frac{d\phi}{dt} = \pm\frac{GM}{R^3}\left(1 \pm \frac{S}{MRc}\sqrt{\frac{GM}{Rc^2}}\right)^{-1}, \tag{65}$$

where $M$ is the mass and $S$ the absolute value of the spin of the Kerr black hole. $R$ denotes the radial coordinate of the particle in Boyer-Lindquist coordinates.

The total mass (energy) of a black hole is given by $M = \sqrt{M_{\text{irr}}^2 + S^2/4M_{\text{irr}}^2}$. $M_{\text{irr}}$ denotes its irreducible mass defined by the area of the event horizon which never can shrink (classically).

The energy connected with the difference $M - M_{\text{irr}}$ can be extracted by e.g., the *Penrose process* in the ergosphere or the *Bardeen-Petterson effect* outside the ergosphere (e.g. Thorne et al. 1986). The Bardeen-Petterson effect around the black hole is a gravitomagnetic-field viscous-disk interrelated effect, where for a thin ring of orbiting matter at radial distance $r$ its orbital angular momentum $\mathbf{L}$ precesses around the frequency vector $\mathbf{\Omega}_{\text{GM}}$ according to the equation

$$\frac{d\mathbf{L}}{dt} = \mathbf{\Omega}_{\text{GM}} \times \mathbf{L}, \qquad \mathbf{\Omega}_{\text{GM}} = \frac{2G\mathbf{J}}{c^2r^3}. \tag{66}$$

$\mathbf{J}$ is the spin of the central black hole. The *Blandford-Znajek process* near the black hole is a magnetic-gravitomagnetic 'surface battery' driven process. Its electro-motoric force reads

$$\text{EMF} \equiv \oint N\mathbf{E}\cdot d\mathbf{r} = -\oint(\mathbf{N}\times\mathbf{B})\cdot d\mathbf{r}, \tag{67}$$

where the magnetic field $\mathbf{B}$ has its origin in the plasma matter surrounding the black hole. For more details see also, e.g. Straumann (2007).

## 8 Higher Order Dynamics of Gravitating Spinning Binaries

In general relativity, for conservative, isolated pole-dipole particle systems or conservative parts of them, the total linear and angular momenta, respectively given by

$$\mathbf{P} = \sum_a \mathbf{p}_a, \quad \mathbf{J} = \sum_a (\mathbf{x}_a \times \mathbf{p}_a + \mathbf{S}_a), \tag{68}$$

are conserved. To the post-Newtonian order $1/c^4$ in the spin dynamics, which quantitatively includes the order $1/c^6$ (3PN) in the point-mass dynamics because the spin magnitudes must be counted of order $1/c$ (notice, for black holes $|\mathbf{S}|c/Gm^2$ can be at most equal to one), the total particle Hamiltonian $H$ and the center-of-mass vector $\mathbf{G} = (G_i)$ read

$$\begin{aligned} H = {} & H_{\mathrm{N}} + H_{1\mathrm{PN}} + H_{2\mathrm{PN}} + H_{3\mathrm{PN}} \\ & + H^{\mathrm{LO}}_{\mathrm{SO}} + H^{\mathrm{LO}}_{\mathrm{S_1S_2}} + H^{\mathrm{LO}}_{\mathrm{S_1^2}} + H^{\mathrm{LO}}_{\mathrm{S_2^2}} \\ & + H^{\mathrm{NLO}}_{\mathrm{SO}} + H^{\mathrm{NLO}}_{\mathrm{S_1S_2}} + H^{\mathrm{NLO}}_{\mathrm{S_1^2}} + H^{\mathrm{NLO}}_{\mathrm{S_2^2}}, \end{aligned} \tag{69}$$

$$\begin{aligned} \mathbf{G} = {} & \mathbf{G}_{\mathrm{N}} + \mathbf{G}_{1\mathrm{PN}} + \mathbf{G}_{2\mathrm{PN}} + \mathbf{G}_{3\mathrm{PN}} \\ & + \mathbf{G}^{\mathrm{LO}}_{\mathrm{SO}} + \mathbf{G}^{\mathrm{NLO}}_{\mathrm{SO}} + \mathbf{G}^{\mathrm{NLO}}_{\mathrm{S_1S_2}} + \mathbf{G}^{\mathrm{NLO}}_{\mathrm{S_1^2}} + \mathbf{G}^{\mathrm{NLO}}_{\mathrm{S_2^2}}. \end{aligned} \tag{70}$$

In the following only the expressions containing spin will be given. For the other expressions see Damour et al. (2000, 2001). To simplify the expressions, we put $G = c = 1$.

In non-center-of-mass frames, the Hamiltonian of leading-order spin-orbit coupling reads,

$$H^{\mathrm{LO}}_{\mathrm{SO}} = \sum_a \sum_{b \neq a} \frac{1}{r_{ab}^2} (\mathbf{S}_a \times \mathbf{n}_{ab}) \cdot \left[ \frac{3m_b}{2m_a} \mathbf{p}_a - 2\mathbf{p}_b \right], \tag{71}$$

where $r_{ab}^2 = (x_a^i - x_b^i)(x_a^i - x_b^i)$, $r_{ab} n_{ab}^i = (x_a^i - x_b^i)$, and the one of leading-order spin($a$)-spin ($b$) coupling, with $a \neq b$, is given by,

$$H^{\mathrm{LO}}_{\mathrm{S_1S_2}} = \sum_a \sum_{b \neq a} \frac{1}{2r_{ab}^3} [3(\mathbf{S}_a \cdot \mathbf{n}_{ab})(\mathbf{S}_b \cdot \mathbf{n}_{ab}) - (\mathbf{S}_a \cdot \mathbf{S}_b)]. \tag{72}$$

The leading-order spin-squared Hamiltonians read,

$$H^{\mathrm{LO}}_{\mathrm{S_1^2}} = \frac{m_2}{2m_1 r_{12}^3} [3(\mathbf{S}_1 \cdot \mathbf{n}_{12})(\mathbf{S}_1 \cdot \mathbf{n}_{12}) - (\mathbf{S}_1 \cdot \mathbf{S}_1)], \qquad H^{\mathrm{LO}}_{\mathrm{S_2^2}} = H^{\mathrm{LO}}_{\mathrm{S_1^2}}(1 \leftrightarrow 2). \tag{73}$$

The leading-order spin-orbit and spin($a$)-spin($b$) center-of-mass vectors $\mathbf{G} = (G_i)$ take the form (Hergt and Schäfer 2008a),

$$\mathbf{G}^{\mathrm{LO}}_{\mathrm{SO}} = \sum_a \frac{1}{2m_a} (\mathbf{P}_a \times \mathbf{S}_a), \qquad \mathbf{G}^{\mathrm{LO}}_{\mathrm{SS}} = 0. \tag{74}$$

The Hamiltonian of next-to-leading-order spin-orbit coupling reads (Damour et al. 2008; Steinhoff et al. 2008a),

$$\begin{aligned}
H_{\mathrm{SO}}^{\mathrm{NLO}} = & -\frac{((\mathbf{p}_1 \times \mathbf{S}_1)\cdot\mathbf{n})}{r^2}\left[\frac{5m_2\mathbf{p}_1^2}{8m_1^3} + \frac{3(\mathbf{p}_1\cdot\mathbf{p}_2)}{4m_1^2} - \frac{3\mathbf{p}_2^2}{4m_1m_2}\right.\\
& \left. + \frac{3(\mathbf{p}_1\cdot\mathbf{n})(\mathbf{p}_2\cdot\mathbf{n})}{4m_1^2} + \frac{3(\mathbf{p}_2\cdot\mathbf{n})^2}{2m_1m_2}\right]\\
& + \frac{((\mathbf{p}_2 \times \mathbf{S}_1)\cdot\mathbf{n})}{r^2}\left[\frac{(\mathbf{p}_1\cdot\mathbf{p}_2)}{m_1m_2} + \frac{3(\mathbf{p}_1\cdot\mathbf{n})(\mathbf{p}_2\cdot\mathbf{n})}{m_1m_2}\right]\\
& + \frac{((\mathbf{p}_1 \times \mathbf{S}_1)\cdot\mathbf{p}_2)}{r^2}\left[\frac{2(\mathbf{p}_2\cdot\mathbf{n})}{m_1m_2} - \frac{3(\mathbf{p}_1\cdot\mathbf{n})}{4m_1^2}\right]\\
& - \frac{((\mathbf{p}_1 \times \mathbf{S}_1)\cdot\mathbf{n})}{r^3}\left[\frac{11m_2}{2} + \frac{5m_2^2}{m_1}\right]\\
& + \frac{((\mathbf{p}_2 \times \mathbf{S}_1)\cdot\mathbf{n})}{r^3}\left[6m_1 + \frac{15m_2}{2}\right] + (1 \leftrightarrow 2),
\end{aligned} \tag{75}$$

$r = r_{12}$, $n^i = n^i_{12}$, and the one of next-to-leading-order spin(1)-spin(2) coupling has been obtained by Steinhoff et al. (2008a),

$$\begin{aligned}
H_{\mathrm{S_1S_2}}^{\mathrm{NLO}} = & \frac{1}{2m_1m_2r^3}\left[\frac{3}{2}((\mathbf{p}_1\times\mathbf{S}_1)\cdot\mathbf{n})((\mathbf{p}_2\times\mathbf{S}_2)\cdot\mathbf{n})\right.\\
& + 6((\mathbf{p}_2\times\mathbf{S}_1)\cdot\mathbf{n})((\mathbf{p}_1\times\mathbf{S}_2)\cdot\mathbf{n}) - 15(\mathbf{S}_1\cdot\mathbf{n})(\mathbf{S}_2\cdot\mathbf{n})(\mathbf{p}_1\cdot\mathbf{n})(\mathbf{p}_2\cdot\mathbf{n})\\
& - 3(\mathbf{S}_1\cdot\mathbf{n})(\mathbf{S}_2\cdot\mathbf{n})(\mathbf{p}_1\cdot\mathbf{p}_2) + 3(\mathbf{S}_1\cdot\mathbf{p}_2)(\mathbf{S}_2\cdot\mathbf{n})(\mathbf{p}_1\cdot\mathbf{n})\\
& + 3(\mathbf{S}_2\cdot\mathbf{p}_1)(\mathbf{S}_1\cdot\mathbf{n})(\mathbf{p}_2\cdot\mathbf{n}) + 3(\mathbf{S}_1\cdot\mathbf{p}_1)(\mathbf{S}_2\cdot\mathbf{n})(\mathbf{p}_2\cdot\mathbf{n})\\
& + 3(\mathbf{S}_2\cdot\mathbf{p}_2)(\mathbf{S}_1\cdot\mathbf{n})(\mathbf{p}_1\cdot\mathbf{n}) - 3(\mathbf{S}_1\cdot\mathbf{S}_2)(\mathbf{p}_1\cdot\mathbf{n})(\mathbf{p}_2\cdot\mathbf{n})\\
& \left. + (\mathbf{S}_1\cdot\mathbf{p}_1)(\mathbf{S}_2\cdot\mathbf{p}_2) - \frac{1}{2}(\mathbf{S}_1\cdot\mathbf{p}_2)(\mathbf{S}_2\cdot\mathbf{p}_1) + \frac{1}{2}(\mathbf{S}_1\cdot\mathbf{S}_2)(\mathbf{p}_1\cdot\mathbf{p}_2)\right]\\
& + \frac{3}{2m_1^2r^3}[-((\mathbf{p}_1\times\mathbf{S}_1)\cdot\mathbf{n})((\mathbf{p}_1\times\mathbf{S}_2)\cdot\mathbf{n})\\
& + (\mathbf{S}_1\cdot\mathbf{S}_2)(\mathbf{p}_1\cdot\mathbf{n})^2 - (\mathbf{S}_1\cdot\mathbf{n})(\mathbf{S}_2\cdot\mathbf{p}_1)(\mathbf{p}_1\cdot\mathbf{n})]\\
& + \frac{3}{2m_2^2r^3}[-((\mathbf{p}_2\times\mathbf{S}_2)\cdot\mathbf{n})((\mathbf{p}_2\times\mathbf{S}_1)\cdot\mathbf{n})\\
& + (\mathbf{S}_1\cdot\mathbf{S}_2)(\mathbf{p}_2\cdot\mathbf{n})^2 - (\mathbf{S}_2\cdot\mathbf{n})(\mathbf{S}_1\cdot\mathbf{p}_2)(\mathbf{p}_2\cdot\mathbf{n})]\\
& + \frac{6(m_1+m_2)}{r^4}[(\mathbf{S}_1\cdot\mathbf{S}_2) - 2(\mathbf{S}_1\cdot\mathbf{n})(\mathbf{S}_2\cdot\mathbf{n})].
\end{aligned} \tag{76}$$

The quite tedious to derive next-to-leading order spin-squared Hamiltonian reads (Steinhoff et al. 2008b),

$$H_{\mathrm{S_1^2}}^{\mathrm{NLO}} = \frac{1}{r^3}\left[\frac{m_2}{4m_1^3}(\mathbf{p}_1\cdot\mathbf{S}_1)^2 - \frac{3}{4m_1m_2}\mathbf{p}_2^2\mathbf{S}_1^2\right.$$

$$+\frac{3m_2}{8m_1^3}(\mathbf{p}_1\cdot\mathbf{n})^2\,\mathbf{S}_1^2-\frac{3m_2}{8m_1^3}\mathbf{p}_1^2\,(\mathbf{S}_1\cdot\mathbf{n})^2$$
$$-\frac{3m_2}{4m_1^3}(\mathbf{p}_1\cdot\mathbf{n})\,(\mathbf{S}_1\cdot\mathbf{n})\,(\mathbf{p}_1\cdot\mathbf{S}_1)-\frac{3}{4m_1m_2}\mathbf{p}_2^2\mathbf{S}_1^2+\frac{9}{4m_1m_2}\mathbf{p}_2^2\,(\mathbf{S}_1\cdot\mathbf{n})^2$$
$$+\frac{3}{4m_1^2}(\mathbf{p}_1\cdot\mathbf{p}_2)\,\mathbf{S}_1^2-\frac{9}{4m_1^2}(\mathbf{p}_1\cdot\mathbf{p}_2)\,(\mathbf{S}_1\cdot\mathbf{n})^2$$
$$-\frac{3}{2m_1^2}(\mathbf{p}_1\cdot\mathbf{n})\,(\mathbf{p}_2\cdot\mathbf{S}_1)\,(\mathbf{S}_1\cdot\mathbf{n})+\frac{3}{m_1^2}(\mathbf{p}_2\cdot\mathbf{n})\,(\mathbf{p}_1\cdot\mathbf{S}_1)\,(\mathbf{S}_1\cdot\mathbf{n})$$
$$+\frac{3}{4m_1^2}(\mathbf{p}_1\cdot\mathbf{n})\,(\mathbf{p}_2\cdot\mathbf{n})\,\mathbf{S}_1^2-\frac{15}{4m_1^2}(\mathbf{p}_1\cdot\mathbf{n})\,(\mathbf{p}_2\cdot\mathbf{n})\,(\mathbf{S}_1\cdot\mathbf{n})^2\Bigg]$$
$$-\frac{m_2}{2r^4}\left[9(\mathbf{S}_1\cdot\mathbf{n})^2-5\mathbf{S}_1^2+\frac{14m_2}{m_1}(\mathbf{S}_1\cdot\mathbf{n})^2-\frac{6m_2}{m_1}\mathbf{S}_1^2\right],\tag{77}$$

$$H^{\rm NLO}_{S_2^2}=H^{\rm NLO}_{S_1^2}(1\leftrightarrow 2).\tag{78}$$

These Hamiltonians contain information about the quadrupole deformation of rotating black holes. They go beyond the structure of (37) but still fit into (36).

The next-to-leading-order spin-orbit and spin($a$)-spin($b$) center-of-mass vectors take the forms (Hergt and Schäfer 2008b),

$$\mathbf{G}^{\rm NLO}_{\rm SO}=-\sum_a\frac{\mathbf{p}_a^2}{8m_a^3}(\mathbf{p}_a\times\mathbf{S}_a)$$
$$+\sum_a\sum_{b\neq a}\frac{m_b}{4m_ar_{ab}}\left[((\mathbf{p}_a\times\mathbf{S}_a)\cdot\mathbf{n}_{ab})\frac{5\mathbf{x}_a+\mathbf{x}_b}{r_{ab}}-5(\mathbf{p}_a\times\mathbf{S}_a)\right]$$
$$+\sum_a\sum_{b\neq a}\frac{1}{r_{ab}}\left[\frac{3}{2}(\mathbf{p}_b\times\mathbf{S}_a)-\frac{1}{2}(\mathbf{n}_{ab}\times\mathbf{S}_a)(\mathbf{p}_b\cdot\mathbf{n}_{ab})\right.$$
$$\left.-((\mathbf{p}_a\times\mathbf{S}_a)\cdot\mathbf{n}_{ab})\frac{\mathbf{x}_a+\mathbf{x}_b}{r_{ab}}\right],\tag{79}$$

$$\mathbf{G}^{\rm NLO}_{S_1S_2}=\frac{1}{2}\sum_a\sum_{b\neq a}\left\{[3(\mathbf{S}_a\cdot\mathbf{n}_{ab})(\mathbf{S}_b\cdot\mathbf{n}_{ab})-(\mathbf{S}_a\cdot\mathbf{S}_b)]\frac{\mathbf{x}_a}{r_{ab}^3}+(\mathbf{S}_b\cdot\mathbf{n}_{ab})\frac{\mathbf{S}_a}{r_{ab}^2}\right\},\tag{80}$$

$$\mathbf{G}^{\rm NLO}_{S_1^2}=\frac{m_2}{m_1}\left[-2\frac{(\mathbf{S}_1\cdot\mathbf{n}_{12})\,\mathbf{S}_1}{r_{12}^2}+\frac{3\,(\mathbf{S}_1\cdot\mathbf{n}_{12})^2}{4r_{12}^3}(\mathbf{x}_1+\mathbf{x}_2)+\frac{\mathbf{S}_1^2}{4r_{12}^3}(3\mathbf{x}_1-5\mathbf{x}_2)\right],\tag{81}$$

$$\mathbf{G}^{\rm NLO}_{S_2^2}=\mathbf{G}^{\rm NLO}_{S_1^2}(1\leftrightarrow 2).\tag{82}$$

Employing standard Hamilton equations of motion calculations, higher order spin effects analogous to those presented in Sect. 6 can easily be obtained. In the future, within gravitational wave astronomy, those effects should become accessible to observations.

**Acknowledgement** Financial support from the International Space Science Institute (ISSI) is thankfully acknowledged.

## References


B. Barker, R. O'Connell, The gravitational interaction: Spin, rotation, and quantum effects—a review. Gen. Relativ. Gravit. **11**, 149–175 (1979)

B. Bertotti, I. Ciufolini, P. Bender, New test of general relativity: Measurement of the de Sitter geodetic precession rate for lunar perigee. Phys. Rev. Lett. **58**, 1062–1065 (1987)

I. Ciufolini, J. Wheeler, *Gravitation and Inertia* (Princeton University Press, Princeton, 1995)

T. Damour, Coalescence of two spinning black holes: An effective one-body approach. Phys. Rev. D **64**, 124013 (2001)

T. Damour, G. Schäfer, Higher-order relativistic periastron advances and binary pulsars. Nuovo Cimento B **101**, 127–176 (1988)

T. Damour, P. Jaranowski, G. Schäfer, Poincaré invariance in the ADM Hamiltonian approach to the general relativistic two-body problem. Phys. Rev. D **62**, 021501 (2000)

T. Damour, P. Jaranowski, G. Schäfer, Erratum: Poincaré invariance in the ADM Hamiltonian approach to the general relativistic two-body problem. Phys. Rev. D **63**, 029903 (2001)

T. Damour, P. Jaranowski, G. Schäfer, Hamiltonian of two spinning compact bodies with next-to-leading order gravitational spin-orbit coupling. Phys. Rev. D **77**, 064032 (2008)

W. de Sitter, On Einstein's theory of gravitation and its astronomical consequences. Mon. Not. R. Astron. Soc. **77**, 155–184 (1916)

B. DeWitt, *Dynamical Theory of Groups and Fields* (Gordon and Breach, New York, 1965)

G. Fleming, Covariant position operators, spin and locality. Phys. Rev. **137**, B188–B197 (1965)

J. Frenkel, Die Elektrodynamik des rotierenden Elektrons. Z. Phys. **37**, 243–262 (1926)

Y. Hagihara, *Celestial Mechanics I. Dynamical Principles and Transformation Theory* (MIT Press, Cambridge, 1970)

A. Hanson, T. Regge, The relativistic spherical top. Ann. Phys. (NY) **87**, 498–566 (1974)

S. Hergt, G. Schäfer, Higher-order-in-spin interaction Hamiltonians for binary black holes from source terms of Kerr geometry in approximate ADM coordinates. Phys. Rev. D **77**, 104001 (2008a)

S. Hergt, G. Schäfer, Higher-order-in-spin interaction Hamiltonians for binary black holes from Poincaré invariance. Phys. Rev. D **78**, 124004 (2008b)

C. Jentsch, T. Müller, E. Rasel, W. Ertmer, HYPER: A satellite mission in fundamental physics based on high precision atom interferometry. Gen. Relativ. Gravit. **36**, 2197–2221 (2004)

T. Klügel, U. Schreiber, W. Schlüter, A. Velikoseltsev, Advances in inertial Earth rotation measurements—new data from the Wettzell G Ring Laser, in *Proceedings of the "Journees Systemes de Reference Spatio-temporels 2007"*, ed. by N. Capitaine, Observatoire de Paris, 2007, pp. 173–176. http://syrte.obspm.fr/journees2007/PDF/s4_08_Klugel.pdf

C. Lämmerzahl, G. Neugebauer, The Lense-Thirring effect: From the basic notions to the observed effects, in *Gyros, Clocks, Interferometers...: Testing Relativity in Space*, ed. by C. Lämmerzahl, C. Everitt, F. Hehl (Springer, Berlin, 2001)

L. Landau, E. Lifschitz, *Mechanik* (Akademie-Verlag, Berlin, 1981)

L. Landau, E. Lifshitz, *The Classical Theory of Fields* (Pergamon, Oxford, 1985)

J. Lense, H. Thirring, Über den Einfluss der Eigenrotation der Zentralkörper auf die Bewegung der Planeten und Monde nach der Einsteinschen Gravitationstheorie. Phys. Z. **19**, 156–163 (1918). See also English translation by B. Mashhoon, F.W. Hehl, D.S. Theis, Gen. Relativ. Gravit. **16**, 711–750 (1984)

B. Mashhoon, Gravitoelectromagnetism, in *Reference Frames and Gravitomagnetism*, ed. by J. Pascual-Sánchez, L. Floría, A. San Miguel, F. Vicente (World Scientific, Singapore, 2001)

C. Misner, K. Thorne, J. Wheeler, *Gravitation* (Freeman, San Francisco, 1973)

C. Møller, *The Theory of Relativity* (Oxford University Press, Oxford, 1969)

G. Pugh, *Proposal for a Satellite Test of the Coriolis Prediction of General Relativity*. Weapons Systems Evaluation group Research Memorandum, vol. 11 (The Pentagon, Washington, 1959)

T. Regge, T. Teitelboim, Role of surface integrals in the Hamiltonian formulation of general relativity. Ann. Phys. (NY) **88**, 286–318 (1974)

M. Ruggiero, A. Tartaglia, Gravitomagnetic effects. Nuovo Cimento B **117**, 743–768 (2002)

G. Schäfer, Gravitomagnetic effects. Gen. Relativ. Gravit. **36**, 2223–2235 (2004)

G. Schäfer, Gravitation: Geometry and dynamics. Ann. Phys. (Leipz.) **14**, 148–164 (2005)

L. Schiff, Motion of a gyroscope according to Einsteins theory of gravitation. Proc. Natl. Acad. Sci. **46**, 871 (1960a)

L. Schiff, Possible new experimental test of general relativity theory. Phys. Rev. Lett. **4**, 215–217 (1960b)

J. Schwinger, *Particles, Sources, and Fields I* (Perseus Books, Reading, 1998)

I. Shapiro, R. Reasenberg, J. Chandler, R. Babcock, Measurement of the de Sitter precession of the Moon: A relativistic three-body effect. Phys. Rev. Lett. **61**, 2643–2646 (1988)

S. Shapiro, S. Teukolsky, *Black Holes, White Dwarfs, and Neutron Stars* (Wiley, New York, 1983)

G. Stedman, Ring-laser tests of fundamental physics and geophysics. Rep. Prog. Phys. **60**, 615–688 (1997)
J. Steinhoff, G. Schäfer, S. Hergt, ADM canonical formalism for gravitating spinning objects. Phys. Rev. D **77**, 104018 (2008a)
J. Steinhoff, S. Hergt, G. Schäfer, Spin-squared Hamiltonian of next-to-leading gravitational interaction. Phys. Rev. D **78**, 101503 (2008b)
N. Straumann, Energy extraction from black holes (2007). arXiv:0709.3895v1
K. Thorne, R. Price, D. Macdonald, *Black Holes: The Membrane Paradigm* (Yale University Press, New Haven, 1986)
K. Yee, M. Bander, Equations of motion for spinning particles in external electromagnetic and gravitational fields. Phys. Rev. D **48**, 2797–2799 (1993)

Space Sci Rev (2009) 148: 53–69
DOI 10.1007/s11214-009-9524-7

# Gravity Probe B Data Analysis

## Status and Potential for Improved Accuracy of Scientific Results

**C.W.F. Everitt · M. Adams · W. Bencze · S. Buchman · B. Clarke · J.W. Conklin · D.B. DeBra · M. Dolphin · M. Heifetz · D. Hipkins · T. Holmes · G.M. Keiser · J. Kolodziejczak · J. Li · J. Lipa · J.M. Lockhart · J.C. Mester · B. Muhlfelder · Y. Ohshima · B.W. Parkinson · M. Salomon · A. Silbergleit · V. Solomonik · K. Stahl · M. Taber · J.P. Turneaure · S. Wang · P.W. Worden Jr.**

Received: 20 April 2009 / Accepted: 26 April 2009 / Published online: 27 June 2009
© Springer Science+Business Media B.V. 2009

**Abstract** This is the first of five connected papers detailing progress on the Gravity Probe B (GP-B) Relativity Mission. GP-B, launched 20 April 2004, is a landmark physics experiment in space to test two fundamental predictions of Einstein's general relativity theory, the geodetic and frame-dragging effects, by means of cryogenic gyroscopes in Earth orbit. Data collection began 28 August 2004 and science operations were completed 29 September 2005. The data analysis has proven deeper than expected as a result of two mutually reinforcing complications in gyroscope performance: (1) a changing polhode path affecting the calibration of the gyroscope scale factor $C_g$ against the aberration of starlight and (2) two larger than expected manifestations of a Newtonian gyro torque due to patch potentials on the rotor and housing. In earlier papers, we reported two methods, 'geometric' and 'algebraic', for identifying and removing the first Newtonian effect ('misalignment torque'), and also a preliminary method of treating the second ('roll-polhode resonance torque'). Central to the progress in both torque modeling and $C_g$ determination has been an extended effort on "Trapped Flux Mapping" commenced in November 2006. A turning point came in August 2008 when it became possible to include a detailed history of the resonance torques into the computation. The East-West (frame-dragging) effect is now plainly visible in the processed data. The current *statistical uncertainty* from an analysis of 155 days of data is 5.4 marc-s/yr ($\sim$ 14% of the predicted effect), though it must be emphasized that this is a preliminary result requiring rigorous investigation of systematics by methods discussed in the accompanying paper by Muhlfelder et al. A covariance analysis incorporating models of the patch effect torques indicates that a 3–5% determination of frame-dragging is possible with more complete, computationally intensive data analysis.

C.W.F. Everitt (✉) · M. Adams · W. Bencze · S. Buchman · B. Clarke · J.W. Conklin · D.B. DeBra · M. Dolphin · M. Heifetz · D. Hipkins · T. Holmes · G.M. Keiser · J. Kolodziejczak · J. Li · J. Lipa · J.M. Lockhart · J.C. Mester · B. Muhlfelder · Y. Ohshima · B.W. Parkinson · M. Salomon · A. Silbergleit · V. Solomonik · K. Stahl · M. Taber · J.P. Turneaure · S. Wang · P.W. Worden Jr.
W.W. Hansen Experimental Physics Laboratory, Stanford University, Stanford, CA 94305-4085, USA
e-mail: francis@relgyro.stanford.edu

J. Kolodziejczak
NASA Marshall Space Flight Center, Huntsville, AL 35812, USA

**Keywords** General Relativity · Frame-dragging · Lense-Thirring

## 1 The NASA-Stanford Gravity Probe B Mission (GP-B)

In 1960, L.I. Schiff (Schiff 1960) showed that an ideal gyroscope in orbit around the Earth or other massive body would undergo two relativistic precessions with respect to a distant undisturbed inertial frame: (1) a geodetic effect in the orbit plane due to the local curvature of space-time; (2) a frame-dragging effect due to the Earth's rotation. The geodetic term is fundamentally equivalent to the curvature precession of the Earth-Moon system around the Sun first given by W. de Sitter in 1916. The Schiff frame-dragging is related to, but not identical with, the dragging of the orbit plane of a satellite around a rotating planet computed in 1918 by J. Lense and H. Thirring (Lense and Thirring 1918). It has, as Thirring noted, connections to the possible relation between General Relativity and Mach's principle.

Schiff's formula for the combined gyroscope precession $\boldsymbol{\Omega}$ in Einstein's theory is:

$$\boldsymbol{\Omega} = \frac{3GM}{2c^2R^3}(\boldsymbol{R}\times\boldsymbol{v}) + \frac{GI}{c^2R^3}\left[\frac{3\boldsymbol{R}}{R^2}(\boldsymbol{\omega}\cdot\boldsymbol{R}) - \boldsymbol{\omega}\right]$$

where the first term is the geodetic, and the second the frame-dragging effect, with $G$ the gravitational constant, $c$ the velocity of light, $M$, $I$, $\boldsymbol{\omega}$, the mass, moment of inertia, and angular velocity of the Earth, and $\boldsymbol{R}$ and $\boldsymbol{v}$ the instantaneous position and velocity of the gyroscope. In the polar orbit selected for Gravity Probe B, the two effects are at right angles. Their magnitudes are found by integrating Schiff's formula around the orbit. For the geodetic effect, it is necessary to take into account the oblateness of the Earth which modifies both the orbit and the term itself: a correction investigated first by D.C. Wilkins, then B.M. Barker and R.F. O'Connell (Barker and O'Connell 1970), and finally, most elegantly, by J.V. Breakwell (Breakwell 1988, see also Adler and Silbergleit 2000). In the 642 km orbit of GP-B, the predicted geodetic effect is 6,606 marc-s/yr and the frame-dragging 39 marc-s/yr. GP-B was designed to measure both to $< 0.5$ marc-s/yr.

Conceptually, Gravity Probe B is simple. All it needs is a star, a telescope, and a spinning sphere (the gyroscope). The difficulty lies in the numbers. To reach the 0.5 marc-s/yr experiment goal calls for:

(1) One or more exceedingly accurate gyroscopes with drift rates $< 10^{-11}$ deg/h, i.e. 6 to 7 orders of magnitude better than the best modeled inertial navigation gyroscopes
(2) A reference telescope $\sim 3$ orders of magnitude better than the best previous star trackers
(3) A sufficiently bright, suitably located guide star (IM Pegasi was chosen) whose proper motion with respect to remote inertial space is known to $< 0.5$ marc-s/yr

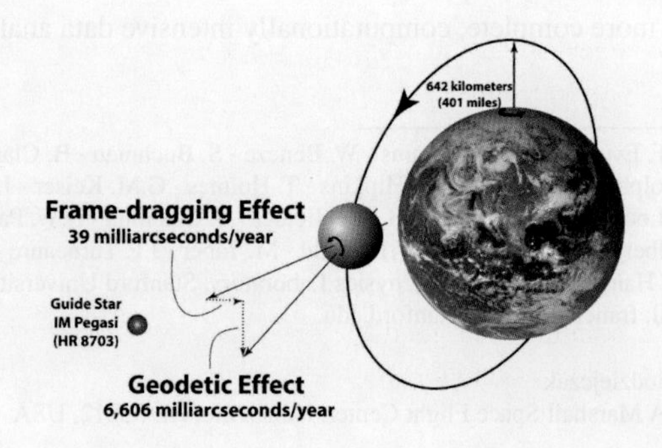


**Fig. 1** Predicted precessions of the GP-B gyroscope (North and West are positive in this coordinate system)

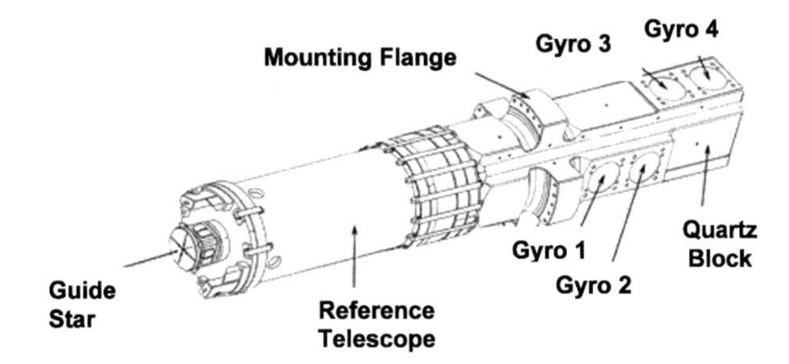


**Fig. 2** Gravity Probe B instrument

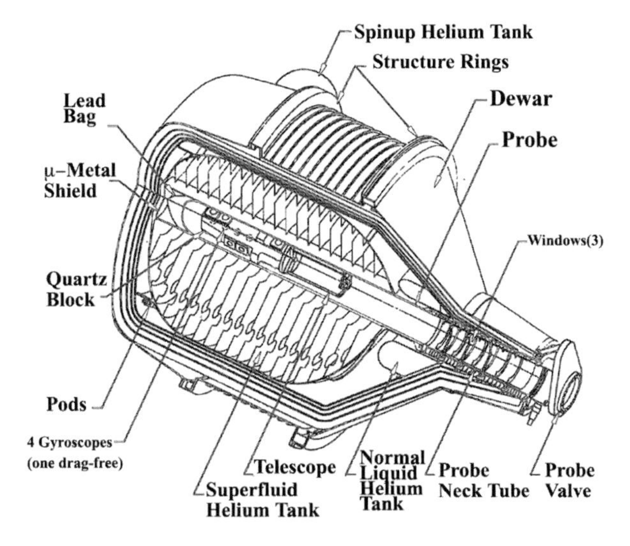


**Fig. 3** Flight dewar

(4) Sufficiently accurate orbit information to calibrate the science signal and calculate the two predicted effects

These four constraints led to 10 fundamental requirements on the design of GP-B, and then to 27 distinct experimental subsystems for executing the mission with 200–300 subrequirements.

The heart of the instrument is the integrated fused quartz structure, operating at cryogenic temperatures, illustrated in Fig. 2. It comprises a 0.14 m aperture, 3.81 m focal length folded Cassegrainian telescope bonded to a quartz block containing four gyroscopes mounted in line to within 0.1 mm of the boresight of the telescope. Two gyroscopes spin clockwise and two counter-clockwise with their axes initially aligned to within a few arc-s of the star. Thus, each gyroscope measures both relativity effects. The spacecraft rolls slowly (77.5 s period) around the line to the star. Since the guide star is occulted for almost half the orbit (see Fig. 1), the telescope must reacquire it every orbit.

The instrument is mounted in an evacuated chamber within a 2440 $\ell$ superfluid helium dewar (Fig. 3) fabricated by Lockheed Martin (Parmley et al. 1986) operating at 1.8 K. The boil-off gas provides thrust for attitude, translational and roll control of the spacecraft. While pointing was good ($\sim 200$ marc-s rms), reaching the $< 0.5$ marc-s/yr experiment goal requires accurate subtraction of the telescope signal $\boldsymbol{T}$ from the gyroscope signal $\boldsymbol{G}$. Thus, in addition to near perfect telescope and gyroscopes, we face the challenge of accu-

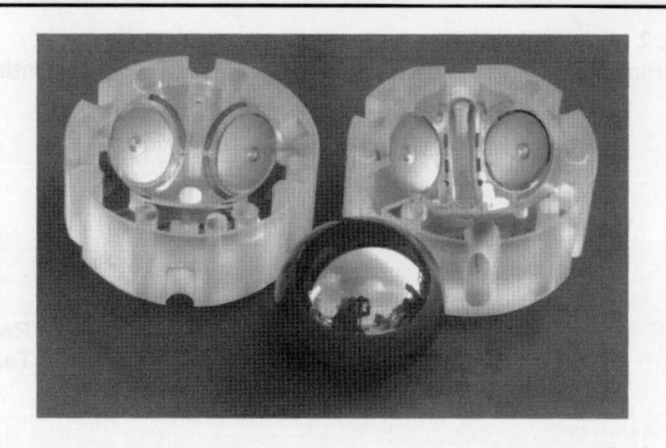

**Fig. 4** Gravity Probe B gyroscope

rately matching the two scale factors to make valid ***G***–***T*** subtractions, and as will appear, an interesting separate process of calibration for ***G***.

The gyroscope (Fig. 4) comprises a niobium-coated fused quartz sphere (3.81 cm dia) electrically suspended about its geometric center in a spherical quartz housing. A simple argument reveals the benefit of performing the experiment in space. Let the maximum allowed Newtonian drift rate be $\Omega_0$ ($\sim 2 \times 10^{-17}$ rad/s for a $< 0.5$ marc-s/yr experiment). Picture a perfectly spherical, but not perfectly homogeneous, rotor with its mass center a distance $\boldsymbol{\delta r}$ from the geometric center. An acceleration $\boldsymbol{f}$ at right angles to the spin axis will cause a torque $\boldsymbol{T} = M(\boldsymbol{\delta r} \times \boldsymbol{f})$ yielding the requirement

$$f\frac{\delta r}{r} < \frac{2}{5} v_s \Omega_0,$$

where $v_s$ is the peripheral velocity of the rotor ($\sim 10$ m/s). Inserting numbers, one finds for an experiment on Earth ($f = 1$ g) the ridiculous homogeneity requirement $6 \times 10^{-18}$. For a satellite, the acceleration level from air drag, solar radiation pressure, etc. might typically be $f \sim 10^{-8}$ g which yields the still exceedingly difficult requirement $\sim 6 \times 10^{-10}$. This brings us to the 'drag-free' concept first suggested in 1959 by G.E. Pugh (Pugh 1959), and then extensively developed at Stanford University (DeBra 1988). The gyroscope, being shielded from drag, tends to follow an ideal gravitational orbit. By sensing its position inside the satellite and applying compensating thrust, the net acceleration is further reduced to $f \sim 10^{-11}$ g. The resulting homogeneity requirement is $6 \times 10^{-7}$; the homogeneity of the actual rotors is $3 \times 10^{-7}$. A parallel argument for suspension torques acting on the out-of-roundness of the gyro rotor is met with an equally satisfactory margin.

Having a rotor spherical and mass-balanced to 10 nm poses a readout problem; for how is one to measure, with marc-s accuracy, the direction of spin of a perfectly round, perfectly homogeneous unmarked sphere? The key is cryogenics. Below 9 K, the rotor's niobium coating is superconducting; it generates a small dipole magnetic moment, the London moment ($\boldsymbol{M}_L$) (London 1954) proportional to spin, aligned with the instantaneous spin axis. Detection is by a highly sensitive SQUID magnetometer coupled to a 4-turn thin film superconducting coil patterned on the flat surface of the left-hand housing half (Fig. 4). Figure 5 indicates how the amplitude and phase of the misalignment angle $\theta$ are determined at the spacecraft's 77.5 s roll period (Muhlfelder et al. 2003). Because the rotor is cooled initially in a finite field, some trapped magnetic flux is superimposed on the London moment. This complicates the readout process. See Sect. 3 below.

Cryogenics helps in three other ways: magnetic shielding, ultra-high vacuum, and mechanical stability.

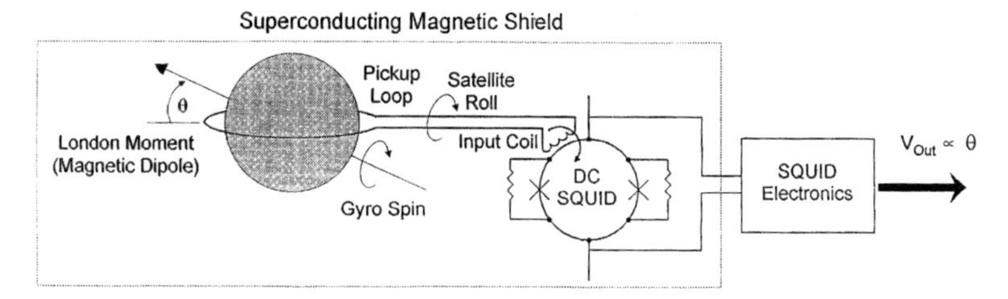


**Fig. 5** London moment readout

- *Magnetic Shielding*: The shielding has two formidable requirements, (i) a DC field level of $10^{-7}$ gauss to reduce trapped flux in the rotor to $\sim 1\%$ of the London moment, and (ii) 240 dB AC attenuation to isolate the gyroscopes from varying external fields. The $10^{-7}$ gauss DC field was obtained by an expanding lead-bag technique devised by B. Cabrera (Cabrera 1982) and applied to GP-B by M. Taber, J. Mester, J. Lockhart, and colleagues (Mester et al. 2000); AC shielding was implemented with a succession of nested Cryoperm and superconducting shields including transverse cylindrical shields around each gyroscope. Two critical engineering challenges were inserting the warm probe into a cold dewar and ensuring that the probe itself was non-magnetic. Insertion was a severely disciplined 24-hour-long process requiring a sophisticated airlock (or rather helium lock). As for the probe itself, this took an even more disciplined process in which every component, tool, and manufacturing procedure were subject to appropriate limits and testing. An engineer in a good position to judge (R. Parmley of Lockheed Martin) has estimated that 30% of the cost came from meeting the magnetic requirements (Parmley et al. 2003).
- *Ultra-high vacuum*: To spin a gyroscope, a torque must be applied. For GP-B, the only viable technique is by helium gas run at sonic velocity through the differentially pumped channel in the right-hand housing half (Fig. 4). The spin-up gas and torque are then switched off. Let the spin torque be $T_s$ and its residual value after spin-up be $T_r$. With spin-up time $t_s$ and the previously stated Newtonian drift limit $\Omega_0$, $T_r/T_s$ has to be $< t_s\Omega_0$, i.e. $< 10^{-14}$, another formidable requirement, and one that makes cryogenic operation seem at first disadvantageous since at a given pressure, gas torques increase as the temperature is lowered. The key is 'low-temperature bakeout' (Turneaure et al. 1982). At 2.5 K, everything except $He^3$ and $He^4$ is frozen out. By raising the temperature from 2.5 K to 6 K for 10 hours, any adsorbed gas is driven off from the exposed surfaces, first allowed to be exhausted to space and then taken up in a large cryopump after the valve to space is closed. Tests on orbit showed that the pressure in the chamber after bakeout was $< 10^{-14}$ torr, perhaps the lowest ever attained in a scientific instrument.
- *Mechanical stability*: Two obvious concerns in the total design of the GP-B instrument have been: (i) possible thermal distortion of the telescope/quartz block structure, and (ii) possible viscoelastic creep in the shape of the gyro rotor under the surprisingly large centrifugal acceleration from spin. Both matter at room temperature; both become vanishingly small at 2.5 K. At the 230 K ambient satellite temperature, sunlight falling on the side of the telescope would shift the star image many arc-s. The greatly reduced expansion coefficient, and greatly improved thermal isolation in the GP-B dewar, together eliminate any significant image motion. Likewise with creep. Metal rotors at room temperature have been known to change shape significantly after 100 days. Fused quartz at 2.5 K has a viscoelastic time well in excess of 1000 years; no detectable change can occur. One further interesting aspect of cryogenic operation is the reduced thermal relaxation time; the GP-B gyro rotor equilibrates thermally in seconds, not minutes.

Issues of mechanical stability are, as H.A. Rowland remarked 120 years ago, the nightmare of precision measurement. In GP-B, not only do cryogenics help; so does space. On Earth, the GP-B telescope, cantilevered under its own weight, sags, causing a 130 marc-s displacement of the image. In space, such distortions vanish.

Turning from mechanical to optical considerations in the design of the telescope, we note first that a focused 0.14 m aperture diffraction-limited telescope has a 1.8 arc-s Airy disk giving, at 3.81 m focal length, an image diameter of $\sim 30$ μm. Measurement to $< 0.5$ marc-s requires image division at the nm level, yielding two main design constraints, one on pointing, the other on the sharpness of the image divider. First, two star images are formed by means of a beam splitter. These then separately fall on the edges of two roof prisms to give reference signals in two orthogonal directions from the intensities of the beams reflected from the prisms. The criteria determining performance are photon noise and the sharpness of the prism edge. For IM Pegasi (5.9 magnitude), the effective telescope accuracy determined by photon noise was $\sim 50$ marc-s. With that, as already remarked, an accurate gyro-telescope subtraction ($\boldsymbol{G}$–$\boldsymbol{T}$) becomes necessary. Two separate methods have been used. In the first, dither signals of known periods ($\sim 30$ s and $\sim 40$ s) and amplitudes $\sim 60$ marc-s were injected into the pointing servo, and then demodulated. In the second, correlation functions were drawn on the natural pointing variations. The on-orbit agreement was excellent.

The subtraction process just described does not require that either the gyro or telescope scale factor be known; only that they be matched. In fact, determining the gyroscope scale factor $\boldsymbol{C}_g$ would seem at first formidably difficult. The London moment readout is a voltage which, though proportional to angle, depends on several experimental parameters and hence is not directly in angular measure. Happily, Nature itself provides the calibration signal. The apparent position of the guide star varies by known amounts through the Earth's motion around the Sun (annual aberration) and the spacecraft's motion around the Earth (orbital aberration). Because the telescope is held pointing at the star, these signals appear in the gyro readout. They provide exact cross-checking calibrations of $\boldsymbol{C}_g$. Recalling that the line of sight to the star is occulted by the Earth for nearly half of each orbit (Fig. 1), we define the orbit in two parts, Guide Star Valid (GSV) and Guide Star Invalid (GSI). Figure 6 illustrates data from one GSV period. The 5 arc-s orbital aberration envelope is clearly visible superimposed on the output of the gyroscope modulated at the 77.5 s roll period.

Gravity Probe B was launched from Vandenberg Air Force Base, California on 20 April 2004. The dewar stayed cold on orbit for 17 months and 9 days, surpassing the 16 1/2 month

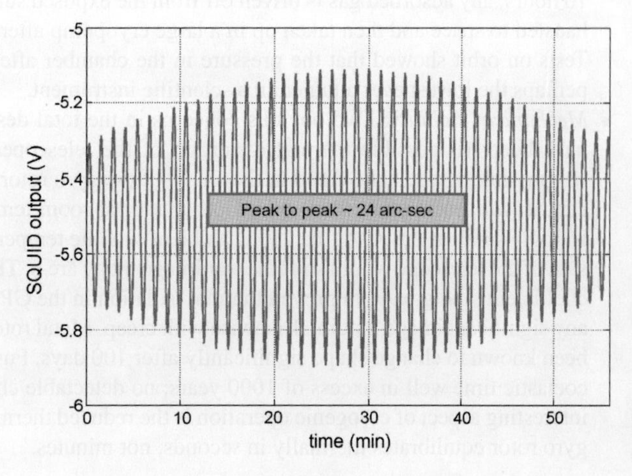


**Fig. 6** Gyroscope London moment data over one Guide Star Valid period

design lifetime. Initial Orbital Checkout (IOC) took 128 days during which time extensive calibration tests re-verified numerous pre-launch performance measures. Spin-up of the four gyroscopes was completed on 14 July 2004; science data collection began 28 August 2004. A final Post-Science Calibration Phase, starting 14 August 2005 and ending with the depletion of the dewar on 29 September 2005, involved tests in which certain potential systematic effects were deliberately enhanced. These included, for example, pointing the spacecraft for limited periods at other real or virtual guide stars to investigate the magnitude of misalignment torques.

## 2 The Actual vs. the Ideal Gravity Probe B

Our aim was an apparatus where all disturbances would be vanishingly small so that all four gyroscopes would agree to within the $< 0.5$ marc-s/yr design requirement. Since the gyros have different mechanical and electrical parameters, and function in different magnetic, electric, and acceleration environments, that four-way agreement, if achieved, would be exceedingly powerful. Other fundamentals included: (1) rolling the spacecraft about the line to the star for torque averaging and removing $1/f$ readout noise, (2) an expected $< 5$ marc-s pointing accuracy at roll period, and (3) the use of drag-free control to reduce the cross-track average acceleration to $< 10^{-12}$ g. With a series of ground-based and IOC tests, known natural variations during the Science Phase, and calibrated increases of possible systematic effects during the Calibration Phase, GP-B would become the most rigorously validated of all tests of Einstein's theory. That aim, though attained by different means in the actual as against the ideal GP-B, is one we still hold to.

Ideally, also, there would be no data interruptions other than the once per orbit occultations of the guide star. With relativity signals increasing as time $t$, and readout noise averaging as $t^{-1/2}$, the measurement accuracy should then advance with the error going down as $t^{-3/2}$, a result slightly modified by the process of calibrating the scale factor against aberration. We knew, however, that interruptions might occur and had a provision for dealing with them.

Most of the ideal conditions were met. The performances of the telescope and dewar met or beat expectation. The launch vehicle delivered GP-B into an almost perfect orbit. The gyroscopes were nearly $10^6$ times better than the best inertial navigation gyroscopes. The UV electric discharge of the rotors proved easier than expected. The low field and trapped flux levels remained rigorously constant. The SQUID noise in the gyro readouts was actually less than the pre-launch expectation, reaching levels corresponding to ideal measurements ranging from 0.14 marc-s/yr for Gyro 3 to 0.35 marc-s/yr for Gyro 4. Nevertheless, some deviations from the ideal did occur, of which the most important were two mutually reinforcing complications in gyro behavior, a *changing polhode period* affecting the determination of the scale factor, and two larger than expected forms of Newtonian torque, known respectively as *misalignment* and *roll-polhode resonance* torques (slightly variant names appear in the accompanying papers by Keiser et al. and Heifetz et al.).

Figure 7 (April 2007) shows what is effectively raw data from the four gyroscopes in the North-South (orbital) plane. The geodetic effect is at once obvious, but so are various unexplained wiggles. The mean $1\sigma$ result at this stage of analysis was $-6{,}638 \pm 97$ marc-s/yr. After subtracting known corrections for the N-S components of guide star proper motion ($+28 \pm 1$ marc-s/yr) and the solar geodetic effect ($+7$ marc-s/yr), the net result was $-6{,}673 \pm 97$ marc-s/yr, to be compared with the $-6{,}606$ marc-s/yr prediction of General Relativity. The E-W results required for frame-dragging were at this stage inconclusive.

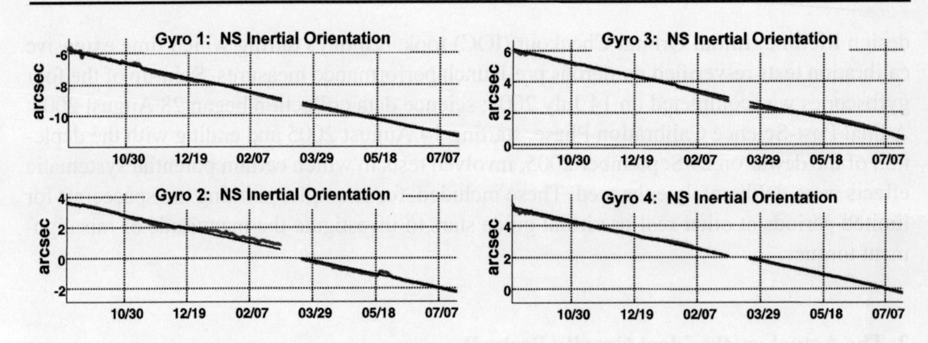


**Fig. 7** Direct measure of the North-South (NS) geodetic precession

**Table 1** The original seven near zeros

| | | |
|---|---|---|
| *Rotor properties* | | |
| • density homogeneity | $< 6 \times 10^{-7}$ | met |
| • sphericity | < 10 nm | met |
| • electric dipole moment | < 0.1 V-m | ***issue*** |
| *Environment* | | |
| • cross-track acceleration | $< 10^{-11}$ g | met |
| • gas pressure | $< 10^{-12}$ torr | met |
| • magnetic field | $< 10^{-6}$ gauss | met |
| *Mixed* | | |
| • rotor electric charge | $< 10^{8}$ electrons | met |

To follow the dramatically improved analysis that has led to the clear observation of frame-dragging reported here, we need to elaborate on the differences between the actual and ideal Gravity Probe B.

The original ideal $< 10^{-11}$ deg/h GP-B gyroscope depended on making seven distinct quantities simultaneously 'near zero'. Three, homogeneity, sphericity, and electric dipole moment, were properties of the rotor. Three, drag-free acceleration, pressure, and magnetic field level, were properties of the environment. The seventh concerned maintaining the rotor's electric charge (due either to lift-off or cosmic ray impacts) at an acceptably low level. Table 1 gives the requirements set before launch. Detailed on-orbit investigations showed that six of the seven were met or surpassed, but with regard to the electric dipole 'near zero', there was an issue, or to be more exact, a double issue due to the patch effect.

The term 'patch effect' refers to contact potential differences between different crystalline or contamination regions on a metal surface (Darling 1989; Speake 1996). Prior to launch, our concern was whether patches on the rotor would create a net dipole moment interacting with the gyro suspension voltages to cause a torque. Kelvin probe measurements on flat samples indicated crystals so minute that any such effect would vanish. This conclusion was wrong as later UV scanning measurements on one flight-quality rotor revealed. More important, it gradually became evident that individual patches on the rotor could interact with nearby patches on the housing, causing significant translational forces and therefore torques, and probably also the changing polhode period. Put simply, while mechanically both rotor and housing are exceedingly spherical, electrically they are not.

*Issue 1: Changing Polhode Period* The discovery a few days into the Science Phase that the polhode periods were changing (Fig. 8) was a surprise. Pre-launch calculations strongly indicated that they would remain fixed throughout the mission. With Gyros 1 and 2, the initial 1.8 and 6 hour periods rapidly increase reaching separatrices with extremely long periods after 20 to 30 days, and then decrease over the rest of the year to asymptotic values of $\sim 1$ and $\sim 3$ hours. They begin by spinning near their minimum moments of inertia, pass through their intermediate moments at the separatrix, and finally approach spinning about their maximum moments. With Gyros 3 and 4 initially spinning more nearly around their maximum moments, there is a steady decrease to asymptotic values of 1.55 and 4.2 hours. Observe that while the body axes go through large motions, the angular momentum of the gyroscope in the inertial frame is unaffected.

Why does a changing polhode period matter? The answer lies in the process of calibrating the gyro scale factor $\boldsymbol{C}_g$ against the aberration of starlight. Performing a $< 0.5$ marc-s/yr measurement in the presence of a 6.6 arc-s/yr geodetic effect, 5 arc-s orbital aberration, and 20 arc-s annual aberration requires a readout scaled to 1–2 parts in $10^5$. Now, the signal for that calibration is the SQUID readout data taken during the half-orbit GSV segments. Reaching $\sim 10^{-5}$ accuracy depends on connecting data from successive half-orbits. Trapped flux in the rotor complicates this process by causing polhode variations in $\boldsymbol{C}_g$ (see Fig. 11). Two methods of resolving the issue are described in the accompanying paper, "Polhode Motion, Trapped Flux, and the GP-B Science Data Analysis" (Silbergleit et al. 2009).

*Issues 2 & 3: The Two Forms of Newtonian Torque* The accompanying paper "Misalignment and Resonance Torques and Their Treatment in the GP-B Data Analysis" (Keiser et al. 2009) details the two forms of Newtonian torque.

- *Misalignment Torque*: During the Calibration Phase, we deliberately pointed the spacecraft off to a series of real and virtual stars to produce large misalignments between the spacecraft roll axis and the gyroscope spin axes. To our surprise, the drift rates were much larger than expected. A detailed investigation revealed that, for misalignments up to 1°, the magnitudes of the torques were proportional to the misalignment angle and produced drift rates in directions at right angles to the misalignment plane (see Fig. 9). The Calibration Phase data provided a valuable initial determination of these torques, but not one sufficiently accurate for the final analysis. More advanced calibration methods are described in Sect. 3.
- *Roll-Polhode Resonance Torque*: This second unexpected torque was discovered by J. Kolodziejczak in April 2007 by exploiting the fact that gyro-to-gyro comparison signals can be processed even during the GSI phase when the guide star is not visible. From time to time, the spin direction of one of the gyroscopes would shift by angles of up to 100 marc-s in 1 to 3 days with no such effect in the other three. Further investigation, reported in the accompanying paper by Keiser et al. revealed that these shifts occur when harmonics of the changing polhode period resonated with the 77.5 s satellite roll period, with the gyro displacement broadly following the Cornu spiral pattern of Fig. 10.

## 3 The Evolving Data Analysis Process

The geodetic effect, being much larger, is easier to separate from Newtonian and other disturbances than the frame-dragging effect. As early as September 2005, it was visible in the raw data, or to be more exact, the data after an initial processing where the aberration signals were used for an approximate calibration, and then removed from the plot of Fig. 7.

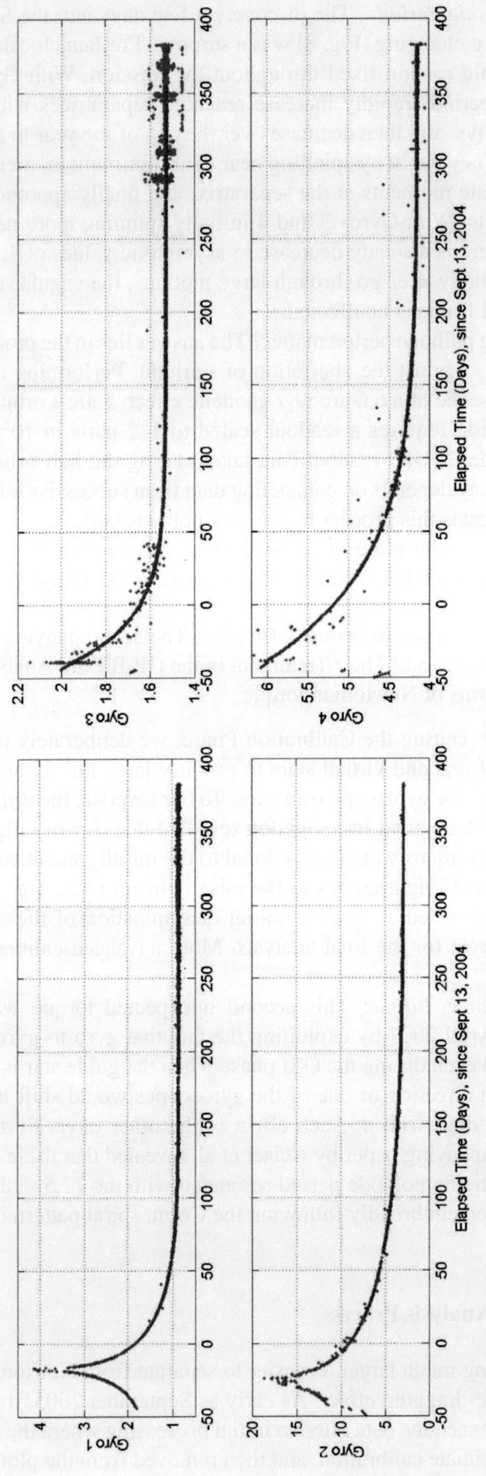


**Fig. 8** Polhode period history of GP-B gyros: *red* from HF FFT, *blue* from snapshots

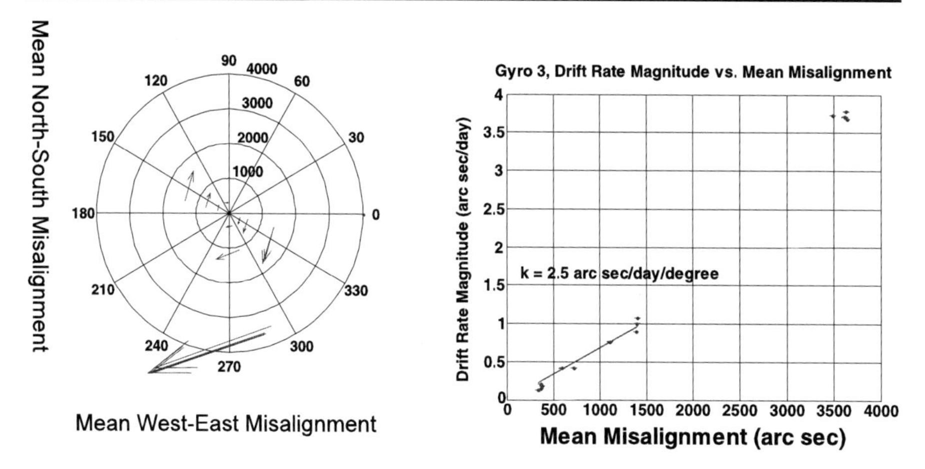


**Fig. 9** Mean drift rate (mas/day) vs. mean misalignment (as) for gyroscope 3 during calibration maneuvers

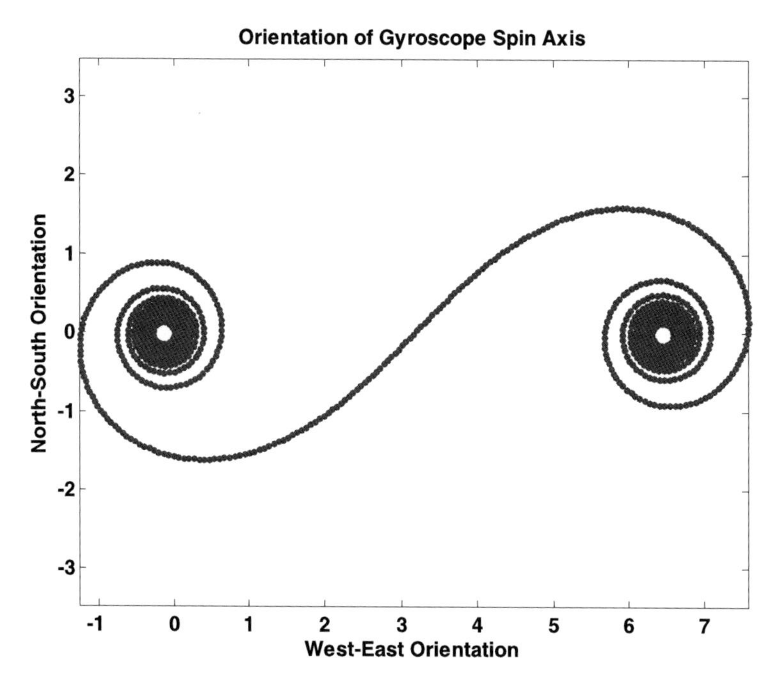


**Fig. 10** Characteristic change in orientation due to a resonance torque. The units are dimensionless but equal in the North-South and West-East directions

Two complementary data processing methods emerged as we proceeded. One, *algebraic*, partially set up before launch, depends on a sophisticated filtering process with explicitly modeled torques. The other, *geometric*, begun in August 2006, has the immediate benefit of illuminating why a clean separation of relativity from the misalignment drift is possible. The relativity signal is in a fixed direction in inertial space; the misalignment drift rate, depending on spacecraft pointing, varies over the year in an exactly known way from the annual aberration of starlight. Define two directions, one 'radial', parallel to the misalignment angle, and the other 'axial', at right angles to it. The signal can be resolved into an axial term where the relativistic and misalignment drift rates are superimposed, and a radial one entirely free of the misalignment term.

The 'geometric' method of separating relativity from the misalignment drifts proved strikingly effective. It has gone through a number of refinements. An essential feature in it and the corresponding 'algebraic' treatment is the need for a continuous history of the misalignment angle through both GSV and GSI periods. An intriguing feature of the experiment, as thus developed, is that the required pointing information during GSI is obtained from the science gyroscopes.

*Trapped Flux Mapping (TFM)* The misalignment torque so elegantly treated by the geometric method is, as already emphasized, only one of three main issues. Progress on the other two, scale factor calibration and treatment of the resonance torques, has depended on a detailed process of Trapped Flux Mapping, begun in November 2006 and now complete.

Figure 11 illustrates the trapped flux and its effect assuming initially that the trapped field is a simple dipole. Superimposed on the London moment is a moment $\boldsymbol{M}_T$ ranging from 0.3%–5% $\boldsymbol{M}_L$ depending on which gyroscope. It has components $\boldsymbol{M}_{T\parallel}$ parallel and $\boldsymbol{M}_{T\perp}$ perpendicular to $\boldsymbol{M}_L$. $\boldsymbol{M}_{T\perp}$ provides a very accurate measurement of the gyro spin speed. $\boldsymbol{M}_{T\parallel}$ adds to $\boldsymbol{M}_L$, evolving with time as the polhode path varies. The SQUID readout signal, $z_i$, is the dot product of the normal, $\boldsymbol{n}$, to the gyro readout loop with the difference between $\boldsymbol{\tau}$ the satellite roll axis direction and $\boldsymbol{s}$ the inertial orientation of the gyroscope, multiplied by $\boldsymbol{C}_g$, which depends on $\boldsymbol{M}_L + \boldsymbol{M}_{T\parallel}$. The $\boldsymbol{n}$ direction changes uniformly with the vehicle roll angle $\phi_r$. The angle $\boldsymbol{\tau}$, measured to high accuracy by the cryogenic telescope, includes both annual (20.4958 arc-s) and orbital (5.1856 arc-s) aberrations, computed respectively from JPL ephemerides data and precise orbit measurements from onboard GPS receivers. Since the measurement noise, $\nu_{SQUID}$, near roll frequency is $\sim 200$ marc-s/Hz$^{1/2}$ ; data from many successive half-orbits must be connected to determine $\boldsymbol{C}_g$ to the necessary $\sim 1$ part in $10^5$ accuracy.

Making reliable orbit-to-orbit connections requires a continuous accurate knowledge of the polhode phase $\phi_p$. Though the basic London Moment readout was not designed to measure $\phi_p$, relevant data was available from engineering telemetry channels in the form of 2 s long 2.2 kHz bursts. From it we obtained 'snapshots' of the trapped flux signal over a num-

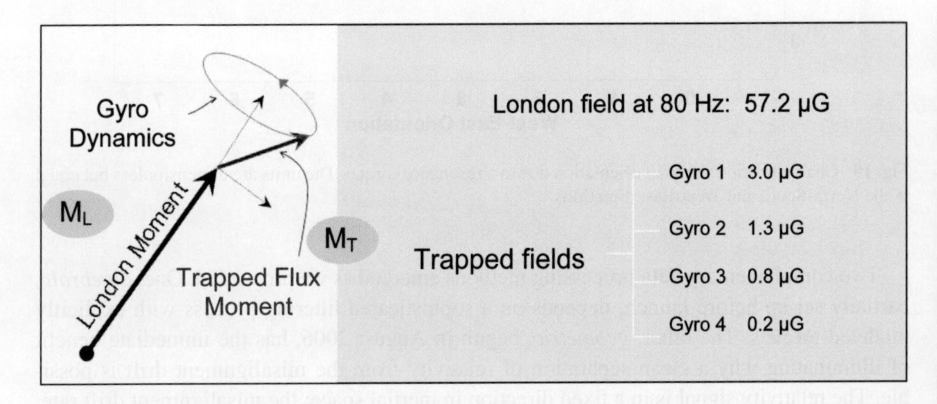


**Fig. 11** Ideal vs. actual London moment readout

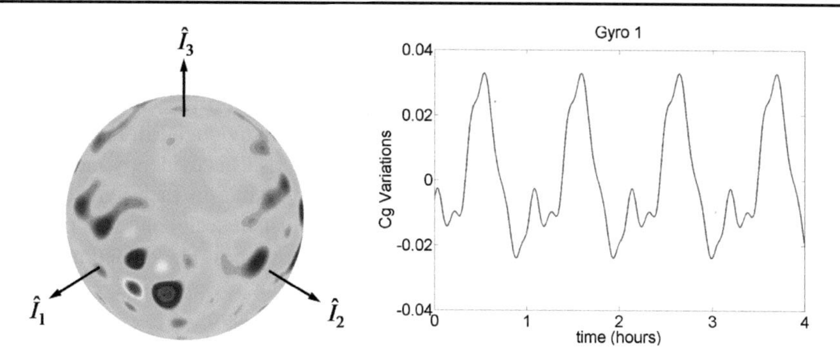


**Fig. 12** Gyro 1 Trapped Flux Map (*left*) and resulting $\boldsymbol{C}_g$ variations scaled to London moment scale factor (*right*)

ber of gyroscope spin cycles, and from the evolving flux pattern thus seen were able to track the motion of the spin axis in the body frame with remarkable precision.

This Trapped Flux Mapping, accomplished by fitting a constant-coefficient rotor-fixed flux map to a dynamic model of the polhode motion, has produced three critical results: (1) a history of the polhode phase good to 1° over the entire 353 days of science, (2) continuous knowledge of rotor spin speeds to $\sim$ 1 part $10^{10}$ and spin-down rates to $\sim$ 1 pHz/s over the entire mission, and (3) a direct estimate of polhode-driven variations in the gyro scale factor $\boldsymbol{C}_g$. Figure 12 shows (left) the flux map for Gyro 1 and (right) the periodic variation in scale factor over a 4-hour time period as the polhoding modulates the direction of the trapped flux. The results for Gyros 2, 3 and 4 had similar forms with successively lower flux levels. Tests on all four gyros demonstrated that the trapped flux distributions remained constant on the surface of the rotors within the measurement limit throughout the mission. For more details, see the accompanying Silbergleit et al. paper.

These advances did not come immediately. Prior to August 2007, the algebraic processing relied on an incompletely known polhode phase derived from the measured polhode period, which yielded batch-to-batch discontinuities in the data. A dramatic moment came in August 2007 when the first TFM data became available. The variations in the estimated parameters was reduced from $100\sigma$ to $6\sigma$, and then with further refinement to $2\sigma$. Data from multiple orbits could be chained together in order to calibrate $\boldsymbol{C}_g$ with enough precision to resolve marc-s changes in gyro orientation.

*TFM and the Resonance Torques* As already remarked, the roll-polhode resonance torques had the effect of producing unexpected (and initially unexplained) shifts in spin direction for individual gyroscopes of up to $\sim$ 100 marc-s in 1–3 days. Coming as they did to one gyroscope at a time, the displacements were in general easily identifiable. With regard to data processing, our first approach was to exclude the resonant period data. A second approach, slightly more advanced, was to bridge across it with a simple ramp function. These two methods were useful as far as they went, but did not get to the heart of the matter, dynamical understanding, nor did they produce the hoped for improvement in results. The spread between the processed signals from the four gyroscopes was substantially larger than the statistical uncertainties recorded for each.

Decisive progress required two things: (1) equations correctly describing the resonance effect (showing *inter alia*) that the displacements follow or nearly follow a Cornu spiral and (2) the accurate continuous time history of polhode phase $\phi_p$ and angle $\gamma_p$ given by Trapped Flux Mapping. The new treatment incorporates additional terms into the basic data processing model with continuously varying functions of $\phi_p$ and $\gamma_p$. Details appear in the accompanying paper "The Gravity Probe B Data Analysis Approach" (Heifetz et al. 2009).

The results are presented in two ways in Figs. 13 and 14. Figure 13, which may usefully be compared with the 'raw data' curves of Fig. 7, shows the individual drift rates in the N-S and E-W directions for the four gyroscopes. The geodetic result is evidently much cleaner than in the raw data; the frame-dragging effect is clearly visible in each gyroscope. Figure 14 shows the same results with the E-W precession (*vertical* scale) and N-S precession (*horizontal* scale) plotted against each other for all four gyroscopes. The reduction in spread from anything previously shown is remarkable, with the four separate 50% error ellipses overlapping for the first time. This result is exceedingly encouraging.

It must be emphasized that this is a preliminary result and does not include either a treatment of systematic error or a model sensitivity analysis. The accompanying paper "GP-B Systematic Error" (Muhlfelder et al. 2009) reviews the current limits. More than a hundred uncertainty sources have been considered using five different classes of evaluation technique: (1) the spread in relativity estimates from the four gyroscopes; (2) studies of different partial data sets at different times of year; (3) comparisons between physics-based models of possible disturbances with appropriate sections of on-orbit data; (4) sensitivity testing by propagating modified values of disturbances (e.g. gyroscope misalignment) through the Science analyses; and (5) comparisons between the results of the 'geometric' and 'algebraic' data analysis methods.

Continuing work on the geometric approach has produced two notable results. Around 11 March, the line to IM Pegasi reaches its closest point to the Sun, with an ecliptic latitude of 22.1°. By using the science gyroscopes as references for the telescope, it is possible to invert GP-B and determine the gravitational deflection of light by the Sun. The accuracy so far is modest $\sim 20\%$, but it serves as a valuable cross-check on the experiment. In addition, IM Pegasi is known to be a binary with period 24.6 days and amplitude 1–2 marc-s. The motion has been determined with considerable accuracy by our colleagues at the Smithsonian Astrophysical Observatory and York University, Canada (Shapiro et al. 2007). Careful investigation has now identified an ellipse of the same period and approximate amplitude in the processed GP-B data. When data processing reaches a more advanced stage, the two independent measurements will be made the basis of a suitable blind comparison test.

Another limitation so far has been computational speed. To bring the processing within the available constraints, we had to average the data once per orbit. We are now actively pursuing high-speed parallel processing techniques to overcome this limitation and allow 2 second processing which, in turn, may provide insight into the fine details of the roll-polhode resonance torque.

## 4 Conclusion

The physics underlying the three major disturbances to the science measurement is well understood, and credible methods for removing them from the measurement now exist. The key synchronizing element for tying together the entire data set is the polhode phase known to within 1° throughout the mission. The Trapped Flux Mapping machinery can accurately model polhode-modulated variations of the readout scale factor so that they can be separated from estimates of gyro drift.

The results of Sect. 3 from the algebraic method for the period 10 December 2004 through 27 May 2005 (155 effective days) give a greatly improved fit to the data model. They yield current *statistical uncertainties* of 12.3 marc-s/yr in the North-South direction

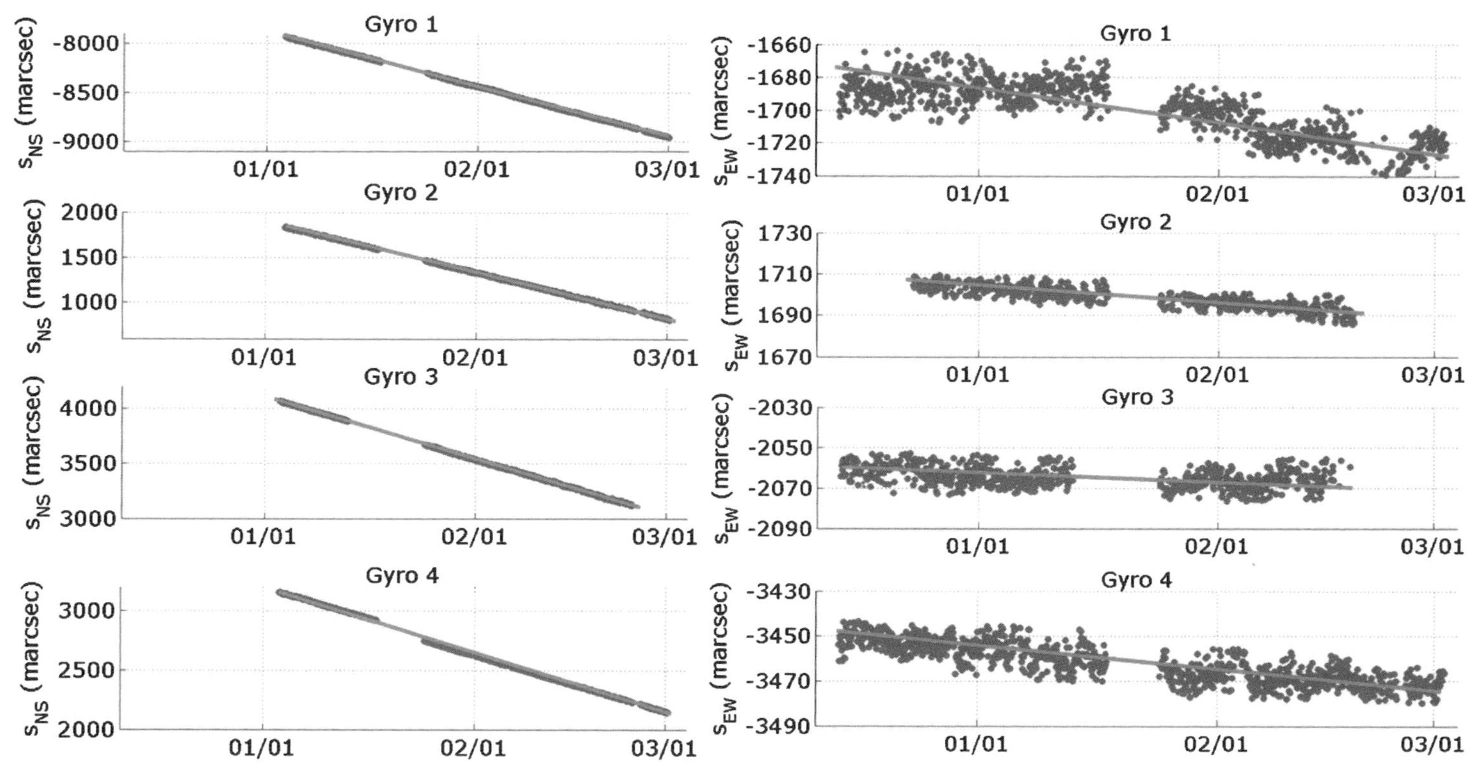


**Fig. 13** Seeing General Relativity directly in all four gyroscopes

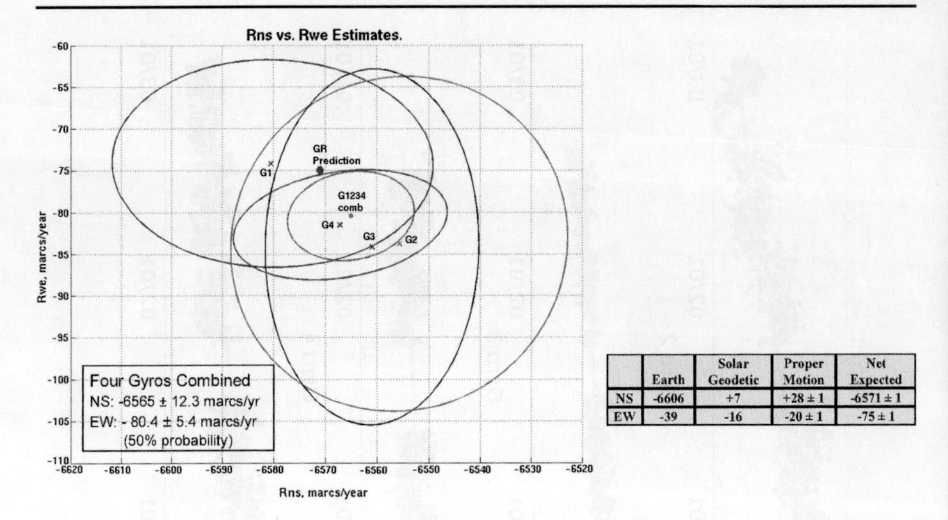


| | Earth | Solar Geodetic | Proper Motion | Net Expected |
|---|---|---|---|---|
| NS | -6606 | +7 | +28 ± 1 | -6571 ± 1 |
| EW | -39 | -16 | -20 ± 1 | -75 ± 1 |

**Fig. 14** Preliminary Relativity Estimates. Gyroscopes 1, 2, 4: Segments 5, 6, and 9 (Gyro 3: Segments 5, 6), not including systematic error or a model sensitivity analysis. Note that the "GR prediction" in the figure is the net expected value of the attached table allowing for solar geodetic and guide star proper motion terms

($<$ 0.5% of the GR prediction) and 5.4 marc-s/yr in the West-East direction ($\sim$ 14% of the GR prediction). These preliminary results require rigorous investigation of systematics by the methods just indicated. With the extension to 279 days or longer of science data and the use of more advanced processing techniques, we hope to approach the 3–5% covariance limit mentioned in Sect. 2.

Analysis through September 2007 provided 20% determinations of the $\sim$ 1 marc-s orbital motion of the guide star and of the bending of starlight due to the Sun.

Reaching the final limit will include completing the work on the geometric and algebraic science analyses and performing an even more extensive investigation of the systematics than is given in the accompanying paper by Muhlfelder et al. The use of high speed computing to decrease the analysis time step from once per orbit (5859 s) to once every two seconds will significantly increase the fidelity of the analysis by providing multiple data points per roll cycle for the modeling of the roll-polhode resonance torque. No fundamental breakthroughs appear to be needed to extend the analysis to these other segments. The necessary techniques are being developed in collaboration with Prof. Charbel Farhat of the Stanford University Department of Aeronautics and Astronautics and his colleagues.

Reviewing progress, we note three dramatic advances in the understanding and treatment of the GP-B science data, with 12-month intervals between them. The first (August 2006) was the geometric insight that the direction of the misalignment torque is varied in an exactly known way over the year as a result of the aberration of starlight. James Bradley's great discovery in 1729 is vital not only for calibration of the gyroscope scale factor, but also for the proper treatment of the misalignment torque. The second advance (August 2007) was the sudden reduction in scatter of the processed SQUID data obtained by applying the greatly improved polhode phase information from Trapped Flux Mapping. The third (August 2008) was the large reduction in spread in the processed signals from the four gyroscopes brought about by including the roll-polhode resonance term into the analysis model.

In the third breakthrough lies a certain irony. Depending as it did on Trapped Flux Mapping, it actually required the presence of trapped magnetic flux in the rotors. With the 'ideal'

London moment readout at one time hoped for, we would not have been able to eliminate the resonance torque from the GP-B data.

**Acknowledgements** This work was supported by NASA Contract NAS8-39225 through the George C. Marshall Space Flight Center (MSFC). From January 15, 2008 through September 30, 2008 critical bridge funding was provided through a generous personal gift by Mr. Richard Fairbank with matching funds from Stanford University and NASA. Acknowledgement is also made for significant funding from KACST in the period since September 30, 2008.

## References

R.J. Adler, A.S. Silbergleit, General treatment of orbiting gyroscope precession. Int. J. Theor. Phys. **39**(5), 1291–1316 (2000)

B.M. Barker, R.F. O'Connell, Phys. Rev. D **2**, 1428 (1970)

J.V. Breakwell, The Stanford relativity gyroscope experiment (F): correction to the predicted geodetic precession of the gyroscope resulting from the Earth's oblateness, in *Near Zero: New Frontiers of Physics*, ed. by J.D. Fairbank, B.S. Deaver, C.W.F. Everitt, P.F. Michelson (Freeman, New York, 1988), pp. 685–690

B. Cabrera, Near zero magnetic fields with superconducting shields, in *Near Zero: New Frontiers in Physics* (Freeman, New York, 1982)

T.W. Darling, Electric Fields on Metal Surfaces at Low Temperatures. *School of Physics*, Parkville, Victoria 3051, Australia, University of Melbourne, p. 88 (1989)

D. DeBra, The Stanford relativity gyroscope experiment (G): translation and orientation control, in *Near Zero: New Frontiers of Physics*, ed. by J.D. Fairbank, B.S. Deaver, C.W.F. Everitt, P.F. Michelson (Freeman, New York, 1988), pp. 691–699

M. Heifetz, W. Bencze et al., The gravity probe B data analysis approach. Space Sci. Rev. (2009)

G.M. Keiser, J. Kolodziejczak, A.S. Silbergleit, Misalignment and resonance torques and their treatment in the GP-B data analysis. Space Sci. Rev. (2009)

J. Lense, H. Thirring On the influence of the proper rotations of central bodies on the motions of planets and moons according to Einstein's theory of gravitation. Z. Phys. **19**, 156 (1918). English translation in B. Mashhoon, F.W. Hehl et al., On the gravitational effects of rotating masses: the Lense-Thirring papers. Gen. Relativ. Gravit. **16**, 711–750 (1984)

F. London, *Superfluids*, vol. II (Wiley, New York, 1954)

F. London, *Superfluids, Macroscopic Theory of Superconductivity* (Dover, New York, 1960)

J.C. Mester, J.M. Lockhart et al., Ultralow magnetic fields and gravity probe B gyroscope readout. Adv. Space Res. **25**(6), 1185–1188 (2000)

B. Muhlfelder, J.M. Lockhart et al., The gravity probe B gyroscope readout system. Adv. Space Res. **32**(7), 1397–1400 (2003)

B. Muhlfelder et al., GP-B systematic error. Space Sci. Rev. (2009)

R.T. Parmley, G.A. Bell et al., Performance of the relativity mission superfluid helium flight dewar. Adv. Space Res. **32**(7), 1407–1416 (2003)

R.T. Parmley, J. Goodman et al., Gravity probe B dewar/probe concept. Proc. SPIE **619**, 126–133 (1986)

G.E. Pugh, *Proposal for a Satellite Test of the Coriolis Prediction of General Relativity*. Washington, DC, Weapons Systems Evaluation Group, The Pentagon (1959)

I.I. Shapiro et al., Bull. Am. Phys. Soc. (3) (2007)

A.S. Silbergleit, J. Conklin et al., Polhode motion, trapped flux, and the GP-B science data analysis. Space Sci. Rev. (2009)

L.I. Schiff, Possible new experimental test of General Relativity Theory. Phys. Rev. Lett. **4**(5), 215–217 (1960)

C.C. Speake, Forces and force gradients due to patch fields and contact-potential differences. Class. Quantum Gravity **13**, A291–A297 (1996)

J.P. Turneaure, E.A. Cornell et al., The Stanford relativity gyroscope experiment (D); ultrahigh vacuum techniques for the experiment, in *Near Zero: New Frontiers of Physics* (Freeman, New York, 1982)

Space Sci Rev (2009) 148: 71–104
DOI 10.1007/s11214-009-9585-7

# Towards a One Percent Measurement of Frame Dragging by Spin with Satellite Laser Ranging to LAGEOS, LAGEOS 2 and LARES and GRACE Gravity Models

**Ignazio Ciufolini · Antonio Paolozzi · Erricos C. Pavlis · John C. Ries · Rolf Koenig · Richard A. Matzner · Giampiero Sindoni · Hans Neumayer**

Received: 26 April 2009 / Accepted: 23 October 2009 / Published online: 18 December 2009
© Springer Science+Business Media B.V. 2009

**Abstract** During the past century Einstein's theory of General Relativity gave rise to an experimental triumph; however, there are still aspects of this theory to be measured or more accurately tested. Today one of the main challenges in experimental gravitation, together with the direct detection of gravitational waves, is the accurate measurement of the gravito-

I. Ciufolini (✉)
University of Salento and INFN Sezione di Lecce, via Monteroni, 73100 Lecce, Italy
e-mail: ignazio.ciufolini@unisalento.it

A. Paolozzi · G. Sindoni
Sapienza University of Rome, Scuola di Ingegneria Aerospaziale, via Salaria 851/881, 00138 Roma, Italy

A. Paolozzi
e-mail: antonio.paolozzi@uniroma1.it

G. Sindoni
e-mail: giampiero.sindoni@uniroma1.it

E.C. Pavlis
University of Maryland, Baltimore County, 1000 Hilltop Circle, Baltimore, MD 21250, USA
e-mail: epavlis@umbc.edu

J.C. Ries
University of Texas at Austin, Center for Space Research, Austin, TX 78759, USA
e-mail: ries@csr.utexas.edu

R. Koenig · H. Neumayer
GFZ German Research Centre for Geosciences, Potsdam, Germany

R. Koenig
e-mail: rolf.koenig@gfz-potsdam.de

H. Neumayer
e-mail: neumayer@gfz-potsdam.de

R.A. Matzner
Center for Relativity, University of Texas at Austin, Austin TX 78712, USA
e-mail: matzner2@physics.utexas.edu

magnetic field generated by the angular momentum of a body. Here, after a brief introduction on frame-dragging, gravitomagnetism and Lunar Laser Ranging tests, we describe the past measurements of frame-dragging by the Earth spin using the satellites LAGEOS, LAGEOS 2 and the Earth's gravity models obtained by the GRACE project. We demonstrate that these measurements have an accuracy of approximately 10%.

We then describe the LARES experiment to be launched in 2010 by the Italian Space Agency for a measurement of frame-dragging with an accuracy of a few percent.

We finally demonstrate that a number of claims by a single individual, that the error budget of the frame-dragging measurements with LAGEOS-LAGEOS 2 and LARES has been underestimated, are indeed ill-founded.

**Keywords** General Relativity · Frame-dragging · Gravitomagnetism · Lense-Thirring effect

## 1 Introduction

A number of experiments have been performed and proposed to accurately measure the gravitomagnetic field (Ciufolini 2007a; Thorne et al. 1986; Ciufolini and Wheeler 1995) generated by the angular momentum of a body, and frame-dragging, from the complex space experiment Gravity Probe B, launched by NASA in 2004 after more than 40 years of preparation (GRAVITY PROBE-B http://einstein.stanford.edu/), to the observations of the LAGEOS and LAGEOS 2 satellites (Ciufolini and Pavlis 2004; Ciufolini et al. 2006, 2010a) and from the LARES satellite, to be launched in 2010 by ASI (Italian Space Agency) (Ciufolini et al. 2010b) using the new launch vehicle VEGA of ESA (European Space Agency), to Lunar Laser Ranging (Williams et al. 2004a), binary pulsars (Stairs et al. 2004) and other astrophysical observations (Nordtvedt 1988; Cui et al. 1998). A number of other space experiments are also currently proposed to various international space agencies.

In Einstein's gravitational theory the local inertial frames have a key role (Misner et al. 1973; Weinberg 1972; Ciufolini and Wheeler 1995). The strong equivalence principle, at the foundations of General Relativity, states that the gravitational field is locally 'unobservable' in the freely falling frames and thus, in these local inertial frames, all the laws of physics are the laws of Special Relativity. The local inertial frames are determined, influenced and dragged by the distribution and flow of mass-energy in the Universe; the axes of these non-rotating, local, inertial frames are determined by torque-free test gyroscopes that are dragged by the motion and rotation of nearby matter, for this reason this phenomenon is called dragging of inertial frames or frame-dragging (Ciufolini and Wheeler 1995; Ciufolini 2007a).

In General Relativity, a torque-free test gyroscope defines an axis non-rotating relative to the local inertial frames; the orbital plane of a test particle is also a kind of gyroscope. The frame-dragging effect on the orbit of a satellite, due to the angular momentum vector **J** of a central body, is known as Lense-Thirring effect:

$$\boldsymbol{\Omega}^{\text{Lense-Thirring}} = \frac{2G\mathbf{J}}{c^2 a^3 (1-e^2)^{3/2}} \tag{1}$$

where $\boldsymbol{\Omega}^{\text{Lense-Thirring}}$ is the rate of change of the longitude of the nodal line of the satellite, that is the intersection of its orbital plane with the equatorial plane of the central body, i.e., it represents the rate of change of the orbital angular momentum vector, $a$ is the semimajor axis of the orbit of the test-particle, $e$ its orbital eccentricity, $G$ the gravitational constant and $c$ the speed of light. Frame-dragging by the Earth spin has been measured using the LAGEOS satellites with about 10 percent accuracy (Ciufolini and Pavlis 2004; Ciufolini et al. 2006, 2010a; Ries et al. 2008) (Sects. 2, 3, 4, 6, 7 and 8, below), might be detected by further Gravity Probe B data analysis (GRAVITY PROBE-B http://einstein.stanford.edu/) and will be measured with an accuracy of a few percent by the LARES satellite (Ciufolini et al. 2010b) (Sects. 5, 6, 7 and 8 below).

In General Relativity there is another type of frame-dragging effect and precession of a gyroscope known as geodetic precession or de Sitter effect (Ciufolini and Wheeler 1995; Ciufolini 2007a). If a gyroscope is at rest with respect to a non-rotating mass, it does not experience any drag. However, if the gyroscope starts to move with respect to the non-rotating mass it acquires a precession that will again disappear when the gyroscope will stop relative to the non-rotating mass. The geodetic precession, due to the velocity $\mathbf{v}$ of a test gyroscope, is $\boldsymbol{\Omega}_{\text{geodetic}} = \frac{3}{2}\frac{GM}{c^2r^3}\mathbf{x} \times \mathbf{v}$, where $M$ is the mass of the central body and $\mathbf{x}$ and $r$ are position vector and radial distance of the gyroscope from the central mass.

A basic difference between frame-dragging by spin and geodetic precession is that in the case of the former (the Lense-Thirring effect) the frame-dragging effect is due to the additional spacetime curvature produced by the rotation of a mass, whereas in the case of the latter (the de Sitter effect) the frame-dragging effect is due to the motion of a test gyroscope on a static background and its motion produces no spacetime curvature, (see below and Sect. 6.11 of Ciufolini and Wheeler 1995; for a discussion on frame-dragging and geodetic precession see Ashby and Shahid-Saless 1990; O'Connell 2005; Ciufolini 2007a, 2009b).

The geodetic precession has been measured on the Moon's orbit by LLR with an accuracy of the order of 0.6 percent (Williams et al. 2004a), see also Bertotti et al. (1987), Williams et al. (1996), by Gravity Probe B with approximately 1 percent accuracy (GRAVITY PROBE-B http://einstein.stanford.edu/) and has been detected on binary pulsars (Weisberg and Taylor 2002; Stairs et al. 2004). Recently, a number of authors have debated whether the gravitomagnetic interaction and frame-dragging by spin have also been accurately measured on the Lunar orbit by Lunar Laser Ranging (Murphy et al. 2007a, 2007b; Kopeikin 2007; Ciufolini 2007a, 2009b). This is a recent chapter of a long debate on the meaning of frame-dragging and gravitomagnetism (Ashby and Shahid-Saless 1990; O'Connell 2005; Barker and O'Connel 1979; Khan and O'Connell 1976; Murphy et al. 2007a, 2007b; Kopeikin 2007; Ciufolini 1994, 2007a, 2009b; Ciufolini and Wheeler 1995); a basic issue treated in Murphy et al. (2007a, 2007b), Kopeikin (2007) is whether the effect detected by LLR is a frame-dependent effect or not.

In order to answer to this question, we have proposed a distinction between gravitomagnetic effects generated by the translational motion of the frame of reference where they are observed, e.g., by the motion of a test gyroscope with respect to a central mass (not necessarily rotating), and those generated by the rotation of a mass or by the motion of two masses (not test-particles) with respect to each other, without any necessary motion of the frame of reference where these effects are observed. The geodetic precession is a translational effect due to the motion of the 'Earth–Moon gyroscope' in the static field of the Sun. The Lense-Thirring effect measured by the LAGEOS satellites, that might also be detected by further Gravity Probe B data analysis and by LARES, is due to the rotation of a mass, i.e., by the rotation of the Earth's mass. In order to distinguish between the Lense-Thirring effect, that we call an 'intrinsic' gravitomagnetic effect, and 'translational' ones, such as the geodetic precession, we have proposed to use spacetime curvature invariants (Ciufolini and Wheeler 1995; Ciufolini 1994, 2009b). For example, in Ciufolini (2009b) we have shown that the phenomena measured by Lunar Laser Ranging are translational gravitomagnetic effects. In general, one cannot derive intrinsic gravitomagnetic effects from translational ones unless making additional theoretical hypotheses, such as the linear superposition of the translational gravitomagnetic effects, i.e., the linear superposition of the terms contained in the non-diagonal part of the metric tensor (the so-called gravitomagnetic potential) in the standard PPN (Post-Newtonian-Parametrized) coordinates; for example, the magnetic field generated by the intrinsic magnetic moment (Bohr magneton) is an intrinsic phenomenon due to the intrinsic spin of a particle that cannot be explained and derived as a translational effect by any Lorentz and frame transformation.

Recently, rotational frame dragging has been derived for all linear cosmological perturbations of Friedmann–Robertson–Walker cosmologies with $k = 0$ (Schmid 2006). Finally, we mention that the uncertainty in proposed determination of the gravitomagnetic field of Mars (Iorio 2006) was shown, by several authors (Krogh 2007; Sindoni et al. 2007; Felici 2007) to be underestimated by a factor of at least ten thousand.

## 2 The Measurement of Gravitomagnetism with LAGEOS, LAGEOS 2 and the GRACE Earth Gravity Models

The orbital plane of a planet, moon or satellite is like a huge gyroscope that feels general relativistic effects, e.g., the Lense-Thirring effect (1), that is the dragging of the whole orbital plane of a test-particle due to the angular momentum $J$ of the central body. The Lense-Thirring effect is extremely small for Solar System objects, so in order to measure its effect on the orbit of a satellite we need to measure the position of the satellite to extremely high accuracy.

Laser-ranging is the most accurate technique to measure distances to the Moon (Bender et al. 1973; Williams et al. 2004b) and to artificial satellites such as LAGEOS (Cohen and Dunn 1985). Short-duration laser pulses are emitted from lasers on Earth and then reflected back to the emitting laser-ranging stations by retro-reflectors on the Moon or on artificial satellites. By measuring the total round-trip travel time we are today able to determine the instantaneous distance of a retro-reflector on the LAGEOS satellites with a few millimeters precision (Noomen et al. 2003).

LAGEOS, LAser GEOdynamics Satellite (Cohen and Dunn 1985), was launched by NASA in 1976 and LAGEOS 2 by the Italian Space Agency (ASI) and NASA in 1992. The LAGEOS satellites are heavy brass and aluminum satellites, of about 406 kg weight, completely passive and covered with retro-reflectors, orbiting at an altitude of about 6000 km above the surface of Earth. LAGEOS and LAGEOS 2 have an essentially identical structure but they have different orbits. The semimajor axis of the orbit of LAGEOS is $a \cong 12270$ km, the period $P \cong 3.758$ hr, the orbital eccentricity $e \cong 0.004$ and the orbital inclination $I \cong 109.9°$. The orbital inclination is the angle between the satellite orbital plane and the Earth equatorial plane. The semimajor axis of LAGEOS 2 is $a_{II} \cong 12163$ km, the orbital eccentricity $e_{II} \cong 0.014$ and the orbital inclination $I_{II} \cong 52.65°$.The LAGEOS satellites position can be predicted, over a 15-day period, with an uncertainty of just a few centimetres (Ciufolini et al. 2006; Noomen et al. 2003).

The Lense-Thirring drag (1) of the orbital plane of LAGEOS and LAGEOS 2 is (Ciufolini 1986, 1996) respectively 31 and 31.5 milliarcsec/yr (a milliarcsec is a thousandth of a second of arc), corresponding at the LAGEOS altitude to approximately 1.9 meters/yr. Since by laser-ranging we can determine their orbits with a few centimeters accuracy, the Lense-Thirring effect can be very accurately measured on the LAGEOS satellites orbit if all other orbital perturbations can be modelled well enough (Ciufolini 1989, 1996, 1998c; Tapley et al. 1989). Indeed, the precession of the node of LAGEOS and LAGEOS 2 can be measured with an accuracy of a fraction of milliarcsec per year (one milliarcsec corresponds to about 6 centimeters at the LAGEOS altitude).

On the other hand, the LAGEOS satellites are very heavy, small cross-sectional area, spherical satellites, therefore atmospheric particles and photons can only slightly perturb their orbit (Rubincam 1990) and especially they can hardly change the orientation of their orbital plane (Ciufolini 1989; Tapley et al. 1989; Rubincam 1990; Lucchesi 2002). Indeed, by far the main perturbation of their orbital plane is due to the Earth's deviations from spherical symmetry. The deviations of the Earth's gravitational potential from spherical symmetry are described by a spherical harmonics expansion of the potential (Kaula 1966). However, the only secular perturbations of the node of a satellite are due to the so-called even zonal harmonics, $J_{2n}$, i.e., the spherical harmonics terms of even degree and zero order (axially symmetric deviations from spherical symmetry of the Earth's gravitational potential that are also symmetric with respect to the Earth's equatorial plane). In particular, the flattening of the Earth's gravitational potential, described by the quadrupole moment, produces a large perturbation of the LAGEOS node (Ciufolini 1986, 1989; Tapley et al. 1989). The rate of change of the nodal longitude of a satellite $\dot{\Omega}$, due to the quadrupole moment, $J_2$, and to the second largest even zonal harmonic of degree four, i.e., $J_4$ (see Fig. 1) are described by

$$\dot{\Omega}^{\text{Class}} = -\frac{3}{2}n\left(\frac{R_\oplus}{a}\right)^2 \frac{\cos I}{(1-e^2)^2} \times \left\{J_2 + J_4\left[\frac{5}{8}\left(\frac{R_\oplus}{a}\right)^2 \times (7\sin^2 I - 4)\frac{(1+\frac{3}{2}e^2)}{(1-e^2)^2}\right] + \cdots\right\} \tag{2}$$

where $n = 2\pi/P$ is the orbital mean motion, $P$ is the orbital period, $R_\oplus$ is the Earth equatorial radius and $J_{2n}$ are the even zonal harmonic coefficients. The orbital parameters $n, a, e$ and $I$ in (2), i.e., mean motion, semimajor axis, orbital eccentricity and orbital inclination are determined with sufficient accuracy via LAGEOS laser ranging (Cohen and Dunn 1985; Noomen et al. 2003; Ciufolini 1989; Tapley et al. 1989; Ciufolini et al. 1989, 2004), see Sects. 7 and 8. Any other quantity in (2) can be determined or is known with sufficient accuracy, apart from the $J_{2n}$.

We stress that what is *critical for the measurement of the Lense-Thirring effect* is that the *modeling* of this classical *node precession* (i.e., the prediction of its behavior on the basis of the available physical models) must be accurate enough (i.e., at the level of a few milliarcsec) compared to the Lense-Thirring effect (of size of about 31 milliarcsec Ciufolini 1986). What is critical is *not* that all the quantities entering this equation, i.e., the Earth parameters and the orbital parameters, and in particular the Earth spherical harmonic coefficients and the semimajor axis and the inclination, must be *predicted* in their variations, but instead what is critical is that they must be *determined* with sufficient accuracy via satellite laser ranging and other techniques (such as GRACE). For example if the variations of the *inclination* and of the *semimajor axis* of LAGEOS are not well modeled because the effect of particle drag (i.e., atmospheric drag) is not known with sufficient accuracy, this is *not* critical for our measurement of the Lense-Thirring effect because we are able to *measure* the variations of inclination and semimajor axis accurately enough with satellite laser ranging (see Sect. 8 on the uncertainty in the determination of the orbital inclination due to the atmospheric refraction modeling errors) and we are thus able to precisely quantify the effect of these variations on the nodal rate, (2) (with the orbital estimators GEODYN, EPOS-OC and UTOPIA). Indeed, in Sect. 8 we show that the average measurement error in the inclinations of LAGEOS

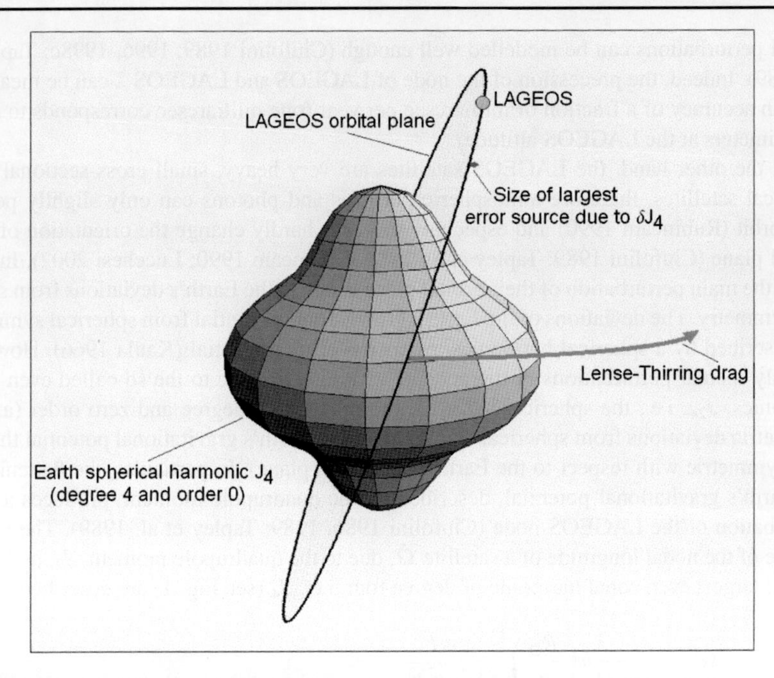


**Fig. 1** (Color online) The Lense-Thirring effect on the orbital plane of a test-particle. The Lense-Thirring precession of the orbital plane of a test-particle by the spin of a central body is represented by the *big red arrow* (*lower thicker arrow*). Also shown is the Earth deviation from spherical symmetry (enhanced, and so not to scale) described by the so-called even zonal harmonic of degree four, $J_4$. The uncertainty in its static part is the largest source of error in the measurement of Earth's frame-dragging using the LAGEOS satellites. The maximum precession of the LAGEOS orbital plane due to the uncertainty in $J_4$ of the EIGEN-GRACE02S model, that is, the nodal precession error due to $\delta J_4$, is represented by the *blue arrow* (*upper smaller arrow*); this error and the Lense-Thirring effect are drawn to scale

and LAGEOS 2 (that are roughly constant) can be estimated to be at the level of a few tens of $\mu$arcsec for LAGEOS and LAGEOS 2, and this error, when propagated in the nodal rate, (2), corresponds to less than one percent of the Lense-Thirring effect only.

In fact, the only quantities in (2) for $\dot{\Omega}$ that are not measured with sufficient accuracy, in order to accurately measure the Lense-Thirring effect, are the even zonal harmonic coefficients $J_{2n}$ and therefore the main uncertainty in the measurement of the Lense-Thirring effect is due to the uncertainty in these $J_{2n}$ coefficients. For example, a relative uncertainty $\frac{\delta J_2}{J_2}$ of the quadrupole coefficient, $J_2$, of the order of Reigber et al. (2005): $\frac{\delta J_2}{J_2} \sim 10^{-7}$, corresponds, from (2) to an uncertainty in the nodal precession of about 45 milliarcsec/year, i.e., to a systematic error in the measurement of the Lense-Thirring effect (that has a predicted size of 31 milliarcsec) of about 150%. In addition, we need to include the uncertainty due to the higher $J_{2n}$ coefficients.

In order to solve the problem of the systematic error due to the uncertainty in the Earth's even zonal harmonics coefficients, such as Earth's flattening, the following main techniques were proposed.

One technique would be to use polar satellites; in fact, from formula (2), for a polar satellite, since $I = 90°$, $\dot{\Omega}^{\text{Class}}$ is equal to zero. Yilmaz proposed the use of polar satellites in 1959 (Yilmaz 1959) and in 1976, Van Patten and Everitt (1976) proposed an experiment with two drag-free, guided, counter–rotating, polar satellites. The reason to propose *two* counter-rotating polar satellites was to avoid the inclination measurement errors.

Another solution (Ciufolini 1984, 1986, 1989; Tapley et al. 1989; Ciufolini et al. 1989, 2004) would be to orbit a new satellite of LAGEOS type (called LAGEOS III), with the same semimajor axis, the same eccentricity, but the inclination supplementary to that of LAGEOS. With this choice, the classical precession $\dot{\Omega}^{\mathrm{Class}}$, (2), would be equal and opposite for the two satellites. By contrast, since the Lense-Thirring precession, (1), is independent of the inclination, $\dot{\Omega}^{\mathrm{Lense\text{-}Thirring}}$ would be the same in magnitude and sign for both satellites. Therefore, by combining the measured nodal precessions of LAGEOS and LAGEOS III we could eliminate the uncertainties due to all the even zonal harmonics in $\dot{\Omega}^{\mathrm{Class}}$ and very accurately measure $\dot{\Omega}^{\mathrm{Lense\text{-}Thirring}}$.

Another technique, that was proposed in Ciufolini (1989), Ciufolini and Wheeler (1995), is to orbit several high-altitude, laser-ranged satellites, similar to LAGEOS; this is the method that we used in Ciufolini and Pavlis (2004), Ciufolini (1996) and in this paper, and that is described below in Sect. 3. A similar technique, proposed in Ciufolini (1996), is the use of three observables: both nodes of LAGEOS and LAGEOS II and the perigee of LAGEOS II, in order to remove the error due to the uncertainties of the Earth's quadrupole moment, $J_2$, and of the next largest even zonal harmonic, $J_4$ (see Fig. 1), and to measure the Lense-Thirring effect, i.e., in order to use three observables for the three unknowns: Lense-Thirring effect, $\delta J_2$ and $\delta J_4$ (Ciufolini 1996). This technique led to the observation of the Lense-Thirring effect in 1996–1998 (Ciufolini et al. 1997a, 1997b, 1998b). However, the accuracy of these earlier observations of the Lense-Thirring effect could not be easily assessed, the two limiting factors were (a) the knowledge of the Earth's gravity field in 1996, indeed the Earth gravity models JGM-3 and EGM96, even though representing the state of the art in 1995–1996, were not accurate enough for a precise measurement of the Lense-Thirring effect, thus forcing to use a third observable, i.e., the perigee; (b) the use of the perigee of LAGEOS 2, in fact the perigee of a satellite is less stable than the node under non-gravitational perturbations; in classical mechanics the node (orbital angular momentum) is conserved under any central force, however, the perigee (Runge-Lenz vector) is conserved only under a central force of the type $\sim 1/r^2$. Thus, the perigee of an Earth satellite such as LAGEOS 2 is affected by a number of non-gravitational perturbations whose impact in the final error budget is not easily assessed.

Therefore, since 1996, in order to use the nodes only, our effort was to find a new observable to replace the perigee of LAGEOS 2 (see the LARES Sect. 5) and to eliminate the perigee from our analysis using a more accurate Earth gravity field model.

Overcoming the problem (Ries et al. 2003a, 2003b) of the Earth's gravity field uncertainties came in March 2002 when NASA's two identical GRACE (Reigber et al. 2002; Tapley 2002, 2004) spacecraft (Gravity Recovery and Climate Experiment) were launched in a polar orbit at an altitude of approximately 450 km and at a mutual distance of about 200–250 km. The spacecraft range to each other via a radar and they are tracked by the GPS satellites. The GRACE satellites have provided dramatic improvements in the knowledge of the Earth's gravitational field. Indeed, by using the two LAGEOS satellites and GRACE Earth gravity models (Reigber et al. 2005), the orbital uncertainties due to the modelling errors in the non-spherical Earth gravitational field are only a few percent of the Lense-Thirring effect (Ciufolini et al. 2006), see Sects. 3, 4 and 6. In 2004, nearly eleven years of laser-ranging data were analyzed. This analysis resulted in a measurement of the Lense-Thirring effect, described in Sects. 3 and 4, with an accuracy (Ciufolini and Pavlis 2004; Ciufolini et al. 2006, 2010a) of approximately 10%; the main error source was the uncertainty in some axially symmetric Earth's departures from sphericity (see Figs. 1, 2).

After 2004, other accurate Earth gravity models have been published using longer periods of GRACE observations. The LAGEOS analyses have been recently repeated with

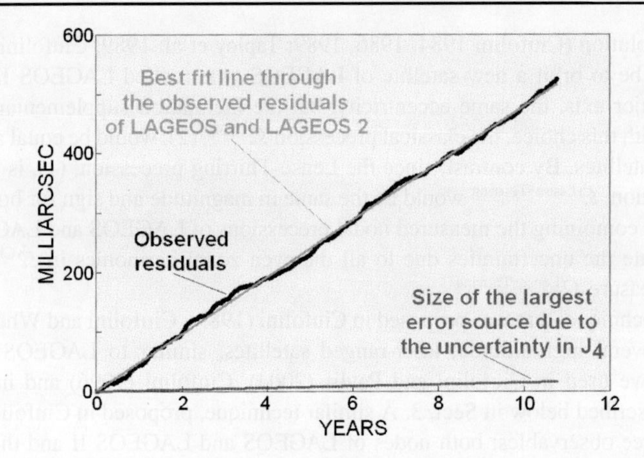


**Fig. 2** The Lense-Thirring effect measured via the LAGEOS satellites in 2004. The *red solid straight line* (*upper straight line*) is the best-fit line through the observed residuals (in *black*, *thicker curve*) and the *blue solid straight line* (*lower straight line*) represents the uncertainty in the combined nodal longitudes of the LAGEOS satellites from the largest error source due to the uncertainty in the Earth's even zonal harmonic of degree four, $J_4$, of the EIGEN-GRACE02S model; see Fig. 1. The observed slope of the *red line* is $0.99 \pm 0.1$, where 1 is the prediction of General Relativity and the $\pm 0.1$ uncertainty is the estimated total systematic error; see Sects. 2, 3, 4, 7 and 8

these models, over a longer period and by using different orbital programs independently developed by NASA Goddard and GFZ of Potsdam/Munich. These recent frame-dragging measurements (Ciufolini et al. 2010a), by a team of the universities of Salento (Lecce), Sapienza (Rome), Maryland BC, NASA-Goddard and GFZ of Potsdam/Munich, have improved the precision of the 2004 LAGEOS determination of the Lense-Thirring effect. No deviations from the predictions of General Relativity have been observed.

In 2008, these measurements of the Lense-Thirring effect were repeated and extended (Ries et al. 2008) by an independent group of CSR-University of Texas at Austin, with an independent orbital estimator called UTOPIA and using more recent and more accurate GRACE Earth gravity field models. Their results, reported in the next section, confirmed our measurements of the Lense-Thirring effect with an accuracy of about 12%. The laser-ranged satellite LARES (ASI) will provide a future improved test of Earth's gravitomagnetism with accuracy of a few percent, see Sects. 5 and 6.

## 3 Method of the 2004 Analysis of LAGEOS and LAGEOS 2 Data Using the GRACE Models

The accurate measurement of the Lense-Thirring effect, obtained in 2004 and described in this paper, has been obtained using the laser–ranging data of the satellites LAGEOS and LAGEOS 2 and the Earth gravity field models EIGEN-GRACE02S (Reigber et al. 2005), EIGEN-GRACE03S and JEM03G. An independent GRACE group of the Center for Space Research (CSR) of the University of Texas at Austin has extended these measurements using GGM02S, EIGEN-CG03C, GIF22a, JEM04G, EIGEN-GL04C, JEM01-RL03B, GGM03S, ITG-GRACE03S, EIGEN-GL05C. The analysis covered an observational period between about 11 years and 14 years, i.e. more than 2.5 times longer than in any previous analysis.

We have analyzed the laser–ranging data using the principles described in International Earth Rotation Service (1996) and adopted the underlying IERS conventions in our mod-

**Table 1** Models used in the orbital analysis with EIGEN-GRACE02S, EIGEN-GRACE03S, JEM03G

| | |
|---|---|
| geopotential (static part) | EIGEN-GRACE02S, EIGEN-GRACE03S, JEM03G |
| geopotential (tides) | Ray GOT99.2 and FES2002, FES2004 |
| lunisolar and planetary perturbations | JPL ephemerides DE-403 |
| general relativistic corrections | PPN except L–T |
| Lense-Thirring effect | set to zero |
| direct solar radiation pressure | cannonball model |
| albedo radiation pressure | Knocke–Rubincam model |
| Yarkovsky–Rubincam effect | GEODYN model |
| spin axis evolution of LAGEOS satellites | LOSSAM 2004 (Andrès et al. 2004) |
| station positions (ITRF) | ITRF2000 |
| ocean loading | Scherneck model with GOT99.2 and FES2002 tides |
| polar motion | estimated |
| Earth rotation | VLBI + GPS |

eling, except that, in the 2004 analysis and following ones, we used the GRACE Earth's static gravity models listed above. Our analysis was performed using 15-day arcs. For each 15-day arc, initial state vector (position and velocity), coefficient of reflectivity ($C_R$) and polar motion were adjusted. Solar radiation pressure, Earth albedo, and anisotropic thermal effects were modeled according to Rubincam (1988, 1990), Rubincam and Mallama (1995), Martin and Rubincam (1996). In modeling the thermal effects, the orientation of the satellite spin axis was obtained from Andrès et al. (2004). We have applied a $\dot{J}_4 = -1.41 \cdot 10^{-11}$ correction (Reigber et al. 2005; Ciufolini and Pavlis 2005). Lunar, solar, and planetary perturbations were also included in the equations of motion, formulated according to Einstein's general theory of relativity with the exception of the Lense-Thirring effect, which was purposely set to zero. Polar motion was adjusted and Earth's rotation was modeled from the very long baseline interferometry-based series SPACE (Gross 1996) which are extended annually. We analyzed the laser-ranging data and the orbits of the LAGEOS satellites using the orbital analysis and data reduction software GEODYN II (NASA Goddard) (Pavlis et al. 1998) and EPOS-OC (GFZ) (the CSR-UT team used UTOPIA). The models used in the GEODYN II and EPOS-OC analysis are listed in Table 1.

As we have pointed out in the previous section, the perigee of an Earth satellite such as LAGEOS 2 is affected by a number of perturbations whose impact in the final error budget is not easily assessed and this was one of the two main points of concern of Ries et al. (2003a). The other point of concern was some favorable correlation of the errors of the Earth's spherical harmonics for the EGM96 model that might lead to some underestimated error budget. However, these points of concern do not exist in our 2004–2009 analyses with the GRACE models; using the previous models JGM-3 and EGM96, we were forced to use three observables to eliminate the $J_4$ uncertainty (Figs. 1, 2) and thus we needed to use the perigee of LAGEOS 2. However, for a measurement of the Lense-Thirring effect with accuracy of the order of 10%, using a number of 2004–2008 GRACE models, thanks to the dramatic improvement in the determination of the Earth's gravity field due to GRACE, it is just enough to eliminate the uncertainty in the Earth's quadrupole moment and thus to use two observables only, i.e. the two nodes of the LAGEOS satellites. Nevertheless, for a measurement of the Lense-Thirring effect with accuracy of a few percent, it is necessary to use one additional observable that will be provided by the node of the LARES satellite (see the LARES Sects. 5 and 6).

To precisely quantify and measure the gravitomagnetic effects we have introduced the parameter $\mu$ that is by definition 1 in general relativity (Ciufolini and Wheeler 1995) and zero in Newtonian theory (thus, our approach is not based on the metric gravitational theories described by the PPN, Post-Newtonian-Parametrized, approximation).

The main error in this measurement is due to the uncertainties in the Earth's even zonal harmonics and their time variations. The unmodeled orbital effects due to some harmonics of lower degree are of order of magnitude comparable to the Lense-Thirring effect (Ciufolini 1996). However, analyzing the GRACE models and their uncertainties in the even zonal harmonics, and propagating these errors on the nodes of LAGEOS and LAGEOS 2, we find that by far the main source of error in the determination of the Lense-Thirring effect is only due to the first even zonal harmonic, $J_2$ (Ciufolini and Pavlis 2004; Ciufolini et al. 2006).

We can therefore use the two observable quantities $\dot{\Omega}_I$ and $\dot{\Omega}_{II}$ to determine $\mu$ (Ciufolini 1989, 1996, 2002), thereby avoiding the largest source of error arising from the uncertainty in $J_2$. We can do this by solving the system of the two equations for $\delta\dot{\Omega}_I$ and $\delta\dot{\Omega}_{II}$ in the two unknowns $\mu$ and $J_2$, obtaining for $\mu$:

$$\begin{aligned}
&\delta\dot{\Omega}^{\text{Exp}}_{\text{LAGEOS I}} + c\,\delta\dot{\Omega}^{\text{Exp}}_{\text{LAGEOS II}} \\
&\quad = \mu(31 + c31.5)\ \text{milliarcsec/yr} + \text{other errors} \\
&\quad \cong \mu(48.2\ \text{milliarcsec/yr}),
\end{aligned} \tag{3}$$

where $c = 0.545$.

The use of the nodes of two laser-ranged satellites of LAGEOS type to measure the Lense-Thirring effect by eliminating in this way the Earth spherical harmonics uncertainties was first proposed and published in Ciufolini (1984, 1989) and further studied in Tapley et al. (1989), Ciufolini et al. (1989, 2004), Ries (1989). The calculation of the standard relativistic perigee precession of LAGEOS was carried out in Rubincam (1977) and the proposal to use laser ranging to artificial satellites to detect relativistic effects, among which the Lense-Thirring effect, was published in Cugusi and Proverbio (1978), in this paper the LAGEOS Lense-Thirring precession was calculated to be 4 arcsec/century, i.e. 40 milliarcsec/yr, instead of the correct 31 milliarcsec/yr figure calculated in Ciufolini (1986) and the problem of the Earth's even zonal harmonics errors was not treated in Cugusi and Proverbio (1978). A solution to the problem of the Earth's spherical harmonics using polar satellites was proposed in Yilmaz (1959) and then in Van Patten and Everitt (1976), see Sect. 2. The use of the nodes of $N$ high-altitude, laser-ranged satellites, similar to LAGEOS, to determine the first $N-1$ even zonal harmonics $J_2$, $J_4$, $J_6, \dots$, and to measure the Lense-Thirring effect was first published in Ciufolini (1989), p. 3102 (see also Ciufolini and Wheeler 1995, p. 336). The use of the nodes of LAGEOS and LAGEOS 2, together with the explicit expression of the LAGEOS satellites nodal equations, was proposed in Ciufolini (1996). A detailed study of the various possibilities to measure the Lense-Thirring effect using LAGEOS and other laser-ranged satellites was presented in Peterson (1997). The use of GRACE-derived gravitational models, when available, to measure the Lense-Thirring effect with accuracy of a few percent was published by Ries et al. (2003a, 2003b) and Pavlis (2000).

Equation (3) for $\mu$ does not depend on $J_2$ nor on its uncertainty, thus, this value of $\mu$ is unaffected by the largest error, due to $\delta J_2$, and it is sensitive only to the smaller uncertainties due to $\delta J_{2n}$, with $2n \geq 4$.

The next largest error source is due to uncertainty in $J_4$ that may be as large as 10% of the Lense-Thirring effect. To eliminate this error source we need a new observable that will indeed be provided by the node of LARES, allowing a measurement of the Lense-Thirring effect with accuracy of a few percent.

The various error sources that can affect the measurement of the Lense-Thirring effect using the nodes of the LAGEOS satellites have been extensively treated in a large number of papers by several authors, see, e.g., Ciufolini (1986, 1989, 1996), Ciufolini et al. (1997a, 1997b, 1998b, 2006, 2010a), Rubincam (1988, 1990), Rubincam and Mallama (1995), Martin and Rubincam (1996), Lucchesi (2001, 2002); the main error sources are treated in details in Ciufolini et al. (2006, 2010a). We refer to these papers for a detailed error analysis and error budget. Here, we only point out that a single author has claimed in a number of papers (Iorio 2005a; Iorio 2008a, 2008b, 2009) that the error analysis and error budget of the frame-dragging measurements with LAGEOS-LAGEOS 2 and LARES have been quite underestimated. Therefore, in Sect. 7 we describe the main misunderstandings and miscalculations that led to these ill-founded claims; any interested reader should study Sects. 2, 7 and 8, and Ciufolini and Pavlis (2005), Ciufolini et al. (2006), Lucchesi (2005), Ries et al. (2008).

## 4 Results of the Measurement of the Lense-Thirring Effect

In this section we present the results in the measurement of the Lense-Thirring effect on the basis of a number of different GRACE Earth gravitational models, through the analysis of the nodal rates of the LAGEOS and LAGEOS 2 satellites. We first report the results we obtained using EIGEN-GRACE02S, EIGEN-GRACE03S and JEM03G (Figs. 2, 3, 4, 5) over a period of about 11 years. We also separately plot the integrated residuals of the node of LAGEOS, of the node of LAGEOS 2 and of their $J_2$-free combination (Fig. 3). In Fig. 4d, we report the result of the measurement of the Lense-Thirring effect with EIGEN-GRACE02S using the orbital estimator EPOS-OC by the German GRACE group of GFZ, this program is independent of GEODYN; thus this independent result confirms our previous analyses obtained with GEODYN. Then, in Fig. 5, we present the results with these three models, including an error bar representing the total estimated error in our measurement including systematic errors. Finally, in Fig. 6 we report the result of Ries et al. (2008) for the measurement of the Lense-Thirring effect using a number of GRACE models, including EIGEN-GRACE02S, GGM02S, EIGEN-CG03C, GIF22a, JEM04G, EIGEN-GL04C, JEM01-RL03B, GGM03S, ITG-GRACE03S and EIGEN-GL05C.

Figure 3 shows (in blue) the integrated residuals of the node of LAGEOS from January 1993 to October 2003 using the model EIGEN-GRACE02S, the integrated node residuals of LAGEOS 2 (in green), using EIGEN-GRACE02S, and the integrated combination of the nodes residuals of LAGEOS and LAGEOS 2 (in red), according to formula (3), using EIGEN-GRACE02S; this figure displays that the large residuals of the node of each satellite (in blue and green) due to the error in $J_2$ are eliminated in the combined residuals (in red), see also Ries et al. (2008).

Figure 4 shows the result of the measurement of the Lense-Thirring effect using the three GRACE Earth models EIGENGRACE02S (Fig. 4a), EIGENGRACE03S (Fig. 4b) and JEM03G (Fig. 4c), by fitting the orbital residuals with a secular trend together with *six* periodic effects. Figure 4d represents the result of the measurement of the Lense-Thirring effect using *the orbital estimator EPOS-OC* of the German GRACE team of GFZ, with EIGEN-GRACE02S and by fitting the orbital residuals with a secular trend together with *six* periodic effects.

In conclusion, by fitting our combined residuals with a secular trend plus 6 periodic signals, using EIGENGRACE02S, we found over an observational period of 11 years:

$$\mu_{\text{EIGENGRACE02S}} = 0.994 \pm 0.10$$

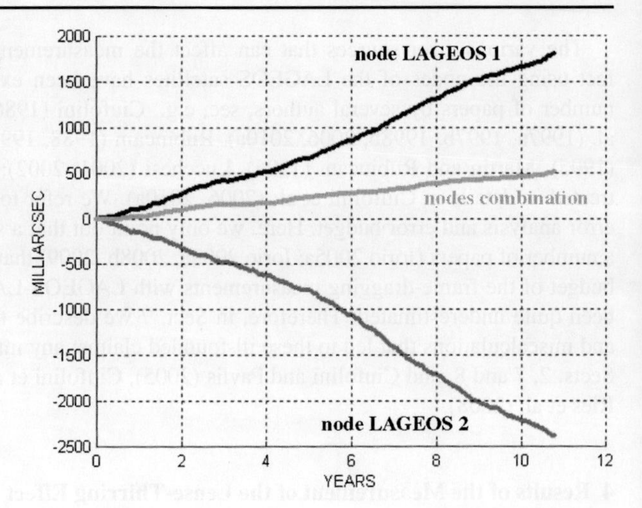


**Fig. 3** (Color online) Residuals of the node of LAGEOS (in *blue*; *upper curve*), node of LAGEOS 2 (in *green*; *lower curve*) and combination of the nodes of LAGEOS and LAGEOS 2 using formula (3) (in *red*; *middle line*), JEM03G is the GRACE Earth gravity model used here

with RMS of the post–fit residuals of 5.98 milliarcsec using GEODYN II however, using EPOS-OC and the corresponding GFZ set-up describing the satellites orbital perturbations, we found: $\mu_{\mathrm{EIGENGRACE02S}} = 1.0 \pm 0.10$ (with RMS of the post–fit residuals of 6.92 milliarcsec); this small 0.6% difference for $\mu$ using the two orbital estimators is due to a different modeling of the orbital perturbations in the two cases.

By fitting our combined residuals with a secular trend plus six periodic signals, using EIGENGRACE03S and JEM03G, we found respectively:

$$\mu_{\mathrm{EIGENGRACE03S}} \simeq 0.93 \pm 0.13$$

and

$$\mu_{\mathrm{JEM03G}} \simeq 0.99 \pm 0.18,$$

where these uncertainties include all systematic errors. The static gravitational errors for the models EIGEN-GRACE02S, EIGEN-GRACE03S and JEM03G have been calculated by simply adding the absolute values of the errors in our combined residuals due to each even zonal harmonics uncertainty and by then multiplying this total error for a factor 2 to take into account possible underestimations of the published uncertainties of each model (see Sect. 7). However, in the case of JEM03G, only the formal uncertainties of the $J_{2n}$ were available to us but *not* their calibrated uncertainties, i.e., the uncertainties including the systematic errors, and we have then tentatively used the uncertainties of the GRACE model GGM02S; then, using the published, calibrated, uncertainties of GGM02S, we obtain a total error budget of the order of 10%, or by doubling the static even zonal harmonics uncertainties, we get a total error of the order of 18%.

In the case of EIGENGRACE02S, by fitting our combined residuals with 2, 6, or 10 periodic terms we practically get the same value for the Lense-Thirring effect and by analyzing the data with the NASA orbital estimator GEODYN II and with the GFZ orbital estimator EPOS-OC with their corresponding different set-up for the orbital perturbations, we practically obtain the same result. Furthermore, these different measurements of the Earth frame-dragging effect obtained with EIGEN-GRACE02S, EIGEN-GRACE03S and JEM03G are in agreement with each other within their uncertainties. Therefore, our measured value of the Lense-Thirring effect with the Earth gravity models EIGEN-GRACE02S,

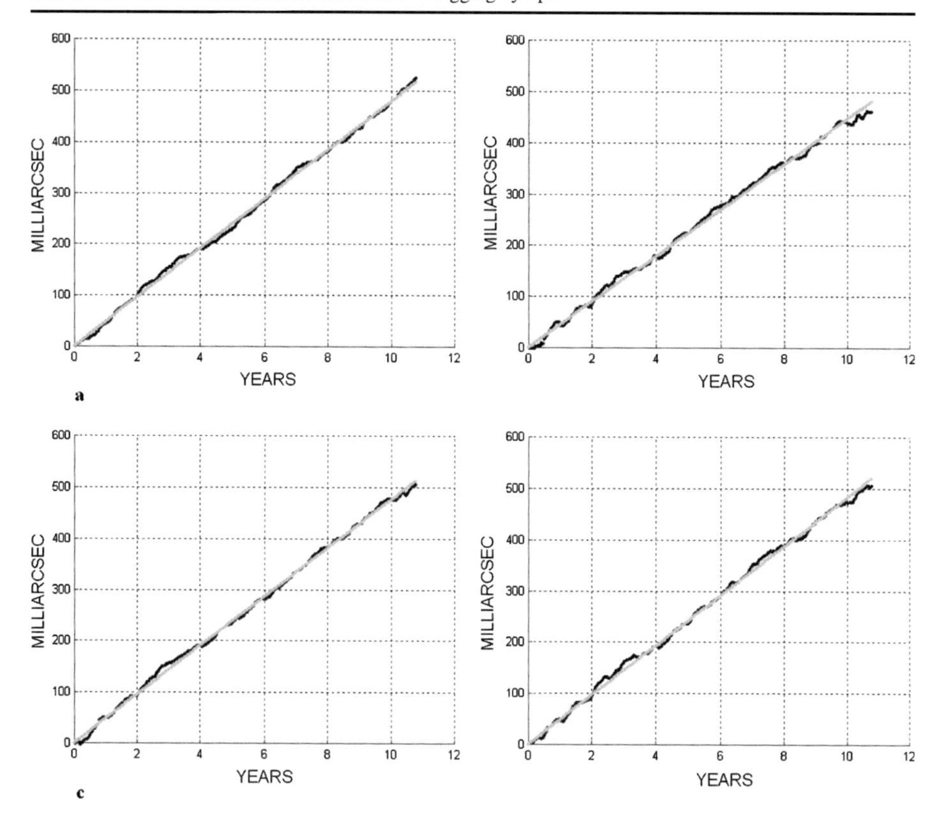


**Fig. 4** Linear fit of the residuals of the nodes of LAGEOS and LAGEOS 2, using the combination (3) with: **a** the model EIGEN-GRACE02S, **b** EIGEN-GRACE03S, **c** JEM03G. In **d** is shown the fit of the nodes of LAGEOS and LAGEOS 2, using *the orbital estimator EPOS-OC* of the German GRACE team of GFZ and the model EIGEN-GRACE02S. The fits in **a** to **d** are with a secular trend plus *six* periodic terms. The slope in **a** is $\mu \simeq 0.99$, in **b** $\mu \simeq 0.93$, in **c** $\mu \simeq 0.99$ and in **d** $\mu \simeq 1.0$. The scale of the axes is different with respect to Fig. 3

EIGEN-GRACE03S and JEM03G agree with the general relativistic prediction and the corresponding uncertainty of our measurement using the more recent GRACE models is of the order of 10% (see Appendix of Ciufolini et al. 2010a).

Ries et al. (2008) have extended the measurement of the Lense-Thirring effect to a number of more recent models, including EIGEN-GRACE02S, GGM02S, EIGEN-CG03C, GIF22a, JEM04G, EIGEN-GL04C, JEM01-RL03B, GGM03S, ITG-GRACE03S and EIGEN-GL05C, using the orbital estimator UTOPIA and the corresponding set-up for the orbital perturbations, the results are presented in Fig. 6. Ries et al. have concluded that the mean value of the Lense-Thirring effect using these models is $0.99\mu$ with a total error budget in the measurement of the Lense-Thirring effect of about 12%.

*In conclusion, the analysis of the University of Salento, Sapienza University of Rome, University of Maryland Baltimore County and GFZ Potsdam/Munich (using the orbital estimators GEODYN and EPOC-OC), and of the Center for Space Research of the University of Texas at Austin (using the orbital estimator UTOPIA) have confirmed the general relativistic prediction for Lense-Thirring effect using the LAGEOS and LAGEOS 2 orbital data with an accuracy of about 10% (see* Figs. 5, 6).

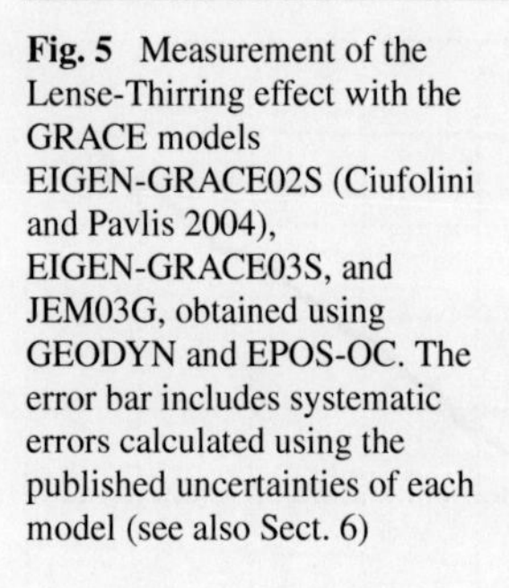
Fig. 5 Measurement of the Lense-Thirring effect with the GRACE models EIGEN-GRACE02S (Ciufolini and Pavlis 2004), EIGEN-GRACE03S, and JEM03G, obtained using GEODYN and EPOS-OC. The error bar includes systematic errors calculated using the published uncertainties of each model (see also Sect. 6)

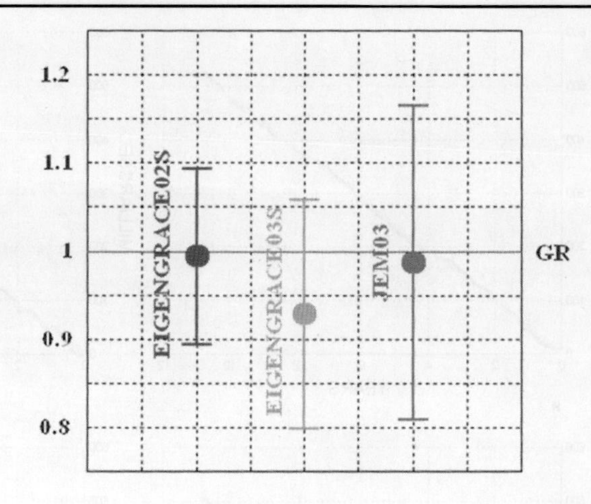


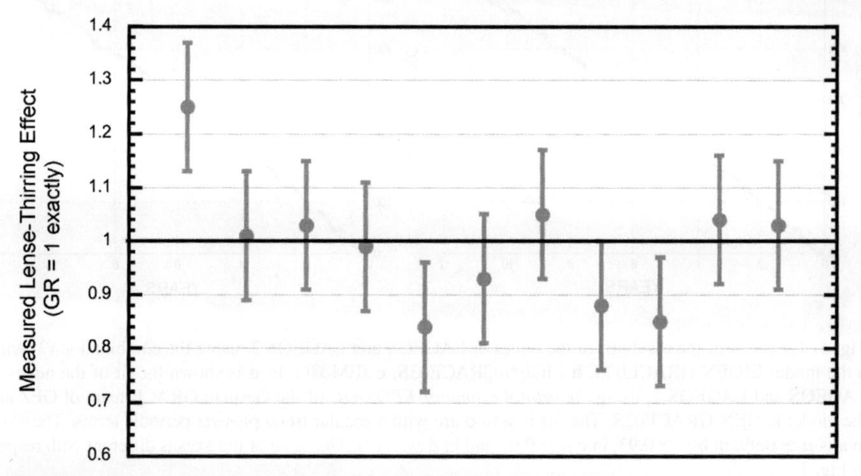


**Fig. 6** Measurement of the Lense-Thirring effect with the GRACE models: EIGEN-GRACE02S, GGM02S, EIGEN-CG03C, GIF22a, JEM04G, EIGEN-GL04C, JEM01-RL03B, GGM03S, ITG-GRACE03S and EIGEN-GL05C, obtained using UTOPIA (Ries et al. 2008)

## 5 LARES

The LARES space experiment, by the Italian Space Agency (ASI), is based on the launch of a new laser-ranged satellite, called LARES (LAser RElativity Satellite), using the new launch vehicle VEGA (ESA-ELV-ASI-AVIO), Fig. 7. LARES will have an altitude of about 1450 km, an orbital inclination of about 71.5 degrees and a nearly zero eccentricity. The LARES satellite together with the LAGEOS (NASA) and LAGEOS 2 (NASA and ASI) satellites and with the GRACE (NASA-CSR and DLR-GFZ) Earth gravity field models will allow a measurement of the Earth gravitomagnetic field and of Lense-Thirring effect with an uncertainty of a few percent.

In this section, after a description of the LARES experiment and of its orbit, we discuss the main error sources affecting the measurement of gravitomagnetism with LARES; these are due to the uncertainties in the Earth's gravitational field, in particular in the Earth's even zonal harmonics and to the time-dependent component of the Earth's gravitational field, in

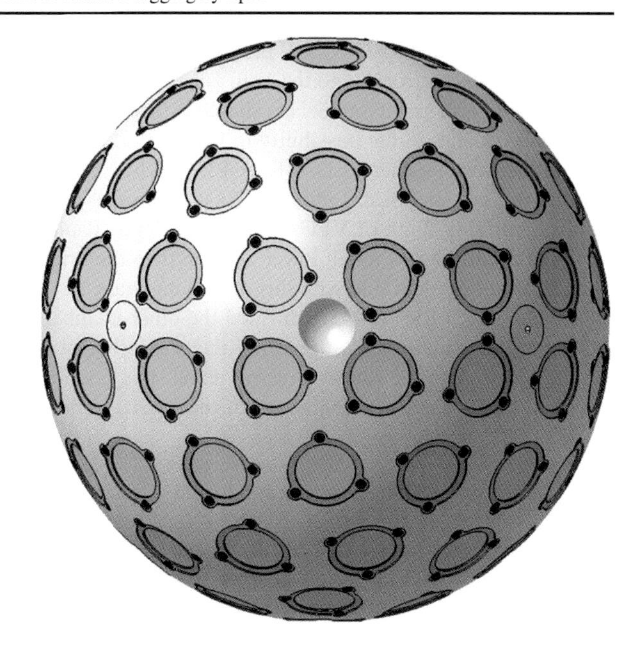

**Fig. 7** Drawing of one version of the LARES satellite

particular $\dot{J}_6$ and the $K_1$ tide. We also briefly discuss the effect of particle drag and the error due to the uncertainties in the measurement of the orbital inclination (see Sect. 8). We finally describe the structure of the LARES satellite that is designed and built in order to minimize the non-gravitational perturbations.

## 5.1 Introduction

In Sect. 2 we briefly reported the LAGEOS III 1984 proposal, i.e., the use of the nodes of two laser-ranged satellites of LAGEOS type to measure the Lense-Thirring effect (Ciufolini 1984, 1986). Several papers (Ciufolini 1989), international studies (Tapley et al. 1989; Ciufolini et al. 1989, 2004), proposals (Ciufolini et al. 1998a) and Ph.D. dissertations (Ries 1989; Peterson 1997), analyzed the LAGEOS III proposal.

Unfortunately, even though such an orbit of LARES would have allowed a complete removal of the static Earth spherical harmonics secular errors, in order to measure the much smaller Lense-Thirring effect, the weight of the proposed LARES satellite, of about 400 kg, and especially the high altitude of its orbit implies an expensive launch vehicle. For this reason a LARES satellite of only about 100 kg of weight was later designed (Ciufolini et al. 1998a), with a non-zero orbital eccentricity to allow some equivalence principle tests as proposed by Nordtvedt (1998).

Nevertheless, three new factors have changed the need of such a high-altitude, expensive, orbit for LARES: (a) the idea to use N laser-ranged satellites to measure the Lense-Thirring effect and to cancel the uncertainty due to the first N-1 even zonal harmonics (Ciufolini 1989; Ciufolini and Wheeler 1995; Ciufolini and Pavlis 2004) (that led to the 1995–1996 observations Ciufolini et al. 1997a, 1997b, 1998b; Ciufolini 2000 and to the 2004–2008 measurements Ciufolini and Pavlis 2004; Ciufolini et al. 2006, 2010a; Ries et al. 2008 of the Lense-Thirring effect, described in Sects. 2–4); (b) the launch of the GRACE spacecraft in 2002 and the publication of a new generation of very accurate Earth's gravity field models using the GRACE observations (Reigber et al. 2002, 2005; Tapley 2002; Watkins et al. 2002) (at the time of the first LAGEOS III proposal, Ciufolini 1989; Tapley et al. 1989; Ciufolini et al. 1989, 2004, the error due to the even zonal harmonics was much larger due to the much less accurate Earth gravity models available at that time and the LAGEOS 2 satellite was not yet launched); and (c) the possibility to launch the LARES satellite using the qualification flight of the new launcher VEGA (ESA-ELV-ASI-AVIO), however at a much lower altitude than the original proposal. Indeed in 2004, one of us (A.P.) discovered the possibility to use the qualifying flight of VEGA to orbit LARES (Paolozzi 2005). However, this launch for LARES will be at a much lower altitude than the originally planned satellite at 12270 km. The altitude achievable with this qualifying launch is of about 1450 km. Nevertheless, one of us (J.C.R.) Ries (2005) informed the other members of the LARES team that CSR had done some simulations supporting the possibility of using a lower orbit laser-ranged satellite to measure the Lense-Thirring effect, this possibility was later on also discussed in Iorio (2005b). Precise calculations of the LARES gravitational errors were analyzed in Ciufolini (2006)

### 5.2 A New Laser-Ranged Satellite at a Lower Altitude than LAGEOS and LAGEOS 2

The simplest conceivable orbit in order to cancel the effect of all the even zonal harmonics on the node of a satellite would be a polar orbit (for such an orbit the effect of the even zonal harmonics on the satellite node would be zero and, however, the node of the satellite would still be perturbed by the Earth gravitomagnetic field, i.e., by the Lense-Thirring effect, Lucchesi and Paolozzi 2001, see Sects. 1 and 2).

Unfortunately, as pointed out in the 1989 LAGEOS III NASA/ASI study (Tapley et al. 1989; Ciufolini et al. 1989, 2004) and as explicitly calculated by Peterson (1997) (Chap. 5 of Peterson 1997), the uncertainty in the $K_1$ tide (tesseral, $m = 1$, tide) would make such an orbit unsuitable for the Lense-Thirring measurement. Indeed, a polar satellite would have a secular precession of its node whose uncertainty would introduce a large error in the Lense-Thirring measurement (in addition, it would be quite demanding to launch LARES with the requirement of a small orbital injection deviation from a polar orbit; in fact, in order to cancel the error due to the uncertainties in the static Earth gravity field, for a very accurate measurement of the Lense-Thirring effect at an altitude lower than the one planned for LAGEOS III, the deviation from a polar orbit should be less than a tenth of a degree).

Nevertheless, a non-polar orbit would have a nodal precession, due to its departure from 90 degrees of inclination, and thus one could simply fit for the effect of the $K_1$ tide using a periodical signal exactly at the nodal frequency. Such signal (with the periods of the LAGEOS and LAGEOS 2 satellites nodes) is indeed observed in the already mentioned LAGEOS and LAGEOS 2 analyses (Ciufolini et al. 1998b; Ciufolini and Pavlis 2004) and is the periodical signal with the largest amplitude observed in the combined residuals.

Furthermore, in regard to the effect of the static even zonal harmonics, by using the technique explained in Ciufolini (1989, 1996) and by using the nodes of the satellites LARES, LAGEOS and LAGEOS 2, we would be able to cancel the uncertainties due to the first two even zonal harmonics, $J_2$ and $J_4$, and our measurement will only be affected by the uncertainties of the even zonal harmonics with degree strictly higher than 4.

By solving the system of the three equations for the nodal precessions of LAGEOS, LAGEOS 2 and LARES in the three unknowns, $J_2$, $J_4$ and Lense-Thirring effect, we have a combination of three observables (the three nodal rates) which determines the Lense-Thirring effect independently of any uncertainty, $\delta J_2$ and $\delta J_4$, in the first two even zonal harmonics. This same technique was applied in Ciufolini et al. (1998b) using the nodes of LAGEOS and LAGEOS 2 and the perigee of LAGEOS 2 and in Ciufolini and Pavlis (2004) using the nodes of LAGEOS and LAGEOS 2 only.

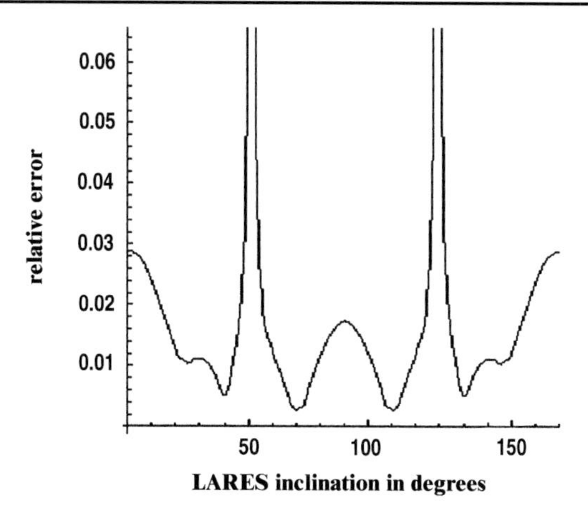


**Fig. 8** Percent uncertainty in the measurement of the Lense-Thirring effect, due to the even zonal harmonics uncertainties, as a function of the inclination of LARES, using LARES, LAGEOS and LAGEOS 2. The altitude of LARES is here 1500 km and the range of the inclination between 0 and 180 degrees

It turns out that some values of the inclination of LARES would minimize the error in the measurement of the Lense-Thirring effect since they would minimize the error due to the uncertainty in the largest (not cancelled using the combination of the three observables) even zonal harmonic $J_6$.

In Fig. 8 we have plotted the relative error in the measurement of the Lense-Thirring effect as a function of the inclination by assuming an altitude of LARES of 1500 km, i.e., a LARES semimajor axis of about 7880 km.

From Fig. 8 we can see that any inclination from 60 degrees to 86 degrees and from 94 to 120 degrees would be suitable for a measurement of the Lense-Thirring effect with an accuracy of a few percent. An inclination of LARES of about 110 degrees or 70 degrees would minimize the error. In deriving this result, we have assumed: (a) zero eccentricity for the LARES orbit, (b) we have only considered the effect of the first 5 even zonal harmonics: $J_2$, $J_4$, $J_6$, $J_8$ and $J_{10}$ and (b) we have considered the uncertainties in the spherical harmonics $J_6$, $J_8$ and $J_{10}$ equal to those of the EIGEN-GRACE02S Earth's gravity model (Reigber et al. 2005), i.e., we have assumed $\delta C_{60} = 0.2049 \cdot 10^{-11}$, $\delta C_{80} = 0.1479 \cdot 10^{-11}$, $\delta C_{10\,0} = 0.2101 \cdot 10^{-11}$, where the $C_{l0}$ are the normalized zonal harmonic coefficients related to the un-normalized zonal harmonic coefficients $J_l$ by: $J_l \equiv -\sqrt{2l+1}C_{l0}$. However, by including higher degree even zonal harmonics, the results of Fig. 8 would only slightly change.

Indeed, in the next section and in Figs. 9, 10, 11, 12, we calculate and show the error in the measurement of the Lense-Thirring effect, with LARES, LAGEOS and LAGEOS 2, due to the even zonal harmonics up to degree 70, corresponding to a LARES orbit of 1450 km of altitude and 71.5 degrees of inclination. Some of the results described in this section and those shown in Fig. 8, are based on the calibrated uncertainties (i.e., including systematic errors) of the EIGEN-GRACE02S model; even though the real errors in these EIGEN-GRACE02S coefficients would probably be about two or three times larger than these published uncertainties, EIGEN-GRACE02S is a 2004 model and by the time of the launch of LARES and of its data analysis (about 2010–2017), much improved Earth's gravity field models based on much longer data set of GRACE observations would be available, with true errors even smaller than the EIGEN-GRACE02S uncertainties that we used in the present analysis.

In regard to the other orbital perturbations that affect the LARES experiment, we briefly discuss here the tidal effects, particle drag and thermal drag; for a detailed treatment of other perturbations we refer to Ciufolini (1989), Tapley et al. (1989), Ciufolini et al.

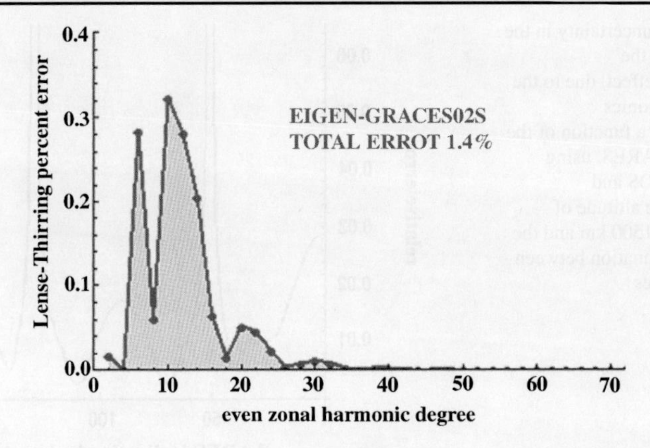


**Fig. 9** Percent error in the measurement of the Lense-Thirring effect using LARES, LAGEOS and LAGEOS 2 as a function of the uncertainties of each even zonal harmonic. The model used is EIGEN-GRACE02S (GFZ Potzdam, 2004) and the uncertainties in this model include systematic errors. Using EIGEN-GRACE02S, the total error in the measurement of the Lense-Thirring effect due to the even zonal harmonics is 1.4%. An improvement by about an order of magnitude is expected at the time of the LARES data analysis

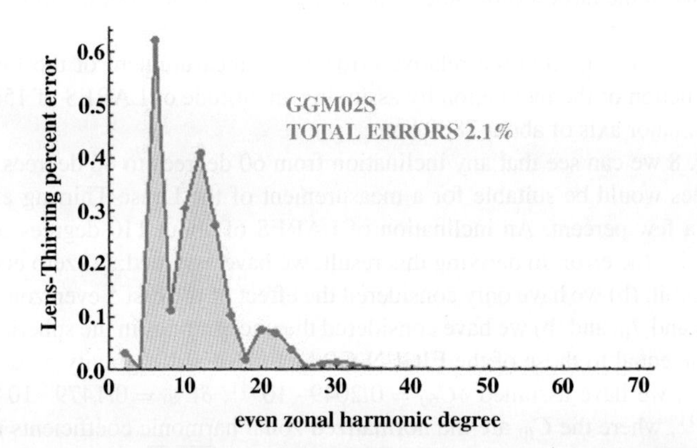


**Fig. 10** Percent error in the measurement of the Lense-Thirring effect using LARES, LAGEOS and LAGEOS 2 as a function of the uncertainties of each even zonal harmonic. The model used is GGM02S (CSR, 2004) and the uncertainties in this model include systematic errors. Using GGM02S, the total error in the measurement of the Lense-Thirring effect due to the even zonal harmonics is 2.1%. An improvement by about an order of magnitude is expected at the time of the LARES data analysis

(1989, 2004, 2006, 2010a). In regard to the orbital perturbations of the LARES experiment due to the time-dependent Earth's gravity field, we observe that the largest tidal signals are due to the zonal tides with $l = 2$ and $m = 0$, due to the Moon node, and to the $K_1$ tide with $l = 2$ and $m = 1$ (tesseral tide). However, the medium and long period zonal tides ($l = 2$ and $m = 0$) will be cancelled using the combination of the three nodes together with the static $J_2$ uncertainty (also the uncertainty in the time-dependent secular variations $\dot{J}_2$, $\dot{J}_4$ will be cancelled using this combination of three observables). Furthermore, the tesseral tide $K_1$ will be fitted for over a period multiple of the LARES nodal period, as explained above (see Tapley et al. 1989 and Chap. 5 of Peterson 1997) and this tide would then introduce a small uncertainty in our combination. In regard to the non-gravitational orbital perturbations, we stress here that the unmodeled thermal drag perturbations on the LARES orbit would be reduced

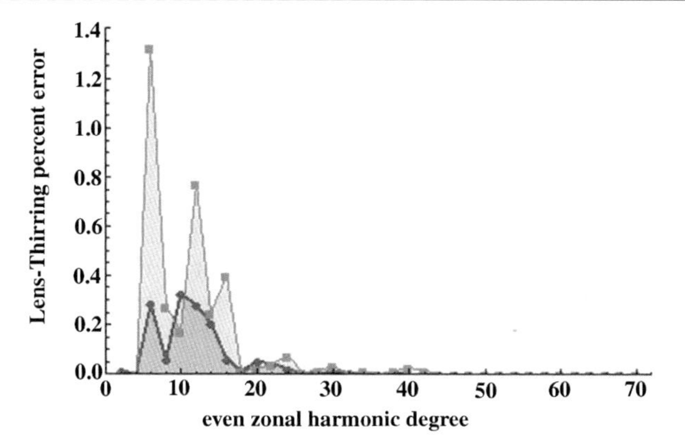


**Fig. 11** (Color online) Percent error in the measurement of the Lense-Thirring effect using LARES, LAGEOS and LAGEOS 2 as a function of the uncertainty due to each even zonal harmonic. The points in *blue* (*lower points* joined by the *solid line* boundary of the darker shadow) are the errors obtained using the model EIGEN-GRACE02S (i.e. Fig. 9 rescaled) and the points in *red* (*upper points* joined by the *solid line* boundary of the lighter shadow) are the errors obtained using, as uncertainty of each coefficient, the difference between the value of this coefficient in the two different models EIGEN-GRACE02S and GGM02S. The total error in the measurement of the Lense-Thirring effect using EIGEN-GRACE02S is 1.4% and by using as uncertainties the differences between the coefficients of the two models is 3.4%. However, at the time of the LARES data analysis an improvement of about one order of magnitude has to be taken into account with respect with these 2004 models that were based on less than 365 days of observations of the GRACE spacecraft

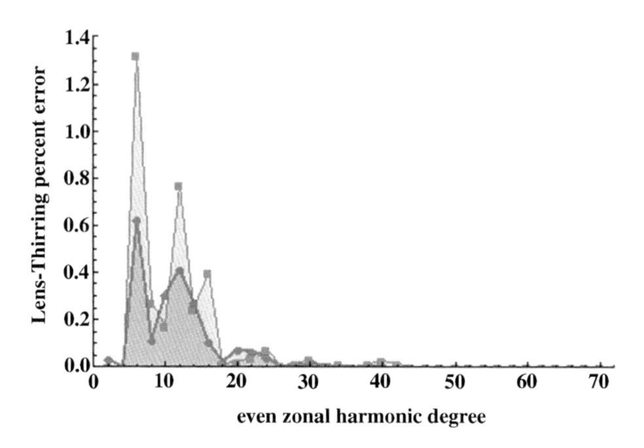


**Fig. 12** (Color online) Percent error in the measurement of the Lense-Thirring effect using LARES, LAGEOS and LAGEOS 2 as a function of the uncertainty due to each even zonal harmonic. The points in *green* (*lower points* joined by the *solid line* boundary of the darker shadow) are the errors obtained using the model GGM02S (i.e. Fig. 10 rescaled) and the points in *red* (*upper points* joined by the *solid line* boundary of the lighter shadow) are the errors obtained using as uncertainty of each coefficient the difference between the value of this coefficient in the two different models EIGEN-GRACE02S and GGM02S. The total error in the measurement of the Lense-Thirring effect using GGM02S is 2.1% and by using as uncertainties the differences between the coefficients of the two models is 3.4%. However, at the time of the LARES data analysis an improvement of about one order of magnitude has to be taken into account with respect with these 2004 models that were based on less than 365 days of observations of the GRACE spacecraft

with respect to the LAGEOS satellites thanks to the much smaller (by a factor of about 0.34) cross-sectional to mass ratio of LARES (see the Sect. 5.3 on the LARES structure); furthermore, accurate measurements of the thermal properties of LARES and of its retro-reflectors should be performed. We finally point out that the neutral and charged particle drag on the LARES node at an altitude of about 1450 km will be a small effect (of a fraction of a percent of the Lense-Thirring effect) for an orbit with very small eccentricity, even by assuming that the exosphere would be co-rotating with Earth at 1450 km of altitude and by considering the exosphere density inhomogeneities at that altitude (Ciufolini et al. 2009c). Indeed, as calculated in Ciufolini (1989), Ciufolini et al. (1990) for the LAGEOS III satellite, in the case of zero orbital eccentricity $e = 0$, the total drag effect on the LARES node would be zero; indeed the nodal rate of a satellite due to particle drag is a function of $\sin\nu \cdot \cos\nu$ ($\nu$ is the true anomaly) and the total nodal shift is then zero over one orbit; in the case of a small orbital eccentricity, the total shift would be proportional to the eccentricity and it would still be a small effect, as calculated in Ciufolini (1989). *In regard to the orbital inclination, as explained in Sects.* 2, 7.1 *and* 8, *we stress that what is critical for the measurement of the Lense-Thirring effect is that the measurement of the inclination be accurate enough but not its modeling (i.e., the prediction of its behavior on the basis of the available physical models).* In Sect. 8, we treat the accuracy in the measurement of the orbital inclination of the LAGEOS satellites and its main limitation due to atmospheric refraction.

### 5.3 Structural Requirements for LARES Satellite

The LARES satellite has been designed in such a way to minimize all the non gravitational perturbations such as particle drag and thermal thrust (induced by the anisotropic thermal radiation from the satellite due to the anisotropic temperature distribution over the satellite surface). Indeed, the orbit of LARES is much lower than that of the two LAGEOS satellites, therefore the minimization of the non-gravitational perturbations, such as particle drag, is especially important for LARES. In the following we briefly report on the relevant aspects of the design of the LARES satellite that are different with respect to those of the LAGEOS satellites.

The minimization of the cross-sectional area-to-mass ratio of a spherical satellite yields: $\frac{A}{M} = \frac{3}{4\rho r}$, where $\rho$ is the mean satellite density and $r$ its radius; this is a critical parameter for the minimization of the non-gravitational perturbations, indeed these accelerations of the satellite are substantially proportional to the satellite cross-sectional area and inversely proportional to its mass. From the above simple relation one deduces that to minimize $A/M$ the satellite should be very large. However, due to launch cost, the size and weight of the satellite is fixed to about 400 kg. The optimal condition will then be obtained by taking the highest value for $\rho$. The best compromise for what concerns cost, density and mechanical properties is obtained with tungsten (density of 19350 kg/m$^3$). However, pure tungsten is not workable and therefore a tungsten alloy has to be considered for LARES. There are a variety of tungsten alloys, some of which can reach a density of 18500 kg/m$^3$. To further reduce the $A/M$ ratio, the cavities housing the CCRs should be as small as possible (Ciufolini et al. 2010b). Another issue addressed in Ciufolini et al. (2010b) is the number of CCRs to be mounted on the satellite surface, their distribution and their orientation, which must be calculated to optimize the average laser return for all attitudes. A higher number of CCRs increases the reflective area of the satellite but on the other hand it increases the $A/M$ ratio. For what concerns the reduction of errors induced by thermal thrust, besides a small $A/M$ ratio, one can try a better modeling of this perturbation that can be achieved by determining the satellite attitude from ground-based observations. This can be performed by photometric measurements of sun glints from the CCRs front faces or from the laser pulses that may contain a spin signature. Both techniques require a suitable CCRs arrangement on the surface (Ciufolini 2009a). However, it is important to stress that the reduced $A/M$ ratio of LARES with respect to LAGEOS (by a factor of about 0.34) will make thermal thrust much smaller on LARES than what it is on LAGEOS.

These aspects and additional ones regarding the design of LARES are discussed in Ciufolini et al. (2010b). The satellite has been designed to minimize the non-gravitational perturbations and to be as nearly as possible a test-particle freely falling in the Earth gravitational field. In conclusion, the orbital parameters chosen for LARES are: inclination about 71.5 degrees, semimajor axis about 7830 km (i.e., altitude of about 1450 km) and eccentricity nearly zero; a relatively large orbital injection error from these values is acceptable for the measurement of gravitomagnetism. The final weight of the LARES satellite will be about 385 kg, its radius about 18 cm and it will have 92 CCRs housed in conical cavities. LARES, together with the LAGEOS satellites and with the GRACE models, will allow a measurement of Earth gravitomagnetism and Lense-Thirring effect with an accuracy of a few percent.

## 6 Gravitational Uncertainties and Even Zonal Harmonics

In the LARES experiment, the error sources of gravitational origin, i.e., those due to the uncertainties in the Newtonian gravitational field, are by far larger than the uncertainties of non-gravitational origin, i.e. radiation pressure, both from Sun and Earth, thermal thrust and particle drag. Indeed the LAGEOS satellites and especially LARES are extremely dense spherical satellites with very small cross-sectional-to-mass ratio (Ciufolini 1989; Ciufolini et al. 2010b); in particular LARES is the densest known single object in the solar system (see the previous Sect. 5). As explained in Sects. 2 and 3, the only secular perturbations affecting the node of an Earth satellite are due to the Earth spherical harmonics of even degree and zero order, e.g., the $J_2$ harmonic describing the well known Earth quadrupole moment.

In this section we report the result of the precise calculation of the uncertainty of each even zonal harmonic, up to order 70, in the measurement of the Lense-Thirring effect using the satellites LAGEOS, LAGEOS 2 and LARES, and the GRACE Earth gravity models, *confirming* a total error budget of a few percent. Figures 9, 10, 11 and 12 clearly display that the uncertainties of the even zonal harmonics of degree higher than 26 are negligible in this measurement.

In Figs. 9, 10, 11 and 12 we display the error in the LARES experiment due to each even zonal harmonic up to degree 70. Once again we stress that the large errors due to the uncertainties of the first two even zonal harmonics, i.e., of degree 2 and 4, are eliminated using the 3 observables, i.e. the 3 nodes. Figures 9, 10, 11 and 12 clearly display that the error due to each even zonal harmonic of degree higher than 4 is considerably less than 1% and in particular that the error is substantially negligible for the even zonal harmonics of degree higher than 26. In Fig. 9 we show the percent errors in the measurement of the Lense-Thirring effect with the LARES experiment due to each individual uncertainty of each even zonal harmonic corresponding to the model EIGEN-GRACE02S (Reigber et al. 2005). In Fig. 10 we show these percent errors for the model GGM02S (CSR, Austin, 2004). In Fig. 11 we display the maximum percent errors, due to each even zonal harmonic, obtained by considering as uncertainty for each harmonic the difference between the value of that harmonic in the EIGEN-GRACE02S model minus its value in the GGM02S model; this is a technique used in space geodesy to estimate the reliability of the published uncertainties of a model. *Of course, in order to use this technique, one must difference models of comparable accuracy, i.e., models that are indeed comparable, or use this technique to only assess the errors of the less accurate model* (*see next Sect.* 7.2).

In conclusion, using EIGEN-GRACE02S and GGM02S, the total error in the measurement of the Lense-Thirring effect due to the even zonal harmonics is respectively 1.4% and 2.1%. However, an improvement by at least an order of magnitude is expected (with respect to EIGEN-GRACE02S) at the time of the LARES data analysis (today we already have a substantial improvement with GGM03S). Indeed, these two models, EIGEN-GRACE02S and GGM02S, have been obtained with a relatively small amount of observations of the GRACE spacecraft (launched in 2002), over less than 365 days, and therefore a substantial factor of improvement of these GRACE models, of about one order of magnitude, should be taken into account at the time of the LARES data analysis (between 2010 and 2016), thanks to longer GRACE observational periods.

## 7 A Reply to the Critical Remarks by Iorio on the Error Analysis and Error Budget of the Gravitomagnetism Measurements with LAGEOS, LAGEOS 2 and LARES

Here we demonstrate that the various claims of a single author (Iorio 2005a, 2008a, 2008b, 2009), that the LAGEOS and LARES error analysis and error budget have been underestimated, are ill-founded. These claims are mainly based on the following four arguments: (a) the effect of the uncertainties in the measured rate of change of $J_4$ and of higher even zonal harmonics, (b) the unmodeled changes in the inclination due to atmospheric drag, (c) the reliability of the published uncertainties of the GRACE gravity field models and (d) in the case of the LARES experiment, the inclusion of the even zonal harmonics of degree higher than 20 in the calculation of the total uncertainty.

We show that these claims are based on, at least, four misunderstandings or miscalculations (see Sects. 2, 7.1, 7.2, 7.3, 7.4 and 8 and Ciufolini and Pavlis 2005; Ciufolini et al. 2006; Lucchesi 2005; Ries et al. 2008). In regard to the claims (a), this author is claiming in Iorio (2005a) that the uncertainties in the measured rate of change of $J_4$ and of higher even zonal harmonics would introduce a bias that can be as large as 45% of the Lense-Thirring effect, however several independent authors (Ciufolini and Pavlis 2005; Ciufolini et al. 2006; Lucchesi 2005; Ries et al. 2008) have proved that this type of error, in the LAGEOS and LAGEOS 2 measurements, is in fact at the level of a few percent only of the Lense-Thirring effect. In regard to the claims (b), in order to measure gravitomagnetism using the nodal rate of an Earth satellite, the orbital inclination of a satellite needs only to be accurately *measured* but *not* accurately *predicted* in its temporal behavior; this has been explained in Sect. 2. Then, in Sects. 7.1 and 8, we show that the orbital inclination of LAGEOS-type satellites is very accurately *measured* by the technique of laser ranging and thus, the corresponding error in the determination of the Lense-Thirring effect due to the *measurement* errors in the inclination of LAGEOS, LAGEOS 2 and LARES is at most at a level of a fraction of 1% of the Lense-Thirring effect. In regard to the claims (c), in Sect. 7.2, we show that different Earth gravity field models with intrinsically different accuracies cannot be compared or they *must* be compared in a proper way, i.e., the accuracy of models derived with a larger GRACE data set or with the standard, accurate, GRACE techniques to derive Earth gravity models *cannot* be estimated by taking their difference with less accurate models, derived with a smaller GRACE data set or based on less accurate techniques or on other less accurate satellite observations. Finally, in regard to the claims (d), in Sect. 6 (Figs. 9, 10, 11 and 12) we have proven that the error due to the even zonal harmonics of degree higher than 26 is negligible in the LARES experiment, contrary to a number of miscalculations by the same author (Iorio 2008b, 2009). Indeed, the contribution of higher degree spherical harmonics of the Earth's gravity field to a satellite orbital motion, and in particular their contributions to the satellite nodal precession, decreases as the inverse power of the semimajor axis to the degree of the even zonal harmonic, and thus quickly *decreases* with the degree and does *not increase* as claimed in various papers by the same author (Iorio 2008b, 2009): the expansion of the Earth's gravity field in spherical harmonics is indeed an expansion with terms that decrease as the inverse power of the semimajor axis to the degree of the harmonics (Kaula 1966).

### 7.1 Orbital Inclination Determination

In Sect. 2 we have discussed and stressed that what is *critical for the measurement of the Lense-Thirring effect* is that the *modeling* of the classical *node* precession (2) (i.e., the prediction of its behavior on the basis of the available physical models) must be accurate, at the level of a few milliarcsec or less (Ciufolini 1986). What is critical is *not* that all the quantities entering this equation, i.e., the Earth parameters and the orbital parameters, and in particular the Earth spherical harmonic coefficients and the semimajor axis and the inclination, must be *predicted* in their variations, but instead what is critical is that they must be *determined* with sufficient accuracy via satellite laser ranging and other techniques (such as the GRACE spacecraft to determine the Earth gravity field). For example if the variations of the *inclination* and of the *semimajor axis* of LAGEOS are not well modeled because the effect of particle drag (i.e., "atmospheric drag") is not known with sufficient accuracy, this is *not* critical for our measurement of the Lense-Thirring effect because we are able to *measure* the variations of inclination and semimajor axis accurately enough with satellite laser ranging (see next Sect. 8 on the uncertainty in the determination of the orbital inclination due to the atmospheric refraction modeling errors) and thus we are able to precisely quantify the effect of these variations on the nodal rate, (2) (with the orbital estimators GEODYN, EPOS-OC and UTOPIA). Indeed, in the next Sect. 8 we estimate that the average measurement error in the inclinations of LAGEOS and LAGEOS 2 (that are roughly constant) is respectively about 30 μarcsec (i.e., 0.03 milliarcsec) for LAGEOS and about 10 μarcsec for LAGEOS 2, that, when propagated in the nodal rate, (2), corresponds to respectively about 0.6% and 0.36% of the Lense-Thirring effect.

The difference between *modeling* and *determining* an orbital element may seem trivial but is critical to understand one of the misunderstandings published in Iorio (2008a) about the error induced in our measurement of the Lense-Thirring effect by the unmodeled inclination variations; indeed this paper (Iorio 2008a) concludes: "The atmospheric drag, both in its neutral and charged components, will induce a non-negligible secular decrease of the inclination of the new spacecraft yielding a correction to the node precession of degree $l = 2$ which amounts to 3–9% $\text{yr}^{-1}$ of the total gravitomagnetic signal. Such a corrupting bias would be very difficult to be modeled". However, as stressed above, these variations in the inclination are very accurately determined with satellite laser ranging (see Sect. 8) even though they are not modeled (i.e., predictable) at a comparable level of accuracy; nevertheless, the measurement of the inclination variations is what is only needed to accurately model the nodal rate, (2).

### 7.2 Comparing Different Earth Gravity Field Models

The GRACE mission provided a leap in the accurate determination of models for the gravitational field and its temporal variations. Nonetheless, not all models derived from GRACE data are of equal or even nearly equal quality, simply because they are based on the highly accurate GRACE observations.

There are several reasons behind this fact which is easily asserted from an examination of the error spectra for nearly every such model available, at the site of the Int. Center for Global Earth Models (ICGEM) at: http://icgem.gfz-potsdam.de/ICGEM. The first and most obvious reason is that most of these models are based on an entirely different set of data, from as little as one month to as much as several years. Some of them are derived rigorously from standard reduction techniques and others use innovative approximation techniques that rely on the accurate position and velocity determination of the orbits from the GPS tracking data. When one additionally considers the fact that GRACE's primary mission is to observe the temporal change of the field, one can realize that all of these models are de facto associated with a mean epoch as far as the representation of the static part of the field and since they span quite diverse time periods, an additional source of diversity is the way that temporal change was handled during their creation, i.e. whether it was accounted for and if so, what rates and for which part of the model they were applied. These models are not referenced to a fixed epoch, and this implies that especially for what concerns the very low degrees, users must apply very carefully the re-referencing (mapping) of these coefficients to a fixed epoch for all of them, prior to attempting any comparisons. This task is difficult to accomplish rigorously for all fields, since there are several other background models used during the data reduction process that affect these rates, each of which could be handled in various ways in the different software and by the different groups that generated these fields. Even for the tightly controlled models produced by the three groups that comprise the GRACE project office, UT/CSR, GFZ/Potsdam and NASA/JPL, the direct comparison at the single coefficient level is problematic, although comparison of the RMS coefficient differences by degree or by order, do seem to provide much more reliable comparison statistics and give a more robust way of assessing the relative accuracy of these models (see Fig. 13).

A recommended approach to compare these diverse models is to examine their "by degree" and "per degree" spectral differences with respect to a single benchmark model and after a careful mapping of the coefficients to the same epoch. Such a comparison is available already online at the ICGEM site. What we can surmise from an inspection of those spectra is that the models are very diverse, even when they come from the same group (e.g. GGM02 and GGM03). In conclusion, we must stress that one should not use each coefficient differences between non-equivalent models as a substitute for the accuracy estimates of any particular GRACE gravity model. The proper calibration of their errors is accomplished with the use of ground truth data sets as for example a global set of GPS-derived geoid undulations over the solid Earth surface and oceanographic observations over the oceans.

Let us further discuss how to correctly compare different Earth gravity field models and some of the previous misleading comparisons. In Iorio (2007), the author is comparing different models, some of which already obsolete in 2007 and obtained with the use of less accurate space missions before GRACE, such as CHAMP. For example the author is comparing two state of the art models in 2004, EIGEN-GRACE02S and GGM02S, with the preliminary, older and obsolete models EIGEN-GRACE01S and GGM01S. Of course, the author is getting a large uncertainty as a result, indeed in Iorio (2007) he concludes "it turns out that the systematic error $\delta\mu$ in the Lense-Thirring measurement is quite larger than in the evaluations so far published based on the use of the sigmas of one model at a time separately, amounting up to 37–43% for the pairs GGM01S/GGM02S and EIGEN-GRACE01S/EIGEN-GRACE02S". However, in order to assess the uncertainty of the newer, much more accurate models, this is simply wrong; for example whereas the uncertainty associated with the measurement using EIGENGRACE02S is about 10% of the Lense-Thirring effect, the uncertainty using the older models is much larger, e.g., for GGM01S was already published (Ciufolini et al. 2006) to be as large as 24%. This explains why the author

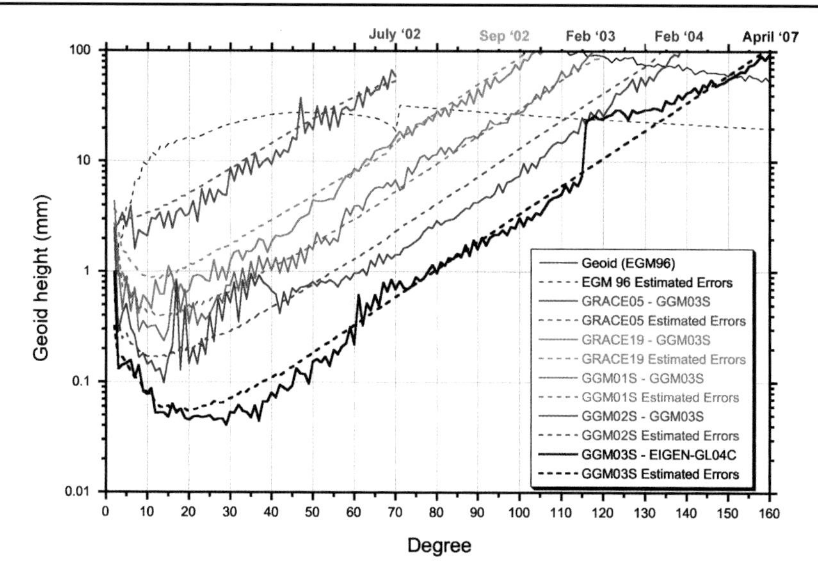


**Fig. 13** Successive static gravity model improvement from GRACE data as a function of the data span used in the solution and correlation of the model errors and the assumed calibrated model errors (Tapley et al. 2007). In this figure different gravity field models are compared by comparing the difference of the corresponding coefficients for each degree (by including the sum over every corresponding *order*, see Sect. 7.2) with their published uncertainties. For example the gravity field model GGM01S, a preliminary GRACE model, less accurate and based on a much shorter period of observation of GRACE, is compared with GGM03S, one of the best models today available. Clearly the difference between GMM01S and GGM03S, the *solid line* in *light blue* (the *solid line in the middle* corresponding to Feb '03) is of comparable size with the accuracy of the less accurate model, i.e., GGM01S, the *dashed line* in *light blue* (the *dashed line in the middle* corresponding to Feb '03). This difference cannot be interpreted as the uncertainty in the much more accurate GGM03S model

gets a large percent "error" in comparing EIGEN-GRACE02S and GGM02S with EIGEN-GRACE01S and GGM01S.

One cannot evaluate the accuracy of the latest GRACE gravity models by comparing them with previous obsolete models obtained using shorter periods of GRACE observations or obtained with different and less accurate techniques. One can see the error spectra for the various models following the links under each model in the tables available at: http://www.gfz-potsdam.de/.

Furthermore, different models cover different periods of time (thus they refer to a different mean epoch). For example, when any of these models state that they have a "zero tide" $J_2$, this means that their quadrupole coefficient $J_2$ is different from the $J_2$ of most of the other models by $4.173 \times 10^{-9}$ in fully normalized coefficient space. If one does not remove that "kind of reference frame bias", taking the difference of $J_2$ for two such models, this would make them look as extremely different and it will generate a bias in the error calibration of the accuracy estimates of the coefficients. Similarly, in regard to other even zonal harmonics such as $J_4$, $J_6$, etc.

In a more recent paper (Iorio 2008b) by the same author, the claims (Iorio 2009) of an error of 1000% in the measurement the Lense-Thirring effect was resized down by almost two orders of magnitude, with a new claim of an error that, in the worst possible case, can be as large as 33% (by comparing the models AIUB-GRACE01S and ITG-Grace03s with EIGEN-GRACE02S). However, even this more recent comparison is affected by at least four misunderstandings and miscalculations: (a) as just explained, one cannot simply take the difference between the corresponding even zonal harmonics coefficient of two different models, because, *among other things*, they refer to different mean epochs and they have been obtained with different $\dot{C}_{lm}$ corrections; (b) in Iorio (2008b) the author is comparing more recent models, however, the author is still taking the difference between models with different intrinsic accuracies, for example in Table 2 he is comparing the model GGM02S with ITG-Grace03s, however, GGM02S has a lower intrinsic accuracy than ITG-Grace03s; (c) the author is often comparing the difference of each even zonal coefficient of two different models with their *formal* error to conclude that the differences "$\Delta C_{l0}$ are always larger than the linearly added sigmas, apart from l = 12 and l = 18", however, it is well known and rather trivial that the formal error, i.e., the error that does *not* include the *systematic* errors, is usually much smaller than the true error (which includes systematic errors); (d) one cannot take the difference between the even zonal coefficients of different models and then sum the *absolute* values of each difference, called "SAV" in Iorio (2008b), (or similarly take the root sum square of these differences) and then compare this "SAV" sum with the Lense-Thirring effect (48.2 milliarcsec/yr) in order to more or less implicitly imply that the results of the measurement of the Lense-Thirring effect with these two models would differ by this "SAV" sum. In fact, one *must* perform a *real* data reduction to compare the results with two different models and in the real data reduction with a model, some even zonal coefficients will be larger and some smaller with respect to the other model, in other words in order to compare the results with two different models, the differences with their plus or minus signs have to be considered. For example, in the more recent paper (Iorio 2008b), in Tables 3 and 10, the author calculates a difference between GGM02S and GGM03S of 24%, however by doing the real data reduction, Ries et al. (2008) obtained a difference between the two models of 13% only.

The appropriate way of comparing different Earth gravity models is manifest from Fig. 13. In this figure different gravity field models are compared by comparing the difference of the corresponding coefficients for each degree (by including the sum over every corresponding *order*, see above) with their published uncertainties. For example the gravity field model GGM01S, a preliminary GRACE model, less accurate and based on a much shorter period of observation of GRACE, is compared with GGM03S, one of the best models today available. Clearly the difference between GMM01S and GGM03S, the solid line in light blue, is of comparable size with the accuracy of the less accurate model, i.e., GGM01S, the dashed line in light blue. This difference cannot be interpreted as the uncertainty in the much more accurate GGM03S model.

We further elucidate this point with a very simple example. Suppose we wish to discuss the accuracy in the measurement of the gravitational constant G, we then take one of its latest measured values, let us say, $6.674215 \times 10^{-11}$ with a relative uncertainty of 0.0015%, whereas the G value measured in the eighteen century was, let us say, about $6.74 \times 10^{-11}$ with a relative uncertainty of a few percent. Now, in order to evaluate the accuracy of the latest value of G (and this is exactly what the author of Iorio (2008b, 2009) is doing by comparing in Iorio (2008b, 2009) even zonal coefficients of different gravity field models with different intrinsic accuracies), we take the difference between these two values. In this way we obtain a relative difference of about 1%. What can one reasonably conclude from this? One may conclude that the older estimate of the uncertainty in G was of the correct order of magnitude but certainly one *cannot* conclude that the 0.0015% uncertainty of the modern measurement of G is wrong by a factor 1000. One cannot seriously difference the older value of the gravitational constant with its newer value to assess the accuracy of its recent measurement. Similarly one cannot seriously compare some obsolete and preliminary Earth gravity models, such as GGM01 and EIGEN-GRACE01S (as this author is doing in Iorio 2009) with some of the most accurate models obtained with GRACE,[1] such as the latest EIGEN-GRACE and GGM models, used for the Lense-Thirring measurement. Furthermore one cannot seriously compare models obtained with GRACE only with models that use GRACE together with other less accurate observations; nevertheless, this is exactly what this author is doing in Iorio (2008b, 2009).

### 7.3 Uncertainties in the Higher Degree Even Zonal Harmonics

In Sect. 6, Figs. 9, 10, 11 and 12, we have shown that the total error in the LARES experiment, due to the uncertainties in the even zonal harmonics of degree strictly higher than four, is of the order of one percent only.

Figures 9, 10, 11 and 12 prove indeed that the claims (published in Iorio 2008b, 2009) that by considering the uncertainties in the even zonal harmonics higher than degree 20, i.e., from degree 20 to degree 70, the error increases by as much as 1000%, are obviously wrong and misleading by at least three orders of magnitude. Indeed, in Iorio (2009), the author concludes "it turns out that, by using the sigmas of the covariance matrices of some of the latest global Earth's gravity solutions based on long data sets of the dedicated GRACE mission, the systematic bias due to the mismodeled even zonal harmonics up to $l = 70$ will amount to ≈100–1000%", nevertheless, in a more recent paper (Iorio 2008b), these claims have been quite weakened and the author now claims an error that can be as large as 26% using the combination LAGEOS, LAGEOS 2 and LARES: "Straightforward calculations up to degree $l = 60$ with the standard Kaula approach yield errors as large as some tens percent".

Indeed, the results shown in Figs. 9, 10, 11 and 12 have been obtained both by precise analytical propagation of the error of each coefficient in the nodal equation (2) (using Mathematica) and confirmed by orbital propagation (with GEODYN). However, these results can also be easily understood and derived, in order of magnitude, in the following simple way: (a) the uncertainties of each even zonal harmonic published with a GRACE gravity field model are very roughly constant, from degree 12 to degree 60; they may at most change by a factor 5,[2] so it is the difference between different Earth gravity models of comparable accuracy (see Fig. 13 that displays the difference between a number of gravity field models as a function of the degree of each harmonic; see also the above discussion about the appropriate way of comparing different Earth gravity field model); furthermore, (b) in the nodal rate (2), the size of each even zonal coefficient of degree $l$ roughly decreases as the inverse, $l + 1.5$, power of the satellite semimajor axis, $a$ (or better of the ratio between the Earth radius, $R_\oplus$, and the satellite semimajor axis, $a$), where the 1.5 power comes from the common coefficient $\frac{n}{a} = \frac{2\pi}{Pa}$ in (2) and $P$ is the satellite orbital period. For example, for $l = 2$ the error corresponding to the term $l = 2$ is proportional to $(R_\oplus/a)^{3.5}$, however for $l = 60$ is proportional to $(R_\oplus/a)^{61.5}$. Now, since the error corresponding to each even zonal harmonic uncertainty can be calculated by multiplying each even zonal uncertainty (very roughly constant from $l = 12$ to $l = 60$) for the corresponding even zonal coefficient in the nodal rate equation (2) (that goes as the inverse, $l + 1.5$, power), the total error due to each

[1] The error spectra of most GRACE Earth's gravity models are available at: http://icgem.gfz-potsdam.de/ICGEM.

[2] See, e.g., the EIGEN-GRACE02S calibrated errors at: http://www-app2.gfz-potsdam.de/pb1/op/grace//results/.

even zonal harmonic uncertainty is roughly proportional to the inverse, $l + 1.5$, power of the satellite semimajor axis $a$. This simply explains the results displayed in Figs. 9, 10, 11 and 12.

### 7.4 Uncertainties in the Rate of Change of $J_4$ and of Higher Even Zonal Harmonics

The error due to the uncertainties in the measured rate of change of $J_4$ and of higher even zonal harmonics, claimed by the same author to be as large as 45%, has been proved by several independent authors (Ciufolini and Pavlis 2005; Ries et al. 2008) to be at the level of a few percent only in the measurement of the Lense-Thirring effect with the satellites LAGEOS and LAGEOS 2. Here, we simply refer to these papers.

## 8 Measurement of the Orbital Inclination of LAGEOS-type Satellites and Atmospheric Delay Modeling Errors in SLR

In this section we analyze the uncertainty in the measurement of the inclination of the LAGEOS satellites. This measurement uncertainty is mainly due to the atmospheric refraction. As discussed in Sects. 2 and 7.1, the corresponding error in the determination of the Lense-Thirring effect using the LAGEOS satellites is induced by the uncertainty in the *measurement* of the inclination and *not* by the uncertainty in the *modeling* of the inclination (i.e., the uncertainty in the prediction of its behavior).

Atmospheric refraction is an important accuracy-limiting factor in the use of Satellite Laser Ranging (SLR) for high-accuracy science applications. In most of these applications, and particularly for the establishment and monitoring of the Terrestrial Reference Frame, of great interest is the stability of its scale and its implied height system. The modeling of atmospheric refraction in the analysis of SLR data is based on the determination of the delay in the zenith direction and on the subsequent projection to a given elevation angle using a mapping function. Mendes et al. (2002) pointed out some limitations in the Marini–Murray model used in SLR since its introduction in 1973, namely, the modeling of the elevation dependence (the mapping function component of the model), and a less than 1 mm bias in the computation of the zenith delay. The mapping functions developed by Mendes et al. (2002) represent a significant improvement over the built-in mapping function of the Marini–Murray model and other known mapping functions.

The new mapping functions can be used in combination with any zenith delay model. Mendes et al. (2002) concluded that current zenith delay models have errors at the millimeter level, which increase significantly at 0.355 micrometers, reflecting inadequacy in the dispersion formulae incorporated in these models. A more accurate zenith delay model was developed, applicable to the range of wavelengths used in modern SLR instrumentation (0.355 to 1.064 micrometers) (Mendes and Pavlis 2004). Using a three-dimensional ray-tracing procedure based on globally distributed satellite data from the Atmospheric Infrared Sounder (AIRS) instrument on NASA's AQUA platform, as well as three-dimensional analysis fields from the European Center for Medium Weather Forecasting (ECMWF), Mendes et al. assessed the new zenith delay models and mapping functions both spatially and temporally. They also looked at the magnitude of the horizontal gradient contribution to the total delay by ray-tracing and using a parametric model. Ray-tracing does not depend on any models or mapping functions, it uses a three-dimensional spherical grid that covers Earth from its surface to the top of the atmosphere and generates the atmospheric delay value by following the (non-planar) path of a light ray from the tracking station to the satellite and back. The path of the ray is governed by the local refractive index computed on the basis of the conditions within each three-dimensional cell of the grid (with horizontal and vertical resolution ~50 km).

Mendes et al. used meteorological data sets from NASA's AIRS in order to validate the new zenith delay and mapping function models, and to develop new models that include variations in horizontal refractive indices. The AIRS Level-2 support product gives profiles of temperature, pressure and water vapor from the surface to the top of the atmosphere in 100 standard pressure levels. The pressure levels extend from 1100 mb up to 0.1 mb.

Mendes et al. developed a ray-tracing algorithm specifically tailored for AIRS and ECMWF data in order to calculate the atmospheric delay by directly integrating all the values through which the ray traverses, independent of any mapping function. They used a new formulation for the group refractivity based on formulas by Ciddor (1996) that include both hydrostatic and non-hydrostatic components of the group refractivity. In order to perform the ray-tracing, the data are first processed and grouped into 10 × 10 degree latitude/longitude grids up to 0.1 mb in order to build three-dimensional atmospheric profiles around each operational ILRS-SLR tracking station. AIRS-based Ray-Tracing (ART) and ECMWF-based Ray-Tracing (ERT) were used to calculate both the total atmospheric delay as well as the delay due to horizontal gradients in refractivity. Given the independent accuracy assessment of the simple, closed-formula models in use today, (e.g. Marini–Murray), the effect of the horizontal gradients (which is not accounted for in these models), is the largest remaining error in refraction delay modeling today. If we can quantify the level of that unaccounted error induced on the orbit of LAGEOS due to neglecting the horizontal gradients, we can then place an upper bound on the maximum error in our Lense-Thirring estimate. Using the three-dimensional ray-tracing approach which *fully* accounts for the *total* atmospheric delay including the gradients, we analyzed a few years of LAGEOS and LAGEOS 2 data as a test case. Comparing the results of these reductions to those obtained using the standard model (Mendes–Pavlis without gradients), we showed that the most significant result was a reduction in the variance of the observation residuals up to 25%, with only random and insignificant differences in the orbits. This was the most significant outcome of implementing the new approach in calculating the atmospheric delay using the three-dimensional ray-tracing methodology. With insignificant, random and no secular differences in the orbits, the effect on the estimated Lense-Thirring parameter is also equally insignificant and of no further concern for the Lense-Thirring experiment.

Since refraction does not enter the dynamical model being a media propagation effect on the tracking data, one would expect a priori that the effects on the orbit would be very small, much smaller than the actual variations seen in the tracking data residuals that absorb the majority of refraction errors.

From Fig. 14 we can estimate that on LAGEOS, on the average, the effect on inclination is at the 30 μarcseconds (0.03 milliarcseconds) level with a comparable standard deviation. For comparison purposes notice that 30 μarcseconds at LAGEOS altitude are only less than 1.5 mm. On the LAGEOS orbit, a 30 μarcseconds measurement uncertainty in the inclination corresponds (from the nodal rate (2) of LAGEOS due to the even zonal harmonics) to a nodal rate of 0.6% of the Lense-Thirring effect. Similar conclusions are reached from inspection of Fig. 15, displaying the results of the analysis of several years of older data taken on LAGEOS 2. We notice here that from the analysis of three years of data we again see a very minor effect at the level of about 10 μarcseconds on the average in either the inclination or the node of LAGEOS 2, with an even smaller standard deviation than the one for LAGEOS. This may be explained by the fact that in recent years (after 2004) the network has allowed tracking at lower elevations below the original 20° minimum and all the way down to 10°.

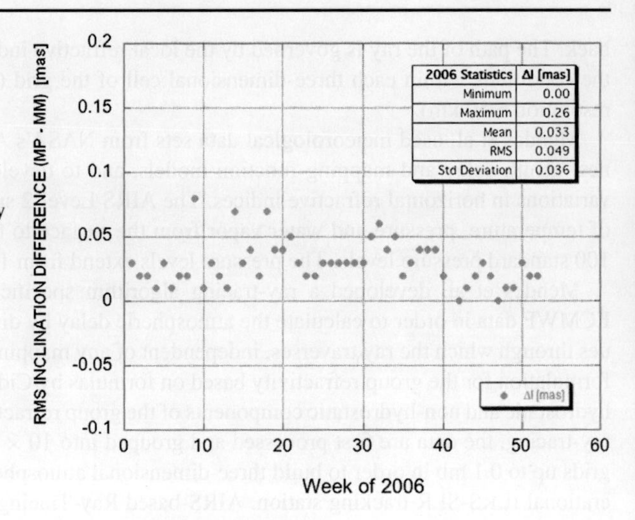


**Fig. 14** Weekly arc RMS variation (RMS Difference, Marini–Murray minus Mendes–Pavlis, in mas (milliarcsec)) in the inclination fits of LAGEOS data reduced with two different atmospheric delay models, the Marini–Murray and the Mendes–Pavlis

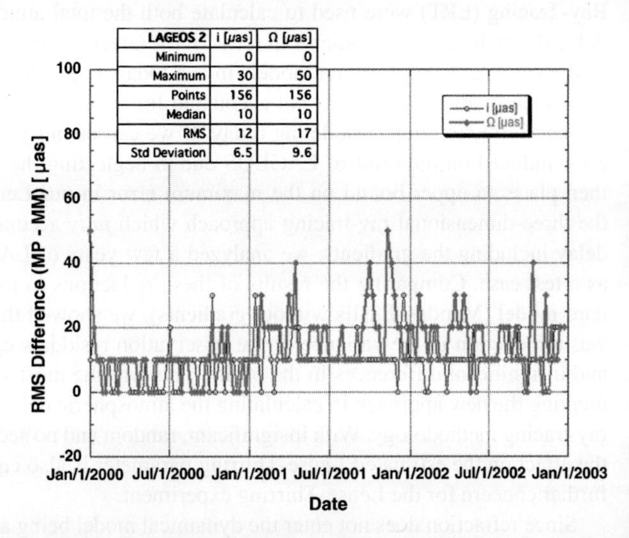


**Fig. 15** Weekly arc RMS variation (RMS Difference, Marini–Murray minus Mendes–Pavlis, in μarcsec (microarcsec)) in the inclination and node fits of LAGEOS 2 data reduced with two different atmospheric delay models, the Marini–Murray and the Mendes–Pavlis

Refraction errors drop significantly at higher elevations and since the preponderance of the LAGEOS 2 data in these early years is at elevations above 20°, the discrepancy between the two orbits is less affected by the much smaller refraction errors. On the LAGEOS 2 orbit, a 10 μarcsecond measurement uncertainty in the inclination corresponds (from the nodal rate (2) of LAGEOS 2 due to the even zonal harmonics) to a nodal rate of 0.36% of the Lense-Thirring effect.

Final tests were performed with the direct application of three-dimensional atmospheric ray-tracing as detailed in Hulley and Pavlis (2007), where two years of data were reduced with atmospheric delay corrections that are obtained with this approach and are thus free of any error in the zenith delay or the mapping function used in the model. Furthermore, three-dimensional AIRS-based Ray-Tracing includes automatically the effect of the horizontal gradients as explained ibidem, and what these tests demonstrated is that indeed, as expected, the errors in the atmospheric delay modeling are absorbed by the residuals of the individual stations. The use of the superior 3D ART approach results in a significant variance reduction in the station residuals, whether the meteorological information comes from the AIRS global fields or the ECMWF assimilation fields, which in recent years begun using the AIRS data as part of their standard input for their assimilation scheme.

In conclusion, on the combination of the nodes of LAGEOS and LAGEOS 2 (to measure the Lense-Thirring effect) the error due to inclination measurement uncertainties is of the order of 0.5% and similarly for the LARES + LAGEOS + LAGEOS 2 experiment.

## 9 Conclusions

We have described the measurements of frame-dragging by the Earth spin with the satellites LAGEOS and LAGEOS 2 using the Earth's gravity models obtained by the GRACE spacecraft. These measurements of Earth gravitomagnetism have confirmed the general relativistic prediction of the Lense-Thirring effect with an accuracy of about 10% and have been recently *independently* repeated by the group of the Center for Space Research of University of Texas at Austin that reported an accuracy of about 12%. We have then introduced the LARES satellite to be launched in 2010 by the Italian Space Agency for a measurement of frame-dragging with an accuracy of a few percent.

Our detailed analyses and the agreement of the independent results of the four groups: University of Salento and University of Rome, University of Maryland, GFZ of Potsdam and Center for Space Research of the University of Texas at Austin, confirm our error budget of the LAGEOS-LAGEOS 2 and LARES frame-dragging measurements.

These independent results demonstrate that the claims of a single individual, that the LAGEOS-LAGEOS 2 and LARES frame-dragging error budget has been underestimated, are based on erroneous analyses; we have then pointed out the misunderstandings and miscalculations on which these claims are based. We have indeed analyzed and precisely quantified a number of conceivable error sources, e.g., the uncertainties in higher Earth's even zonal harmonics, up to degree 70, and the uncertainties in the orbital elements of the LAGEOS satellites, including the orbital inclination. We have also shown the correct way to compare different Earth's gravity field models obtained with GRACE and discussed the reliability of the published uncertainties of the GRACE models.

**Acknowledgements** We acknowledge the support of the Italian Space Agency (ASI), grant I/043/08/0. We thank the organizers of the ISSI workshop, Rudolf Treumann and Martin C.E. Huber, and Claudio Paris, Cristian Vendittozzi, Marcello Ramigoni, Giuliano Battaglia of the LARES team of Sapienza University of Rome.

## References

J.I. Andrès et al., Spin axis behavior of the LAGEOS satellites. J. Geophys. Res. **109**, B06403–1-12 (2004)

N. Ashby, B. Shahid-Saless, Geodetic precession or dragging of inertial frames? Phys. Rev. D **42**, 1118–1122 (1990)

B.M. Barker, R.F. O'Connel, The gravitational interaction: Spin, rotation, and quantum effects – A review. Gen. Relativ. Gravit. **11**, 149–175 (1979)

P.L. Bender et al., The lunar laser ranging experiment. Science **182**, 229–238 (1973)

B. Bertotti, I. Ciufolini, P.L. Bender, New test of general relativity: measurement of de Sitter geodetic precession rate for lunar perigee. Phys. Rev. Lett. **58**, 1062–1065 (1987)

P.E. Ciddor, Refractive index of air: New equations for the visible and near infrared. Appl. Opt. **35**(9), 1566–1573 (1996)

I. Ciufolini, Theory and experiments in General Relativity and other metric theories, Ph.D. Dissertation, Univ. of Texas, Austin, Pub. Ann Arbor, Michigan, 1984

I. Ciufolini, Measurement of the Lense-Thirring drag on high-altitude laser-ranged artificial satellites. Phys. Rev. Lett. **56**, 278–281 (1986)
I. Ciufolini, A comprehensive introduction to the Lageos gravitomagnetic experiment: from the importance of the gravitomagnetic field in physics to preliminary error analysis and error budget. Int. J. Mod. Phys. A **4**, 3083–3145 (1989); see also Tapley et al. (1989)
I. Ciufolini, Gravitomagnetism and status of the LAGEOS III experiment. Class. Quantum Gravity **11**, A73–A81 (1994)
I. Ciufolini, On a new method to measure the gravitomagnetic field using two orbiting satellites. Nuovo Cim. A **109**, 1709–1720 (1996)
I. Ciufolini, The 1995–99 measurements of the Lense-Thirring effect using laser-ranged satellites. Class. Quantum Gravity **17**, 2369–2380 (2000)
I. Ciufolini, in *Proceedings of the I SIGRAV School on General Relativity and Gravitation*, Frascati (Rome), September 2002, IOP, Bristol (2005), pp. 27–69
I. Ciufolini, On the orbit of the LARES satellite (2006). arXiv:gr-qc/0609081v1
I. Ciufolini, Dragging of inertial frames. Nature **449**, 41–48 (2007a)
I. Ciufolini, A. Paolozzi, G. Sindoni, E.C. Pavlis, A. Gabrielli, Scientific Aspects of LARES mission, *Proc. International Astronautical Congress 09.B4.2.9*, Daejeon, Republic of Korea, 12–16 October 2009 (2009a)
I. Ciufolini, Frame-dragging, gravitomagnetism and lunar laser ranging. New Astron. (2009b). doi:10.1016/j.newast.2009.08.004
I. Ciufolini, E.C. Pavlis, A confirmation of the general relativistic prediction of the Lense-Thirring effect. Nature **431**, 958–960 (2004)
I. Ciufolini, E.C. Pavlis, On the measurement of the Lense-Thirring effect using the nodes of the LAGEOS satellites, in reply to "On the reliability of the so-far performed tests for measuring the Lense-Thirring effect with the LAGEOS satellites" by L. Iorio. New Astron. **10**(8), 636–651 (2005)
I. Ciufolini, J.A. Wheeler, *Gravitation and Inertia* (Princeton Univ. Press, Princeton, 1995)
I. Ciufolini et al., ASI-NASA Study on LAGEOS III, CNR, Rome, Italy, 1989
I. Ciufolini et al., Effect of particle drag on the LAGEOS node and measurement of the gravitomagnetic field. Nuovo Cim. B **105**, 573–588 (1990)
I. Ciufolini, F. Chieppa, D. Lucchesi, F. Vespe, Test of Lense-Thirring orbital shift due to spin. Class. Quantum Gravity **14**, 2701–2726 (1997a)
I. Ciufolini, D. Lucchesi, F. Vespe, F. Chieppa, Measurement of gravitomagnetism. Europhys. Lett. **39**, 359–364 (1997b)
I. Ciufolini, A. Paolozzi et al., LARES phase. A study for the Italian Space Agency, 1998a
I. Ciufolini, E.C. Pavlis, F. Chieppa, E. Fernandes-Vieira, J. Perez-Mercader, Test of general relativity and measurement of the Lense-Thirring effect with two Earth satellites. Science **279**, 2100–2103 (1998b)
I. Ciufolini et al., Italian Space Agency Phase A Report on LARES, Italian Space Agency, ASI, Rome, 1998c
I. Ciufolini et al., INFN study on LARES/WEBER-SAT, 2004
I. Ciufolini, E.C. Pavlis, R. Peron, Determination of frame-dragging using Earth gravity models from CHAMP and GRACE. New Astron. **11**, 527–550 (2006)
I. Ciufolini et al., Gravitomagnetism and its measurement with laser ranging to the LAGEOS satellites and GRACE Earth gravity models, in *General Relativity and John Archibald Wheeler: Frame-Dragging, Gravitational-Waves and Gravitational Tests*, ed. by I. Ciufolini, R. Matzner (Springer, Berlin, 2010a)
I. Ciufolini et al., The LARES space experiment: LARES orbit, error analysis and satellite structure, in *General Relativity and John Archibald Wheeler: Frame-Dragging, Gravitational-Waves and Gravitational Tests*, ed. by I. Ciufolini, R. Matzner (Springer, Berlin, 2010b)
I. Ciufolini, A. Anselmo et al., Effect of particle drag in the LARES experiment (2009c, to be published)
S.C. Cohen, P.J. Dunn, LAGEOS scientific results. J. Geophys. Res. B **90**, 9215–9438 (1985)
L. Cugusi, E. Proverbio, Relativistic effects on the motion of Earth's artificial satellites. Astron. Astrophys. **69**, 321–325 (1978)
W. Cui, S.N. Zhang, W. Chen, Evidence for frame dragging around spinning black holes in X-ray binaries. Astrophys. J. **492**, L53–L58 (1998)
G. Felici, The meaning of systematic errors, a comment to "Reply to On the systematic errors in the Detection of the Lense-Thirring effect with a Mars orbiter", by L. Iorio (2007). arXiv:gr-qc/0703020v1
R.S. Gross, Combinations of Earth orientation measurements: SPACE94, COMB94, and POLE94. J. Geophys. Res. **101**(B4), 8729–8740 (1996)
G.C. Hulley, E.C. Pavlis, A ray-tracing technique for improving Satellite Laser Ranging atmospheric delay corrections, including the effects of horizontal refractivity gradients. J. Geophys. Res. **112**, B06417-1-19 (2007). doi:10.1029/2006JB004834
International Earth Rotation Service (IERS) Annual Report, 1996. Observatoire de Paris, Paris, July 1997

L. Iorio, On the reliability of the so-far performed tests for measuring the Lense-Thirring effect with the LAGEOS satellites. New Astron. **10**, 603–615 (2005a)
L. Iorio, The impact of the new Earth gravity models on the measurement of the Lense-Thirring effect with a new satellite. New Astron. **10**, 616–635 (2005b)
L. Iorio, Evidence of the gravitomagnetic field of Mars. Class. Quantum Gravity **23**, 5451–5454 (2006)
L. Iorio, On some critical issues of the LAGEOS/LAGEOS II Lense-Thirring experiment (2007). arXiv:0710.1022v1 [gr-qc]
L. Iorio, On the impact of the atmospheric drag on the LARES mission. arXiv:0809.3564v2 (2008a); see also: arXiv:0809.3564v1
L. Iorio, An assessment of the systematic uncertainty in present and future tests of the Lense-Thirring effect with satellite laser ranging (2008b). arXiv:0809.1373v2 [gr-qc]
L. Iorio, Will the recently approved LARES mission be able to measure the Lense-Thirring effect at 1%? Gen. Relativ. Gravit. **41**, 1717–1724 (2009). doi:10.1007/s10714-008-0742-1; see also: arXiv:0803.3278v5 [gr-qc]
W.M. Kaula, *Theory of Satellite Geodesy* (Blaisdell, Waltham, 1966)
A.R. Khan, R.F. O'Connell, Gravitational analogue of magnetic force. Nature **261**, 480–481 (1976)
S.M. Kopeikin, Comment on "Gravitomagnetic Influence on Gyroscopes and on the Lunar Orbit". Phys. Rev. Lett. **98**, 229001–1 (2007)
K. Krogh, Iorio's "high-precision measurement" of frame-dragging with the Mars Global Surveyor. Class. Quantum Gravity **24**, 5709–5715 (2007)
D.M. Lucchesi, Reassessment of the error modelling of non-gravitational perturbations on LAGEOS 2 and their impact in the Lense-Thirring determination. Part I. Planet. Space Sci. **49**, 447–463 (2001)
D.M. Lucchesi, Reassessment of the error modelling of non-gravitational perturbations on LAGEOS II and their impact in the Lense-Thirring determination. Part II. Planet. Space Sci. **50**, 1067–1100 (2002)
D.M. Lucchesi, The impact of the even zonal harmonics secular variations on the Lense-Thirring effect measurement with the two Lageos satellites. Int. J. Mod. Phys. D **14**, 1989–2023 (2005)
D.M. Lucchesi, A. Paolozzi, A cost effective approach for LARES satellite, in *XVI AIDAA*, Palermo, September 2001, pp. 1–14. Paper no. 111
C.F. Martin, D.P. Rubincam, Effects of Earth albedo on the LAGEOS I satellite. J. Geophys. Res. **101**(B2), 3215–3226 (1996)
V.B. Mendes, E.C. Pavlis, High-accuracy zenith delay prediction at optical wavelengths. Geophys. Res. Lett. **31**, L14602–1-5 (2004). doi:10.1029/2004GL020308
V.B. Mendes, G. Prates, E.C. Pavlis, D.E. Pavlis, R.B. Langley, Improved mapping functions for atmospheric re-fraction correction in SLR. Geophys. Res. Lett. **29**, 1414–1-4 (2002). doi:10.1029/2001GL014394
C.W. Misner, K.S. Thorne, J.A. Wheeler, *Gravitation* (Freeman, San Francisco, 1973)
T.W. Murphy Jr., K. Nordtvedt, S.G. Turyshev, Phys. Rev. Lett. **98**, 071102–1-4 (2007a)
T.W. Murphy Jr., K. Nordtvedt, S.G. Turyshev, Phys. Rev. Lett. **98**, 229002–1 (2007b)
R. Noomen, S. Klosko, C. Noll, M. Pearlman (eds.), *Toward Millimeter Accuracy in NASA CP 2003-212248. Proc. 13th Int. Laser Ranging Workshop* (NASA Goddard, Greenbelt, 2003)
K. Nordtvedt, Existence of the gravitomagnetic interaction. Int. J. Theor. Phys. **27**, 1395–1404 (1988)
K. Nordtvedt, LARES and tests on new long ranges forces, in *LARES phase. A study for the Italian Space Agency* (1998), pp. 34–38
R.F. O'Connell, A note on frame dragging. Class. Quantum Gravity **22**, 3815–3816 (2005)
A. Paolozzi, private communication (2005)
E.C. Pavlis, Geodetic contributions to gravitational experiments in space, in *Recent Developments in General Relativity*, ed. by R. Cianci et al., Genoa, 2000 (Springer, Berlin), pp. 217–233
D.E. Pavlis et al., GEODYN Operations Manuals, Contractor Report, Raytheon, ITSS, Landover, MD, 1998
G.E. Peterson, Estimation of the Lense-Thirring precession using laser-ranged satellites, Ph.D. Dissertation, Univ. of Texas, Austin, 1997
Ch. Reigber, F. Flechtner, R. Koenig, U. Meyer, K. Neumayer, R. Schmidt, P. Schwintzer, S. Zhu, GRACE orbit and gravity field recovery at GFZ Potsdam—first experiences and perspectives. Eos. Trans. AGU **83**(47) (2002). Fall Meet. Suppl., Abstract G12B-03
C. Reigber, R. Schmidt, F. Flechtner, R. Konig, U. Meyer, K.H. Neumayer, P. Schwintzer, S.Y. Zhu, J. Geodyn. **39**, 1–10 (2005)
J.C. Ries, private communication (2005)
J.C. Ries, Simulation of an experiment to measure the Lense-Thirring precession using a second LAGEOS satellite, Ph.D. dissertation, The University of Texas, Austin, 1989
J.C. Ries, R.J. Eanes, B.D. Tapley, Lense-Thirring precession determination from laser ranging to artificial satellites, in *Nonlinear Gravitodynamics, the Lense-Thirring Effect, Proc. III William Fairbank Meeting* (World Scientific, Singapore, 2003a), pp. 201–211

J.C. Ries, R.J. Eanes, B.D. Tapley, G.E. Peterson, Prospects for an improved Lense-Thirring test with SLR and the GRACE gravity mission, in *Toward Millimeter Accuracy, Proc. 13th Int. Laser Ranging Workshop*, Report NASA CP 2003-212248, NASA Goddard, Greenbelt, Maryland, 2003b
J.C. Ries, R.J. Eanes, M.M. Watkins, Confirming the frame-dragging effect with satellite laser ranging, in *16th International Workshop on Laser Ranging*, Poznan, Poland, 13–17 October 2008
D.P. Rubincam, General relativity and satellite orbits: the motion of a test particle in the Schwarzschild metric. Celest. Mech. **15**, 21–33 (1977)
D.P. Rubincam, Yarkovsky Thermal Drag on LAGEOS. J. Geophys. Res. **93**(B11), 13805–13810 (1988)
D.P. Rubincam, Drag on the LAGEOS satellite. J. Geophys. Res. B **95**, 4881–4886 (1990)
D.P. Rubincam, A. Mallama, J. Geophys. Res. **100**(B10), 20285–20990 (1995)
C. Schmid, Cosmological gravitomagnetism and Mach's principle. Phys. Rev. D **74**, 044031-1-18 (2006)
G. Sindoni, C. Paris, P. Ialongo, On the systematic errors in the detection of the Lense-Thirring effect with a Mars orbiter (2007). arXiv:gr-qc/0701141
I.H. Stairs, S.E. Thorsett, Z. Arzoumanian, Measurement of gravitational spin-orbit coupling in a binary-pulsar system. Phys. Rev. Lett. **93**, 141101-1-4 (2004)
B.D. Tapley, The GRACE mission: status and performance assessment. Eos. Trans. AGU **83**(47) (2002). Fall Meet. Suppl., Abstract G12B-01
B.D. Tapley, J.C. Ries, R.J. Eanes, M.M. Watkins, *NASA-ASI Study on LAGEOS III*, CSR-UT publication n. CSR-89-3, Austin, Texas, 1989
B.D. Tapley, S. Bettadpur, M. Watkins, C. Reigber, The gravity recovery and climate experiment: Mission overview and early results. Geophys. Res. Lett. **31**, L09607-1-4 (2004)
B.D. Tapley, J. Ries, S. Bettadpur, D. Chambers, M. Cheng, F. Condi, S. Poole, The GGM03 mean Earth gravity model from GRACE, Eos Trans. AGU **88**(52) (2007). Fall Meet.Suppl., Abstract G42A-03
K.S. Thorne, R.H. Price, D.A. Macdonald, *The Membrane Paradigm* (Yale Univ. Press, New Haven, 1986)
R.A. Van Patten, C.W.F. Everitt, Possible experiment with two counter-orbiting drag-free satellites to obtain a new test of Einstein's General Theory of Relativity and improved measurements in geodesy. Phys. Rev. Lett. **36**, 629–632 (1976)
M. Watkins, D. Yuan, W. Bertiger, G. Kruizinga, L. Romans, S. Wu, GRACE gravity field results from JPL. Eos. Trans. AGU **83**(47) (2002). Fall Meet. Suppl., Abstract G12B-02
S. Weinberg, *Gravitation and Cosmology: Principles and Applications of the General Theory of Relativity* (Wiley, New York, 1972)
J.M. Weisberg, J.H. Taylor, General relativistic geodetic spin precession in binary pulsar B1913+16: mapping the emission beam in two dimensions. Astrophys. J. **576**, 942–949 (2002)
J.G. Williams, X.X. Newhall, J.O. Dickey, Relativity parameters determined from lunar laser ranging. Phys. Rev. D **53**, 6730–6739 (1996)
J.G. Williams, S.G. Turyshev, D.H. Boggs, Progress in lunar laser ranging tests of relativistic gravity. Phys. Rev. Lett. **93**, 261101-1-4 (2004a)
J.G. Williams, S.G. Turyshev, T.W. Jr. Murphy, Improving LLR tests of gravitational theory. Int. J. Mod. Phys. D **13**, 567–582 (2004b)
H. Yilmaz, Proposed test of the nature of gravitational interaction. Bull. Am. Phys. Soc. **4**, 65 (1959)

Space Sci Rev (2009) 148: 105–121
DOI 10.1007/s11214-009-9627-1

# Lense-Thirring Precession in the Astrophysical Context

**Luigi Stella · Andrea Possenti**

Received: 14 August 2009 / Accepted: 21 December 2009 / Published online: 16 January 2010
© Springer Science+Business Media B.V. 2010

**Abstract** This paper surveys some of the astrophysical environments in which the effects of Lense-Thirring precession and, more generally, frame dragging are expected to be important. We concentrate on phenomena that can probe *in situ* the very strong gravitational field and single out Lense-Thirring precession in the close vicinity of accreting neutron stars and black holes: these are the fast quasi periodic oscillations in the X-ray flux of accreting compact objects. We emphasise that the expected magnitude of Lense-Thirring/frame dragging effects in the regions where these signals originate are large and thus their detection does not pose a challenge; rather it is the interpretation of these phenomena that needs to be corroborated through deeper studies. Relativistic precession in the spin axis of radio pulsars hosted in binary systems hosting another neutron star has also been measured. The remarkable properties of the double pulsar PSR J0737–3039 has opened a new perspective for testing the predictions of general relativity also in relation to the precession of spinning bodies.

**Keywords** Relativity and gravitation · Neutron stars · Black holes · X-ray binaries · Pulsars

## 1 Introduction

General Relativity predicts that the *mass currents* associated e.g. to a rotating body alter spacetime in such a way that the local inertial frames are dragged by them. Both the motion of test particles and the direction in space of gyroscopes is affected by this *frame-dragging*. In an electromagnetic analogy these and other phenomena resulting from mass currents are qualified as *gravitomagnetic*. Lense-Thirring precession is probably the best-known effect

L. Stella (✉)
INAF-Osservatorio Astronomico di Roma, Via Frascati, 33, 00040 Monteporzio Catone, Rome, Italy
e-mail: stella@mporzio.astro.it

A. Possenti
INAF-Osservatorio Astronomico di Cagliari, Loc. Poggio dei Pini, 09012 Capoterra, CA, Italy
e-mail: possenti@ca.astro.it

of this type (Lense and Thirring 1918). Predicted only a few years after the formulation of general relativity, this effect demonstrates the embodiment of Mach's principle in the theory of general relativity (Pfister 2007, but see Rindler 1994). As a result of this effect, the line of nodes of a particle orbiting a spinning star is expected to drift in the direction of the star rotation, leading to *prograde nodal precession*. Similarly the spin axis of a gyroscope orbiting a rotating star will precess around the star's angular momentum (for a monograph see Ruffini and Sigismondi 2002).

The Lense-Thirring precession rate is extremely slow around the Earth: that is why Earth satellite experiments[1] must achieve exceptionally high accuracy in order to single out Lense-Thirring effects against a number of much larger effects. The strong gravitational field regions in the vicinity of rotating collapsed objects, i.e. neutron stars and black holes, provide an alternative and largely complementary opportunity to observe and single out Lense-Thirring and frame dragging effects. Since these effects can take place at a very fast pace around collapsed objects, the reliability of the interpretation of the phenomena that are attributed to them (rather than the magnitude of the effects themselves) is usually the key factor.

Matter accretion towards rotating neutron stars and black holes has been studied extensively in last decades. In the deep gravitational fields close to these objects, the inflowing matter moves at a sizable fraction of the speed of light $c$ and its gravitational energy is converted into radiation (mainly X and gamma-rays) usually with high efficiency. Since neutron stars and black holes are small objects, matter endowed with angular momentum (albeit small) will settle in an disk and gradually spiral toward the collapsed object as a result of viscous coupling between adjacent rings of matter at different radii. The (near-Keplerian) orbits of matter in accretion disks can extend down to the radius of the marginally stable orbit. Therefore the motion of matter in the innermost regions of such disks holds crucial information on the properties of spacetime in the close vicinity of neutron stars and black holes, including Lense-Thirring and other frame dragging effects. Over the last two decades diagnostics have been discovered and studied which appear to the especially promising in this respect. Among these are the quasi periodic oscillations in the flux of a number of X-ray binary systems, which are very likely related to the fundamental frequencies of matter motion in the innermost disk regions; we summarise the main results in this field of research in Sects. 2 and 3, with special emphasis on phenomena which are interpreted on the basis of Lense-Thirring precession.

The motion of matter around neutron stars and black holes is also investigated through X-ray spectral diagnostics. The Fe K$\alpha$ emission line which are observed in a number of accreting collapsed objects, ranging from neutron stars to very massive black holes in active galactic nuclei, is probably the best studied of them. The very broad and often markedly redshift profiles of these lines are interpreted in terms of a combination of relativistic effects in the motion of disk matter in the vicinity of these objects (Fabian et al. 1989; Laor 1991). Strong evidence has been obtained that the accretion disk extends well inside the marginally stable orbit of a non-rotating black hole ($r_{\rm ms} = 6$ GM/c$^2$), implying that the accreting black hole must be rotating. The first clear case emerged with the X-ray CCD observations of the Fe K$\alpha$ line profile from the active nucleus of MGC 6-30-15, a well studied Seyfert galaxy. The red wing from this line extended shortwards of the lowest energies than can be produced in a disk orbiting a Schwarzschild black hole, implying that the disk extended as far in as $\sim 2 \div 3$ GM/c$^2$. This requires a Kerr black hole with $a/M \sim 0.93$ (Tanaka et al. 1995;

[1] These experiments are described in detail elsewhere in this issue.

Bromley et al. 1997; Fabian et al. 2002; Reynolds et al. 2005). Insomuch as frame dragging is responsible for the smaller radius of the marginally stable orbit around Kerr black holes (as compared to the case of Schwarzschild black holes), the Fe K$\alpha$ lines of these X-ray sources yield information on gravitomagnetic effects in the very strong field regime (e.g. Fabian 2008; Miller 2006 and references therein; see Titarchuk et al. 2009 for an alternative interpretation). A similar conclusion has been reached through detailed modeling of the soft X-ray spectral continuum in several accreting stellar mass black holes hosted in transient X-ray binaries (see e.g. McClintock et al. 2007 and references therein). Since this review concentrates on relativistic precession, we will not discuss these X-ray spectral diagnostics further.

Another line of research is based on relativistic binary pulsars, systems comprising two neutron stars, of which at least one is a radio pulsar. These binaries have yielded remarkably accurate gravity tests and confirmed to very good accuracy some key predictions of general relativity. Their pulses are very accurate clocks which provide high precision information on the motion of the two stars and probe directly the spacetime properties over distances of order the orbital separation. Even though these regions are characterised by comparatively weak gravitational fields, nevertheless the fact that relativistic binary pulsars host objects (neutron stars) which possess a *strong self-field* provide a unique opportunity to test general relativity against alternative theories of gravity in the strong field regime. In Sects. 4–6 we survey the results that have been recently obtained in the study of relativistic precession from the double pulsar PSR J0737–3039.

Finally we remark that the possible role of frame dragging has been considered and investigated in a variety of other astrophysical models. Examples of these include: the physics of warped accretion disks (Bardeen and Petterson 1975; Armitage and Natarajan 1999; Martin et al. 2007); the launching of jets by accreting black holes and neutron stars (Martin and Rees 1979; Rees 1984; Martin et al. 2008); the spin evolution in active galactic nuclei (King et al. 2008); the kinematic of galactic center stars (Kannan and Saha 2009); precession in the crust of isolated radio pulsars (Blandford 1995); curvature emission in millisecond radio pulsars (Venter and de Jager 2008). The fact that frame dragging effects are considered nearly routinely in these and many other calculations stems from people's increasing trust in general relativity and its predictions. However prospects for detecting unambiguously these effects through observations do not appear to be for the near future.

## 2 Quasi Periodic Oscillations in Accreting Neutron Stars and Black Holes

Low mass X-ray binaries, LMXRBs, are systems containing a compact object, either an old neutron star or a stellar mass black hole, which accretes mass from its companion, usually a low mass main sequence star. The intense X-ray flux produced by the accretion process frequently displays quasi-periodic oscillations, QPOs, with different modes often excited at the same time. Presently these QPOs can be studied only through the peaks in the power spectra of a source's X-ray light curve, since current instrumentation does not achieve a sufficient signal to noise ratio to study individual QPO cycles (or even just trains of cycles). The relevant power spectrum peaks are broad (usually between $1/3$ and $1/100$ of their centroid frequency), testifying that a range of frequencies is excited in each QPO and thus their signal has only poor coherence.

The *low frequency* QPOs ($\sim$1–100 Hz) were discovered from LMXRBs in the mid-eighties. They are classified into horizontal, normal and flaring branch oscillations (HBOs, NBOs and FBOs, respectively), depending on the energy spectrum of the source at a given time. This, in turn, is most commonly determined through the instantaneous position occupied by a given source in the X-ray colour-colour diagram (the above terminology for the different branches is based on the properties of the most luminous LMXRBs, the so-called *Z-sources*, for a review see van der Klis 1995).

Much faster QPOs ($\sim$0.2 to $\sim$1.4 kHz) were detected and studied in a number of NS LMXRBs since the mid nineties (see e.g. van der Klis 2000 and references therein); these *kHz QPOs* have periods close to the dynamical timescales in the vicinity of the neutron star surface. Two kHz QPOs (centroid frequencies of $\nu_1$ and $\nu_2$) are often present at the same time. The kHz QPO drift in frequency while maintaining their separation $\Delta\nu \equiv \nu_2 - \nu_1 \approx$ 250–360 Hz approximately constant, while these sources move in the colour-colour diagram (over time intervals ranging from minutes to weeks). However, significant variations of $\Delta\nu$ by up to $\sim$100 Hz have been seen in several sources. These QPOs possess similar properties across neutron stars in LMXRBs that differ substantially in some of their other properties. For example the luminosity of Z-sources is on average a factor of $\sim$10 higher than in the so-called *Atoll* sources, the largest class of neutron star LMXRBs.

Many old accreting neutron stars in LMXRBs emit type I X-ray bursts, sharp increases of luminosity that last typically for tens of second and recur irregularly over hours. These bursts originate in thermonuclear flashes which burn the hydrogen and/or helium accumulated on the neutron star surface by accretion. In about a dozen Atoll sources, a nearly coherent signal at a frequency of $\nu_{\mathrm{burst}} \sim$ 250–600 Hz has been detected during type I bursts (for a review see Strohmayer 2001). These *burst oscillations* have frequencies very close to the spin frequency $\nu_s$ of the neutron star, as directly inferred from those few transient systems that possess a magnetosphere and show both burst oscillations and coherent pulsations resulting (like in standard X-ray pulsars) from a lighthouse effect due to beamed emission from accretion onto the magnetic poles (for reviews see Chakrabarty 2005 and Di Salvo et al. 2007). Therefore burst oscillation provide a powerful means of measuring the neutron star spin.

HBOs are often excited in sources of both the Atoll and Z classes. The changes in HBO frequency, $\nu_{\mathrm{HBO}}$ ($\sim$15 to $\sim$60 Hz), of individual sources displays an approximately quadratic dependence on the corresponding higher kHz QPO frequency ($\sim\nu_2^2$). Several diagnostics indicate that the frequency changes of the kHz QPOs and HBOs in individual sources correlate with the instantaneous mass accretion rate. HBOs show also a higher harmonic content in their signal, sometimes with 3 power spectrum peaks separated by a factor of 2 in frequency.

Two high frequency QPO peaks have also been detected in several LMXRBs hosting a black hole candidate. For instance QPOs at $\sim$300 Hz and $\sim$450 Hz were seen in the X-ray transient GRO J1655-40 (Remillard et al. 2000). Similarly XTE J1550-564 displayed simultaneous QPOs at $\nu_1 \simeq 187$ Hz $\nu_2 \simeq 268$ Hz during its April-May 2000 outburst (Miller et al. 2001). In these and other instances the frequency ratio of high frequency QPOs from BHCs was consistent with $3/2$. Low frequency QPOs (tens of Hz) are often present simultaneously with the high frequency QPOs also in accreting black hole candidates in LMXRBs.

The centroid frequency of different QPO modes (or peaked noise components) in LMXRBs display remarkable correlations among them (Psaltis et al. 1999a, PBV). One of these correlations extends over nearly 3 decades in frequency and encompasses both neutron stars and black hole candidate systems (see the points in Fig. 2); this correlation involves the lower frequency kHz QPOs, $\nu_1$, and the low frequency, HBO or HBO-like QPOs, $\nu_{\mathrm{HBO}}$. A nearly linear relationship ($\nu_{\mathrm{HBO}} \sim \nu_1^{0.95}$) holds.

Besides the QPOs whose properties are outlined above, other QPO and noise components have been identified and characterised, especially for the power spectra of accreting neutron star systems (see van der Klis 2006 and references therein); these will not be discussed here. The frequencies of QPOs, in spite their limited coherence, provide some of the most accurately measured observables of LMXRBs. Their frequency range and dependence offers a unique opportunity to study the deep gravitational fields in the close vicinity of neutron stars and black holes, where most of the high energy radiation is emitted.

## 3 QPO Models and Lense-Thirring Precession

Two main classes of QPO models have been proposed: models that apply only to accreting neutron stars *or* black holes, and models that apply to both accreting neutron stars *and* black holes. Models applicable only to neutron stars include: the magnetospheric and sonic point beat frequency models (Alpar and Shaham 1985; Lamb et al. 1985; Miller et al. 1998), the two-oscillator disk model (Osherovich and Titarchuk 1999), the photon bubble model (Klein et al. 1996) and neutron star oscillation models (Bildsten 1998; Bildsten and Cumming 1998). In this class of models, the similarities and correlations of QPOs across the neutron star and black hole systems result from mere coincidence.

QPO models that apply to both neutron stars and black holes involve: disk oscillations and trapped modes (Nowak and Wagoner 1991, 1992; Nowak et al. 1997; Milsom and Taam 1997) or disk warping and (nodal) precession (Cui et al. 1998; Markovic and Lamb 1998; Fragile et al. 2001; Merloni et al. 1999; Karas 1999) or the parametric epicyclic resonance (Abramowicz and Kluzniak 2001, 2005). The relativistic (periastron and nodal) precession model, RPM (Stella and Vietri 1998a; 1999; Vietri and Stella 1999; Stella et al. 1999; Psaltis and Norman 2000; Stella 2001; Psaltis 2001) that is outlined below also belongs to the latter class of models.

A variety of models associate the QPO signals to the fundamental frequencies of particle motion in the immediate vicinity of the accreting compact object. In particular the higher frequency kHz QPOs at $\nu_2$ are believed to correspond to the Keplerian frequency of matter motion at the innermost radius of the accretion disk that feeds matters to the compact object.

Let us consider a circular equatorial geodesic around a Kerr black hole of mass $M$ and specific angular momentum $a$. The coordinate $\phi$-frequency measured by a static observer at infinity is

$$\nu_\phi = \pm M^{1/2} r^{-3/2} [2\pi(1 \pm a M^{1/2} r^{-3/2})]^{-1} \tag{1}$$

(we use $G = c = 1$ units; the upper sign refers to prograde orbits). If we slightly perturb the circular orbit in the $r$ and $\theta$ directions, the small amplitude oscillations on the equatorial plane (with radial epicyclic frequency $\nu_r$) and perpendicular to it (with vertical frequency $\nu_\theta$) have coordinate frequencies

$$\nu_r^2 = \nu_\phi^2 (1 - 6Mr^{-1} \pm 8aM^{1/2} r^{-3/2} - 3a^2 r^{-2}), \tag{2}$$

$$\nu_\theta^2 = \nu_\phi^2 (1 \mp 4aM^{1/2} r^{-3/2} + 3a^2 r^{-2}). \tag{3}$$

The difference $\nu_\phi - \nu_\theta$ gives the nodal precession frequency, i.e. the frequency at which the particle orbital plane precesses around the spin axis of the black hole. In the limit of a non-rotating (Schwarzschild) black hole ($a = 0$) $\nu_\theta$ is equal to $\nu_\phi$ (as expected in a spherically symmetric spacetime) and thus the nodal precession frequency is identically zero; $\nu_r$ initially increases with increasing $\nu_\phi$ (or, equivalently, decreasing $r$), reaches a maximum for $r = 8M$ and then decreases, reaching zero at $r_{\rm ms} = 6M$ (the radius of the marginally stable orbit). This qualitative behaviour of $\nu_r$ is preserved in the Kerr field ($a \neq 0$). By comparing (1)–(2) it is easy to see that the periastron precession frequency $\nu_{\rm per} \equiv \nu_\phi - \nu_r$ is usually dominated by the "Schwarzschild" terms (i.e. those that do not contain $a$).

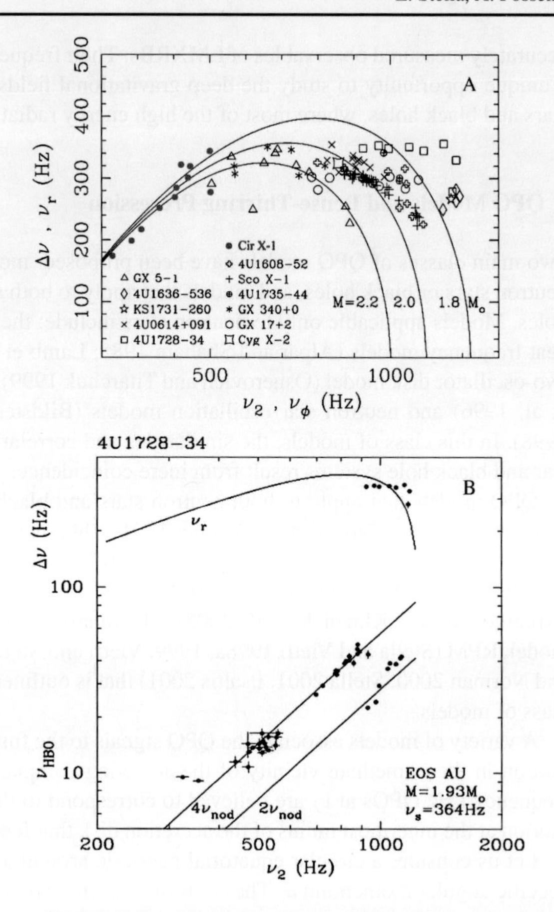


**Fig. 1** (**A**) kHz QPO frequency difference $\Delta\nu$ versus higher QPO frequency $\nu_2$ for a dozen LMXRBs containing an old neutron star (Stella and Vietri 1999). *Error bars* are not plotted for the sake of clarity. The curves give the $r$- and $\phi$-frequencies of matter in nearly circular orbit around a non-rotating neutron star, of mass 2.2, 2.0 and 1.8 $M_\odot$. The Cir X-1 data (*filled circles*) confirmed the low frequency behaviour of $\Delta\nu$ predicted by the RPM (Boutloukos et al. 2006). (**B**) kHz QPO frequency difference $\Delta\nu$ and (double-branched) HBO frequency versus higher QPO frequency $\nu_2$ in the Atoll LMXRB 4U1728-34 (Strohmayer et al. 1996; Ford and van der Klis 1998; Mendez and van der Klis 1999). The *solid lines* give the $r$-frequency and the 2nd and 4th harmonics of the nodal precession frequency $\nu_{\rm nod}$ as a function of the $\phi$-frequency for infinitesimally eccentric and tilted orbits in the spacetime of a 1.93 $M_\odot$ neutron star spinning at 364 Hz with EOS AU

In the RPM the higher and lower frequency kHz QPOs are identified with $\nu_2 = \nu_\phi$ and $\nu_1 = \nu_{\rm per}$, respectively. Therefore the frequency separation of the kHz QPOs, $\Delta\nu \equiv \nu_2 - \nu_1 = \nu_\phi - (\nu_\phi - \nu_r) = \nu_r$, is the radial epicyclic frequency. For $a = 0$, (1)–(2) give

$$\nu_r = \nu_\phi(1 - 6M/r)^{1/2} = \nu_\phi[1 - 6(2\pi\nu_\phi M)^{2/3}]^{1/2}. \quad (4)$$

Figure 1A shows different curves for $\nu_r$ vs. $\nu_\phi$ for $a = 0$ and selected values of $M$, the only free parameter in (4). The measured $\Delta\nu$ vs. $\nu_2$ for a dozen neutron star LMXRBs showing kHz QPOs is also plotted. For neutron star masses of $\sim$2 $M_\odot$, the simple model above is in pretty good agreement with measured values, including the decrease of $\Delta\nu$ for increasing $\nu_2$ seen in Sco X-1, 4U1608-52, 4U1735-44 and 4U1728-34. Remarkably, most points are close to the maximum of the radial epicyclic frequency. The rising behaviour of $\nu_r$ at low frequencies, a clear prediction of the RPM, was recently confirmed through the QPO separation in Cir X-1 which was found to increase for increasing QPO frequencies (Boutloukos et al. 2006) (see Fig. 1A).

It should be emphasised that the model above and the curves in Fig. 1A are approximate in that (4) retains only the strong-field Schwarzschild terms, which are dominant over a very wide range of parameters. Lower order expansions of these terms would not reproduce the observed $\Delta\nu - \nu_2$ relation. In this sense the RPM holds the potential to test the strong-field properties of the Schwarzschild spacetime. However, in general rotation of the compact object must be taken into account. In the case of a rotating black hole, the Kerr formulae suffice. Instead, the spacetime around rotating neutron stars is different from a Kerr spacetime and, in general, must be calculated numerically. In particular, the star's oblateness induced by rotation gives rise to additional quadrupole and higher multipole terms in the gravitational field.

Analytical formulae to approximate the spacetime around a rotating neutron star are given in Stella and Vietri (1999). At the first order of the angular momentum in the Kerr metric, $\nu_\phi$ and $\nu_r$ contain a Lense-Thirring term given by $\nu_{\mathrm{LT},\phi} = -\nu_{\mathrm{LT,nod}}/2$ and $\nu_{\mathrm{LT},r} = 3\nu_{\mathrm{LT,nod}}/2$. Here $\nu_{\mathrm{LT,nod}}$ is the Lense-Thirring nodal precession frequency (see (5) below). These Lense-Thirring correction terms amount to ~1–2% and ~5–10% of the values $\nu_\phi$ and $\nu_r$, for the angular momentum of neutron stars in LMXRBs; singling them out through a comparison between measured value of the kHz QPOs and the RPM has proven unfeasible so far.

In the RPM, the maximum value of $\nu_r = \Delta\nu$ depends (mainly) on the mass of the compact object. The neutron star masses inferred from the curves in Figs. 1–2 are ~1.8–2.0 $\mathrm{M}_\odot$. While somewhat higher than most neutron star mass measurements, these values are compatible with masses obtained from spectro-photometric optical observations of a few X-ray binaries (e.g. Cyg X-2, $M = 1.78 \pm 0.23\ \mathrm{M}_\odot$, Orosz and Kuulkers 1999, and Vela X-1, $M = 1.86 \pm 0.16\ \mathrm{M}_\odot$, Barziv et al. 2001), and from pulse timing studies of PSR J0437-4715 ($M = 1.76 \pm 0.20\ \mathrm{M}_\odot$, Verbiest et al. 2008) and some radio pulsars in globular clusters (e.g. Freire et al. 2008).

In the RPM the HBO frequency is related to the nodal precession frequency, $\nu_{\mathrm{nod}}$, at the same radius where the signals at $\nu_\phi$ and $\nu_{\mathrm{per}}$ are produced. In the slow rotation case ($a/M \ll 1$), i.e. the Lense-Thirring limit, the nodal precession frequency $\nu_{\mathrm{nod}}$ is approximated by (see (1) and (3))

$$\begin{aligned}\nu_{\mathrm{LT,nod}} &= aMr^{-3}/\pi = 4\pi a\nu_\phi^2 \simeq 6.2\times10^{-5}(a/M)m\nu_\phi^2\ \mathrm{Hz}\\ &\simeq 4.4\times10^{-8}\ I_{45}\nu_\phi^2\nu_s/m\ \mathrm{Hz},\end{aligned}\tag{5}$$

where $M = m\ \mathrm{M}_\odot$. This is the well known Lense-Thirring formula for the nodal precession of the orbital plane of a test particle around the spin axis of the compact object. The latter expression on the right hand side is for a rotating neutron star, where $aM = 2\pi\nu_s I$, with $I = 10^{45} I_{45}\ \mathrm{g\,cm^2}$ its moment of inertia. In general, the nodal precession frequency is considerably slower than the other fundamental frequencies of motion and close to the frequency range of HBOs; this led (Stella and Vietri 1998a) to propose that HBOs are due to the nodal precession of matter in the innermost disk region.

If $\nu_\phi$ and $\nu_s$ are known (through $\nu_2$ and $\nu_{\mathrm{burst}}$), then $I_{45}m^{-1}$ is the only free parameter in (5). Neutron star models show that $0.5 < I_{45}m^{-1} < 2$, for virtually any mass and equation of state (EOS, see e.g. the rotating NS models of Friedman et al. 1986; Cook et al. 1992). The stellar oblateness induced by the star rotation (and thus the "classical" quadrupole and higher multipole terms in the gravitational potential), alters the nodal precession frequency somewhat (Stella and Vietri 1998a). A detailed post-Newtonian treatment for the classical and Lense-Thirring contributions to the nodal precession frequency around realistic neutron star models is given in Morsink and Stella (1999).

The relative importance of the "classical" correction increases for higher values of $\nu_s$ and $\nu_2$. The quadrupole component causes a retrograde nodal precession that scales like $\sim r^{-7/2}$ (as opposed to the $\sim r^{-3}$ dependence of $\nu_{\mathrm{LT,nod}}$). The "classical" correction amounts to few tens percent over a wide range of parameters, such that the nodal precession frequency is well approximated by the Lense-Thirring scaling $\nu_{\rm nod} \propto \nu_\phi^2$.[2] An approximately quadratic dependence of $\nu_{\rm HBO}$ on $\nu_2$ has been measured in a number of LMXRBs (Stella and Vietri 1998a, 1998b; Ford and van der Klis 1998; Psaltis et al. 1999b) in agreement with one of the basic features of the RPM. Concerning the absolute values of $\nu_{\rm HBO}$ a good agreement with the RPM is obtained if the HBO frequency is identified with the 2nd harmonics of $\nu_{\rm nod}$ (i.e. $2\nu_{\rm nod}$) (Stella and Vietri 1998a; Morsink and Stella 1999).

In sources displaying simultaneous kHz QPOs and HBOs and with known spin period the predictions of the RPM can be tested directly.

4U1728-34 is one such source (Strohmayer et al. 1996; Ford and van der Klis 1998; Mendez and van der Klis 1999). The neutron star spin frequency is $\nu_s \simeq 364$ Hz, as inferred from burst oscillations. In the application of the RPM to the QPOs of 4U1728-34, a numerical approach was adopted in order to compute accurately the spacetime and geodetic motion around the star (Stergioulas and Friedman 1995); $\nu_{\rm nod}$ was calculated following the prescription of Morsink and Stella (1999), whereas $\nu_r$ was obtained based the second radial derivative of the effective potential (Stella and Vietri 1999).

Figure 1B shows the measured values of $\Delta\nu$ and $\nu_{\rm HBO}$ versus $\nu_2$ in 4U1728-34. The solid lines represent the RPM frequencies for a 1.93 $M_\odot$ NS with EOS AU and $\nu_s = 364$ Hz. Also in this case a good matching is obtained if the HBO frequency, that is the lower of the two branches detected in 4U1728-34, is interpreted in terms of the 2nd harmonics of $\nu_{\rm nod}$ (i.e. $2\nu_{\rm nod}$); correspondingly the upper HBO branch is well fit by $4\nu_{\rm nod}$. The behaviour and values of the epicyclic frequency $\nu_r$ calculated for the model above are also in general agreement with the $\Delta\nu$ measurement (Stella 2001).

In summary, the RPM is capable of reproducing the salient features of both the $\Delta\nu$ versus $\nu_2$ and $\nu_{\rm HBO}$ versus $\nu_2$ relationships in 4U1728-34. We note that only two free parameters are involved, $M$, which can assume only a very limited range of values, and the EOS, which is varied in a discrete manner to give different neutron star models. Concerning Z-sources and all other Atoll sources for which no burst oscillations nor coherent pulsations have yet been detected, the neutron star spin is to be regarded as another free parameter in applications of the RPM.

The RPM reproduces accurately also the PBV correlation, without resorting to additional assumptions (Stella et al. 1999). Figure 2A shows $2\nu_{\rm nod}$ and $\nu_\phi$ (see (1)–(3)), as a function of $\nu_{\rm per}$ for corotating orbits and selected values of $M$ and $a/M$ in a Kerr spacetime. The high-frequency end of each line corresponds to the orbital radius reaching the marginally stable orbit; as customary for the RPM $\nu_{\rm HBO}$ is identified with $2\nu_{\rm nod}$. By taking weak field ($M/r \ll 1$) and slow rotation ($a/M \ll 1$) limit in (1)–(3), the first order dependence between the relevant RPM frequencies is made explicit

$$\nu_\phi \simeq (2\pi)^{-2/5} 3^{-3/5} M^{-2/5} \nu_{\rm per}^{3/5} \simeq 33\, m^{-2/5} \nu_{\rm per}^{3/5}\ \text{Hz}, \tag{6}$$

$$\nu_{\rm nod} \simeq (2/3)^{6/5} \pi^{1/5} (a/M) M^{1/5} \nu_{\rm per}^{6/5} \simeq 6.7 \times 10^{-2}\, (a/M) m^{1/5} \nu_{\rm per}^{6/5}\ \text{Hz}. \tag{7}$$

The lines plotted in Fig. 2A show that $\nu_{\rm nod}$ depends weakly on $M$ and strongly on $a/M$; the opposite holds for $\nu_\phi$. Therefore the black hole mass and angular momentum can be inferred from $\nu_1$, $\nu_2$ and $\nu_{\rm HBO}$. Note that relatively small values of $a/M \sim 0.1$–$0.3$ are required for the RPM to match the QPO data from black hole candidates.

[2] A slightly flatter dependence is expected for high values of $\nu_s$ and $\nu_\phi$, due to the increasingly important role of classical precession.

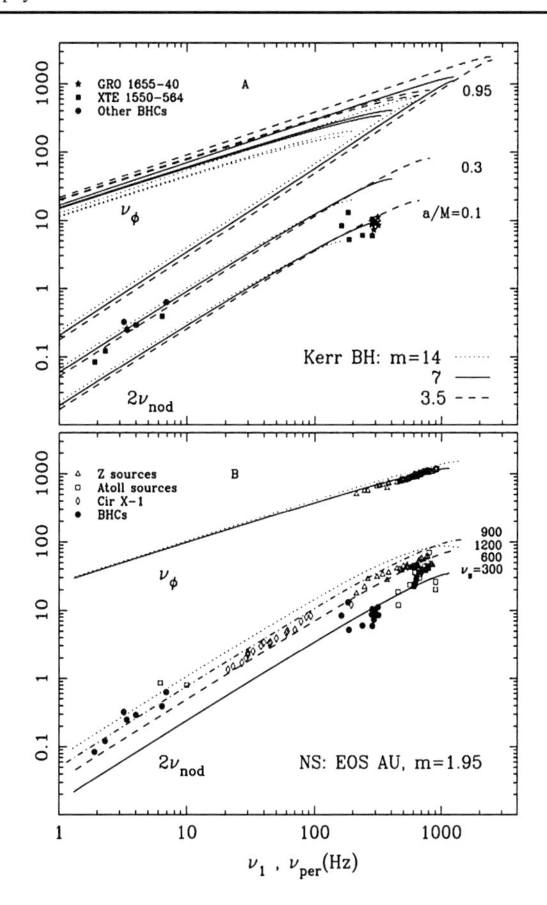


**Fig. 2** Twice the nodal precession frequency, $2\nu_{\rm nod} = \nu_{\rm HBO}$, and $\phi$-frequency, $\nu_\phi = \nu_2$, vs. periastron precession frequency, $\nu_{\rm per} = \nu_1$, for black hole candidates of various masses and angular momenta (panel **A**) and rotating neutron star models (EOS AU, $m = 1.95$) with selected spin frequencies (panel **B**) (Stella and Vietri 1999). The measured QPO (or peaked noise) frequencies $\nu_1$, $\nu_2$ and $\nu_{\rm HBO}$ giving rise to the PBV correlation are also shown in panel **B** for both BHC and NS LMXRBs and in panel **A** for BHC LMXRBs only; *errors bars* are not plotted (see Psaltis et al. 1999a for a complete list of references). We included only those cases in which QPOs at $\nu_1$ were unambiguously detected. NBO and FBO frequencies are not plotted

For the case of rotating NSs the numerical approach summarised above is adopted also. The lines in Fig. 2B are for a NS mass of 1.95 $M_\odot$, EOS AU and $\nu_s$ = 300, 600, 900 and 1200 Hz. The measured QPO and peaked-noise frequencies giving rise to the PBV correlation are also plotted in Fig. 2B. The approximately linear dependence of $\nu_{\rm nod}$ on $\nu_{\rm per}$, nicely matches the PBV correlation. Note that the approximate scalings in (6)–(7) remain valid over a wide range of frequencies. Only for the largest values of $\nu_{\rm per}$ and $\nu_s$, $\nu_{\rm nod}$ departs substantially from the $\sim\nu_{\rm per}^{6/5}$ dependence. The fact that the RPM matches nicely the observed $\nu_{\rm HBO} - \nu_{\rm l}$ correlation over $\sim$3 decades in frequency (down to $\nu_{\rm l}$ of a few Hz) provides additional support in favor of the model and testifies that also the scaling of $\nu_{\rm nod}$ on $\nu_{\rm per}$ (besides that of $\nu_{\rm nod}$ on $\nu_\phi$, see above) is in agreement with the Lense-Thirring interpretation of the HBOs (which implicitly confirms the radial dependence of the Lense-Thirring nodal precession frequency).

We have summarised above the salient characteristics and merits of the RPM and emphasised that the model holds the potential to single out and test several general relativistic effects in the motion of matter orbiting very close to a collapsed object. In such an environment these effects are large: for instance there is no weak-field expansion that could approximate the behaviour of periastron precession in this regime. Moreover close to the fast rotating neutron stars of LMXRBs, nodal precession is expected to be dominated by the Lense-Thirring effect, with classical nodal precession contributing only at a level of a few

**Table 1** Approximate frequency identification in some QPO models applicable to neutron stars and black holes

| Model | $\nu_{HBO}$ | $\nu_1$ | $\nu_2$ |
|---|---|---|---|
| Generic | – | – | $\nu_\phi$ |
| Disk oscillations | – | $\nu_{nod}$ | $\nu_r$ |
| Disk warping | – | – | $\nu_{nod}$ |
| Parametric epicyclic resonance | – | $\nu_r$ | $\nu_\phi$ |
| Relativistic precession | $\nu_{nod}$ | $\nu_{per}$ | $\nu_\phi$ |

tens percent.[3] Therefore the magnitude of general relativistic effects is not an issue. Rather, in this context (and the astrophysical context in general) the key issue is making sure that the relevant diagnostics are interpreted correctly and that there are no ambiguities in singling out the effects.

For instance the interpretation of the kHz QPOs and HBOs in LMXRBs is still open to other possibilities. Table 1 gives a schematic summary the frequency interpretations proposed in different classes of QPO models, which resort also to the fundamental frequencies of motion in a strong gravitational field. We will not attempt to discuss pros and cons of various models here, though, to our knowledge, no other model can presently match the QPO frequency scaling and correlations at a level comparable to the RPM (especially in neutron star systems). We simply stress that further model development and detailed comparison with the QPO properties in accreting neutron stars and black hole candidates are required to determine without ambiguity the origin of these signals.

## 4 Relativistic Precession in Binary Pulsars

Radio pulsars hosted in binary systems containing another compact star (a neutron star or a white dwarf) very often behave as very stable clocks, and thus allow scientists to accurately determine their orbital motion, by using pulse arrival time techniques (see e.g. Lorimer and Kramer 2005 for a detailed description of this procedure).

When the orbital separation is still relatively large, well before coalescence, each component of the binary system moves in the relatively weak gravitational field of the companion. In most alternative theories of gravity, e.g. scalar-tensor theories, the stars' motion and gravitational radiation damping depend on the gravitational binding energy (*i.e. self-field energy*) of the stars (see e.g. Esposito-Farèse 2005 and Will 2006). Since the gravitational binding energy of the neutron stars is very large (i.e. their compactness parameter is close to unity), significant deviations in the motion of the bodies are expected to occur due to *strong self-field effects*. This provides the observational and theoretical basis for using binary pulsars as laboratory for performing tests of the gravity theories in the *strong-field* regime. Note that in this context the expression *strong field* is used in a somewhat different meaning than the case of particle motion in the vicinity of collapsed object, where the *strong field* is that experienced directly *in situ* by the accreting matter.

[3] For comparison LAGEOS-type satellite around the earth undergo classical nodal precession at a rate about 7 orders of magnitude faster than Lense-Thirring precession.

## 5 The Outstanding Case of the Double Pulsar

The PSR J0737–3039 system (also known as the Double Pulsar) represents a discovery paramount importance in pulsar science and stands as a unique object among the nine (confirmed and likely) double neutron star binaries known to date in the galactic plane (e.g. Lorimer 2008). The system comprises two radio pulsars: PSR J0737–3039A (hereafter pulsar A; Burgay et al. 2003), with a spin period of ∼23 ms, and PSR 0737–3039B (hereafter pulsar B; Lyne et al. 2004), with a spin period of ∼2.7 s. The system has a ∼2.4 hr orbital period and a $e \sim 0.09$ eccentric orbit. At least four factors make the Double Pulsar an unsurpassed cosmic laboratory for studying general relativistic effects: (i) the large orbital velocity (more than 0.001 $c$); (ii) the small orbital separation ($8.5 \times 10^{10}$ cm on average); (iii) the high orbital inclination (>88 deg); (iv) the presence of pulsations from both stars, which can be used as independent clocks. The first two factors are essential to measure relativistic effects, while the third and fourth factors significantly improve the detectability of the effects and the precision with which they can be measured.

A few days after the discovery of J0737−3039, it was already possible to measure the relativistic advance of the periastron $\dot{\omega}$ of pulsar A (Burgay et al. 2003): about 16.9 degree/yr, four times larger than in any other binary pulsar known. After 6 more months of observation, a measurement of 3 additional post-Keplerian parameters was obtained (i.e. the combined time dilation and gravitational redshift parameter $\gamma$, the range $r$ and the shape $s \equiv \sin i$ of the Shapiro delay, where $i$ is the inclination of the orbital plane), as well as a *direct* measurement (never achieved before as a post-Keplerian parameter) of the mass ratio $R$ in the system (Lyne et al. 2004). With another year of data, a fifth post-Keplerian parameter (i.e. the orbital decay $\dot{P}_b$) was also detected, confirming that the system shrinks and will coalesce in ∼85 million years, owing to the emission of gravitational waves.

Damour and Deruelle (1985, 1986) and Damour and Taylor (1992) provided the original definition of the aforementioned post-Keplerian parameters and introduced a powerful framework for testing gravity theories in binary systems. An illustration of these tests is given in the so-called mass-mass diagram for the two stars in the binary, where each measured post-Keplerian parameter (with its uncertainties) is associated to a pair of lines. The functional dependence on the masses and (usually well determined) Keplerian parameters is specific to a given gravity theory. The region in the diagram enclosed between a pair of lines is the only region allowed by that theory. Therefore, if there is no intersection between the regions defined by the various pairs of lines, then that specific gravity theory is to be rejected. In fact, if such a common region of overlap does not exist, there must be an observational constraint (corresponding to a pair of lines in the mass-mass diagram) which is not satisfied by any allowed set of values for the masses of the two stars; this in turn leads to an intrinsic incongruence in the gravity theory. The mass-mass diagram for the case of the Double Pulsar is shown in Fig. 3, where the pairs of lines related to the post-Keplerian parameters $\dot{\omega}$, $\gamma$, $r$, $s$, and $\dot{P}_b$ are drawn based on general relativity. The pair of lines related to the mass ratio $R$ holds to first post-Newtonian order for general relativity as well as a number of alternative gravity theories (e.g. Damour and Taylor 1992). The diagram (see also the zoomed inset) shows a common (albeit very small) region of overlap for the 6 pairs of lines representing the parameters listed above (plus a 7th pair of lines which is related to the relativistic precession of the spin axis of pulsar B, as explained in next section), implying that general relativity passes 5 independent tests with full marks. In fact the area of intersection between two pairs of lines (e.g. resulting from the observation of two Post-Keplerian parameters) constrains the allowed values for the masses of the two neutron stars in the system. Any additional pair of lines (there are $7-2=5$ of them in the case of

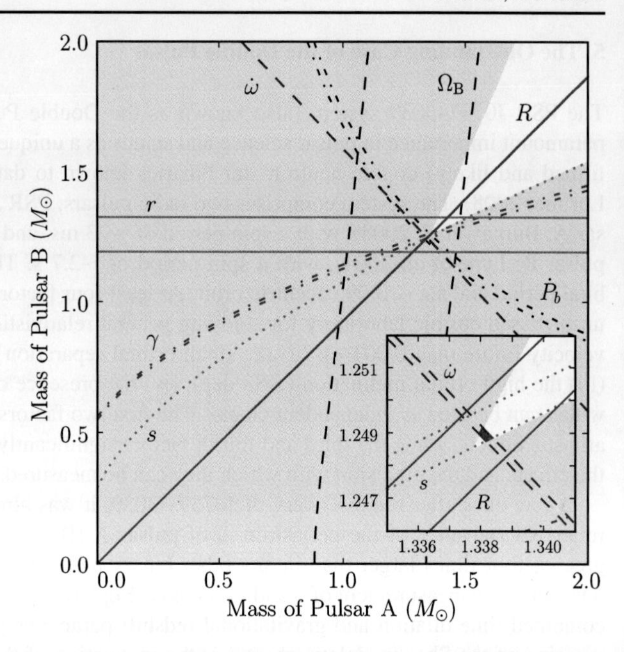


**Fig. 3** Mass-mass diagram for the Double Pulsar. The *blueish colored regions* are those excluded by the Keplerian mass function of the two pulsars. The additional constraints are depicted as *pairs of lines* labeled by the corresponding parameter, as described in the text. Courtesy of René Breton 2009

Fig. 3) over-determines the masses of the two stars, thus providing an independent test of general relativity.

Incidentally we note that after ~3 years of observations, the shape $s$ of the Shapiro delay provided the most stringent astrophysical test of general relativity to date, with the prediction matching the observed value to within 0.05% (Kramer et al. 2006). Better levels of agreement have so far been obtained only through some tests in the much weaker field of the Solar System (see e.g. Bertotti et al. 2003).

## 6 Relativistic Precession of the Spin of PSR J0737–3039B

A variety of remarkable phenomena takes place in the Double Pulsar system. The radio emission from pulsar B is strong during two short and secularly varying orbital phases (Burgay et al. 2005), alternated with large portions of the orbits where the signal is very weak or undetectable (Lyne et al. 2004). Furthermore, the shape of the pulse profile of pulsar B changes along the orbit (Lyne et al. 2004) and a fraction of the emission from pulsar B (at least at suitable orbital phases) is modulated at the beat frequency between the spin frequency of the two pulsars (McLaughlin et al. 2004a). These properties testify that there is an unprecedented interaction between the flux emitted by pulsar A (whose spin-down luminosity is about 3600 larger than that of pulsar B) and the magnetosphere of pulsar B.

Due to the high orbital inclination, the radio signal from pulsar A (at least for wavelengths longer than ~20 cm) also experiences a very short (~30 s) eclipse when the pulsar transits at superior conjunction. What makes the latter phenomenon especially interesting is that the residual signal from pulsar A during the eclipse is modulated by the rotation of pulsar B, indicating that the absorbing material responsible for the eclipse corotates with pulsar B (McLaughlin et al. 2004b). According to the model by Lyutikov and Thompson (2005), absorption is due to synchrotron resonance with relativistic electrons confined within the closed field lines of the magnetic dipole associated with pulsar B. This very simple model reproduces very accurately the complex eclipse phenomenology, once suitable parameters for the magnetosphere of pulsar B and the orientation of its spin and magnetic axes are adopted. A remarkable results is that the orientation in space of the magnetic and spin axes of pulsar B was found to vary in a continuous fashion. Besides yielding unprecedented information on some physical characteristics of pulsar magnetospheres (e.g. Lyutikov and Thompson 2005), these findings provide the observational basis for the first very accurate measurement of an additional relativistic effect, i.e. the precession of the spin axis of pulsar B. Below we describe this in greater detail.

The eclipses of pulsar A were intensively monitored for over 4 years at the Green Bank Radio Telescope at 820 MHz (Breton et al. 2008). First, the parameters of each eclipse were determined. Then, a search for changes in the orientation of the spin and magnetic axes was carried out. No significant secular variation of the angle between the magnetic and the spin axis was found; moreover, the angle between the spin axis and the total angular momentum of the system was consistent with being constant. On the contrary, a clear evolution in the azimuthal spin axis angle (i.e. the angle describing the precession of the spin vector around the total angular momentum vector) was detected. The rate of precession determined by fitting a linear variation of the azimuthal angle in time is $\Omega_B = 4.77(+0.66/-0.65)\ \mathrm{deg\,yr^{-1}}$. The agreement with the value predicted by general relativity, $5.0734 \pm 0.0007\ \mathrm{deg\,yr^{-1}}$ (Barker and O'Connell 1975), is at 13% level,[4] much better than the only other measurement of this effect available so far in a binary pulsar system (Stairs et al. 2004).

The measurement of $\Omega_B$ in the Double Pulsar can also constrain the relativistic spin precession predicted in alternate gravity theories. In particular for the class of the Lorentz-invariant gravity theories based on a Lagrangian (Damour and Taylor 1992), a class which of course comprises general relativity, the general expression for the precession rate of the spin axis is

$$\Omega_B = \left[\frac{x_A x_B}{s^2}\right]\left[\frac{8\pi^3}{(1-e^2)P_b^3}\right]\left[\frac{c^2\sigma_B}{G_{AB}}\right], \tag{8}$$

where $x_A, x_B, P_b$ and $e$ are Keplerian parameters (the projected semi-major axis of the orbit of pulsar A and pulsar B, the orbital period and the eccentricy, respectively), $s$ is the post-Keplerian shape of the Shapiro delay, $\sigma_B$ is a (theory dependent) strong-field spin-orbit coupling constant and $G_{AB}$ is the (theory dependent) gravitational constant for the gravity interaction between the two pulsars. Inspection of (8) shows that an almost edge-on orbit (required for a precise determination of $s$) and two active pulsars in the system (essential for determining $x_A$ and $x_B$ separately) are required to solve for the rightmost factor. In the so far unique case of the Double Pulsar, (8) yields $[c^2\sigma_B/G_{AB}] = 3.38(+0.49/-0.46)$. This value must be matched by the predictions of any successful gravity theory belonging to the large aforementioned class. Since in general relativity $[c^2\sigma_B/G_{AB}]_{GR} = 2 + \frac{3m_A}{2m_B}$, the mass ratio of pulsar A and B gives $[c^2\sigma_B/G_{AB}]_{GR} = 3.60677 \pm 0.00035$, which lies well inside the range given above. Therefore, as anticipated in Sect. 5 and clearly illustrated in Fig. 3 (where a pair of lines are depicted for the new constraint from $\Omega_B$), general relativity passes also this test. This is a results which goes beyond any weak-field test of relativistic spin precession, in that it supports, to within uncertainties, the *effacement* property of gravity in general relativity (Damour 1987) also for spinning bodies, i.e. the fact that the internal

[4]Despite extensive searches, no spin precession has yet been detected in pulsar A (Manchester et al. 2005). This indicates that the misalignment angle between the orbital angular momentum and the spin axis of pulsar A is small, likely <15 deg (Ferdman et al. 2008).

structure of the neutron star (and related strong self-field) does not prevent the star from behaving like a spinning test particle in a weak external field.

What is generically dubbed above (following Breton et al. 2008) as relativistic spin precession is commonly referred to as *geodetic* (*or de Sitter*) precession in other papers related to binary pulsar science. This reflects the fact that, for the Double Pulsar as well as for all the other known binary pulsars, the contribution of the spin of the two objects to the curvature invariant (really a pseudo-invariant for coordinate reflections) $^{*}\mathbf{RR}$, constructed from the Riemann tensor $\mathbf{R}$ and its dual $^{*}\mathbf{R}$, is negligible and thus it follows $^{*}\mathbf{RR} = 0$. In fact, Ciufolini and Wheeler (1995) and, more recently, Ciufolini (2008) stated that *geodetic* (*or de Sitter*) (de Sitter 1916a, 1916b) and *frame-dragging* (*or Lense-Thirring*) effects (Lense and Thirring 1918), are physically different phenomena, which can be unambiguously discriminated on the basis of the aforementioned invariant: a genuine frame-dragging effect (i.e. due to the *mass currents* of a rotating body) occurs only when $^{*}\mathbf{RR} \neq 0$.

When the mass of the body whose spin axis precesses is much smaller than the mass of the body around which it orbits, relativistic spin precession (at first post-Newtonian approximation) is commonly interpreted (see e.g. Misner et al. 1973) as a combination of the two aforementioned terms.[5] In particular, the *de Sitter* precession of the spin axis results from the spacetime curvature produced by the more massive body. Lense-Thirring precession, instead, is due to the rotation of the more massive body. However, the physical interpretation of the various pieces entering the relativistic precession formulae is more involved in the case of a general two-body problem (i.e. when the masses of the two orbiting objects are comparable, like for binary pulsars). Moreover it has been suggested (see e.g. the works of Ashby and Shahid-Saless 1990 or O'Connell 2005) that the adoption of some peculiar frame of reference could lead (even for the simpler one-body problem) to an apparent mixture, or even a reversal, of the nominal contribution of the geodetic and frame-dragging effects to the total precession.

As explained above, using of the value of the pseudo-invariant $^{*}\mathbf{RR}$ (Ciufolini and Wheeler 1995) establishes a physically sound framework for dealing with the different types precession of the spin axis. On the other hand, alternative approaches have been proposed in order to disentangle the various contributions to relativistic precession and identify their physical origin. One of them is borrowed from quantum electrodynamics and uses the concepts of spin-orbit (S-L) and spin-spin (S-S) coupling, to describe the gravitational effects due to the rotation of the two bodies (e.g. O'Connell 2008 and reference therein). This approach directly relates the various coupling terms to the corresponding parts in the Lagrangian (or Hamiltonian) of the system, which in turn depends on the specific gravity theory (at least for the very large class of theories already mentioned above, Damour and Taylor 1992). In general the Lagrangian naturally includes $\mathbf{S}^{(i)} \cdot \mathbf{L}$ and $\mathbf{S}^{(i)} \cdot \mathbf{S}^{(j)}$ terms (plus other more complex terms), where $\mathbf{S}^{(i)}$ is the spin of the $i$-th body and $\mathbf{L}$ the total orbital angular momentum; however, for the sample of binary pulsars discovered so far the S-S coupling is always negligible and we are left with precession effects arising from the spin-orbit coupling only. In this perspective, the de Sitter and Lense-Thirring effects could be regarded as different manifestations of the same basic spin-orbit interaction (O'Connell 2008).

Indeed a thorough analysis of the various contributions resulting from spin-orbit coupling has been conducted by Damour and Taylor (1992). These authors investigated the most general type of L-S contribution to the Lagrangian of the system (thus including terms also depending on any acceleration experienced by the bodies). It turns out that, besides a *Thomas*

[5] A third term, the so-called Thomas precession does not come into play for an orbiting (i.e. freely falling) body.

*precession* (i.e. acceleration related) effect, independent of the specific type of interaction, there is a *gravity-related* precession term, which depends on the chosen gravity theory and may be largely affected by strong-field effects. This further highlights the importance of the strong-field test of relativistic spin precession (the only one available so far) performed with the Double Pulsar (Breton et al. 2008).

## References


M.A. Abramowicz, W. Kluzniak, Astron. Astrophys. **374**, L19 (2001)

M.A. Abramowicz, W. Kluzniak, Astrophys. Space Sci. **300**, 143 (2005)

M.A. Alpar, J. Shaham, Nature **316**, 239 (1985)

P. Armitage, P. Natarajan, Astrophys. J. **525**, 909 (1999)

N. Ashby, B. Shahid-Saless, Phys. Rev. D **42**, 1118 (1990)

J.M. Bardeen, J.A. Petterson, Astrophys. J. **195**, L65 (1975)

B.M. Barker, R.F. O'Connell, Phys. Rev. D **12**, 329 (1975)

O. Barziv et al., Astron. Astrophys. **377**, 925 (2001)

R.D. Blandford, J. Astrophys. Astron. **16**, 191 (1995)

B. Bertotti, L. Iess, P. Tortora, Nature **425**, 374 (2003)

L. Bildsten, Astrophys. J. **501**, 89 (1998)

L. Bildsten, A. Cumming, Astrophys. J. **506**, 842 (1998)

S. Boutloukos et al., Astrophys. J. **653**, 1435 (2006)

R.P. Breton, V.M. Kaspi, M. Kramer, M.A. McLaughlin, M. Lyutikov, S.M. Ransom, I.H. Stairs, R.D. Ferdman, F. Camilo, A. Possenti, Science **321**, 104 (2008)

B.C. Bromley, K. Chen, W.A. Miller, A. Warner, Astrophys. J. **475**, 57 (1997)

M. Burgay et al., Nature **426**, 531 (2003)

M. Burgay, A. Possenti, R.N. Manchester, M. Kramer, M.A. McLaughlin, D.R. Lorimer, I.H. Stairs, B.C. Joshi, A.G. Lyne, F. Camilo, N. D'Amico, P.C.C. Freire, J.M. Sarkissian, A.W. Hotan, G.B. Hobbs, Astrophys. J. **624**, 113 (2005)

D. Chakrabarty, in *Interacting Binaries: Accretion, Evolution, and Outcomes*. AIP Conf. Proc., vol. 797 (AIP, New York, 2005), p. 71

I. Ciufolini, arXiv:0809.3219 [gr-qc] (2008)

I. Ciufolini, J.A. Wheeler, *Gravitation and Inertia* (Princeton University Press, Princeton, 1995)

G.B. Cook, S.L. Shapiro, S.A. Teukolsky, Astrophys. J. **398**, 20 (1992)

W. Cui, S.N. Zhang, W. Chen, Astrophys. J. **492**, L53 (1998)

T. Damour, in *Three Hundred Years of Gravitation*, ed. by S.W. Hawking, W. Israel (Cambridge Univ. Press, Cambridge, 1987), p. 128

T. Damour, N. Deruelle, Ann. Inst. H. Poincaré (Phys. Théor.) **43**, 107 (1985)

T. Damour, N. Deruelle, Ann. Inst. H. Poincaré (Phys. Théor.) **44**, 263 (1986)

T. Damour, J.H. Taylor, Phys. Rev. D **45**, 1840 (1992)

W. de Sitter, Mon. Not. R. Soc. **72**, 155 (1916a)

W. de Sitter, Mon. Not. R. Soc. **76**, 699 (1916b)

T. Di Salvo et al., in *The Multicolored Landscape of Compact Objects*. AIP Conf. Proc., vol. 924 (AIP, New York, 2007), p. 613

G. Esposito-Farèse, in *The Tenth Marcel Grossmann Meeting: On Recent Developments in Theoretical and Experimental General Relativity, Gravitation and Relativistic Field Theories*, ed. by M. Novello, S. Perez Bergliaffa, R. Ruffini (World Sci., Singapore, 2005), p. 647

A.C. Fabian, Astron. Nachr. **329**, 155 (2008)

A.C. Fabian, M.J. Rees, L. Stella, N.E. White, Mon. Not. R. Astron. Soc. **238**, 729 (1989)

A.C. Fabian et al., Mon. Not. R. Astron. Soc. **335**, L1 (2002)

R.D. Ferdman, I.H. Stairs, M. Kramer, R.N. Manchester, A.G. Lyne, R.P. Breton, M.A. McLaughlin, A. Possenti, M. Burgay, in *40 Years of Pulsars: Millisecond Pulsars, Magnetars and More*. AIP Conf. Proc., vol. 983 (AIP, New York, 2008), p. 474

E.C. Ford, M. van der Klis, Astrophys. J. **506**, L39 (1998)

P.C. Fragile, G.J. Mathews, J.R. Wilson, Astrophys. J. **553**, 955 (2001)

P.C.C. Freire et al., Astrophys. J. **675**, 670 (2008)

J.L. Friedman, J.R. Ipser, L. Parker, Astrophys. J. **304**, 115 (1986)

R. Kannan, P. Saha, Astrophys. J. **690**, 1553 (2009)

V. Karas, Astrophys. J. **526**, 953 (1999)

A.R. King, J.E. Pringle, J.A. Hofmann, Mon. Not. R. Astron. Soc. **385**, 1621 (2008)
R.I. Klein et al., Astrophys. J. **469**, L119 (1996)
M. Kramer et al., Science **314**, 97 (2006)
F.K. Lamb, N. Shibazaki, A. Alpar, J. Shaham, Nature **317**, 681 (1985)
A. Laor, Astrophys. J. **376**, 90 (1991)
J. Lense, H.Ü. Thirring, Phys. Z. **19**, 156 (1918)
D.R. Lorimer, in *Living Reviews in Relativity*, vol. 11 (Max Planck Society, Munich, 2008), p. 8
D.R. Lorimer, M. Kramer, *Handbook of Pulsar Astronomy* (Cambridge University Press, Cambridge, 2005)
A.G. Lyne et al., Science **303**, 1153 (2004)
M. Lyutikov, C. Thompson, Astrophys. J. **634**, 1223 (2005)
R.N. Manchester, M. Kramer, A. Possenti, A.G. Lyne, M. Burgay, I.H. Stairs, A.W. Hotan, M.A. McLaughlin, D.R. Lorimer, G.B. Hobbs, J.M. Sarkissian, N. D'Amico, F. Camilo, B.C. Joshi, P.C.C. Freire, Astrophys. J. **621**, 49 (2005)
D. Markovic, F.K. Lamb, Astrophys. J. **507**, 316 (1998)
P.G. Martin, M.J. Rees, Mon. Not. R. Astron. Soc. **189**, 19P (1979)
R.G. Martin, J.E. Pringle, C.A. Tout, Mon. Not. R. Astron. Soc. **381**, 1617 (2007)
R.G. Martin, R.C. Reis, J.E. Pringle, Mon. Not. R. Astron. Soc. **391**, L15 (2008)
J.E. McClintock, R. Narayan, R. Shafee, in *Black Holes*, ed. by M. Livio, A. Koekemoer (Cambridge University Press, Cambridge, 2007). arXiv:0707.4492
M.A. McLaughlin, M. Kramer, A.G. Lyne, D.R. Lorimer, I.H. Stairs, A. Possenti, R.N. Manchester, F. Camilo, I.H. Stairs, M. Kramer, M. Burgay, N. D'Amico, P.C.C. Freire, B.C. Joshi, N.D.R. Bhat, Astrophys. J. **613**, 57 (2004a)
M.A. McLaughlin, A.G. Lyne, D.R. Lorimer, A. Possenti, R.N. Manchester, P.C.C. Freire, B.C. Joshi, M. Burgay, F. Camilo, N. D'Amico, Astrophys. J. **616**, 131 (2004b)
M. Mendez, M. van der Klis, Astrophys. J. **517**, L51 (1999)
A. Merloni, M. Vietri, L. Stella, D. Bini, Mon. Not. R. Astron. Soc. **304**, 155 (1999)
J.M. Miller, Astron. Nachr. **327**, 997 (2006)
M.C. Miller, F.K. Lamb, D. Psaltis, Astrophys. J. **508**, 791 (1998)
J.M. Miller et al., Astrophys. J. **563**, 928 (2001)
J.A. Milsom, R.E. Taam, Mon. Not. R. Astron. Soc. **286**, 358 (1997)
C.W. Misner, K.S. Thorne, J.A. Wheeler, *Gravitation* (Freeman & Co., San Francisco, 1973)
S.M. Morsink, L. Stella, Astrophys. J. **513**, 827 (1999)
M.A. Nowak, R.V. Wagoner, Astrophys. J. **378**, 656 (1991)
M.A. Nowak, R.V. Wagoner, Astrophys. J. **393**, 697 (1992)
M.A. Nowak, R.V. Wagoner, M.C. Begelman, D.E. Lehr, Astrophys. J. **477**, L91 (1997)
R.F. O'Connell, Class. Quantum Gravity **22**, 3815 (2005)
R.F. O'Connell, in *Atom Optics and Space Physics*, ed. by E. Arimondo, W. Ertmer, W. Schleich. Proc. of CLXVIII Course of the International School of Physics "Enrico Fermi", Varenna, Italy (2008). arXiv:0804.3806
J.A. Orosz, E. Kuulkers, Mon. Not. R. Astron. Soc. **305**, 132 (1999)
V. Osherovich, L. Titarchuk, Astrophys. J. **518**, L95 (1999)
H. Pfister, Gen. Relativ. Gravit. **39**, 1735 (2007)
D. Psaltis, Adv. Space Res. **28**, 481 (2001)
D. Psaltis, C. Norman, astro-ph/0001391 (2000)
D. Psaltis, T. Belloni, M. van der Klis, Astrophys. J. **520**, 262 (1999a)
D. Psaltis et al., Astrophys. J. **520**, 763 (1999b)
M.J. Rees, Ann. Rev. Astron. Astrophys. **22**, 471 (1984)
R.A. Remillard, M.P. Muno, J.E. McClintock, J.A. Orosz, Astrophys. J. **580**, 1030 (2000)
C.S. Reynolds, L.W. Brenneman, D. Garofalo, Astrophys. Space Sci. **300**, 7 (2005)
W. Rindler, Phys. Lett. A **187**, 236 (1994)
R. Ruffini, C. Sigismondi, *Nonlinear Gravitodynamics: The Lense-Thirring Effect* (World Scientific, Singapore, 2002)
I.H. Stairs, S.E. Thorsett, Z. Arzourmanian, Phys. Red. Let. **93**, 141101 (2004)
L. Stella, in *X-ray Astronomy 2000*, ed. by R. Giacconi, S. Serio, L. Stella. ASP Conf. Proc., vol. 234 (Astron. Soc. Pac., San Francisco, 2001), p. 213
L. Stella, M. Vietri, Astrophys. J. **492**, L59 (1998a)
L. Stella, M. Vietri, Nucl. Phys. B (Proc. Suppl.) **69**, 135 (1998b)
L. Stella, M. Vietri, Phys. Red. Let. **82**, 17 (1999)
L. Stella, M. Vietri, S.M. Morsink, Astrophys. J. **524**, L63 (1999)
N. Stergioulas, J.L. Friedman, Astrophys. J. **444**, 306 (1995)
T.E. Strohmayer, Adv. Space Res. **28**, 511 (2001)

T.E. Strohmayer et al., Astrophys. J. **469**, L9 (1996)
Y. Tanaka et al., Nature **375**, 659 (1995)
T. Titarchuk, P. Laurent, N. Shaposhnikov, Astrophys. J. **700**, 1831 (2009)
M. van der Klis, in *X-ray Binaries*, ed. by W.H.G. Lewin, J. van Paradijs, E.P.J. van den Heuvel (Cambridge University Press, Cambridge, 1995), p. 252
M. van der Klis, Ann. Rev. Astron. Astrophys. **38**, 717 (2000)
M. van der Klis, in *Compact Stellar X-ray Sources*, ed. by W.H.G. Lewin, M. van der Klis. Cambridge Ap. Ser., vol. 39 (Cambridge Univ. Press, Cambridge, 2006), p. 39
C. Venter, O.C. de Jager, Astrophys. J. **680**, L125 (2008)
J.P.W. Verbiest et al., Astrophys. J. **679**, 675 (2008)
M. Vietri, L. Stella, Astrophys. J. **503**, 350 (1999)
C. Will, in *Living Reviews in Relativity*, vol. 9 (Max Planck Society, Munich, 2006), p. 3

Space Sci Rev (2009) 148: 123–147
DOI 10.1007/s11214-009-9613-7

# New Frontiers at the Interface of General Relativity and Quantum Optics

**C. Feiler · M. Buser · E. Kajari · W.P. Schleich · E.M. Rasel · R.F. O'Connell**

Received: 27 August 2009 / Accepted: 19 November 2009 / Published online: 6 January 2010
© Springer Science+Business Media B.V. 2009

**Abstract** In the present paper we follow three major themes: (i) concepts of rotation in general relativity, (ii) effects induced by these generalized rotations, and (iii) their measurement using interferometry. Our journey takes us from the Foucault pendulum via the Sagnac interferometer to manifestations of gravito-magnetism in double binary pulsars and in Gödel's Universe. Throughout our article we emphasize the emerging role of matter wave interferometry based on cold atoms or Bose–Einstein condensates leading to superior inertial sensors. In particular, we advertise recent activities directed towards the operation of a coherent matter wave interferometer in an extended free fall.

**Keywords** General relativity · Sagnac interferometer · Atom interferometer · Sagnac time delay · Gödel Universe · Gravito-magnetism

## 1 Introduction

When Faust is about to sign his pact with the devil, he is asked to use his own blood as ink, see Goethe (1808). The devil explains:

*Blut ist ein ganz besondrer Saft.*

*Blood is a quite peculiar juice.*

In the same vein time has always played and still continues to play a rather peculiar role in physics. The concept of an absolute time prevalent in the mechanics of Newton ultimately

C. Feiler (✉) · M. Buser · E. Kajari · W.P. Schleich
Institut für Quantenphysik, Universität Ulm, Albert-Einstein-Allee 11, 89069 Ulm, Germany
e-mail: cornelia.feiler@uni-ulm.de

E.M. Rasel
Institut für Quantenoptik, Leibniz Universität Hannover, Welfengarten 1, 30167 Hannover, Germany

R.F. O'Connell
Department of Physics and Astronomy, Louisiana State University, Baton Rouge, LA 70803-4001, USA

had to give way to the distinction between coordinate and proper time—a trademark of special as well as general relativity. The fact that time depends on the observer had already been anticipated by William Shakespeare who writes in Act 3 of *As you like it*, Bartlett and Kaplan (2002):

*Time travels in diverse paces with diverse persons.*
*I'll tell you who Time ambles withal, who time trots withal,*
*Who Time gallops withal, and who he stands still withal.*

Nowhere clearer do we see the difference between the past, the present and the future than in the light cone description of special relativity. Nevertheless, the problem of the arrow of time has not found a convincing solution yet. As Edward FitzGerald points out in his poem *The Rubiáyát of Omar Khayyám*:

*The Moving Finger writes; and, having writ,*
*Moves on: Nor all your Piety nor Wit*
*Shall lure it back to cancel half a Line*
*Nor all your Tears wash out a Word of it.*

On first sight one might think that general relativity supports this notion of an arrow of time and that it rules out the possibility of time travel; however, in 1949 Kurt Gödel showed that using an exact solution of Einstein's field equations "[...] it is theoretically possible [...] to travel into the past". This unusual feature is closely linked to the concept of rotation in general relativity. In the present paper we address various such concepts starting from the rotation of a coordinate system. We then turn to gravitational fields induced by a rotating mass distribution; finally we address the Gödel solution where the whole universe rotates. We also analyze methods of measuring the effects of the rotation using light and matter interferometers.

Our paper is organized as follows: In Sect. 2 we highlight mechanical and optical rotation sensors by focusing on the Foucault pendulum and the Sagnac interferometer. We then briefly discuss the state of the art of atom interferometers with Bose–Einstein condensates (BEC) and advertise a novel experimental apparatus designed by the QUANTUS Group for creating a BEC in free fall. In this way the concept of a freely falling elevator which was so important in the development of general relativity has been promoted from a "Gedanken" experiment into a real experiment. We dedicate Sect. 3 to a brief discussion of the effects of gravito-magnetism such as the Lense–Thirring effect starting from Einstein's field equations in the weak-field limit. Here, we only allude to the recent tests of the Lense–Thirring effect using Gravity Probe B and LAGEOS and LAGEOS II since this work is already covered by other articles in the book. Instead, we focus on the observation of gravito-magnetic effects in two-body systems such as double binaries. We then turn to the discussion of Gödel's Universe and illustrate by help of light cone diagrams how to perform time travel. Moreover, we visualize light propagation in this unusual metric. Needless to say we do not live in a Gödel Universe; nevertheless, it would be still interesting to create a laboratory system to study the propagation of light in this unusual and time travel allowing spacetime metric. Such an idea might become reality due to an analogy between Maxwell's equations in a dielectric medium and Maxwell's equations in a metric field. We conclude in Sect. 5 by summarizing our main results and give one more historical perspective on the central role of the weak equivalence principle.

Needless to say in a review such as the present one, i.e. space limitations make it impossible for us to present detailed derivations. As a consequence we can only but refer to the literature for more information.

On a lighter note it is amusing but not too surprising that in a paper dedicated to the discussion of rotation a great variety of spirals make their appearance in unexpected places. For example, atom interferometry can be described most conveniently with the help of the Feynman path integral. An essential ingredient of this formulation is the Cornu spiral. Moreover, in the Gravity Probe B experiment geared towards the measurement of the Lense–Thirring effect the loxodrome appears. Finally the null geodesics in Gödel's Universe are helices giving rise to unusual views of objects.

## 2 Rotation Sensors

In the present section, we provide a historic overview over rotation sensors. We start by discussing the Foucault pendulum, briefly mention the original Sagnac effect and close with a concise introduction into the modern field of atom interferometers and BECs.

### 2.1 Foucault Pendulum

In 1851 Jean Bernard Léon Foucault (1819–1868) demonstrated to a stunned public the rotation of the Earth using a pendulum hanging from the ceiling of the Pantheon, see Foucault (1851, 1878). However, his experiment in the Pantheon was only the repetition of one performed already on January 6, 1851 at 2 a.m. in the basement of his mother's house at the corner of rue de Vaugirard and rue d'Assas in Paris. Foucault noted the event in his diary, see Aczel (2003):

> *Two o'clock in the morning, the pendulum has moved in the direction of the diurnal motion of the heavenly sphere.*

Foucault had quite a remarkable career, despite the fact that he had no formal scientific training in natural sciences. He had studied medicine, but since he could not see blood turned to theoretical medicine. He made his living as an assistant to Alfred Donné who was a professor of medicine specializing in microscopy. One of the many problems they were facing was to illuminate plants underneath a microscope in order to take photographs. At that time, this task was a non-trivial endeavor, since photography had just been developed independently by Henry Fox Talbot in England and by Louis-Jacques-Mandé Daguerre in France. Foucault solved the illumination problem by developing a carbon-arc electric lamp which later on was even used in the Paris opera for a sun rise in Giacomo Meyerbeer's "Le Prophète".

Another problem of the Daguerre method of photography was the fact that it was impossible to take pictures of people due to exposure times up to half an hour. Together with Armand Hippolyte Louis Fizeau, Foucault found a method to reduce the exposure time to twenty seconds. In this way it becomes possible for the first time to take pictures of human beings.

Obviously the year 1851 was an extremely successful one for Foucault. Not only did he achieve instant fame with his pendulum experiment but he was also awarded a permanent research position which he held till his death. Indeed, he was appointed by the emperor Napoléon III as *Physicist Attached to the Imperial Observatory* in Paris. Moreover, in the same year Foucault also invented the mechanical gyroscope. Hence, he can be considered as an important contributor to the Gravity Probe B satellite launched in 2004 and discussed in Sect. 3.

## 2.2 Sagnac Interferometer

In 1913 Georges Sagnac performed an experiment based on a light interferometer to measure the effect of rotation on the arrival times of two counter-propagating light rays. A sketch of his apparatus is shown in Fig. 1. In this device, light emitted by a mercury-arc lamp (O) is split by a prism (J) into two counter-propagating light beams. After successive reflections at the mirrors (M), the beams are recombined by the same prism (J) such that interference fringes on the photographic plate (PP′) emerge.

Sagnac found that when he started to rotate the table with a constant rotation rate, these fringes moved. His discovery is a consequence of the fact that the light which propagates against the rotation returns to the starting point before the light that has to catch up with the rotation.

The fact stands out most clearly for the case of a circular light path of length $P \equiv 2\pi r$. In the absence of rotation the light with velocity $c$ takes the time

$$T = \frac{P}{c} = \frac{2\pi r}{c} \tag{1}$$

to complete the round. When the table rotates with the rate $\Omega$ the paths of the two light beams get extended or shortened by an amount $\delta P = \Omega T r = \Omega T \cdot cT/(2\pi)$ leading to the path length difference

$$2 \cdot \delta P = \frac{c}{\pi} \Omega T^2 \tag{2}$$

experienced by the two counter-propagating beams.

With the wave vector $k \equiv \omega / c$ we find from (2) for the phase shift $\Delta\varphi_{\mathrm{L}} \equiv k \cdot 2 \cdot \delta P$ the expression

$$\Delta\varphi_{\mathrm{L}} = \frac{1}{\pi} \omega \Omega T^2. \tag{3}$$

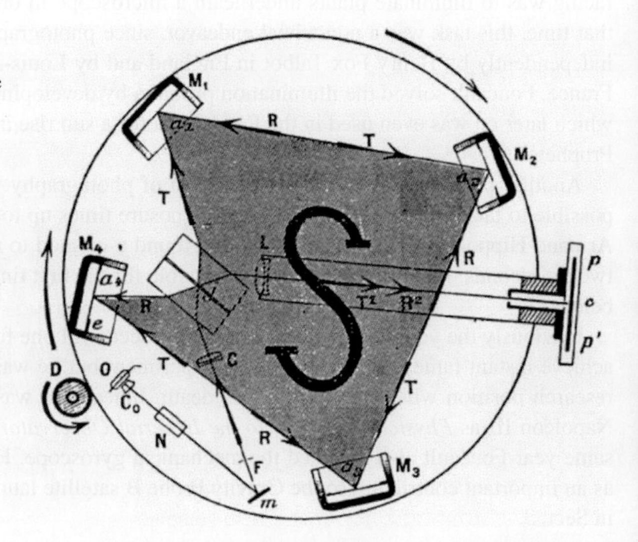


**Fig. 1** Sketch of the original light interferometer Sagnac presented in his second paper, see Sagnac (1913a, 1913b)

The frequency of light $\omega = kc \equiv c/\lambda\!\!\!^{-}$ and (1) allow us to cast $\Delta\varphi_{\mathrm{L}}$ in the more familiar form

$$\Delta\varphi_{\mathrm{L}} = 4\frac{\Omega S}{\lambda\!\!\!^{-} c}, \tag{4}$$

see e.g. Post (1967) and Chow et al. (1985). Here, $\lambda\!\!\!^{-} \equiv \lambda/(2\pi)$ denotes the reduced wave length and $S \equiv \pi r^2$ is the area of the circle enclosed by the two counter-propagating beams. It is interesting to note that although (4) was only derived for a circular path it holds true for a closed loop of arbitrary shape.

The expression (3) for the Sagnac phase shift of light shows that $\Delta\varphi_{\mathrm{L}}$ increases quadratically with the time $T$ the light spends in the interferometer. Two methods to increase $T$ offer themselves: (i) to increase the path length by circulating the light many times, and (ii) to use slow light, see Scully and Welch (2004). The first method is utilized in modern optical fiber gyroscopes, which are valuable tools to measure rotation rates to a high precision. Such gyroscopes are sometimes called *passive* Sagnac interferometers since the light source is located outside of the interferometer. The second approach is based on the fact that in appropriately prepared media the speed of light can be made arbitrarily small. Indeed, light can be (almost) stopped. This phenomenon which has been verified in many spectacular experiments can be used to build a gyroscope, see Zimmer and Fleischhauer (2004, 2006).

Another modern off-spring of Sagnac interferometry is the ring laser gyroscope. In such a device, the lasing medium providing the light source sits inside the interferometer. As a result the clockwise and counter-clockwise propagating waves in this so-called *active* Sagnac interferometer experience different resonance conditions, giving rise to a beat frequency

$$\Delta\omega \cong \omega\frac{\delta P}{P} = 4\frac{\Omega S}{\lambda\!\!\!^{-} P}.$$

Here we have made use of (1) and (2). For a detailed discussion of ring laser gyroscopes we refer to the review article by Stedman (1997).

In a seminal experiment, a huge ring laser gyroscope with an area of 16 $\mathrm{m}^2$ was used by Stedman and Schreiber et al. (2004), to measure variations in the location of the rotational pole of the Earth to an accuracy of a few centimeters. Moreover, there are strong indications that general relativistic effects connected to gravito-magnetism as discussed in the next section might be detected by these devices, see Scully (1979, 1981) and Schleich and Scully (1984). We are confident that future improvements of active as well as passive Sagnac interferometers will open new avenues for tests of general relativity.

### 2.3 Atom Interferometer

Another interesting way to detect rotation consists of replacing in an interferometer light waves by matter waves. The emerging field of matter wave optics is still a young, but rapidly progressing branch of quantum optics, see e.g. Baklanov et al. (1976), Bordé (1989), Berman (1997) and Cronin et al. (2009). It has recently led to sensational discoveries such as laser cooling and atom lasers which were awarded by several Nobel Prizes.

Atomic quantum sensors rely on the wave nature of the center-of-mass motion of an atom, which implies that a single atom moves at the same time along different trajectories, see Rasel et al. (1995). Interference of trajectories is the crucial ingredient of quantum mechanics summarized most beautifully in Young's double-slit experiment and expressed in Feynman's path integral.

During the last decades matter wave optics has turned into an important tool in the ultra-precise monitoring of accelerations and rotations, see Kasevich and Chu (1991), Riehle et al. (1991), Gustavson et al. (2000), Gauguet et al. (2006) and Le Gouët et al. (2008). The potential of atom interferometers and atom lasers can be compared with Superconducting Quantum Interference Device (SQUID) sensors, but without the need for cryogenic equipment. Atomic inertial quantum sensors function similarly to atomic clocks, which are today the most-accurate standards for time and frequency measurement. Like atomic clocks, which revolutionized frequency metrology, inertial and rotational sensors using atom interferometers display a high potential for replacing classic state-of-the-art sensors, see Pereira Dos Santos and Landragin (2007) and Lamporesi et al. (2008).

### 2.3.1 *Enhancement Factor*

The main advantage of using de Broglie rather than light waves in rotation sensors originates from the fact that their wavelength is much shorter, which in principle allows for a much larger Sagnac phase shift. This claim can easily be motivated by the following argument: When we define the mass $m$ of light by the familiar energy relation

$$mc^2 \equiv \hbar\omega = \hbar\frac{c}{\lambda},$$

we can cast the Sagnac phase shift for light given by (4) into the form

$$\Delta\varphi_{\mathrm{L}} = 4\frac{m}{\hbar}\Omega S.$$

Moreover, it can be shown, see e.g. Staudemann et al. (1980) and Rauch and Werner (2000), that in complete analogy the corresponding Sagnac phase shift $\Delta\varphi_{\mathrm{M}}$ for matter waves after one full circulation around the enclosed area $S$ reads

$$\Delta\varphi_{\mathrm{M}} = 4\frac{M}{\hbar}\Omega S,$$

with $M$ being the mass of the massive particle.

Thus, when we compare the phase shift obtained for a light ray, say a helium–neon laser with typical wavelength $\lambda = 632.8$ nm, to a matter wave of rubidium-87 atoms with mass $M = 86.91$ u, we obtain for their ratio

$$\frac{\Delta\varphi_{\mathrm{M}}}{\Delta\varphi_{\mathrm{L}}} = \frac{M}{m} = \frac{Mc^2}{\hbar\omega} \approx 4\times 10^{10}.$$

Here, we have assumed that the helium–neon laser and the rubidium atoms enclose the same area $S$. In this case, we gain a factor of $10^{10}$ in resolution taking advantage of matter instead of light waves. However, the problem with matter waves is that one has not yet a source of comparable large particle fluxes as in the case of the laser. Hence, the signal to noise ratio is much worse for matter waves than for light waves, see Scully and Dowling (1993).

### 2.3.2 *Cold Atoms and the Need for Microgravity*

The fact that the Sagnac phase shift increases quadratically with $T$ is not limited to light waves but also holds true for matter waves. Hence, atoms with an extremely small translational kinetic energy, that is cold atoms, allow us to achieve unusually long transit times $T$. Unfortunately, now we face a problem: Due to the non-vanishing mass atoms are attracted to the Earth. For this reason fountain experiments where the atoms are launched against the gravitational field of the Earth and the measurement is performed at the turning point of the motion are of interest. Another method to avoid the grip of gravity on the atoms consists of performing atom interferometer experiments in a microgravity environment as provided for example by a freely falling elevator or satellite, an air plane in parabolic flight or in the International Space Station.

An example for a space version of a quantum test of the equivalence principle is the matter wave explorer of gravity (MWXG) which will employ atomic Mach–Zehnder-type interferometers, see Ertmer et al. (2009). Here, atomic wave packets made out of cold atoms are coherently split, redirect and recombine to observe matter wave interferences, see Rasel et al. (1995). The splitting is achieved by the interaction between the atoms and pulsed light fields generated by two counter-propagating laser beams. An atom absorbs a photon from one of the laser beams and is stimulated by the other laser beam to re-emit another one. In this way twice the recoil of a photon is transferred coherently to the atomic wave (rather than atoms) to generate a new spatial mode of a matter wave, see Müller et al. (2009).

Due to their high symmetry Mach–Zehnder atom interferometers are perfect tools for tests of gravity. For example, the phase shift

$$\Delta\varphi = \mathbf{k}_{\mathrm{eff}} \cdot \mathbf{a}T^2 = \mathbf{k}_{\mathrm{eff}} \cdot \mathbf{a}\frac{L^2}{v^2} \tag{5}$$

due to an acceleration **a** is independent of the atomic mass of the particular species. Therefore, this formula implies the equality of the inertial and the gravitational mass.

Moreover, the phase shift $\Delta\varphi$ given by (5) depends only on the square of the drift time $T$ of the atom inside the interferometer and on the momentum difference $\hbar\mathbf{k}_{\mathrm{eff}}$ of the two interfering matter waves. The latter is determined by the photon recoil of the light beam interacting with the atom.

The simplicity of the expression (5) for the phase shift $\Delta\varphi$ suggests the use of atom interferometers as inertial references for measuring absolute accelerations. However, according to (5) the Mach–Zehnder interferometer senses accelerations in only one particular direction defined by the coherently scattered photons.

The sensitivity of the atomic accelerometer increases for a given length scale $L$ of the detection volume and effective photon recoil of the interferometer with the square of the atomic drift time $T$ or the inverse square of the atomic velocity. Thus, for an atomic accelerometer the sensitivity as well as the precision by using ultra-cold atoms in a space-bound experiment would increase by several orders ($10^3$ to $10^4$) of magnitude allowing longer measurement times and providing a more stable environment, see Zoest et al. (2008).

### *2.3.3 Present Activities and Goals*

The recent improvement of sensor performance is mainly due to the rapid progress in the development of lasers and in the manipulation of atoms. World wide groups try to develop transportable variants of matter wave gyroscopes, gravimeters and gradiometers and reach for stationary large scale devices and experiments in space, see Nyman et al. (2006), Lamporesi et al. (2008) and De Angelis et al. (2009). Activities aim for both, increasing the sensitivity and, at the same time, improving the sensor's transportability, as these devices should be used in navigation, in Earth observation, metrology and fundamental physics, see Le Gouët et al. (2008). First prototypes of compact one- and three-axes sensors were tested with the gyroscopes and improved variants are under development. They should provide an enhanced short-term sensitivity aiming for a reduction of the integration time by several orders of magnitudes, using for example the so-called butterfly configuration. This interferometer was demonstrated recently in a proof-of-principle experiment. Moreover, novel stationary devices operated in large towers are under development in order to investigate their ultimate sensitivity. Another goal is to transfer the knowledge acquired during experiments in extended free fall in drop towers, parabolic flights or in space to improve current ground based precision.

## 2.4 Bose–Einstein Condensates

As mentioned in the preceding subsection the sensitivity of an atom interferometer is determined by the transit time of the atom. Quantum gases at ultralow temperatures permit to extend this time since the atomic cloud expands rather slowly. In this way many novel devices and effects, such as atom interferometers and atomic clocks with unprecedented accuracy, phase transitions or atom lasers have been achieved. In this section we present a historical perspective on the phenomenon of Bose–Einstein condensation and conclude by briefly summarizing recent activities on BEC in microgravity.

### *2.4.1 A Brief History*

The unusual state of matter in which all atoms are in the ground state of their center-of-mass motion originates from a paper by Satyendra Nath Bose, in which he derived Planck's radiation law by taking advantage of the statistical properties of light, see Bose (1924). He sent the paper to Einstein and asked for his opinion. Einstein was quite impressed by the work, translated the English manuscript into German and submitted the paper to *Zeitschrift für Physik* where all articles at that time appeared in German. Einstein also generalized Bose's work in a subsequent paper entitled "Quantentheorie des einatomigen idealen Gases" (Quantum theory of the monoatomic ideal gas), Einstein (1924). At the end of it Einstein concludes:

> *Die Klammer drückt den Quanteneinfluß auf das Maxwellsche Verteilungsgesetz aus. Man sieht, daß die langsamen Moleküle gegenüber den raschen häufiger sind, als es gemäß Maxwells Gesetz der Fall wäre.*
>
> *The bracket expresses the influence of the quanta on Maxwell's distribution law. One sees that the slower molecules occur more frequently, as compared to the fast ones, than they would by virtue of Maxwell's law.*

In a second paper also published in the Proceedings of the Prussian Academy of Science, Einstein (1925) talks about the possibility of condensation of atoms. Here most of the atoms are in a ground state of a box, very much like the condensation of steam.

> *Ich behaupte, daß in diesem Falle eine mit der Gesamtdichte stets wachsende Zahl von Molekülen in den 1. Quantenzustand (Zustand ohne kinetische Energie) übergeht, während die übrigen Moleküle sich gemäß dem Parameterwert $\lambda = 1$ verteilen. Die Behauptung geht also dahin, daß etwas Ähnliches eintritt wie beim isothermen Komprimieren eines Dampfes über das Sättigungsvolumen. Es tritt eine Scheidung ein; ein Teil »kondensiert«, der Rest bleibt ein »gesättigtes ideales Gas«.*
>
> *I claim that in this case a number of molecules which always grows with the total density makes a transition to the 1. quantum state (state without kinetic energy), whereas the remaining molecules distribute themselves according to the parameter*

*value $\lambda = 1$. The claim thus asserts that something similar happens as when isothermally compressing a vapor beyond the volume of saturation. A separation occurs; a part condenses", the rest remains a "saturated ideal gas".*

We note that already in 1911 Ladislas Natanson derived in his paper entitled "On the Statistical Theory of Radiation" the Planck radiation formula using the concept of particle indistinguishability, Natanson (1911). According to Spalek (2006) Natanson visited Einstein in Berlin in 1914. However, it is not clear if on this occasion he mentioned his own work to him, since Natanson was a shy person. He became an eminent intellectual after the rebirth of Poland in 1918 and even *Rector Magnificus* of the Jagiellonian University in Cracow. It is amusing that he also received highest awards in literature for his essays on Shakespeare and the ancient Greek dramas.

### *2.4.2 BEC and Microgravity*

It is interesting to speculate if quantum degenerate gases will be of advantage in metrological applications. The ground state of a harmonic trap is a minimum uncertainty state determined by the Heisenberg uncertainty relation between position and momentum. Lowering the trap will adiabatically release the atoms forming an ideal wave packet. The perfect control of the external degrees of freedom is mandatory for minimizing systematic errors such as the Coriolis force on the atoms or wave front curvatures of the light fields acting as beamsplitters for the matter waves. The control of these errors becomes even more critical when we increase the sensitivity of an atom interferometer by extending the transit time. The gain in sensitivity and in accuracy is compensating for the loss in atom number, which only enters as a square root into the sensitivity at the shot noise limit.

In this regard microgravity can be beneficial in several respects. Indeed, it allows us to study quantum gases at temperatures below pico-Kelvins, to obtain de Broglie wavelength of macroscopic dimensions, and to achieve unperturbed evolutions of these distinguished quantum objects for an unusually long duration. Hence, microgravity sets the stage for the physics of ultra-dilute gases and giant matter waves and the control of these macroscopic quantum objects and mixtures in an environment unbiased by gravity. In particular, microgravity is of high relevance for matter waves as it permits the extension the unperturbed free fall of these test particles in a low-noise environment. This is a prerequisite for fundamental tests in the quantum domain such as the equivalence principle or the realization of ideal reference systems.

The QUANTUS team, formed by a consortium of the Leibniz University of Hanover, the Universities of Hamburg, Berlin and Ulm together with ZARM as well as the Max-Planck Institute for Quantum Optics and ENS, realized a compact facility to study a Rubidium BEC in the extended free fall at the drop tower in Bremen and during parabolic flights. These environments will permit the QUANTUS team to investigate the generation of a BEC and its release from the trap in microgravity, as well as a detailed experimental analysis of decoherence. The remote controlled and miniaturized BEC trap is under test in the drop tower since November 2007.

## 3 Gravito-magnetism

After this overview of seminal experiments dealing with rotation and modern issues related to matter waves, we now turn to the discussion of experimental tests of gravito-magnetism. For this purpose, we first address the notion of rotation in general relativity and motivate

**Table 1** Basic equations of tensor calculus and general relativity including the linearized field equations. The gravitational constant $G$ and the speed of light appear in the definition of the coupling constant $\kappa = (8\pi G)/c^4$. Moreover, $\Lambda$ and $T^{\mu\nu}$ denote the cosmological constant and the contravariant components of the stress-energy tensor, respectively

| | |
|---|---|
| Minkowski metric | $(\eta_{\mu\nu}) \equiv \mathrm{diag}(1, -1, -1, -1)$ |
| Christoffel symbols | $\Gamma^{\mu}{}_{\alpha\beta} \equiv \frac{1}{2} g^{\mu\nu}(g_{\nu\alpha,\beta} + g_{\nu\beta,\alpha} - g_{\alpha\beta,\nu})$ |
| Curvature tensor | $R^{\mu}{}_{\alpha\beta\gamma} \equiv \Gamma^{\mu}{}_{\alpha\gamma,\beta} - \Gamma^{\mu}{}_{\alpha\beta,\gamma} + \Gamma^{\mu}{}_{\rho\beta}\Gamma^{\rho}{}_{\alpha\gamma} - \Gamma^{\mu}{}_{\rho\gamma}\Gamma^{\rho}{}_{\alpha\beta}$ |
| Ricci tensor and scalar curvature | $R_{\alpha\beta} \equiv R^{\mu}{}_{\alpha\mu\beta}$ and $R \equiv R^{\mu}{}_{\mu}$ |
| Einstein's field equations | $R_{\mu\nu} - \frac{1}{2} g_{\mu\nu} R = \kappa T_{\mu\nu} + \Lambda g_{\mu\nu}$ |
| Weak-field approximation | $g_{\mu\nu} = \eta_{\mu\nu} + h_{\mu\nu}$ with $\lvert h_{\mu\nu}\rvert \ll 1$ |
| Hilbert gauge | $\phi^{\mu\nu}{}_{,\nu} = 0$ with $\phi_{\mu\nu} \equiv h_{\mu\nu} - \frac{1}{2}\eta_{\mu\nu} h^{\alpha}{}_{\alpha}$ |
| Linearized field equations | $\Box\phi_{\mu\nu} = -2\kappa T_{\mu\nu}$ |

the Lense–Thirring metric. We then briefly summarize two recent experiments on gravito-magnetism: (i) the Gravity Probe B experiment (GP-B) and (ii) the laser ranging of the satellites LAGEOS and LAGEOS II. We conclude with an introduction to two-body effects in general relativity and their recent experimental verification by astronomical observations of the double binary pulsar PSR J0737-3039 A/B.

There exist many introductions into the physics of gravito-magnetism. Here we only refer to the text book by Ciufolini and Wheeler (1995) and the articles by Lämmerzahl and Neugebauer (2001) and Mashhoon et al. (2001). The fact that all gravito-magnetic effects give rise to non-geodesic motion and the corresponding relationship between spin-dependent coordinate transformations and spin supplementary conditions is discussed in Barker and O'Connell (1974). Up-to date reviews appear in Ciufolini (2007) and O'Connell (2009a, 2009b).

In order to clarify the sign conventions used throughout this article, we summarize several fundamental equations of tensor calculus and general relativity in Table 1.

## 3.1 The Relativistic Notion of Rotation

In general relativity spacetime is no longer presumed to be flat. Instead, the metric is the basic quantity which incorporates gravity and provides the link between theoretical predictions and experimental results. In the description of experiments in general relativity, one distinguishes between local effects which are determined by the curvature of spacetime in the local neighborhood of an observer, and global effects which involve the coordinate dependence of the metric in an extended region of spacetime. As a result two concepts of rotation exist in general relativity. They are based on the *inertial compass* and the *stellar compass*, see Weyl (1924).

Let us first elucidate the term *inertial compass*. For this purpose we consider a freely falling observer who travels along a timelike geodesic in curved spacetime. He carries with him three spatial coordinate axes (spatial tetrad vectors) which allow him to identify certain directions in space. Moreover, the observer is endowed with three gyroscopes, which are aligned mutually orthogonal with respect to each other and which define the *inertial compass*. Only when the spatial coordinate axes of the observer do not rotate relative to the three gyroscopes, the observer is said to be freely falling and non-rotating along his world line. In this case, his local reference frame corresponds to a local inertial frame. However, when his spatial coordinate axes rotate relative to the gyroscopes, the observer experiences Coriolis-

and centrifugal like forces in his local neighborhood: The observer rotates relative to the inertial compass. Hence, the inertial compass provides us with an absolute characterization of the state of motion of the observer's spatial coordinate axes based on a *local* criterion along his trajectory.

In contrast to this local method to quantify rotation by means of gyroscopes, one could likewise use celestial light sources such as catalog stars as spatial reference frame, see e.g. Soffel (1989). In this way, one takes advantage of the stellar compass, which is defined by the directions of the incident light rays of the "fixed stars" along the world line of the observer. This method represents an alternative way to quantify the rotation of the observer's spatial coordinate axes. Obviously, the tangent vectors of the incident light rays crucially depend on the global aspects of the metric and thus, the stellar compass represents a *non-local* criterion of rotation in general relativity.

The stellar and inertial compass represent two different ways to determine the rotation of an observer. They give different results for stationary but non-static spacetimes when the "fixed stars" and the observer follow the integral curves of the Killing vector field, see Straumann (2004).

### 3.2 Lense–Thirring and Geodetic Precession

In this subsection, we briefly sketch the so-called Lense–Thirring metric as well as the resulting Lense–Thirring and geodetic precession, see Lense and Thirring (1918) and Thirring (1918) and e.g. Schleich and Scully (1984) and Straumann (2004).

This metric is the starting point for the verification of gravito-magnetism, a notion which most prominently reflects the striking analogy between the linearized field equations of gravitation and electromagnetism, see Table 1. A direct consequence of this analogy is the so-called Lense–Thirring metric whose line element reads in the small-velocity limit

$$\begin{aligned} \mathrm{d}s^2 = &\left(1-\frac{2Gm_2}{c^2 r}\right)c^2\,\mathrm{d}t^2-\left(1+\frac{2Gm_2}{c^2 r}\right)\left(\mathrm{d}x^2+\mathrm{d}y^2+\mathrm{d}z^2\right)\\ &-\frac{4GS_z^{(2)}}{c^3}\frac{y}{r^3}c\,\mathrm{d}t\,\mathrm{d}x+\frac{4GS_z^{(2)}}{c^3}\frac{x}{r^3}c\,\mathrm{d}t\,\mathrm{d}y. \end{aligned} \tag{6}$$

It follows from the special stress-energy tensor of incoherent dust $T^{\mu\nu}=\rho u^\mu u^\nu$ for a spherically symmetric and homogeneous mass distribution $\rho$ rotating around the $z$-axis with a constant rotation rate. The angular momentum and the mass of the rotating body are denoted by $\mathbf{S}^{(2)}=(0,0,S_z^{(2)})$ and $m_2$, respectively.

The Lense–Thirring precession of a drag-free gyroscope ($a^\mu=0$) around a massive, spinning body originates from the non-diagonal elements of the Lense–Thirring metric. In this spirit, the Lense–Thirring precession is the general relativistic analogy of the electrodynamic precession of a magnetic needle in an external magnetic field.

On its path $\mathbf{r}(t)$ the angular momentum of the gyroscope $\mathbf{S}^{(1)}$ experiences a precession according to

$$\frac{\mathrm{d}\mathbf{S}^{(1)}}{\mathrm{d}t}=\boldsymbol{\Omega}^{(1)}(t)\times\mathbf{S}^{(1)}(t),$$

where the instantaneous precession rate $\boldsymbol{\Omega}^{(1)}(t)$ contains beside the much smaller Lense–Thirring precession $\boldsymbol{\Omega}_{LT}$ the geodetic precession $\boldsymbol{\Omega}_{GP}$. We note that the latter is also present for a non-rotating body. For the case of the spherical symmetric, massive body the total precession rate reads

$$\boldsymbol{\Omega}^{(1)}(t) = \boldsymbol{\Omega}_{GP}(t) + \boldsymbol{\Omega}_{LT}(t) = \frac{3}{2}\frac{Gm_2}{c^2}\frac{\mathbf{r}(t)\times\mathbf{v}(t)}{r^3(t)} + \frac{G}{c^2}\left(\frac{3\left(\mathbf{S}^{(2)}\cdot\mathbf{r}(t)\right)\mathbf{r}(t) - r^2(t)\mathbf{S}^{(2)}}{r^5(t)}\right).$$

Here, the vectors $\mathbf{r}$ and $\mathbf{v}$ denote position and velocity of the gyroscope with respect to the center of the massive body.

## 3.3 Experiments on Gravito-magnetism

We now turn briefly to the two most prominent measurements of the Lense–Thirring effect in one-body systems: Gravity Probe B and the LAGEOS satellite experiment, after which we discuss the generalization to two-body systems, especially the double binary system.

### *3.3.1 Gravity Probe B*

In order to observe the Lense–Thirring precession, NASA has launched in April 2004 the Gravity Probe B satellite (GP-B), see e.g. Keiser (2009). This satellite travels on a polar orbit around the Earth and carries four mechanical gyroscopes and one telescope. The telescope, which points at the guide star IM Pegasi (HR8703), is used to measure the relativistic precession of the gyroscopes with respect to the “fixed stars”. Based on the data, collected between August 2004 and September 2005, the predicted geodetic effect was verified to better than 1%. Unfortunately, due to patch effects on the balls of the gyroscopes the much smaller Lense–Thirring precession has not been confirmed to the expected accuracy so far.

Fortunately, the additional precession of the gyroscopes caused by the patch effects and described by a loxodromic curve can be substracted from the data. As a result a confirmation of the Lense–Thirring effect to 15% is achieved. However, for a more detailed discussion of all systematic effects we refer to the literature.

According to the Merriam–Websters Collegiate Dictionary (1999) the word *loxodrome* originates from the Greek words *Loxos* meaning *oblique*, and *Dromos* meaning *course*. It dates back to ca 1795 and is also referred to as the rhumb line whose definition is:

> *a line on the surface of the earth that follows a single compass bearing and makes equal oblique angles with all meridians.*

When projected onto a plane the loxodrome, that is, the rhumb line becomes a spiral. It is interesting to note that another spiral, namely the Cornu spiral, plays a decisive role in the foundations of quantum mechanics. Indeed, the convergence of the Feyman path integral rests on the Cornu spiral. Hence, as it stands right now, the measurement of the Lense–Thirring effect due to Gravity Probe B again relies on a spiral.

### *3.3.2 LAGEOS Satellites*

Another experiment testing the Lense–Thirring effect is due to Ciufolini and Pavlis (2004), Ciufolini (2007). In contrast to the GP-B mission, here the nodal precession of the two satellites LAGEOS and LAGEOS2 due to the spinning Earth is measured. Each satellite consists of retro-reflectors mounted to a heavy sphere of aluminum-covered brass. The reflectors allow the determination of the Earth-satellite distance with a precision of a few millimeters using laser ranging. This experiment has confirmed the Lense–Thirring effect to an accuracy of approximately 10%. We note that another satellite, called LARES, is currently under construction and will be probably launched in 2010. LARES is designed to increase the accuracy to approximately 1%.

### 3.4 Observation of Gravito-magnetic Effects in a Two-Body System

So far, we only considered gravito-magnetic effects on test-masses. The purpose of the present subsection is to briefly outline the necessary generalizations encountered for two-body systems.

In the previous subsections, we saw that a single rotating body gives rise to a metric, the so-called Lense–Thirring metric, which incorporates the effect of the rotation of a massive body. For a much smaller mass orbiting around this body we have the conventional one-body system. More generally, in the two-body case, we define the mass and rotation (spin) of the larger (smaller) body by $m_2$ and $\mathbf{S}^{(2)}$ ($m_1$ and $\mathbf{S}^{(1)}$), see Fig. 2.

Turning to the situation where $m_1$ and $m_2$ are comparable, we know that even in the absence of rotation the problem becomes more difficult so that it was non-trivial to calculate the periastron precession, see Robertson (1938) and Einstein et al. (1938). However, in the weak-field limit, even with inclusion of spin, it turns out that an exact analysis is feasible; see Barker and O'Connell (1975b, 1977). In particular, a Hamiltonian $H$ was obtained which is completely analogous to the corresponding problem in quantum electrodynamics (QED). Thus we find it convenient to refer to use the term "spin" in the generic sense of meaning "rotational angular momentum" in the case of a macroscopic body (and "internal spin" in the case of an elementary particle, which is not of interest in the present context). In fact, $H$ contains spin–orbit and spin–spin coupling terms which compare term by term with the corresponding results for, say, the QED interaction of an $e^-$ and $\mu^-$ if we let

$$e^2 \longrightarrow (\ldots)Gm_1m_2,$$

where $(\ldots)$ refers to small (but important) coefficients, see O'Connell (1974, 2005). In other words, the spin–orbit coupling in the weak-field limit of the two-body system in general relativity (GR) corresponds to the fine structure in QED, whereas the spin–spin coupling in GR corresponds to the hyperfine structure in QED.

The advantage of starting with a Hamiltonian is that it provides analogous paths to the calculation of spin–orbit and spin–spin contributions to both spin precession and orbital precession. With regard to the latter, we find after averaging over a period of the orbit defined by the relative coordinate $\mathbf{r}(t) = \mathbf{r}_2(t) - \mathbf{r}_1(t)$, that the Kepler ellipse precesses as a whole with angular velocity $\boldsymbol{\Omega}^*$ such that

$$\frac{\mathrm{d}\mathbf{L}}{\mathrm{d}t} = \boldsymbol{\Omega}^* \times \mathbf{L}$$

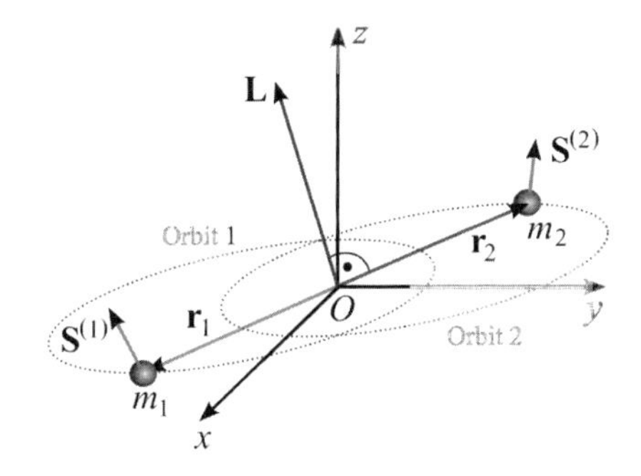


**Fig. 2** Sketch of a two-body system consisting of two comparable masses $m_1$ and $m_2$ with center of mass $O$. The orbital angular momentum $\mathbf{L}$ is perpendicular to the orbits 1 and 2, which are defined by the position vectors $\mathbf{r}_1(t)$ and $\mathbf{r}_2(t)$ of the two-body system. The spin of each mass is denoted by $\mathbf{S}^{(1)}$ and $\mathbf{S}^{(2)}$, respectively

and

$$\frac{\mathrm{d}\mathbf{A}}{\mathrm{d}t} = \boldsymbol{\Omega}^* \times \mathbf{A},$$

where **L** denotes the orbital angular momentum, and **A** is the Runge–Lenz vector pointing towards the periastron of the Kepler ellipse. In particular, we note that

$$\frac{\mathrm{d}}{\mathrm{d}t}(\mathbf{A} \cdot \mathbf{L}) = 0,$$

as it should, and that $\boldsymbol{\Omega}^*$ has contributions from a variety of sources viz.

$$\boldsymbol{\Omega}^* = \boldsymbol{\Omega}^{*(E)} + \boldsymbol{\Omega}^{*(1)} + \boldsymbol{\Omega}^{*(2)} + \boldsymbol{\Omega}^{*(1,2)},$$

where $\boldsymbol{\Omega}^{*(E)}$ is the Einstein two-body (no spin) contribution, and $\boldsymbol{\Omega}^{*(1)}$, $\boldsymbol{\Omega}^{*(2)}$ and $\boldsymbol{\Omega}^{*(1,2)}$ refer to the spin 1, spin 2 and spin–spin contribution whose explicit form can be found in Barker and O'Connell (1975b) and O'Connell (2009a). There are also contributions from the quadrupole moments of both bodies, which are important in some cases such as the GP-B experiment, see O'Connell (1969), Barker and O'Connell (1975a, 1975b), but we will not consider these here. With regard to spin precession of the individual bodies, we find that

$$\frac{\mathrm{d}\mathbf{S}^{(1)}}{\mathrm{d}t} = \boldsymbol{\Omega}^{(1)} \times \mathbf{S}^{(1)}$$

and

$$\frac{\mathrm{d}\mathbf{S}^{(2)}}{\mathrm{d}t} = \boldsymbol{\Omega}^{(2)} \times \mathbf{S}^{(2)},$$

where

$$\boldsymbol{\Omega}^{(1)} = \boldsymbol{\Omega}^{(1)}_{\mathrm{SO}} + \boldsymbol{\Omega}^{(1)}_{\mathrm{SS}}$$

and similarly for $1 \to 2$. Here, $\boldsymbol{\Omega}^{(1)}_{\mathrm{SO}}$ and $\boldsymbol{\Omega}^{(1)}_{\mathrm{SS}}$ are the two-body generalizations of the one-body de Sitter geodetic precession and the Lense–Thirring precession, respectively.

It should also be noted that

$$\frac{\mathrm{d}}{\mathrm{d}t}\left(\mathbf{L} + \mathbf{S}^{(1)} + \mathbf{S}^{(2)}\right) \equiv \frac{\mathrm{d}\mathbf{J}}{\mathrm{d}t} = 0.$$

In other words, the total angular momentum **J** is conserved.

In summary, we are dealing with three main precession rates $\boldsymbol{\Omega}^*$, $\boldsymbol{\Omega}^{(1)}$ and $\boldsymbol{\Omega}^{(2)}$, all of which are derived from three main terms in $H$. Moreover, following Barker and O'Connell (1975b), it is useful to write

$$\boldsymbol{\Omega}^* = \frac{\mathrm{d}\Omega}{\mathrm{d}t}\mathbf{n}_0 + \frac{\mathrm{d}\omega}{\mathrm{d}t}\mathbf{n} + \frac{\mathrm{d}i}{\mathrm{d}t}\frac{\mathbf{n}_0 \times \mathbf{n}}{|\mathbf{n}_0 \times \mathbf{n}|},$$

where $\Omega$, $\omega$ and $i$ denote the longitude of the ascending node, the argument of the periastron and the inclination of the orbit, respectively, in the reference system of the plane of the sky (the tangent plane to the celestial sphere at the center of mass of the binary system), see Fig. 3. In addition, $\mathbf{n}_0$ is a unit vector normal to the plane of the sky directed from the center of mass of the binary system towards the Earth in $z$-direction. The angle between $\mathbf{n}_0$ and $\mathbf{n} = \mathbf{L}/|\mathbf{L}|$ is the inclination $i$. In the absence of spin, only the periastron precession is present.

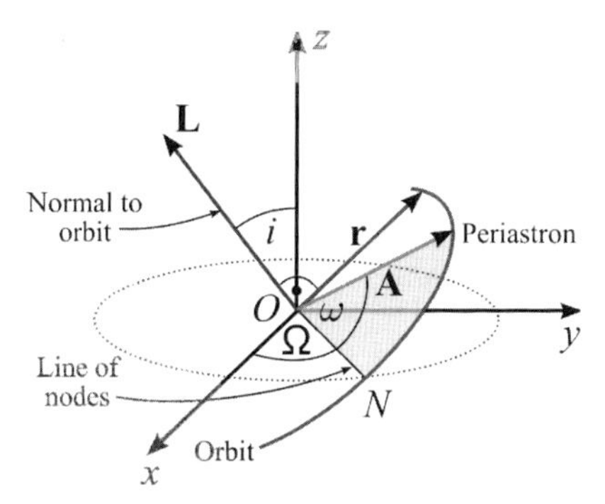


**Fig. 3** Illustration of orbital coordinates with the center of mass of the two-body system denoted by $O$. The angles $\Omega$, $\omega$ and $i$ represent the longitude of the ascending node, the argument of the periastron and the inclination of the orbit in the reference system of the plane of the sky represented by the $x$-$y$-plane. The relative coordinate of the two-body system $\mathbf{r}(t) = \mathbf{r}_2 - \mathbf{r}_1$ defines the orbit which lies in the plane orthogonal to the orbital angular momentum $\mathbf{L}$. Moreover, the node $N$ is the intersection point of the orbit with the plane of the sky, whereas $\mathbf{A}$ stands for the Runge–Lenz vector which points toward the periastron

We turn next to a discussion of specific attempts to measure or observe gravito-magnetic effects. In this context, exact expressions for the various precession angles have been given in the literature, see Barker and O'Connell (1975b) and O'Connell (2009a). As it turns out, the GP-B experiment measures $\boldsymbol{\Omega}^{(1)}$, whereas current astronomical observations attempt to measure both, $\boldsymbol{\Omega}^{(1)}$ and $\boldsymbol{\Omega}^{(2)}$. By contrast, the Ciufolini–Pavlis experiment and the JPL lunar lasing experiment measure $\boldsymbol{\Omega}^{*}$. Moreover, as emphasized by O'Connell (2009a), for a particular system, the magnitude of the periastron and spin–orbit precession depends significantly on the product of the averaged orbital angular velocity

$$\bar{\omega} = \frac{2\pi}{T} = \sqrt{\frac{GM}{a^3}}$$

and the gravitational coupling constant

$$\alpha_g = \frac{GM}{c^2 a},$$

where $T$ is the period, $a$ is the semi-major axis of the Kepler ellipse and $M = m_1 + m_2$. For illustration we provide some calculated values of $\bar{\omega}$ and $\alpha_g$ in Table 2.

In particular, we note that for the double binary pulsar the product $\alpha_g\bar{\omega}$ has the largest value (5.6°/yr); this difference amounting to nearly 4 orders of magnitude, is reflected in the significantly larger precession (both orbital and spin) obtained for the double binary pulsar, compared to Earth related systems. Thus, we are motivated to present more details on the latter system, particularly because the new two-body spin precession effect has already been measured; see Breton et al. (2008). The double binary system, see Burgay et al. (2003), Lyne et al. (2004) and Kramer et al. (2006), is the only system observed so far in which both components are neutron stars which are pulsing. Its distance from the Earth is about $2.2 \times 10^{16}$ miles or $\frac{1}{22}$ times the milky way diameter. It has a semi-major axis of $1.25R_{\odot} = 6.7 \times 10^5$ km and thus the entire binary could fit within our sun, whose radius we denoted by $R_{\odot}$. It has a very small eccentricity $e = 0.088$ and its angle of inclination is 88.69° so that the system is observed nearly perfectly edge-on. Also it has a very small orbital period of 2.45 hours. Other useful numbers appear in Table 3.

**Table 2** Periastron precession $\frac{d\omega}{dt} = \frac{3}{1-e^2}(\alpha_g \bar{\omega})$ and spin–orbit precession $\Omega_{SO} = \frac{m_2 + \frac{\mu}{3}}{2M}\frac{d\omega}{dt}$ of $m_1$ is calculated for different one-body and two-body systems. The gravitational coupling constant is given by $\alpha_g = \frac{GM}{c^2 a}$, whereas the average orbital angular velocity reads $\bar{\omega} = \frac{2\pi}{T} = \sqrt{\frac{GM}{a^3}}$. Also $M_\odot$ is the solar mass, $\mu = \frac{m_1 m_2}{m_1+m_2}$, $M = m_1 + m_2$ and $e$ is the eccentricity of the Kepler ellipse. Note the huge values of the periastron precession $\frac{d\omega}{dt}$ and the spin–orbit precession $\Omega_{SO}$ for the Hulse–Taylor pulsar PRS 1913+16 and the double binary pulsar PSR J0737-3039 A/B

| System | Sun–(Earth & Moon) Gyro | Sun–Mercury | Earth–Gyro (GP-B) | PSR 1913+16 | PSR J0737 -3039 A/B |
|---|---|---|---|---|---|
| $a$ [km] | $1.5 \times 10^8$ | $6 \times 10^7$ | $6 \times 10^3$ | $2 \times 10^6$ | $9 \times 10^5$ |
| $M/M_\odot$ | 1 | 1 | $3 \times 10^{-6}$ | 1.44 (Pulsar)<br>1.39 (Com.) | 1.34 (A) $\to m_2$<br>1.25 (B) $\to m_1$ |
| $e$ | 0.017 | 0.206 | 0.0014 | 0.6 | 0.088 |
| $\alpha_g$ | $9.8 \times 10^{-9}$ | $3 \times 10^{-8}$ | $7 \times 10^{-10}$ | $2 \times 10^{-6}$ | $4.4 \times 10^{-6}$ |
| $\bar{\omega}[\frac{\text{rad}}{\text{s}}]$ | $2 \times 10^{-7}$ | $8.3 \times 10^{-7}$ | $10^{-3}$ | $2.2 \times 10^{-4}$ | $7 \times 10^{-4}$ |
| $\alpha_g\bar{\omega}$ $[\frac{\circ}{\text{yr}}]$ | $3.3 \times 10^{-6}$ | $4.5 \times 10^{-5}$ | $1.4 \times 10^{-3}$ | 0.79 | 5.6 |
| $\frac{d\omega}{dt}$ $[\frac{\circ}{\text{yr}}]$ | $10^{-5}$ | $1.2 \times 10^{-4}$ | $3.5 \times 10^{-3}$ | 4.2 | 16.9 |
| $\Omega_{SO}$ $[\frac{\circ}{\text{yr}}]$ | $5.3 \times 10^{-6}$ | $5.8 \times 10^{-5}$ | $1.8 \times 10^{-3}$ | 1.1 | 4.8 (A)<br>5.1 (B) |

**Table 3** Some observed properties of the double binary pulsar PSR J0737-3039 A/B

| Pulsar | Mass $[m/M_\odot]$ | Pulse frequency [Hz] | Pulse period [ms] | Surface $B$-field [G] | Spin–orbit precession [°/yr] | Time for 360° revolution of spin [yr] |
|---|---|---|---|---|---|---|
| $A$ | $m_2 = 1.3381$ | 44 | 22.7 | $6 \times 10^9$ | | 75 |
| $B$ | $m_1 = 1.2489$ | 0.36 | 2773.5 | $2 \times 10^{12}$ | $4.77 \pm 0.66$ | 71 |

Its periastron precession has been measured to an enormous accuracy of

$$\frac{d\omega}{dt} = 16.899949(68)°/\text{yr} \approx 43000 \cdot \left(\frac{d\omega}{dt}\right)_{\text{Mercury}}.$$

This enables $M$ to be determined with great accuracy

$$M = m_1 + m_2 = 2.58708(16) \cdot M_\odot,$$

where $M_\odot$ denotes the solar mass. However, in order to determine the spin–orbit precession, it is necessary to know both $m_1$ and $m_2$. In particular, $\frac{m_1}{m_2}$ is obtainable from $a_1 = \frac{m_2}{M}a$ and $a_2 = \frac{m_1}{M}a$ and the measured projected semi-major values $\frac{a_1 \sin i}{c}$ and $\frac{a_2 \sin i}{c}$. In addition, $m_1$ and $m_2$ are obtainable from the Shapiro time delay (delay of 6.2 μs due to the effect of pulsar $A$ on the light from pulsar $B$). As a result, the spin precession of the axis of pulsar $B$ has been observed to be $4.77 \pm 0.66°/\text{yr}$, to an accuracy of 13%, in agreement with the two-body spin–orbit calculations of Barker and O'Connell (1975b), as given in the last row of Table 2. Spin–spin contributions are negligible in an astronomical context (because neutron stars have very small radii compared to $a$). Even for terrestrial systems, they are not as interesting as spin–orbit effects since they are relatively small and do not involve motion. However, they play a role in the conservation of total angular momentum $\mathbf{J}$.

Finally, we note that further observations on the double binary system will undoubtedly lead to improved values for the spin precession.

## 4 Gödel's Universe

So far, we have discussed recent observations of gravito-magnetism manifesting itself in one- and two-body systems. These systems share the property that the metric is asymptotically flat and follows from the weak-field approximation. However, there exist other solutions of Einstein's field equations which exhibit gravito-magnetic effects and are not asymptotically flat. A prominent and illuminative example of such a solution is the so-called Gödel metric which represents a cosmological solution.

In the present section, we briefly introduce Gödel's metric and discuss time travel using a light cone diagram. Moreover, we illustrate the propagation of light in Gödel's Universe by showing a simulation of a particular visualization scenario. We conclude by summarizing an analogy between Maxwell's equations in a curved spacetime and in the vacuum, and Maxwell's equations in flat spacetime but within a gyrotropic medium. This bridge might open the door for a laboratory approximation of Gödel's Universe.

### 4.1 Basic Features

On the occasion of Einstein's 70th birthday, Gödel proposed an exact solution of Einstein's field equations which contains closed timelike world lines, see Gödel (1949a). Due to the peculiar metric expressed by the line element

$$\mathrm{d}s^2 = c^2\mathrm{d}t^2 - \frac{\mathrm{d}r^2}{1+(\frac{r}{2a})^2} - r^2\left(1-\left(\frac{r}{2a}\right)^2\right)\mathrm{d}\phi^2 - \mathrm{d}z^2 + 2r^2\frac{c}{\sqrt{2}a}\,\mathrm{d}t\,\mathrm{d}\phi, \tag{7}$$

it is in principle possible to travel back in time. Here the parameter $a > 0$ has the dimension of a length and characterizes the curvature of Gödel's Universe, see e.g. Schücking and Ozsváth (2003) and Hawking and Ellis (2006).

Before we address the startling opportunity of time travel we briefly list some properties of Gödel's metric (7). We start by noting that it is a cosmological solution of Einstein's field equations, see Table 1, with the stress-energy tensor of an ideal fluid

$$T_{\mu\nu} \equiv \left(\rho + \frac{p}{c^2}\right)u_\mu u_\nu - p g_{\mu\nu}$$

as source. The line element (7) is expressed in coordinates comoving with the ideal fluid, so that $u^\mu = (c, 0, 0, 0)$. According to the field equations, see Table 1, the mass density $\rho$ and the pressure $p$ of the ideal fluid are coupled to the cosmological constant $\Lambda$ and to the parameter $a$ by the relations

$$\kappa\left(\rho + \frac{p}{c^2}\right) = \frac{1}{a^2c^2} \quad \text{and} \quad \kappa p = \Lambda + \frac{1}{2a^2}.$$

From the characterization of timelike vector fields by Ehlers (1961), one concludes that the volume expansion and the shear tensor vanish for the velocity field $u^\mu$ of the ideal fluid in Gödel's Universe, see e.g. Hawking and Ellis (2006). In contrast the rotation tensor has non-zero components, which lead to the rotation scalar

$$\Omega_{\mathrm{G}} \equiv \frac{c}{\sqrt{2}a} > 0. \tag{8}$$

We note that $\Omega_{\mathrm{G}}$ is inversely proportional to the parameter $a$ and vanishes in the limit $a \to \infty$ for which (7) reduces to the line element of flat spacetime.

Finally, we emphasize that Gödel's metric allows for five independent Killing vector fields. It is rotational symmetric and its spacetime is homogeneous. Moreover, Gödel's spacetime is stationary but not static which results in gravito-magnetic effects.

In particular, the inertial compass of an observer comoving with the ideal fluid rotates relative to the stellar compass defined by the "fixed stars" of the ideal fluid. This feature stands out most clearly when we consider Gödel's metric for a small rotation scalar $\Omega_{\mathrm{G}} \ll 1$. In this limit we obtain in first order the line element

$$\mathrm{d}s^2 = c^2\,\mathrm{d}t^2 - \mathrm{d}r^2 - r^2\,\mathrm{d}\phi^2 - \mathrm{d}z^2 + 2r^2\Omega_{\mathrm{G}}\,\mathrm{d}t\,\mathrm{d}\phi \tag{9}$$

of flat space time in a frame of reference rotating with rate $\Omega_{\mathrm{G}}$. This rotation gives rise to the Sagnac effect of Gödel's Universe, see Delgado et al. (2002) and Kajari et al. (2004).

We conclude by emphasizing that the metric (7) is by no means a realistic model of our universe, since it cannot explain the red-shift observed for the distant stars. Gödel was well aware of this fact and later on published a metric which contains both expansion as well as rotation, see Gödel (1950).

### 4.2 Time Travel

The most puzzling property of Gödel's Universe is the existence of closed timelike world lines. Gödel (1949a) points out in his original article that

> [...] it is theoretically possible in these worlds to travel into the past, or otherwise influence the past.

This statement is illustrated most clearly by the light cone diagram of Fig. 4 where quasi "infinitesimal" light cones are attached to several points of spacetime. The light cones are depicted on two planes of constant coordinate time $t$ for three typical radii $r$. On the inner circle, the light cones are only slightly tilted and rather reminiscent of the ones in a rotating reference frame in flat spacetime. However, already the light cones on the next circle with radius $r = 2a$ (critical Gödel radius) are substantially different from the ones in a rotating reference frame in flat spacetime. Whereas in the rotating frame the opening angle of the cones decreases with increasing radius, the light cones in Gödel's Universe are still wide open and touch the plane of constant coordinate time. Hence, the middle circle corresponds to a closed null curve, provided the $z$-component of the curve is constant.

Finally, on the outer radius the light cones have fallen through the plane of constant coordinate time. For this reason the curves

$$x^{\mu}(\tau) = (t_0, r_0, \alpha\tau, z_0) \quad \text{with} \quad \alpha = \frac{c}{r_0\sqrt{\left(\frac{r_0}{2a}\right)^2 - 1}}$$

and $2a < r_0$ represent closed timelike world lines. Since it is possible to follow a timelike world line and return to the same spacetime point at a later proper time, Gödel's Universe in principle allows for time travel.

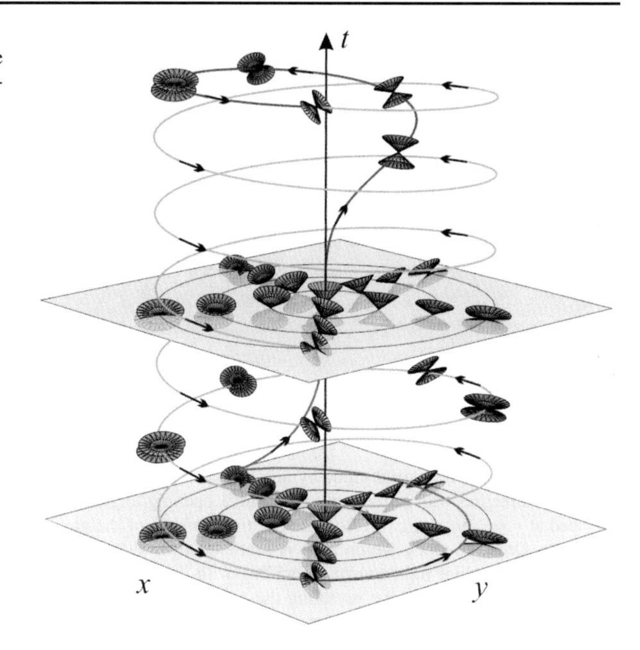


**Fig. 4** Light cone diagram of Gödel's Universe illustrating the possibility of time travel. The $x$- and $y$-coordinates are given by $x = r\cos\phi$ and $y = r\sin\phi$, whereas the trivial $z$-coordinate has been suppressed in this figure. Coordinate time flows vertically

In order to bring out this opportunity of time travel in the most vivid way we start our journey at the origin $r = 0$ of the upper plane of constant coordinate time shown in Fig. 4 and seek to return to the very same spacetime point at a later proper time. For this purpose we first follow a timelike world line leading from the origin to a point beyond the critical Gödel radius. The coordinate time along this world line will of course increase during the first part of our journey. After crossing the critical Gödel radius, we take a sharp turn to the "left" in the spacetime diagram Fig. 4 and follow the spiral curve down to earlier coordinate times $t$. Having completed our trip "back into coordinate time", we take another "left" turn onto a world line with increasing coordinate time which brings us back to our initial spacetime point.

In conclusion, at every point of Gödel's Universe there exist closed timelike world lines which pass through it. For more discussion about time travel and Gödel's view on it see Gödel (1949b, 1995).

## 4.3 Visualization of Light Propagation

The rotational structure of Gödel's Universe also manifests itself in the way how an observer visually experiences his surroundings. Due to the non-vanishing rotation tensor light is constantly deflected by Coriolis-like forces and can even return to the point where it was originally emitted. In several articles, see Grave and Buser (2008) and Kajari et al. (2009), we have visualized Gödel's Universe emphasizing effects due to its optical horizon which acts as a mirror and produces multiple images. In the present section we add a novel feature to our visualizations which goes beyond the results presented in Kajari et al. (2009): the appearance of shadows.

Visualizations of relativistic phenomena have become rather popular, see Ertl et al. (1989) and Weiskopf (2000). However, so far shadows have not been considered in this context. Our approach is made possible by exploiting the intrinsic symmetries of the Gödel metric expressed by Killing vector fields, see Grave and Buser (2008).

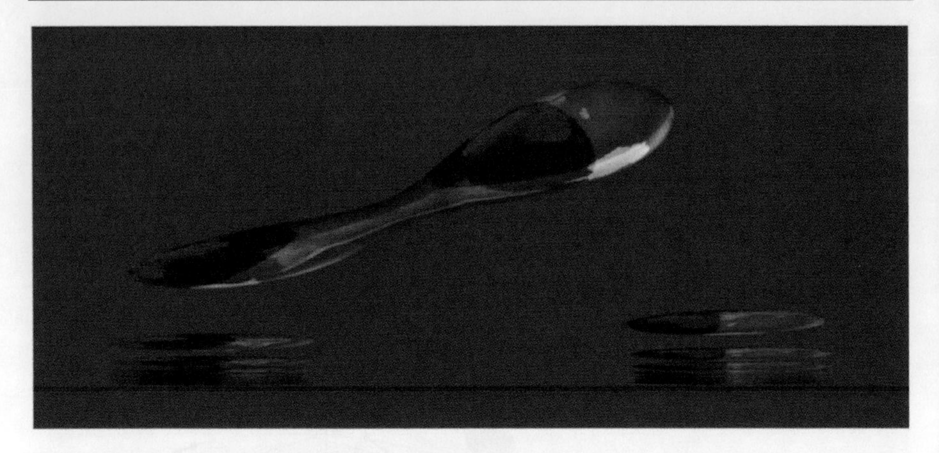

**Fig. 5** View on the terrestrial globe by an observer located at the origin of Gödel's Universe. The observer looks in $r$-direction with horizontal and vertical opening angles of 110° and 90°, respectively. The globe is placed at $r = 1.5a$ and $z = 1.2a$. The multiple images of the globe are a result of distinct paths existing in Gödel's Universe along which light is scattered from the globe to the observer. The whole scenario is illuminated by a point light source coinciding with the observer. Unlike in flat space, in Gödel's Universe shadows are cast on the globe revealing the spots which are illuminated by the light source. The *red line* marks the $z = 0$ plane. We have suppressed the lower half image plane since the globe only appears on the upper one

Kajari et al. (2009) provides a detailed discussion of null geodesics in Gödel's Universe and finds two types of light rays: (i) those which initially propagate towards the observer, and (ii) those which initially propagate away from him. The latter can be redirected to him due to a reflection from the horizon. These two classes of rays create two images of the object, which may even fuse into a single one. Whereas the direct rays display the front of the object, the indirect ones yield the mirror image of its back.

Moreover, Kajari et al. (2009) showed that all null geodesics are of the form of a helix with an axis parallel to the $z$-axis. Since a helix can have an arbitrary number of revolutions there exist a large variety of null geodesics along which light emitted from an object can reach the observer. Each of these paths generates a separate image of an object as exemplified in Fig. 5 by the view on the terrestrial globe. In this scenario the observer is located at the origin of the coordinate system looking in $r$-direction. The globe is positioned at a radial distance $r = 1.5a$ and at the height $z = 1.2a$. All images of the globe are arranged in two different clusters due to the spiral nature of the null geodesics.

In order to obtain a visualization which is more realistic than the ones in Kajari et al. (2009) we have illuminated in Fig. 5 the globe by a light source located at the position of the observer. In flat space no shadows appear in a scenery in which the location of the light source coincides with that of the observer. The paths of the rays from the light source illuminating the scenery and the paths of the rays scattered back to the observer are identical. In contrast, light rays in Gödel's Universe are not invertible due to the intrinsic rotation; light which illuminates the globe takes a different route than light which returns to the observer. This effect leads in Fig. 5 to a superposition of shadows on the surface of the globe.

## 4.4 Possible Analogies

Recently, analogies between light propagation in appropriately prepared media and in curved spacetime have received a lot of attention, see Geoff (2008) and Leonhardt (2009). Indeed, three routes to draw these analogies offer themselves: (i) arbitrarily moving media as generalizations of the famous Fizeau experiment, (ii) sound wave propagation in superfluids, and finally (iii) electrodynamics in a gyrotropic medium.

In the present subsection we illuminate the phenomenon of light propagating in a gravitational field from the point of view of light propagating in a gyrotropic medium. Indeed, it was proposed by Tamm (1924) and Skrotskii (1957), that Maxwell's equations in curved spacetime can be formulated in complete analogy to Maxwell's equations in flat spacetime but in the present of a gyrotropic medium, see also Plebanski (1960). We now briefly summarize this formulation, Schleich and Scully (1984).

We start by introducing the spacetime coordinate $x^{\mu} \equiv (ct, x, y, z)$ and the four current $J^{\mu} \equiv (c\rho, J_x, J_y, J_z)$. In this notation Maxwell's equations for the field tensor $F_{\mu\nu}$ take the relativistic form

$$\frac{1}{\sqrt{-g}}\frac{\partial}{\partial x^{\nu}}\left(\sqrt{-g}g^{\mu\rho}F_{\rho\beta}g^{\beta\nu}\right) = -\mu_0 J^{\mu} \tag{10}$$

and

$$F_{\alpha\beta,\gamma} + F_{\beta\gamma,\alpha} + F_{\gamma\alpha,\beta} = 0 \tag{11}$$

where $g \equiv \det g_{\mu\nu}$.

Next, we identify the components of the field tensor to the three-dimensional fields $\mathbf{E}$, $\mathbf{D}$, $\mathbf{B}$ and $\mathbf{H}$ via

$$\begin{aligned} \frac{1}{c}E_j &\equiv F_{0j}, \qquad & D_j &\equiv -c\sqrt{-g}F^{0j}, \\ B_j &\equiv -\frac{1}{2}\bar{\varepsilon}_{jlm}F_{lm}, \qquad & H_j &\equiv -\frac{1}{2}\bar{\varepsilon}_{jlm}\sqrt{-g}F^{lm}, \end{aligned} \tag{12}$$

with the completely antisymmetric Levi–Civita symbol $\bar{\varepsilon}_{jlm}$. We note that we only use lower indices for $\mathbf{E}$, $\mathbf{D}$, $\mathbf{B}$ and $\mathbf{H}$, since there are no co- or contravariant components for these fields according to their definition. However, we still follow the Einstein convention and sum over double indices.

With the identification (12) and the relation $F^{\mu\nu} = g^{\mu\alpha}g^{\nu\beta}F_{\alpha\beta}$, we can establish the constitutive relations

$$D_j = \mu_{js}E_s - c\bar{\varepsilon}_{jlm}g_l H_m \tag{13}$$

and

$$B_j = \mu_{js}H_s + \frac{1}{c}\bar{\varepsilon}_{jlm}g_l E_m \tag{14}$$

of a gyrotropic medium. Here we have introduced the abbreviations

$$\mu_{js} \equiv -\frac{\sqrt{-g}}{g_{00}}g^{js} \quad \text{and} \quad g_j \equiv -\frac{g_{0j}}{g_{00}}. \tag{15}$$

Moreover, we find by substitution of the identities (12) into Maxwell's equations (10) and (11) their familiar formulation

$$\operatorname{div}\mathbf{B} = 0, \qquad \operatorname{rot}\mathbf{E} = -\frac{\partial \mathbf{B}}{\partial t} \tag{16}$$

and

$$\mathrm{div}\mathbf{D} = \frac{\rho}{\epsilon_0}, \qquad \mathrm{rot}\,\mathbf{H} = \mu_0 \mathbf{j} + \frac{1}{c^2}\frac{\partial \mathbf{D}}{\partial t}, \tag{17}$$

with the charge density

$$\rho \equiv \sqrt{-g}\,\frac{J^0}{c}$$

and the current density

$$\mathbf{j} \equiv \sqrt{-g}\mathbf{J}.$$

The electric field vector is like the other vectors defined by $\mathbf{E} = (E_1, E_2, E_3)$.

So far, we have interpreted the gravitational field as a medium. However, we can also take the opposite point of view and consider the light propagation in a medium as simulating light propagation in a gravitational field. Hence, our strategy for simulating Gödel's metric by a gyrotropic medium is obvious: (i) find from Gödel's line element (7) the metric coefficients $g_{\mu\nu}$, (ii) substitute them into the definitions of $\mu_{js}$ and $g_j$, (15), and (iii) obtain the constitutive relations from (13) and (14).

Obviously this approach can only provide us with the connections between all four fields **E**, **B**, **D** and **H**. Unfortunately, it is unable to tell us how to implement them in an experiment. The answer to this question goes beyond the scope of the present paper and will be addressed in a future publication. It suffices to say that BECs, meta-materials, see Smith et al. (2004), and electromagnetically induced chirality, see Kästel et al. (2007), represent excellent candidates for such unusual media.

## 5 Summary

In the present article, we have followed the changes in the concept of rotation and its manifestations through several centuries starting from Foucault and Sagnac via Lense–Thirring and Gödel and have finally reached today's rotation sensors based on cold atoms and quantum degenerate gases. In particular, we have discussed several manifestations of gravito-magnetism in the weak-field limit focusing mainly on the relativistic two-body system. In contrast, Gödel's metric serves as an example of a cosmological solution of Einstein's field equations showing gravito-magnetic features. In this context we have pointed out an analogy between electrodynamics in curved spacetime and in a gyrotropic medium, which might open up an avenue for laboratory simulations of light propagation in Gödel's Universe.

During this journey, we have appreciated the fact that atom optics represents a lively field of research at the frontier of experimental general relativity. Moreover, we have emphasized the important role of microgravity in order to achieve long transit times in atom interferometry. This observation serves as one of the many motivations of the QUANTUS project where a BEC is created and observed in a freely falling elevator. Such a system allows us to study the behavior of quantum objects in an environment central to relativity. Indeed, the Gedanken experiment of a freely falling elevator to eliminate locally the effects of gravity is at the very heart of the weak equivalence principle assuming the identity of gravitational and inertial mass.

John Archibald Wheeler frequently told the story that Einstein, during his work on general relativity, had learned about a roofer who had survived his fall off a building. Einstein visited him in the hospital to find out what he felt while he was in free fall. The man explained to Einstein that he had experienced no sensation commonly considered as the effect of gravity.

Einstein (1982) himself describes the birth of this corner stone of general relativity as follows:

> *I was sitting on a chair in my patent office in Bern. Suddenly a thought struck me: If a man falls freely, he would not feel his weight. I was taken aback. This simple thought experiment made a deep impression on me. This led to the theory of gravity.*

This paradigmatic change of the concept of gravity stands out most clearly in the poem:

> *What's the fault of the force on my feet?*
> *What pushes my feet down on the street?*
> *Says Newton, the fault's with the earth's core.*
> *Einstein says, the fault's with the floor;*
> *Remove that and gravity's beat.*

(Frances Ruml, 1978, see Wheeler (1983))

The QUANTUS project bears testimony to the fact that atom optics is well on its way to space.

**Acknowledgements** We thank W. Ertmer, C. Lämmerzahl, H. J. Dittus, I. Ciufolini, J. Ehlers, J. Frauendiener, D. Giulini and H. Pfister for many enlightening discussions, as well as F. Grave, T. Müller, H. Ruder and G. Wunner for the fruitful collaboration. In particular we appreciate conversations with S. Varro, D. Wharam and M. Weidemüller concerning the role of L. Natanson in the creation of the Bose–Einstein statistics. As part of the QUANTUS collaboration, we are grateful for the financial support of the German Space Agency DLR with funds provided by the Federal Ministry of Economics and Technology (BMWi) under grant number DLR 50 WM 0837. We would also like to thank the DFG (German Research Foundation) for the financial support of the project "Visualisierung geschlossener zeitartiger Kurven in der Allgemeinen Relativitätstheorie". The work of R. F. O'C. on this project was carried out in Ulm and he would like to thank the Institute for Quantum Physics for the very gracious hospitality. W. P. S. appreciated the hospitality of the International Space Science Institute (ISSI) during this most exciting workshop in Bern. In particular, he would like to thank R. Treumann and M. Huber for the excellent organization.

## References

A.D. Aczel, *Pendulum—Léon Foucault and the Triumph of Science* (Washington Square Press, New York, 2003)

Ye.V. Baklanov, B.Ya. Dubetsky, V.P. Chebotayev, Appl. Phys. **9**, 171 (1976)

B.M. Barker, R.F. O'Connell, Gen. Relativ. Gravit. **5**, 539 (1974)

B.M. Barker, R.F. O'Connell, Phys. Rev. D **11**, 711 (1975a)

B.M. Barker, R.F. O'Connell, Phys. Rev. D **12**, 329 (1975b)

B.M. Barker, R.F. O'Connell, General relativistic effects in binary systems, in *Physics and Astrophysics of Neutron Stars and Black Holes*, ed. by R. Giacconi, R. Ruffini. Proceedings of the International School of Physics "Enrico Fermi", Course LXV (North-Holland, Amsterdam, 1977)

J. Bartlett, J. Kaplan, *Bartlett's Familiar Quotations*, 17th edn. (Little, Brown and Company, Boston, 2002)

P.R. Berman (ed.), *Atom Interferometry* (Academic Press, New York, 1997)

Ch.J. Bordé, Phys. Lett. A **140**, 10 (1989)

S.N. Bose, Zeitschrift für Physik **26**, 178 (1924). Translation: Am. J. Phys. **44**, 1057 (1976)

R.P. Breton, V.M. Kaspi, M. Kramer, M.A. McLaughlin, M. Lyutikov, S.M. Ransom, I.H. Stairs, R.D. Ferdman, F. Camilo, A. Possenti, Science **321**, 104 (2008)

M. Burgay, N. D'Amico, A. Possenti, R.N. Manchester, A.G. Lyne, B.C. Joshi, M.A. McLaughlin, M. Kramer, J.M. Sarkissian, F. Camilo, V. Kalogera, C. Kim, D.R. Lorimer, Nature **426**, 531 (2003)

W.W. Chow, J. Gea-Banacloche, L.M. Pedrotti, V.E. Sanders, W.P. Schleich, M.O. Scully, Rev. Mod. Phys. **57**, 61 (1985)
I. Ciufolini, Nature **449**, 41 (2007)
I. Ciufolini, E.C. Pavlis, Nature **431**, 958 (2004)
I. Ciufolini, J.A. Wheeler, *Gravitation and Inertia* (Princeton University Press, Princeton, 1995)
A.D. Cronin, J. Schmiedmayer, D.E. Pritchard, Rev. Mod. Phys. **81**, 1051 (2009)
M. De Angelis, A. Bertoldi, L. Cacciapuoti, A. Giorgini, G. Lamporesi, M. Prevedelli, G. Saccorotti, F. Sorrentino, G.M. Tino, Meas. Sci. Technol. **20**, 1 (2009)
A. Delgado, W.P. Schleich, G. Süssmann, New J. Phys. **4**, 37 (2002)
J. Ehlers, Beiträge zur relativistischen Mechanik kontinuierlicher Medien, in *Abhandlungen der Mathematisch-Naturwissenschaftlichen Klasse*, vol. 11 (Verlag der Akademie der Wissenschaften und der Literatur, Mainz, 1961)
A. Einstein, How I created the theory of relativity, speech in Kyoto 1922. Phys. Today **35**, 45–47 (1982)
A. Einstein, Sitz. ber. Preuss. Akad. Wiss. **22**, 261 (1924)
A. Einstein, Sitz. ber. Preuss. Akad. Wiss. 3 (1925)
A. Einstein, L. Infeld, B. Hoffmann, Ann. Math. **39**, 65 (1938)
T. Ertl et al., Visualization in astrophysics, in *Eurographics'89: Proceedings of the European Computer Graphics Conference and Exhibition*, ed. by W. Hansmann, F.R.A. Hopgood, W. Strasser (Elsevier/North-Holland, Amsterdam, 1989)
W. Ertmer, C. Schubert, T. Wendrich, M. Gilowski, M. Zaiser, T.V. Zoest, E. Rasel, Ch.J. Bordé, A. Clairon, A. Landragin, P. Laurent, P. Lemonde, G. Santarelli et al., Exp. Astron. **23**, 611 (2009)
J.B.L. Foucault, Compte-Rendu des séances de l'Académie Des Sci. **32**, 135 (1851)
V. Foucault, *Recueil des Travaux Scientifiques de Léon Foucault*, ed. by C.-M. Gariel (Gauthier, Paris, 1878)
A. Gauguet, B. Canuel, F. Leduc, D. Holleville, N. Dimarcq, A. Clairon, A. Landragin, J. Phys. IV **135**, 357 (2006)
B. Geoff, Nature **451**, 236 (2008)
K. Gödel, Rev. Mod. Phys. **21**, 447 (1949a), reprinted in Gen. Relativ. Gravit. **32**, 1409 (2000)
K. Gödel, A remark about the relationship between relativity theory and idealistic philosophy, in *Albert Einstein: Philosopher-Scientist*, ed. by P.A. Schilpp. The Library of Living Philosophers, vol. VII (Evanston, Illinois, 1949b)
K. Gödel, Proc. Int. Cong. Math. **1**, 175 (1950), reprinted in Gen. Relativ. Gravit. **32**, 1419 (2000)
K. Gödel, Lecture on rotating universes, in *Kurt Gödel: Collected Works*, ed. by S. Feferman, J.W. Dawson Jr., W. Goldfarb, C. Parsons, R.N. Solovay. Unpublished Essays and Lectures, vol. III (Oxford University Press, Oxford, 1995)
J.W. Goethe, *Faust I* (1808), English version: Project Gutenberg, translated by G.M. Priest. http://www.einam.com/faust/index.html
J. Le Gouët, T.E. Mehlstäubler, J. Kim, S. Merlet, A. Clairon, A. Landragin, F. Pereira Dos Santos, Appl. Phys. B **92**, 133 (2008)
F. Grave, M. Buser, Visiting the Gödel Universe. IEEE Trans. Vis. Comput. Graph. **14**, 1563 (2008)
T.L. Gustavson, A. Landragin, M.A. Kasevich, Class. Quantum Gravity **17**, 2385 (2000)
S.W. Hawking, G.F.R. Ellis, *The Large Scale Structure of Space-Time* (Cambridge University Press, Cambridge, 2006)
J. Kästel, M. Fleischhauer, S.F. Yelin, R.L. Walsworth, Phys. Rev. Lett. **99**, 073602 (2007)
E. Kajari, R. Walser, W.P. Schleich, A. Delgado, Gen. Relativ. Gravit. **36**, 2289 (2004)
E. Kajari, M. Buser, C. Feiler, W.P. Schleich, Rotation in relativity and the propagation of light, in *Atom Optics and Space Physics*, ed. by E. Arimondo, W. Ertmer, E.M. Rasel, W.P. Schleich. Proceedings of the International School of Physics "Enrico Fermi", Course CLXVIII (IOS, Amsterdam, 2009)
M. Kasevich, S. Chu, Phys. Rev. Lett. **67**, 181 (1991)
G.M. Keiser, Gravity probe B, in *Atom Optics and Space Physics*, ed. by E. Arimondo, W. Ertmer, E.M. Rasel, W.P. Schleich. Proceedings of the International School of Physics "Enrico Fermi", Course CLXVIII (IOS, Amsterdam, 2009)
M. Kramer, I.H. Stairs, R.N. Manchester, M.A. McLaughlin, A.G. Lyne, R.D. Ferdman, M. Burgay, D.R. Lorimer, A. Possenti, N. D'Amico, J.M. Sarkissian, G.B. Hobbs, J.E. Reynolds, P.C.C. Freire, F. Camilo, Science **314**, 97 (2006)
C. Lämmerzahl, G. Neugebauer, The Lense–Thirring effect: from the basic notations to the observed effects, in *Gyros, Clocks, Interferometers...: Testing Relativistic Gravity in Space*, ed. by C. Lämmerzahl, C.W.F. Everitt, F.W. Hehl (Springer, Berlin, 2001)
G. Lamporesi, A. Bertoldi, L. Cacciapuoti, M. Prevedelli, G.M. Tino, Phys. Rev. Lett. **100**, 050801 (2008)
J. Lense, H. Thirring, Phys. Z. **19**, 156 (1918). Translation in Gen. Relativ. Gravit. **16**, 727 (1984)
U. Leonhardt, *Essential Quantum Optics: From Quantum Measurements to Black Holes* (Cambridge University Press, Cambridge, 2009)

A.G. Lyne, M. Burgay, M. Kramer, A. Possenti, R.N. Manchester, F. Camilo, M.A. McLaughlin, D.R. Lorimer, N. D'Amico, B.C. Joshi, J. Reynolds, P.C.C. Freire, Science **303**, 1153 (2004)
B. Mashhoon, F. Gronwald, H.I.M. Lichtenegger, Gravitomagnetism and the Clock effect, in *Gyros, Clocks, Interferometers...: Testing Relativistic Gravity in Space*, ed. by C. Lämmerzahl, C.W.F. Everitt, F.W. Hehl (Springer, Berlin, 2001)
*Merriam–Websters Collegiate Dictionary* (Merriam–Webster Incorporated, Springfield, 1999)
T. Müller, M. Gilowski, M. Zaiser, T. Wendrich, W. Ertmer, E.M. Rasel (2009). doi:10.1140/epjd/e2009-00139-0
L. Natanson, Extraits du Bulletin de l'Academie des Sciences de Cracovie. Série A, 134 (1911). German version: Phys. Z. **12**, 659 (1911)
R.A. Nyman, G. Varoquaux, F. Lienhart, D. Chambon, S. Boussen, J.-F. Clément, T. Müller, G. Santarelli, F. Pereira Dos Santos, A. Clairon, A. Bresson, A. Landragin, P. Bouyer, Appl. Phys. B **84**, 673 (2006)
R.F. O'Connell, Astrophys. Space Sci. **4**, 119 (1969)
R.F. O'Connell, Spin, rotation and C, P and T effects in the gravitational interaction and related experiments, in *Experimental Gravitation*, ed. by B. Bertotti. Proceedings of the International School of Physics "Enrico Fermi", Course 56 (Academic Press, San Diego, 1974), p. 496
R.F. O'Connell, Class. Quantum Gravity **22**, 3815 (2005)
R.F. O'Connell, Gravito-magnetism in one-body and two-body systems: Theory and experiments, in *Atom Optics and Space Physics*, ed. by E. Arimondo, W. Ertmer, E.M. Rasel, W.P. Schleich. Proceedings of the International School of Physics "Enrico Fermi", Course CLXVIII (IOS, Amsterdam, 2009a)
R.F. O'Connell, Rotation and spin in physics, in *Frame-Dragging, Gravitational-Waves and Gravitational Tests*, ed. by I. Ciufolini, R. Matzner (Springer, Berlin, 2009b, in press)
F. Pereira Dos Santos, A. Landragin, Phys. World **20**(11), 32 (2007)
J. Plebanski, Phys. Rev. **118**, 1396 (1960)
E.J. Post, Rev. Mod. Phys. **39**, 475 (1967)
E.M. Rasel, M.K. Oberthaler, H. Batelaan, J. Schmiedmayer, A. Zeilinger, Phys. Rev. Lett. **75**, 2633 (1995)
H. Rauch, S.A. Werner, *Neutron Interferometry: Lessons in Experimental Quantum Mechanics* (Oxford University Press, Oxford, 2000)
F. Riehle, T. Kisters, A. Witte, J. Helmcke, C.J. Bordé, Phys. Rev. Lett. **67**, 177 (1991)
H.R. Robertson, Ann. Math. **39**, 101 (1938)
G. Sagnac, C. R. Acad. Sci., Paris **157**, 708 (1913a)
G. Sagnac, C. R. Acad. Sci., Paris **157**, 1410 (1913b)
W.P. Schleich, M.O. Scully, General relativity and modern optics, in *New Trends in Atomic Physics, Les Houches 1982, Session XXXVIII*, ed. by G. Grynberg, R. Stora (North-Holland, Amsterdam, 1984)
K.U. Schreiber, A. Velikoseltsev, M. Rothacher, T. Klugel, G.E. Stedman, D.L. Wiltshire, J. Geophys. Res. **109**, B06405 (2004)
E. Schücking, I. Ozsváth, Am. J. Phys. **71**, 801 (2003)
M.O. Scully, Suggestion and analysis for a new optical test of general relativity, in *Laser Spectroscopy IV*, ed. by H. Walther, K.W. Rothe (Springer, Berlin, 1979)
M.O. Scully, J.P. Dowling, Phys. Rev. A **48**, 3186 (1993)
M.O. Scully, G.R. Welch, Phys. World **17**(10), 31 (2004)
M.O. Scully, M.S. Zubairy, M.P. Haugan, Phys. Rev. A **24**, 2009 (1981)
G.B. Skrotskii, Dokl. Akad. Nauk SSSR **114**, 73 (1957). Translation: Soviet Phys.-Dokl. **2**, 226 (1957)
D.R. Smith, J.B. Pendry, M.C.K. Wiltshire, Science **305**, 788 (2004)
M.H. Soffel, *Relativity in Astrometry, Celestial Mechanics and Geodesy* (Springer, Berlin, 1989)
J. Spalek, Letter to Physics Today (2006). http://th-www.if.uj.edu.pl/ztms/jspalek_pl.htm
J.L. Staudemann, S.A. Werner, R. Colella, A.W. Overhauser, Phys. Rev. A **21**, 1419 (1980)
G.E. Stedman, Rep. Prog. Phys. **60**, 615 (1997)
N. Straumann, *General Relativity with Applications to Astrophysics* (Springer, Berlin, 2004)
J.E. Tamm, J. Russ. Phys.-Chem. Soc. **56**(2–3), 284 (1924)
H. Thirring, Phys. Z. **19**, 33 (1918). Erratum: **22**, 29 (1921), Translation in Gen. Relativ. Gravit. **16**, 712 (1984)
D. Weiskopf, Four-dimensional non-linear ray tracing as a visualization tool for gravitational physics, in *Proceedings of the Conference on Visualization '00* (IEEE Computer Society Press, Los Alamitos, 2000)
H. Weyl, Naturwissenschaften **12**, 197 (1924)
J.A. Wheeler, Introduction to general relativity, in *Quantum Optics, Experimental Gravitation and Measurement Theory*, ed. by P. Meystre, M.O. Scully (Plenum, New York, 1983)
F. Zimmer, M. Fleischhauer, Phys. Rev. Lett. **92**, 253201 (2004)
F.E. Zimmer, M. Fleischhauer, Phys. Rev. A **74**, 063609 (2006)
T. Zoest, T. Müller, T. Wendrich, M. Gilowski, E.M. Rasel, W. Ertmer, T. Könemann, C. Lämmerzahl, H. Dittus, A. Vogel, K. Bongs, K. Sengstock, W. Lewczko, A. Peters, T. Steinmetz, J. Reichelt, G. Nandi, W.P. Schleich, R. Walser, Int. J. Modern Phys. D **16**, 2421 (2008)

Space Sci Rev (2009) 148: 149–167
DOI 10.1007/s11214-009-9543-4

# The Pioneer Anomaly in the Light of New Data

**Slava G. Turyshev · Viktor T. Toth**

Received: 24 May 2009 / Accepted: 1 June 2009 / Published online: 13 June 2009
© Springer Science+Business Media B.V. 2009

**Abstract** The radio-metric tracking data received from the Pioneer 10 and 11 spacecraft from the distances between 20–70 astronomical units from the Sun has consistently indicated the presence of a small, anomalous, blue-shifted Doppler frequency drift that limited the accuracy of the orbit reconstruction for these vehicles. This drift was interpreted as a sunward acceleration of $a_P = (8.74 \pm 1.33) \times 10^{-10}$ m/s$^2$ for each particular spacecraft. This signal has become known as the Pioneer anomaly; the nature of this anomaly is still being investigated.

Recently new Pioneer 10 and 11 radio-metric Doppler and flight telemetry data became available. The newly available Doppler data set is much larger when compared to the data used in previous investigations and is the primary source for new investigation of the anomaly. In addition, the flight telemetry files, original project documentation, and newly developed software tools are now used to reconstruct the engineering history of spacecraft. With the help of this information, a thermal model of the Pioneers was developed to study possible contribution of thermal recoil force acting on the spacecraft. The goal of the ongoing efforts is to evaluate the effect of on-board systems on the spacecrafts' trajectories and possibly identify the nature of this anomaly.

Techniques developed for the investigation of the Pioneer anomaly are applicable to the New Horizons mission. Analysis shows that anisotropic thermal radiation from on-board sources will accelerate this spacecraft by $\sim 41 \times 10^{-10}$ m/s$^2$. We discuss the lessons learned from the study of the Pioneer anomaly for the New Horizons spacecraft.

**Keywords** Pioneer anomaly · Gravitational experiments · Deep-space navigation · Thermal modeling

S.G. Turyshev (✉)
Jet Propulsion Laboratory, California Institute of Technology,
4800 Oak Grove Drive, Pasadena, CA 91109, USA
e-mail: turyshev@jpl.nasa.gov

V.T. Toth
Ottawa, ON K1N 9H5, Canada
e-mail: vttoth@vttoth.com

## 1 Introduction

The first spacecraft to leave the inner solar system (Anderson et al. 1998, 2002; Turyshev et al. 2000, 2005), Pioneer 10 and 11 were designed to conduct an exploration of the interplanetary medium beyond the orbit of Mars and perform close-up observations of Jupiter during the 1972–1973 Jovian opportunities.

The spacecraft were launched in March 1972 (Pioneer 10) and April 1973 (Pioneer 11) on top of identical three-stage Atlas-Centaur launch vehicles. After passing through the asteroid belt, Pioneer 10 reached Jupiter in December 1973. The trajectory of its sister craft, Pioneer 11, in addition to visiting Jupiter in 1974, also included an encounter with Saturn in 1979 (Anderson et al. 2002; Turyshev et al. 2006a).

After the planetary encounters and successful completion of their primary missions, both Pioneers continued to explore the outer solar system. Due to their excellent health and navigational capabilities, the Pioneers were used to search for trans-Neptunian objects and to establish limits on the presence of low-frequency gravitational radiation (NASA Ames Research Center 1971).

Eventually, Pioneer 10 became the first man-made object to leave the solar system, with its official mission ending in March 1997. Since then, NASA's Deep Space Network (DSN) made occasional contact with the spacecraft. The last successful communication from Pioneer 10 was received by the DSN on 27 April 2002. Pioneer 11 sent its last coherent Doppler data in October 1990; the last scientific observations were returned by Pioneer 11 in September 1995.

The orbits of Pioneer 10 and 11 were reconstructed primarily on the basis of radio-metric Doppler tracking data. The reconstruction between heliocentric distances of 20–70 AU yielded a persistent small discrepancy between observed and computed values (Anderson et al. 1998, 2002; Turyshev et al. 2000, 2005). After accounting for known systematic effects, the unmodeled change in the Doppler residual for Pioneer 10 and 11 is equivalent to an approximately sunward constant acceleration of

$$a_P = (8.74 \pm 1.33) \times 10^{-10}\ \mathrm{m/s^2}. \tag{1}$$

The magnitude of this effect, measured between heliocentric distances of 40–70 AU, remains approximately constant within the 3 dB gain bandwidth of the high-gain antenna (Turyshev et al. 2005, 2006a). The nature of this anomalous acceleration remains unexplained; this signal has become known as the Pioneer anomaly.

There were numerous attempts in recent years to provide an explanation for the anomalous acceleration of Pioneer 10 and 11. These can be broadly categorized as either invoking conventional mechanisms or utilizing principles of "new physics".

Initial efforts to explain the Pioneer anomaly focused on the possibility of on-board systematic forces. While these cannot be conclusively excluded (Anderson et al. 2002; Turyshev et al. 2005), the evidence to date did not support these mechanisms: it was found that the magnitude of the anomaly exceeds the acceleration that these mechanisms would likely produce, and the temporal evolution of the anomaly differs from that which one would expect, for instance, if the anomaly were due to thermal radiation of a decaying nuclear power source.

Conventional mechanisms external to the spacecraft were also considered. First among these was the possibility that the anomaly may be due to perturbations of the spacecrafts' orbits by as yet unknown mass distributions in the Kuiper belt. Another possibility is that dust in the solar system may exert a drag force, or it may cause a frequency shift, proportional to distance, in the radio signal. These proposals could not produce a model that is consistent with the known properties of the Pioneer anomaly, and may also be in contradiction with the known properties of planetary orbits.

The value of the Pioneer anomaly happens to be approximately $cH_0$, where $c$ is the speed of light and $H_0$ is the Hubble constant at the present epoch. Attempts were made to exploit this numerical coincidence to provide a cosmological explanation for the anomaly, but it has been demonstrated that this approach would likely produce an effect with the opposite sign (Anderson et al. 2002; Turyshev et al. 2006a).

As the search for a conventional explanation for the anomaly appeared unsuccessful, this provided a motivation to seek an explanation in "new physics". No such attempt to date produced a clearly viable mechanism for the anomaly (Turyshev et al. 2006a).

Here we report on the status of the recovery of the Pioneers' radiometric Doppler data and flight telemetry and their usefulness for the analysis of the Pioneer anomaly.

## 2 New Doppler Data and Their Preliminary Analysis

The inability to explain the anomalous behavior of the Pioneers with conventional physics has resulted in a growing discussion about the origin of the detected signal. The limited size of the previously analyzed data set also limits our current knowledge of the anomaly. To determine the origin of $a_P$ and especially before any serious discussion of new physics can take place, one must analyze the entire set of radio-metric Doppler data received from the Pioneers.

Since 2002, multiple on-going efforts have contributed significantly to our ability to explore and comprehend the nature of the Pioneer anomaly (Turyshev et al. 2005, 2006a, 2006b; Toth and Turyshev 2006, 2008). Most notable among these are: (i) the availability of an extended Doppler data set; (ii) the recovery of spacecraft telemetry; (iii) the recovery of Pioneer project documentation; (iv) the development of a comprehensive thermal model of the spacecraft; (v) the development of new methods to incorporate thermal telemetry into orbit determination; and (vi) several independent confirmations of the Pioneer anomaly. These developments led to the formulation of a comprehensive strategy to establish reliably the temporal dependence and direction of the anomalous acceleration, and correlate it with revised estimates of the thermal recoil force.

### 2.1 The Extended Pioneer Doppler Data Set

Immediately after the results of the first major study of the anomaly were announced (Anderson et al. 2002), a focused effort began at JPL to recover as much Doppler data as possible. Initially, it was hoped that nearly all the Doppler record of the Pioneer 10 and 11 spacecraft from 1972 until the end of their respective missions can be recovered. Unfortunately, this proved much more difficult than anyone anticipated at the time.

Recovery of radio-metric data for a mission operating for more then 30 years is an effort that was never attempted before. Indeed, 30 years is a long time, presenting many unique challenges, including changes in the data formats, navigational software, as well as supporting hardware (Turyshev et al. 2006a). Even the DSN configuration had changes—new stations were built, and some stations were moved, upgraded and reassigned. By 2005 all the DSN data formats, navigational software used to support Pioneers, all the hardware used to read, write and maintain the data have become obsolete and are no longer operationally supported by existing NASA protocols. The main asset of the entire mission support—its people—changed the most, as personnel with the necessary expertise to answer questions or shed light on obscure details are either retired or no longer with us.

Despite these multiple complexities (Turyshev et al. 2006a; Toth and Turyshev 2006, 2008), the transfer of the available Pioneer Doppler data to modern media formats has been completed. However, as a result of these issues, less data is available for analysis than what was initially hoped. Nonetheless, we now have a significantly expanded data set that is available for study.

For Pioneer 10, good quality Doppler data are available covering the period between 1980 and 1995. Some additional data from the year 2000 also appear usable. Yet more data files, from 1996–1997, may be usable if absent ramp information can be recovered from the transmitting DSN stations.

For Pioneer 11, good quality Doppler data are available from the periods 1983–1985, and 1987–1990. Data from 1980–1982 and from 1986, while recovered, appear unusable.

Additional data covering the Jupiter encounter of Pioneer 10, and the Saturn encounter of Pioneer 11, are also available and appear to be of good quality. Modeling encounters with gas giants is fraught with additional difficulties, as both gravitational and nongravitational effects of the complex planetary environment must be modeled with precision. On the other hand, the rapid changes in the spacecrafts' velocities can help uncover small systematic effects that are otherwise difficult to observe.

## 2.2 A Strategy to Find the Origin of the Pioneer Anomaly

The primary objective of the new investigation is to determine the origin of the Pioneer anomaly. Specifically, the investigation intends to accomplish the following three major objectives:

I. By using the early mission Doppler data we aim: (i) to determine the true direction of anomalous acceleration by discriminating between the four possible directions: sunward, Earth-pointing, along the velocity vector, or along the spin-axis, and (ii) to study the physics of the planetary encounters and to learn more about the onset of the anomaly (see Figs. 6–7 in Anderson et al. 2002);

II. Analysis of the entire set of Doppler data: (iii) to study the temporal behavior of the signal, and (iv) to perform a comparative analysis of the individual anomalous accelerations of the two Pioneers with data taken from similar heliocentric distances; and finally

III. The newly recovered telemetry information from the spacecrafts' thermal, electrical, power, propulsion, and communication subsystems will be used (v) to investigate contributions of known sources of on-board systematics and study their effects on $a_P$. This investigation of the on-board small forces will be done in conjunction with the analysis of the Doppler data, thereby strengthening the ultimate outcome.

The new extended data set enables us to investigate the Pioneer anomaly with the entire available Pioneer 10 and 11 radiometric Doppler data (Turyshev et al. 2006a; Toth and Turyshev 2008). In particular, it is used to build a thermal/electrical/dynamical model of the Pioneer vehicles and verify it with the actual data from the telemetry; the goal here is the development of a model that can be used to calibrate the Doppler anomaly with respect to the on-board sources of dynamical noise.

### *2.2.1 Direction of the Pioneer Anomaly*

Analysis of the earlier data is critical in establishing a precise time history of the effect. With the radiation pattern of the Pioneer antennae and the lack of precise navigation in the plane of the sky, the determination of the exact direction of the anomaly was a difficult task (Anderson et al. 2002). While in deep space, for standard antennae without good 3-D navigation, the directions: (i) towards the Sun, (ii) towards the Earth, (iii) along the direction of motion of the craft, or (iv) along the spin axis, are all observationally synonymous (Nieto and Turyshev 2004).

The four possible directions for the anomaly all indicate a different physical mechanism. Specifically, if the anomaly is: (i) in the direction *towards the Sun*, this would indicate a force originating from the Sun, likely signifying a need for gravity modification; (ii) in the direction *towards the Earth*, this would indicate an effect on signal propagation or a time signal anomaly impacting the design of the DSN hardware and space flight-control methods; (iii) in the direction *of the velocity vector*, this would indicate an inertial force or a drag force providing support for a media-dependent origin; or, finally, (iv) in *the spin-axis direction*, this would indicate an on-board systematic, which is the most plausible explanation for the effect. The corresponding directional signatures of these four directions are distinct and could be extracted from the data (Nieto and Turyshev 2004; Turyshev et al. 2005; Toth and Turyshev 2006).

The increased data span, and especially the earlier data segment, is crucial in our ability to determine the direction of the signal and its true nature. If the anomaly is due to on-board effects, direct along the spin axis, the solution for the anomaly will be different after each re-pointing of the spacecraft, namely, every re-pointing will produce a step-function discontinuity in the solution for the $a_P$. Earlier in the mission, there were many of these re-pointing maneuvers; understanding their impact on the anomaly would be a very important activity of the proposed analysis of the earlier data (Turyshev et al. 2005).

With the early data, we expect to improve the sensitivity of the solutions in the directions perpendicular to the line-of-sight by at least an order of magnitude. We started to analyze the early parts of the trajectories of the Pioneers with the goal of determining the true direction of the Pioneer anomaly and possibly its origin (Nieto and Turyshev 2004; Turyshev et al. 2005, 2006b); results will be reported elsewhere.

### 2.2.2 *Study of the Planetary Encounters*

It is alarming is that the early Pioneer 10 and 11 data (before 1987) was never analyzed in detail, especially in regards to the effect of on-board systematics. For instance, a nearly constant anomalous acceleration seems to exist in the data of Pioneer 10 as close as 27 AU from the Sun (Anderson et al. 2002; Turyshev et al. 2005). Pioneer 11, beginning just after Jupiter flyby, shows a small value for the anomaly during the Jupiter-Saturn cruise phase in the interior of the solar system. However, right at Saturn encounter, when the craft passed into a hyperbolic escape orbit, analysis of early navigation results tentatively shows a fast increase in the anomaly to its canonical value (Nieto and Turyshev 2004; Turyshev et al. 2006b).

We first study the Saturn encounter for Pioneer 11. We will use the data for nearly two years surrounding this event. If successful, we should be able to learn the mechanism that led to the onset of the anomaly during the flyby. The Jovian encounters are also of interest; however, they were in the region much too close to the Sun. One expects large contributions from the standard sources of acceleration noise that exist at heliocentric distances $\sim$5 AU. We use a similar strategy as with the Saturn encounter and will attempt to make full use of the data available.

### *2.2.3 Study of the Temporal Evolution of the Anomaly*

The same investigation as above is used to revisit the question of collimated thermal emission by studying the temporal evolution of the anomaly. If the anomaly is due to the on-board nuclear fuel inventory ($^{238}$Pu) present on the vehicles and the related heat recoil force, one expects that a decrease in the anomaly's magnitude will be correlated with $^{238}$Pu decay with a half-life of 87.74 years. The previous analysis of 11.5 years of data (Anderson et al. 2002) found no support for a thermal mechanism. However, Markwardt (2002) and Toth (2009) did not rule out this possibility finding an appropriate trend in the solution for $a_P$. The now available 30-year interval of data (20 years for Pioneer 11) may demonstrate the effect of a $\sim$21% reduction in the heat contribution to the anomaly. This behavior would strongly support a thermal origin of the effect (Toth and Turyshev 2008).

Asymmetrically radiated heat due to available on-board thermal inventory, if appropriately directed, may result in a recoil force with properties similar to those of the Pioneer anomaly. This is the "primary suspect" for the origin of the effect. Therefore, an investigation of heat exchange mechanisms that may lead to a thermal thrust on the craft is an important part our effort. This investigation is done in two ways: (i) to use the longest possible set of Doppler data to study the temporal evolution of $a_P$ and (ii) to build a model of thermal thrust of the Pioneer craft and use available telemetry data to derive the resulted acceleration. If the anomaly is due to a thermal mechanism, the two methods should agree.

The much extended data span, augmented by all the ancillary spacecraft information, will help to study a signature of the exponential decay of the on-board power source (Markwardt 2002; Toth 2009). The wealth of the recently acquired data presents an exciting opportunity to learn more about the anomaly in various regimes and will help to determine the nature of this anomalous signal.

### *2.2.4 Analysis of the Individual Trajectories for Both Pioneers*

The much larger newly recovered sets of the Pioneer 10 and 11 Doppler data make it possible to study the properties of the individual acceleration solutions for both Pioneers obtained with the data collected from similar heliocentric distances. The limited data set used in the previous analysis (Anderson et al. 1998, 2002) precluded such a comparison; however, the now available data allows such an investigation.

Previously, even though we had individual solutions from the two craft, the fact remained that $a_{P10}$ and $a_{P11}$ were obtained from data segments that not only were very different in length (11.5 and 3.75 years), but they were also taken from different heliocentric distances. We anticipate that analysis of their data from similar heliocentric regions will help us better understand the properties of the anomaly, especially if it were to be attributed to an on-board systematic source. Analysis of the individual data would also help to calibrate the final solution for the anomaly by properly accounting for the individual properties of the spacecraft.

### *2.2.5 Investigation of On-Board Systematics*

The availability of telemetry information makes it possible to conduct a detailed investigation of the on-board systematic forces as a source of the anomalous acceleration. Here we consider forces that are generated by on-board spacecraft systems and that are thought to contribute to the constant acceleration seen in the analysis of the Pioneer Doppler data (see Sect. 3 for discussion).

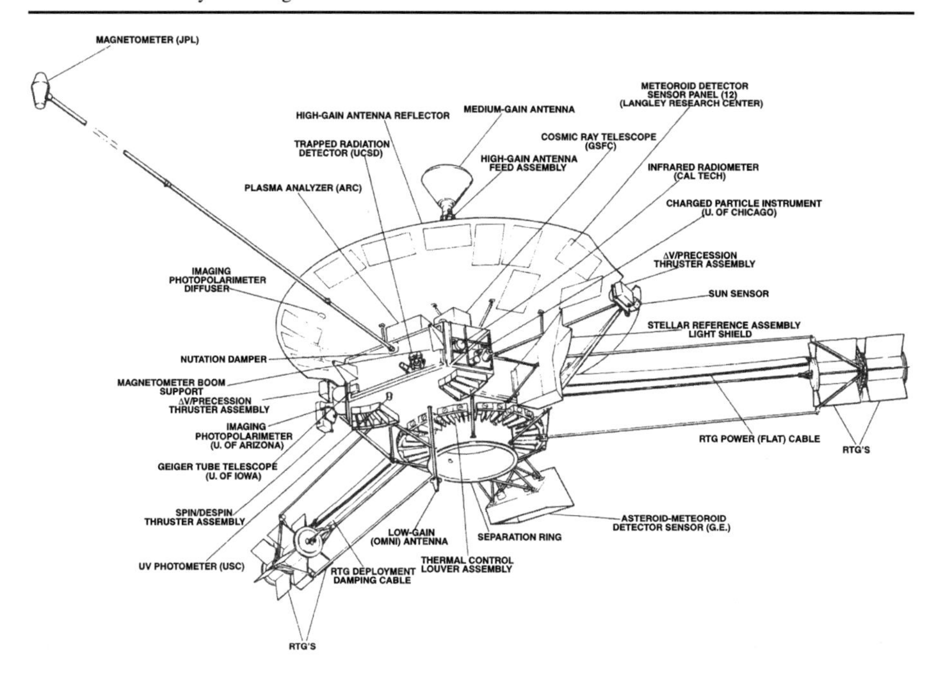


**Fig. 1** A drawing of the Pioneer spacecraft

## 3 Using Flight Telemetry to Study the Spacecrafts' Behavior

All transmissions of both Pioneer spacecraft, including all engineering telemetry, were archived (Turyshev et al. 2006a) in the form of files containing Master Data Records (MDRs). Originally, MDRs were scheduled for limited retention. Fortunately, the Pioneers' mission records avoided this fate: with the exception of a few gaps in the data (Turyshev et al. 2006a) the entire mission record has been saved. These recently recovered telemetry readings are important in reconstructing a complete history of the thermal, electrical, and propulsion systems for both spacecraft. This, may data lead to a better determination of the crafts' acceleration due to on-board systematic effects.

### 3.1 The Pioneer Spacecraft

As evident from Fig. 1, the appearance of the Pioneer spacecraft is dominated by the 2.74 m diameter high gain antenna (HGA). The spacecraft body, located behind the HGA, consists of a larger, regular hexagonal compartment housing the propellant tank and spacecraft electronics; an adjacent, smaller compartment housed science instruments. The spacecraft body is covered by multilayer thermal insulating blankets, except for a louver system located on the side opposite the HGA, which was activated by bimetallic springs to expel excess heat from the spacecraft.

Each spacecraft was powered by four radioisotope thermoelectric generators (RTGs) mounted in pairs at the end of two booms, approximately three meters in length, extended from two sides of the spacecraft body at an angle of 120°. A third boom, approximately 6 m long, held a magnetometer.

The total (design) mass of the spacecraft was ∼250 kg at launch, of which 27 kg was propellant (NASA Ames Research Center 1971).

For the purposes of attitude control, the spacecraft were designed to spin at the nominal rate of 4.8 rpm. Six small monopropellant (hydrazine) thrusters, mounted in three thruster cluster assemblies, were used for spin correction, attitude control, and trajectory correction maneuvers (see Fig. 3).

### 3.2 Compartment Temperatures and Thermal Radiation

If the Pioneer anomaly is due to anisotropically emitted thermal radiation, one expects to find a near constant supply of heat radiated off the back of the spacecraft that would produce a thermal recoil force with the well established properties (Turyshev et al. 2005). The lack of constancy of heat dissipated during the longest Doppler segment analyses (i.e. 11.5 years of the Pioneer 10 data (Anderson et al. 2002)) appears to have invalidated the hypothesis.

The newly acquired data (both Doppler and telemetry) is very valuable for the investigation as it contributes to addressing this possibility. We also have a much larger segment of Doppler data that will be used to analyze these heat dissipation processes on the vehicles. In addition, we now have the actual design, fabrication, testing, pre- and in-flight calibration data that characterize the Pioneer craft performance for the duration of their missions. Finally, we have all the detailed information on properties of the spacecraft and the data needed to reconstruct the behavior of its major components, including electrical power and thermal subsystems.

This data tells precisely at what time the louvers were open and closed, when a certain instrument was powered “on” and “off”, what was the performance of the battery, shunt current and all electric parts of the spacecraft. This information is being used in the development of a model of the Pioneers that is needed to establish the true thermal and electrical power dissipation history of the vehicles and also to correlate major events on the Pioneers (such as powering “on” or “off” certain instruments or performing a maneuver) with the analysis of the available Doppler data.

### 3.3 Telemetry Overview

Telemetry formats can be broadly categorized as science formats versus engineering formats. Telemetry words included both analog and digital values. Digital values were used to represent sensor states, switch states, counters, timers, and logic states. Analog readings, from sensors measuring temperatures, voltages, currents and more, were encoded using 6-bit words. This necessarily limited the sensor resolution and introduced a significant amount of quantization noise. Furthermore, the analog-to-digital conversion was not necessarily linear; prior to launch, analog sensors were calibrated using a fifth-order polynomial. Calibration ranges were also established; outside these ranges, the calibration polynomials are known to yield nonsensical results.

With the help of the information contained in these words, it is possible to reconstruct the history of RTG temperatures and power, radio beam power, electrically generated heat inside the spacecraft, spacecraft temperatures, and propulsion system history.

Relevant on-board telemetry falls into two categories: temperature and electrical measurements. In the first category, we have data from several temperature sensors on-board, most notably the fin root temperature readings for all four RTGs. Figure 5 shows the location of most temperature sensors on board for which readings are available. Other temperature sensors are located at the RTGs and inside the propellant tank.

The electrical power profile of the spacecraft can be reconstructed to a reasonable degree of accuracy using electrical telemetry measurements. Available are the individual voltage and current readings for the RTGs, the readings on the main bus voltage and current, as well as the shunt current; known are the power on/off state of most spacecraft subsystems. From these and other readings, one can calculate the complete electrical profile of the spacecraft.

### 3.4 RTG Temperatures and Anisotropic Heat Reflection

It has been argued that the anomalous acceleration may be due to anisotropic reflection of the heat coming from the RTGs off the back of the spacecraft high gain antennae. Note that only ~65 W of directed constant heat is required to explain the anomaly, which certainly is not a great deal of power when the craft has heat sources capable of producing almost 2.5 kW of heat at the beginning of the missions. However, using available information on the spacecraft and RTG designs, Anderson et al. (2002) estimated that only 4 W of directed power could be produced by this mechanism. Adding an uncertainty of the same size, they estimated a contribution to the anomalous acceleration from heat reflection to be $a_{\rm hr} = (-0.55 \pm 0.55) \times 10^{-10}$ m/s$^2$. Furthermore, if this mechanism were the cause of the anomaly, ultimately an unambiguous decrease in the size of $a_P$ should be observed, because the RTGs' radioactively produced radiant heat is decreasing. In fact, one would expect a decrease of about $0.75 \times 10^{-10}$ m/s$^2$ in $a_P$ over the 11.5 year Pioneer 10 data interval if this mechanism were the origin of $a_P$.

Alternative estimates presented put the magnitude of this effect at ~24 W or the corresponding value for $a_{\rm hr}$ at the level of $a_{\rm hr} \sim -3.3 \times 10^{-10}$ m/s$^2$ (see discussion in Scheffer 2003). We comment on the fact that both groups acknowledged that a thermal model for the Pioneer spacecraft is hard to build. However, this seems to be exactly what one would have to do in order to reconcile the differences in analyzing the role of the thermal heat in the formation of the Pioneer anomaly.

It is clear that any thermal explanation should clarify why either the radioactive decay (if the heat is directly from the RTGs (Katz 1999)) or electrical power decay (if the heat is from the instrument compartment (Murphy 1999)) is not seen. One reason could be that previous analyses used only a limited data set of only 11.5 years when the thermal signature was hard to disentangle from the Doppler residuals or the fact that the actual data on the performance of the thermal and electrical systems was not complete or unavailable at the time the analyses were performed.

The present situation is very different. Not only do we have a much longer Doppler data segment for both spacecraft, we also have the actual telemetry data on the thermal and electric power subsystems for both Pioneers for the entire lengths of their missions. The electrical power profile of the spacecraft can be reconstructed to a reasonable degree of accuracy using electrical telemetry measurements. We have individual voltage and current readings for the RTGs; we have readings on the main bus voltage and current, as well as the shunt current; we know the power on/off state of most spacecraft subsystems. From these and other readings, the complete electrical profile of the spacecraft can be calculated, as discussed in Turyshev et al. (2006a).

To utilize this data for the upcoming investigation, it is possible to reconstruct the direction of heat flow (with the help of a thermal model built for this purpose) and study the absorption and re-emission by, and reflection off the craft surfaces (Turyshev et al. 2006a; Toth and Turyshev 2008).

In the following discussion, telemetry words are labeled using identifiers in the form of $C_n$, where $n$ is a number indicating the word position in the fixed format telemetry frames.

The exterior temperatures of the RTGs were measured by one sensor on each of the four RTGs: the so-called "fin root temperature" sensor. Telemetry words $C_{201}$ through $C_{204}$

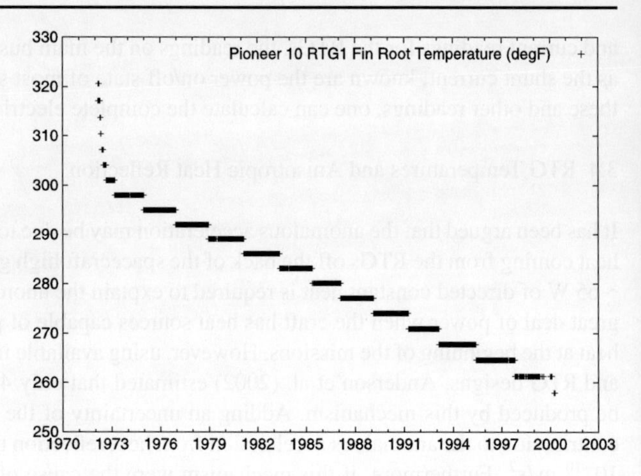


**Fig. 2** RTG 1 fin root temperatures (telemetry word $C_{201}$; in °F) for Pioneer 10

contain the fin root temperature sensor readings for RTGs 1 through 4, respectively. Figure 2 depicts the evolution of the RTG 1 fin root temperature for Pioneer 10 (other fin root sensors exhibited similar behavior (Turyshev et al. 2006a)).

A best fit analysis confirms that the RTG temperature indeed evolves in a manner consistent with the radioactive decay of the nuclear fuel on board. The results for all the other RTGs on both spacecraft are similar, confirming that the RTGs were performing thermally in accordance with design expectations.

## 3.5 RTG Power Electrically Generated Heat

RTG electrical power can be estimated using two sensor readings per RTG, measuring RTG current and voltage. Currents for RTGs 1 through 4 appear as telemetry words $C_{127}$, $C_{105}$, $C_{114}$, and $C_{123}$, respectively; voltages are in telemetry words $C_{110}$, $C_{125}$, $C_{131}$, and $C_{113}$. Combined, these words yield the total amount of electrical power available on board:

$$P_E = C_{110}C_{127} + C_{125}C_{105} + C_{131}C_{114} + C_{113}C_{123}. \tag{2}$$

All this electrical power is eventually converted to waste heat by the spacecrafts' instruments, with the exception of power radiated away by transmitters.

Whatever remains of electrical energy (Fig. 3) after accounting for the power of the transmitted radio beam is converted to heat on-board. Some of it is converted to heat outside the spacecraft body by externally mounted components.

The Pioneer electrical system is designed to maximize the lifetime of the RTG thermocouples by ensuring that the current draw from the RTGs is always optimal. This means that power supplied by the RTGs may be more than that required for spacecraft operations. Excess electrical energy is absorbed by a shunt circuit that includes an externally mounted radiator plate. Especially early in the mission, when plenty of RTG power was still available, this radiator plate was the most significant component external to the spacecraft body that radiated heat. Telemetry word $C_{122}$ tells us the shunt circuit current, from which the amount of power dissipated by the external radiator can be computed using the known ohmic resistance ($\sim$5.25 $\Omega$) of the radiator plate.

Other externally mounted components that consume electrical power are the Plasma Analyzer ($P_{\mathrm{PA}} = 4.2$ W, telemetry word $C_{108}$ bit 2), the Cosmic Ray Telescope ($P_{\mathrm{CRT}} = 2.2$ W, telemetry word $C_{108}$, bit 6), and the Asteroid/Meteoroid Detector ($P_{\mathrm{AMD}} = 2$ W, telemetry

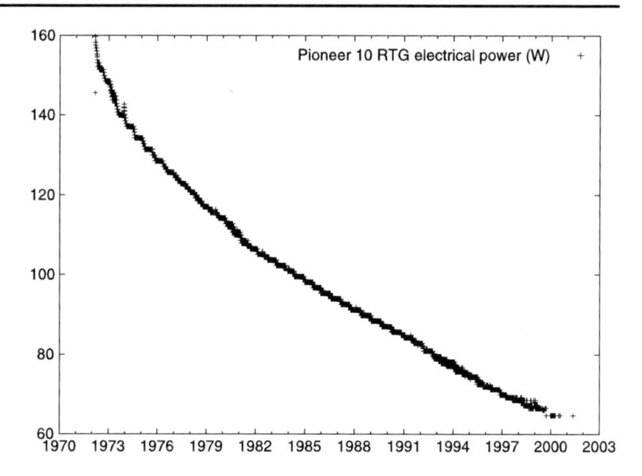


**Fig. 3** Changes in total RTG electrical output (in W) on board Pioneer, as computed using the mission's on-board telemetry

word $C_{124}$, bit 5). Though these instruments' exact power consumption is not telemetered, we know their average power consumption from design documentation, and the telemetry bits tell us when these instruments were powered.

Two additional external loads are the battery heater and the propellant line heaters. These represent a load of $P_{\rm LH} = P_{\rm BH} = 2$ W (nominal) each. The power state of these loads is not telemetered. According to mission logs, the battery heater was commanded off on both spacecraft on 12 May 1993.

Yet a further external load is the set of cables connecting the RTGs to the inverters. The resistance of these cables is known: it is 0.017 Ω for the inner RTGs (RTG 3 and 4), and 0.021 Ω for the outer RTGs (RTG 1 and 2). Using the RTG current readings it is possible to accurately determine the amount of power dissipated by these cables in the form of heat:

$$P_{\rm cable} = 0.017(C_{114}^2 + C_{123}^2) + 0.021(C_{127}^2 + C_{105}^2). \tag{3}$$

After accounting for all these external loads, whatever remains of the available electrical power on board is converted to heat inside the spacecraft. So long as the body of the spacecraft is in equilibrium with its surroundings, heat dissipated through its walls has to be equal to the heat generated inside:

$$P_{\rm body} = P_E - P_{\rm cable} - P_{\rm PA} - P_{\rm CRT} - P_{\rm AMD} - P_{\rm LH} - P_{\rm BH}, \tag{4}$$

with all the terms defined above.

### 3.6 The Thermal Control Subsystem

The passive thermal control system consisted of a series of spring-activated louvers (see Fig. 4). The springs were bimetallic, and thermally (radiatively) coupled to the electronics platform beneath the louvers. The louver blades were highly reflective in the infrared. The assembly was designed so that the louvers fully open when temperatures reach 30°C, and fully close when temperatures drop below 5°C.

The effective emissivity of the thermal blankets used on the Pioneers is $\varepsilon_{\rm sides} = 0.085$ (Toth and Turyshev 2006). The total exterior area of the spacecraft body is $A_{\rm walls} = 4.92$ m$^2$. The front side of the spacecraft body that faces the HGA has an area of $A_{\rm front} = 1.53$ m$^2$, and its effective emissivity, accounting for the fact that most thermal radiation this side emits is

New Horizons might provide a means to further investigate the Pioneer anomaly (Anderson et al. 2002).

There are some differences between the two spacecraft. New Horizons uses X-band frequencies for communication, which allow for better quality radiometric data. Also, during the New Horizons cruise phase to Pluto, the spacecraft is in "hibernation" mode, with low operational activity and minimal tracking.

The study of the anomalous acceleration of the Pioneer 10 and 11 spacecraft (Anderson et al. 2002; Toth and Turyshev 2006) demonstrates the importance of taking into account very small forces that might affect the accuracy of spacecraft navigation. We developed an approach to investigate this claim, thereby establishing a foundation for the study of the Pioneer anomaly using recently recovered extended set of radio-metric Doppler and flight telemetry data (Toth and Turyshev 2006, 2008).

During the course of our work, we learned some lessons that are directly applicable to New Horizons. Specifically, we found that the two main sources of thermal acceleration are asymmetrically reflected heat from the RTGs, and electrical heat generated inside the spacecraft's body.

Based solely on publicly available information sources, we can develop a preliminary estimate of these heat sources on the trajectory of New Horizons. We start with electrically generated heat.

### 4.1 Electrically Generated Heat

The New Horizons power source is a GPHS-RTG fueled with 61 fuel elements. The nominal electrical power of a GPHS-RTG fueled with the maximum of 72 fuel elements is 285 W, and its total power is ~4.3 kW. Thus, we calculate the electrical power of the New Horizons RTG at the time of launch as $P_E = 240$ W, and its total power, 3.36 kW. These values agree with values published in New Horizons Launch Press Kit (2006).

To maximize the lifetime of the thermocouple elements inside the RTG, the spacecraft is designed to always draw the optimal amount of current from the RTG; excess electrical power is directed to a shunt circuit. Shunted power may be dissipated in the form of heat either inside the spacecraft body or outside, depending on thermal conditions. This design suggests that the amount of electrical power converted to heat inside the craft body remains approximately constant throughout the mission, at least after it leaves the inner solar system and is no longer significantly heated by the Sun.

Accordingly, we assume that the amount of electrically generated heat inside the spacecraft body will be the nominal figure of 150 W, constant throughout the mission except for the very early phases, when solar heating is significant, and the very late phases, when RTG power drops to a low level.

Heat will leave the interior of the spacecraft body through its walls. The walls are generally covered with multilayer insulation, whose typical effective emissivity is on the order of $\varepsilon_w = 0.01$ (Stimpson and Jaworski 1972). One exception is the area covered by the louvers of the thermal control system. On Pioneer 10 and 11, the effective emissivity of the closed louvers is approximately $\varepsilon_l = 0.04$ (NASA Ames Research Center 1971). As we do not have a separate value for New Horizons, we shall use the Pioneer figure. The emissivity changes as the louvers open partially, but the system also becomes highly non-Lambertian, as the angled louvers preferentially reflect internal heat in one particular direction. Due to the complexities of this scenario, we ignore the case when louvers are partially open, i.e., when the spacecraft is still near the Sun.

We assume that the spacecraft walls and the louver blades are Lambertian emitters. The side walls will not contribute to acceleration; heat radiated by these walls will, on average, be emitted in a direction perpendicular to the spin axis. Similarly, the wall facing the back of the HGA will not contribute to acceleration either; as it is facing the highly (∼99%) reflective backside of the HGA, very little heat will leave the spacecraft body in this direction. Therefore, the amount of heat contributing to acceleration can be calculated as the ratio of heat leaving through the bottom of the spacecraft (which is partially covered by louvers) to the total heat leaving through the bottom and sides:

$$P_{E_{\rm accel}} = \frac{2}{3}\frac{\varepsilon_w(A_b - A_l) + \varepsilon_l A_l}{\varepsilon_w(A_b - A_l + A_s) + \varepsilon_l A_l} P_E, \tag{6}$$

where $P_E$ is the total electrically generated heat inside the spacecraft body, $P_{E_{\rm accel}}$ is the amount of electrically generated heat contributing to acceleration, $\varepsilon$ is the emissivity, and $A$ is the area of a surface ($b$: bottom, $s$: sides). The factor of $2/3$ is due to the assumption that the surfaces are Lambertian emitters.

No published data appears to exist on the actual size of the New Horizons louvers, but published "artist's renderings" appear to depict at least four louver assemblies with ten louver blades each. On Pioneer 10 and 11, a set of 30 louver blades covered an area of $\sim 0.3\ \mathrm{m}^2$; therefore, we shall assume that the area of the louver system on New Horizons is $A_l \simeq 0.4\ \mathrm{m}^2$. The surface area of the rest of the craft body can be calculated using low-resolution, but dimensionally correct drawings, leading to the estimate

$$P_{E_{\rm accel}} = 0.46 P_E = 68.4\ \mathrm{W}. \tag{7}$$

## 4.2 Heat from the RTG

Most of the power from the RTG is not converted to electricity but radiated away into space in the form of heat. The radiation pattern of the RTG is fore-aft symmetric, resulting in no net (average) acceleration force on a spinning spacecraft. Some of the heat, however, is reflected by the back of the HGA, resulting in an acceleration force.

A quick look at published drawings of the New Horizons spacecraft makes it clear that no parts of the spacecraft body stand in the way of thermal radiation emitted from the RTG in the direction of the back of the HGA. However, an approximately circular heat shield at the base of the RTG blocks some thermal radiation.

Any radiation that impacts the back of the HGA will transfer momentum to the spacecraft. Additionally, radiation that is reflected by the HGA transfers further momentum. To estimate the total amount of momentum transferred to the spacecraft this way, one needs to enumerate the following:

(i) $P_{\rm incident}$, the spin-axis component of thermal radiation incident on the back of the HGA, as a function of the angle between the ray of radiation and the spin axis;
(ii) $P_{\rm specular}$, the spin-axis component of thermal radiation specularly reflected by the back of the HGA, as a function of the angle of incidence at the HGA and the angle of the HGA surface normal at the point of incidence relative to the spin axis;
(iii) $P_{\rm diffuse}$, the spin-axis component of thermal radiation diffusely reflected by the back of the HGA, as a function of the angle between the HGA surface normal at the point of incidence relative to the spin axis.

To compute these quantities, we begin with the equation of heat transfer between two Lambertian surfaces (Toth and Turyshev 2009):

$$P_{1\to 2} = \iint \frac{\cos\theta_1 \cos\theta_2 \mathcal{P}_1}{4\pi r^2} dA_2 dA_1, \tag{8}$$

where $A_1$ and $A_2$ represent the surface of the emitting and the absorbing body, respectively; $r$ is the distance between the surface elements $dA_1$ and $dA_2$; $\theta_1$ and $\theta_2$ are the angles between the line connecting the two surface elements and their respective normals; and $\mathcal{P}_1$ is the emitted power density (power per unit area) at the surface $A_1$.

We can eliminate one of the double integrals by noting that the RTG is approximately isothermal along its length, and it is approximately cylindrically symmetrical (indeed, the hexagonal fin arrangement means that one can substitute a cylindrical body of a diameter that is the mean of unity and $\cos 30° = \sqrt{3}/2$, introducing an error no larger than ~7%; in actuality, as the fins are likely colder than the core of the RTG, the error will be even smaller). Put together, these considerations help us reduce (8) to

$$P_{\text{RTG}\to\text{antenna}} = P_{\text{RTG}} \iint \frac{\sin\beta\cos\theta}{4\pi r^2} dL dA, \tag{9}$$

where $L$ represents the length of the RTG and $A$ represents the portion of the back of the RTG illuminated by that point of the RTG. As the RTG is assumed to be lengthwise isothermal, we could also move $P_{\text{RTG}}$, denoting the RTG thermal power (total power minus power removed by the thermocouples in the form of electricity; ~3360 W at the beginning of mission), outside the integration sign. $\beta$ now denotes the angle between the ray of radiation and the lengthwise RTG axis, and $\theta$ is the angle of incidence relative to the RTG surface normal.

To obtain the spin-axis component of this incident ray of radiation, we need to further multiply by $\sin\beta$:

$$P_{\text{incident}} = P_{\text{RTG}} \iint \frac{\sin^2\beta\cos\theta}{4\pi r^2} dL dA. \tag{10}$$

For diffusely reflected radiation, we note that for a Lambertian reflector, momentum transferred will be in a direction perpendicular to the surface element, and it will be proportional to the incident radiation times $2\rho/3$, where $\rho$ denotes the reflectance. To calculate the spin-axis component of diffusely reflected radiation, we need to therefore compute

$$P_{\text{diffuse}} = \frac{2}{3}\rho(1-\sigma) P_{\text{RTG}} \iint \frac{\sin\beta\cos\gamma\cos\theta}{4\pi r^2} dL dA, \tag{11}$$

where $\gamma$ is the angle between the normal of the surface element $dA$ and the spin axis, and $\sigma$ is the ratio of specular vs. total reflected radiation.

Radiation that is specularly reflected will be emitted in accordance with the laws of geometric optics. Denoting the angle of specularly reflected radiation at surface element $dA$ by $\delta$, we can compute the spin axis component of specularly reflected radiation as

$$P_{\text{specular}} = \rho\sigma P_{\text{RTG}} \iint \frac{\sin\beta\cos\delta\cos\theta}{4\pi r^2} dL dA. \tag{12}$$

The values of $\beta$, $\gamma$, $\delta$, and $\theta$, as well as the integration limits for the surface integrals (i.e., the boundaries of the HGA area illuminated by the RTG) can be computed using geometric considerations. An examination of the geometry also tells us that approximately 20% of reflected radiation will be intercepted by the spacecraft body, and thus not contribute to thrust. The total amount of radiation that contributes to thrust in the direction of the spin axis can be summed as:

$$P_{\text{thrust}} = P_{\text{incident}} + 0.8(P_{\text{diffuse}} + P_{\text{specular}}). \tag{13}$$

Numerical evaluation of these integrals yields

$$P_{\text{thrust}} = 0.15 P_{\text{RTG}}, \tag{14}$$

as the amount of thermal power, as a function of total RTG power $P_{\text{RTG}}$, that will contribute to acceleration along the spin axis. At the beginning of mission, this translates into approximately 500 W of power contributing to acceleration.

While the approximations used in this section are clearly no substitute for the evaluation of a detailed finite element model of the spacecraft, the result indicates that anisotropic thermal radiation is a potentially significant source of acceleration for New Horizons during its long interplanetary cruise.

### 4.3 Acceleration Due to Emitted Heat

The relationship between the momentum and energy of electromagnetic radiation is well known: $p = E/c$. To calculate the acceleration due to collimated electromagnetic radiation of power $P$ emitted by a body of mass $m$, one can use the formula $a = P/(mc)$. The nominal mass at launch of the New Horizons spacecraft is about 465 kg (NHF 2005). Thus, we can calculate an anomalous acceleration for the New Horizons spacecraft due to thermal radiation from the RTG and electrical equipment as

$$a_{\text{NH}} = 41 \times 10^{-10}\,\text{m/s}^2 \simeq 4.7\,a_P. \tag{15}$$

When the spacecraft's 15 W transmitter is operating, the power of the radio beam (emitted in the direction opposite to the direction of thermal effects), which translates into an acceleration of $1 \times 10^{-10}\,\text{m/s}^2$, would have to be subtracted from our result.

The availability of the entire history of the Pioneer spacecraft makes it possible for us to calculate acceleration due to on-board forces to a significantly greater precision, and as a function of spacecraft parameters that evolve with time. The contribution of on-board forces to acceleration can be significant. Therefore, preserving and making available raw (engineering) telemetry of New Horizons to researchers is of great importance if accurate orbit determination is desired, and especially if conclusions are to be derived from any discrepancies between computed and observed orbits.

## 5 Conclusions

By 2009, the existence of the Pioneer anomaly is no longer in doubt. Our continuing effort to process and analyze Pioneer radio-metric and telemetry data is part of a broader strategy (Turyshev et al. 2005, 2006a).

Based on the information provided by the telemetry, we were able to develop a high accuracy thermal, electrical, and dynamical model of the Pioneer spacecraft. This model is used to investigate the anomalous acceleration and especially to study the contribution from the on-board thermal environment to the anomaly.

The available thermal model for the Pioneer spacecraft accounts for all heat radiation produced by the spacecraft. In fact, we use telemetry information to accurately estimate the amount of heat produced by the spacecrafts' major components. We are in the process of evaluating the amount of heat radiated in various directions.

This entails, on the one hand, an analysis of all available radio-metric data, to characterize the anomalous acceleration beyond the periods that were examined in previous studies.

Telemetry, on the other hand, enables us to reconstruct a thermal, electrical, and propulsion system profile of the spacecraft. Soon, we should be able to estimate effects on the motion of the spacecraft due to on-board systematic acceleration sources, expressed as a function of telemetry readings. This provides a new and unique way to refine orbital predictions and may also lead to an unambiguous determination of the origin of the Pioneer anomaly.

Concluding, we mention that before Pioneer 10 and 11, Newtonian gravity had never been measured—and was therefore never confirmed—with great precision over great distances. The unique "built-in" navigation capabilities of Pioneer 10 and 11 allowed them to reach the levels of $\sim 10^{-10}$ m/s$^2$ in acceleration sensitivity. Such an exceptional sensitivity means that Pioneer 10 and 11 represent the largest-scale experiment to test the gravitational inverse square law ever conducted. However, the experiment failed to confirm the validity of this fundamental law of Newtonian gravity in the outer regions of the solar system. One can demonstrate, beyond 15 AU the difference between the predictions of Newton and Einstein is negligible. So, at the moment, two forces seem to be at play in deep space: Newton's laws of gravity and the mysterious Pioneer anomaly. Until the anomaly is thoroughly accounted for by natural causes, and can therefore be eliminated from consideration, the validity of Newton's laws in the outer solar system will remain unconfirmed. This fact justifies the importance of the investigation of the nature of the Pioneer anomaly.

**Acknowledgements** This work was partially performed at the International Space Science Institute (ISSI), Bern, Switzerland, when both of us visited ISSI as part of an International Team program. In this respect we thank Roger M. Bonnet, Vittorio Manno, Brigitte Schutte and Saliba F. Saliba of ISSI for their hospitality and support. We especially thank The Planetary Society for support and, in particular, Louis D. Freidman, Charlene M. Anderson, and Bruce Betts for their interest, stimulating conversations and encouragement.

The work of SGT was carried out at the Jet Propulsion Laboratory, California Institute of Technology, under a contract with the National Aeronautics and Space Administration.

## References

J.D. Anderson, P.A. Laing, E.L. Lau, A.S. Liu, M.M. Nieto, S.G. Turyshev, Phys. Rev. Lett. **81**, 2858 (1998). gr-qc/9808081

J.D. Anderson, P.A. Laing, E.L. Lau, A.S. Liu, M.M. Nieto, S.G. Turyshev, Phys. Rev. D **65**, 082004/1-50 (2002). gr-qc/0104064

Final Environmental Impact Statement for the New Horizons Mission. NASA, 2005

J.I. Katz, Phys. Rev. Lett. **83**, 1892 (1999). gr-qc/9809070

C.B. Markwardt, Independent confirmation of the Pioneer 10 anomalous acceleration. gr-qc/0208046

E.M. Murphy, Phys. Rev. Lett. **83**, 1890 (1999). gr-qc/9810015

NASA Ames Research Center, Pioneer F/G: Spacecraft Operational Characteristics. PC-202 (1971)

New Horizons Launch Press Kit January 2006, NASA, 2006, for details consult http://www.nasa.gov/pdf/139889main_PressKit12_05.pdf

M.M. Nieto, S.G. Turyshev, Class. Quantum Gravity **21**, 4005 (2004). gr-qc/0308017

L.K. Scheffer, Phys. Rev. D **67**, 084021 (2003). gr-qc/0107092

L. Stimpson, W. Jaworski, in *AIAA 7th Thermophysics Conference, San Antonio, Texas*, AIAA paper # 72–285 (1972)

V.T. Toth, S.G. Turyshev, Can. J. Phys. **84**(12), 1063 (2006). gr-qc/0603016

V.T. Toth, S.G. Turyshev, AIP Conf. Proc. **977**, 264 (2008). arXiv:0710.2656 [gr-qc]

V.T. Toth, S.G. Turyshev, Phys. Rev. D **79**, 043011 (2009). arXiv:0901.4597 [physics.space-ph]

V.T. Toth, Independent analysis of the orbits of Pioneer 10 and 11, to be published in Int. J. Mod. Phys. D (2009). arXiv:0901.3466 [physics.space-ph]

S.G. Turyshev, J.D. Anderson, P.A. Laing, E.L. Lau, A.S. Liu, M.M. Nieto, The apparent anomalous, weak, long-range acceleration of Pioneer 10 and 11, in *Gravitational Waves and Experimental Gravity, Proc. of the XVIIIth Workshop of the Rencontres de Moriond*, ed. by J. Dumarchez, J. Tran Thanh Van. Les Arcs, Savoi, France, 23–30 January 1999 (World Publishers, Hanoi, 2000), pp. 481–486. gr-qc/9903024

S.G. Turyshev, M.M. Nieto, J.D. Anderson, Am. J. Phys. **73**, 1033 (2005). physics/0502123

S.G. Turyshev, M.M. Nieto, J.D. Anderson, A route to understanding of the Pioneer anomaly, in *The XXII Texas Symposium on Relativistic Astrophysics*, ed. by P. Chen, E. Bloom, G. Madejski, V. Petrosian. Stanford University, 13–17 December 2004, SLAC-R-752, Stanford e-Conf #C041213, paper #0310. eprint: http://www.slac.stanford.edu/econf/C041213/. arXiv:gr-qc/0503021
S.G. Turyshev, V.T. Toth, L.R. Kellogg, E.L. Lau, K.J. Lee, Int. J. Mod. Phys. D **15**(1), 1 (2006a). gr-qc/0512121
S.G. Turyshev, M.M. Nieto, J.D. Anderson, EAS Publ. Ser. **20**, 243 (2006b). gr-qc/0510081

Space Sci Rev (2009) 148: 169–174
DOI 10.1007/s11214-009-9571-0

# The Puzzle of the Flyby Anomaly

**Slava G. Turyshev · Viktor T. Toth**

Received: 20 July 2009 / Accepted: 23 July 2009 / Published online: 5 August 2009
© Springer Science+Business Media B.V. 2009

**Abstract** Close planetary flybys are frequently employed as a technique to place spacecraft on extreme solar system trajectories that would otherwise require much larger booster vehicles or may not even be feasible when relying solely on chemical propulsion. The theoretical description of the flybys, referred to as gravity assists, is well established. However, there seems to be a lack of understanding of the physical processes occurring during these dynamical events. Radio-metric tracking data received from a number of spacecraft that experienced an Earth gravity assist indicate the presence of an unexpected energy change that happened during the flyby and cannot be explained by the standard methods of modern astrodynamics. This puzzling behavior of several spacecraft has become known as the flyby anomaly. We present the summary of the recent anomalous observations and discuss possible ways to resolve this puzzle.

**Keywords** Flyby anomaly · Gravitational experiments · Spacecraft navigation

## 1 Introduction

Significant changes to a spacecraft's trajectory require a substantial mass of propellant. In particular, placing a spacecraft on a highly elliptical or hyperbolic orbit, such as the orbit required for an encounter with another planet, requires the use of a large booster vehicle, substantially increasing mission costs. An alternative approach is to utilize a gravitational assist from an intermediate planet that can change the direction of the velocity vector. Although such an indirect trajectory can increase the duration of the cruise phase of a mission,

S.G. Turyshev (✉)
Jet Propulsion Laboratory, California Institute of Technology, 4800 Oak Grove Drive, Pasadena, CA 91109, USA
e-mail: turyshev@jpl.nasa.gov

V.T. Toth
Ottawa, ON K1N 9H5, Canada
e-mail: vttoth@vttoth.com

the technique nevertheless allowed several interplanetary spacecraft to reach their target destinations economically (Anderson 1997; Van Allen 2003).

Notable missions[1] that relied on an Earth gravity assist maneuver and are relevant to the main topic of this paper include Galileo,[2] which had two encounters with the Earth and one each with Venus and an asteroid to reach Jupiter more quickly; the Near Earth Asteroid Rendezvous[3] (NEAR Shoemaker) mission; the Cassini mission[4] with encounters with Venus, Earth, and Jupiter to speed it on its way to Saturn; and the European Space Agency's Rosetta mission[5] en route to an encounter with the comet 67 P/Churyumov-Gerasimenko.

However, during the Earth flybys, these missions experienced an unexpected navigational anomaly. In the following, we discuss the nature of gravity assist maneuvers, characterize the flyby anomalies experienced by these spacecraft, and discuss the challenges that one faces in attempting to find an explanation of this effect.

## 2 Gravity Assist Maneuvers

A gravity assist maneuver is a specific application of the restricted three body problem, in which an effectively massless test particle (such as a spacecraft) moves in the combined gravitational field of two larger bodies. When the larger bodies move in circular orbits, the problem is known as the circular restricted three-body problem, or Euler's three-body problem (Euler 1760a, 1760b, 1760c), among other names. In this problem, the energy and momentum of the test particle are not conserved, although other conserved quantities exist. The energy gain or loss by the test particle is offset by a corresponding loss (gain) in energy by the two larger bodies in the system, however, due to the differences in mass, the corresponding changes in the larger bodies' velocities are not perceptible.

The circular restricted three-body problem is exactly solvable: after a suitable set of generalized coordinates are chosen, the solution can be expressed in the form of elliptic integrals (Whittaker 1937). The problem has also been analyzed by use of the method of patched conics (Breakwell et al. 1961; Battin 1987), which notionally patches two conics together at the trajectory's intersection with the sphere of influence surrounding the smaller mass. As well, the problem can be addressed by numerically integrating the equations of motion using a suitably chosen integration method of sufficient accuracy.

However, precision calculation of the trajectory of a spacecraft in the vicinity of a planet requires detailed analysis that takes into account all the effects including a complicated gravitational potential (usually represented in the form of spherical harmonics), perturbations due to the gravitational influence of the planet's moons, if any, the pressure of light and thermal radiation received from the Sun and the planet, drag forces that may be present

[1] Several additional missions have used planetary assists to reach their target destinations, including Mariner 10 (Venus and Mercury), Pioneer 10 and 11 (Jupiter and Saturn), and also Voyager 1 and 2, which used gravity assists from Jupiter to reach Saturn. Voyager 2 continued to Uranus and Neptune, using the gravity assist of each planetary encounter to target the spacecraft to the next planet. The most feasible plans of space missions inward toward the Sun (such as the Ulysses mission that used a Jupiter flyby to form a trajectory outside the ecliptic plane) and outward to Pluto (such as the New Horizons mission that used a Jupiter flyby to increase significantly the craft's velocity) depend on gravitational assists from Jupiter.

[2] http://www2.jpl.nasa.gov/galileo/.

[3] http://near.jhuapl.edu/.

[4] http://saturn.jpl.nasa.gov/.

[5] http://sci.esa.int/science-e/www/area/index.cfm?fareaid=13.

in the planet's upper atmosphere, and on-board events such as thruster firings. The precision with which a trajectory can be computed is determined by the accuracy with which all these effects can be accounted for. When sufficient accuracy is achieved, this results in a diminishing difference between computed an observed values of tracking observables (e.g., radio-metric Doppler and ranging data.) The methods and tools used for high-precision navigation have seen major improvements (Moyer 2003).

In particular, when a spacecraft receives a gravity assist from the Earth, its trajectory must be computed after taking into account the gravitational effects of the Earth and Moon, and nongravitational effects in the near Earth environment, including the upper atmosphere, thermal recoil forces generated on-board due to heat dissipation processes, etc. The dynamics of an Earth flyby are well understood, and are a subject of continuous study thanks to the large number of Earth-orbiting satellites, in particular as a result of the enormous success of the GRACE mission, which led to the construction of the GRACE Gravity Model[6] (Tapley et al. 2005).

Nonetheless, during the Earth flyby of several spacecraft, a small, anomalous increase in velocity was observed (Anderson et al. 2008). Although these anomalies were not mission critical events and did not prevent the spacecraft to proceed to their ultimate destinations, the puzzle of the flyby anomaly, as this effect became known, remains unresolved.

## 3 The Earth Flyby Anomaly

Several spacecraft, including Galileo, NEAR, Cassini, and Rosetta, utilized Earth flybys to achieve their desired trajectories. For all four of these flybys an anomalous change in the modeled flyby velocity is required (Anderson et al. 2007) in order to fit the Deep Space Network (DSN) Doppler and ranging data. This increase can be represented by a fictitious trajectory maneuver at perigee, or it can be demonstrated by fitting the pre-encounter data, and subsequently using the resulting trajectory to predict the post-encounter data. The difference between the actual post-encounter data and the predicted data is consistent with the velocity change determined from the fictitious maneuver.

This anomalous trajectory behavior was first noted after the first Earth gravity assist for the Galileo spacecraft on 8 December 1990 (Fig. 1). The increase in the asymptotic velocity $V_\infty$ of its hyperbolic trajectory was $\Delta V_\infty = 3.92 \pm 0.08$ mm/s. The efforts by the Galileo navigation team and the Galileo radio science team to find a cause for this anomaly were unsuccessful. Two years later, during Galieo's second Earth flyby on 8 December 1992, the Tracking and Data Relay Satellite (TDRS) system was scheduled to track the spacecraft (Edwards et al. 1993). This time the perigee altitude was lower, 303 km vs. 960 km. Consequently, any anomalous velocity increase was masked by atmospheric drag. However, results were published for the two Galileo flyby navigational anomalies (Edwards et al. 1994).

After the Earth flyby by the NEAR spacecraft on 23 January 1998, at an altitude of 539 km, the anomalous velocity increase was observed once again, with a magnitude of $\Delta V_\infty = 13.46 \pm 0.13$ mm/s. Results were presented at a spaceflight conference (Antreasian and Guinn 1998), including a reanalysis of the two Galileo flybys, and with all three flybys based on the best Earth gravity field available in August 1998.

Subsequently, the Earth flybys of additional spacecraft were studied, in order to establish if an anomalous velocity change is present in the tracking data. When Cassini flew by

[6]For details on the GRACE mission and the GRACE Gravity Model 02 (GGM02), please consult http://www.csr.utexas.edu/grace/gravity/.

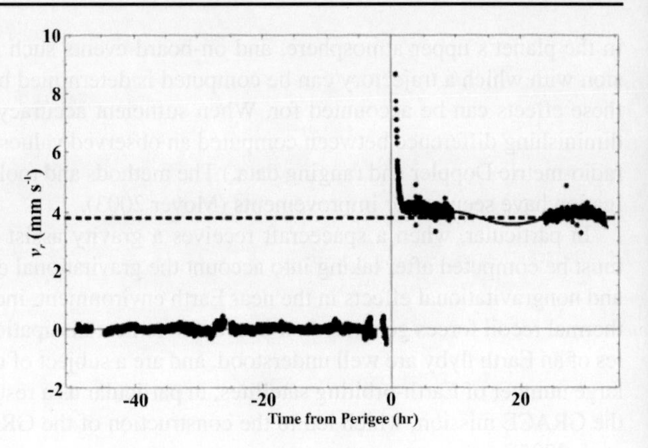


**Fig. 1** Doppler residuals (observed minus computed Doppler frequency, converted into units of line-of-sight velocity) during Galileo's December 8, 1990 Earth flyby (Anderson and Nieto 2009). Residuals are based on a pre-perigee fit to radio-metric Doppler data, and demonstrate the anomalous post-perigee velocity increase

**Table 1** Summary of Earth flyby spacecrafts, based on Anderson et al. (2008)

| Quantity | Galileo | NEAR | Cassini | Rosetta | Messenger |
|---|---|---|---|---|---|
| Speed at infinity, km/s | 8.949 | 6.851 | 16.01 | 3.863 | 4.056 |
| Minimal altitude, km | 956.053 | 532.485 | 1171.505 | 1954.303 | 2336.059 |
| Spacecraft mass, kg | 2497.1 | 730.40 | 4612.1 | 2895.2 | 1085.6 |
| $\Delta V_\infty$, mm/s | $3.92 \pm 0.08$ | $13.46 \pm 0.13$ | – | $1.82 \pm 0.05$ | – |
| Gained energy, J/kg | $35.1 \pm 0.7$ | $92.2 \pm 0.9$ | – | $7.03 \pm 0.19$ | – |
| Inclination to equator | 142.9° | 108.8° | 25.4° | 144.9° | 133.1° |
| Deflection angle | 47.46° | 66.92° | 19.66° | 99.396° | 94.7° |

the Earth on August 18, 1999 at an altitude of 1175 km, thruster maneuvers prevented the immediate detection of any anomalous behavior. However, subsequent analysis showed an anomalous velocity change of $2 \pm 1$ mm/s in magnitude.

The navigation team of the European Space Agency's Rosetta spacecraft, analyzing the spacecraft's trajectory during its Earth flyby on 4 March 2005, also observed an anomalous change in that spacecraft's velocity (Morley and Budnik 2006), at $\Delta V_\infty = 1.80 \pm 0.03$ mm/s, as that spacecraft flew by the Earth at an altitude of 1956 km.

However, during the subsequent flyby by the MESSENGER spacecraft on 2 August 2005, at an altitude of 2347 km, no anomalous velocity change was seen.

Attempts to find this anomalous orbital energy increase for flybys of other planets have failed so far because the gravity fields are not known nearly as well as for the Earth. Perhaps in the future, models of the gravity fields of Venus and Mars, determined from orbiter data, could become sufficiently accurate that DSN Mars and Venus flyby data from earlier NASA missions could be used to search for an anomalous $\Delta V_\infty$. These old DSN data can be retrieved from NASA's National Space Science Data Center (NSSDC) archive, where they were deposited by various radio science teams.

The results from Galileo, NEAR, Cassini, and Rosetta, as well as the null result from MESSENGER, summarized in Table 1, suggest that a relationship may exist between the magnitude of the flyby anomaly and the altitude and geometry of the flyby trajectory. Indeed, an empirical relationship was proposed by Anderson et al. (2008). While the reported anomalous energy changes were claimed to be consistent with a simple empirical formula that fits most of the data in Table 1, the physical origin of this expression is yet unclear.

Furthermore, the data seem to indicate that the search for a phenomenological model is yet unfinished.

## 4 Conclusions

Multiple experiments performed by different teams using a variety of spacecraft have firmly established the existence of an anomalous effect during Earth flybys. The cause of this effect remains unknown. Lämmerzahl et al. (2006) have considered, and dismissed, a variety of possible conventional causes, including effects due to the atmosphere, tides, electric charge and magnetic moment of the spacecraft, radiation pressure due to the Earth's albedo, solar wind, and spin-rotation coupling. Antreasian and Guinn (1998) have earlier dismissed explanations involving the Moon, the Sun, and relativistic effects.

Another possibility is the incorrect modeling of the flybys and their realization in navigational software. We must emphasize the need for a consistent treatment of the gravitational assist problem from the standpoint of the modern theory of astronomical reference frames. Of particular concern is the theoretical treatment and practical realization of the concept of spheres of influence in the software codes used to navigate the spacecraft (Moyer 2003). In this respect, an appropriate treatment must properly describe the dynamical and signal propagation effects relevant to the transition between various reference frames (Soffel et al. 2003) in the overlapping region. For instance, in the case of the Earth flyby problem these frames are the solar system barycentric and geocentric frames. Appropriate values for astronomical constants must be used to model the spacecraft motion when it transits from the sphere of influence of one body to that of another. Finally, one must also make sure that all the ancillary data derived by external means (such as VLBI, lunar laser ranging, spacecraft radar etc.) and used to describe solid-Earth effects (namely Earth's precession, nutation, sidereal rotation, polar motion, tidal effects, tectonic plates drift, etc.) have consistent definitions within the overlapping region between the frames used to treat the flyby problem, which currently is not the case.

It is tempting to consider the flyby anomaly in conjunction with other spacecraft navigation anomalies, including the Pioneer anomaly (Anderson et al. 2002), and the anomalous increases in the Astronomical Unit (AU) and the eccentricity of the Moon's orbit (Anderson and Nieto 2009). However, there is no direct evidence that these anomalies are related, and at least in the case of the Pioneer anomaly, the possibility of a conventional explanation cannot be dismissed (Turyshev et al. 2005; Toth and Turyshev 2009; Anderson and Nieto 2009).

Nonetheless, the study of the Pioneer anomaly teaches a very valuable lesson concerning the preservation of old navigational and other spacecraft data. As discussed above, it may be possible to re-analyze radio tracking data from past Mars and Venus flybys once more accurate models of the gravitational fields of these planets become available. Such a re-analysis will only be possible if the archived tracking data is still available. So long as the possibility exists that the flyby anomaly has an origin in on-board systematics, it is also advisable to preserve the engineering telemetry record of the spacecraft in question.

The work outlined above is important but rather tedious. Yet it must be done before any discussion of new physics as an explanation for the flyby anomaly can seriously take place. To that extent, a continuing effort is needed to better characterize the anomaly, using both past and future flyby data, similar to the on-going efforts to study the Pioneer anomaly (Turyshev et al. 2005). Some of this work has already begun, for example in the context of the International Flyby Collaboration[7] at the International Space Science Institute (ISSI). However, significant additional effort is required. This paper intends to motivate initiation of such a work in the near future.

By 2009, the existence of the flyby anomaly has been confirmed by a variety of teams with several spacecraft. While it is unlikely that the origin of the flyby anomaly will prove to be anything other conventional in nature, because of its importance for high-precision navigation further investigations into the physics of this puzzling effect are certainly justified.

**Acknowledgements** The work of SGT was carried out at the Jet Propulsion Laboratory, California Institute of Technology, under a contract with the National Aeronautics and Space Administration.

## References

J.D. Anderson, in *Encyclopedia of Planetary Sciences*, ed. by J.H. Shirley, R.W. Fairbridge (Chapman and Hall, New York, 1997), pp. 287–289

J.D. Anderson, M.M. Nieto, in *Proc. of the IAU Symposium 261* (2009, to be published). arXiv:0907.2469

J.D. Anderson, P.A. Laing, E.L. Lau, A.S. Liu, M.M. Nieto, S.G. Turyshev, Phys. Rev. D **65**, 082004/1-50 (2002). gr-qc/0104064

J.D. Anderson, J.K. Campbell, M.M. Nieto, New Astron. Rev. **12**, 383 (2007)

J.D. Anderson, J.K. Campbell, J.E. Ekelund, J. Ellis, J.F. Jordan, Phys. Rev. Lett. **100**, 091102 (2008)

P.G. Antreasian, J.R. Guinn, AIAA Paper No 98-4287 presented at the AIAA/AAS Astrodynamics Specialist Conference and Exhibit, Boston, 10–12 August 1998, and available at http://www.issibern.ch/teams/Pioneer/pa-literature.htm

R.H. Battin, *An introduction to the Mathematics and Methods of Astrodynamics* (American Institute of Aeronautics and Astronautics (AIAA), Washington, 1987)

J.V. Breakwell, R.W. Gillespie, S. Ross, ARS J. **31**, 201–208 (1961)

C. Edwards et al., *Tracking Galileo at Earth-2 Perigee Using the Tracking and Data Relay Satellite System. Publication 93-0165* (Jet Propulsion Laboratory, Pasadena, 1993). Available at http://hdl.handle.net/2014/34792

C. Edwards et al., *Astrodynamics 1993*. Advances in the Astronautical Sciences, ed. by A.K. Misra et al. (Univelt, San Diego, 1994), p. 1609

L. Euler, Nov. Commun. Acad. Imp. Petropolitanae **10**, 207–242 (1760a)

L. Euler, Nov. Commun. Acad. Imp. Petropolitanae **11**, 152–184 (1760b)

L. Euler, Mém. Acad. Berl. **11**, 228–249 (1760c)

C. Lämmerzahl, O. Preuss, H. Dittus, in *Lasers, Clocks, and Drag-Free Control: Exploration of Relativistic Gravity in Space*, ed. by H. Dittus, C. Laemmerzahl, S. Turyshev (Springer, Berlin, 2006), pp. 75–101. gr-qc/0604052

T. Morley, F. Budnik, in *Proceedings of the 19th International Symposium on Space Flight Dynamics 2006*, Paper ISTS 2006-d-52

T.D. Moyer, *Formulation for Observed and Computed Values of Deep Space Network Data Types for Navigation*. JPL Deep-Space Communications and Navigation Series (Wiley, Hoboken, 2003)

M. Soffel, S.A. Klioner, G. Petit, P. Wolf, S.M. Kopeikin, P. Bretagnon, V.A. Brumberg, N. Capitaine, T. Damour, T. Fukushima, B. Guinot, T.-Y. Huang, L. Lindegren, C. Ma, K. Nordtvedt, J.C. Ries, P.K. Seidelmann, D. Vokrouhlický, C.M. Will, C. Xu, Astron. J. **126**, 2687–2706 (2003)

B. Tapley, J. Ries, S. Bettadpur, D. Chambers, M. Cheng, F. Condi, B. Gunter, Z. Kang, P. Nagel, R. Pastor, T. Pekker, S. Poole, F. Wang, J. Geod. **79**(8), 467–478 (2005)

V.T. Toth, S.G. Turyshev, Phys. Rev. D **79**, 043011 (2009). arXiv:0901.4597 [physics.space-ph]

S.G. Turyshev, V.T. Toth, L.R. Kellogg, E.L. Lau, K.J. Lee, Int. J. Mod. Phys. D **15**, 1–55 (2005). gr-qc/0512121

J.A. Van Allen, Am. J. Phys. **71**(5), 448–451 (2003)

E.T. Whittaker, in *A Treatise on the Analytical Dynamics of Particles and Rigid Bodies, with an Introduction to the Problem of Three Bodies*, 4th edn. (Dover, New York, 1937), pp. 97–99

[7]For more information on the International Flyby Collaboration at ISSI, please visit team's website at http://www.issibern.ch/teams/investflyby/.

Space Sci Rev (2009) 145: 285–335
DOI 10.1007/s11214-009-9518-5

# Time-Dependent Nuclear Decay Parameters: New Evidence for New Forces?

**E. Fischbach · J.B. Buncher · J.T. Gruenwald · J.H. Jenkins · D.E. Krause · J.J. Mattes · J.R. Newport**

Received: 14 April 2009 / Accepted: 29 April 2009 / Published online: 28 May 2009
© Springer Science+Business Media B.V. 2009

**Abstract** This paper presents an overview of recent research dealing with the question of whether nuclear decay rates (or half-lives) are time-independent constants of nature, as opposed to being parameters which can be altered by an external perturbation. If the latter is the case, this may imply the existence of some new interaction(s) which would be responsible for any observed time variation. Interest in this question has been renewed recently by evidence for a correlation between nuclear decay rates and Earth–Sun distance, and by the observation of a dip in the decay rate for $^{54}$Mn coincident in time with the solar flare of 2006 December 13. We discuss these observations in detail, along with other hints in the literature for time-varying decay parameters, in the framework of a general phenomenology that we develop. One consequence of this phenomenology is that it is possible for different experimental groups to infer discrepant (yet technically correct) results for a half-life depending on where and how their data were taken and analyzed. A considerable amount of attention is devoted to possible mechanisms which might give rise to the reported effects, including fluctuations in the flux of solar neutrinos, and possible variations in the magnitudes of fundamental parameters, such as the fine structure constant and the electron-to-proton mass ratio. We also discuss ongoing and future experiments, along with some implications of our work for cancer treatments, $^{14}$C dating, and for the possibility of detecting the relic neutrino background.

**Keywords** Solar physics: Particle emission · Solar physics: Flare · Nuclear reactions: Fluctuation phenomena

E. Fischbach (✉) · J.B. Buncher · J.T. Gruenwald · J.H. Jenkins · J.J. Mattes · J.R. Newport
Physics Department, Purdue University, West Lafayette, IN 47906, USA
e-mail: ephraim@physics.purdue.edu

D.E. Krause
Physics Department, Wabash College, Crawfordswille, IN 47933, USA

## 1 Introduction

Few issues frame the history of natural radioactivity as fundamentally as the question of whether the decay rates of nuclides are constants of nature, unaffected by the external environment. Following the discovery of radioactivity by Becquerel (1896), an intense effort was mounted to ascertain whether the decay rates of nuclides could be affected by external influences including temperature, pressure, chemical composition, concentration, and magnetic fields. One example of this effort is an experiment carried out by Rutherford (1913, p. 505) who contained a quantity of radon gas in a high pressure bomb along with the explosive cordite. The authors estimated that when the cordite was detonated the maximum temperature in the bomb reached 2500°C, and the pressure $\sim$1000 atm, and yet the $\gamma$-ray activity from the radon was unchanged. By 1930 Rutherford et al. (1930, p. 161) concluded that "The rate of transformation of an element has been found to be a constant under all conditions." The object of the present paper is to revisit this question in light of recent suggestions that some radioactive decays may in fact be affected by solar activity and/or by other external influences.

In order to define the question at hand more precisely it is useful to distinguish among the following modes of nuclear decay, which will be the focus of our discussion; $\alpha$, $\beta^{\pm}$, and $\epsilon$ (electron capture). It is not surprising to find that electron capture rates depend on the environment of the atom: these rates depend on the overlap of the wavefunction of an orbital electron with the nucleus (Wu and Moszkowski 1966, p. 195ff) which can be affected by the local chemical or physical environment of the decaying atom (Emery 1972; Hahn et al. 1976; Norman et al. 2001; Ohtsuki et al. 2004). Similarly, $\beta$-decay rates in stars can be influenced by very strong magnetic fields which alter the wavefunctions of the emitted $\beta$-particles, and hence the phase space available for the decays (Matese and O'Connell 1969; Fassio-Canuto 1969). In the present paper we will not deal with either of these cases. In the ensuing discussion we will, however, consider in some detail the electron capture process $^{54}\mathrm{Mn} + e^- \rightarrow {}^{54}\mathrm{Cr} + \nu_e$, but only in connection with data we obtained for this decay during the solar flare of 2006 December 13.

This work is an outgrowth of an effort to apply the GRIP randomness test (Tu and Fischbach 2003; Tu and Fischbach 2005) to nuclear decays. Although it is almost universally assumed that nuclear decays are random, experimental tests of this assumption are relatively scant (Anderson and Spangler 1973; Silverman et al. 1999; Silverman et al. 2000). In the course of designing an appropriate experiment based on the GRIP formalism, we came upon a paper by Alburger et al. (1986) which revealed an unexpected annual variation in the decay rates of $^{32}$Si and $^{36}$Cl (see Sect. 2 below). This experiment, which was carried out at Brookhaven National Laboratory (BNL) between 1982 and 1986, questioned our understanding of nuclear decays, particularly our belief that these decays are uncorrelated in time with any other influence. Further exploration of the literature revealed yet another data set which exhibited an annual variation in decay rates, this from an experiment studying $^{152}$Eu, $^{154}$Eu, and $^{226}$Ra at the Phyikalische-Technische Bundesanstalt (PTB) in Germany. Other researchers have also reported time-varying count-rates, and Table 1 provides a guide to some of these experiments. As we discuss in Sect. 2, the annual variation of the BNL and PTB data, along with the correlation of these data sets with each other, raised the question of whether these decay rates (and possibly others as well) were in fact being influenced by an external source, or whether they simply represented some poorly understood instrumental effects.

The unexpected correlations observed in the BNL and PTB data have served as the motivation for an ongoing series of experiments at Purdue. In one of these, data taken

**Table 1** Summary of experiments suggesting a possible time-dependence of nuclear decay rates. For each entry the observed nuclides and their dominant decay modes are exhibited along with the approximate dates when the corresponding experiments were carried out. The decay modes are indicated by $\alpha$ (alpha-decay), $\beta$ (beta-decay), and $\epsilon$ (electron-capture). Observed periodicities in the decay rates are noted. The interested reader is referred to the references for further details

| Source | Mode | Duration | Periodicity (d) | Reference |
|---|---|---|---|---|
| $^{3}$H | $\beta^-$ | | 0.5 y | (Lobashev et al. 1999)[#] |
| $^{3}$H | $\beta^-$ | Fall 1980–Spring 1982 | 365 | (Falkenberg 2001)* |
| $^{32}$Si | $\beta^-$ | 1982–1986 | 365 | (Alburger et al. 1986) |
| $^{60}$Co | $\beta^-$ | 1999.06–2001.12 | 1, ~30, 365 | (Parkhomov 2005) |
| $^{60}$Co | $\beta^-$ | 1998.12–1999.04 | 1, 27 | (Baurov et al. 2007) |
| $^{60}$Co | $\beta^-$ | 2000.03–2000.04 | | (Baurov et al. 2001) |
| $^{137}$Cs | $\beta^-$ | 1998.12–1999.04 | 1, 27 | (Baurov et al. 2007) |
| $^{137}$Cs | $\beta^-$ | 2000.03–2000.04 | | (Baurov et al. 2001) |
| $^{226}$Ra | $\alpha$ | 1981–1996 | 365 | (Siegert et al. 1998) |
| $^{238}$Pu | $\alpha$ | 1978–1980; 1982–7 | 365 | (Ellis 1990) |
| Various | $\alpha, \beta, \epsilon$ | | 1, 27, 365 | (Shnoll et al. 1998, 2000)[†] |

*Indicates comments on this paper by Bruhn (2002) and Falkenberg (2002), and the †indicates comments by Derbin et al. (2000) and Kushnirenko and Pogozhev (2000). The #denotes that Lobashev et al. (1999) report a periodic effect ("Troitsk anomaly") in the determination of $m_{\nu}^{2}$ from $^{3}$H $\beta$-decay. A recent paper by Kostenko and Yuriev (2008) discusses a possible connection between some of the time-dependence of half-lives reported in this table and magnetic monopoles.

during the solar flare of 2006 December 13 exhibited a dip in the counting rate of a $^{54}$Mn sample which coincided in time with the flare. These data are presented and analyzed in Sect. 3. They serve as part of the motivation for the remainder of this paper, where we explore the phenomenology of a time-dependent $\lambda$, which is defined by the decay law $\dot{N}(t) \equiv dN(t)/dt = -\lambda(t)N(t)$. Note that the unperturbed half-life $T_{1/2}$ is given by $T_{1/2} = \ln 2/\lambda$. We will henceforth avoid the awkward oxymoronic construction "time-dependent decay constant" by referring to $\lambda$ as the "decay parameter".

It must be emphasized that even if the correlations reported in the BNL and PTB data prove to be reproducible, and similarly for the solar flare data, this does not necessarily imply that they arise from a time-dependent modification of $\lambda$, as described by the phenomenology in Sect. 4 below. The observed effects could arise from a field or particles emanating from the Sun which modify the experimental apparatus rather than $\lambda$, or from some conventional, but overlooked, influence on the apparatus arising from local fluctuations in temperature, pressure, humidity, etc. As we will discuss below, all of these alternatives remain viable at present.

The suggestion that the decay parameter $\lambda$ may be time-dependent derives not only from the BNL, PTB, and solar flare data cited above (and discussed below), but also from a number of earlier experiments which are summarized in Table 1. In addition to the experiments cited there, which report direct evidence for time-varying decay parameters, there may be indirect indications of varying decay parameters if one takes seriously apparent discrepancies arising from half-life determinations of some nuclides by different groups. Although it would be reasonable to attribute these discrepancies to unspecified systematic effects, the possibility that these arise from a common source is suggested by the fact that many such discrepancies are cited in the literature in experiments carried out by experienced, well-known groups (Alburger et al. 1986; Begemann et al. 2001; Chiu et al. 2007; Pommé 2007; Pommé et al. 2008).

We conclude by outlining the present paper. As noted above, Sect. 2 contains the discussion of the BNL and PTB correlations, and Sect. 3 presents the data from the solar flare of 2006 December 13. In Sect. 4 we develop the phenomenology of a time-varying decay parameter $\lambda(t)$, and discuss some experimental implications. One of the main objectives of the present paper is to discuss possible mechanisms through which solar activity could affect nuclear decays, and this discussion for $\beta$-decays is presented in Sect. 5, along with a discussion of the relevant $\beta$-decay phenomenology. This section also discusses some of the constraints on mechanisms which aim to explain the data in Sects. 2 and 3 in terms of new interactions. In Sect. 6 we present a brief discussion of constraints on possible mechanisms to explain time-varying decay parameters in $\alpha$-decay. Following the appearance of our original papers on the BNL/PTB data (Jenkins et al. 2008), and on the solar flare data (Jenkins and Fischbach 2008), additional data came to light which both questioned and supported our results, and these are discussed in Sect. 7. In Sect. 8 we comment on the implications of this work to $^{14}$C dating. In Sect. 9 we present a summary of present and future experiments testing the constancy of nuclear decay parameters, and in Sect. 10 we discuss the implications of this work for possibly detecting the relic neutrino background. We conclude by summarizing our results in Sect. 11.

## 2 Evidence for Correlations Between Nuclear Decay Rates and Earth–Sun Distance

As noted in the Introduction, our interest in the possibility that nuclear decay parameters $\lambda$ might vary with time arose from an unrelated effort to apply a new randomness test (Tu and Fischbach 2003, 2005) to nuclear decays. This effort eventually led us to a number of interesting data sets which indicated an apparent time-dependence of $\lambda$, most notably the BNL data of Alburger et al. (1986), and the PTB data of Siegert et al. (1998), Schrader (2008). (We note that a time-dependence of $\lambda$ does not necessarily imply any deviation from randomness, but only a deviation in the probability distribution which governs the decays.) In this Section we present a summary and analysis of these two data sets.

Between 1982 and 1986, Alburger et al. (1986) measured the half-life of $^{32}$Si at BNL via a direct measurement of the counting rate as a function of time. As is typical in such counting experiments, the counting rate for $^{32}$Si was continually monitored in the same detector against a long-lived comparison standard, which in the BNL experiment was $^{36}$Cl ($T_{1/2} = 301{,}000$ y). Counts were taken with a precision sample changer (Harbottle et al. 1973), where each of the $^{32}$Si and $^{36}$Cl sources was counted for 30 minutes alternately, 10 times each. The 10 counts were then summed for each source, and the $^{32}$Si/$^{36}$Cl ratio was generated. Three days of counting were done for each reported week, and all of the data points were used in our analysis. Since the fractional change in the $^{36}$Cl counting rate over the four year duration of the experiment was only $\mathcal{O}(10^{-5})$, which was considerably smaller than the overall uncertainty of the final result, $T_{1/2}(^{32}\mathrm{Si}) = 172(4)$ y, the $^{36}$Cl decay rate was assumed to be constant. Any time dependence for $^{36}$Cl beyond the expected statistical fluctuations was then presumed to arise from various systematic effects, such as drift in the electronics. By computing the ratio $^{32}\mathrm{Si}/^{36}\mathrm{Cl} \equiv \dot{N}(^{32}\mathrm{Si})/\dot{N}(^{36}\mathrm{Cl})$, these apparatus-dependent systematic effects should have largely cancelled (see discussion below), and hence this ratio was used to obtain the half-life of $^{32}$Si.

If there was any residual unexpected behavior of the ratio $^{32}\mathrm{Si}/^{36}\mathrm{Cl}$, one would naturally seek to explain this behavior in terms of the differential response of the two nuclides' individual measured counting rates to possible variations in the measuring apparatus. If such an explanation did not quantitatively explain the anomalous residual behavior, one could then reasonably consider the possibility that the residual unexpected behavior was in fact due to variations in the individual decay parameters themselves. Since we generally expect different nuclides to exhibit different time-dependent fractional variations, it follows that such an effect could account for any residual unexplained behavior in the ratio $^{32}$Si/$^{36}$Cl. We note, however, that two different nuclides might indeed experience similar variations in their intrinsic decay parameters, perhaps as a consequence of the details of the mechanism, or perhaps just by chance. Hence, the failure to see any unexpected behavior in the ratio of the nuclides' count-rates does not necessarily imply that the interesting behavior of the nuclides individually is entirely due to systematic effects in the measuring apparatus. As such, if the behavior of the nuclides' individual measured counting rates seems difficult to explain in terms of systematic effects of the apparatus, one could tentatively consider the possibility that the experiment is suggesting an intrinsic variation in individual decay parameters. This latter comment will be seen to be relevent to the PTB data, to be discussed shortly.

The BNL data for the ratio $^{32}$Si/$^{36}$Cl revealed an unexpected annual variation which led Alburger et al. (1986) to study the sensitivity of their detection system to changes in various experimental parameters, including counter voltage and gas flow, box pressure and temperature, and discriminator level. In addition, backgrounds were measured and found to be negligible. In the end, Alburger et al. (1986) concluded that "... systematic periodic variations are present but that they cannot be fully accounted for by our tests or estimates."

Since several annually varying effects happen to closely track the varying distance $R$ between the Earth and the Sun (e.g. local temperatures), it is reasonable to ask whether the BNL data correlate directly with either $1/R^2$ or $1/R$ (we note that our existing data cannot distinguish between these two possibilities). When comparing the results from experiments on different nuclides, it is convenient to study the function $U(t) \equiv [\dot{N}(t)/\dot{N}(0)]\exp(+\lambda t)$ rather than $\dot{N}(t)$ itself, since $U(t)$ should be time-independent for all nuclides. For $^{32}$Si, we used $\lambda = 4.0299 \times 10^{-3}$ y$^{-1}$ from Alburger et al. (1986). Figure 1 exhibits $U(t)$ for the $^{32}$Si/$^{36}$Cl BNL data, along with a plot of $1/R^2$. An annual modulation of the $^{32}$Si/$^{36}$Cl ratio is clearly evident, as was first reported in Alburger et al. (1986). The Pearson correlation coefficient, $r$ (Taylor 1997, p. 217), between the raw BNL data and $1/R^2$ (or $1/R$) is $r = 0.52$ for $N = 239$ data points, which translates to a formal probability of $6 \times 10^{-18}$ that this correlation would arise from two data sets which were uncorrelated. There is also a suggestion in Fig. 1 of a phase shift between $1/R^2$ and the BNL data, which we discuss in greater detail below.

The annual variation observed in the BNL decay data raised the question of whether similar effects could be present in other decays as well. Although there are hundreds of potentially useful nuclides whose half-lives have been measured, the data from many of the experiments we examined were generally not useful, most often because data were not acquired continuously over sufficiently long time periods. However, we were able to obtain the raw data from an experiment carried out at the PTB in Germany (Schrader 2008; Siegert et al. 1998) measuring the half-lives for $^{152}$Eu and $^{154}$Eu, in which $^{226}$Ra was the long-lived comparison standard. The data were collected with a high-pressure $4\pi\gamma$ ionization chamber, measured as a current, and each measurement was corrected for background, as described by the PTB experimental protocol: "The ionization chamber was equipped with an automated sample changer that allowed the sources under study, background, and a radium reference source to be successively measured" (Schrader 2008; Siegert et al. 1998). Each data point in Fig. 2 is then an average of approximately 30 individual measurements of the $^{226}$Ra current corrected for background. The contribution to the $^{226}$Ra counting rate from such background radiation as $^{222}$Rn, which is known to fluctuate seasonally (Wissman 2006), is subtracted

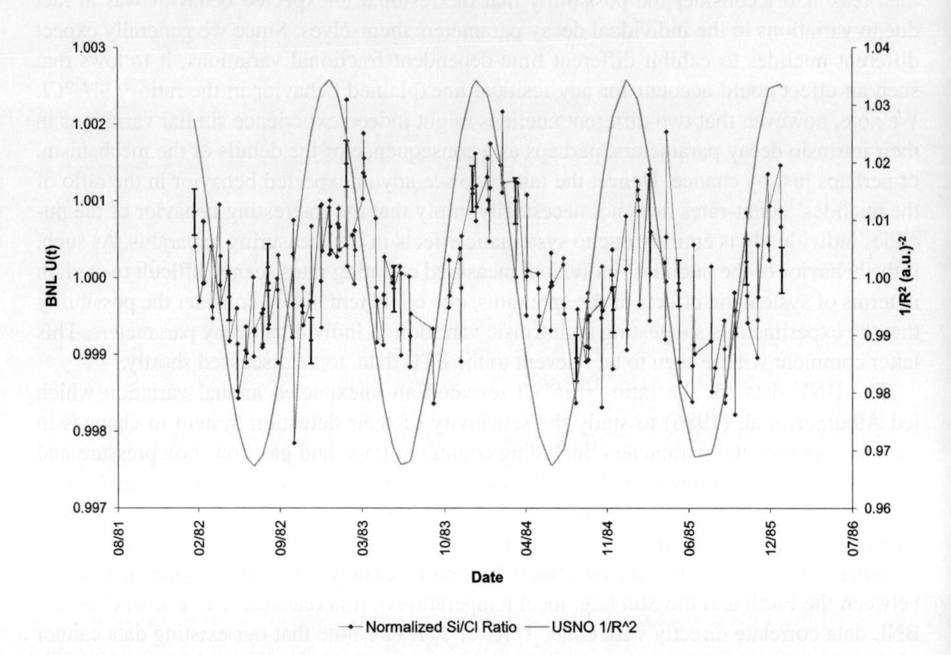


**Fig. 1** Plot of $U(t)$ for the raw BNL $^{32}$Si/$^{36}$Cl ratio along with $1/R^2$ where $R$ is the Earth–Sun distance. $U(t)$ is obtained by multiplying each data point by $\exp(+\lambda t)$ where $\lambda = \ln(2)/T_{1/2}$ and $T_{1/2} = 172$ y for $^{32}$Si. The *left axis* gives the scale for the normalized $U(t)$, and the *right axis* denotes the values of $1/R^2$ in units of $1/(\text{a.u.})^2$ obtained from the U.S. Naval Observatory (USNO). The fractional change in $^{32}$Si counting rates between perihelion and aphelion is approximately $3 \times 10^{-3}$. As noted in the text, the correlation coefficient between the BNL data and $1/R^2$ (or $1/R$) is $r = 0.52$ for $N = 239$ points. The formal probability that the indicated correlation could have arisen from uncorrelated data sets is $6 \times 10^{-18}$

out for each individual data point, thus suppressing any seasonal contribution. The same subtraction applies to other potential backgrounds, such as cosmic rays. Other environmental factors such as ambient temperature, pressure, and relative humidity would lie within what is typical for a laboratory environment, and are not expected to have a significant effect on the detection system. Any changes within the normal laboratory setting, however, should be mitigated by the fact that the system was a high-pressure (2 MPa) ionization chamber, detecting photons traversing a very short range within the $4\pi\gamma$ setup. The PTB experiment, which extended over 15 years, exhibited annual fluctuations in the $^{226}$Ra data similar to those seen at BNL.

We note that the data plotted in Fig. 2 are the raw $^{226}$Ra data rather than the ratio $^{152}$Eu/$^{226}$Ra as might be expected from an analogy to Fig. 1. There are two reasons for this: (a) In contrast to the BNL data, where we were given copies of the original notebooks containing the raw $^{32}$Si and $^{36}$Cl data, in the PTB case the data we received had already been processed in a way that would have been difficult for us to undo. (b) Secondly, $^{152}$Eu is an interesting nuclide for present purposes because it decays by both electron capture (72%) and $\beta$-decay (28%). Since these two decay modes could very well exhibit different time-dependent responses to a given external perturbation, the time-dependence of $^{152}$Eu could be more complicated than that of the other nuclides whose data we have analyzed.

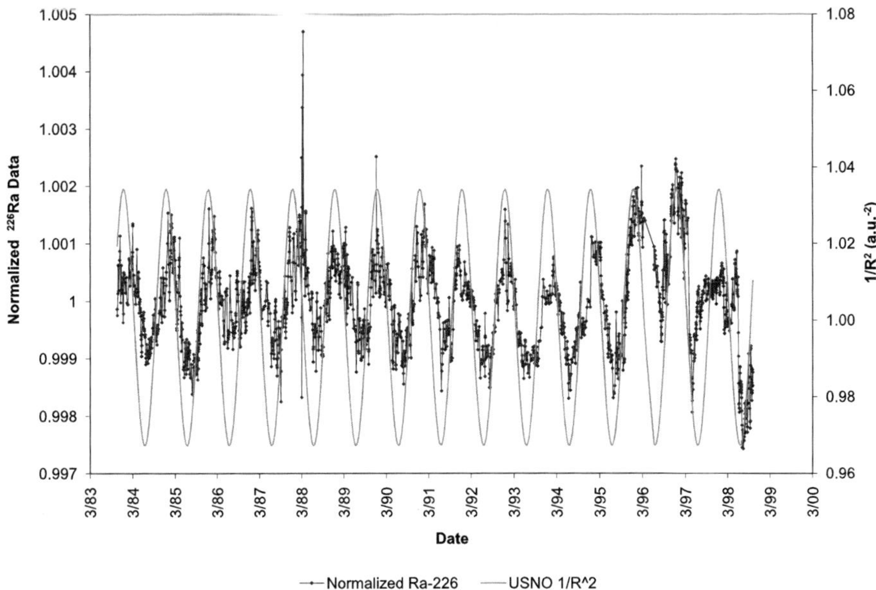


**Fig. 2** Plot of $U(t)$ for the raw PTB $^{226}$Ra data (using $T_{1/2} = 1600$ y) along with $1/R^2$. As noted in the text, the correlation coefficient between the PTB data and $1/R^2$ is $r = 0.62$ for $N = 1974$ points. The formal probability that the indicated correlation could have arisen from uncorrelated data sets is $5 \times 10^{-210}$. See caption to Fig. 1 for additional details

As in the case of the BNL data, it is again reasonable to ask whether these data correlate with $1/R^2$. The Pearson correlation coefficient $r$ for the data in Fig. 2 is $r = 0.62$ for $N = 1974$ data points corresponding to a formal probability of $5 \times 10^{-210}$ that the two data sets were in fact uncorrelated. There is also a suggestion of a phase shift between $1/R^2$ and the PTB data, as in the BNL data.

Given that the BNL and PTB experiments overlapped for ~2 years, we can also calculate the correlation coefficient between the BNL and PTB data. For the weeks during which the BNL and PTB data sets had concurrent measurements, averages for each week's measurements (the $^{32}$Si/$^{36}$Cl ratios for BNL and the currents for PTB) were taken in each data set in order to maintain consistency, and the resulting correlation is exhibited in Figs. 3 and 4. The Pearson correlation coefficient for the raw BNL and PTB data is $r = 0.66$ for $N = 39$ points, which corresponds to a formal probability of $6 \times 10^{-6}$ that this correlation could have arisen from two uncorrelated data sets.

Notwithstanding the possible implications of the correlations between the BNL and PTB data, some words of caution are appropriate relating to our use of the $^{226}$Ra data. When the ratio is taken between the counts of $^{154}$Eu and $^{226}$Ra, a periodic signal is no longer evident (see Fig. 3 in Siegert et al. 1998), in contrast to what is seen in the BNL data. Referring to the previous discussion, the absence of a periodic signal in the $^{154}$Eu/$^{226}$Ra ratio could be attributed to a similar response of the individual $^{154}$Eu and $^{226}$Ra count-rates to a variation in the measuring apparatus. We note that, unlike the PTB europium data, the $^{226}$Ra data we have analyzed are raw data, having no corrections other than background subtraction. It would thus appear that the periodic signal present in the $^{226}$Ra data is either a manifestation

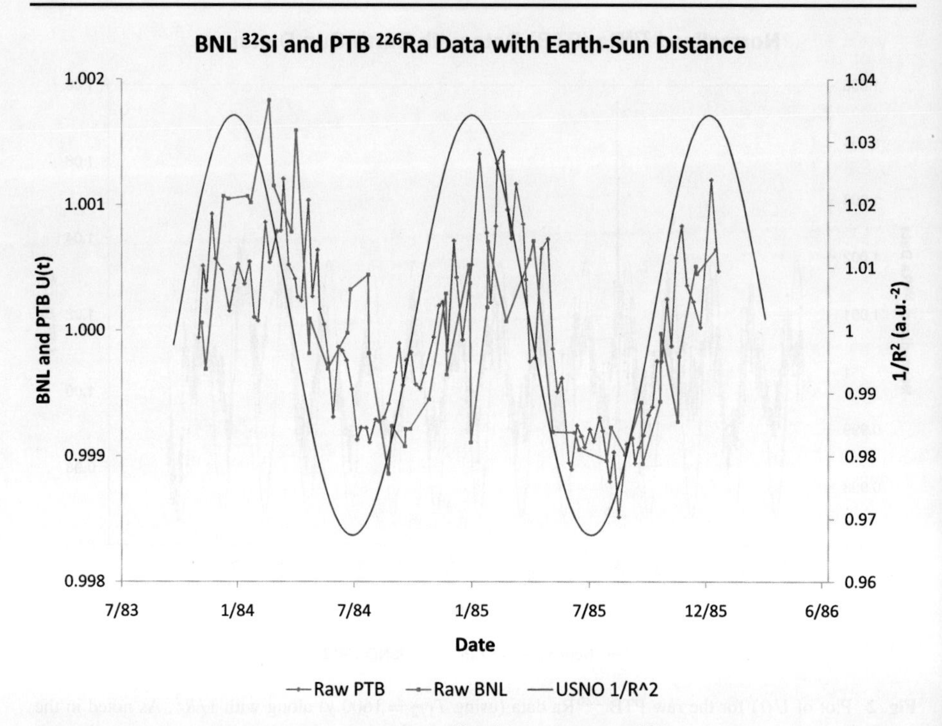


**Fig. 3** Correlation between the raw decay rates of $^{32}Si/^{36}Cl$ at BNL and $^{226}Ra$ at PTB. The BNL and PTB data for $U(t)$ have been averaged in common weekly bins for purposes of comparison. The correlation coefficient between the BNL and PTB data is $r = 0.66$ for $N = 39$ points, which corresponds to a probability of $6 \times 10^{-6}$ that the BNL/PTB correlation could have arisen from uncorrelated data sets as a result of statistical fluctuation. *Error bars* are shown for representative BNL data points, and the *error bars* for the PTB data lie within the points themselves

of fluctuations arising from the measuring apparatus, or else results from fluctuations in the $^{226}Ra$ decay parameter itself. One would naturally seek to ascertain whether any variations in the measuring instrumentation would be capable of producing the observed periodic signal in the $^{226}Ra$ data. Siegert et al. (1998) have proposed a qualitative mechanism whereby background radioactivity due to radon and daughter products could affect the apparatus, but no quantitative analysis was given. It therefore seems reasonable to regard the origin of the periodic signal in the $^{226}Ra$ measured count-rate as an open question, and we have thus suggested the possibility that the PTB data are indicating variations in the decay parameters of $^{226}Ra$ or its daughters. This seems to imply that the fractional variations in the decay parameters of $^{154}Eu$ and $^{226}Ra$ are very similar, which one generally would not expect, and this could be a clue to the nature of a possible physical mechanism.

The correlations of the BNL and PTB data with $1/R^2$, as well as with each other, do not in and of themselves point to an origin for these effects. Not only are there several potential influences which could depend on $R$, but additionally there are seasonal variations that roughly track with $R$ even though the Earth–Sun distance is not their primary cause (e.g. local temperature variations). Having previously addressed the possibility that these correlations arise from seasonal fluctuations in the detectors used in the BNL and PTB experiments, we next consider the possibility that the time-dependence of the $^{32}Si/^{36}Cl$ ratio and the $^{226}Ra$ decay rate are being modulated by an annually varying flux or field originating

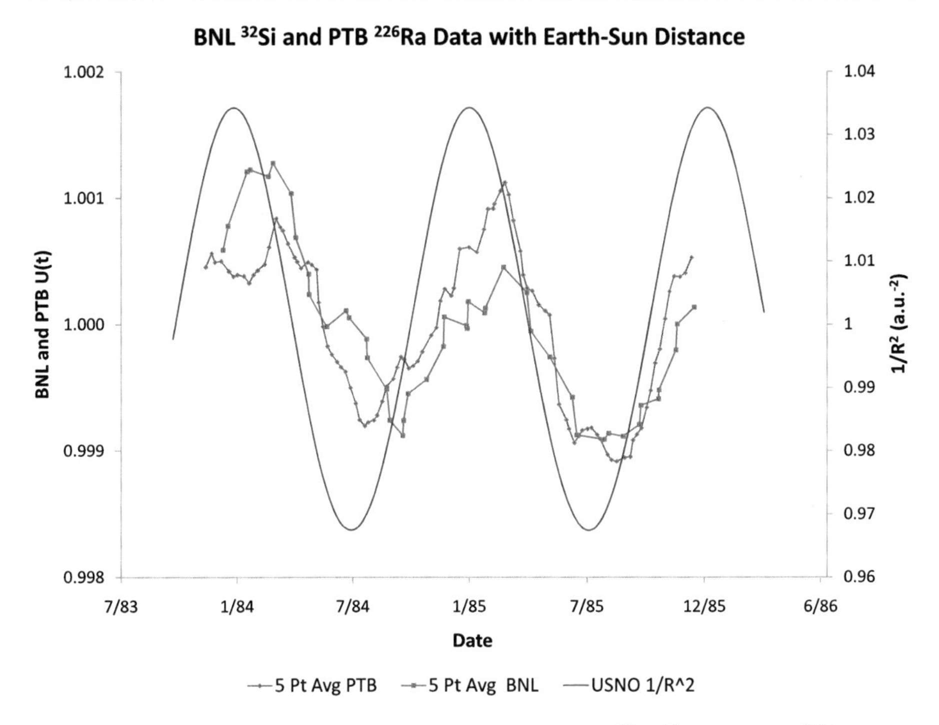


**Fig. 4** Correlation between the centered 5 point averaged decay rates of $^{32}Si/^{36}Cl$ at BNL and $^{226}Ra$ at PTB. The centered 5 point average illustrates the effect of smoothing out short-term fluctuations in both data sets. The BNL and PTB data for $U(t)$ have been averaged in common weekly bins for purposes of comparison

from the Sun. The fact that the two decay processes are very different ($\alpha$-decay for $^{226}Ra$ and $\beta$-decay for $^{32}Si$) would seem to preclude a common mechanism for both. We note, however, that even though $^{226}Ra$ decays via $\alpha$-emission, its daughter products include $^{214}Bi$ and $^{214}Pb$, which are $\beta$-decays and which rapidly reach equilibrium with the parent $^{226}Ra$ (see Sect. 7 below). It is thus possible that a single mechanism could explain the $^{32}Si$ and $^{226}Ra$ data, provided that its dominant effects appeared in the $\beta$-decays. However, there are also mechanisms which could affect both $\beta$- and $\alpha$-decays, such as proposed in the recent papers by Barrow and Shaw (Barrow and Shaw 2008; Shaw 2007) which we discuss in more detail in Sect. 6 below.

Returning to Figs. 1–4, we briefly explore the suggestion noted above of a possible phase shift of $1/R^2$ relative to both the BNL and PTB data. One possibility could be additional contributions to periodic variations in neutrino flux, which might arise from a small neutrino magnetic dipole moment. Sturrock (2008) has recently shown that data on solar neutrino flux and solar irradiance exhibit a common modulation at 11.85 $y^{-1}$ (period = 31 d), which is thought to arise from the rotation of the solar core. It is thus possible that the ~1 month phase lag evident in Figs. 1–4 could be related in some way to Sturrock's observations, although this would require more than a simple superposition of a monthly and an annual sine curve. Yet another possibility for the apparent phase shift is that the effect depends on the Earth's velocity relative to some fixed direction in space. Such an effect might share similarities with the mechanism proposed by the DAMA/LIBRA collaboration (Bernabei et al. 2008), though we note that their particular model involves the Earth's motion through the galactic rest frame, and would thus be roughly 2 months out of phase with the BNL/PTB data. However, we note in passing that the superposition of two annually varying periodic effects of comparable strength, one of which is in phase with $1/R^2$, and the other with our speed through the galactic rest frame, could give rise to a phase shift of $\sim 30$ days, as present in the BNL/PTB data.

## 3 Implications of the Solar Flare of 2006 December 13

As noted in the Introduction, data taken during the solar flare of 2009 December 13 lend support to the inference from the BNL and PTB data that solar activity can influence nuclear decay rates. In this Section we summarize this experiment and refer the reader to Jenkins and Fischbach (2008) for a more detailed discussion.

The apparatus that was in operation during the solar flare was a $\sim$1 μCi sample of $^{54}$Mn attached to the front of a Bicron 2 × 2 inch NaI(Tl) crystal detector, which was connected to an Ortec photomultiplier (PMT) base with pre-amplifier. An Ortec 276 spectroscopy amplifier was used to analyze the pre-amplifier signal, and this was connected to an Ortec Trump® PCI card running Ortec's Maestro32® MCA software. The system recorded the 834.8 keV $\gamma$-ray emitted from the de-excitation of $^{54}$Cr produced from the electron-capture process $^{54}\mathrm{Mn} + e^- \rightarrow {}^{54}\mathrm{Cr} + \nu_e$. The detector and the $^{54}$Mn sample were shielded on all sides by several inches of lead, except at the end of the PMT base where a space was left to accommodate cables. The apparatus was located in a windowless, air-conditioned interior 1st floor room in the Physics building at Purdue in which the temperature was maintained at a constant 19.5(5)°C.

During the course of the data collection, which extended from 2006 December 2 to 2007 January 2, a solar flare was detected on 2006 December 13 at 02:37 UT (21:37 EST on December 12) by the Geostationary Operational Environmental Satellites (GOES-10 and GOES-11). Spikes in the X-ray and particle fluxes were recorded on all of the GOES satellites (NOAA 2009) during the course of the four days following the initial flare on December 13. The X-ray data from this X-3 class solar flare are shown in Figs. 5 and 6 along with the $^{54}$Mn counting rates: in each 4 hour live-time period ($\sim$4.25 hours real-time) we recorded $\sim 2.5 \times 10^7$ counts of 834.8 keV $\gamma$-rays with a fractional $1/\sqrt{N}$ statistical uncertainty of $\sim 2 \times 10^{-4}$. Each $^{54}$Mn data point in Figs. 5 and 6 then represents the number of counts in the subsequent 4 hour period, which are normalized in Fig. 6 by the number of counts $N(t)$ expected from a monotonic exponential decay $N(t) = N_0 \exp(-\lambda t)$, with $\lambda = 0.002347(2)\ \mathrm{d}^{-1}$ determined from our December data. We see from Figs. 5 and 6 that, to within the time resolution offered by the 4 hour width of our bins, the $^{54}$Mn count rates exhibit a dip which is coincident in time with the spike in the X-ray flux which signalled the onset of the solar flare. Although a second X-ray peak on December 14 at 17:15 EST corresponds to a relatively small dip in the $^{54}$Mn count-rate, a third peak on December 17 at 12:40 EST is again accompanied by an obvious dip in the $^{54}$Mn counting rate, as seen in Figs. 5 and 6. The fact that some X-ray spikes in these and other data sets are not accompanied by correspondingly prominent dips in the $^{54}$Mn data may provide clues to the underlying mechanisms that produce these solar events. Additionally, the X-ray flare that occurred earlier in the month on December 5 was not Earth-directed, which may explain the relatively small change in the observed $^{54}$Mn count-rate. Conversely, peaks or dips in the $^{54}$Mn data not accompanied by visible X-ray spikes may correspond to other types of solar events, or to events on the opposite side of the Sun, which are possibly being detected via neutrinos. In particular, the dip on 2006 December 22 (09:04 EST) was coincident in time with a severe solar storm (NOAA 2006), but did not have an associated X-ray spike.

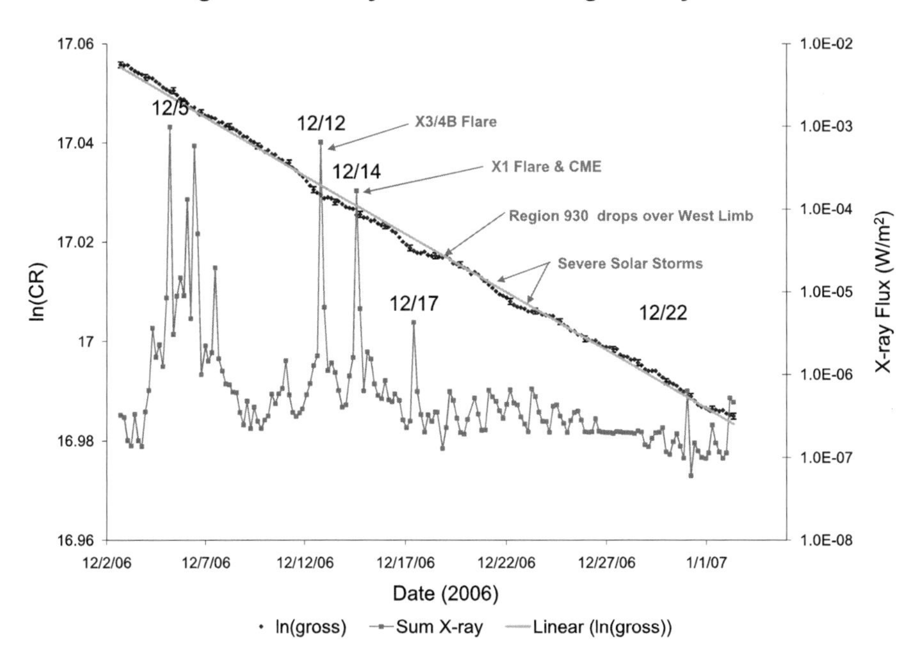


**Fig. 5** December 2006 $^{54}$Mn data, and GOES-11 X-ray data, both plotted on a logarithmic scale. For $^{54}$Mn, each point represents the natural logarithm of the number of counts $\sim 2.5 \times 10^7$ in the subsequent 4 hour period, and has a $3 \times 1/\sqrt{N}$ statistical error shown by the indicated (small) error bars. For the GOES-11 X-ray data, each point is the solar X-ray flux in $\mathrm{W/m^2}$ summed over the same real time intervals as the corresponding decay data. The *solid line* is a fit to the $^{54}$Mn data, and deviations from this line coincident with the X-ray spikes are clearly visible on 12/12 and 12/17. As noted in the text, the deviation on 12/22 was coincident with a severe solar storm, with no associated flare activity (NOAA 2009). The dates for other solar events are also shown by *arrows*

Before considering more detailed arguments supporting our inference that the $^{54}$Mn count-rate dips are due to changes in the flux of solar neutrinos, we address the question of whether the coincident fluctuations in the decay data and the solar flare data could simply arise from statistical fluctuations in each data set. Referring to Fig. 5, we define the dip region in the decay data as the 84 hour period (encompassing our runs 51–71 inclusive) extending between 2006 December 11 (17:52 EST) and 2006 December 15 (06:59 EST). The measured number of decays $N_m$ in this region can then be compared to the number of events $N_e$ expected in the absence of the observed fluctuations, assuming a monotonic exponential decrease in the counting rate. Since the systematic errors in $N_e$ and $N_m$ are small compared to the statistical uncertainties in each, only the latter are retained and we find

$$N_e - N_m = (7.51 \pm 1.07) \times 10^5, \tag{1}$$

where the dominant contributions to the overall uncertainty arise from the $\sqrt{N}$ fluctuations in the counting rates. If we interpret (1) in the conventional manner as a $\sim 7\sigma$ effect, then the formal probability of such a statistical fluctuation in this 84 hour period is $\sim 3 \times 10^{-12}$. This conclusion is not altered by including additional small systematic corrections.

We next estimate the probability that a solar flare would have occurred during the same 84 hour period shown in Fig. 5. The frequency of solar radiation storms varies with their

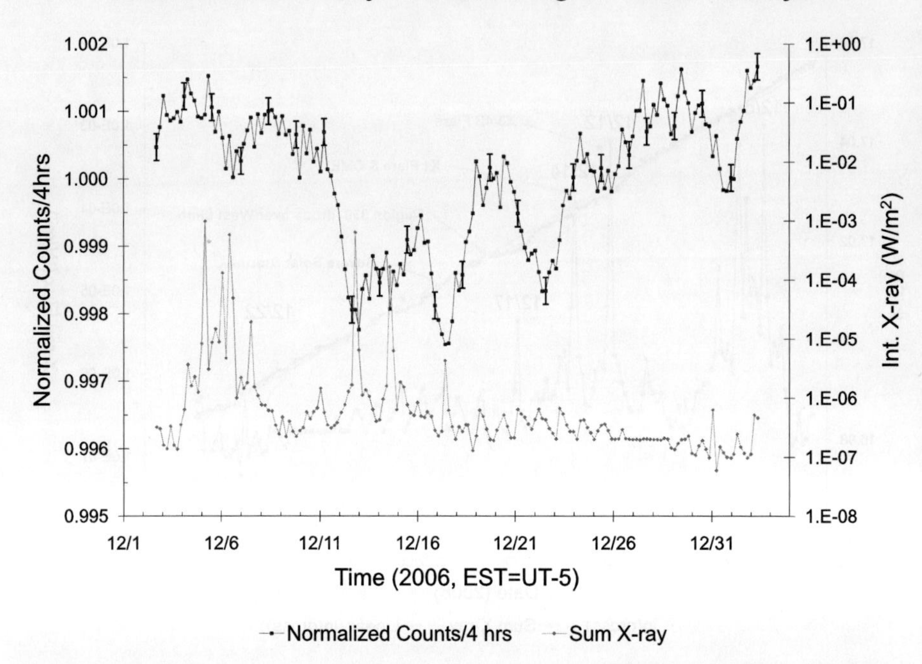


**Fig. 6** Normalized December 2006 $^{54}$Mn decay data along with GOES-11 X-ray data on a logarithmic scale. For $^{54}$Mn, each point represents the number of counts in the subsequent four hour period normalized to the average decay rate (see text), and has a fractional $1/\sqrt{N}$ statistical uncertainty of $\sim 2 \times 10^{-4}$. For the GOES-11 X-ray data, each point is the solar flux in W/m$^2$ summed over the same real-time intervals. The 2006 December 12 spike in solar flux occurred at $\sim$21:37 EST

intensities, which are rated on a scale from S1(Minor) to S5(Extreme) (NOAA 2005). The 2006 December 13 event was rated as S2 (Moderate), and S2 storms occur with an average frequency of 25 per 11 year solar cycle (NOAA 2009). In total, the frequency of storms with intensity $\geq$S2 is $\sim$39 per 11 year solar cycle, or $9.7 \times 10^{-3}$ d$^{-1}$, and hence the probability of a storm occurring at any time during the 84 hour window in Fig. 5 is $\sim 3.4 \times 10^{-2}$. Evidently, if the X-ray and decay peaks were uncorrelated, the probability that they would happen to coincide as they do over the short time interval of the solar flare would be smaller still, and hence a conservative upper bound on such a statistical coincidence occurring in any 84 hour period is $\sim (3 \times 10^{-12})(3 \times 10^{-2}) = 1 \times 10^{-13}$. Since a similar analysis would apply to the coincident peak and dip at 12:40 EST on December 17, the probability that random fluctuations would produce two sets of coincidences several days apart is negligibly small, and hence we turn to consider other possible explanations for the data in Figs. 5 and 6.

Solar flares are known to produce a variety of electromagnetic effects on Earth, including changes in the Earth's magnetic field, and power surges in the electric grids. It is thus conceivable that the observed dips in the $^{54}$Mn counting rate could have arisen from the response of our detection system (rather than the $^{54}$Mn atoms themselves) to the solar flare. The most compelling argument against this explanation of the $^{54}$Mn data is that the $^{54}$Mn decay rate began to decrease more than one day *before* any signal was detected in X-rays by the GOES satellites (see Figs. 5 and 6 in text). Since it is unlikely that any other electromagnetic signal would reach the Earth earlier than the X-rays, we can reasonably exclude any explanation of

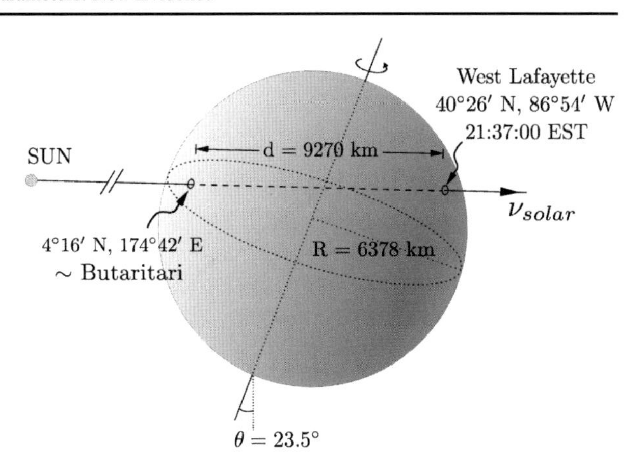


**Fig. 7** Trajectory of neutrinos from the solar flare of 2006 December 12 21:37 EST. The neutrinos would have entered the Earth near Butaritari, in the Pacific Ocean, and travelled ~9270 km through the Earth before the coincident minimum in the count-rate was detected in West Lafayette, Indiana

the $^{54}$Mn data in terms of a conventional electromagnetic effect arising from the solar flare. This is particularly true since the most significant impact on the geomagnetic field occurs with the arrival of the charged particle flux, several hours after the arrival of the X-rays. Figure 8 exhibits the $A_p$ index for the Earth's magnetic field during 2006 December,[1] along with the $^{54}$Mn counting rate. We see immediately that the sharp spike in the $A_p$ index at approximately 00:00 EST on 2006 December 15 occurred more than two days *after* the solar flare and the accompanying dip in the $^{54}$Mn counting rate, and hence was presumably not the cause of this dip. This conclusion was further strengthened by the results of a series of measurements carried out in our laboratory, which established that our detection system was insensitive to applied magnetic fields that were more than 100 times stronger than the spike exhibited in Fig. 8.

The response of our detection system to fluctuations in line voltages was also studied. No unusual behavior was detected by either the Purdue power plant, or by the Midwest Independent Systems Operator (MISO) which also supplies power to Purdue. MISO did in fact receive notification on 14 December 2006 of a "Geo-magnetic disturbance of K-7 magnitude" at 02:46 UT, but noted that there were no reported occurrences of excessive neutral currents during the time-frame of 10–18 December 2006. At Purdue, an alert would have been triggered had the line voltage strayed out of the range 115–126 V, and hence we can infer that the voltage remained within this range during the solar flare. Moreover, since the main effect of a power surge would have been to shift the $^{54}$Mn peak slightly out of the nominal region of interest (ROI) for the 834.8 keV $\gamma$-ray, this would have been noted and corrected for in the routine course of our data acquisition. No significant changes to either the peak shape or location were noted during this period. Additional post-experiment testing showed that the line voltage at Purdue is confined to the range $120 \pm 5$ V, and within this range, no apparent effects were seen in similar counting experiments.

We therefore turn to the possibility that this dip was a response to a change in the flux of solar neutrinos during the flare, as implied by the analysis of Jenkins et al. (2008). To begin we note that the X-ray spike coincident with the maximum deviation of the $^{54}$Mn count-rate from the expected rate occurred at ~21:40 EST, approximately 4 hours after local sunset, which was at ~17:21 EST on 2006 December 12. As can be seen from Fig. 7, the neutrinos (or whatever other agent produced this dip) had to travel ~9,270 km through the Earth

[1] http://www.swpc.noaa.gov.

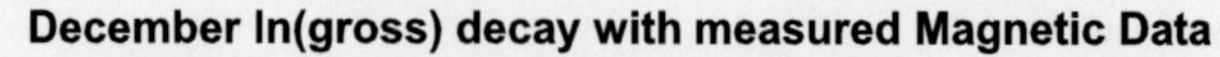


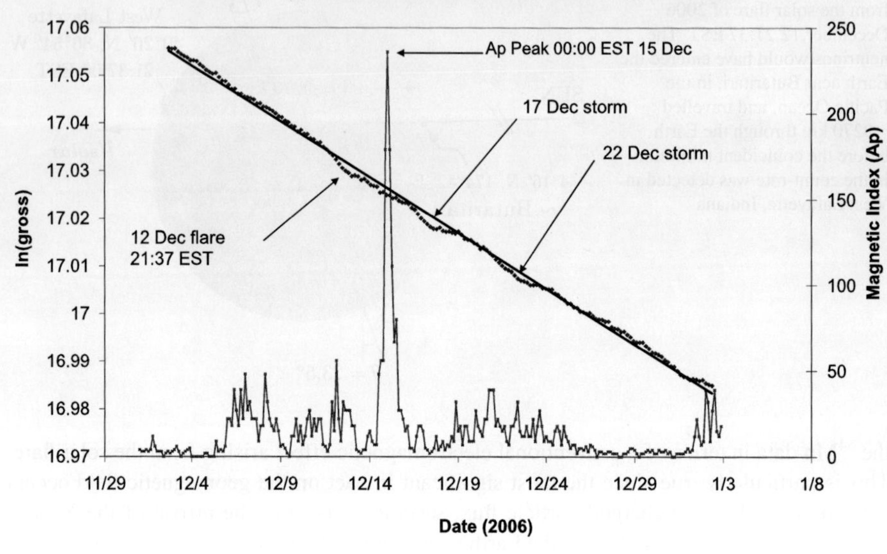


**Fig. 8** Fluctuations in the Earth's magnetic field in December 2006. The magnetic field fluctuations, which are characterized by the $A_p$ index, are plotted along with the natural logarithm of the $^{54}$Mn count-rate. We note that the spike in the magnetic data on 2006 December 15 occurred ~2 days *after* the dip in the $^{54}$Mn count-rate at 21:37 EST on 2006 December 12

before reaching the $^{54}$Mn source, and yet produced a dip in the counting rate coincident in time with the peak of the X-ray burst. Significantly, the monotonic decline of the counting rate in the 40 hours preceding the dip occurred while the Earth went through 1.7 revolutions, and yet there are no obvious diurnal or other periodic effects. These observations support our inference that this effect may have arisen from neutrinos, some neutrino-like particles, or some field emanating from the Sun, and not from any conventionally known electromagnetic effect or other source, such as known charged particles.

## 4 Phenomenology of a Time-Dependent Decay Parameter $\lambda(t)$

As noted in the previous sections, one implication of the BNL/PTB data correlations and the solar flare data is that nuclear decay parameters may be influenced by the external environment, and hence may not actually be constant. In this section we outline some of the phenomenological implications of a decay parameter $\lambda = \lambda(t)$ which is time-dependent.

If $P(t)$ denotes the probability that a nucleus has not decayed after a time $t$, then the probability that it has still not decayed after a time $(t + dt)$ is given by

$$P(t + dt) = P(t)[1 - \lambda(t)dt]. \tag{2}$$

In (2) $\lambda(t)dt$ is the probability of decay in the time interval $t$ to $(t + dt)$ and hence $[1 - \lambda(t)dt]$ is the corresponding survival probability. Rearranging (2) we have

$$\dot{P}(t) = \frac{P(t + dt) - P(t)}{dt} = -\lambda(t)P(t), \tag{3}$$

**Fig. 9** Schematic illustration of a flat region in the decay curve. As shown in Fig. 10 flat regions are seen experimentally in the $^{54}$Mn data. See text for additional details

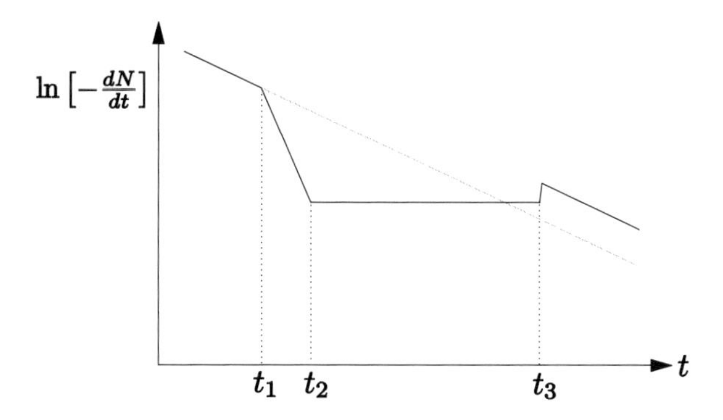


and integrating (3) we find

$$P(t) = P(0)\exp\left[-\int_0^t dt'\lambda(t')\right]. \tag{4}$$

When $\lambda(t)$ is time-independent $P(t)$ reduces to the familiar exponential decay law with $P(0)$ set to unity. For a sample of $N_0$ atoms at $t = 0$ experiencing a common external interaction, the number $N(t)$ surviving at a time $t$ is then given by

$$N(t) = N_0\exp\left[-\int_0^t dt'\lambda(t')\right]. \tag{5}$$

In principle one can extract from (5) a wide range of expressions for $N(t)$ by an appropriate choice of $\lambda(t)$. Any "anomalous slope region", i.e. a deviation of the experimental data for $N(t)$ and $dN(t)/dt$ from the expected forms (with constant $\lambda$)

$$N(t) = N_0 e^{-\lambda t}, \qquad \frac{dN}{dt} = -\lambda N_0 e^{-\lambda t}, \tag{6}$$

could then be taken as evidence that $\lambda$ is itself time-dependent. Among the possible functional dependencies for $\lambda(t)$ that one might consider, two are of special interest, since they lead to variations in $dN/dt$ that are suggested by existing data. One is a sinusoidal variation of $dN/dt$ as we have already discussed in Sect. 2. The other is a "flat region", which is an extended period of time during which $dN/dt$ remains approximately constant, notwithstanding the depletion of the sample population. Although in principle a "flat region" is no more fundamental than any other deviation from the expected behavior of $dN/dt$, it is both visually striking and straightforward to describe. In what follows we consider this case first, and discuss some potential medical implications of anomalous slope regions.

To see how we may obtain a flat region mathematically, we start with (5) and, taking a derivative, we find

$$\frac{dN}{dt} = -N_0\lambda(t)\exp\left[-\int_0^t \lambda(t')dt'\right] \cong -N_0\lambda(t)\cdot\left[1 - \int_0^t \lambda(t')dt'\right], \tag{7}$$

where we have assumed that $t$ is sufficiently small that the exponent may be expanded to lowest order. It is straightforward to show that if

$$\lambda(t) = \frac{\lambda(0)}{\sqrt{1 - 2\lambda(0)t}}, \tag{8}$$

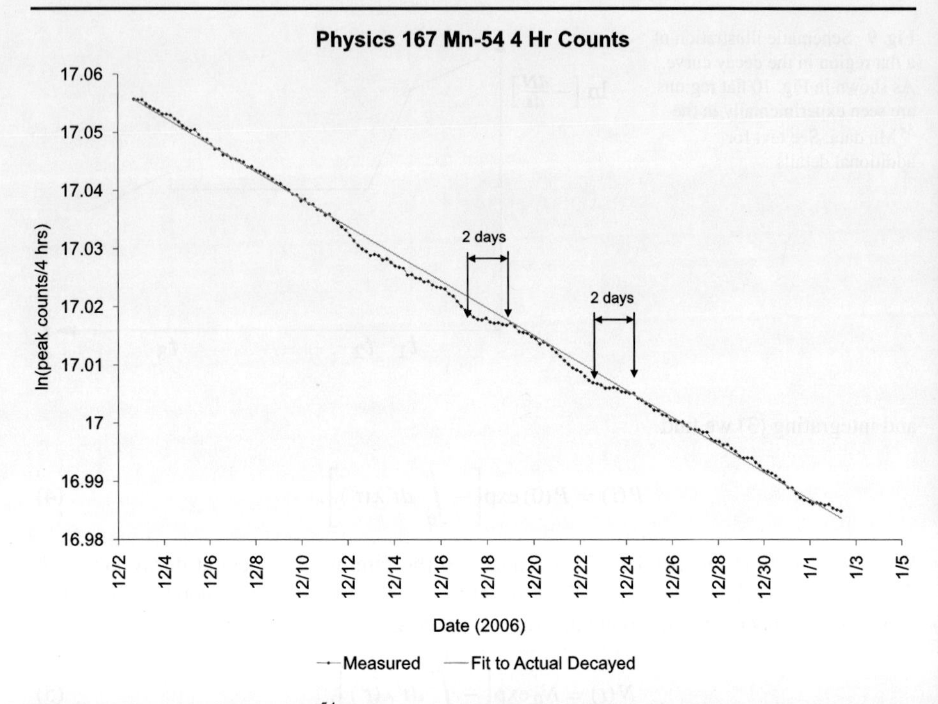


**Fig. 10** Flat regions in the Purdue $^{54}$Mn decay data. During the periods shown the count-rates were approximately constant

then (8) gives a flat region with constant decay-rate given by $N_0\lambda(0)$ when $\lambda(0)t \ll 1$. One can see in this way that (7) can produce a decay curve of basically any shape, given the appropriate $\lambda(t)$, though in the general case it may be difficult to find analytic forms for $\lambda(t)$. One can also write $\lambda(t)$ as a power series, which is valid for larger values of $\lambda(0)t$, by writing

$$\lambda(t) = \lambda(0) + \sum_{n=1}^{\infty} \lambda_n t^n, \tag{9}$$

where $\lambda_n$ are appropriate dimensional constants. Setting $\lambda_n = \lambda^{n+1}(0)$ eliminates the $t$-dependence in $dN/dt$ through $\mathcal{O}[(\lambda(0)t)^n]$, and hence creates a "flat region" of this order. The leading $t$-dependent term is then $\lambda_1 t = \lambda^2(0)t$, in agreement with (8). We note from (8) that $\lambda(t)$ is an increasing function of $t$ as we expect on physical grounds: since the number of atoms $N(t)$ available to decay is always a decreasing function of time, the only way for the actual decay-rate $|dN/dt|$ to remain approximately constant is if the effect of decreasing $N(t)$ is offset by an increasing $\lambda(t)$. As we note from Fig. 10, regions where $dN/dt \sim$ constant are seen in the $^{54}$Mn data we have taken at Purdue. (There are also suggestions of flat regions in the $^{152}$Eu data from PTB.)

Returning to Fig. 9 we describe what is happening physically in the different time intervals as follows: at $t_1$ the decay parameter begins to decrease, and therefore so does the decay rate $|dN/dt|$. At $t_2$ the decay parameter $\lambda(t)$ starts to increase in such a way that the decay rate remains constant. By $t_3$, the decay parameter has increased back to its original value $\lambda(t_1)$ and thereafter remains constant. Note that the specified variation in $\lambda$ implies

author further comments that this result is in accordance with the fact that, in the absence of the field, the square of the matrix element, averaged over spins and neutrino directions, does not depend on the electron momentum and neutrino energy. Thus, when the magnetic field is present, no field dependence is exhibited by the matrix element. However, the electron phase-space volume is modified as a result of two influences. It is altered by the presence of the interaction energy of the electron with the magnetic field, $V = -\boldsymbol{\mu}_e \cdot \mathbf{B}$, which may either increase or decrease the energy available to the decay products, depending upon the spin orientation of the emitted electron relative to the field. The phase-space volume is also modified by virtue of quantization of the electron-orbits in the external field, which in general diminishes the number of final states accessible to the electron.

In the case of electromagnetism, the behavior of particles in external fields is well understood. For other types of interactions, it may be difficult or impossible to use Dirac's equation to obtain the wavefunctions needed in Furry's formalism. Furthermore, as mentioned in Sect. 5.1, an exact calculation using Feynman diagrams is also difficult. Thus, rather than directly apply Furry's formalism to decays in general external fields, we seek to apply the prescription described above: that external fields are ultimately accounted for by modifying the phase-space volume available to the decay products. Since we generally do not know what the wavefunction will look like in an external field whose properties are not fully understood, it is difficult (if not impossible) to calculate the phase-space modifications due to energy-level quantization in the external field. As such, we seek to account for the effect of the external field by computing phase-space modifications due solely to the presence of the field's interaction energy with the decay particles. We shall henceforth refer to this as the "phase-space prescription", or the PSP. In Sect. 5.3.2 we investigate the accuracy of the PSP in the specific case of $\beta$-decay in external magnetic fields, and find that it predicts decay-rate modifications that are accurate to within an order of magnitude for a wide range of external field strengths.

### *5.2.3 Extension of Furry's Formalism to Time-Dependent Fields*

An extension of this formalism has been developed to deal with time-varying external electromagnetic fields (Nikishov and Ritus 1964; Ritus 1969; Lyul'ka 1975; Ternov et al. 1978; Ternov et al. 1984; Reiss 1983), where again the effect of the external field is solely accounted for by the wavefunctions, which are now time-dependent by virtue of the time-dependence of the electromagnetic background. Reiss (1983) notes that since the external field is time-dependent, the asymptotic states will be time-dependent also, and hence Fermi's Golden Rule cannot be immediately applied. In general, we expect that Fermi's Golden Rule may not apply if variations in the external field cause the asymptotic states to vary on a timescale which is short compared to the Heisenberg uncertainty time of the decay interaction.

## 5.3 Decay Modifications by Constant Fields

### *5.3.1 $\beta$-Decay in a Static, Homogeneous, Isotropic Field*

By virtue of its computational simplicity, we will first apply the PSP to the case of $\beta$-decay in a static field which is isotropic and homogeneous. Practically speaking, such a field would need to be homogeneous and isotropic only over a length scale which is very large compared to the interatomic spacing in a decay sample, and to the de Broglie wavelengths of the decay particles. In such an equipotential region, there is no energy-level quantization, and hence we expect the PSP to be most accurate in this case. We start by considering the decay of a free neutron at rest. In the absence of an external field, with all form factors approximated as unity, and with the neutrino mass set to zero, the differential transition rate is given by (Griffiths 1987)

$$\frac{d\Gamma}{dE} = \frac{G_F^2}{2\pi^3}\left[\frac{1}{2}(m_n^2 - m_p^2 - m^2)\cdot(E_+^2 - E_-^2) - \frac{2m_n}{3}(E_+^3 - E_-^3)\right], \tag{18}$$

where we have set $\hbar = c = 1$. In (18),

$$E_\pm = \frac{\frac{1}{2}(m_n^2 - m_p^2 + m^2) - m_n E}{m_n - E \mp \sqrt{E^2 - m^2}}, \tag{19}$$

$G_F = 1.166 \times 10^{-5}$ GeV$^{-2}$ is the Fermi constant, $m_n$, $m_p$, $m \equiv m_e$ are the neutron, proton, and electron masses, respectively, and $E$ is the energy of the emitted electron. One can simplify this expression by noting that the following quantities are all much smaller than unity:

$$\frac{m_n - m_p}{m_n} \ll 1, \qquad \frac{m}{m_n} \ll 1, \qquad \frac{E}{m_n} \ll 1, \qquad \frac{\sqrt{E^2 - m^2}}{m_n} \ll 1. \tag{20}$$

Expanding (18) in these small parameters and dropping higher-order terms, one obtains

$$\frac{d\Gamma}{dE} \cong \frac{2G_F^2}{\pi^3} E\sqrt{E^2 - m^2}[(m_n - m_p) - E]^2. \tag{21}$$

Note that in this approximation, the parameters $m_n$ and $m_p$ occur only in the combination $m_n - m_p$, and thus rather than regarding $d\Gamma/dE$ as a function of $m_n$ and $m_p$ separately, we may now regard $d\Gamma/dE$ as a function simply of $E_0 \equiv m_n - m_p$, which is the energy available to the decay products:

$$\frac{d\Gamma}{dE} = \frac{2G_F^2}{\pi^3} E\sqrt{E^2 - m^2}(E_0 - E)^2. \tag{22}$$

Before proceeding we note that allowed $\beta$-decays are generally similar to free-neutron decay, insofar as these decays are approximately described by (22) with the appropriate $E_0$. For example, $^3H \rightarrow {}^3He$ is approximately described by (22), with $E_0$ set equal to the difference between the $^3H$ and $^3He$ nuclear masses. The fact that the constant prefactor in such decays may differ from $2G_F^2/\pi^3$ is irrelevant in our analysis, since we will be considering fractional changes in transition-rates, in which case the prefactor will divide out. We will thus use (22) to analyze $\beta$-decays, with particular emphasis on $^3H$ decay. For forbidden $\beta$-decays, the differential transition rate will differ in form from (22) by more than just a prefactor. Thus, for more complicated $\beta$-decays such as $^{32}Si \rightarrow {}^{32}P$, the analysis presented here is unlikely to be quantitatively applicable, though some of the methods and ideas may be of qualitative value.

We next calculate the effect of the static potential field on the free-neutron decay-rate by applying the PSP to (22). In the presence of an external potential field, the energy balance for the process is given by

$$m_n + V_n = (E_p + V_p) + (E + V_e) + (E_{\bar{\nu}} + V_{\bar{\nu}}), \tag{23}$$

where $V_i$ denotes the interaction energy between particle $i$ and the external field. As we will see in (30) below, and in Fig. 12(a), for almost all values of $E_0$, energies at least on

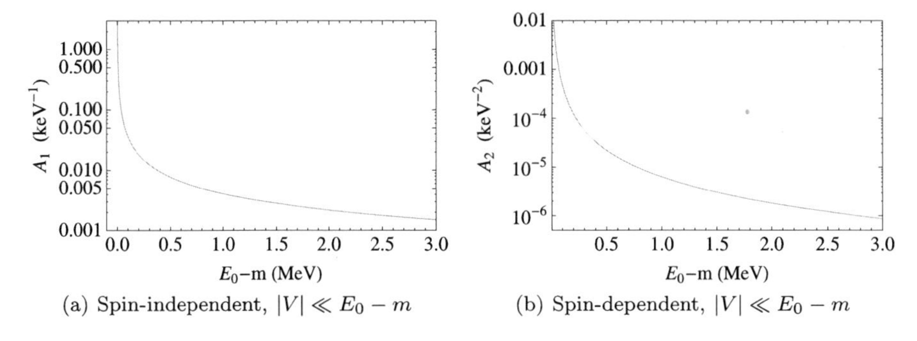


**Fig. 12** (**a**) Gives $A_1$ (in keV$^{-1}$) versus $E_0 - m$, where $A_1$ is defined by $\delta\Gamma/\Gamma_0 = A_1 \cdot V$, for $\delta\Gamma/\Gamma_0$ given by (29). (**b**) Gives $A_2$ (in keV$^{-2}$) versus $E_0 - m$, where $A_2$ is defined by $\delta\Gamma/\Gamma_0 = A_2 \cdot V^2$, for $\delta\Gamma/\Gamma_0$ given by (36). In both cases $|V|$ is assumed to be spatially constant

the order of $\gtrsim$1 eV are needed to explain variations on the order of 0.1% (which is the rough scale indicated by the data in Sects. 2 and 3). Since such an energy is substantial on the scale of atomic and low-temperature physics, it may be that the presence of such an external field would be in contradiction with a large body of experiments if it coupled in a spin-independent manner to the proton, neutron, or electron. Although this inference may turn out to be incorrect, we will assume for the present that a static, isotropic, and homogeneous field would couple predominantly to the emitted antineutrino. Setting $V \equiv V_{\bar{\nu}}$ and $V_p = V_e = V_n = 0$, we then have

$$E + E_{\bar{\nu}} = m_n - m_p - V = E_0 - V. \tag{24}$$

It follows that the energy available to the decay products is altered by the presence of the external field such that

$$E_0 \to E_0 - V, \tag{25}$$

and hence the PSP in this case amounts to substituting (25) into (22):

$$\frac{d\Gamma}{dE} = \frac{2G_F^2}{\pi^3} E\sqrt{E^2 - m^2}(E_0 - V - E)^2. \tag{26}$$

We see that, in the case where the interaction affects only the decay products, a negative interaction energy $V$ increases the available phase-space energy, and hence tends to increase the transition rate, while a positive $V$ decreases the available phase-space energy, and tends to decrease the transition rate. Before proceeding, we note that electron-capture transition rates also depend on phase-space factors containing $E_0$, and the PSP applies straightforwardly in this case.

It is of interest to examine the Kurie function $K(E)$, defined by

$$K(E) \equiv \sqrt{\frac{d\Gamma}{dE} \cdot \frac{1}{E\sqrt{E^2 - m^2}}} = \frac{2G_F^2}{\pi^3}(E_0 - V - E). \tag{27}$$

From (27), we see that when $V \neq 0$, $K(E)$ is obtained by the replacement $E \to E + V$, as noted in Horvat (1998). He further notes that if (27) is modified to include a finite neutrino mass, the Kurie plot is altered in a way that might explain the $^3$H end-point anomaly, to be discussed below. Another possible explanation, which we explore in Sect. 5.3.2, is that the neutrino mass is in fact negligible, but that the $^3$H end-point anomaly may be explained by the presence of a spin-dependent external field.

We now calculate the fractional change in the total transition rate as a result of the external field by integrating equation (26). Using the notation $\tilde{E}_0 = E_0 - V$, we have

$$\Gamma = \int_m^{\tilde{E}_0} dE \frac{d\Gamma}{dE} = \frac{G_F^2}{30\pi^2}\Bigg[\sqrt{\tilde{E}_0^2 - m^2}(2\tilde{E}_0^4 - 9\tilde{E}_0^2 m^2 - 8m^4) + 15\tilde{E}_0 m^4 \ln\left(\frac{\tilde{E}_0 + \sqrt{\tilde{E}_0^2 - m^2}}{m}\right)\Bigg]. \quad (28)$$

Here it is understood that the potential $V$ is always restricted to $V < E_0 - m$, as otherwise the decay is energetically forbidden. There is no energy-balance restriction on negative values of $V$. We denote the total transition rate in the absence of the field by $\Gamma_0$, which is obtained from (28) by setting $\tilde{E}_0 = E_0$. The fractional change in the transition rate in the presence of the field is given by

$$\frac{\delta\Gamma}{\Gamma_0} \equiv \frac{\Gamma - \Gamma_0}{\Gamma_0}. \quad (29)$$

For $|V| \ll E_0 - m$, it can be shown that (29) is approximately linear in $V$, and hence one may write

$$\delta\Gamma/\Gamma_0 = A_1(E_0) \cdot V, \quad (30)$$

where $A_1(E_0)$ is a function of $E_0$. A plot of $A_1$ versus $E_0 - m$ is shown in Fig. 12(a). The results for $^3$H and free-neutron decay are

$$\left.\frac{\delta\Gamma}{\Gamma_0}\right|_{^3\mathrm{H}} \approx 0.2 \times \left(\frac{V}{1\ \mathrm{keV}}\right), \qquad |V| \ll E_0 - m \approx 18.6\ \mathrm{keV}, \quad (31)$$

$$\left.\frac{\delta\Gamma}{\Gamma_0}\right|_{n} \approx 0.005 \times \left(\frac{V}{1\ \mathrm{keV}}\right), \qquad |V| \ll E_0 - m \approx 780\ \mathrm{keV}. \quad (32)$$

For a given $|V|$, we expect decays with smaller $E_0$ to experience larger fractional changes in decay-rate. Thus, $V = \pm 200$ eV would give a fractional change on the order of $\pm 0.1\%$ in the transition rate for free-neutron decay, and a fractional change on the order of $\pm 4\%$ in the $^3$H transition rate. It follows that decay-rates can be very sensitive to small changes in the available phase-space energy. As an example, the matrix elements for free-neutron decay and $^3$H decay are very similar (neglecting small meson-exchange corrections in $^3$H), and thus the difference in the respective decay rates is primarily due to the phase-space available to the decay products. Even though $E_0$ in free-neutron decay is larger than $E_0$ in $^3$H decay by only a factor of 2.5, the free-neutron decay rate is larger than the $^3$H decay rate by a factor of $\approx 6 \times 10^5$. Thus, even potentials that are quite small compared to $E_0$ are capable of producing observable changes in decay rates.

In the case of a static, spatially constant, isotropic field, the existence of spin-independent potentials $V$ substantially larger than 1 keV is perhaps implausible. However, in Sect. 5.4 we consider the case of time-varying external fields due to solar-neutrinos passing through a sample of radioactive material. In such a circumstance, the interaction energy may become very large for a solar neutrino which passes close to a specific nucleus. Thus, because it will

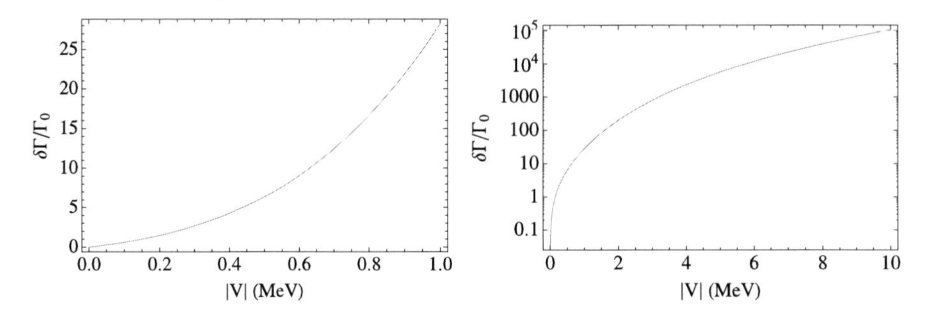


**Fig. 13** The fractional change in the free-neutron decay rate, $\delta\Gamma/\Gamma_0$, predicted by the PSP formalism for a static, homogeneous, isotropic, spin-dependent potential field which has an interaction energy $V=-|V|$ with the emitted antineutrino. These plots were generated using (35), which uses the proper relativistic integration limits. Note that the same function is exhibited in both plots, but the vertical scales are linear (*left*) and log (*right*)

have some relevance in Sect. 5.4, the exact fractional change in transition rate as a function of $|V|$ up to 1 GeV is plotted in Fig. 13. Nonetheless, it very well may be that the current formalism breaks down at such large interaction energies.

We note here that for sufficiently large negative values of $V$, some care is needed with regard to the upper integration limit in (28). The limit $E_{\rm max}=E_0-V$ assumes that the parent nucleus is left with negligible kinetic energy. However, since the additional amount of energy $-V$ is available to all the decay products, at sufficiently high negative $V$, the proton can acquire a significant amount of kinetic energy. The electron attains its maximum energy when its 3-momentum $\mathbf{p}$ is equal and opposite to that of the daughter nucleus, while the neutrino is emitted at rest. For a parent P which $\beta$-decays to a daughter D, the energy balance is

$$m_{\rm P}-V=E_{\rm D}+E_{\rm max}=\sqrt{m_{\rm D}+p^2}+\sqrt{m^2+p^2}. \tag{33}$$

Solving for $p\equiv|\mathbf{p}|$ gives the maximum electron energy

$$E_{\rm max}(V)=\sqrt{m^2+p^2}=\frac{1}{2}\sqrt{\frac{m^2+(m_{\rm P}-V)^2-m_{\rm D}^2}{m_{\rm P}-V}}, \tag{34}$$

where again we restrict the potential to $V<m_{\rm P}-m_{\rm D}-m$. For free-neutron decay, the maximum electron energy given by (34) is very close to $E_0-V$ up to $V\sim 25$ MeV, after which point they start to differ significantly. Since different nuclei will have differing points of divergence, it is generally more accurate to use

$$\Gamma=\Gamma(V)=\int_m^{E_{\rm max}(V)}dE\frac{d\Gamma}{dE}=\int_m^{E_{\rm max}(V)}dE\frac{2G_F^2}{\pi^3}E\sqrt{E^2-m^2}(E_0-V-E)^2, \tag{35}$$

with $E_{\rm max}(V)$ given by (34), and hence Fig. 13 exhibits plots derived from (35), and not (28).

We now proceed to discuss a point related to the interpretation of the PSP used in (25). In (21) we rightly identify the factor $E_0\equiv m_n-m_p$ as the energy available to the decay products, assuming zero proton recoil kinetic energy. However, the constants $m_n$ and $m_p$ that appear in (21) arise from factors of $m_n$ and $m_p$ that appear in the invariant amplitude, as well as in the energy-balance-derived integration limits used at earlier stages in the calculation.

Hence, it is not completely clear that in modifying $E_0$ in (21), one is modifying only the phase-space. Consistent with this concern is the observation that the functional dependence of (18) on $m_n$ and $m_p$ is not simply through the combination $E_0 = m_n - m_p$. It is perhaps more accurate to regard the PSP in this context as describing the decay of a neutron of mass $m_n - V$. Although this makes sense for a neutron at rest, whose entire mass becomes energy available to the decay products, (just as $-V$ is the energy available to the decay products), if the neutron mass is significantly altered, it is not clear that the approximations in (20) continue to hold. Furthermore, it may not make sense to regard the interaction energy $V$ as modifying the masses of the proton, electron, or the antineutrino: not only because one could easily end up with negative masses and various kinematic absurdities, but also because in such a case the energy $-V$ is no longer energy that is available to all the decay products in the same fashion as the neutron rest energy. There is thus some degree of ambiguity in how to accurately account for the presence of $V$ within the PSP formalism. However, these concerns are mitigated by the fact that for potentials $|V| \lesssim 1$ GeV, the fractional variation in transition rate computed using Eqs. (18) and (19), with a neutron mass $m_n - V$, and with the integration limit given by (34), agrees with the fractional variation given by (35) to within an order of magnitude. Thus, while there may be other reasons why the calculation becomes invalid for such large $|V|$, it appears that there is some level of internal consistency in using (35) for $|V| \lesssim 1$ GeV.

#### 5.3.2 *β-Decay in a Static, Homogeneous, Spin-Dependent Field*

We now consider $\beta$-decays in the presence of a static, homogeneous, spin-dependent field, which is of some interest for a number of reasons. First, as discussed in Sect. 5.2.2, it is a case that has been well investigated in the context of $\beta$-decay in constant magnetic fields. As such, this allows for a comparison between our PSP formalism and the exact calculations found in the literature. Secondly, for an external field coupling to the proton, neutron, or electron, it may be more likely that the field is spin-dependent, rather than spin-independent, given that the experimental limits on spin-dependent interactions are much less stringent (Fischbach and Talmadge 1999; Fischbach and Krause 1999a, 1999b; Adelberger et al. 2003). Finally, as we will see shortly, such an interaction may offer a novel explanation of the $^3$H end-point anomaly.

We consider the $\beta$-decay of a sample of unpolarized nuclei, and for definiteness we assume that the external field interacts significantly with only one of the decay particles. When comparing our PSP calculation with earlier calculations in the literature for $\beta$-decay in magnetic fields, we will assume that this interaction is with the emitted electron. For all other calculations in this section we shall continue to assume that the interaction is with the emitted antineutrino, and in both cases we will denote the interaction energy with the field by $V$.

In the case of a spin-dependent field we cannot regard all nuclei in the sample as experiencing the same potential. (As we will discuss below, it is precisely this feature of spin-dependent interactions which may explain the $^3$H end-point anomaly, specifically the fact that determinations of $m_\nu^2$ from the $^3$H end-point give negative values (Amsler et al. 2008).) For the case of the emitted electron coupling to a magnetic field, the interaction energy $V = -\boldsymbol{\mu}_e \cdot \mathbf{B}$ (where $\boldsymbol{\mu}_e$ is the electron magnetic dipole moment) depends upon the orientation of the magnetic moment $\boldsymbol{\mu}_e$ of the emitted electron relative to $\mathbf{B}$. We note that a sufficiently strong magnetic field can significantly influence the likelihood of the electron being emitted with a certain spin orientation. For a strong enough field, decays in which an electron is emitted with its spin (magnetic moment) in a direction parallel (antiparallel) with the magnetic field are energetically blocked.

While it may be possible to implement the PSP by not performing the angular integrations in the formulae leading to (18) until an angular-dependent $V$ is accounted for in the phase space, we simplify our discussion by adopting the following model. To approximate a sample of unpolarized decaying nuclei, we imagine that half of the decays produce electrons which experience $V = +\mu_e B$, while the other half of the decays produce electrons which experience $V = -\mu_e B$. The average transition rate for a sample of decaying nuclei then becomes

$$\langle \Gamma \rangle \equiv \lambda = \frac{1}{2}\Gamma(-|V|) + \frac{1}{2}\Gamma(+|V|)\Theta(E_0 - m - |V|) \tag{36}$$

where $\Gamma(V)$ is given by (35), and $|V| = |\mu_e B|$. The step-function, $\Theta(x) = 1$ for $x \geq 0$, $\Theta(x) = 0$ for $x < 0$, accounts for the possible energy-blocking of decays with electron spin emitted antiparallel to the magnetic field. Here we introduce the symbol $\lambda$ to denote the overall, average transition rate for a sample comprised of nuclei which may individually have differing transition rates. A more accurate model would involve an actual averaging over the spin directions of the emitted electrons. Furthermore, it is clear that for sufficiently strong fields, practically all of the decays will experience $V = -|V|$, so that (36) may be in error by a factor of $\sim$2. Given that we are primarily interested in order-of-magnitude estimates for an unpolarized sample, (36) should be adequate for present purposes. We note, however, that the above model could give inaccurate fractional decay-rate variations if there happens to exist a range of $|V|$ where decays with $V = +|V|$ are significantly blocked (so that the above factors of 1/2 become inaccurate), while at the same time $\delta\Gamma(-|V|)/\Gamma_0 \lesssim 1$. For $^3$H decay, we note that $\delta\Gamma(-|V|)/\Gamma_0 \lesssim 1$ occurs only for $|V| < 500$ eV, which is an energy sufficiently small compared to $E_0 - m \approx 18$ keV that decays with $V = +|V|$ should not be significantly blocked. For an allowed $\beta$-decay with $E_0 - m \gtrsim 500$ eV, which includes free-neutron decay, there is no problematic range of $|V|$, although for decays where this is not the case, a more accurate analysis is necessary.

For $|V| \ll E_0 - m$, it can be shown that (29) is approximately quadratic in $V$, and hence one may write $\delta\Gamma/\Gamma_0 \equiv (\langle\Gamma\rangle - \Gamma_0)/\Gamma_0 = A_2(E_0) \cdot V^2$, where $A_2(E_0)$ is a function of $E_0$. A plot of $A_2$ versus $E_0 - m$ is shown in Fig. 12(b). The results for $^3$H and free-neutron decay are

$$\left.\frac{\delta\Gamma}{\Gamma_0}\right|_{^3\mathrm{H}} \approx 0.013 \times \left(\frac{V}{1\ \mathrm{keV}}\right)^2, \quad |V| \ll E_0 - m = 18.6\ \mathrm{keV}, \tag{37}$$

$$\left.\frac{\delta\Gamma}{\Gamma_0}\right|_{n} \approx 10^{-5} \times \left(\frac{V}{1\ \mathrm{keV}}\right)^2, \quad |V| \ll E_0 - m = 780\ \mathrm{keV}. \tag{38}$$

Comparing (37) and (38) with (31) and (32), we see that for a given $|V| \lesssim 1$ keV, the fractional change in decay rate is smaller for the spin-dependent potential. This is expected, given that half of the nuclei decay with $V = +|V|$ and experience a decrease in transition rate which largely cancels the increase in rate experienced by the half of the nuclei which decay with $V = -|V|$. As $|V|$ grows larger, eventually the $V = +|V|$ decays become significantly blocked in the spin-dependent field, and by this point the fractional modifications to decay rates are almost identical for the spin-dependent and the spin-independent fields.

Before proceeding to discuss the modification to the $\beta$-decay spectrum, we pause to compare our PSP calculation, applied in the simplified model of (36), to exact calculations performed in the literature for the case of free-neutron decay in strong magnetic fields. We have computed the ratio of the fractional change in decay-rates predicted by our calculation to those of the exact calculation (Matese and O'Connell (1969), (18)). This ratio is less than 10 for $|V| \lesssim 1$ MeV, corresponding to magnetic fields $B \lesssim 3 \times 10^{14}$ G, although the discrepancy between the two methods grows quite large for stronger magnetic fields. Such a discrepancy is not surprising given that the gap between electron energy levels increases with increasing magnetic field strength, while our calculation ignores phase-space modifications from quantization effects due to the external field. Thus, the PSP formalism may be inapplicable for sufficiently strong static homogeneous spin-dependent fields (or any field where quantization effects become important).

We now discuss the modification to the $\beta$-decay spectrum due to the spin-dependent field. To simplify the discussion, we will assume that $|V| \ll E_0 - m$, so that factors of 1/2 in (36) are fairly accurate. Using (26), the differential transition rate is

$$\left\langle \frac{d\Gamma}{dE} \right\rangle = \frac{1}{2}\frac{d\Gamma}{dE}\bigg|_{V=-|V|} + \frac{1}{2}\frac{d\Gamma}{dE}\bigg|_{V=+|V|}$$

$$= \begin{cases} 2\frac{G_F^2}{\pi^3} E\sqrt{E^2 - m^2}((E_0 - E)^2 + V^2) & \text{for } E \le E_0 - |V|, \\ \frac{G_F^2}{\pi^3} E\sqrt{E^2 - m^2}(E_0 + |V| - E)^2 & \text{for } E_0 - |V| < E \le E_0 + |V|, \end{cases} \tag{39}$$

where the piecewise structure arises from the fact that the $V = +|V|$ and $V = -|V|$ nuclei have different electron end-point energies. We note that (39) bears an interesting resemblance to the expression for the electron spectrum in the absence of an external field, but where the neutrino mass $m_\nu$ is included (Fukugita and Yanagida 2003, p. 263):

$$\frac{d\Gamma}{dE} = \frac{2G_F^2}{\pi^3} E\sqrt{E^2 - m^2} \cdot (E_0 - E)\sqrt{(E_0 - E)^2 - m_\nu^2}. \tag{40}$$

For electron energies $E$ sufficiently far from the end-point such that $m_\nu^2 \ll (E_0 - E)^2$, we may expand (40) in powers of $m_\nu^2/(E_0 - E)^2$ and, dropping higher order terms, we obtain

$$\frac{d\Gamma}{dE} \cong \frac{2G_F^2}{\pi^3} E\sqrt{E^2 - m^2}\left((E_0 - E)^2 - \frac{1}{2}m_\nu^2\right). \tag{41}$$

Comparing (41) and (39) for $E \le E_0 - |V|$, we see that in the region $V^2 \ll (E_0 - E)^2$, *if we set* $V^2 = -m_\nu^2/2$, *then a massless neutrino spectrum in the presence of a spatially constant, spin-dependent real potential* $\pm V$ *is identical to a spectrum with no external field but with negative* $m_\nu^2$. This is quite interesting given that most experiments which fit $^3$H end-point data to a function of the form (40) obtain a negative best-fit value of $m_\nu^2$ (Amsler et al. 2008). Consider, for example, the experiment of Stoeffl and Decman (1995) who find $m_\nu^2 = -130$ eV$^2$. From the previous discussion, a possible explanation of their data is the presence of a long-range spin-dependent field which couples to the final-state neutrino with an interaction energy of $|V| \approx \sqrt{65}$ eV. While these two spectra do look different very close to the end-point, where $(E_0 - E)^2 \lesssim V^2$, counting statistics are generally poor in this region, and the difference may be difficult to resolve in an experiment. Such a mechanism predicts that the electron spectrum should extend past the expected $E_0$ by $\sim|V|$, though again it may be difficult to resolve this in practice. Figure 14 shows Kurie plots for $^3$H for $V^2 = 0$, $V^2 = 65$ eV$^2$, and $m_\nu^2 = \pm 130$ eV$^2$. Returning to the discussion at the beginning of this subsection, we see from (36) that what allows an external potential $V$ to simulate a negative value of $m_\nu^2$ is precisely the fact that $V$ is spin-dependent. In the present simplified model of an unpolarized sample, this spin-dependence leads to two distinct populations of decaying nuclei, those whose interactions are $\pm|V|$. It is straightforward to show that for

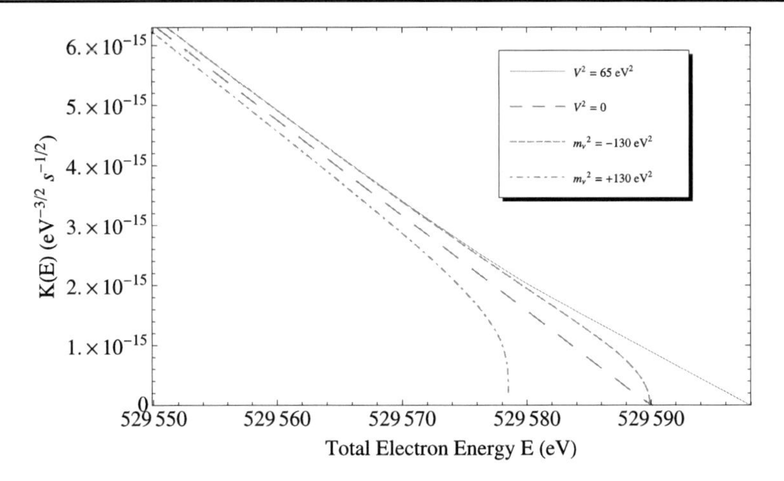


**Fig. 14** The Kurie plot for $^3$H decay, demonstrating how a static, spatially-dependent and spin-dependent potential $V$ produces a Kurie function which mimics the experimentally observed anomalous end-point behavior. The *straight dashed line* is the Kurie function for $V^2 = 0$, $m_\nu^2 = 0$, and intersects the *horizontal axis* at $E = E_0 \approx 529$ keV. The *dotted-dashed curve* which bends down from the straight line is the Kurie function corresponding to (40) for $m_\nu^2 = +130$ eV$^2$, while the *dashed line* which curves down at $E_0$ is the Kurie function for (40) with $m_\nu^2 = -130$ eV$^2$. The *solid line* is the Kurie function for (39) with $V^2 = 65$ eV$^2$. Note that the *solid line* tracks the $m_\nu^2 = -130$ eV$^2$ curve quite well except for $E$ within ~10 eV of the end-point, and that outside this region they both curve upward relative to the expected straight-line Kurie function

a spin-independent interaction, where all nuclei would see the same potential $V$, the Kurie plot would remain a straight line (rather than being curved as in the case $m_\nu^2 \neq 0$), with the only change being $E_0 \to \tilde{E}_0 = E_0 - V$.

It is important to point out that the anomalous $^3$H end-point behavior does not depend upon a *variation* in the external field, but only on the *presence* of an external spin-dependent field. This is to be contrasted with the data described in Sect. 2, where the oscillatory signal is presumably a manifestation of a $\sim 7\%$ *variation* of the field seen by the nuclear sample. These two situations can provide complementary information on the mechanism in question. In the former case one can roughly infer the absolute scale of the potential interaction in question (within the PSP model described above), while in the latter case one can roughly infer the absolute scale of the fractional change in decay rate due to the presence of the mean external field.

## 5.4 Decay Modifications from Time-Dependent External Fields

The correlation of the BNL and PTB data with the Earth–Sun distance, combined with the solar-flare data for $^{54}$Mn, suggests the possibility that nuclear decay rates are being modified via an interaction with solar neutrinos. As reported in Jenkins and Fischbach (2008), this picture is suggested by the coincidence in time between the occurrence of the solar flare of 2006 December 13, and a dip in the $^{54}$Mn count-rate which was detected on the dark side of the Earth. Given that the density of solar neutrinos $\nu_s$ at the Earth is $\rho_{\nu_s} \approx 1$ cm$^{-3}$ (Fukugita and Yanagida 2003), an interaction range $\gg 1$ cm would result in a potential field which is sufficiently homogeneous and static that the results of Sect. 5.3 should be applicable. For an interaction range $\ll 1$ cm, the external field experienced by an individual nucleus could vary rather rapidly in time. Hence, we require a more general formalism that is capable of describing decay processes in an external field with significant time-dependence. We have developed a formalism which partially addresses some of the issues involved, but which does not completely resolve these issues. We nonetheless present the formalism we have developed, in the hope that it offers a starting point for further work which will result in more robust calculational schemes.

As discussed in Sect. 5.2.3, $\beta$-decay in a time-varying electromagnetic field has been previously investigated (Nikishov and Ritus 1964; Ritus 1969; Lyul'ka 1975; Ternov et al. 1978; Ternov et al. 1984; Reiss 1983). Such a case allows for an extension of Furry's formalism, as one can solve for the now time-dependent electron wavefunction in the presence of the external electromagnetic field. However, for more general background fields, one typically cannot use the Dirac equation to solve for the wavefunctions, and thus we will again be resorting to the phase-space prescription (PSP).

We note immediately that in the case of the decay particles interacting with a single solar neutrino, one can no longer regard the source of the external field as experiencing negligible back-reaction. Our calculation will nonetheless neglect changes in the solar neutrino phase space, though this could be a large source of error, especially for the case when the interaction energy $V$ is comparable to the solar neutrino single-particle energy. Perhaps a complete computation in this case would involve the evaluation of Feynman diagrams similar to Fig. 11. With this significant caveat in mind, we shall nonetheless proceed with the naive PSP calculation, as it may turn out to give a rough order-of-magnitude estimate of the effect.

We begin by considering the general case of a particle which experiences a time-dependent transition rate $\Gamma(t)$. At this stage the origin of the time variation of $\Gamma$ is not specified, though we may imagine it arises from a time-dependent potential. From (4), the general solution for the survival probability $P(T)$ is

$$P(T) = \exp\left[-\int_0^T dt\,\Gamma(t)\right], \tag{42}$$

where we have set $P(0) = 1$. For a sample which initially contains $N_0$ active nuclei, each with a different transition rate $\Gamma_i(t)$ (and hence a different probability of survival $P_i(t)$), the expected number of nuclei remaining at a time $T$ is then

$$N(T) = \sum_{i=1}^{N_0} P_i(T) = \sum_{i=1}^{N_0} \exp\left[-\int_0^T dt\,\Gamma_i(t)\right]. \tag{43}$$

The instantaneous decay rate $dN(T)/dt$ and electron energy spectrum of the sample $d\Gamma(T)/dE$ at time $T$ are then given by

$$\frac{dN(T)}{dT} = -\sum_{i=1}^{N_0} \Gamma_i(T)\exp\left[-\int_0^T dt\,\Gamma_i(t)\right], \tag{44}$$

$$\frac{d\Gamma(T)}{dE} = \sum_{i=1}^{N_0} \frac{d\Gamma_i(T)}{dE}\exp\left[-\int_0^T dt\,\Gamma_i(t)\right]. \tag{45}$$

Depending upon the nature of the time-variation of the individual transition rates, an exact analytical evaluation of these formulae could be quite difficult. Hence, in the general case, a Monte-Carlo computer simulation of the system may be the only way to obtain accurate results. Here we will present a simplified analysis which is valid when (42) is close to unity for all nuclei in the sample. While this requirement does allow for an analytic calculation, it ultimately seems to yield unobservably small fractional sample decay rate modifications.

We can cast our physical problem into a more manageable form as follows. According to (42), the probability of decay $P_D$ for the nucleus in a time-interval $T$ is given by

$$P_D = 1 - \exp\left[-\int_0^T dt\, \Gamma(t)\right]. \tag{46}$$

Assuming that the nucleus is unlikely to decay during the time $T$, we have

$$\int_0^T dt\, \Gamma(t) \ll 1, \tag{47}$$

and so to leading order

$$P_D \simeq \int_0^T dt\, \Gamma(t). \tag{48}$$

We now approximate the integral by considering a discrete sum over time intervals $\Delta t = T/n$, where $n$ is sufficiently large that $\Gamma(t)$ is roughly constant over any such time interval $\Delta t$. We label these time-intervals by the integer-valued index $j$, so that the transition rate during the time-interval $t = \Delta t \cdot j \to \Delta t \cdot (j+1)$ is denoted $\Gamma_j$. We then have

$$\int_0^T dt\, \Gamma(t) = \sum_{j=0}^{n-1} \Gamma_j \Delta t. \tag{49}$$

Thus, at least as regards the statistically expected number of decays, whenever (47) is satisfied, *observing a single nucleus with transition rate $\Gamma(t)$ for a time $T$ is equivalent to observing for a time $\Delta t = T/n$ an ensemble of $n$ nuclei, each with a generally different time-independent transition rate $\Gamma_j = \Gamma(j \cdot \Delta t)$*. If (47) is not satisfied, the expected number of decays is much larger in the ensemble than for the single nucleus. We may easily extend this picture to a collection of $N$ nuclei, which are labeled by the index $i$. The expected number of decays during a time $T$ is then

$$\sum_{i=1}^{N} \int_0^T dt\, \Gamma_i(t) = \sum_{i=1}^{N} \sum_{j=0}^{n-1} \Gamma_{ij} \Delta t, \tag{50}$$

where $\Gamma_{ij}$ is the transition rate for nucleus $i$ during the $j^{\text{th}}$ time-interval $\Delta t$. The right hand side effectively describes an ensemble of $N \cdot n$ different nuclei which are observed for a time $\Delta t$.

We can now define an overall, effective transition rate $\lambda$ for the sample of $N$ nuclei by

$$N \cdot \lambda \cdot T \equiv \sum_{i=1}^{N} \int_0^T dt\, \Gamma_i(t) = \sum_{i=1}^{N} \sum_{j=0}^{n-1} \Gamma_{ij} \Delta t, \tag{51}$$

so that

$$\lambda = \frac{1}{N \cdot T} \sum_{i=1}^{N} \sum_{j=0}^{n-1} \Delta t\, \Gamma_{ij} = \frac{1}{N \cdot n} \sum_{i=1}^{N} \sum_{j=0}^{n-1} \Gamma_{ij} \equiv \langle \Gamma \rangle, \tag{52}$$

where the right-hand side of (52) is simply the arithmetic ensemble average of $\Gamma_{ij}$. Given that we require (47) to hold for all nuclei, $\lambda \cdot T \ll 1$ is automatically satisfied, and hence this formalism demands that $(N(T) - N)/N \ll 1$, where $N$ is the initial number of active nuclei.

For the case of $\beta$-decay, we follow (52) in defining the overall, effective differential transition rate $d\lambda/dE$, which gives the observed electron spectrum:

$$\frac{d\lambda}{dE} = \frac{1}{N \cdot n} \sum_{i=1}^{N} \sum_{j=0}^{n-1} \frac{d\Gamma_{ij}}{dE} \equiv \left\langle \frac{d\Gamma}{dE} \right\rangle, \tag{53}$$

where the right-hand side is simply the ensemble average of $d\Gamma_{ij}/dE$ for a given value of the electron-energy $E$. In doing numerical computations of (53), especially for $E > E_0$, one must either carefully control integration limits or use step-functions, because the analytic formulae for $d\Gamma/dE$ are defined mathematically even for non-physical ranges of $E$.

Having established some general results, we now consider the specific case of nuclei interacting with solar neutrinos. In the case of a homogeneous, isotropic, and static potential field, the discussion following (23) suggests that the solar neutrinos are likely not interacting with either the neutrons, protons, or electrons, given that the energies involved would probably give rise to effects that contradict existing experiments. For example, for a sufficiently high solar neutrino flux a spin–spin coupling to nucleons or electrons could lead to a polarization of ordinary matter at a level incompatible with observation. However, in the case of a short-range interaction, the *intermittence* of significant potential interactions between a solar neutrino and any of the decay particles may very well give rise to effects that are not detectable in many types of traditional experiments. Thus, it is possible that a short-ranged solar neutrino interaction could allow for a coupling to not only the emitted antineutrino, but also to the nucleons and the electrons. For concreteness, and also for reasons of continuity and comparison with Sect. 5.3, we shall (arbitrarily) assume that the solar neutrinos are interacting with the emitted antineutrino; however, the formalism we present extends in a straightforward way to interactions with the nucleons or electrons. In this regard, we note that in the event of a spin-dependent solar-neutrino coupling to the nucleons, the possibility of preparing a polarized sample of parent nuclei may allow the proposed mechanism to be amplified to a level that would not be attainable for an interaction with the emitted antineutrino (whose unperturbed spin-distribution is isotropic).

We now consider the specific case of nuclei interacting with solar neutrinos via a spin-independent potential, so that $V$ is a function of the nucleus-neutrino separation only. If the interaction potential has a range much shorter than the average distance between two solar neutrinos, each nucleus can be considered to be interacting significantly with at most one neutrino at a given time. As solar neutrinos stream through the nuclear sample, the distance $r(t)$ between a given nucleus $i$ and the closest solar neutrino is changing with time, and hence the interaction potential $V_i(r(t))$ experienced by the nucleus is also changing with time. As a result, according to the phase-space prescription, each nucleus $i$ is characterized by a time-dependent transition rate $\Gamma_i(t) = \Gamma(V_i(r(t)))$. Recalling that the ensemble of (52) is composed of a system of $N \cdot n$ nuclei, observed for a time $\Delta t = T/n$, each with a time-independent $\Gamma_{ij}$, we see that different members of the ensemble are distinguished solely by the $V(r)$ which they experience. Thus, one could compute $\langle \Gamma \rangle$ if there were a method for computing the fraction $f(r)dr$ of the ensemble which experiences a potential in the range

$V(r) \to V(r+dr)$:

$$\lambda = \langle \Gamma \rangle = \frac{1}{N \cdot n} \sum_{i=1}^{N} \sum_{j=0}^{n-1} \Gamma_{ij} = \int_0^\infty f(r)dr \cdot \Gamma(V(r)). \tag{54}$$

Similarly, in the case of $\beta$-decay, the differential transition rate is given by

$$\frac{d\lambda}{dE} = \left\langle \frac{d\Gamma}{dE} \right\rangle = \int_0^\infty f(r)dr \cdot \frac{d\Gamma(V(r))}{dE} \cdot \Theta(E_0 - E - V(r)), \tag{55}$$

where the step-function $\Theta(x) = 1$ for $x \geq 0$, $\Theta(x) = 0$ for $x < 0$, accounts for the fact that decays with electron energy $E > E_0 - V$ are energetically blocked.

If the potential is spin-dependent, we adopt the simplified model of Sect. 5.3.2. Suppressing the explicit spin-orientation dependence of the potential, and defining $|V(r)|$ to be the maximum positive potential attainable at $r$, we then have that

$$\lambda = \langle \Gamma \rangle = \int_0^\infty f(r)dr \cdot \frac{1}{2}[\Gamma(+|V(r)|) + \Gamma(-|V(r)|)], \tag{56}$$

and

$$\frac{d\lambda}{dE} = \left\langle \frac{d\Gamma}{dE} \right\rangle = \int_0^\infty f(r)dr \cdot \frac{1}{2} \cdot \left[ \frac{d\Gamma(+|V(r)|)}{dE} \cdot \Theta(E_0 - |V(r)| - E) + \frac{d\Gamma(-|V(r)|)}{dE} \cdot \Theta(E_0 + |V(r)| - E) \right]. \tag{57}$$

We further note here that neglecting back-reaction on the solar neutrinos implies that they traverse the sample in a straight line at constant speed.

If the density of solar neutrinos is $\rho_\nu$, then the mean separation between solar neutrinos is $l \sim \rho_\nu^{-1/3}$, and it is unlikely that the closest solar neutrino to a nucleus would be much farther away than $l$. Thus, in computing (54), one can replace the upper integration limit of $\infty$ by a distance on the order of $l$, and approximately calculate $f(r)$ as follows. Consider each of the $N \cdot n$ nuclei in the ensemble described in (52) to be placed at the center of a sphere of volume $1/\rho_\nu$. Each such spherical volume $ij$ is expected to contain roughly one solar neutrino, which is the neutrino closest to the nucleus $i$ during the time step $j$. If one looks at the larger ensemble, a very large number of different solar neutrinos and nuclei are present, and hence the solar neutrino distribution appears approximately random. Therefore, the fraction of the $N \cdot n$ ensemble nuclei which experience a potential in the range $V(r) \to V(r+dr)$ is given by the probability $P(r)dr$ of finding a neutrino inside a spherical shell of radius $r$ and thickness $dr$. We thus have

$$f(r)dr = P(r)dr = \frac{\text{volume of shell}}{\text{volume of sphere}} = \frac{4\pi r^2 dr}{1/\rho_\nu} = 4\pi \rho_\nu r^2 dr, \tag{58}$$

and we set the upper integration limit in (54) to be the sphere radius

$$R = \left( \frac{4}{3} \pi \rho_\nu \right)^{-1/3}. \tag{59}$$

Using (54)–(59), we find that for the short-range spin-independent potential, the overall effective transition rate for the sample of $N$ nuclei is given by

$$\lambda = \langle \Gamma \rangle = 4\pi \rho_\nu \int_0^R r^2 \cdot \Gamma(V(r)) dr. \tag{60}$$

Hence the overall effective differential transition rate, which yields the electron spectrum, is given by

$$\frac{d\lambda}{dE} = \left\langle \frac{d\Gamma}{dE} \right\rangle = 4\pi \rho_\nu \int_0^R r^2 \cdot \frac{d\Gamma(V(r))}{dE} \cdot \Theta(E_0 - E - V(r)) \cdot dr. \tag{61}$$

For a spin-dependent potential,

$$\lambda = \langle \Gamma \rangle = 4\pi \rho_\nu \int_0^R r^2 \cdot \frac{1}{2} \big[ \Gamma(V(r)) + \Gamma(-V(r)) \big] dr, \tag{62}$$

and

$$\frac{d\lambda}{dE} = \left\langle \frac{d\Gamma}{dE} \right\rangle = 4\pi \rho_\nu \int_0^R r^2 \cdot \frac{1}{2} \cdot \left[ \frac{d\Gamma(+|V(r)|)}{dE} \cdot \Theta(E_0 - |V(r)| - E) + \frac{d\Gamma(-|V(r)|)}{dE} \cdot \Theta(E_0 + |V(r)| - E) \right] dr. \tag{63}$$

We now demonstrate that the perturbation in a sample's decay parameter is proportional to $\rho_\nu$ when the range of the interaction is much smaller than $R \sim \rho_\nu^{-1/3}$. We consider a sample composed of $N$ identical nuclei. Let $\lambda = \lambda_0 + \delta\lambda$, where $\lambda_0$ is the decay parameter in the absence of solar neutrinos, and write $\Gamma(V(r)) = \Gamma_0 + \delta\Gamma(V(r))$, where $\Gamma_0 = \lambda_0$ is the transition rate for a single nucleus in the absence of solar neutrinos. Considering the spin-independent potential for simplicity, (60) becomes

$$\lambda_0 + \delta\lambda = 4\pi \rho_\nu \int_0^R r^2 \cdot \big[ \Gamma_0 + \delta\Gamma(V(r)) \big] dr. \tag{64}$$

By virtue of (59) we then have

$$4\pi \rho_\nu \int_0^R \Gamma_0 r^2 dr = \Gamma_0 = \lambda_0, \tag{65}$$

so that (64) becomes

$$\delta\lambda = 4\pi \rho_\nu \int_0^R r^2 \cdot \delta\Gamma(V(r)) dr. \tag{66}$$

Other than the factor of $4\pi\rho_\nu$ multiplying the integral, the only $\rho_\nu$ dependence of (66) is through the upper integration limit $R$. However, if the range (denoted $r_0$) of the potential $V$ is much less than $R \sim \rho_\nu^{-1/3}$, then we expect that the integrand in (66) is negligible near the upper integration limit $R$, and in fact is only significant for $r \lesssim r_0$. Hence, the only $\rho_\nu$ dependence of (66) is through the factor $4\pi\rho_\nu$, so that $\delta\lambda \propto \rho_\nu$.

### *5.4.1 Modeling The Interaction Potential*

Before presenting numerical results, we first discuss how we are modeling the interaction potential $V(r)$. It is well known that in the center-of-mass frame, the non-relativistic 2-body interaction potential is given by the Fourier-transform of the invariant amplitude computed using Feynman rules. We shall assume that such a prescription holds true for the case of relativistic solar neutrinos interacting with nucleons in the lab frame. There is some precedent for this type of treatment: the MSW effect, describing solar neutrino oscillations in matter, can be treated in a non-relativistic potential picture, as was originally done (Bethe 1986); these results were later confirmed using the full machinery of relativistic quantum field theory (Nieves 1989).

One possible choice of potential would be the pseudoscalar exchange potential (Frauenfelder and Henley 1974):

$$V_{12} = \frac{A}{r^3} \cdot [3(\boldsymbol{\sigma}_1 \cdot \hat{\mathbf{r}})(\boldsymbol{\sigma}_2 \cdot \hat{\mathbf{r}}) - \boldsymbol{\sigma}_1 \cdot \boldsymbol{\sigma}_2], \tag{67}$$

where $A$ is a constant depending upon the pseudoscalar mass and its couplings to the two particles, $\mathbf{r}$ is the separation vector between the particles, and $\boldsymbol{\sigma}_1, \boldsymbol{\sigma}_2$ are the spin orientations of the two particles. When considering spin-dependent interactions between solar neutrinos and nuclear decay products, we have chosen to adopt a simplified model in which the spin-orientation dependence of the potential is suppressed, and the nuclei in the sample are considered to be experiencing a potential

$$V(r) = \pm \frac{B}{r^3}, \tag{68}$$

where $B$ is a spin-independent constant roughly equal to $A$, and $r$ is the separation between the particles. In such a picture half of the nuclei experience a potential $V = +|V|$ and half experience a potential $V = -|V|$. We note that while the semi-classical potential (68) formally diverges as $r \to 0$, one should not use such a formula for $r$ smaller than the distance at which higher-order corrections start to dramatically modify the interaction in question. This distance is likely to be much smaller than any distance scales in our problem, and specifically this distance is not related to the quantity $r_{\min}$ to be discussed shortly.

As discussed in Sect. 5.3.2, the PSP formalism breaks down at sufficiently high energies. Energy-level quantization and the solar neutrino phase space are also more important at higher energies, and hence it is likely incorrect to use the potential (68) for arbitrarily small values of $r$. For the interaction of a solar neutrino with an electron, it would be reasonable to suppose that the functional form of $V(r)$ in (68) could remain valid down to distances as small as $\sim 10^{-16}$ cm $\approx 1/M_W$, assuming the semi-classical picture still holds at this energy scale, since existing data support the view that an electron behaves "point-like" down to this scale (Odom et al. 2006). However, since $|V(r)|$ would be sufficiently large at such small distances to violate the approximations we have made in calculating its effects, we will cut off the potential at some value of $r$, hereafter called $r_{\min}$, defining the potential to be

$$V(r) = \pm \begin{cases} V_{\max} & \text{for } r \leq r_{\min}, \\ V_{\max} \cdot (\frac{r_{\min}}{r})^3 & \text{for } r \geq r_{\min}. \end{cases} \tag{69}$$

We stress here that $r_{\min}$ has no physical significance, as it represents only the likely breakdown of our calculational formalism, and hence has no relationship to the range of the potential. Unfortunately, any "leverage" to be gained by the $r$-dependence of the potential is thus lost by using (69), which is essentially a step-function as regards order-of-magnitude calculations. Regrettably, then, the specific $r$-dependence of the potential does not significantly affect the results of our calculation. It follows from the preceding discussion that the numerical results to be presented below, which suggest that the solar neutrinos produce too small an effect to explain existing data, are only valid within the framework of the present approximations. It is thus possible that a more general formalism capable of dealing with smaller values of $r$, and hence with larger $|V(r)|$ could in fact account for existing data in terms of a spin-dependent $V(r)$.

In any numerical calculation, both $V_{\max}$ and $r_{\min}$ must be set. As mentioned in our previous discussion, in order for the PSP to give accurate values of $\Gamma$ we are restricted on the range of $V_{\max}$ we may use. We thus choose $V_{\max} = 1$ MeV, which is the rough energy scale of solar neutrinos, noting that the error in neglecting the solar neutrino phase-space redistribution is likely increasingly significant after this point. We also note that this is the energy past which the PSP calculation starts to differ significantly from the exact calculation of Matese and O'Connell (1969), though perhaps this has no bearing on more general interactions. With $V_{\max}$ set, $r_{\min}$ is then essentially a free parameter. We also give results for $V_{\max} = 400$ MeV, where the PSP prescription is still internally consistent, as discussed in Sect. 5.3.1. We regard these results very cautiously, however, given that at these energies the issue of the solar neutrino phase space may become significant and, additionally, the PSP prescription diverges significantly from the results of Matese and O'Connell (1969) in the case of a constant external magnetic field. We note in passing that even for $V_{\max} = 400$ MeV we satisfy the condition in (47) which allows us to generate the ensemble discussed in Sect. 5.4.

We note that, with energies significantly higher than $E_0$, the use of spin-independent rather than spin-dependent potentials does not result in large changes to the predicted modification of the decay rate. This is in contrast to what was shown in Sect. 5.3.2 where the predicted modification to the decay rates of free neutrons and $^3$H were significantly smaller for a spin-dependent potential when $|V| \ll E_0$. When $|V| \ll E_0$, the $+|V|$ and $-|V|$ contributions largely cancel, assuming that half of the emitted particles experience $+|V|$ and the other half experience $-|V|$. When $|V| \gg E_0$, a spin-dependent potential energetically forbids decays experiencing $+|V|$, so all decays will experience the $-|V|$ interaction, which is the same as if the potential were spin-independent. Thus, spin-independent potentials essentially give the same results as spin-dependent potentials in the PSP when $|V| \gg E_0$.

The preceding considerations allow us to address the implications of the present formalism for solar dynamics. Given the fact that the density of solar neutrinos in the Sun is far greater than it is on Earth, could the mechanism presented here lead to an unacceptably large decay rate for nuclei undergoing radioactive decay in the Sun? (A similar question arises regarding nucleosynthesis in the early universe.) One possibility is that in the case of a spin–spin interaction the isotropy of the solar neutrinos in the Sun might sum to a largely null interaction energy. Note that this would assume that the spin–spin interaction range is much larger than the mean inter-neutrino spacing, so that the interactions of solar neutrinos could be described by a coherent field rather than as the "ballistic" interaction of individual neutrinos. It may also be the case that the neutrino–nucleus interaction energy is a sufficiently small contribution to the ambient energy density in the Sun, that its presence would not materially affect nuclear decay rates in the Sun. Finally, increasing the neutrino density or flux does not necessarily lead to a monotonic increase in nuclear decay rates, as might naively be thought: increasing $\lambda$ eventually leads to a more rapid depletion of the population of atoms available to decay, and this is the principle that underlies the SID effect which we discuss in Sect. 9 below. Taken together, these observations suggest that the role of neutrinos in influencing decay rates of nuclei in the Sun requires a more careful analysis.

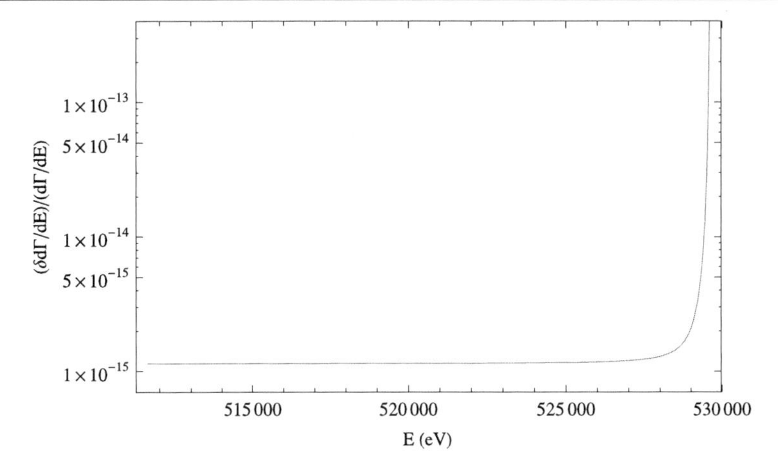


**Fig. 15** The fractional change in the spectrum for $^3$H $\beta$-decay for $V_{\max} = 1$ MeV and $r_{\min} = 4.5$ Å. Note that near $E_0$ where the spectrum is the smallest, the fractional change is the largest. This is expected given that we are now allowing more electrons to decay with energies in that region (and higher)

## 5.5 Calculational Results

For a spin-dependent potential with $V_{\max} = 1$ MeV, we find that for $^3$H

$$\frac{\delta\lambda}{\lambda_0} = \frac{\langle\Gamma\rangle - \Gamma_0}{\Gamma_0} = \frac{\delta\acute{\Gamma}}{\Gamma_0}\bigg|_{^3\mathrm{H}} \approx 1.9 \cdot 10^{-17} \cdot \left(\frac{r_{\min}}{1\ \text{Å}}\right)^3, \tag{70}$$

where $r_{\min} \ll 1$ cm. We study $^3$H since this is the simplest superallowed decay for which the PSP formalism is applicable. The fractional changes in $\Gamma(^3\mathrm{H})$ are exceedingly small, roughly 12 orders of magnitude smaller than the observed changes in the count-rate in the BNL/PTB data for the nuclides studied by these groups (see Sect. 2). That $\delta\Gamma/\Gamma$ is proportional to $r_{\min}^3$ is expected based on (68). This is because, as mentioned in Sect. 5.4.1, our potential approximates a step-function, and thus $\delta\Gamma/\Gamma$ is proportional to the number of affected nuclei, which is proportional to $r_{\min}^3$. We note that to get a value consistent with the $\sim 0.1\%$ scale characteristic of the BNL/PTB data with $V_{\max} = 1$ MeV, $r_{\min}$ must be increased to $\sim 10^{-6}$ m. This would imply extremely large values of the potential even at relatively large distances (Angstroms), and as such may be an indication that the chosen potential is unphysical. We again emphasize that the large disparity between our calculated results for $\Gamma(^3\mathrm{H})$ and the general order-of-magnitude effects suggested by the BNL/PTB data may simply reflect the limitations of our formalism in dealing with small values of $r$, where $|V(r)|$ could be large enough to account for the data. It is also possible that the PSP formalism combined with the semi-classical potential picture developed here may not be appropriate. In either case, it is our calculational scheme which is inadequate, and solar neutrino interactions may still be the responsible mechanism.

It is instructive to investigate the electron spectrum to determine the energy range in which the majority of the additional decays are occurring. Figure 15 shows the fractional change in the $^3$H spectrum for $V_{\max} = 1$ MeV, $r_{\min} = 4.5$ Å. For these values, $\delta\Gamma/\Gamma = 1.7 \times 10^{-15}$. In Fig. 15, we see that for electron energies $m < E < \sim E_0$, the fractional change is one part in $10^{15}$, so that the extra decays are spread somewhat evenly throughout the traditional range of electron energies. We also see that near $E_0$ the fractional change in

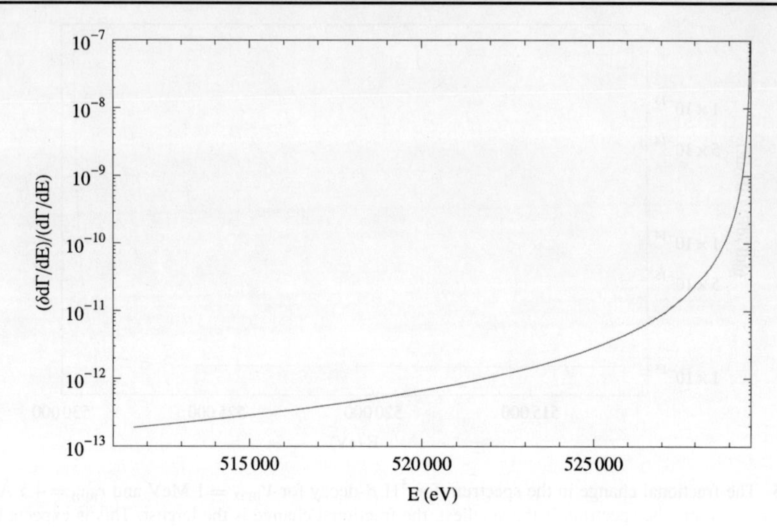


**Fig. 16** The fractional change in the spectrum for $^3$H $\beta$-decay for $V_{max} = 400$ MeV and $r_{min} = 4.5$ Å. Note that near $E_0$ where the spectrum is the smallest, the fractional change is the largest. This is expected given that we are now allowing more electrons to decay with energies in that region (and higher)

the spectrum increases by several orders of magnitude. This is expected, as in the absence of an external potential the number of electrons that are emitted with energies near $E_0$ is relatively small, but in the presence of such a potential there may be many electrons emitted with those energies. There is also a portion of the spectrum that extends beyond $E_0$, as expected in the presence of the additional energy $|V|$ available to the decay products. For the chosen values of $r_{min}$ and $V_{max}$, the fraction of electrons emitted with energies greater than $E_0$ is $\sim 10^{-15}$, which is on the order of $\delta\Gamma/\Gamma$. We also note that fractional changes on the order of $10^{-15}$ to $10^{-13}$ are far too small to explain the $^3$H end-point, within the present formalism.

In the absence of any restrictions on the range of $V_{max}$ in the PSP formalism, one can ask what value would be needed to obtain an observable change in the decay rate on the order of 0.1%. Keeping $r_{min}$ at 4.5 Å, $V_{max}$ must be increased to $\sim$400 MeV in order to obtain $\delta\Gamma/\Gamma \sim 0.1\%$. A plot of the fractional change in the spectrum for $V_{max} = 400$ MeV is given in Fig. 16. While the fractional change over energies $m < E < E_0$ is certainly greater than for the 1 MeV case shown in Fig. 15, the greatest contribution to the change in the decay rate comes from electrons with energies greater than $E_0$, as these make up $\sim$0.1% of the electrons emitted. This would not necessarily be easily seen in the spectrum, however, since the peak beyond $E_0$ is roughly 7 orders of magnitude smaller than the peak in the range from $m$ to $E_0$. If this calculation were to accurately model the physical mechanism responsible, the change in the decay rate might not even be detectable in some experiments. For example, an experiment that directly counted $\beta$-particles may have discriminators set up to screen out energies greater than $E_0$, in which case most of the extra signal would be lost. Assuming that the presence of the potential does not significantly alter the state of the daughter products, however, an experiment set up to detect the de-excitation of a daughter state at a fixed energy would be able to detect the modification to the decay rate. We also note that even with $V_{max}$ at 400 MeV, the modifications to the spectrum near $E_0$ are far too small to explain the $^3$H end-point anomaly, within our present calculational framework.

In summary, using values of $r_{\min}$ and $V_{\max}$ most compatible with the PSP formalism as applied to a short-ranged intermittent potential generated by an interaction with solar neutrinos, the resulting changes in decay rates are too small to explain either the PTB/BNL data or the $^{3}$H end-point anomaly. While increasing $r_{\min}$ and $V_{\max}$ can give changes in decay rates on the order of the PTB/BNL data, we do not currently know how to justify the required values in the context of the PSP formalism. However, our results certainly do not disprove the hypothesis that solar neutrinos could be influencing decay rates. Leaving aside the fact that $^{3}$H was not one of the isotopes used in the BNL/PTB experiments (and may be affected very differently than those used), and that the PSP formalism may break down in the general case, in order to overcome calculational difficulties and limitations some potentially severe additional approximations and assumptions were made. The first was that the scattering of the solar neutrino was negligible, which is almost certainly not true for interaction energies as high as 1 MeV. Thus correctly including the effects of the solar neutrino phase space may give more promising results. The other significant approximation was the implementation of $V_{\max}$, which neglects the increasing strength of the potential at short distances. A calculation which properly addresses these issues would help in understanding the modification of decay rates by solar neutrinos, and thus the possibility that such a mechanism could account for the observed decay data. We finally note that if the solar neutrino interaction in question has a range of $\gg 1$ cm, then the formalism and mechanism of Sect. 5.3 appear to provide for a viable explanation of the observed decay data.

## 6 Seasonal Variation of the Fine Structure Constant and $\alpha$-Decay

In this section we discuss a different model which characterizes a class of theories that could lead to a seasonal variation of radioactive decay rates. All decay processes depend on fundamental constants, including the strengths of the electromagnetic, weak, and strong interactions. Recently, several authors have discussed the possibility that the electromagnetic fine structure constant $\alpha_{\mathrm{EM}}$ might exhibit a seasonal time dependence (Flambaum and Shuryak 2008; Shaw 2007; Barrow and Shaw 2008). Since both $\alpha$- and $\beta$-decay processes depend on $\alpha_{\mathrm{EM}}$ (Uzan 2003), any time variation in $\alpha_{\mathrm{EM}}$ could lead to a modulation of the these decay rates.

A seasonal variation in the fine structure constant would occur if the magnitude of $\alpha_{\mathrm{EM}}$ depended on a scalar field $\phi$ produced by the Sun. Since a light scalar field would look like gravity, one can write the relative variation in $\alpha_{\mathrm{EM}}$ due to $\phi$ as Flambaum and Shuryak (2008)

$$\frac{\delta\alpha_{\mathrm{EM}}}{\alpha_{\mathrm{EM}}} = \frac{k_{\alpha}\delta U_{g}}{c^{2}}, \tag{71}$$

where $k_{\alpha}$ is a dimensionless parameter, $c$ is the speed of light, and $U_{g}$ is the gravitational potential. (In this section, we reinstate factors of $c$.)

For an experiment conducted on the Earth, the greatest variation in the gravitational potential is due to the Earth's motion around the Sun (Flambaum and Shuryak 2008). The eccentricity of the Earth's orbit is small ($\epsilon \simeq 0.0167$), so the Earth–Sun separation can be written as

$$r(t) = r_{\oplus} + \epsilon r_{\oplus}\cos\left[\frac{2\pi(t-t_{0})}{T}\right] + \mathcal{O}(\epsilon^{2}), \tag{72}$$

where $r_\oplus$ is the mean Earth–Sun separation, $T = 1$ y is the orbital period, and $t = t_0$ when the Earth is at its aphelion. The Sun's gravitational potential experienced by the Earth is then

$$U_g(r) = -\frac{GM_\odot}{r} \cong -\frac{GM_\odot}{r_\oplus}\left\{1 - \epsilon \cos\left[\frac{2\pi(t-t_0)}{T}\right] + \mathcal{O}(\epsilon^2)\right\}, \tag{73}$$

where $M_\odot$ is the mass of the Sun. Substituting (73) into (71) leads to an expression for the expected seasonal variation in $\alpha_{\mathrm{EM}}$:

$$\alpha_{EM}(t) \simeq \alpha_{EM}^\infty + \alpha_{EM}^\infty \frac{k_\alpha \epsilon GM_\odot}{r_\oplus c^2} \cos\left[\frac{2\pi(t-t_0)}{T}\right], \tag{74}$$

where $\alpha_{EM}^\infty \simeq 1/137$ is the value of the fine structure constant when $\phi = 0$, which occurs at $r \to \infty$.

These variations in $\alpha_{\mathrm{EM}}$ will affect the $\alpha$-decay rate of unstable nuclei since the $\alpha$-particles must tunnel through the Coulomb barrier (for a review, see Uzan 2003). For a nucleus with atomic number $Z$, one can write its $\alpha$-decay rate as (Uzan 2003; Hodgson et al. 1997)

$$\Gamma \simeq \Gamma_0(\alpha_{\mathrm{EM}}, v) \exp\left[-4\pi Z \alpha_{\mathrm{EM}} \frac{c}{v}\right], \tag{75}$$

where $v/c = \sqrt{2Q/m_\alpha c^2}$ is the speed of the emitted $\alpha$-particle, and $m_\alpha$ is the mass of the $\alpha$-particle. $Q$ is the total change in the system's binding energy,

$$Q = B(Z-2, A-4) + B(Z=2, A=4) - B(Z, A), \tag{76}$$

where $Z$ $(A)$ is the atomic (mass) number of the nucleus, and $B(Z, A)$ is the associated binding energy. The factor $\Gamma_0(\alpha_{\mathrm{EM}}, v)$ has a slow dependence on $\alpha_{\mathrm{EM}}$ and $v$ compared to the exponential term and so will be treated as constant.

To determine the effect of the scalar field $\phi$ on the $\alpha$-decay rate, we substitute (74) into (75), which gives

$$\begin{aligned}\Gamma(t) &\simeq \tilde{\Gamma}_0 \exp\left\{-\alpha_{EM}^\infty \frac{4\pi k_\alpha \epsilon (c/v) ZGM_\odot}{r_\oplus c^2} \cos\left[\frac{2\pi(t-t_0)}{T}\right]\right\}, \\ &\simeq \tilde{\Gamma}_0 - \tilde{\Gamma}_0 \left\{\alpha_{EM}^\infty \frac{4\pi k_\alpha \epsilon (c/v) ZGM_\odot}{r_\oplus c^2} \cos\left[\frac{2\pi(t-t_0)}{T}\right]\right\}\end{aligned} \tag{77}$$

where

$$\tilde{\Gamma}_0 \equiv \Gamma_0(\alpha_{\mathrm{EM}}, v) \exp\left[-4\pi Z \alpha_{\mathrm{EM}}^\infty \frac{c}{v}\right], \tag{78}$$

is independent of time. The relative change in the decay rate is then

$$\frac{\delta\Gamma(t)}{\Gamma} \equiv \frac{\Gamma(t) - \tilde{\Gamma}_0}{\tilde{\Gamma}_0} = \left(\frac{\delta\Gamma_0}{\Gamma}\right)_{k_\alpha} \cos\left[\frac{2\pi(t-t_0)}{T}\right], \tag{79}$$

where the amplitude of the relative decay oscillation is

$$\left(\frac{\delta\Gamma_0}{\Gamma}\right)_{k_\alpha} \equiv -\alpha_{EM}^\infty \frac{4\pi k_\alpha \epsilon (c/v) ZGM_\odot}{r_\oplus c^2}. \tag{80}$$

Thus, the effect of $\phi$ on alpha decay would be a sinusoidal variation in the decay rate in phase (or exactly out of phase, depending on the sign of $k_\alpha$) with the Earth–Sun separation.

To explain the seasonal variation of the $\alpha$-decay rate of $^{226}$Ra observed recently by the PTB (Siegert et al. 1998), which has an amplitude (Jenkins et al. 2008)

$$\left(\frac{\delta\Gamma_0}{\Gamma}\right)_{\text{Ra}} \sim 3\times 10^{-3}, \tag{81}$$

(80) requires that

$$k_\alpha = \frac{r_\oplus c^2}{4\pi\alpha_{EM}^{\infty}\epsilon(c/v)ZGM_\odot}\left(\frac{\delta\Gamma_0}{\Gamma}\right)_{\text{Ra}}. \tag{82}$$

Using $Q = 4.9$ MeV for the $\alpha$-decay of $^{226}$Ra (NNDC 2008), $v/c \simeq 0.051$, together with $GM_\odot/r_\oplus c^2 \simeq 9.8\times 10^{-9}$, (82) then gives

$$k_\alpha \simeq 4.5\times 10^7, \tag{83}$$

which is significantly larger than the limit $k_\alpha = (-5.4\pm 5.1)\times 10^{-8}$ obtained by Barrow and Shaw from a recent atomic physics experiment (Rosenband et al. 2008). Therefore it is highly unlikely that the seasonal variation of the $^{226}$Ra decay rate seen by the PTB group can be attributed to effects of a single scalar field $\phi$. However, this model illustrates how a seasonal variation of a fundamental constant (e.g. $\alpha_{EM}$ or $m_e/m_p$) affects nuclear decay rates, and it is conceivable that a more elaborate model taking into account variations in more than one fundamental constant could lead to observable effects that would have escaped detection in other experiments.

## 7 Discussion of Critical Papers

The appearance of our original papers (Jenkins et al. 2008; Jenkins and Fischbach 2008) motivated a search for archived data sets which might contain useful results. Norman et al. (2009) examined data for $^{22}$Na, $^{44}$Ti, $^{108}$Ag$^{\text{m}}$, $^{121}$Sn$^{\text{m}}$, $^{133}$Ba, and $^{241}$Am and found no evidence for an annual variation of these decay rates at a level below that detected in both the BNL and PTB data sets. It is possible that the BNL/PTB data experienced some overlooked (but not fundamental) systematic effect which was not present in the data of Norman et al. (2009). However, since it is likely that different nuclides would be sensitive in different degrees to any external influence, it is also possible that $^{32}$Si, $^{36}$Cl, and $^{226}$Ra are simply more sensitive "detectors" than the nuclides studied by Norman et al. (2009). In any case, although the data in Norman et al. (2009) are quite interesting, they do not necessarily contradict the BNL/PTB data presented in Jenkins et al. (2008), or the solar flare data of Jenkins and Fischbach (2008).

We turn next to the analysis by Cooper (2009) of data from the radioisotope thermoelectric generators (RTGs) obtained from the Cassini mission, searching for a variation of the decay rate of $^{238}$Pu as a function of the distance of Cassini from the Sun. The premise of this analysis is that since the energy produced by the RTG comes from $^{238}$Pu $\alpha$-decay, the indication that the PTB data from $^{226}$Ra (which is also an $\alpha$-emitter) shows a variation with the Earth–Sun distance suggests that the Cassini data should as well. Cooper analyzed two years of Cassini data and set a limit of $0.84\times 10^{-4}$ on a contribution varying as $1/R^2$ in the $^{238}$Pu decay rate. In what follows we temporarily set aside questions about the reliability and interpretations of the RTG data, and assume that Cooper's analysis is correct.

Notwithstanding the fact that $^{238}$Pu decays via $\alpha$-emission, as does $^{226}$Ra, there is a difference in their respective decay chains which may play a role in understanding the RTG data. $^{238}$Pu ($T_{1/2} = 87.7$ y) decays to $^{234}$U ($T_{1/2} = 246{,}000$ y), and hence for practical purposes only $\alpha$-decays contribute significantly to the Cassini RTG. By way of contrast, the dominant decay chain for $^{226}$Ra is more complicated (Chisté et al. 2008): $^{226}\mathrm{Ra} \xrightarrow{1600\,\mathrm{y}} {}^{222}\mathrm{Rn} \xrightarrow{3.8\,\mathrm{d}} {}^{218}\mathrm{Po} \xrightarrow{3.1\,\mathrm{m}} {}^{214}\mathrm{Pb} \xrightarrow{27\,\mathrm{m}} {}^{214}\mathrm{Bi} \xrightarrow{20\,\mathrm{m}} {}^{214}\mathrm{Po} \xrightarrow{160\,\mu\mathrm{s}} {}^{210}\mathrm{Pb} \xrightarrow{22\,\mathrm{y}} {}^{210}\mathrm{Bi} \xrightarrow{5\,\mathrm{d}} {}^{210}\mathrm{Po} \xrightarrow{140\,\mathrm{d}} {}^{206}\mathrm{Pb}$. Thus a sample of $^{226}$Ra quickly comes to equilibrium with daughters $^{214}$Pb and $^{214}$Bi undergoing $\beta$-decay (Christmas et al. 1983; Siegert et al. 1998). As noted in Sect. 2, it then follows that if we accept the Cooper analysis at face value, a possible explanation of the $^{226}$Ra data (which evidence a time-varying signal), and the Cassini RTG data (which do not) is that the dominant effects of the underlying mechanism appear in $\beta$-decays.

Support for this conjecture comes from data reported by Ellis (1990) who used a $^{238}$PuBe neutron irradiator to carry out *in vivo* neutron activation analysis in humans. Ellis observed that when the $^{238}$PuBe irradiator was used to initiate the reaction $^{55}\mathrm{Mn}(n,\gamma)^{56}\mathrm{Mn}$, the resulting $\gamma$ counts in the 847 keV photopeak exhibited a "… seasonal difference of approximately 0.5% … between the winter and summer months." Given that the higher counts occurred during the winter months, Ellis' data agree in both magnitude and phase with the BNL and PTB data. Since both the Cooper and Ellis data depend on $^{238}$Pu decay, it is not clear at present what the origin is of the (apparent) discrepancy in the reported data, but one possibility may be problems with modeling the behavior of RTGs in a space environment, or some unknown systematic effect in the Ellis experiment.

## 8 Discussion of $^{14}$C Decay

Following the appearance of Jenkins and Fischbach (2008) and Jenkins et al. (2008), a paper appeared (Sanders 2008) discussing the possible relevance of our work to $^{14}$C dating. Here we add a few comments on $^{14}$C dating.

Chiu et al. (2007) analyzed data on $^{14}$C atmospheric concentrations (denoted by $\Delta^{14}$C) in an effort to understand the causes of $\Delta^{14}$C fluctuations. Although a precise knowledge of $T_{1/2}(^{14}\mathrm{C})$ is not necessary for some dating applications, if $^{14}$C dates are calibrated against tree-ring dates, $T_{1/2}(^{14}\mathrm{C})$ is needed to understand and interpret the causes of $\Delta^{14}$C fluctuations. Chiu et al. (2007) give an extensive discussion of the different determinations of $T_{1/2}(^{14}\mathrm{C})$ that have been reported in the literature, along with various discrepancies among the published values.

Returning to the discussions of Sect. 4 and Sect. 5.3, we note that if $\lambda$ is in fact modified by external fields, some of the reported discrepancies could have arisen from measurements carried out at different times (when the external field had different values), or perhaps through the use of different measurement techniques (see Sect. 4). As an example of the latter effect, consider a comparison of $T_{1/2}(^{14}\mathrm{C})$ obtained from a calorimeter experiment versus a determination of $T_{1/2}(^{14}\mathrm{C})$ by a direct counting experiment conducted at the same time. In the presence of a constant potential interaction $V$ of the type described in Sect. 5.3, the decay parameter $\lambda$ will differ from its unperturbed value $\lambda_0$. The exact modified decay parameter would presumably be measured in a direct counting experiment, and is given by

$$\lambda = \lambda_{\mathrm{count}} = \lambda_0 \cdot \left[1 + \frac{1}{\lambda_0}\int_m^{E_0+|V|} \delta\left(\frac{d\lambda}{dE}\right) \cdot dE\right], \tag{84}$$

where $\delta(d\lambda/dE)$ is the modification to the spectrum due to the potential interaction. However, as we now demonstrate, the calorimetric method may not yield the exact modified decay parameter. If one knows the number of active nuclei $N$ in the calorimeter, the decay parameter $\lambda_{\rm cal}$ is determined by measuring the collected power $P$ and setting

$$P = N \cdot \overline{K} \cdot \lambda_{\rm cal}, \tag{85}$$

where $\overline{K}$ is the expected mean electron kinetic energy $K = E - m$. *It follows that one can obtain incorrect values of $\lambda_{\rm cal}$ if an incorrect value of $\overline{K}$ is used.* This may occur if the assumed value of $\overline{K}$ was measured at a time when the external field differed, or if it is computed under the assumption that there is no perturbing external field. For example, in the absence of a perturbing field, $\overline{K}$ for an allowed $\beta$-decay with endpoint energy $E_0$ is given by

$$\overline{K}_0 \equiv \overline{K}|_{V=0} = \frac{1}{\lambda_0}\int_m^{E_0} dE \frac{d\lambda_0}{dE} \cdot K = \frac{\int_m^{E_0} dE\sqrt{E^2 - m^2}E(E_0 - E)^2 \cdot K}{\int_m^{E_0} dE\sqrt{E^2 - m^2}E(E_0 - E)^2}, \tag{86}$$

where $d\lambda_0/dE$ is the differential transition rate in the absence of external fields. In the presence of a potential $V = -|V|$ (chosen negative for definiteness), the actual measured power would be given by

$$P = N \cdot \left[\int_m^{E_0} dE \frac{d\lambda_0}{dE} \cdot K + \int_m^{E_0+|V|} dE\delta\left(\frac{d\lambda}{dE}\right) \cdot K\right] \tag{87}$$

$$= N \cdot \lambda_0 \cdot \left[\overline{K}_0 + \frac{1}{\lambda_0}\int_m^{E_0+|V|} dE\delta\left(\frac{d\lambda}{dE}\right) \cdot K\right]. \tag{88}$$

Note that (86) was used in obtaining (88) from (87). Using $\overline{K} = \overline{K}_0$ in (84), combining it with (85) and (88), and then expanding to first order in $(\lambda_{\rm count} - \lambda_0)/\lambda_0$, we find that

$$\frac{\lambda_{\rm cal}}{\lambda_{\rm count}} = 1 + \frac{1}{\overline{K}_0\lambda_0}\int_m^{E_0+|V|} dE\delta\left(\frac{d\lambda}{dE}\right) \cdot (K - \overline{K}_0). \tag{89}$$

A similar result holds for a positive potential $V = +|V|$. We see that the calorimetric method yields a decay parameter which could differ from the true value $\lambda = \lambda_{\rm count}$. This discrepancy is solely due to the use of $\overline{K}_0$ in (85), which is no longer valid if the decay parameter is modified by an external field. While fractional count-rate modifications significantly larger than those indicated in Sects. 2 and 3 would be required to explain the $\sim 5\%$ discrepancies present in the $^{14}$C literature, these considerations nonetheless underscore the point made earlier in Sect. 4: it is necessary to exercise care when comparing half-life measurements using different techniques, when the decay parameter $\lambda$ is influenced by an external source.

$^{14}$C decays are particularly interesting from the present point of view because of the extensive literature comparing $^{14}$C dates and tree-ring dates. Here we note that the tree-ring record presents a cumulative memory of not only the time-dependent production rate of $^{14}$C, but also of its possible time-dependent decay if $\lambda(t)$ is not a constant. Specifically, an event such as the 1859 solar storm (Odenwald and Green 2008) could in principle influence the decay rate of the $^{14}$C atoms already embedded in a tree ring sample, but not more recent atoms. Given improving $^{14}$C dating techniques it may be of interest to search for evidence of major storms by looking for possible shifts in $^{14}$C trend lines, as discussed in Sect. 4.

## 9 Ongoing and Future Experiments

We present in this Section a brief discussion of several experiments, both ongoing and future, with which our group is presently involved. We will defer for the present a discussion of experiments by other groups which are in progress or being planned. The experiments we discuss include generic searches for a time-dependence of the decay parameter $\lambda(t)$, as well as experiments which depend specifically on the suggestion that solar neutrinos may be the source of time-variation in the count-rates seen in the BNL/PTB data.

1. A long-term study of the decay of $^{54}$Mn is underway at Purdue. The results of Sect. 3 suggest that $^{54}$Mn decay may be sensitive to perturbations originating from the Sun, and this suggestion is qualitatively supported by the data we have acquired since December 2006. In contrast to Dec. 2006, which was a period of considerable solar activity, the period since then has been unusually quiet (Phillips 2009). No solar flares of equal or greater magnitude have been detected since then. Interestingly, our $^{54}$Mn data have largely tracked solar behavior: although we recorded a number of dips in the December 2006 $^{54}$Mn data (see Fig. 8), the recent count-rate in this experiment has been devoid of any significant fluctuations. At present a second experiment studying $^{54}$Mn is running at a different location at Purdue with the aim of studying possible correlations between the two detectors.
2. As we have discussed above, one possible explanation of the dip in our $^{54}$Mn count-rate during the solar flare of 2006 December 13 is that it arose from a fluctuation in the flux of solar neutrinos during the flare. In January 2007 we attempted to test this hypothesis by carrying out an experiment at the Brezeale Triga Reactor located at Pennsylvania State University. Although it is natural to suppose that exposing a radioactive sample to the flux of $\bar{\nu}_e$ from a reactor would be a straightforward test of the proposed neutrino mechanism, the practicalities of carrying out such an experiment proved to be more complex. In order to achieve a sufficiently high neutrino flux to provide a meaningful test, the radioactive samples had to be positioned as close to the core as possible. This produced a larger-than-expected $\gamma$-background, so the results of this run were inconclusive. We plan to repeat this experiment with a substantial increase in shielding in the near future.

   Another way to circumvent this problem is to have the sample of $^{54}$Mn directly adjacent to the core, but without the associated electronics. The idea here is to irradiate the $^{54}$Mn sample long enough to produce a sufficient suppression of the decay rate that the shift in decay curves shown in Fig. 9 becomes detectable when the sample is removed and counted. We are in the process of estimating how long an exposure would be required to produce a detectable effect. As noted above, the data collected during the flare of 2006 December 13 indicated that the fractional change in counting rate expected would have been only $\sim 10^{-5}$, which is below our detection sensitivity. To suppress the effects of neutrons transmuting the $^{54}$Mn sample in such an experiment, the sample could be contained in borated polyethylene (HDPE) surrounded by cadmium to absorb neutrons. Since transmutation would in any case decrease the $^{54}$Mn population and hence the count-rate, whereas the neutrino flux might work in the opposite direction, these two effects could presumably be disentangled. We are helped by the fact that the sequence $n + {}^{54}\text{Mn} \rightarrow {}^{55}\text{Mn}$ followed by $n + {}^{55}\text{Mn} \rightarrow {}^{56}\text{Mn}$ produces $^{56}$Mn with $T_{1/2} = 2.58$ h. The decay of $^{56}$Mn produces photons at 846.8 keV and 1810.8 keV and these can be used to estimate the neutron flux into the sample. The neutron flux can also be determined by placing gold foils near the sample and then measuring their neutron-induced activity. Photonuclear reactions involving the $^{54}$Mn sample will also be present but with a low cross-section.

To summarize, the search for displaced decay curves is an attractive experiment because if an effect were seen this could in principle provide a quantitative estimate of the contribution of changing neutrino flux on decay rates.

3. We are in the process of carrying out a potentially cleaner test of the neutrino hypothesis in which the source of neutrinos is the decaying sample itself (Lindstrom et al. 2009; Fischbach et al. 2009). In this experiment the decay rates of two samples of $^{198}$Au($T_{1/2} =$ 2.7 d) are compared, where one sample is a thin foil and the other is a sphere of the same mass and approximate activity. It can be shown that the ratio of the sphere and foil decay rates from this "Self-Induced Decay" (SID) effect is given by ($x = t \ln 2/T_{1/2}$)

$$f(x,\alpha) = \frac{(dN/dt)_{\text{sphere}}}{(dN/dt)_{\text{foil}}} = \frac{1+\alpha}{[1+\alpha(1-\mathrm{e}^{-x})]^2}\frac{N_s}{N_f}. \tag{90}$$

   Here $N_s$ and $N_f$ are the numbers of $^{198}$Au atoms initially present in the sphere and foil respectively, and $\alpha = \delta\Gamma/\Gamma|_{t=0}$. As noted in Fischbach et al. (2009), the functional form of (90) can be understood as follows: initially the decay rate in the sphere exceeds that in the foil. However, by $t \approx T_{1/2}$ the resulting increase in the depletion of the population of decaying atoms in the sphere leads to an overall suppression of the decay rate in the sphere compared to that in the foil. An experiment to search for a nonzero contribution proportional to $\alpha$ in (90) is presently underway at the National Institute of Standards and Technology (Lindstrom et al. 2009), and this will eventually set a limit on $\alpha$ for incident $\bar{\nu}_e$.
4. Since the SID effect described above, and the reactor experiment described previously, set limits on the couplings on $\bar{\nu}_e$ but not on other neutrino species, it is of interest to consider experiments which might provide information on the contributions from $\nu_\mu$ and $\nu_\tau$ and/or their antiparticles. One possibility is to search for a day/night variation in count-rates which could arise from flavor oscillations as neutrinos pass through the Earth. To date there is no evidence for a day/night effect in our data. Another possibility is to look for a fluctuation in count-rates during a solar eclipse. Interestingly data on $^{137}$Cs decay that we obtained during the solar eclipse of 2005 April 08 hinted at an effect, and this served as part of the motivation for a series of experiments we carried out at the Thule Air Base in Greenland during the eclipse of 2008 August 01. The Solar Eclipse at Thule (SEAT) collaboration was a joint effort of the U.S. Air Force Academy, Purdue Air Force ROTC, and the Purdue Physics Department, and collected an extensive set of data on a number of nuclides, which are presently being analyzed.
5. From the discussion in Sect. 4, the presence of extended "flat regions" is evidence for a time-varying decay parameter. Although there is nothing inherently more important about a "flat region" compared to any other "anomalous slope region", flat regions represent a potentially interesting confluence between an effect which is easy to spot, and a theoretical formalism for describing the data. From our $^{54}$Mn data, flat regions may appear more frequently than solar flares. We propose to have several identical experiments running at different locations, and if flat regions were observed over the same time interval in two or more widely separated experiments this could be strong evidence for the presence of a time-dependence of $\lambda(t)$.

## 10 Implications for Detecting the Relic Neutrino Background

If the BNL/PTB data discussed in Sect. 2, and the solar flare data presented in Sect. 3, are in fact due to variations in the flux of solar neutrinos, then one implication of the present work is that radioactive nuclides could serve as real-time neutrino detectors for some purposes. In principle, such "radionuclide neutrino detectors" (RNDs) could be combined with existing detectors, such as Super Kamiokande, to significantly expand our understanding of both neutrino physics and solar dynamics.

One potentially interesting application of such an RND would be to the detection of the predicted 1.95 K background of relic neutrinos which decoupled around 1 second after the Big Bang. This would be exciting, given that all previous proposals to detect the relic neutrino sea seem to be impractical at present (Duda et al. 2001). It is promising that the density of relic neutrinos is expected to be roughly 56 $\text{cm}^{-3}$ per species for a total of 330 $\text{cm}^{-3}$, which is larger than the solar neutrino density at Earth. If we assume that the average speed of the Earth relative to the relic neutrino sea (or equivalently the cosmic microwave background) is $\sim$370 $\text{km}\cdot\text{s}^{-1}$, then the resulting flux of neutrinos incident on an RND would be $\sim 1 \times 10^{10}\ \text{cm}^{-2}\,\text{s}^{-1}$. This is comparable to the estimated solar flux, $\sim 6 \times 10^{10}\ \text{cm}^{-2}\,\text{s}^{-1}$, and potentially comparable to the fluctuation in the neutrino flux detected during the solar flare period discussed in Sect. 3.

The Earth's motion around the Sun results in the sinusoidal variation of its speed through the cosmic microwave background. Using the results of Kogut et al. (1993) and Gelmini and Gondolo (2001), one can show that the amplitude of this variation is $\sim$ 29 km/s, and that the Earth's peak speed occurs at around December 10. It is possible that this modulation could be detected in the signal of a decaying nuclear sample.

There are, however, important differences between the Earth's motion through the solar neutrinos compared to the relic neutrinos. Even if the relic neutrinos are galactically clustered, their density across the solar system should be almost completely uniform. Thus, unlike the case of our annual motion through the solar neutrinos, there would be no annual modulation in the density or flux of relic neutrinos. (The difference in number density due to differential Lorentz contraction throughout the year is completely negligible.) It follows that unless the interaction in question depends on the Earth's velocity through the sea of relic neutrinos, perhaps in a manner similar to the Stodolsky effect (Stodolsky 1975), there may be no annual modulation in decay rates due to such neutrinos. If the interaction in question is significantly energy-dependent, then the slight annual variation in relic neutrino energy seen by a nuclear sample could potentially affect the decay rate. However, the expected mean relic neutrino momentum is approximately $10^{-4}$ eV, compared to roughly 1 MeV for solar neutrinos. Thus, the relic-neutrino signal may be unobservably small in such a case.

Another difference is that solar neutrinos all have the same spin-orientation as detected on the Earth, whereas the relic neutrino background is unpolarized in the rest frame of the cosmic microwave background, and only very slightly polarized in the frame of the Earth. While this should make little difference for a short-ranged interaction, if the interaction in question is spin-dependent and sufficiently long-ranged, the nucleus (or any of its decay products) may experience a large number of simultaneous interactions which largely cancel. Furthermore, the relic neutrino background is expected to have roughly equal numbers of particles and antiparticles, and if these give potentials of opposite sign, a long-ranged interaction may result in an effect which largely cancels.

Nonetheless, in view of the current state of ignorance regarding the nature of the $\nu$-nucleon interaction in question, it is perhaps not unreasonable to consider performing an experiment. Given that a signal due to relic neutrinos is likely to be quite small, one might need a very well controlled experiment with extremely high statistics. A modulation in count-rate data in phase with $\sim$ December 10/June 10 might indeed be a signature of the relic neutrino background. However, we note that our maximum speed through the relic neutrino background is coincident with our minimum speed through the galactic rest frame to within

$\sim$1 week. Thus, unless the experimental data allow for a phase resolution on the order of several days or less, it may be difficult to uniquely ascribe an observed signal to the relic neutrino background, as opposed to some other galactic effect, such as the WIMP interaction sought in the DAMA experiment (Bernabei et al. 2008). That said, a detection of any signal in this manner, whether due to neutrinos or WIMPs, would represent an exciting experimental result.

## 11 Summary and Conclusions

Our objective in this paper has been to present as complete a picture as possible of previous and current research dealing with the question of whether nuclear decay rates are being perturbed by some as yet unknown mechanism or new force. By "unknown" we mean to exclude phenomena associated with the more-or-less well-understood response of electron-capture processes to changes in the external environment, as discussed in Sect. 1. The motivation for embarking on this investigation comes from the following observations:

1. Evidence for a correlation between nuclear decay rates and Earth–Sun distance, as reported by the BNL and PTB groups (Sect. 2).
2. Detection of dips in the $^{54}$Mn count-rate coincident in time with the solar flare of 2006 December 13 (Sect. 3).
3. Observation of "anomalous slope regions", particularly "flat regions" in the $^{54}$Mn data taken at Purdue, and perhaps in the PTB data as well.
4. A history of reports of periodic effects in nuclear decays as reported by other groups (Table 1).
5. The existence of a significant number of discrepancies in half-life determinations by different groups, using either the same or different techniques (Sects. 2 and 4).
6. The possibility that the same mechanism which could account for the BNL/PTB data and the 2006 December 13 flare data, could also account for the apparently negative value of $m_{\nu}^2$ inferred from $^3$H decay (39)–(41).

Although it may eventuate that some or all of these effects arise from conventional (but presumably not-understood) systematic effects, the fact that so many effects have been reported in the literature by well-known and respected groups, suggests that we treat this problem seriously at present.

Even if we assume that most or all of the reported data are correct, it does not necessarily follow that nuclear decay rates themselves are being affected, as distinguished from the observed count-rates. Stated otherwise, it is possible (for example) that solar activity is affecting our instrumentation so as to simulate a time-varying decay-rate. Even if this were the case, for some purposes this could be quite interesting and useful: as we note in Sect. 3, there was a precursor signal in the $^{54}$Mn count-rate that preceded the actual flare of 2006 December 13. From a practical point of view the possibility of using such a signal to warn of an impending flare is interesting irrespective of what the underlying mechanism might be.

Much of the present paper has been devoted to understanding how solar activity could in fact provide a non-trivial mechanism to account for the data, as we discuss in detail in Sects. 5 and 6. The conceptually simpler of the two mechanisms, which we discuss in Sect. 6, is one based on a variation of fundamental dimensionless constants such as $\alpha_{\text{EM}}$ and $m_e/m_p$ arising from some new field emanating from the Sun. Although we do not have a specific version of such a model at present which would be compatible with existing data, there is also no compelling argument precluding such a model.

The mechanism to which we have devoted most of our attention is a fluctuation in the flux of solar neutrinos, as discussed in Sect. 5. Perhaps the most compelling argument for such a mechanism are the data in Sect. 3 indicating dips in the $^{54}$Mn count-rate coincident in time with a solar flare. In particular, the flare of 2006 December 13, which occurred at 21:37 EST where the data were taken, appear to require a mechanism in which a signal from the Sun travels through the Earth at approximately the speed of light in order to reach our detectors coincident in time with the flare. Since this would be compatible with a change in the solar neutrino flux, but not with some other mechanisms, this observation has served as part of the motivation for our proposed mechanism. Regrettably, in the event of a short-range neutrino interaction, the technical problems we have encountered in trying to implement this model have not yet been solved, and so in this case we do not have a completely formulated mechanism at present. However, for a sufficiently long-ranged solar-neutrino interaction, a potentially viable mechanism to explain the experimental data is presented in Sect. 5.3.

Even in the absence of a theoretical mechanism, there is nothing precluding us from studying time-varying nuclear decay parameters phenomenologically, along the lines described in Sects. 4 and 9, and we are in the process of doing this through a variety of experiments. In designing new experiments, and/or analyzing earlier experiments, it is important to bear in mind one of the important lessons from Sect. 4: in the presence of a time-varying decay parameter $\lambda(t)$ different experiments which have been carried out correctly can nonetheless legitimately infer different values for $T_{1/2}$ depending on the experimental technique and on the time interval over which the data were acquired. This observation could conceivably explain some of the discrepancies noted in point 5 above. Some of the experiments that are in progress are variants of existing experiments measuring half-lives of different nuclides, while others such as the SID effect and the SEAT experiment are conceptually new. Given the likelihood of increased solar activity in the next few years, the possibility of detecting solar flares via the effects they produce in decay rates of nuclei is quite exciting, and would be all the more exciting if several experiments around the world were running at the same time.

**Acknowledgements** The authors are deeply indebted to D. Alburger and G. Harbottle for supplying us with the raw data from the BNL experiment, to H. Schrader for providing to us the raw PTB data, and to A. Hall and M. Fischbach for their many contributions during the early stages of this collaboration. We also express our appreciation to D. Anderson, V. Barnes, P. Belli, R. Bernabei, B. Budick, R. Chrien, B. Craig, T. Downar, D. Elmore, A. Fentiman, V. Flambaum, Y. Fujii, G. Greene, J. Heim, M. Jones, A. Karam, A. Lasenby, R. Lindstrom, L. Lobashev, A. Longman, E. Merritt, T. Mohsinally, D. Mundy, P. Muzikar, A. Overhauser, L. Polsky, R. Reifenberger, S. Revankar, B. Revis, J. Schweitzer, M. Slothaak, S. Soloway, P. Sturrock, R. Sutter, N. Titov, A. Treacher and J. Uzan, for their assistance and for many helpful conversations. The work of E.F. was supported in part by the U.S. Department of Energy under Contract No.DE-AC02-76ER071428.

## References

E.G. Adelberger, E. Fischbach, D.E. Krause, R.D. Newman, Constraining the couplings of massive pseudoscalars using gravity and optical experiments. Phys. Rev. D **68**(6), 062002 (2003). doi:10.1103/PhysRevD.68.062002

D.E. Alburger, G. Harbottle, E.F. Norton, Half-life of $^{32}$Si. Earth Planet. Sci. Lett. **78**, 168–176 (1986)

C. Amsler et al., Particle data group. Phys. Lett. B **667**(1) (2008)

J. Anderson, G. Spangler, Serial statistics: is radioactive decay random? J. Phys. Chem. **77**(26) (1973)

J.D. Barrow, D.J. Shaw, Varying alpha: New constraints from seasonal variations. Phys. Rev. D **78**(6), 067304 (2008). arXiv:0806.4317v1 [hep-ph]

Y.A. Baurov et al., Experimental investigation of changes in $\beta$-decay rate of $^{60}$Co and $^{137}$Cs. Mod. Phys. Lett. A **16**, 2089–2101 (2001)

Y.A. Baurov et al., Experimental investigation of changes in beta-decay count rate of radioactive elements. Phys. At. Nucl. **70**(11), 1825–1835 (2007)

H. Becquerel, On the invisible rays emitted by phosphorescent bodies. Comptes Rendus **122**, 501–503 (1896)

F. Begemann, K. Ludwig, G. Lugmair, K. Min, L. Nyquist, P. Patchett, P. Renne, C.Y. Shih, I. Villa, R. Walker, Call for an improved set of decay constants for geochronological use. Geochim. Cosmochim. Acta **65**(1), 111–121 (2001)

V. Berestetskii, E. Lifshitz, L. Pitaevskii, *Relativistic Quantum Theory, Part 2* (Pergamon, Oxford, 1979), pp. 402–408

R. Bernabei et al., First results from DAMA/LIBRA and the combined results with DAMA/NaI. Eur. Phys. J. C **56**, 333–355 (2008)

H. Bethe, Possible explanation of the solar neutrino puzzle. Phys. Rev. Lett. **56**(12), 1305–1308 (1986)

N. Bogoliubov, D. Shirkov, *Introduction To The Theory of Quantized Fields* (Wiley, New York, 1959), pp. 260–262

G.W. Bruhn, Does radioactivity correlate with the annual orbit of Earth around Sun? Aperion **9**(2), 28–40 (2002)

V. Chisté, M. Be, C. Dulieu, Evaluation of decay data of radium-226 and its daughters, in *Proceedings of the International Conference on Nuclear Data for Science and Technology, April 22–27, 2007, Nice, France*, ed. by O. Bersillon, F. Gunsing, E. Bauge, R. Jacqmin, S. Leray. (EDP Sciences, France, 2008). doi:10.1051/ndata:07122. http://nd2007.edpsciences.org/

T.C. Chiu, R. Fairbanks, L. Cao, R. Mortlock, Analysis of the atmospheric $^{14}$C record spanning the past 50,000 years derived from high-precision $^{230}$Th/$^{234}$U/$^{238}$U, $^{231}$Pa/$^{235}$U and $^{14}$C dates on fossil corals. Quart. Sci. Rev. **26**, 18–36 (2007)

P. Christmas, R.A. Mercer, M.J. Woods, S.M. Judge, $^{210}$Bi and the apparent half-life of a sealed $^{226}$Ra source. Int. J. Appl. Radiat. Isot. **34**(11), 1555 (1983)

P.S. Cooper, Searching for modifications to the exponential radioactive decay law with the Cassini spacecraft. Astropart. Phys. **31**(4), 267–269 (2009). arXiv:0809.4248v1 [astro-ph]

A. Derbin et al., Comment on the paper "realization of discrete states during fluctuations in macroscopic processes". Phys. Usp. **43**(2), 199–202 (2000)

G. Duda, G. Gelmini, S. Nussinov, Expected signals in relic neutrino detectors. Phys. Rev. D **64**, 122001 (2001)

K.J. Ellis, The effective half-life of a broad beam $^{238}$PuBe total body neutron irradiator. Phys. Med. Biol. **35**(8), 1079–1088 (1990)

G.T. Emery, Perturbation of nuclear decay rates. Ann. Rev. Nucl. Phys. **22**, 165–202 (1972)

E.D. Falkenberg, Radioactive decay caused by neutrinos? Aperion **8**(2), 32–45 (2001)

E.D. Falkenberg, Reply to "Does radioactivity correlate with the annual orbit of Earth around Sun?" by G.W. Bruhn. Aperion **9**(2), 41–43 (2002)

L. Fassio-Canuto, Neutron beta decay in a strong magnetic field. Phys. Rev. **187**(5), 2141 (1969)

E. Fischbach, C. Talmadge, *The Search for Non-Newtonian Gravity* (Springer, New York, 1999)

E. Fischbach, D.E. Krause, New limits on the couplings of light pseudoscalars from equivalence principle experiments. Phys. Rev. Lett. **82**(24), 4753–4756 (1999a). doi:10.1103/PhysRevLett.82.4753

E. Fischbach, D.E. Krause, Constraints on light pseudoscalars implied by tests of the gravitational inverse-square law. Phys. Rev. Lett. **83**(18), 3593–3596 (1999b). doi:10.1103/PhysRevLett.83.3593

E. Fischbach et al., *Possibility of a Self-induced Contribution to Nuclear Decays*, 2009 ANS Proceedings (2009, in preparation)

V.V. Flambaum, E.V. Shuryak, How changing physical constants and violation of local position invariance may occur? in *AIP Conf. Proc.*, vol. 995, 2008, pp. 1–11

H. Frauenfelder, E.M. Henley, *Subatomic Physics*, 1st edn. (Prentice-Hall, New Jersey, 1974)

M. Fukugita, T. Yanagida, *Physics of Neutrinos and Applications to Astrophysics* (Springer, Berlin, 2003), p. 181

W. Furry, On bound states and scattering in positron theory. Phys. Rev. **81**(1), 115–124 (1951)

G. Gelmini, P. Gondolo, Weakly interacting massive particle annual modulation with opposite phase in late-infall halo models. Phys. Rev. D **64**, 023504 (2001)

D. Griffiths, *Introduction to Elementary Particles* (Wiley, New York, 1987)

H.P. Hahn, H.J. Born, J. Kim, Survey on the rate of perturbation of nuclear decay. Radiochim. Acta **23**, 23–37 (1976)

G. Harbottle, C. Koehler, R. Withnell, A differential counter for the determination of small differences in decay rates. Rev. Sci. Inst. **44**(1), 55–59 (1973)

P.E. Hodgson, E. Gadioli, E.G. Erba, *Introductory Nuclear Physics* (Clarendon, Oxford, 1997), p. 375

R. Horvat, Recent results of the neutrino mass squared measurements and the coherent neutrino–cold-dark-matter interaction. Phys. Rev. D **57**(8), 5236 (1998)

J. Jenkins, E. Fischbach, Perturbation of nuclear decay rates during the solar flare of 13 December 2006. arXiv:0808.3156v1 [astro-ph] (2008)

J. Jenkins et al., Evidence for correlations between nuclear decay rates and Earth–Sun distance. arXiv:0808.3283v1 [astro-ph] (2008)

V.R. Khalilov, Electroweak nucleon decays in a superstrong magnetic field. Theor. Math. Phys. **145**(1), 1462 (2005)

A. Kogut et al., Dipole anisotropy in the COBE differential microwave radiometers first-year sky maps. Apstrophys. J. **419**, 1–6 (1993)

B.F. Kostenko, M.Z. Yuriev, Possibility of a modification of life time of radioactive elements by magnetic monopoles. Ann. Found. L. de Broglie **33**(1–2), 93–106 (2008). arXiv:0709.1052v1 hep-ph

E.A. Kushnirenko, I.B. Pogozhev, Comment on the paper by S.E. Shnoll et al. Phys. Usp. **43**(2), 203–204 (2000)

R. Lindstrom et al., *Does the Half-Life of a Radioactive Sample Depend on Its Shape*? 2009 ANS Proceedings (2009, in preparation)

V. Lobashev et al., Direct search for mass of neutrino and anomaly in the tritium beta-spectrum. Phys. Lett. B **460**, 227–235 (1999)

V. Lyul'ka, Elementary particle decays in the field of an intense electromagnetic wave. Sov. Phys.-JETP **42**(3), 408–412 (1975)

F. Mandl, G. Shaw, *Quantum Field Theory–Revised Edition* (Wiley, New York, 1993), pp. 159–165

J.J. Matese, R.F. O'Connell, Neutron beta decay in a uniform constant magnetic field. Phys. Rev. **180**(5), 1289 (1969)

J. Nieves, Neutrinos in a medium. Phys. Rev. D. **40**(3), 866–872 (1989)

A. Nikishov, V. Ritus, Quantum processes in the field of a plane electromagnetic wave and in a constant field. Sov. Phys.-JETP **19**(5), 1191–1199 (1964)

NNDC, 2008. http://www.nndc.bnl.gov/qcalc

NOAA, 2005. http://www.swpc.noaa.gov/NOAAscales/index.html

NOAA, NOAA Space Environment Center, SEC PRF 1634, 2006

NOAA, 2009. http://www.ngdc.noaa.gov/stp/

E. Norman et al., Influence of physical and chemical environments on the decay rates of $^7$Be and $^{40}$K. Phys. Lett. B **519**, 15–22 (2001)

E.B. Norman et al., Evidence against correlations between nuclear decay rates and Earth–Sun distance. Astropart. Phys. **31**, 135–137 (2009)

S.F. Odenwald, J.L. Green, Bracing for a solar superstorm. Sci. Am. **299**(2) (2008)

B. Odom, D. Hanneke, B. D'Urso, G. Gabrielse, New measurement of the electron magnetic moment using a one-electron quantum cyclotron. Phys. Rev. Lett. **97**, 030801 (2006)

T. Ohtsuki, H. Yuki, M. Muto, J. Kasagi, K. Ohno, Enhanced electron-capture decay rate of $^7$Be encapsulated in $C_{60}$ cages. Phys. Rev. Lett. **93**, 112501 (2004)

A.G. Parkhomov, Bursts of count rate of beta radioactive sources during long-term measurements. Int. J. Pure Appl. Phys. **1**, 119–128 (2005)

T. Phillips, 2009. http://science.nasa.gov/headlines/y2009/01apr~deepsolarminimum.htm

S. Pommé, Problems with the uncertainty budget of half-life measurements. Am. Chem. Soc. Symp. Ser. **945**, 282–292 (2007)

S. Pommé, J. Camps, R. Van Ammel, J. Paepen, Protocol for uncertainty assessment of half-lives. J. Rad. Nucl. Chem. **276**(2), 335–339 (2008)

H.R. Reiss, Nuclear beta decay induced by intense electromagnetic fields: Basic theory. Phys. Rev. C **27**(3), 1199 (1983)

V. Ritus, Effect of an electromagnetic field on decays of elementary particles. Sov. Phys.-JETP **29**(3), 532–541 (1969)

T. Rosenband et al., Frequency ratio of $Al^+$ and $Hg^+$ single-ion optical clocks; metrology at the 17th decimal place. Science **319**, 1808 (2008)

E. Rutherford, *Radioactive Substances and Their Radiations* (Cambridge University Press, New York, 1913)

E. Rutherford, J. Chadwick, C. Ellis, *Radiations from Radioactive Substances* (Cambridge University Press, Cambridge, 1930), p. 167

A.J. Sanders, Implications for C-14 dating of the Jenkins-Fischbach effect and possible fluctuation of the solar fusion rate (2008). arXiv:0808.3986v2 [astro-ph]

H. Schrader, Private communication, 2008

D.J. Shaw, Detecting seasonal changes in the fundamental constants (2007). arXiv:gr-qc/0702090v1

S.E. Shnoll et al., Realization of discrete states during fluctuations in macroscopic processes. Phys. Usp. **41**, 1025–1035 (1998)

S.E. Shnoll et al., Regular variation of the fine structure of statistical distributions as a consequence of cosmophysical agents. Phys. Usp. **43**(2), 205–209 (2000)

H. Siegert, H. Schrader, U. Schötzig, Half-life measurements of europium radionuclides and the long-term stability of detectors. Appl. Radiat. Isot. **49**(9–11), 1397 (1998)
M. Silverman, W. Strange, C. Silverman, T. Lipscombe, On the run: Unexpected outcomes of random events. Phys. Teach. **37**, 218–225 (1999)
M. Silverman, W. Strange, C. Silverman, T. Lipscombe, Tests for randomness of spontaneous quantum decay. Phys. Rev. A **61**(042106) (2000)
L. Stodolsky, Speculation on detection of the "neutrino sea". Phys. Rev. Lett. **34**(2), 110–112 (1975)
W. Stoeffl, D.J. Decman, Anomalous structure in the beta decay of gaseous molecular tritium. Phys. Rev. Lett. **75**(18), 3237 (1995)
P.A. Sturrock, Solar neutrino variability and its implications for solar physics and neutrino physics. Astrophys. J. **688**, 53 (2008)
J. Taylor, *An Introduction to Error Analysis* (University Science Books, Sausalito, 1997)
I.M. Ternov et al., $\beta$-decay polarization effects in an intense electromagnetic field. Sov. J. Nucl. Phys. **28**(6), 747 (1978)
I.M. Ternov et al., Polarization effects and electron spectrum in the nuclear $\beta$ decay in the field of an intense electromagnetic wave. Sov. J. Nucl. Phys. **39**(5), 710 (1984)
S.J. Tu, E. Fischbach, Geometric random inner products: A family of tests for random number generators. Phys. Rev. E **67**, 016113 (2003)
S.J. Tu, E. Fischbach, A study on the randomness of the digits of $\pi$. Int. J. Mod. Phys. C **16**(2), 281–294 (2005)
J. Uzan, The fundamental constants and their variation: observational and theoretical status. Rev. Mod. Phys. **75**, 403–455 (2003)
F. Wissman, Variation observed in environmental radiation at ground level. Rad. Prot. Dos. **118**, 3–10 (2006)
C.S. Wu, S.A. Moszkowski, *Beta Decay* (Interscience, New York, 1966)

Space Sci Rev (2009) 148: 175–190
DOI 10.1007/s11214-009-9540-7

# Tests of the Gravitational Inverse Square Law at Short Ranges

**R.D. Newman · E.C. Berg · P.E. Boynton**

Received: 16 March 2009 / Accepted: 12 May 2009 / Published online: 4 June 2009
© The Author(s) 2009

**Abstract** A laboratory test and several geophysical tests conducted in the last decades of the 20th century suggested deviations from the inverse square distance dependence of Newton's law of gravity. While further work has failed to substantiate these results, renewed interest in inverse square law tests of increased sensitivity has been stimulated by a wide range of new theoretical ideas. Of particular interest are tests at submillimeter ranges, which could reveal the existence of compact new dimensions accessible only to gravity. This paper reviews the current status of inverse square law tests, with emphasis on present and proposed experimental techniques.

**Keywords** Inverse square law · Newton's law of gravity

## 1 Introduction

In 1974 Daniel Long published a paper *Why do we believe Newtonian gravitation at laboratory dimensions?* (Long 1974), comparing measurements of $G$ made since 1894 with various mass separations $r$. His plot of $G$ values as a function of $r$ strongly suggested a dependence on mass separation. Two years later, Long reported an experiment of his own (Long 1976) which used a torsion balance to compare the forces produced by source masses at distances of 4.5 cm and 29.9 cm. Long's experiment used ring-shaped source masses, exploiting the fact that the force on a test mass at a certain point on the axis of a ring source mass is at an extremum and thus is quite insensitive to error in its position relative to the ring. Daniel Long reported that the ratio of the torque produced by the more distant ring to that produced by the nearer ring exceeded the Newtonian prediction by $(0.37 \pm 0.07)\%$, a result consistent with the distance dependence found in his analysis of $G$ measurements.

R.D. Newman (✉) · E.C. Berg
Department of Physics and Astronomy, University of California, Irvine, CA 92697, USA
e-mail: rdnewman@uci.edu

P.E. Boynton
Department of Physics, University of Washington, Seattle, WA 98195, USA

This result precipitated a flurry of efforts by others to search for inverse square law violation (ISLV). The early ISLV tests probed ranges greater than one millimeter, and quite conclusively ruled out the violation apparently found by Long. More recently attention has focused on searches for ISLV at sub-millimeter ranges, motivated by a wide variety of theoretical speculations. An excellent review of the 2003 status of these searches and their theoretical motivations has been presented by Adelberger et al. (2003). A recent paper by Adelberger and co-authors (Adelberger et al. 2009) updates this review for tests conducted with torsion balances. Here we review the status of ISLV tests with emphasis on progress since 2003, prospects for improved tests, and the principles of instrumentation employed in this area of research.

*Detecting Anomalous Forces* A natural assumption is that a new force in nature or an anomaly in classical gravity can be characterized by an interaction of Yukawa form:

$$U(r) = \frac{-G m_1 m_2}{r}(1 + \alpha e^{-r/\lambda}) \tag{1}$$

with strength $\alpha$ relative to Newtonian gravity, and range $\lambda = \frac{\hbar}{mc}$ related to the mass of a mediating particle. A power law dependence could also naturally arise in some cases, and is considered briefly later in this review. Here we assume an interaction of Yukawa form (1). The strength parameter $\alpha$ may be a function of the composition of the interacting masses, leading to a real (or apparent) violation of Einstein's Weak Equivalence Principle. There are then two complementary approaches to the search for the anomalous interaction:

1. One may search directly for deviation from inverse square law behavior. This has the advantage that it works even if composition dependence is very weak or nonexistent. It's disadvantage is that for useful sensitivity the scale of the experiment needs to be on the order of $\lambda$. This is evident from the force law following from (1):

$$F(r) = \frac{G m_1 m_2}{r^2}\left(1 + \alpha\left(1 + \frac{r}{\lambda}\right)e^{-r/\lambda}\right) = \frac{G(r) m_1 m_2}{r^2} \tag{2}$$

   where the anomalous $r$ dependence is absorbed in an effective $G(r)$. A plot (Fig. 1) of $G(r)$ shows that to have optimum sensitivity an ISLV test must compare forces at ranges roughly straddling $\lambda$. Thus a thorough test for inverse square law violation requires an extensive set of experiments, each tailored for a different region in $\lambda$ parameter space.
2. Alternatively, one may search for composition dependence of the interaction strength $\alpha$. This has the advantage of retaining sensitivity for all mass separations not much greater than $\lambda$, but suffers from the fact that the difference in composition of suitable test bodies (by such measures as neutron/proton ratio or baryon number/mass ratio) is generally very small. Thus the two approaches are complementary and ideally are both to be pursued.

## 2 Laboratory ISLV Tests at Ranges Greater than a Millimeter

The period between Daniel Long's (1976) paper and 1993 yielded many searches for ISLV testing ranges from a few millimeters to a few meters. Current constraints in $\alpha$–$\lambda$ parameter space established by the more sensitive tests are displayed in Fig. 2, along with the region of parameter space allowed by Long's positive result.

The first experiment (Spero et al. 1980) to apparently exclude Long's effect yielded the constraint labeled "Irvine I" in Fig. 2. This was a "near null" experiment (Fig. 3) conducted

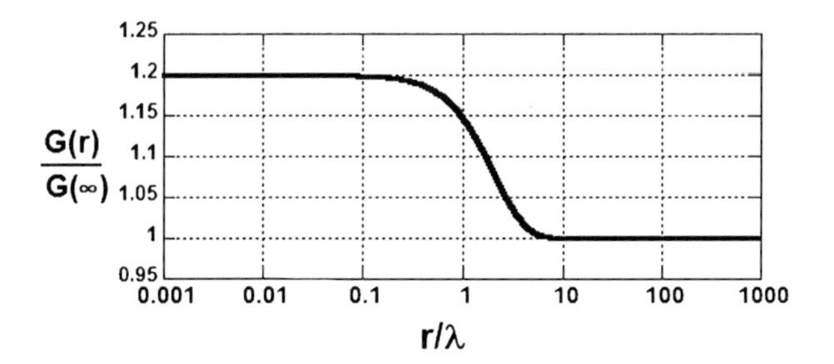


**Fig. 1** Plot of $G(r)$ given by (2) for $\alpha = 0.2$, demonstrating that detection of inverse square law violation by comparing forces at two mass separations requires that the chosen separations roughly flank $\lambda$

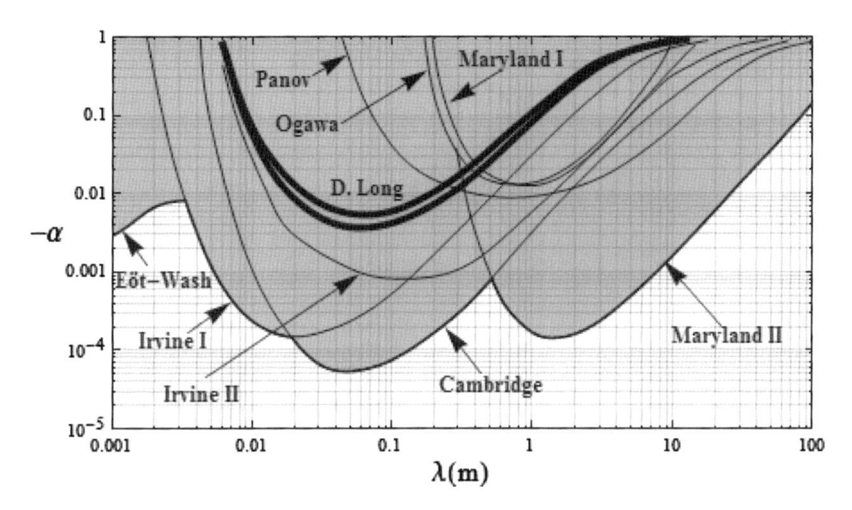


**Fig. 2** Experimental limits on an interaction of the form of (1) with $\lambda > 1$ mm and negative $\alpha$, with the $1\sigma$ limits (*dark lines*) of the parameter space implied by Long's experiment (Long 1976). Other lines are $1\sigma$ limits, except the Eöt-Wash line which is a $2\sigma$ limit on $|\alpha|$. References are as follows: Irvine I and II: (Spero et al. 1980; Hoskins et al. 1985), Eöt-Wash: (Kapner et al. 2007), Cambridge: (Chen et al. 1984), Maryland I and II: (Chan et al. 1982; Moody and Paik 1993), Panov: (Panov and Frontov 1979), Ogawa: (Ogawa et al. 1982). A corresponding plot for positive $\alpha$ is presented in Fig. 6

at UC Irvine, which exploited the fact that for an inverse square force law the gravitational g field vanishes within an infinite mass tube just as within a spherical mass shell. The suspended test mass was used to probe the change in gravitational field near the center of a stainless steel tube as the tube was moved laterally. The small ($\approx$1%) end effect from the finite length of the tube was largely cancelled by a thin copper ring that was moved from one side of the test mass to the other as the tube was translated.

Daniel Long then suggested (Long 1980) that the anomaly he found might arise from a vacuum polarization effect analogous to that producing a logarithmic deviation from $1/r^2$ behavior in the electric force between charges at very short distances. Such an effect might not be observable in a null experiment such as the Irvine test, Long argued, because of the lack of a polarizing field in the region probed by the test mass. However, a number of subsequent experiments that were not of null type, including one at Irvine (Hoskins et al. 1985), gave results consistent with Newton's law but not with the violation reported by Long. It is likely that Long's experiment suffered from an undetected systematic error from apparatus tilt correlated with source mass position—an insidious and common problem in torsion pendulum gravity experiments.

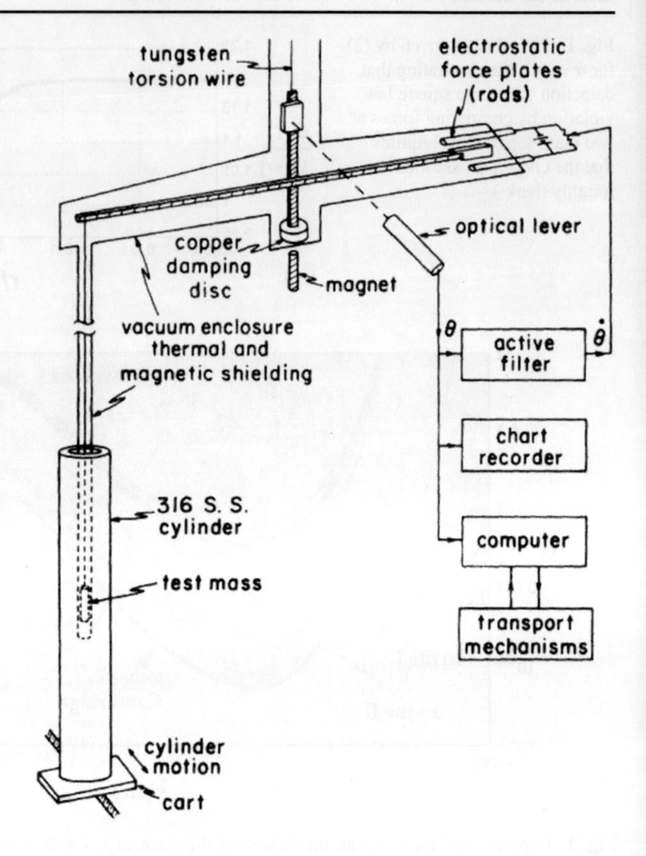


**Fig. 3** Apparatus of the first Irvine ISLV test. A test mass probes the gravitational field within a SS tube of height 60 cm ID 6 cm and OD 8 cm

Laboratory tests of ISLV at ranges above a few millimeters have taken a number of interesting forms. Several have used a torsion pendulum (Chen et al. 1984; Panov and Frontov 1979; Spero et al. 1980; Hoskins et al. 1985). A test by Goodkind et al. (1993) at UC San Diego measured the change of force exerted on a levitated superconducting niobium ball when a spherical mass was placed below it at a distance ranging from 0.4 to 1.4 meters. A test by Ogawa et al. (1982) at the University of Tokyo measured the resonant response of a 1400 kg aluminum quadrupole gravity wave antenna to a 401 kg steel bar rotating at 30.4 Hz and placed at distances ranging from 2.6 to 10.6 meters. A test by Moody and Paik (1993) at the University of Maryland used a three-axis superconducting gradiometer to detect a possible non-zero Laplacian of the gravitational field generated by a 1500 kg lead pendulum (Fig. 4). As the Laplacian of the Newtonian gravitational field ($\nabla^2\Phi$) must vanish, this was an elegant null experiment. Each axis of the gradiometer instrument was a pair of superconducting accelerometers read out by a differential SQUID circuit—a technique that Paik is now applying to a sub-mm ISLV test described here in Sect. 3.

Finally, we describe a torsion pendulum ISLV test being conducted by the authors together with Michael Moore and Ricco Bonicalzi in a U. Washington/UC Irvine (UW/UCI) collaboration. This experiment (Boynton et al. 2007) is designed to have maximum sensitivity in $\alpha$–$\lambda$ parameter space at $\lambda \sim 12$ cm. The test, like that of Moody and Paik (1993), exploits the fact that ISLV is signaled by a non-vanishing Laplacian of the gravitational field. In our test, the signal is the gradient of the Laplacian, manifest as an $m = 1$ variation in the pendulum's torsional frequency and anharmonic behavior as a source mass re-

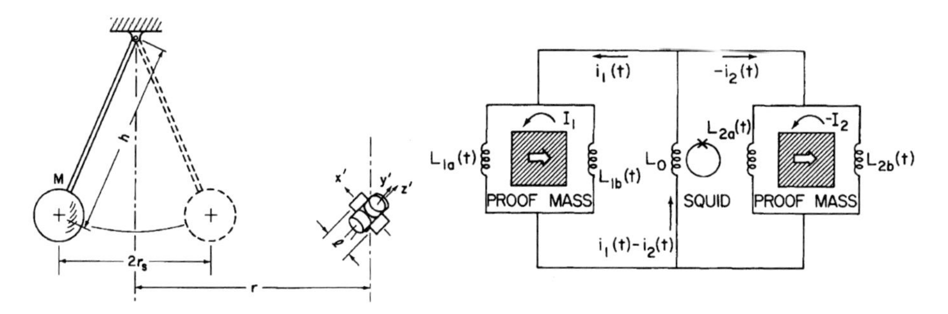


**Fig. 4** Principle of Ho Jung Paik's (1993) inverse square law test. Signals from a cryogenic three-axis gradiometer (*central figure*) are summed to yield the Laplacian of the gravitational field $\Phi$ generated by a swinging pendulum. Each of the three orthogonal gradiometers ($i = 1, 2, 3$) uses the circuit shown to measure the differential displacement of two test masses separated by a distance $\ell$. This yields the differential force $\Delta F_i$ on the test masses $m$, and thus yields $\partial g_i/\partial x_i = \Delta F_i/(m\Delta\ell)$ which are summed to yield $\nabla^2\Phi$

volves around the pendulum in a sequence of stationary positions. The experiment has two key design features: a pendulum that couples to the gradient of the Laplacian, but has no $m = 1$ mass moments that couple to Newtonian gravitational fields through $\ell = 6$, and a source mass configuration that would exaggerate the gradient of the Laplacian ($\nabla\nabla^2\Phi$) of a Yukawa potential, but generates no $m = 1$ Newtonian field moments through $\ell = 8$. Either of these design features, if achieved, would suffice to eliminate significant bias from Newtonian field couplings in this experiment. The combination of these features renders the experiment extremely insensitive to small fabrication errors in either pendulum or source mass. The experiment's pendulum and source mass are shown in Fig. 5.

The experimental design and analysis (Moore 2000) generalizes the standard multipole formalism to allow for non-Newtonian fields such as $V(r) \propto \frac{1}{r}e^{-r/\lambda}$. One finds that the general interaction energy $U = \int \rho V d^3r$ between a pendulum of density distribution $\rho(r)$ and the gravitational field $V$ produced by a source mass may be expanded:

$$U = \sum_{n\ell m} V_{n\ell m} M_{n\ell m} \tag{3}$$

where $M_{n\ell m}$ is the generalized multipole mass moment of the pendulum

$$M_{n\ell m} \propto \int \rho r^n Y_{\ell m} d^3r \tag{4}$$

and $V_{n\ell m}$ is constructed from $n$th-order derivatives of the field $V$ at the origin of the pendulum. This generalized multipole formalism includes non-Newtonian multipole terms for which $\ell$ is smaller than n by increments of 2.

The horizontal gradient of the Laplacian would manifest itself specifically in the multipole term

$$U_{311} = V_{311} M_{311} \propto \frac{\partial}{\partial x}(\nabla^2 V) \int \rho r^3 Y_{11} dx dy dz \tag{5}$$

This experiment is hosted by the Pacific Northwest National Laboratory, and operates in a former Nike missile bunker on a remote arid lands preserve near Richland, Washington. The initial phase of this work is conducted at room temperature, using the second harmonic amplitude of the pendulum's oscillation as a measure of the ISL-violating torque that would

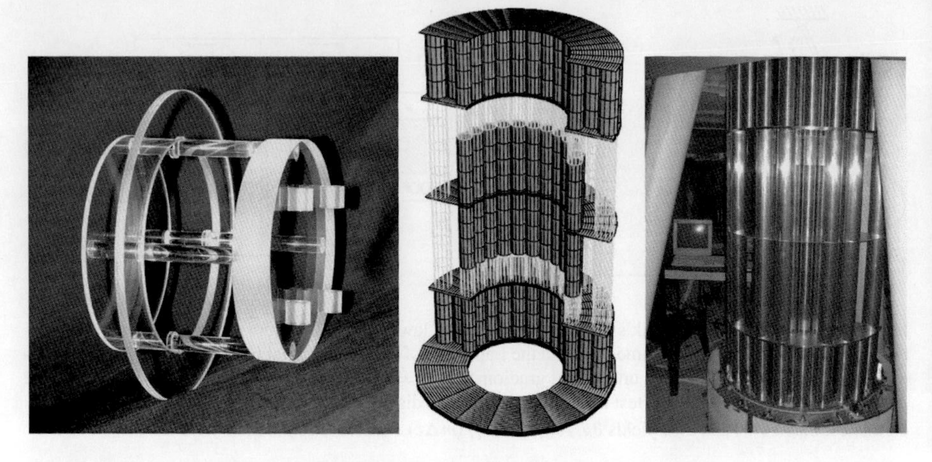

**Fig. 5** The pendulum and source mass of the UW/UCI ISLV test. The pendulum, shown before gold coating, is made of fused silica and has a diameter of 11.5 cm. Not shown are a set of four small gold-plated titanium trim masses. The 1.4 tonne 1.6 m high source mass, constructed from stainless steel solid cylinders and hollow spacer tubes, is shown as a rendition and as realized. The source mass rotates at intervals on an air bearing base. Various alternate mass stacking configurations may be constructed to deliberately exaggerate Newtonian field gradients for diagnostic purposes

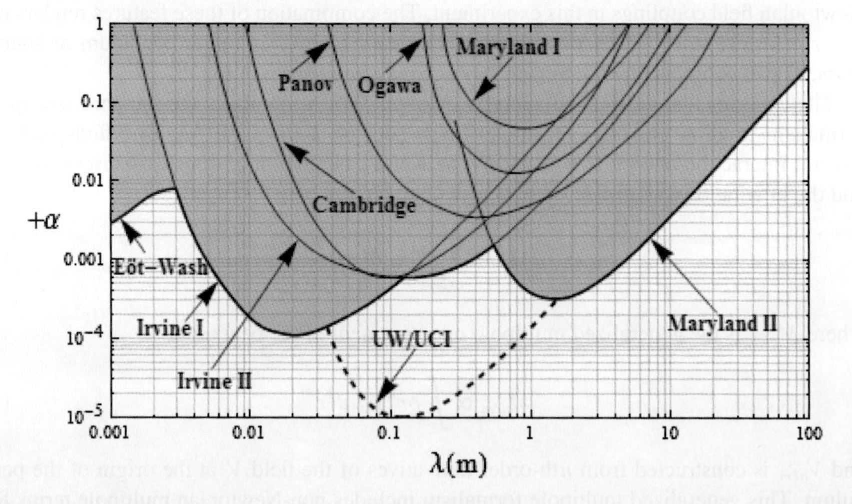


**Fig. 6** Experimental limits on an interaction of the form of (1) with $\lambda > 1$ mm and positive $\alpha$. Limits shown are $1\sigma$, except the Eöt-Wash line which is a $2\sigma$ limit on $|\alpha|$. Plot labels correspond to references as in Fig. 2. The *dashed line* indicates the projected sensitivity of the UW/UCI experiment in progress. A corresponding plot for negative $\alpha$ is presented in Fig. 2

arise from $U_{311}$. In the second experiment phase, the pendulum is in a cryogenic environment at ~4 K, using the variation of torsion frequency as a measure of $U_{311}$. A $1\sigma$ sensitivity at a level $\alpha = 10^{-5}$ is anticipated, with the projected constraint in $\alpha$–$\lambda$ parameter space indicated in Fig. 6.

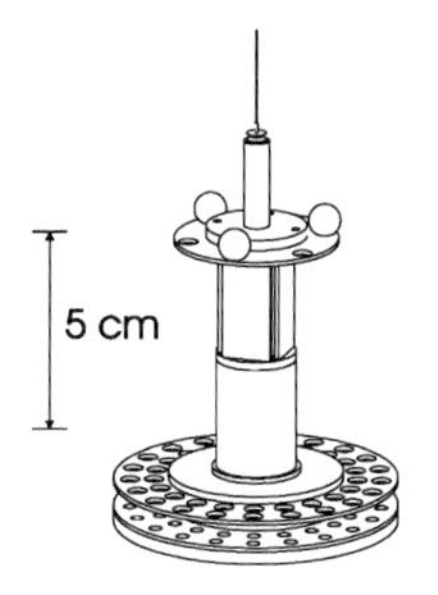


**Fig. 7** Scale drawing of the Eöt-Wash instrument. The three spheres near the top of the detector served for a gravitational calibration of the torque scale. The detector's electrical shield is not shown

## 3 ISLV Tests from 1 μm to 1 mm

Published constraints in $\alpha$–$\lambda$ parameter space and projected improvement in constraints from the experiments discussed in this section are displayed in Figs. 13 and 14.

### 3.1 Torsion Pendulum Sub-mm Experiments

*The University of Washington "Eöt-Wash" group* is conducting a series of ISLV tests (Hoyle et al. 2001, 2004; Kapner et al. 2007). The most recent of these tests probed mass separations ranging from 9.53 mm to 55 μm. These place the best current limits on ISLV below about 3 mm, and put a 95% confidence limit of unity on the magnitude of $\alpha$ (the strength of a Yukawa term relative to gravity, (1)) down to a length scale $\lambda = 56$ μm. The instrument used (Fig. 7) was a torsion pendulum detector suspended above an "attractor" which rotated at a constant velocity $\omega$ lower than the pendulum's resonant frequency by a factor of 28 or 7.5. The detector's 42 "negative mass" test bodies were 5-mm-diameter holes in a 1-mm-thick molybdenum detector ring. The attractor was a 1-mm-thick molybdenum disk with 42 3.2-mm-diameter holes mounted above a thicker tantalum disk containing 21 6.4-mm-diameter holes. A gold-coated 10-μm-thick CuBe shield was stretched between the attractor and pendulum to block electrostatic interaction. The gravitational interaction between the missing masses of attractor and detector holes produced a torque with harmonics at multiples of $21\omega$, with resulting angular displacement read out by an autocollimator. The two disks of the attractor were displaced azimuthally and were designed to nearly cancel the $21\omega$ signal in the absence of ISLV. Capacitive and micrometer techniques served to measure relative position and parallelism of the attractor and pendulum disks. An important advantage of the experiment's multi-hole design is that the signal frequency is far removed from frequencies of possible periodic disturbances. Experiments currently in progress by this group include one using a variant of the apparatus shown in Fig. 7 having 120 radial wedge-shaped 100 μm wide holes, and one using a torsion pendulum with flat plates suspended in a vertical plane near a plane source mass.

*Jun Luo's Huazhong University group in Wuhan China* (Tu et al. 2007) used a torsion pendulum (Fig. 8) to probe the force between a planar gold source and test masses for mass separations from 176 to 341 μm, putting a 95% confidence limit on a gravitational-strength ($|\alpha| = 1$) interaction down to a length scale $\lambda = 66$ μm. Between source and test masses was a 56 μm glass membrane with a 300-nm gold coating on each side. Like the Eöt-Wash instrument, the Wuhan instrument is designed to give a nearly null Newtonian signal over the range of separations explored.

*Ramanath Cowsik at Washington University in Saint Louis* is developing an ISLV test in which a torsion ribbon suspends a 1-mm-thick 100-mm-diameter metal disk in a vertical

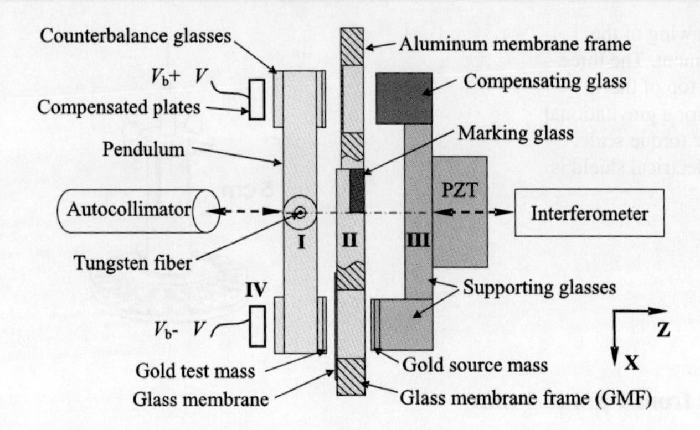


**Fig. 8** Schematic top view of the Wuhan apparatus, showing: (I) the pendulum, carrying the gold test mass and a counterbalance mass, (II) a frame carrying gold-coated electrostatic shield membranes, (III) a PZT-driven moving frame carrying the gold source mass with a compensating mass to largely null the Newtonian interaction, and (IV) electrostatic force plates to null torsional displacement. Translation of element III and rotation of the pendulum I are read by the indicated interferometer and autocollimator

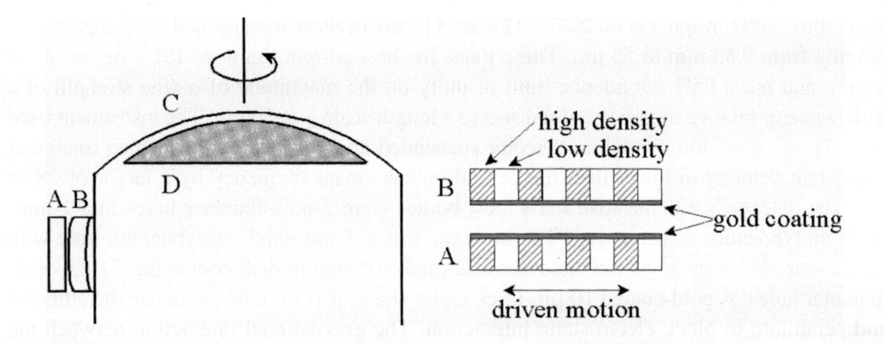


**Fig. 9** Schematic cross-section view of the Birmingham superconducting-suspension torsion balance applied to a sub-mm ISLV test. A superconducting niobium spherical-surface cap (C) is levitated by current in a spiral niobium wire winding embedded in a MACOR spherical-surface bearing (D) of 43 mm radius. Mounted on the levitated "balance" is an array (B) of metal bars of alternately high and low density. A similar array (A) is driven with frequency $\omega$ in the indicated direction, producing a torque on the levitated balance at frequency multiples of $\omega$. The resulting angular displacement of the balance is read out with a capacitive or SQUID-based technique. Mass arrays A and B each have a gold coating, but no intervening shield

plane, parallel to a much larger diameter iridium source mass disk. Between the disks is a thin electrostatic shield. The position and angular orientation of the source mass disk are to be independently modulated at carefully selected frequencies, thereby modulating both the mean and differential distance between the source disk and regions of the test mass on opposite sides of the torsion ribbon. Cowsik has developed a remarkably low-noise autocollimator ($\approx 2 \times 10^{-10}$ rad/$\sqrt{\text{Hz}}$ at 10 mHz) for angular readout of the test mass pendulum.

*Clive Speake and collaborators at the University of Birmingham* are developing a sub-mm ISLV test based on a novel superconducting torsion balance (Hammond et al. 2008), illustrated schematically in Fig. 9.

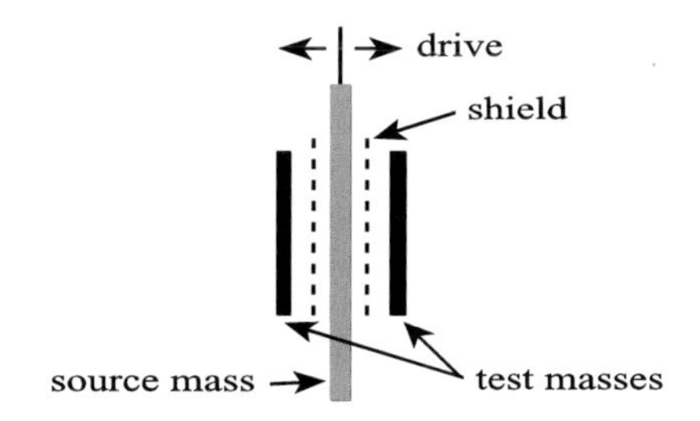


**Fig. 10** Principle of the Maryland ISLV cryogenic instrument. The source mass disk is driven at frequency $\omega$ as shown. The spring-loaded test mass disks experience a differential force at $2\omega$, which would be zero for a Newtonian force law and infinite disk diameter. The detected differential force, corrected for finite disk sizes, is thus a measure of ISLV. The differential displacement of the test masses is measured with a SQUID-based technique like that used in the gradiometers in the 1993 Maryland ISLV test at longer ranges (Fig. 4)

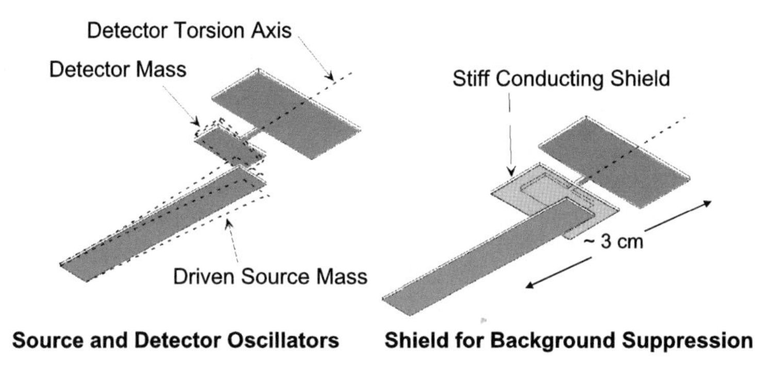


**Fig. 11** Principle of Josh Long's ISLV test. A PZT bimorph drives a tungsten source mass at the ~1 kHz resonant frequency of a tungsten double torsional oscillator that serves as test mass. A thin sapphire plate with 100 nm gold coating on both sides is interposed between source and test masses. A capacitive transducer measures the amplitude of the torsional oscillation. The source and test mass systems are mounted on independent vibration isolation stacks, each of which attenuates vibration by a factor of $10^{10}$ at 1 kHz

## 3.2 Non-Torsion-Pendulum Sub-mm ISLV Tests

*Ho Jung Paik and collaborators at the University of Maryland* are conducting a cryogenic ISLV test based on measurement of the differential motion of a pair of parallel circular test mass disks when an interposed source mass disk is moved away from one test mass and toward the other (Fig. 10).

*Josh Long at Indiana University* is continuing an ISLV test he developed with John Price at the University of Colorado (Long et al. 2003). This experiment measures the resonant response of a test-mass-carrying cantilever to a source mass driven up and down below it (Fig. 11).

*Aharon Kapitulnik and collaborators at Stanford University* have been conducting a series of ISLV tests of steadily increasing sensitivity using a test mass mounted on a resonant cantilever above an array of elements of alternating high and low density material moving below it. In most of these tests a short source mass array is moved back and forth by a PZT driver at a frequency which is a sub-multiple of the cantilever resonant frequency, causing the alternating source mass elements to drive the cantilever at resonance. In this way danger of a spurious signal from mechanical coupling of the drive mechanism to the detector is

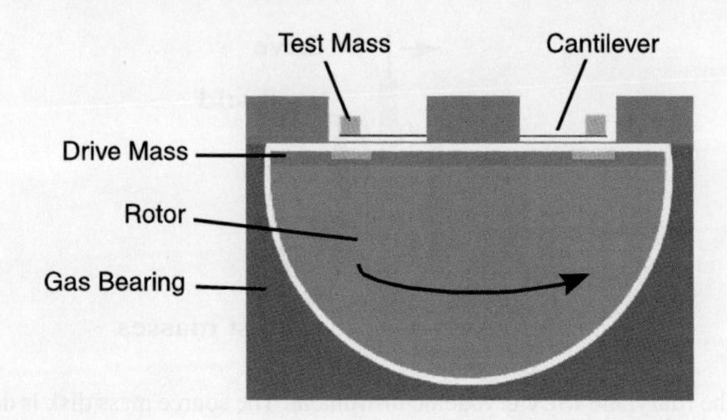


**Fig. 12** Cross-sectional diagram (not to scale) of the Stanford ISLV apparatus of Weld et al. (2008). The indicated drive mass is a disk of low-density material (tungsten or brass) in which trenches have been etched or machined and filled with another, less-dense material for purposes of planarization, yielding a circular array of 100 "source masses". The drive mass is bonded to a 5 cm diameter cryogenic fused quartz helium gas bearing which rotates at about 3.5 Hz. Above the source mass plane are test masses mounted on cantilever detectors resonant at about 350 Hz, whose displacement is read out with fiber interferometers. On the top surface of the source mass array is a uniform flat coating of gold. A gold-coated silicon nitride shield membrane separates the rotating assembly from the test masses. The face-to-face separation between source and test masses can be as small as 29 μm, and may be accurately varied as much as 5 μm by varying the bearing gas flow to raise or lower the spinning rotor

minimized. Preliminary results have been reported from a new experiment design in which a circular array of 100 alternating mass elements mounted on a cryogenic gas bearing passes continuously below the sensor cantilever (Fig. 12). This promising approach has a number of advantages, including ease of stable alignment and ability to employ relatively large area source masses. A problem has been vibrational noise generated by the gas bearing, which the researchers expect to reduce in future work.

### 3.3 General Considerations for Very Short Range ISLV Tests

Among the ISLV tests described above that explore mass separations below 1 mm, most measure a force normal to the surface of parallel mass plates. This has the advantage that the full area of the mass plates contribute to the measured signal force. In contrast, the multi-hole torsion pendulum of the Eöt-Wash group and the spherical superconducting torsion balance of the Birmingham group measure a force tangential to planar arrays of facing mass elements as one plane moves laterally with respect to the other. This approach has the disadvantage that, for an anomalous force with range $\lambda$, only a portion of the interacting masses of width $\lambda$ on their leading and trailing edges contribute significantly to the measured signal. This is an increasing disadvantage as the force range probed becomes smaller. However, this disadvantage is largely compensated by the extraordinary mechanical compliance of a torsion fiber. It is interesting to compare instruments in terms of an effective "spring constant" relating force to displacement. The Eöt-Wash torsion pendulum has a $3 \times 10^{-9}$ Nm/rad torsion constant, or $2.4 \times 10^{-6}$ N/m effective linear spring constant. This may be compared with Decca's MEMS torsion oscillator (following section, Fig. 15), which has a smaller torsion constant ($8.6 \times 10^{-10}$ Nm/rad), but because of its short (0.2 mm) lever arm has a linear spring constant $2.2 \times 10^{-4}$ N/m, two orders of magnitude stiffer. The Stanford cantilever has a linear spring constant of about $10^{-2}$ N/m, nearly four orders of magnitude greater than that of the torsion pendulum.

Gravitation ISLV tests must contend with competing magnetic, electrostatic, and Casimir forces. For tests at ranges greater than a few millimeters, electric and magnetic shielding is

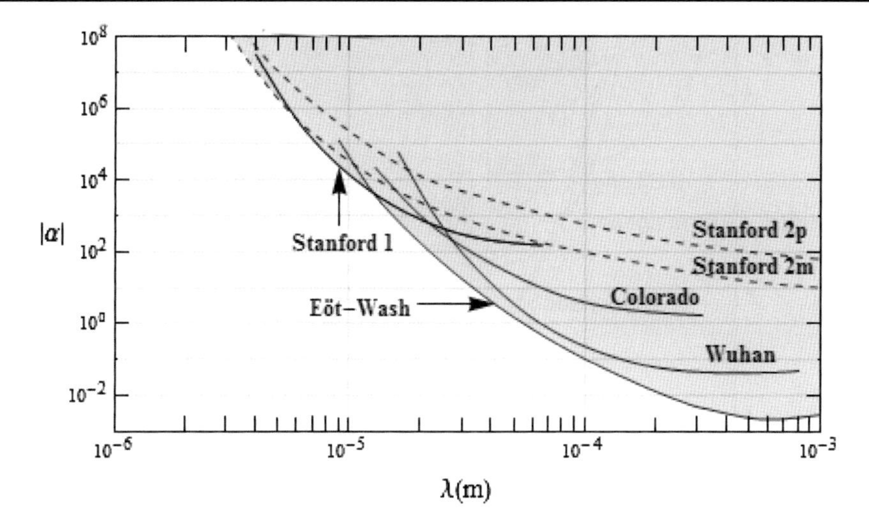


**Fig. 13** Limits at 95% confidence level, as of January 2009, on an anomalous interaction of Yukawa form with range between $\lambda \sim 1$ μm and 1 mm. Curve labels and references correspond as follows: Stanford 1 (Geraci et al. 2008), Stanford 2p and 2m (Weld et al. 2008), Colorado (Long et al. 2003), Wuhan (Tu et al. 2007), Eöt-Wash (Kapner et al. 2007). Stanford 2p and 2m represent $|\alpha|$ limits assuming $\alpha$ greater or less than zero respectively

easily achieved, but at shorter ranges becomes difficult or impossible. Conventional wisdom is that Casimir forces are adequately shielded by a gold layer thicker than the ~140 nm plasma wavelength of gold. Gold-coated shields as thin as 4 μm have been employed between source and test masses, but care must be taken that a shield is sufficiently stiff that no distortion correlated with source mass position is generated by pressure, vibration, or electrostatic forces. One approach, illustrated in the Birmingham experiment, is to dispense with a free-standing shield and rely on "isoelectronic" facing surfaces that are either uniformly gold-coated or possibly made of differing isotopes of the same material. In either case the Casimir force is the same for source mass regions of different density, allowing a difference in gravitational attraction to be discerned. In this approach extraordinary care must be taken to ensure that the surface quality is identical for the differing density mass regions, and is extremely smooth.

Figure 13 shows current limits in $\alpha$–$\lambda$ parameter space on a new force of range ~1 μm to 1 mm. Figure 14 shows targeted improved limits projected by various groups.

## 4 ISLV Tests Below ~1 μm

To explore mass separations below a few microns one has little choice but to accept a distance-dependent Casimir force directly, extracting gravitational anomalies by comparing a measured force with the Casimir prediction. This is complicated by uncertainties in Casimir force calculation, associated with surface quality, temperature, etc. A recent publication (Mostepanenko et al. 2008) derives 95% confidence limits on a new force in $\alpha$–$\lambda$ parameter space from several Casimir force experiments, see Fig. 17. The most recent such experiment (Decca et al. 2005), Fig. 15, used a MEMS torsion oscillator to measure the Casimir force between gold surfaces at separations from 160 to 750 nm. This experiment

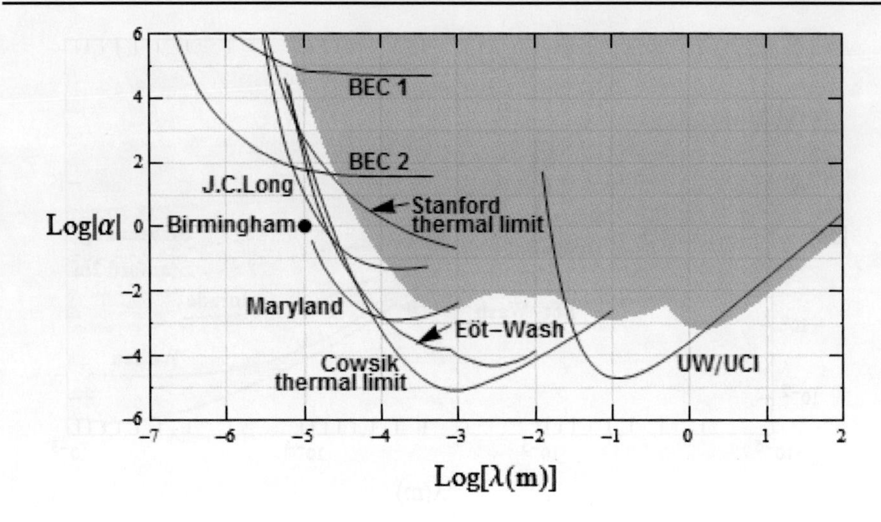


**Fig. 14** Possible future sensitivities projected by various groups, for searches for an anomalous interaction of Yukawa form (1). *Lines* BEC 1 and 2 are from Dimopoulos and Geraci (2003), *line* Stanford is from Weld et al. (2008); the other projections derive from unpublished work

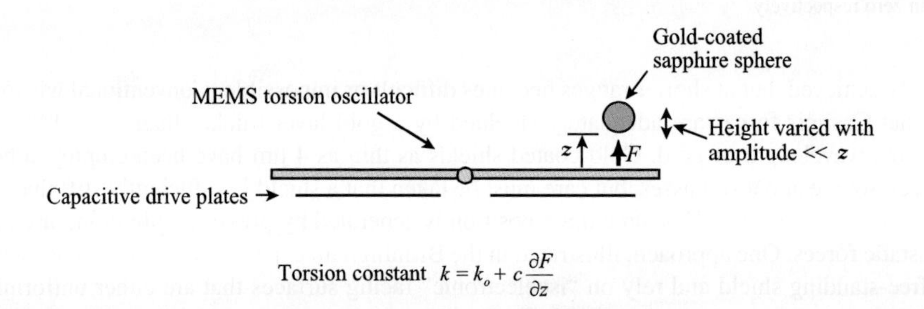


**Fig. 15** Principle of the Casimir measurements of Decca et al. (2005). A micromechanical torsion oscillator operates at ~700 Hz. At height $z$ above the gold-coated oscillator plate a 0.3-mm-diameter gold-coated sapphire sphere exerts a force whose gradient $\partial F/\partial z$ contributes to the oscillator's effective torsion constant. Capacitor plates below the oscillator are used to drive it at resonance, while the height of the sphere is varied harmonically with time. The force $F(z)$ between sphere and oscillator plate is derived from the variation in resonant frequency as a function of $z$. The sphere-plate separation was varied from 160 to 750 nm, measuring forces with relative error varying from 0.19% at 160 nm to 9.0% at 750 nm

produced extremely reproducible results, with Yukawa force limit certainty apparently limited only by uncertainty in Casimir force calculation.

*Dimopoulos and Geraci* (2003) *in consultation with Mark Kasevich and others*, have proposed a probe of submicron-range forces using interferometry of Bose-Einstein condensed atoms. This approach, illustrated in Fig. 16, would load two laser-trap regions at differing distance from a source mass with Bose-Einstein condensates of well-defined initial relative phase. The difference in phase evolution rate of the two condensates would yield a measure of the different potentials in which they sit. After an appropriate interval the condensates would be combined, and their relative phases read out with a fluorescence technique. The contribution to the relative phase shift due to the source mass potential would

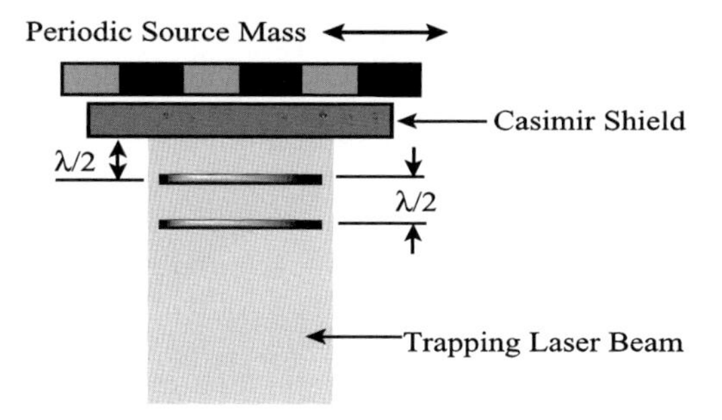


**Fig. 16** The experimental concept of Dimopoulos and Geraci for probing ultra-short distances with an atom interferometry technique. On one side of a gold fixed Casimir shield, an array of tiny source masses of alternating density would move laterally as indicated. On the other side of the gold shield, BEC atom clusters would be trapped in two potential wells a distance $\lambda/2$ apart in a standing wave cavity formed by reflection of a $\lambda = 840$ nm laser from the shield. The BEC clusters would be loaded with an initial well-defined relative phase, whose evolution in time would reflect the different potentials in which they sit. Comparison of relative phase evolution rates for different source mass positions then serves to separate the source-mass-related contribution to the relative potentials at the two regions from the contribution from Casimir effects

be isolated from that due to Casimir and other potentials, by modulating the source mass position.

Measurements of neutron phenomena have been used to put limits on a new Yukawa force; see (Nesvizhevsky et al. 2008; Nesvizhevsky and Protasov 2004) and references therein. Such studies include analyses of the dependence of neutron scattering length on target atomic number, and measurements of quantized neutron energy levels in the Earth's gravitational field. While these limits on $\alpha$ are weak, they extend to ranges below $10^{-11}$ meters. New techniques that could improve such limits using neutrons have been proposed by Greene and Gudkov (2007), Nesvizhevsky et al. (2008) and others. A limit on $\alpha$ extending to extremely short ranges comes from analysis by Nesvizhevsky and Protasov (2004) of data on transition frequencies in antiprotonic atoms ($\bar{p}\,{}^3\mathrm{He}^+$ and $\bar{p}\,{}^4\mathrm{He}^+$). This limit derives from the fact that a new Yukawa interaction between the antiproton and its partner nucleus, with range $\lambda$ greater than the extremely small Bohr radius of the antiprotonic atom, would change the apparent value of the electric binding energy of the system, with a corresponding change in transition frequencies. A $1\sigma$ limit $|\alpha| \leq 1.3 \times 10^{28}$ is obtained, valid for ranges down to $10^{-13}$ m or less.

Constraints in $\alpha$–$\lambda$ space from Casimir, neutron, and atomic studies are shown in Fig. 17.

## 5 Theoretical Ideas Suggesting Inverse Square Law Violation

A wide variety of theoretical ideas suggest a breakdown of the gravitational inverse square law in experimentally accessible regions. Such ideas as of 2003 are reviewed by Adelberger, Heckel and Nelson (AHN) (Adelberger et al. 2003), while more recent ideas (chameleons, unparticles, "fat" gravitons) and their constraints are discussed by Adelberger and colleagues in Adelberger et al. (2009). Here we outline some of these ideas.

One much-referenced idea is the suggestion (Arkani-Hamed et al. 1998) (ADD) that the hierarchy problem presented by the vast difference between Planck and electroweak energy scales could be resolved if there are large extra compact dimensions in which gravitons (but not Standard Model particles) propagate. In the ADD scenario with n equal size extra

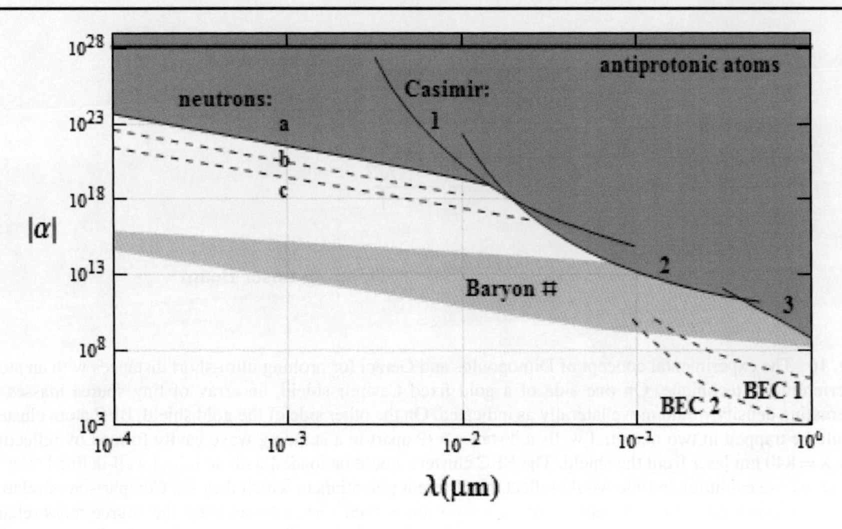


**Fig. 17** Constraints in $\alpha$–$\lambda$ parameter space based on Casimir, neutron, and antiprotonic atom experiments. *Lines 1* and *2* are 95% confidence limits derived by Mostepanenko et al. (2008) from the Casimir experiments of Harris and Chen (2000) and Decca et al. (2005) respectively. *Line 3* is based on the Casimir experiment of Lamoreaux (1997). *Line* a is a constraint obtained by Nesvizhevsky et al. (2008) from neutron scattering data; *lines* b and c are respectively constraint improvements that might be achieved using neutron experiments suggested by Nesvizhevsky et al. (2008) and Greene and Gudkov (2007). The antiprotonic line is the constraint from antiprotonic atom transition frequencies. A region is displayed in which a Baryon gauge particle discussed by Dimopoulos and Geraci (2003) might manifest itself, while *dashed lines* BEC 1 and 2 indicate potential sensitivity of the experimental technique these authors suggest

dimensions of radius $R$, the fundamental energy scale for gravity, $M_*$, is comparable to the electroweak scale $m_{EW}$ and is related to the Planck mass by $M_{Pl}^2 \sim M_*^{2+n} R^n$. Putting $M_* \sim m_{EW}$ then yields:

$$R \sim 10^{\frac{30}{n}-17}\ \mathrm{cm} \times \left(\frac{1\ \mathrm{TeV}}{m_{EW}}\right)^{1+\frac{2}{n}} \tag{6}$$

In this scheme, Kaluza-Klein (KK) excitations in the extra dimensions give rise to an ISL-violating interaction energy $V(r) = -\alpha \frac{Gm_1m_2}{r} e^{-r/\lambda}$, where $\lambda \sim R$ and $\alpha$ (for $n = 2$) is between 3 and ~20, depending on the topology of the extra dimensions (Kehagias and Sfetsos 2000). Taking $m_{EW} = 1$ TeV and $n = 1$ leads to $R \sim 10^{13}$ cm, which is empirically excluded, while for $n \geq 3$, $R$ would be less than a nanometer. For $n = 2$, $R$ would be in an experimentally accessible range, ~100 μm–1 mm. $R$ in this range is excluded by present experiments (Kapner et al. 2007), and as the AHN review paper notes, is also excluded by arguments based on neutrino emission from supernova 1987A, big-bang nucleosynthesis, and non-observation of a diffuse background of cosmological gamma rays from the decay of KK modes. However, AHN discuss other extra-dimension scenarios which can still viably suggest ISLV in a similar distance range, while addressing the hierarchy problem. These scenarios include a model constructed with one large extra dimension (~1 mm) along with several much smaller extra dimensions.

Other possible sources of ISLV are various moduli which arise in superstring theory. Originally massless, such moduli can acquire mass through a variety of possible mechanisms, on various mass scales. One of these moduli is the dilaton, which plays a role in controlling the strength of all interactions. Its mass is unknown, but it would

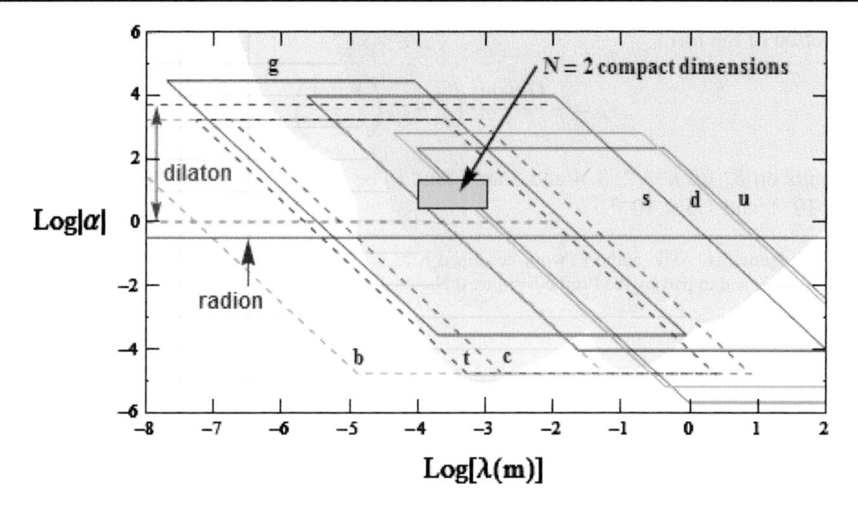


**Fig. 18** Theoretical suggestions of regions in $\alpha$–$\lambda$ parameter space where anomalous forces might be revealed. *Regions* in which Dimopoulos and Geraci (2003) suggest moduli arising in superstring theory might manifest themselves are roughly bounded by parallelograms labeled according to the particle to which the modulus would couple: u, d, s, c, t, or b quark, or gluon (g). A *region* in which two equal-size extra dimensions might be manifested in the ADD scenario (Kehagias and Sfetsos 2000) is indicated, along with regions that might manifest a dilaton (Kaplan and Wise 2000) or radion (Antoniadis et al. 1998). The *two shaded regions* indicate best current experimental limits and potentially improved limits on an interaction of the form of (1)

be expected to produce an interaction between atoms with strength $\alpha$ relative to gravity in a range 1 to $\sim$5000 (Kaplan and Wise 2000). In addition there may be a number of other moduli, with couplings similar to but more specific than that of the dilaton. A possibility explored by Dimopoulos and collaborators (Dimopoulos and Giudice 1996; Dimopoulos and Geraci 2003) is that such moduli acquire mass as a result of supersymmetry breaking at a low energy scale $F$, acquiring mass proportional to $\sim F/M_{Pl}$, with $\sqrt{F}$ in a range 10 TeV to 2000 TeV. These moduli include a gluon modulus that would couple only to Standard Model gluons, and Yukawa moduli that would determine quark and lepton masses and, if sufficiently light, mediate additional interactions. Another possibility is one or more "radions", which can modify effective couplings through their role in fixing the size of compact extra dimensions. A radion would mediate an ISL-violating force with strength $\sim$1/3 of gravity (Antoniadis et al. 1998). Yet another possibility is a bulk gauge particle, such as a baryon gauge particle that couples to baryon number but is free to move off the brane in an extra-dimensional scenario. This possibility is of particular interest to Dimopoulos and his collaborators, because it might manifest itself as ISLV in a small-$\lambda$, large-$\alpha$ region in $\alpha$–$\lambda$ parameter space which could be accessible with the novel BEC experimental technique described in the preceding section and in (Dimopoulos and Geraci 2003). Figure 18 displays some of the regions in $\alpha$–$\lambda$ parameter space in which theories discussed above suggest possible inverse square law breakdown.

Up to this point, this paper has considered only constraints on anomalous interactions of the Yukawa form represented in (1). However inverse square law deviations of power law form could arise from the exchange of multiple bosons. The best limits on such interactions come from the work of the UW Eöt-Wash group (Adelberger et al. 2007). Assuming an interaction of the form

$$V_n = -\frac{Gm_1m_2}{r}\left(1+\beta_n\left(\frac{1\,mm}{r}\right)^{n-1}\right) \tag{7}$$

$1\sigma$ limits on $\beta_n$ for $n = 2$, 3, 4 and 5 are found to be respectively $4.5 \times 10^{-4}$, $1.3 \times 10^{-4}$, $4.9 \times 10^{-5}$, and $1.5 \times 10^{-5}$.

**Acknowledgements** The authors' work described here is supported by NSF grants PHY-0701707 and PHY-0701923, and in part by the Pacific Northwest National Laboratory.

## References


E.G. Adelberger, B.R. Heckel, A.E. Nelson, Ann. Rev. Nucl. Part. Sci. **53** 77, (2003)
E.G. Adelberger, B.R. Heckel, S. Hoedl, C.D. Hoyle, A. Upadhye, Phys. Rev. Lett. **98**, 131104 (2007)
E.G. Adelberger, J.H. Gundlach, B.R. Heckel, S. Hoedl, S. Schlamminger, Prog. Part. Nucl. Phys. **72**, 102 (2009)
I. Antoniadis, S. Dimopoulos, G. Dvali, Nucl. Phys. B **516**, 70 (1998)
N. Arkani-Hamed, S. Dimopoulos, G. Dvali, Phys. Lett. B **429**, 263 (1998)
P.E. Boynton, R.M. Bonicalzi, A.M. Kalet, A.M. Kleczewski, J.K. Lingwood, K.J. McKenney, M.W. Moore, J.H. Steffen, E.C. Berg, W.D. Cross, R.D. Newman, R.E. Gephart, New Astron. Rev. **51**, 334 (2007)
H.A. Chan, M.V. Moody, H.J. Paik, Phys. Rev. Lett. **49**, 1745 (1982)
Y.T. Chen, A.H. Cook, A.J.F. Metherell, Proc. R. Soc. Lond. A **394**, 47 (1984)
R.S. Decca, D. López, H.B. Chan, E. Fischbach, D.E. Krause, C.R. Jamell, Phys. Rev. Lett. **94**, 240401 (2005)
S. Dimopoulos, G.F. Giudice, Phys. Lett. B **379**, 105 (1996)
S. Dimopoulos, A.A. Geraci, Phys. Rev. D **68**, 124021 (2003)
A.A. Geraci, S.J. Smullin, D.M. Weld, J. Chiaverini, A. Kapitulnik, Phys. Rev. D **78**, 022002 (2008)
J.M. Goodkind, P.V. Czipott, A.P. Mills, M. Murakami, P.M. Platzman, C.W. Young, D.M. Zuckerman, Phys. Rev. D **47**, 1290 (1993)
G.L. Greene, V. Gudkov, Phys. Rev. C **75**, 015501 (2007)
G.D. Hammond, C.C. Speake, A.J. Matthews, E. Rocco, F. Peña-Arellano, Rev. Sci. Instrum. **79**, 025103 (2008)
B.W. Harris, F. Chen, U. Mohideen, Phys. Rev. A **62**, 052109 (2000)
J.K. Hoskins, R.D. Newman, R. Spero, J. Schultz, Phys. Rev. D **32**, 3084 (1985)
C.D. Hoyle, U. Schmidt, B.R. Heckel, E.G. Adelberger, J.H. Gundlach, D.J. Kapner, H.E. Swanson, Phys. Rev. Lett. **86**, 1418 (2001)
C.D. Hoyle, D.J. Kapner, B.R. Heckel, E.G. Adelberger, J.H. Gundlach, U. Schmidt, H.E. Swanson, Phys. Rev. D **70**, 042004 (2004)
D.B. Kaplan, M.B. Wise, J. High Energy Phys. (2000)
D.J. Kapner, T.S. Cook, E.G. Adelberger, J.H. Gundlach, B.R. Heckel, C.D. Hoyle, H.E. Swanson, Phys. Rev. Lett. **98**, 021101 (2007)
A. Kehagias, K. Sfetsos, Phys. Lett. B **472**, 39 (2000)
S.K. Lamoreaux, Phys. Rev. Lett. **78**, 5 (1997)
D.R. Long, Phys. Rev. D **9**, 850 (1974)
D.R. Long, Nature **260**, 417 (1976)
D.R. Long, Nuovo Cimento B **55**, 252 (1980)
J.C. Long, H.W. Chan, A.B. Churnside, E.A. Gulbis, M.C.M. Varney, J.C. Price, Nature **421**, 922 (2003)
M.V. Moody, H.J. Paik, Phys. Rev. Lett. **70**, 1195 (1993)
M.W. Moore, Ph.D. Thesis, UMI #9964273, University of Washington (2000)
V.M. Mostepanenko, R.S. Decca, E. Fischbach, G.L. Klimchitskaya, D.E. Krause, D. López, J. Phys. A: Math. Theor. **41**, 164054 (2008)
V.V. Nesvizhevsky, K.V. Protasov, Class. Quantum Gravity **21**, 4557 (2004)
V.V. Nesvizhevsky, G. Pignol, K.V. Protasov, Phys. Rev. D **77**, 034020 (2008)
Y. Ogawa, K. Tsubono, H. Hirakawa, Phys. Rev. D **26**, 729 (1982)
V.I. Panov, V.N. Frontov, Sov. Phys. JETP **50**, 852 (1979)
R. Spero, J.K. Hoskins, R. Newman, J. Pellam, J. Schultz, Phys. Rev. Lett. **44**, 1645 (1980)
L.-C. Tu, S.-G. Guan, J. Luo, C.-G. Shao, L.-X. Liu, Phys. Rev. Lett. **98**, 201101 (2007)
D.M. Weld, J. Xia, B. Cabrera, A. Kapitulnik, Phys. Rev. D **77**, 062006 (2008)

Space Sci Rev (2009) 148: 191–199
DOI 10.1007/s11214-009-9533-6

# The Equivalence Principle and the Constants of Nature

**Thibault Damour**

Received: 29 April 2009 / Accepted: 11 May 2009 / Published online: 30 May 2009
© Springer Science+Business Media B.V. 2009

**Abstract** We briefly review the various contexts within which one might address the issue of "why" the dimensionless constants of Nature have the particular values that they are observed to have. Both the general historical trend, in physics, of replacing a-priori-given, absolute structures by dynamical entities, and anthropic considerations, suggest that coupling "constants" have a dynamical nature. This hints at the existence of observable violations of the Equivalence Principle at some level, and motivates the need for improved tests of the Equivalence Principle.

**Keywords** Gravitation · Equivalence Principle · Constants

## 1 Introduction

The currently known laws of physics contain about twenty independent dimensionless "constants". For instance, some of the most important ones, for determining the world around us, are:

$$\alpha_{\mathrm{EM}} = \frac{e^2}{\hbar c} \simeq \frac{1}{137.035\,999\,7}, \tag{1}$$

$$\frac{m_p}{m_e} \simeq 1836.152\,672, \tag{2}$$

$$\frac{G\,m_e\,m_p}{\hbar c} \simeq 3.216 \times 10^{-42}. \tag{3}$$

An important question is: What determines the values of these constants? According to Leibniz, one of the basic principles of rational thinking is the *Principle of Reason*: "Nihil est sine ratione" ("Nothing is without a reason"). What could be the "reasons" behind the very specific numbers quoted in (1)–(3) above? We do not have any firm answer to this

T. Damour (✉)
Institut des Hautes Etudes Scientifiques, 35, route de Chartres, 91440 Bures-sur-Yvette, France
e-mail: damour@ihes.fr

question. The aim of this note is to recall the various contexts and scenarios within which this question might be addressed. The main conclusion of our discussion will be that it is important to perform improved tests of the Equivalence Principle because these tests are one of our few windows on the physics which is possibly at work for selecting the constants of Nature.

## 2 Are the Constants Constant?

Einstein's theory of General Relativity has deeply transformed one aspect of the general framework of physics. Before 1915, both the structure of spacetime and the laws of local matter interactions were supposed to be "rigid", i.e. given once for all, as absolute structures, independently of the material content of the world. General Relativity introduced the idea that the structure of spacetime might be "soft", i.e. influenced by its material content. On the other hand, one of the basic principles of General Relativity, the Equivalence Principle (EP), postulates that the laws of local physics, and notably the values of all the dimensionless coupling constants, such as $\alpha_{\rm EM}$ or $m_p/m_e$, must be kept "rigidly fixed". General Relativity thereby introduces an *asymmetry* between a soft, dynamical spacetime structure and a rigid, non-dynamical set of coupling constants.

This asymmetry was questioned by Dirac (1937) and Jordan (1937, 1939). Dirac phenomenologically assumed that the small dimensionless coupling $G\,m_e\,m_p/\hbar c$ of (3) varied proportionally to the inverse of the age of the universe, while Jordan (reviving generalizations of General Relativity à la Kaluza-Klein) essentially assumed that both $\alpha_{\rm EM}$ and $G$ could become spacetime fields $\varphi(t,\boldsymbol{x})$. Actually, the first author to clearly realize that Jordan's original theory implied that the fine-structure constant $\alpha_{\rm EM}$ had become replaced by a field $\varphi(t,\boldsymbol{x})$ was Fierz (1956). Fierz then pointed out that astronomical data (line spectra of galaxies) were putting rather strong constraints on the spacetime variability of $\alpha_{\rm EM}$, and suggested to restrict the original, two-parameter class of Jordan's "varying constant" theories to the special one-parameter class where the fine-structure constant $\alpha_{\rm EM}$ remains constant, but where the gravitational coupling $G$ is allowed to become a spacetime field. [This EP-respecting one-parameter Jordan-Fierz theory coincides with the tensor-scalar theory later studied by Brans and Dicke.]

## 3 Varying Constants and Equivalence Principle Violations

The considerations of Jordan and Fierz on field-theory models of varying constants attracted the attention of Dicke. In particular, Dicke realized the important fact that any theory in which the local coupling constants are spatially dependent will entail some violation of the (weak) Equivalence Principle (EP), namely some non-universality in the free-fall acceleration of bodies embedded in an external gravitational field. Dicke's general argument (Dicke 1964) is that the mass $m_i$ of a body, which is made (in view of $mc^2 = E_{\rm tot} = \sum_\alpha E_\alpha$) of many contributions, related to various interaction energies (strong, weak, electromagnetic; to which we can now add the Higgs interactions, responsible for the "rest masses" of the quarks and the leptons), is a certain, complicated function of various coupling constants, notably the gauge and Yukawa coupling constants: $m_i = m_i[\alpha_{\rm EM},\ldots]$. If the coupling constants are spatially dependent, the free-fall acceleration deduced from the action of a point particle embedded in a (general relativistic) gravitational field $g_{\mu\nu}(x)$,

$$S_{m_i} = -\int m_i[\alpha_{\rm EM}(x),\ldots]\sqrt{-g_{\mu\nu}(x)\,dx^\mu\,dx^\nu}, \tag{4}$$

will read (in the slow-velocity limit)

$$\begin{aligned} \boldsymbol{a}_i &= \boldsymbol{g} - \nabla \ln m_i[\alpha_{\rm EM}(x), \ldots] \\ &= \boldsymbol{g} - \frac{\partial \ln m_i[\alpha_{\rm EM}, \ldots]}{\partial \alpha_{\rm EM}} \nabla \alpha_{\rm EM} - \cdots. \end{aligned} \tag{5}$$

The coefficients associated to the spatial gradients of the various coupling constants in (5) are expected not to be universal, so that $\boldsymbol{a}_i \neq \boldsymbol{a}_j$ if the composition of body $i$ differs from that of body $j$.

To turn the result (5) into an explicit prediction for the composition dependence of the EP-violation parameter

$$\eta_{ij} \equiv \frac{a_i - a_j}{\langle a \rangle} \tag{6}$$

one needs: (i) an explicit, Jordan-type, field model of the spacetime variability of coupling constants (predicting both the dynamics of the field $\varphi(t, \boldsymbol{x})$, and the dependence of $\alpha_{\rm EM}$, $m_p/m_e$, $G m_e m_p, \ldots,$ on the field $\varphi$), and (ii) an estimate of the dependence of the mass of a body $i$ (say a chunk of Beryllium) on the various coupling constants of particle physics. Concerning the point (i), several models have been considered in the literature: the original (Kaluza-Klein-)Jordan-type scalar field, coupling (in the "Einstein frame") *only* to the electromagnetic action, and thereby affecting only $\alpha_{\rm EM}$, has been revived by Bekenstein (1982). The properties of this model have been studied by several authors (e.g. Sandvik et al. 2002). Other authors have focussed on the more general type of field models suggested by String Theory, i.e. on "dilaton models" where a scalar field $\varphi(t, \boldsymbol{x})$ monitors, in a correlated manner, the spacetime variability of, essentially, all the coupling constants: gauge couplings, Yukawa couplings, gravitational coupling, . . . . In these models, because of the complex dependence of mass on the various couplings (point (ii)), the EP-violation parameter (6) has, in general, a complicated dependence on the nuclear composition of bodies $i$ and $j$ (see Damour and Polyakov 1994). The dependence of the mass $m_i$ on quark masses, via nuclear interactions, is especially difficult to estimate, see Flambaum and Shuryak (2003), Donoghue (2006), Damour and Donoghue (2008), Dent (2008a). The complexity of the composition-dependence of the EP-violation $\eta_{ij}$ in dilaton models is a phenomenologically interesting fact which might allow, in principle, to experimentally probe the existence of a long-range dilaton-like field, via EP tests comparing several different pairs of bodies. Correlatively, the predicted general structure of the composition dependence of $\eta_{ij}$ can be used to optimize the choice of materials in EP experiments (Damour and Blaser 1994; Damour 1996).

For reviews of the observational and theoretical status of "varying constants" see Uzan (2003), Martins (2003), Dent (2008b), and the popular books (Barrow 2002; Fritzsch 2009). See also the contributions of J.P. Uzan and N. Kolachevsky to these proceedings. Let us only mention the recent progress in the comparison of single-ion optical clocks which has allowed one to constrain the variability of $\alpha_{\rm EM}$ at the $10^{-17}$ yr$^{-1}$ level (Rosenband et al. 2008) (thereby "beating" the indirect constraint on $d \ln \alpha_{\rm EM}/dt$ coming from the analysis of the Oklo phenomenon (Shlyakhter 1976; Damour and Dyson 1996; Fujii et al. 2000)).

## 4 Scenarios for the Selection of Coupling Constants

Let us come back to the main issue of concern: what could be the "reason" behind the specific values taken by the coupling constants of Nature, such as (1)–(3)? We have recalled above that, in General Relativity, the EP represents an obstacle towards finding such a "reason". Indeed, if the coupling constants are "God-given", absolute constants, there is no hope of finding, within Nature, a rational ground for selecting specific values for them. On the other hand, the history of physics suggests that there are no *absolute structures* in physics. Einstein taught us that the concepts of absolute space and absolute time were just approximations to a particular, cosmologically selected, *dynamical* spacetime $g_{\mu\nu}(x)$. Kaluza, Klein, Dirac and Jordan generalized the message of General Relativity by suggesting that both the fine-structure constant $\alpha_{\rm EM}$ and the gravitational constant $G$ might be, like $g_{\mu\nu}(x)$, *dynamical* entities, determined, or at least influenced, by the material content of the universe, and, in particular, by its cosmological evolution. Pauli realized (in his 1953 work which generalized the Kaluza-Klein mechanism to a higher-dimensional "gauge symmetry group"; see Straumann 2008) that $SU(n)$ gauge couplings $g_n^2$ could, like $\alpha_{\rm EM}$, be dynamical entities. Brout-Englert, Guralnik-Hagen-Kibble, and Higgs taught us that the masses of leptons and quarks could be dynamically determined by a (spontaneous symmetry-breaking) mechanism in which a certain auxiliary ("Higgs") field $\phi(x)$ settles to the bottom of its potential well: $\phi(x) \to \phi_0$ (modulo some symmetry operation). String theory appears to be a vast generalization of the Einstein-Kaluza-Klein-Jordan-Pauli idea in which *all* the coupling parameters of the world around us (gauge couplings, Yukawa couplings, gravitational coupling, ...) are *dynamical* entities, related to the mean values (in the "vacuum" around us) of certain "moduli fields" $\varphi_A(x)$.

This historical trend suggests that all the numbers in (1)–(3), and similar equations dealing with other couplings and ratios, might be determined by some dynamical mechanism in which some fields $\varphi_A(x)$ (which determine the local values of the coupling constants) acquire some, approximately constant and uniform, values in the "vacuum" around us. This opens up the hope of finding the "reason" why the numbers in (1)–(3) have the values they take.

Initially, string theorists hoped that the stringent consistency requirements of string theories would somehow select a unique, stable "vacuum", in which consistency requirements and energy minimization would oblige the moduli fields $\varphi_A(x)$ determining the coupling constants of low-energy physics to take particular values $\langle\varphi_A(x)\rangle = \varphi_A^0$. This would be a striking vindication of Leibniz's Principle of Reason. So far it has not been possible to uncover such stringent vacuum-selecting consistency requirements. As a substitute to this grand hope of finding a *unique* consistent vacuum, many string theorists hope that there exists a "*discretuum*" of consistent string vacua, i.e. a discrete set of vacua, in each of which the moduli fields take particular values $\varphi_A^0$, corresponding to some discrete, local minimum of the total energy (for recent reviews see Douglas and Kachru 2007; Denef 2008). If that is the case, this would predict that the coupling constants do not have any temporal or spatial variability because, like in the Higgs mechanism, a fluctuation $\delta\varphi_A(x) = \varphi_A(x) - \varphi_A^0$ has an energy cost $\delta V(\varphi_A) \simeq \frac{1}{2}(\partial^2 V/\partial\varphi_A^0\partial\varphi_B^0)\,\delta\varphi_A\,\delta\varphi_B$ which implies that $\delta\varphi_A(x)$ is a massive, short-ranged field (with Yukawa-type, exponentially suppressed effects). Though such a mechanism might entail observable short-range modifications of gravity (Antoniadis 2008), it predicts the absence of any long-range EP violations. Note that, far from providing no motivation for EP tests, the current majority view of string theorists does imply that EP tests are important: indeed, they represent tests of a widespread theoretical assumption, that any EP-violation observation would refute, thereby teaching us a lot about fundamental issues.[1]

[1] I thank Mike Douglas for suggesting this positive way of formulating the potential theoretical impact of EP tests within the current string-theory majority view.

On the other hand, as the current attempts at stabilizing all the string-theory moduli fields (see, e.g., Denef 2008) are extremely complex and look rather unnatural, one cannot help thinking that there might exist other ways in which string theory (or whatever theory reconciles General Relativity with Particle Physics) connects itself with the world as we observe it. In particular, we know that one of the (generalized) "moduli fields", namely the Einsteinian gravitational field $g_{\mu\nu}(x)$, plays a crucial role in determining the structure of the particle physics interactions via the fact that, in a local laboratory, one can approximate, to a high accuracy, a spacetime varying $g_{\mu\nu}(x)$ by a constant Poincaré-Minkowski metric $\eta_{\mu\nu}$. In other words, when listing the dimensionless coupling constants (1)–(3),..., of particle physics one should include $\eta_{\mu\nu} = \mathrm{diag}(-1, +1, +1, +1)$ in the list, and remember that it comes from a long-range, cosmologically evolving field $g_{\mu\nu}(x)$. In this connection, let us further recall that the "dilaton", $\Phi(x)$, i.e. the moduli field which determines the value of the basic, ten-dimensional string coupling constant $g_s$ can be viewed (à la Kaluza-Klein) as an additional metric component $g_{11\,11}(x)$, measuring the size of a compactified eleventh dimension (Witten 1995). This family likeness between the dilaton $\Phi(x)$ and the metric $g_{\mu\nu}(x)$ (which entails a correlated likeness, say in heterotic string theory, between $g_{\mu\nu}(x)$ and the gauge couplings $g_a^2(x)$, as well as the string-frame gravitational coupling $G(x)$) suggests that there might exist consistent string vacua where some of the moduli fields are not stabilized, but retain their long-range, spacetime-dependent character. As recalled above, such a situation would entail long-range violations of the EP. How come such violations have not yet been observed, given the exquisite accuracy of current tests of the universality of free fall (at the $10^{-13}$ level Schlamminger et al. 2008) and of current tests of the variability of coupling constants (Rosenband et al. 2008)? A possible mechanism for reconciling a long-range, spacetime varying dilaton (or, more generally, moduli) field $\Phi(x)$ with the strong current constraints on the time or space variability of coupling constants is the *cosmological attractor* mechanism (Damour and Nordtvedt 1993; Damour and Polyakov 1994; Damour et al. 2002a) (for other attempts at using cosmological dynamics to stabilize the moduli fields see Greene et al. 2007 and references therein). A simple realization of this mechanism is obtained by assuming that all the coupling functions $B_A(\Phi)$ of $\Phi$ to the fields describing the sub-Planckian particle physics (inflaton, gauge fields, Higgs field, leptons, quarks,...) admit a limit as $\Phi \to +\infty$ ("infinite bare strong coupling") (Veneziano 2002). Under this very general, technically simple (but physically highly non trivial) assumption, one finds that the inflationary stage of cosmological expansion has the effect of naturally driving $\Phi$ towards values so large that the present observational deviations from General Relativity are compatible with all the current tests of Einstein's theory (Damour et al. 2002a, 2002b). This "runaway dilaton" mechanism also yields an interesting connection between the deviations from General Relativity and the amplitude of large-scale cosmological density fluctuations coming out of inflation. In particular, the level of EP violation is predicted to be

$$\eta \equiv \frac{\Delta a}{a} \sim 5 \times 10^{-4} k \left(\frac{\delta\rho}{\rho}\right)^{\frac{8}{n+2}}, \tag{7}$$

where $k = (b_F/(c\, b_\lambda))^2$ is a combination of unknown dimensionless parameters expected to be of order unity, and where $\delta\rho/\rho$ denotes the amplitude of large-scale cosmological density fluctuations, while $n$ denotes the exponent of the inflationary potential $V(\chi) \propto \chi^n$. Inserting the value observed in our universe, $\delta\rho/\rho \sim 5 \times 10^{-5}$, and the value $n = 2$ corresponding to the simplest chaotic inflationary potential ($V(\chi) = \frac{1}{2} m_\chi^2 \chi^2$), the rough prediction (7) yields $\eta \sim k \times 10^{-12}$ which, given that $k$ is only constrained "to be of order unity", is compatible with current EP tests. Note that this runaway dilaton mechanism then predicts (if $n = 2$) that a modest increase in the accuracy of EP tests might detect a non zero violation. Note also the rationally pleasing aspect (reminiscent of Dirac's large number hypothesis Dirac 1937) of (7) which connects the level of variability of the coupling constants to cosmological features (see Damour et al. 2002b for further discussion of this aspect), thereby explaining "why" it is so small without invoking the presence of unnaturally small dimensionless numbers in the fundamental Lagrangian.

The "runaway dilaton" mechanism just mentioned was formulated as a possible way of reconciling, within a string-inspired phenomenological framework, a "cosmologically running" massless[2] dilaton with observational tests of General Relativity. Let us note that some authors (Wetterich 2008; Rabinovici 2008) have suggested that the puzzle of having an extremely small vacuum energy $\rho_{\text{vac}} \lesssim 10^{-123}\,(m_{\text{Planck}})^4$ might be solved by a mechanism of spontaneous breaking of scale invariance of some (unknown) underlying scale-invariant theory. Under the assumption that scale-invariance is re-established only when a certain "dilaton field"[3] $\varphi \sim \ln \chi \to \infty$, it seems (Wetterich 2008) that a "$\varphi$-dilaton runaway" behaviour (technically similar to the $\Phi$-dilaton runaway mentioned above) might take place and entail similar observational violations of the EP.

## 5 Dynamics versus Anthropics

We have mentioned above various visions of the "reason" behind the selection of the observed values (1)–(3),... of the coupling constants. The intellectually most satisfactory one (given the historical pregnancy of the Principle of Reason Heidegger 1991) would be the discovery of subtle consistency requirements which would select an essentially unique physico-mathematical scheme describing the only possible physical laws. In this vision, all the dimensionless numbers of (1)–(3),... would be uniquely determined. [Note that the discovery of asymptotic freedom and dimensional transmutation, see Gross and Wilczek (1973), Politzer (1973), has opened the way to a conceivable rational explanation of very small dimensionless numbers, such as (3) (which baffled Dirac): they might be exponentially related to smallish coupling constants, along the model $\Lambda_{\text{QCD}}/\Lambda \sim \exp(-8\pi b/g^2)$ where $g^2$ is a gauge coupling constant considered at the (high-energy, cut-off) scale $\Lambda$.]

In absence of precise clues for realizing this vision, we are left with two types of less satisfactory visions. In one, the extremely vast "landscape of string vacua" can dynamically channel the coupling constants towards a discretuum of possible "locally special values". This leaves, however, open the problem of finding the "reason" why our world has selected one particular set of such, energy-minimizing locally special values. In the other, all (or some of[4]) the coupling constants are, like the metric of spacetime around us, dynamically determined by some global aspects of our universe. Both visions contain a partial dynamical "reason" behind the selection of the coupling constants, but both visions leave also a lot of room to contingency (or environmental influences). Many authors have

[2] Let us note in passing that an interesting generalization of the cosmological attractor mechanism is obtained by combining the attraction due to the coupling of $\Phi$ (via $B_A(\Phi)$) to the matter density, to the effect of a quintessence-like potential $V(\Phi)$ (Khoury and Weltman 2004).

[3] Beware that, here, the name "dilaton field" refers to a field, say $\varphi$, connected to scale invariance. Such a field $\varphi$ is, a priori, quite different from the "dilaton field" $\Phi$ of string theory. Indeed, string theory, as we currently know it, contains a basic mass (and length) scale, even in the limits $\Phi \to -\infty$ ($m_s^{(D=10)} = 1/\sqrt{\alpha'}$) or $\Phi \to +\infty$ ($m_{\text{Planck}}^{(D=11)} = 1/\ell_{\text{Planck}}^{(D=11)}$).

[4] Indeed, one can evidently mix the two different scenarios.

suggested that a complementary "reason" behind the selection of the coupling constants that we observe, might be the (weak) "Anthropic Principle", i.e. the tautological requirement that the physical laws and conditions around us must be compatible with the existence of information processing organisms able to wonder "why" the world around them is as it is. In other words, this is essentially an issue of Bayesian statistics: one should consider only a posteriori questions, rather than a priori ones. Though the appeal to such an a posteriori consistency requirement is intellectually less thrilling than the demand of a stringent a priori consistency requirement, it might have satisfied Leibniz. Indeed, Leibniz was one of the enthusiastic historical proponents of the "*Principle of Plenitude*" (Lovejoy 1936) which considers that all logically possible "things" (be they objects, beings or, even, worlds) have a tendency to (and therefore *must*, if one does not want contingency—be it God's whim—to reign) *exist*. In addition, in spite of its tautological character, the anthropic consistency requirement does lead to some well-defined, and scientifically interesting (as well as challenging) questions. Indeed, the general scientific question it raises is: what would change in the world around us if the values of the coupling constants (1)–(3), etc., would be different? In its generality, this is a very difficult question to address. Let us mention here some scientifically interesting partial answers. [For the fascinating issue of what happens when one varies the vacuum energy density (or cosmological constant) see Weinberg (1987), Vilenkin (1995), which predicted that one should observe a non zero $\rho_{\mathrm{vac}}$ *before* any data had solidly suggested it.] The "Atomic Principle" refers to the scientific study of the range of coupling parameters compatible with the existence of the periodic table of atoms, as we know it. In particular, one might ask what happens when one changes the ratio $m_q/\Lambda_{\mathrm{QCD}}$ of light quark masses (or the Higgs vacuum expectation value which monitors the quark masses) to the QCD energy scale. This issue has been particularly studied by Donoghue and collaborators (Agrawal et al. 1998a, 1998b; Donoghue 2007; Donoghue et al. 2009). Recent progress (Damour and Donoghue 2008) has shown that the existence of heavy atoms is quite sensitive to the light quark masses. If one were to increase the mass ratio $(m_u + m_d)/\Lambda_{\mathrm{QCD}}$ by about 40%, all heavy nuclei would unbind, and the world would not contain any non trivial chemistry.

Coming back to the issue of EP violation, one might use the idea of a (partially) anthropic selection of coupling constants to predict that the Equivalence Principle should be violated at some level. Indeed, as in the case of the vacuum energy mentioned above, the observed values of the coupling constants (as well as that of their temporal and/or spatial gradients or variability) should only be required to fall within some life-compatible range, and one should not expect that they take any special, more constrained value, except if this is anthropically necessary. When one thinks about it, one can see some reasons why too strong a violation of the universality of free fall might drastically change the world as we know it, but, at the same time, one cannot see any reason why the EP should be rigorously satisfied. Therefore, one should expect to observe

$$\frac{\Delta a}{a} \sim \eta_* \neq 0, \tag{8}$$

where $\eta_*$ is the maximum value of $\eta \equiv \Delta a/a$ tolerable for life (Damour and Donoghue in preparation). It is a challenge to give a precise estimate of $\eta_*$, but the prediction (8) gives an additional motivation for EP tests.

## 6 Conclusions

Despite its name, the "Equivalence Principle" (EP) is not one of the basic principles of physics. There is nothing taboo about having an observable violation of the EP. On the contrary, one can argue (notably on the basis of the central message of Einstein's theory of General Relativity) that the historical tendency of physics is to discard any, a priori given, absolute structure. The EP gives to the set of coupling constants (such as $\alpha_{\rm EM} \simeq 1/137.0359997$) the status of such an, a priori given, absolute structure. It is to be expected that this absolute, rigid nature of the coupling constants is only an approximation. Many theoretical extensions of General Relativity (from Kaluza-Klein to String Theory) suggest observable EP violations in the sense that the set of coupling constants become related to spacetime varying fields.

However, there is no firm prediction for the observable level of EP violation. Actually, the current majority view about the "moduli stabilization" issue in String Theory is to assume that, in each string vacuum, the coupling constants are fixed by an energy-minimizing mechanism which is generically expected to forbid any long-range violation of the EP. This, however, makes EP tests quite important: indeed, they represent crucial tests of a widespread key assumption of string-theory model building. This exemplifies how EP tests are intimately connected with some of the basic aspects of modern attempts at unifying gravity with particle physics.

Some phenomenological models (inspired by string-theory structures, or attempting to understand the cosmological-constant issue) give examples where the observable EP violations would (without fine-tuning parameters) be just below the currently tested level. Such (runaway dilaton) models comprise many different, correlated modifications of Einsteinian gravity ($\Delta a/a \neq 0$, $\dot{\alpha}_{\rm EM} \neq 0$, $\gamma_{\rm PPN} - 1 \neq 0, \ldots$), but EP tests stand out as our deepest possible probe of new physics. Anthropic arguments also suggest that the EP is likely to be violated at some (life-tolerable) level. Let us hope that the refined EP tests which are in preparation (see the contributions of F. Everitt, P. Touboul and M. Kasevich to these proceedings) will open a window on the mysterious physics behind the selection of the coupling constants observed in our world.

## References


V. Agrawal, S.M. Barr, J.F. Donoghue, D. Seckel, The anthropic principle and the mass scale of the standard model. Phys. Rev. D **57**, 5480 (1998a). arXiv:hep-ph/9707380

V. Agrawal, S.M. Barr, J.F. Donoghue, D. Seckel, Anthropic considerations in multiple-domain theories and the scale of electroweak symmetry breaking. Phys. Rev. Lett. **80**, 1822 (1998b). arXiv:hep-ph/9801253

I. Antoniadis, Topics in string phenomenology, in *String Theory and the Real World: From Particle Physics to Astrophysics*, ed. by C. Bachas et al., Les Houches Summer School in Theoretical Physics, Session 87, 2–27 July 2007 (Elsevier, Amsterdam, 2008), pp. 1–44. arXiv:0710.4267 [hep-th]

J.D. Barrow, *The Constants of Nature: From Alpha to Omega – the Numbers that Encode the Deepest Secrets of the Universe* (Jonathan Cape, London, 2002)

J.D. Bekenstein, Fine structure constant: Is it really a constant? Phys. Rev. D **25**, 1527 (1982)

T. Damour, Testing the equivalence principle: Why and how? Class. Quant. Grav. **13**, A33 (1996). arXiv:gr-qc/9606080

T. Damour, J.P. Blaser, Optimizing the choice of materials in equivalence principle experiments, in *Particle Astrophysics, Atomic Physics and Gravitation*, ed. by J. Tran Than Van, G. Fontaine, E. Hinds. Proceedings of the XIVth Moriond Workshop (Editions Frontières, Gif-sur-Yvette, 1994), pp. 433–440

T. Damour, J.F. Donoghue, Constraints on the variability of quark masses from nuclear binding. Phys. Rev. D **78**, 014014 (2008). arXiv:0712.2968 [hep-ph]

T. Damour, J.F. Donoghue, in preparation

T. Damour, F. Dyson, The Oklo bound on the time variation of the fine-structure constant revisited. Nucl. Phys. B **480**, 37 (1996). arXiv:hep-ph/9606486

T. Damour, K. Nordtvedt, General relativity as a cosmological attractor of tensor scalar theories. Phys. Rev. Lett. **70**, 2217 (1993)

T. Damour, A.M. Polyakov, The string dilaton and a least coupling principle. Nucl. Phys. B **423**, 532–558 (1994)

T. Damour, F. Piazza, G. Veneziano, Runaway dilaton and equivalence principle violations. Phys. Rev. Lett. **89**, 081601 (2002a). arXiv:gr-qc/0204094

T. Damour, F. Piazza, G. Veneziano, Violations of the equivalence principle in a dilaton-runaway scenario. Phys. Rev. D **66**, 046007 (2002b). arXiv:hep-th/0205111

F. Denef, Les Houches lectures on constructing string vacua, in *String Theory and the Real World: From Particle Physics to Astrophysics*, ed. by C. Bachas et al., Les Houches Summer School in Theoretical Physics, Session 87, 2–27 July 2007 (Elsevier, Amsterdam, 2008), pp. 483–610. arXiv:0803.1194 [hep-th]

T. Dent, Eotvos bounds on couplings of fundamental parameters to gravity. Phys. Rev. Lett. **101**, 041102 (2008a). arXiv:0805.0318 [hep-ph]

T. Dent, Fundamental constants and their variability in theories of High Energy Physics. Eur. Phys. J. ST **163**, 297 (2008b). arXiv:0802.1725 [hep-ph]

R.H. Dicke, in *Relativity, Groups and Topology*, ed. by C. DeWitt, B. DeWitt (Gordon and Breach, New York, 1964), pp. 163–313

P.A.M. Dirac, The Cosmological constants. Nature **139**, 323 (1937)

J.F. Donoghue, The nuclear central force in the chiral limit. Phys. Rev. C **74**, 024002 (2006). arXiv:nucl-th/0603016

J.F. Donoghue, The fine-tuning problems of particle physics and anthropic mechanisms, in *Universe or Multiverse*, ed. by B. Carr (Cambridge University Press, Cambridge, 2007), pp. 231–246. arXiv:0710.4080 [hep-ph]

J.F. Donoghue, K. Dutta, A. Ross, M. Tegmark, Likely values of the Higgs vev (2009). arXiv:0903.1024 [hep-ph]

M.R. Douglas, S. Kachru, Flux compactification. Rev. Mod. Phys. **79**, 733 (2007). arXiv:hep-th/0610102

M. Fierz, Helv. Phys. Acta **29**, 128 (1956)

V.V. Flambaum, E.V. Shuryak, Dependence of hadronic properties on quark masses and constraints on their cosmological variation. Phys. Rev. D **67**, 083507 (2003). arXiv:hep-ph/0212403

H. Fritzsch, *The Fundamental Constants, a Mystery of Physics* (World Scientific, Singapore, 2009)

Y. Fujii et al., The nuclear interaction at Oklo 2 billion years ago. Nucl. Phys. B **573**, 377 (2000). arXiv:hep-ph/9809549

B. Greene, S. Judes, J. Levin, S. Watson, A. Weltman, Cosmological moduli dynamics. J. High Energy Phys. **0707**, 060 (2007). arXiv:hep-th/0702220

D.J. Gross, F. Wilczek, Ultraviolet behavior of non-Abelian Gauge theories. Phys. Rev. Lett. **30**, 1343 (1973)

M. Heidegger, *The Principle of Reason* (Indiana University Press, Bloomington, 1991)

P. Jordan, Naturwissenschaften **25**, 513 (1937)

P. Jordan, Z. Phys. **113**, 660 (1939)

J. Khoury, A. Weltman, Chameleon cosmology. Phys. Rev. D **69**, 044026 (2004). arXiv:astro-ph/0309411

A.O. Lovejoy, *The Great Chain of Being* (Harvard University Press, Cambridge, 1936) and 1964

C.J.A.P. Martins, ed., *The Cosmology of Extra Dimensions and Varying Fundamental Constants*, A JENAM 2002 Workshop, Astrophys. Space Sci. **283**(4) (2003)

H.D. Politzer, Reliable perturbative results for strong interactions? Phys. Rev. Lett. **30**, 1346 (1973)

E. Rabinovici, Spontaneous breaking of space–time symmetries. Lect. Notes Phys. **737**, 573 (2008). arXiv:0708.1952 [hep-th]

T. Rosenband et al., Science **319**, 1808–1812 (2008)

H.B. Sandvik, J.D. Barrow, J. Magueijo, A simple varying-alpha cosmology. Phys. Rev. Lett. **88**, 031302 (2002). arXiv:astro-ph/0107512

S. Schlamminger, K.Y. Choi, T.A. Wagner, J.H. Gundlach, E.G. Adelberger, Test of the equivalence principle using a rotating torsion balance. Phys. Rev. Lett. **100**, 041101 (2008). arXiv:0712.0607 [gr-qc]

A.I. Shlyakhter, Nature **260**, 340 (1976)

N. Straumann, Wolfgang Pauli and modern physics (2008). arXiv:0810.2213 [physics.hist-ph]

J.P. Uzan, The fundamental constants and their variation: Observational status and theoretical motivations. Rev. Mod. Phys. **75**, 403 (2003). arXiv:hep-ph/0205340

G. Veneziano, Large-N bounds on, and compositeness limit of, gauge and gravitational interactions. J. High Energy Phys. **0206**, 051 (2002). arXiv:hep-th/0110129

A. Vilenkin, Predictions from quantum cosmology. Phys. Rev. Lett. **74**, 846 (1995). arXiv:gr-qc/9406010

S. Weinberg, Anthropic bound on the cosmological constant. Phys. Rev. Lett. **59**, 2607 (1987)

C. Wetterich, Naturalness of exponential cosmon potentials and the cosmological constant problem. Phys. Rev. D **77**, 103505 (2008). arXiv:0801.3208 [hep-th]

E. Witten, String theory dynamics in various dimensions. Nucl. Phys. B **443**, 85 (1995). arXiv:hep-th/9503124

Space Sci Rev (2009) 148: 201–216
DOI 10.1007/s11214-009-9609-3

# Laboratory Tests of the Equivalence Principle at the University of Washington

**Jens H. Gundlach · Stephan Schlamminger · Todd Wagner**

Received: 14 September 2009 / Accepted: 12 November 2009 / Published online: 5 December 2009
© Springer Science+Business Media B.V. 2009

**Abstract** Equivalence principle tests are probes for new fundamental physics. We have conducted several laboratory experiments to test the equivalence principle with unprecedented precision and generality using a torsion balance placed on a continuously rotating turntable. We present preliminary results comparing the differential accelerations of Ti, Be, and Al towards local masses, the earth and the center of our galaxy.

**Keywords** Equivalence principle · Torsion balance · Gravity · Unification · Fundamental interactions · Dark matter

## 1 Introduction, Motivation

The equivalence principle is one of the, if not the, most fundamental symmetries of nature. It basically states that all types of mass-energy fall identically in a uniform gravitational field. In the language of general relativity it means that all test particles must follow geodesics. Local position invariance and Lorentz invariance together with the weak equivalence principle form the Einstein equivalence principle (EEP), establishing the basis of general relativity. These invariance principles, including the weak equivalence principle, are empirical findings and are taken as assumptions. Practically all theories attempting to unify gravity with the quantum world, which is mostly described by the Standard Model, invoke some form of equivalence principle violation. Conversely, it can be conjectured that if the equivalence principle holds at all levels that no connection exists between the two axiomatic models of fundamental physics, General Relativity and the Standard Model of particle physics. In this case, a unification cannot be achieved but rather some sort of dualism would have to be involved. Since the discovery of an equivalence principle violation would have phenomenal consequences, its validity must be tested with the highest possible precision and the widest generality.

J.H. Gundlach (✉) · S. Schlamminger · T. Wagner
Department of Physics, CENPA, University of Washington, Box 354290, Seattle, WA 98195, USA
e-mail: jens@phys.washington.edu

The importance of the equivalence principle has been realized since antiquity, with the most cherished anecdotal test conducted by Galileo dropping spheres from the Leaning Tower of Pisa. Nowadays, tests of the equivalence principle are often interpreted as searches for new interactions. Such new forces in the form of new scalar or vector fields would couple to "charges" resident in the interacting bodies.

A direct consequence of the EEP is the universality of free fall (UFF), which states that any point particle experiences exactly the same acceleration at the same location in a gravitational field. Here the emphasis is on *any,* meaning that the acceleration is not a function of the composition. In the famous UFF experiments by Braginsky (Braginsky and Panov 1971, 1972) and Dicke (Roll et al. 1964) in the 1960s and 1970s, it was assumed that the length scale of the equivalence principle violation is equal to the scale of Newtonian gravity, namely that it also has infinite range. This assumption, referred to as the Newtonian equivalence principle, states that gravitational mass is exactly equal to inertial mass. Braginsky's as well as Dicke's Newtonian equivalence principle tests were carried out by comparing accelerations towards the Sun, far away.

For a more general test of the equivalence principle, it must be assumed that the violation or new interaction may not have an infinite range. It is most general and natural to assume a Yukawa potential for violations of the equivalence principle, implying that the new scalar (spin $=0$) or vector (spin $=1$) fields have a finite range. The potential between two point particles is

$$V_{1,2}(r) = \mp \frac{g^2}{4\pi}(q_1)(q_2)\frac{e^{-r/\lambda}}{r}.$$

Here $q_i$ are the new charges, $\lambda = \hbar/m_b c$ is the range of the interaction, inversely proportional to the mass of the mediating particle, and $g$ is the new interaction's fundamental coupling constant with the potential being attractive between scalar interactions and repulsive between vector charges. Since the observable potential is weak, it is convenient to write it in a way that compares it to gravity:

$$V_{1,2}(r) = \alpha G m_1 m_2 \left(\frac{q}{\mu}\right)_1 \left(\frac{q}{\mu}\right)_2 \frac{e^{-r/\lambda}}{r}. \tag{1}$$

The ratio $(\frac{q_i}{\mu})$ is the charge-to-mass ratio of the interacting objects and is convenient to calculate if the new charge is a property of the elementary constituents, e.g., atoms, of the interacting masses, in which case $\mu$ becomes the atomic mass in atomic mass units $u$. The coupling constant $\alpha$ is dimensionless,

$$\alpha = \mp \frac{g^2}{4\pi G u^2}.$$

The new charge can in principle be anything, but it would probably be a conserved quantity as found in the known interactions. The differences between atomic materials are attributable to their baryon $(B)$ and lepton $(L)$ number content. To preserve generality we assume the new charge to be a linear combination of these:

$$q = B\cos\theta_5 + L\sin\theta_5,$$

with $\theta_5$ being a parameter. $\theta_5 = 0$ or $\pi/2$ are clearly of interest as is $\theta_5 = -\pi/4$, where $q$ is proportional to $B - L$, which is conserved in grand unification. At $\theta_5 = 18.4°$, $q$ is proportional to the total number of fermions $3B + L$.

## 2 The Apparatus

At the University of Washington, we have developed sensitive torsion balance experiments to test the equivalence principle. Torsion balances are exquisitely sensitive detectors for differential accelerations that act perpendicularly to the torsion fiber axis. The instruments have been refined such that they can now resolve horizontal force differences that are $10^{16}$ times smaller than the weight of the test bodies used. To test the equivalence principle, two test masses (each made from different materials) are attached to opposite sides of the torsion balance. A small difference in the attraction of the test masses towards a "source mass" located sideways from the pendulum causes the pendulum to twist. In order to resolve the extraordinarily tiny twist, the equivalence-principle-violating force must be modulated. This can be achieved either by rotating the source about the torsion balance or by rotating the torsion balance apparatus. Here we describe our most precise equivalence principle test in which we have placed the torsion balance on a continuously rotating turntable (Fig. 1).

### 2.1 Torsion Balance

The torsion pendulum consists of a pendulum frame holding eight test bodies in a composition dipole configuration. The test bodies have the same outside shape, and each has a mass of 4.84 g. The test bodies can be individually exchanged, allowing the inversion of the composition dipole with respect to the pendulum frame.

The choice of materials is set by both scientific and practical considerations. To be practical, the materials should be machinable solids, non-magnetic, and conducting. In our first test, we optimized the test-body pair to have a large difference in baryon number per mass. Since the mass of all usable materials is mostly due to their baryon content, the small difference in baryon number is predominantly due to their nuclear binding energy content. As practical materials we chose titanium, near iron, and beryllium, a light element. The specific difference in baryon content is $\Delta(B/\mu)(\mathrm{Be}-\mathrm{Ti}) = -0.0024$. We conducted another test using aluminum and beryllium as the material pair, which was designed to have better sensitivity to $L$, the lepton number, and to $B-L$. Table 1 lists the test masses' new-charge properties. The aluminum and titanium test bodies were made from two halves glued together to create a hollow shell, while each beryllium test body is made from one solid piece.

The pendulum is shown in Fig. 1. Eight test bodies are held onto an aluminum frame. It is up-down symmetric and has fourfold azimuthal symmetry. The total mass of the pendulum is 70 g, about half of it due to the frame. The entire pendulum, i.e., the frame and the test bodies, is coated with $\approx$300 nm of gold to minimize electrostatic interactions with nearby surfaces, which are also gold coated. Titanium screws fasten the test bodies into conical seats on the frame, allowing for reproducible test mass interchanges. Four gold-coated glass pieces are glued onto the pendulum body to reflect the beam of an optical readout system. A copper crimp holds the torsion fiber, and it is, in turn, glued to a screw at the top of the pendulum. At the top and bottom of the pendulum are small set screws which are used to eliminate small mass asymmetries.

A 20 μm diameter, 1.0 m long tungsten fiber suspends the pendulum and provides the restoring torque. At the upper end of the fiber, it is crimped to a 2 cm long, thicker fiber. This "pre-hanger" fiber reduces tilt effects as described in Sect. 4.4. The top of the fiber is attached to a magnetic damper. The damper is designed to reduce the swing and bounce oscillations of the pendulum without affecting the torsional motion. A copper cylinder supported from three beryllium–copper leaf springs is held between two ring magnets contained

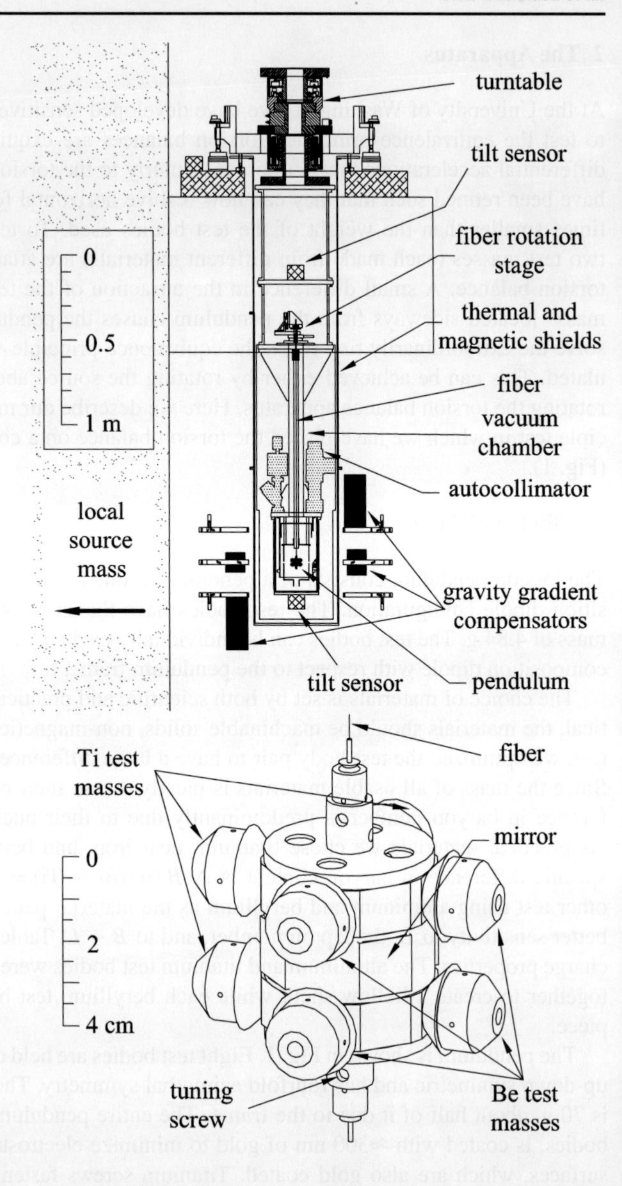


**Fig. 1** Cross-sectional drawing of the apparatus (*top*) and a drawing of the pendulum (*below*). Either titanium or aluminum test bodies can be installed with the beryllium test bodies

in a flux returning housing. The three leaf springs allow for swing and up-and-down motion. The inhomogeneous magnetic field damps these modes.

## 2.2 Torsion Angle Readout

An autocollimator is used to measure the angular deflection of the pendulum. A beam from an infrared diode laser passes through a collimator, a polarizing beam splitter, and a 1/4-wave plate and then passes through a 30 cm focal length lens to form a parallel beam. This beam is reflected off the pendulum twice and returns to the autocollimator, where it is focused to a narrow spot on a position-sensitive photo detector. The laser diode is modulated for lock-in detection.

**Table 1** For each particular test-mass pair there exists a particular $\theta_5^0$ for which $\Delta q = 0$, rendering the test-mass pair insensitive for equivalence principle tests

| | Be | Ti | Al | Be–Ti | Be–Al |
|---|---|---|---|---|---|
| $Z/\mu$ | 0.44384 | 0.45961 | 0.48181 | $-1.58 \times 10^{-2}$ | $-3.80 \times 10^{-2}$ |
| $B/\mu$ | 0.99865 | 1.00108 | 1.00068 | $-2.43 \times 10^{-3}$ | $-2.04 \times 10^{-3}$ |
| $\theta_5^0$ (deg) | −66.0 | −65.3 | −64.3 | −8.7 | −3.1 |

### 2.3 Vacuum Vessel and Shields

The torsion balance is inside an aluminum vacuum vessel at pressures $< 5 \times 10^{-5}$ Pa to minimize gas damping. The pressure is maintained with an ion pump.

The pendulum is surrounded by four layers of $\mu$-metal. Two layers are inside the vacuum vessel. The innermost shield is gold coated to minimize electrostatic effects. One of the external $\mu$-metal shields rotates with the vacuum vessel, while the outermost magnetic shield is stationary.

In addition to magnetic shields, two layers of thermal shields were employed in this experiment. Aluminum shields mounted outside the vacuum vessel enclose the ion pump and autocollimator and were wrapped in super-insulation. The outermost thermal shield is combined with the stationary magnetic shield and is insulated with foam. The entire apparatus is contained in a foam box inside a temperature-controlled room.

### 2.4 Turntable

The entire torsion balance vacuum vessel is suspended from a custom-built air-bearing turntable. An eddy-current motor drives the turntable. Its angular position is measured with a 36,000 line optical angle encoder using two read heads 180° apart. A digital signal processor (DSP) controls the eddy-current motor in a feedback loop to maintain a constant rotation rate of about 20 minutes per revolution.

Small angular acceleration non-uniformities at integer multiples of the rotation frequency were found using the torsion pendulum itself while operating the turntable at various speeds. These irregularities were due to the imperfections in the angle encoder and were reduced by introducing an angle correction map in the DSP code.

Ideally, the turntable's rotation axis should be aligned with the local vertical to reduce effects caused by a small bending of the fiber at the top attachment point; see Sect. 4.4. The tilt of the turntable is inferred from two pairs of orthogonal tilt sensors mounted on the turntable. The difference between the rotation axis and the local vertical can be measured extremely well, since this signal is modulated at the turntable rotation rate and is therefore independent of the zero-point drift of the tilt sensors. One pair of tilt sensors is located 1.7 m above the pendulum, and the other is 0.2 m below the pendulum. The upper pair is used to level the turntable using a feedback loop. The feedback changes the temperature of two turntable support legs, using thermal expansion to adjust the rotation axis parallel to the local vertical. During operation, the measured tilt of the apparatus remained constant within ±3 nrad.

## 3 Data Taking

Four sets of equivalence principle data were taken between September 2005 and June 2007 (see Table 2). Before and after each set, the sensitivities to systematic effects, such as gravity

**Table 2** Four data sets contribute to this measurement. The table shows the beginning and the end of the data taking with the science pendulum. The fourth column gives the composition dipole. The difference between Be–Ti and Ti–Be is the location of the dipole with respect to the pendulum body. Two different fibers, labeled A and B, were used to take the data

| Set | Start date | End date | Dipole | Fiber |
|---|---|---|---|---|
| 1 | 24th November 2005 | 16th January 2006 | Ti–Be | A |
| 2 | 16th May 2006 | 29th June 2006 | Be–Ti | A |
| 3 | 27th September 2006 | 19th October 2006 | Be–Al | B |
| 4 | 13th March 2007 | 27th April 2007 | Al–Be | B |

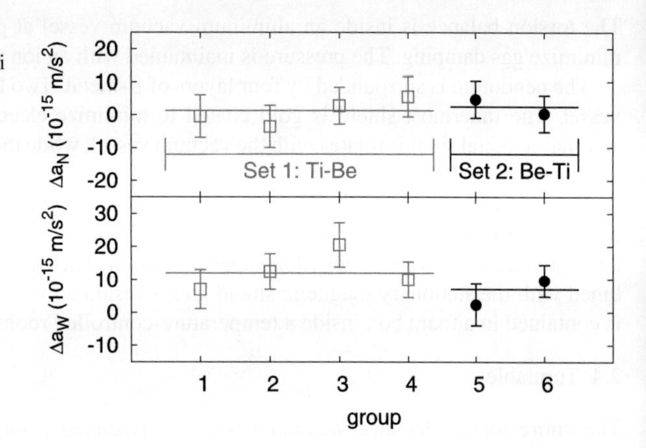


**Fig. 2** Differential acceleration measurements with the Be and Ti test bodies along north and west. The difference between the two composition dipole orientations contains the equivalence-principle-violating signal. The data are preliminary

gradients, tilt, and temperature, were measured. For each test-body pair, we recorded data sets with the composition dipole reversed on the pendulum to suppress systematics associated only with the pendulum frame.

A total of 50 days of data with the Be–Ti test-mass pair were taken in six groups, each consisting of 12 to 14 individual measurements lasting about a day. Groups 1 to 4 were taken with the Be–Ti configuration and groups 5 and 6 with the Ti–Be configuration. Between the about-one-day-long measurements the pendulum was rotated by 180° at the top of the fiber. The rotation and subsequent damping of the pendulum motion was performed under computer control. For groups 1, 2, 4, and 6 this flip alternated between the autocollimator beam reflecting from the 0° and 180° mirrors, aligned with the composition dipole, and for groups 3 and 5 between the 90° and 270° mirrors.

The measurement with the Be and Al test bodies proceeded in a similar manner to the Be and Ti test-body measurement and was divided into nine groups. Groups 1–5 used the Be–Al configuration and groups 6–9 used Al–Be. Groups 1, 5, 7, and 9 used the 90° and 270° mirrors, and groups 2, 3, 4, 6, and 8 used the 0° and 180° mirrors.

### 3.1 Data Reduction

The angle of the pendulum $\theta(t)$, the angle of the turntable $\phi_{tt}(t)$, and 28 other signals, including pressure, temperature, and tilt sensors, were recorded at 2.7 s intervals. These signals were collected in data measurements, called “runs,” lasting a few hours longer than a day. The torsion angle and the turntable angle were reduced to differential acceleration amplitude and direction, while most of the other parameters were used to monitor the data quality

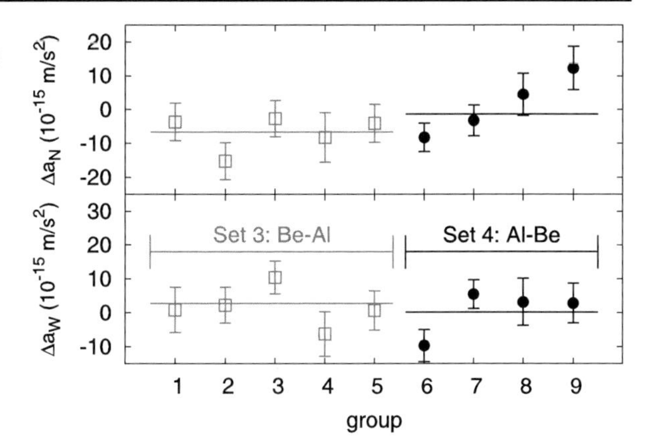


**Fig. 3** Differential acceleration measurements with the Be and Al test bodies along north and west. The data are preliminary

and to measure the differential acceleration sensitivity to these parameters. The sampling interval was chosen such that there was an integer number of samples in a quarter pendulum period, $T_0/4$.

Calibration constants for $\theta(t)$ were obtained by abruptly changing the turntable rotation rate and fitting the response to a damped harmonic oscillator. This procedure also yielded the quality factor, $Q$, of the oscillator and its period.

The pendulum angle data were filtered by adding the two data points half an oscillation period apart:

$$\theta_f(t) = \frac{1}{2}\big(\theta(t - T_0/4) + \theta(t + T_0/4)\big). \tag{2}$$

This filter attenuates the free oscillation of the pendulum by >20 dB. The filtered data is dominated by the fiber drift and residual turntable rotation rate variations.

The data is divided into segments lasting two turntable rotations, i.e., ≈2400 s, resulting in about 38 segments for a typical run. Each data segment is decomposed in Fourier coefficients $a_m^S$ and $a_m^C$ by fitting $\theta_f(t)$ to

$$\theta_f(t) = \sum_{m=1}^{9}\big(a_m^S \sin(-m\phi_{tt}(t)) + a_m^C \cos(m\phi_{tt}(t)) + d_0 P_0(\tilde{t}) + d_1 P_1(\tilde{t}) + d_2 P_2(\tilde{t})\big). \tag{3}$$

The Legendre polynomials, $P_i$, are included to account for a slow fiber drift.

To infer the torque on the pendulum from its position, the response function was taken into account:

$$\theta(\omega) = \frac{1/\kappa}{1 - \omega^2/\omega_0^2 + i/Q}\tau(\omega). \tag{4}$$

After correction for additional attenuation and phase shifts caused by digital and analog filters, the Fourier series of the torques is

$$\tau(\phi_{tt}) = \sum_{m=1}^{9}\big(\tau_m^S \sin(-m\phi_{tt}) + \tau_m^C \cos(m\phi_{tt})\big). \tag{5}$$

The components of the $m = 1$ torques are projected along north and west and converted into differential accelerations:

$$\Delta a_N = \frac{1}{M_{CD} d} \tau_1^S \quad \text{and} \quad \Delta a_W = \frac{1}{M_{CD} d} \tau_1^C, \tag{6}$$

where $M_{CD} = 19.36$ g is the mass of four test bodies with the same composition and $d = 1.9$ cm is their moment arm. The differential acceleration measurements for Be–Ti and Be–Al are shown in Figs. 2 and 3, respectively. Error bars are statistical only. The difference between the two sets in each plot contains any equivalence-principle-violating signal.

Approximately 8% of the data segments were excluded from the data analysis because small disturbances transferred an excessive amount of energy into the torsional mode. The most common disturbances were due to small gas pressure bursts correlated with ion pump current spikes. Additionally, 1% of the remaining segments with exceptionally large $\chi^2$ were excluded if a clear transient event was identified in the pendulum motion.

## 4 Systematic Effects

A host of effects must be considered when looking for extremely small forces in the presence of other intrinsically much larger forces. The leading systematic effects for this experiment are: (1) variations in the turntable rotation rate, (2) coupling of the pendulum to gravitational gradients, (3) tilts of the rotation axis, (4) magnetic coupling of the pendulum to fields in the laboratory, and (5) effects caused by temperature gradients across the torsion balance. For the first three effects we applied corrections to the data, and for the others statistical uncertainties were derived.

### 4.1 Gravity Gradients

Residual gravitational coupling between the pendulum and its environment induces torques at the rotation frequency and harmonics thereof.

The gravitational torque on the pendulum as a function of turntable angle expressed in a multipole expansion is (Su et al. 1994)

$$\tau(\phi_{tt}) = -4\pi G \sum_{l=0}^{\infty} \frac{1}{2l+1} \sum_{m=-l}^{l} i m \bar{q}_{lm} Q_{lm} e^{-im\phi_{tt}}.$$

The multipole moments of the gravity gradient field, $Q_{lm}$, and the moments of the pendulum, $\bar{q}_{lm}$, are

$$Q_{lm} = \int \rho_s(\vec{r}) r^{-l-1} Y_{lm}(\hat{r})\, \mathrm{d}^3 r \tag{7}$$

and

$$\bar{q}_{lm} = \int \rho_p(\vec{r}) r^l Y_{lm}^*(\hat{r})\, \mathrm{d}^3 r. \tag{8}$$

The couplings with $m = 1$ can lead to systematic effects, as they occur at the same frequency as the science signal. Since the gravity gradient fields fall off as $1/r^{l+1}$, the largest contributors to the systematic effects are the $Q_{21}$, $Q_{31}$ and $Q_{41}$ fields.

## 4.2 Gravitational Gradient Compensators

We used an arrangement of masses called field compensators to minimize gravity gradient fields. The masses were installed on turntables, so that they could be rotated to enhance or oppose the ambient gradients. The environmental $Q_{21}$ gradient at the pendulum location was reduced by using two approximately semi-annular lead 444 kg masses positioned about the fiber axis. The magnitude of the $Q_{21}$ was reduced from its uncompensated value of $1.78\ \mathrm{g\,cm^{-3}}$ to less than $0.02\ \mathrm{g\,cm^{-3}}$. Smaller aluminum masses were placed on a separate turntable to compensate the ambient $Q_{31}$ field.

## 4.3 Pendulum Gravitational Moments

To measure seasonal changes in $Q_{21}$ and $Q_{31}$, we constructed a gradiometer pendulum with large, well-determined multipole mass moments, but no composition dipole. The pendulum can be configured separately to have large $q_{21}$, $q_{31}$ or $q_{41}$ moments. It consists of four gold-coated aluminum disks stacked vertically. Each disk contains eight recesses in which to place small titanium spheres. Figure 4 shows a schematic drawing of the gradiometer pendulum with 16 titanium spheres placed to generate a $q_{21}$ moment.

The gravity gradient field varied somewhat during the year, mostly due to changes in the moisture content of the soil surrounding our laboratory. We measured the $Q_{21}$ and the $Q_{31}$ field five times in the course of the reported experiments. The values of the residual gravity gradient are shown in Fig. 5 together with the time of their measurement.

The equivalence principle pendulum was designed to have vanishing gravitational moments for $m = 1$ and $l < 7$. Furthermore, the pendulum $q_{20}$, $q_{30}$ and $q_{40}$ moments were minimized, so that a small tip of the pendulum body does not translate into an $m = 1$ moment. Due to machining imperfections, the pendulum had measurable, unwanted $m = 1$ moments.

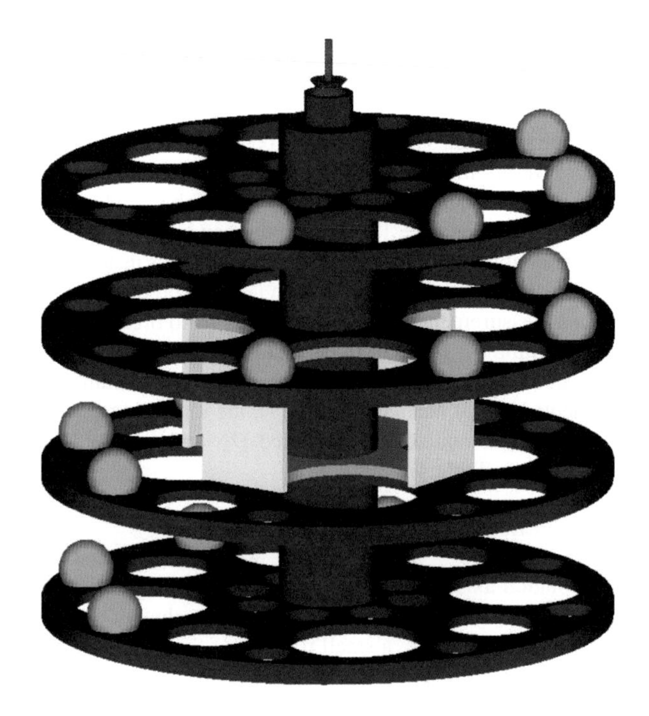

**Fig. 4** A schematic drawing of the gradiometer pendulum. The mass distribution of the 16 titanium balls shown here exhibits a large $q_{21}$, but small $q_{31}$ and $q_{41}$ moments

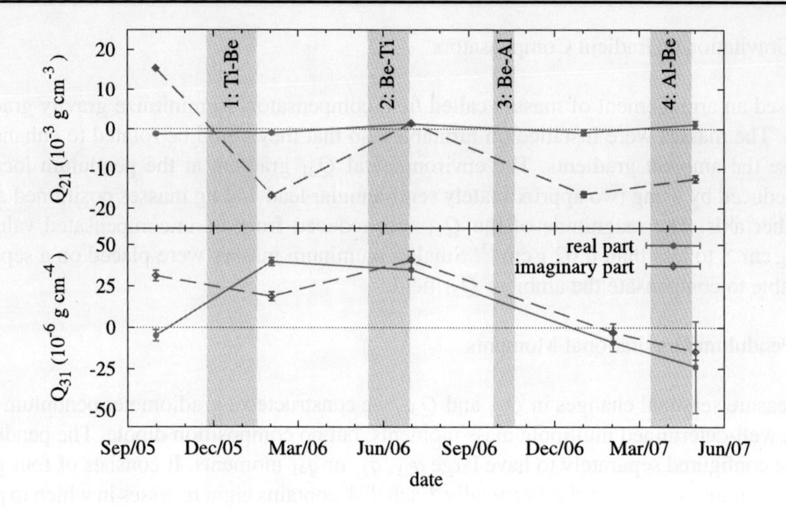


**Fig. 5** The residual gravity gradient field moments. The real part of the field is generated by a mass located north or south of the pendulum. The *shaded areas* indicate the times when data was taken with the composition-dipole pendulum. The *labels* indicate the number of the set and the configuration of the composition dipole

**Table 3** The measured absolute values of the $q_{21}$ and $q_{31}$ moments of our science pendulum. All but one measurement were performed at the beginning and the end of each data set

| | Number of the set | | | |
|---|---|---|---|---|
| | 1 | 2 | 3 | 4 |
| $\lvert q_{21}\rvert$ ($10^{-3}$ g cm$^2$) before | 1.5 | 1.1 | 1.0 | 1.1 |
| $\lvert q_{21}\rvert$ ($10^{-3}$ g cm$^2$) after | 1.7 | 1.1 | 0.9 | 1.0 |
| $\lvert q_{31}\rvert$ ($10^{-1}$ g cm$^3$) before | 1.0 | 1.9 | 3.9 | 3.7 |
| $\lvert q_{31}\rvert$ ($10^{-1}$ g cm$^3$) after | – | 1.8 | 4.1 | 3.7 |

Four pairs of small set screws in the pendulum frame allowed for compensation of these moments. We minimized the $q_{21}$ moments of the pendulum every time an operation, such as a test-body interchange, was performed (see Table 3). The $q_{21}$ of the pendulum was measured by comparing the $m = 1$ torque on the pendulum with the compensator masses rotated by 180° to double the ambient gravity gradient fields instead of canceling them. Minimizing the pendulum's gravity moments was an iterative process. The tuning screws were moved based upon a measurement of the $q_{21}$, requiring us to break the vacuum for each adjustment, until we found the $q_{21}$ moment to be small enough. After taking a complete equivalence principle data set, the $q_{21}$ and $q_{31}$ moments of the pendulum were remeasured to verify that the moments had not changed. Using (7), the remaining $q_{21}$ and $q_{31}$ moments were multiplied with the mean value of the measured $Q_{21}$ and $Q_{31}$ fields to form a small correction to the raw data (see Table 4).

## 4.4 Tilt Effects

The rotation axis of the turntable must be perfectly aligned with the local vertical to avoid systematic effects. A tilt at the fiber's top attachment point bends the fiber. Since the fiber is not perfectly isotropic, a small rotation of the pendulum will occur. A tilt of the apparatus will additionally displace the optical readout beam on the pendulum mirror. A small parasitic

**Table 4** The estimates of the gravitational torques acting on the science pendulum during the four data sets

| Set | 21-Torque ($10^{-19}$ N m) | | 31-Torque ($10^{-19}$ N m) | |
|---|---|---|---|---|
| | $\tau_1^S$ | $\tau_1^C$ | $\tau_1^S$ | $\tau_1^C$ |
| 1 | $-0.5 \pm 0.7$ | $-0.5 \pm 9.4$ | $0.2 \pm 0.5$ | $0.7 \pm 0.5$ |
| 2 | $-0.5 \pm 0.5$ | $3.0 \pm 3.2$ | $2.2 \pm 0.5$ | $0.7 \pm 0.5$ |
| 3 | $0.7 \pm 1.5$ | $2.3 \pm 2.7$ | $-0.5 \pm 1.5$ | $-0.7 \pm 1.8$ |
| 4 | $-0.9 \pm 1.6$ | $-4.6 \pm 1.8$ | $0.9 \pm 1.0$ | $0.6 \pm 1.0$ |

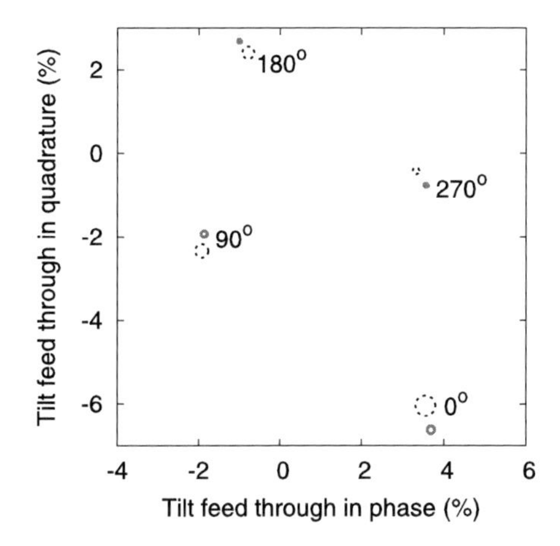


**Fig. 6** Measurements of the tilt feed-through for data set 3 (*solid lines*) and data set 4 (*dashed lines*). The *labels* indicate the mirror on the pendulum that was used for the optical angle readout. Each of the circles indicates a region in which the value can be found with 68.2% probability

curvature of the mirror will lead to an apparent pendulum rotation. Because the apparatus rotates, these effects will occur at the frequency of the science signal.

In order to decrease these tilt effects, several hardware measures were taken. First, the actual torsion fiber is suspended from a short, torsionally stiff but still flexible and more isotropic "pre-hanger" fiber. Second, we used the active leveling feedback described earlier to align the rotation axis with the local vertical.

To quantify and correct for the remaining tilts, a tilt matrix was measured by deliberately tilting the turntable. Figure 6 shows the tilt feed-through as a percentage of tilt. The local vertical changes slightly over the height of the apparatus, so despite the leveling feedback, a tilt of ≈50 nrad persists at the pendulum position.

## 4.5 Temperature Effects

Torsion balances are very sensitive to both temperature changes and temperature gradients. Slow changes of the mean temperature affect the fiber drift rate. This enhanced the noise, but is not expected to cause a torque modulation at the signal frequency. In contrast, horizontal temperature gradients are likely to induce a signal at the turntable frequency.

The temperature of the apparatus is maintained by active temperature control and passive insulation. The temperature of the foam box surrounding the apparatus is typically maintained constant within ≈300 mK peak-to-peak over the 2 weeks of a measurement cycle. Multiple layers of insulation and thermal shields surround the apparatus to reduce the effects of temperature fluctuations due to convection in the room and to attenuate temperature gradients across the apparatus.

**Table 5** The raw differential accelerations for Be–Ti toward north (N) and west (W) are shown in line 1. Lines 2 to 5 list corrections that were applied, and the bottom line gives our corrected results. Uncertainties are $1\sigma$

| Differential acceleration in | $\Delta a_{N,\text{Be–Ti}}$ ($10^{-15}$ m/s$^2$) | $\Delta a_{W,\text{Be–Ti}}$ ($10^{-15}$ m/s$^2$) |
|---|---|---|
| as measured (statistical) | $3.3 \pm 2.5$ | $-2.4 \pm 2.4$ |
| residual gravity gradients | $1.6 \pm 0.2$ | $0.3 \pm 1.7$ |
| tilt | $1.2 \pm 0.6$ | $-0.2 \pm 0.7$ |
| magnetic | $0.0 \pm 0.3$ | $0.0 \pm 0.3$ |
| temperature gradients | $0.0 \pm 1.7$ | $0.0 \pm 1.7$ |
| corrected | $0.6 \pm 3.1$ | $-2.5 \pm 3.5$ |

Horizontal temperature gradients generate the most troubling temperature-related systematic for our instrument. To measure the sensitivity, the stationary outer thermal and magnetic shielding was replaced with two copper panels, one on each side of the apparatus. These were each maintained at temperatures differing by 10°C, inducing a large temperature gradient. Pairs of thermistors on the rotating apparatus were used to measure the temperature gradients and the thermal feed-through. We found that the response to the thermal gradient was independent of the pendulum orientation within the apparatus.

### 4.6 Magnetic Effects

We tested our instrument's sensitivity to magnetic fields using a strong permanent magnet 25 cm from the pendulum with the turntable rotating and the stationary magnetic shields removed. In a second test, the stationary shields were installed and the turntable was stopped, but a magnetic field varying at the signal frequency was applied. In both methods the response of the pendulum was unresolved. Since the permanent magnet generated a stronger field at the pendulum position, we derived magnetic uncertainties from it.

## 5 Preliminary Results

We analyzed the differential acceleration data towards earth-fixed sources and towards space-fixed sources. After corrections for systematic effects, our preliminary values for the $1\sigma$-differential accelerations towards north and towards west are

$$a_N(\text{Be}) - a_N(\text{Ti}) = (0.6 \pm 3.1) \times 10^{-15}\ \text{m/s}^2,$$
$$a_W(\text{Be}) - a_W(\text{Ti}) = (-2.5 \pm 3.5) \times 10^{-15}\ \text{m/s}^2,$$
$$a_N(\text{Be}) - a_N(\text{Al}) = (-2.6 \pm 2.5) \times 10^{-15}\ \text{m/s}^2 \quad \text{and}$$
$$a_W(\text{Be}) - a_W(\text{Al}) = (0.7 \pm 2.5) \times 10^{-15}\ \text{m/s}^2.$$

Table 5 lists the raw accelerations together with the corrections for the Be–Ti data. The Be–Al data had similar systematic uncertainties, but smaller statistical uncertainty.

The classical range independent equivalence principle parameter (Will 1993) $\eta$ can be calculated using $\eta = \Delta a_N / a_\perp^g$, where $a_\perp^g$ is the locally horizontal component of the gravitational acceleration. This component arises because of the centripetal acceleration due to the rotation of the Earth. The latitude of Seattle is 47.6°, near the maximum of this component

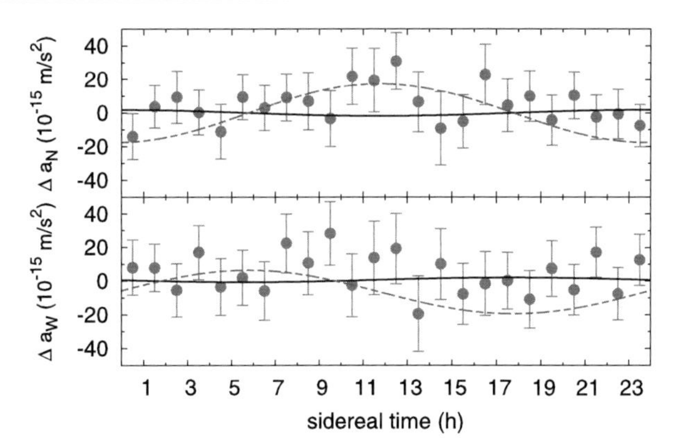


**Fig. 7** The measured differential acceleration to west and north for the Be–Ti dipole. The *solid line* gives the best fit to the data. The *dashed line* is calculated with a hypothetical signal of $20 \times 10^{-15}$ m/s$^2$ pointing towards the center of the galaxy

at latitude 45°. Note that, in contrast to the results of Dicke and Braginsky, our best limits on $\eta$ are derived using the earth and not the Sun as a source.

$$\eta(\text{Be–Ti}) = (0.3 \pm 1.8) \times 10^{-13} \quad \text{and}$$
$$\eta(\text{Be–Al}) = (-1.5 \pm 1.5) \times 10^{-13}.$$

## 5.1 Differential Acceleration Towards Astronomical Sources

Due to the apparent motion of the Sun and the center of our galaxy, the differential acceleration to these sources is additionally modulated with a period of a solar and a sidereal day, respectively. We simultaneously fit the amplitudes $a_N(t)$ and $a_W(t)$ to obtain the four unknowns ($\Delta a$, $\Delta\tilde{a}$, $o_N$ and $o_W$):

$$a_N(t) = \cos\theta(t)\big(-\Delta a \cos(\phi(t) - \phi_0) - \Delta\tilde{a}\sin(\phi(t) - \phi_0)\big) + o_N,$$
$$a_W(t) = \cos\theta(t)\big(\Delta a \sin(\phi(t) - \phi_0) - \Delta\tilde{a}\cos(\phi(t) - \phi_0)\big) + o_W,$$

where $\theta(t)$ and $\phi(t)$ denote the altitude and azimuth of the source, $\Delta a$ the differential acceleration towards the source, and $\Delta\tilde{a}$ its quadrature component; $o_N$ and $o_W$ account for instrumental offsets. In total, 2575 and 2982 data pairs were used for the Be–Ti and the Be–Al data, respectively. Figure 7 shows the data for the Be–Ti dipole binned by sidereal hours.

The preliminary differential accelerations towards the Sun and the center of the Milky Way are

$$\begin{aligned} a_\odot(\text{Be}) - a_\odot(\text{Ti}) &= (-1.8 \pm 2.8) \times 10^{-15}\ \text{m/s}^2, \\ a_\odot(\text{Be}) - a_\odot(\text{Al}) &= (0.0 \pm 2.5) \times 10^{-15}\ \text{m/s}^2, \\ a_g(\text{Be}) - a_g(\text{Ti}) &= (-2.1 \pm 3.1) \times 10^{-15}\ \text{m/s}^2, \quad \text{and} \\ a_g(\text{Be}) - a_g(\text{Al}) &= (1.6 \pm 2.8) \times 10^{-15}\ \text{m/s}^2. \end{aligned} \tag{9}$$

All astronomical source differential accelerations are consistent with zero. The quadrature components were also found to be consistent with zero, but are not detailed here.

Using the values from (9) and the absolute acceleration towards the Sun $a_\odot = 5.9 \times 10^{-3}$ m/s$^2$, our preliminary values for $\eta_\odot$ are

$$\begin{aligned} \eta_\odot(\text{Be} - \text{Ti}) &= (-3.1 \pm 4.7) \times 10^{-13}, \\ \eta_\odot(\text{Be} - \text{Al}) &= (0.0 \pm 4.2) \times 10^{-13}. \end{aligned} \tag{10}$$

The visible galactic disk is embedded in a spherical halo of dark matter. Approximately 25% of the acceleration of the solar system towards the center of the galaxy is due to dark matter residing inside the radius of the Sun's motion in the galaxy: $a_{DM} = 4.8 \times 10^{-11}$ m/s$^2$ (Stubbs 1993). The preliminary value for the equivalence principle parameter for dark matter-ordinary matter interactions is

$$\begin{aligned} \eta_{DM}(\text{Be–Ti}) &= (-4.4 \pm 6.5) \times 10^{-5}, \quad \text{and} \\ \eta_{DM}(\text{Be–Al}) &= (+3.4 \pm 5.8) \times 10^{-5}. \end{aligned} \tag{11}$$

Presently, the composition of dark matter is entirely unknown; however, our tests indicate that gravity is by far the dominant interaction between ordinary matter and galactic dark matter, or, more precisely, that the dark matter-ordinary matter long-range interaction does not distinguish between Be and Ti or Al.

### 5.2 Constraints on Long-Range Yukawa Interactions

In order to interpret our measurements in the framework of a range-dependent new interaction, e.g., a Yukawa interaction, we modeled the mass distribution and composition of the source mass. For short ranges from a meter to hundreds of meters, we performed a mass integration based on measurements, technical drawings, architectural drawings, and local maps. The integration grid is dense for objects near the apparatus and becomes coarser for objects farther away. The model was verified by finding the calculated $Q_{21}$ gradient in agreement with the measured ambient $Q_{21}$ gradient. Between 1 m and 10 m the source integral is dominated by the nearby soil behind the wall of the underground laboratory. For 100 m ranges the east sloping hillside on which the laboratory is built dominates the source strength. For kilometer ranges topological maps were used in conjunction with models of the strata under Seattle. For ranges greater than 1000 km we used an elliptical layered earth model (Dziewonsky and Anderson 1981; Morgan and Anders 1980). For the range region from 10 km to 1000 km we used very coarse data from density maps; however, the knowledge and interpretation of the source strength in this region bears a fair amount of uncertainty.

Using $a_N$, $a_W$ and the range-dependent source integral, constraints on the strength of a long-range Yukawa interaction were deduced. Using (1) the preliminary limits on the strength, $\alpha(\lambda)$, are presented in Fig. 8 in the form of 95 %-confidence exclusion plots for the charges $L$ and $B$.

For baryon number, $B$, as a charge, the Be–Ti and the Be–Al measurements produce approximately equal constraints. Independent of the assumed charge, our preliminary constraints are about an order of magnitude tighter than previous measurement (Su et al. 1994).

Figure 9 shows the constraints as a function of $\theta_5$ for $\lambda = \infty$. For this constraint the measured differential acceleration towards laboratory-fixed sources and towards the Sun have been combined to prevent insensitivity for certain charges.

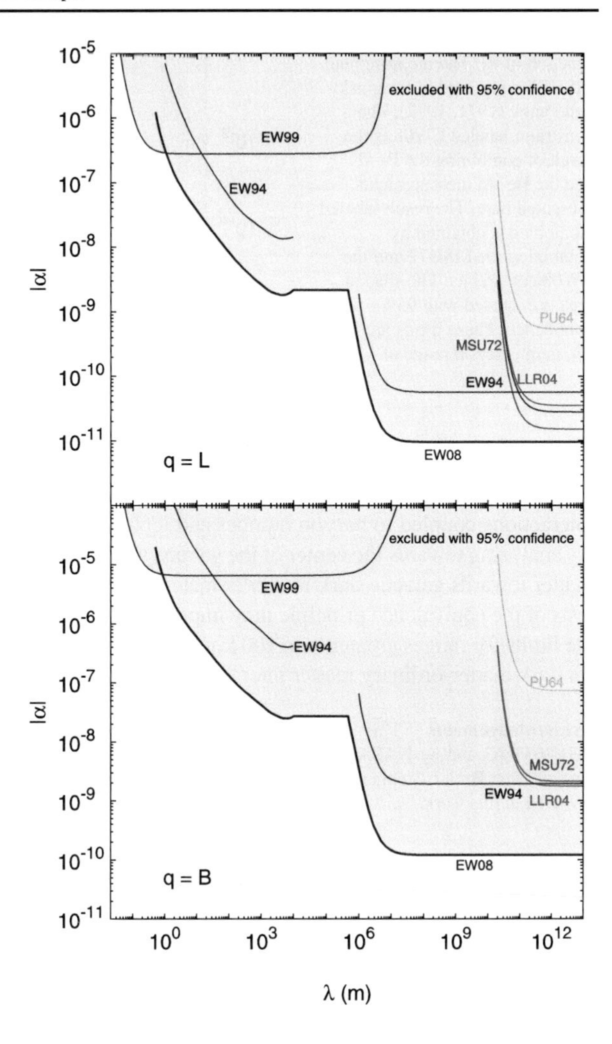


**Fig. 8** Preliminary limits on the strength $\alpha$ of a Yukawa interaction as a function of range $\lambda$. The *upper graph* shows the limit for an interaction coupling to lepton number, $L$, and the lower to baryon number, $B$. The *shaded area* indicates the region of the parameter space that is excluded with 95% confidence. The *curves labeled EW08* show the limits obtained from the results of the Be–Ti and Be–Al measurements. The *other labels* indicate the following references: PU64—Roll et al. (1964), MSU72—Braginsky and Panov (1971, 1972), EW94—Su et al. (1994), EW99—Smith et al. (2000), LLR04—Williams et al. (2004)

## 6 Conclusions

We have tested the equivalence principle using a continuously rotating torsion balance instrument. The rotation transfers the differential acceleration signal to a higher frequency, reducing statistical uncertainties dominated by $1/f$-noise. However, it brings with it systematic uncertainties, such as sensitivity to gravitational gradients and tilt sensitivity that can also generate a signal at the rotation frequency. We have developed methods and protocols to directly minimize the sensitivity to these systematics, such as ambient gravity gradient compensation together with gravity moment zeroing and automatic leveling of the turntable. We also employed additional modulations of an equivalence principle signal by rotating the pendulum within the vacuum chamber and by reversing the composition dipole on the pendulum frame. As a result, our experiment is more than a factor of 10 more sensitive to equivalence principle violations and in detecting new interactions than previous instruments. Our preliminary results presented here do not indicate that the equivalence principle is violated for Be, Ti, and Al test-mass pairs at the level of $\eta \approx 10^{-13}$. We set strict new limits on new

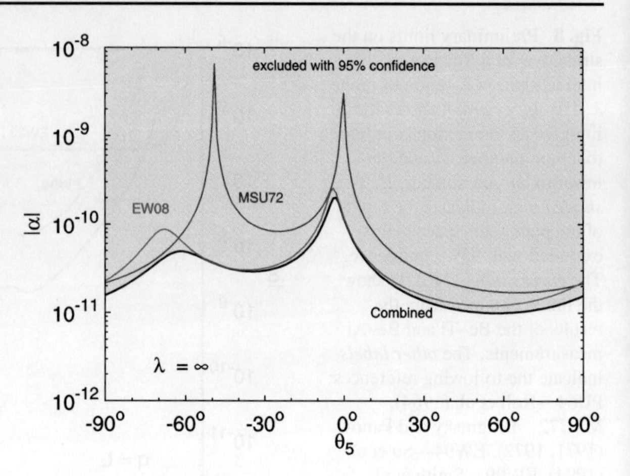


**Fig. 9** Current limits on $\alpha$ as a function of $\theta_5$. The measurement MSU72 is reported in Braginsky and Panov (1971, 1972). The constraint labeled EW08 is the result of combining the Be–Ti and the Be–Al measurements presented here. The result labeled Combined is obtained by combining the MSU72 and the EW08 constraints. The shaded area is excluded with 95% confidence. These limits are valid for an interaction range of $\lambda = \infty$

interactions coupled to baryon number and lepton number for ranges between 1 m and $\infty$. By analyzing towards the center of the galaxy, we can deduce that the acceleration of normal matter towards galactic dark matter is material independent at the $10^{-4}$ level. Space-based tests of the equivalence principle may improve upon the new interaction/equivalence principle limits for ranges greater than 100 km but are unlikely to set significantly improved limits for dark matter-ordinary matter interactions.

**Acknowledgements** This work was funded by grants from the NSF, grant Nos. PHY0355012, PHY0653863 and by NASA, grant No. NNC04GB03G. We would like to thank Eric Adelberger, Blayne Heckel, Tom Butler, Stephen Merkowitz, Ulrich Schmidt, Erik Swanson, and Phil Williams who have been involved in this work.

## References

V.B. Braginsky, V.I. Panov, Zh. Eksp Teor. Fiz. **61**, 873 (1971)
V.B. Braginsky, V.I. Panov, Sov. Phys. JETP **34**, 463 (1972)
A.M. Dziewonsky, D.C. Anderson, Phys. Earth Planet. Inter. **25**, 297 (1981)
J.W. Morgan, E. Anders, Proc. Natl. Acad. Sci. USA **77**, 6973 (1980)
P.G. Roll, R. Krotkov, R.H. Dicke, Ann. Phys. **26**, 442 (1964)
G.L. Smith, C.D. Hoyle, J.H. Gundlach, E.G. Adelberger, B.R. Heckel, H.E. Swanson, Phys. Rev. D **61**, 022001 (2000)
C.W. Stubbs, Phys. Rev. Lett. **70**, 119 (1993)
Y. Su, B.R. Heckel, E.G. Adelberger, J.H. Gundlach, M. Harris, G.H. Smith, H.E. Swanson, Phys. Rev. D **50**, 3614 (1994)
C.M. Will, *Theory and Experiment in Gravitational Physics* (Cambridge University Press, Cambridge, 1993)
J.G. Williams, S.G. Turyshev, D.H. Boggs, Phys. Rev. Lett. **93**, 261101 (2004)

Space Sci Rev (2009) 148: 217–223
DOI 10.1007/s11214-009-9491-z

# Lunar Ranging, Gravitomagnetism, and APOLLO

**T.W. Murphy Jr.**

Received: 19 January 2009 / Accepted: 26 January 2009 / Published online: 20 February 2009
© The Author(s) 2009. This article is published with open access at Springerlink.com 2009

**Abstract** The technique of lunar laser ranging (LLR) has for many decades contributed to cutting-edge tests of the fundamental nature of gravity. These include the best tests to date of the strong equivalence principle, the time-rate-of-change of the gravitational constant, gravitomagnetism, the inverse square law, and preferred frame effects. The phenomenologies of each are briefly discussed, followed by an extended discussion of gravitomagnetism. Finally, the new APOLLO project is summarized, which achieves range precision as low as one millimeter.

**Keywords** Gravitomagnetism · Lunar laser ranging

## 1 Introduction

Lunar laser ranging (LLR) is a technique in which short pulses of laser light are sent from the earth to the moon, reflecting off of arrays of corner cube prisms placed on the moon's surface by astronauts or unmanned missions (Bender et al. 1973; Dickey et al. 1994). The round-trip time is accurately measured, from which the earth-moon distance may be deduced. Comparison to a sophisticated model containing not only gravitational dynamics of the solar system, but also body torque effects, earth tides, surface loading effects, atmospheric propagation delay, etc. allows one to test whether general relativity can adequately describe the lunar orbit, and parameterize any necessary correction.

Because the "Nature of Gravity" conference focused much attention on the gravitomagnetic phenomenon, this paper is largely devoted to that topic, with comparatively little treatment of LLR science in general or the APOLLO project. External references provide ample coverage of these topics.

T.W. Murphy Jr. (✉)
University of California, San Diego, 9500 Gilman Dr., La Jolla, CA 92093-0424, USA
e-mail: tmurphy@physics.ucsd.edu

## 2 LLR Science

A continuous record of laser range measurements between Earth and Moon dating back to 1969 have provided an unprecedented set of data by which to understand dynamics within the solar system and test the fundamental nature of gravity. A model containing all anticipated physical processes impacting the measurement is constructed to simulate the measurements (see Williams et al. 1996 and references within). Differences are minimized by parameter adjustment, many of which represent initial conditions for the numerical integration of solar system bodies. Because each planet and each effect has a unique signature in frequency space—via harmonic distortions on the lunar orbit—a long time span of data allows one to separate effects and isolate influences. Part of the model describes gravity, which is generally formulated in a parameterized post-Newtonian (PPN) framework, described in Will and Nordtvedt (1972). Fits may be configured to test specific violations of general relativity by freezing or releasing various parameters in any combination. Thus far, no departures from Einstein's prescription for gravity have been identified. Such tests include:

- Weak Equivalence principle to $\Delta a/a \approx 1.3 \times 10^{-13}$ (Williams et al. 2004)
- Strong Equivalence Principle to $\eta < 4.5 \times 10^{-4}$ (Williams et al. 2004)
- Evolution of the gravitational constant to $\dot{G}/G < 9 \times 10^{-13}$ (Williams et al. 2004)
- Gravitomagnetism to $\approx 0.1\%$ (Murphy et al. 2007; Soffel et al. 2008)
- Geodetic precession to $< 0.6\%$ (Williams et al. 2004)
- Inverse square law good to $< 10^{-10}$ times strength of gravity at $\sim 10^8$ m scales.

The phenomenologies related to each of the above measurements vary. For instance, the equivalence principle violations (weak or strong) manifest as a displacement, or *polarization* of the lunar orbit, showing up as a $\cos D$ perturbation, where $D$ is the lunar phase angle with respect to the Earth–Sun line (Nordtvedt 1968). A change in $G$ looks like an anomalous evolution of the phase of the large ($2 \times 10^7$ m) elliptical signature of the lunar orbit, in which case the phase anomaly evolves quadratically for a constant $\dot{G}$. Gravitomagnetism imposes both $\cos D$ and $\cos 2D$ perturbations on the orbit (further discussion below). An anomalous precession rate could be interpreted multiple ways, and so is less clear than other perturbations at singling out a specific failure in the theory. For instance, anomalous precession could point to a failure in geodetic precession (predicted at 19.2 mas yr$^{-1}$) or as a "fifth force" modifying the inverse square law—potentially connected to brane-world ideas of Dvali et al. (2003) and Lue and Starkman (2003). LLR can also provide tests of preferred frames (Müller et al. 1996; Nordtvedt 1987) and of Newton's third law (Nordtvedt 2001).

In general, the lunar orbit provides probes to a wide variety of gravitational phenomena, with a rich set of (mostly) unique "fingerprints" by which to distinguish among them. We now turn attention to the topic of gravitomagnetism, and the role it plays in LLR.

## 3 Gravitomagnetism

### 3.1 Introduction to Gravitomagnetism

A covariant theory of gravity must allow the transformation from one coordinate frame to another while maintaining the ability to describe the same phenomenology in a consistent manner. Such transformations in general produce non-zero space-time components in the metric ($g_{0i} \neq 0$), introducing frame-dependent terms in the equations of motion. Such effects are termed gravitomagnetic, in analogy to the magnetic field, which plays a similar role in producing a covariant framework for electromagnetism. Schutz (2003) gives a well-constructed phenomenological derivation of the need for gravitomagnetism in order to satisfy frame independence, for a system of two oppositely directed linear mass currents. Gravitomagnetic phenomena are not readily apparent in human experience due to a $v^2/c^2$ suppression factor, amounting to $\sim 10^{-8}$ times the gravito-electric (Newtonian) influence in the case of earth's orbit about the sun. Precision measurements within the solar system, however, are sensitive to such small effects.

For a point mass, or for a mass element within a system of masses, one may always transform into the frame of the mass and eliminate the gravitomagnetic field generated by that mass—much as one may move along with an electric charge to "kill" any magnetic field, leaving only the electric field. As soon as two masses are in motion with two different velocities, it is impossible to find a non-rotating frame in which the gravitomagnetic field is zero. Such is the case for rotating bodies, where the mass currents of the individual mass elements combine to produce a net gravitomagnetic field that cannot be transformed away by rectilinear boosts—though it could be eliminated in a frame rotating with the object, incurring many practical difficulties in formulating the external universe in such a frame.

### 3.2 Familiar Examples

Well-known examples of gravitomagnetic phenomenology are the Lense-Thirring precession and the Schiff precession, both involving rotating massive bodies. In the first case, the inclined orbital plane of a satellite moving around the rotating mass will be seen to rotate with respect to the fixed stars. In the latter case, a gyroscope will precess due to the massive body's rotation, the direction and magnitude depending on the latitude of the gyroscope relative to the rotation axis. Both phenomena are commonly called "frame-dragging," and are effects sought by the LAGEOS and the Gravity Probe-B experiments, with some success (these may ultimately achieve $\sim$5% and $\sim$10%-level confirmation, respectively).

It is tempting to refer to these phenomena as spin-orbit and spin-spin coupling, respectively, in analogy with quantum mechanics. But this is potentially misleading, since massive bodies do not possess intrinsic gravitational "spin," but rather simply have a net angular momentum due to the superposition of many mass elements in instantaneous rectilinear motion, each contributing infinitesimally and coherently to the total angular momentum. We will now look at the gravitomagnetic influence from individual masses, which—when superimposed—constitute the familiar composite gravitomagnetic field from a rotating body.

### 3.3 Generalized Gravitomagnetism and Application to the Moon's Orbit

The Earth–Moon system also exhibits a gravitomagnetic dependence, in a form that one might be tempted to call orbit–orbit coupling in the misleading jargon of the previous paragraph. In this case, the earth moving around the sun produces gravitomagnetic field (when assessed in the solar system barycenter frame—or any frame not moving with the earth, for that matter). It is important to note that this gravitomagnetic field *does not* arise from rotation, but rather from rectilinear motion. The moon moves through this field, experiencing a Lorentz-like force perpendicular to both its velocity and the gravitomagnetic field. The gravitomagnetic acceleration is given by:

$$\mathbf{a}_{\mathrm{GM}} = \frac{2(1+\gamma+\alpha_1/4)}{c^2}\mathbf{v}_m \times \mathbf{v}_e \times \mathbf{g}_{me}, \tag{1}$$

where subscripts $e$ and $m$ denote earth and moon, $\mathbf{v}$ is a velocity vector, and $\mathbf{g}$ is the familiar gravito-electric acceleration vector acting on the moon from the earth. The pre-factor evaluates to 4 in general relativity, with PPN parameters $\gamma = 1$, and $\alpha_1 = 0$. The gravitomagnetic field is then $\mathbf{B} = 4(\mathbf{v}_e \times \mathbf{g}_{me})/c^2$, so that the Lorentz acceleration has the familiar $\mathbf{v} \times \mathbf{B}$ form. In Murphy et al. (2007), it is shown that this term in the equations of motion, when integrated over individual mass elements in rotating bodies and applied to the geometry of a gyroscope in a polar orbit around the earth, reproduces exactly the 42 mas yr$^{-1}$ precession rate sought by GP-B. It also produces $\sim 6$ m amplitude terms in the moon's orbit as evaluated in the solar system barycenter (SSB) frame, in both $\cos D$ and $\cos 2D$ functions, where $D$ is the phase angle of the moon ($D = 0$ is new moon, $D = 180°$ is full moon). Having determined these amplitudes in fitting LLR data to sub-centimeter accuracy, we may conclude that gravitomagnetism has been checked in the solar system to ~0.1% precision using LLR.

### 3.4 More than a Coordinate Effect?

A natural objection to this statement, as articulated by Kopeikin (2007), is that the gravitomagnetic phenomenology in the lunar orbit is dependent on the frame of evaluation—evaluating in a geocentric frame would null the rectilinear gravitomagnetic influence of earth on the moon. The implication is that, in transforming to the SSB frame, one inserts phenomenological corrections that are trivially recovered in the fit to the data. Such a procedure would indeed be vacuous, and would do nothing to corroborate the existence of gravitomagnetism. But aside from the *practical* difficulties of formulating solar system dynamics in a geocentric frame, we must understand whether the *choice* to execute a coordinate transformation indeed renders the gravitomagnetic sensitivity of LLR meaningless. Note that if the LAGEOS or GP-B were evaluated in a frame rotating with the earth, the "frame dragging" from the rotating earth would disappear, replaced by a host of rotating frame "effects." But does such a coordinate system *choice* also deprive these precession phenomena of value, if measured? To illuminate this issue, we must detail the manner in which the lunar range model is formulated. In doing so, we will expose the notion that the gravitomagnetic phenomenon is intimately connected with time transformation.

### 3.5 Inside LLR Analysis

Consider the fundamental measurement constituted by lunar ranging. A clock on earth, measuring proper time at some defined gravitational potential, is used to timestamp both outgoing pulses and incoming pulses from the moon. The time at the moon bounce is never measured—nor is an actual distance measured. Two times constitute the measurement, each with an absolute uncertainty of about $10^{-8}$ s (0.3 mm at Earth's velocity of 30 km s$^{-1}$), and a differential uncertainty of about $10^{-11}$ s (3 mm at the speed of light). Thus each "range" measurement fundamentally consists of two time measurements on the clock. To transform to the SSB frame, the earth-clock proper times, $\tau$, are converted to SSB frame coordinate times, $t$, according to the prescription formulated by Moyer (1981). In its most basic form:

$$dt \approx \left(1 - \Delta\phi + \frac{1}{2}\dot{s}^2\right)d\tau, \tag{2}$$

where $\Delta\phi$ represents the potential difference between the two locations, and $\dot{s}$ represents the velocity of the clock with respect to the SSB frame. Thus the complete time transformation amounts to gravitational redshift and time dilation corrections. The velocity of the earth in the SSB frame is modulated at the $\cos D$ frequency due to its orbit around the Earth–Moon barycenter. But this amounts to only 12 m s$^{-1}$, imposing only a 1.5 mm $\cos D$ modulation on top of the roughly constant 1.9 m $\dot{s}^2$ correction, modulated annually by $\sim 60$ mm due to earth's eccentric orbit. Following the time transformation, a light propagation correction is applied to account for the fact that the propagation path transits a varying gravitational potential. Called the Shapiro delay, it is computed by the formula:

$$\Delta t = \frac{(1+\gamma)GM_s}{c^3} \ln\left(\frac{R_e + R_m + R_{em}}{R_e + R_m - R_{em}}\right), \tag{3}$$

where the $R_x$ values are the radial coordinates of the bodies from the sun, and $R_{em}$ is the Earth–Moon distance. This term amounts to a nearly constant 25 ns (7.5 m) correction, modulated by only $\sim 10$ mm at the $\cos D$ frequency (29.53 d), $\sim 0.8$ m at the anomalistic month frequency (27.55 d), and $\sim 0.25$ m at an annual frequency. Finally, the earth body figure is Lorentz contracted (31 mm effect) along its velocity vector. To summarize these steps:

1. Obtain launch and return times on a clock at the surface of the Earth;
2. Transform these times to the SSB frame according to the potential energy and velocity of the clock;
3. Compute the Shapiro propagation correction for consideration in the fit;
4. Lorentz contract the earth by a few centimeters;
5. Attempt to fit the observations with a physical model resting on the relativistic equations of motion.

For the last step, with the SSB-corrected times and propagation-correction in hand, one asks the question: are there any world lines for the solar system bodies (Earth and Moon foremost) following the dynamics set out by the equations of motion that allow the time measurements to be consistent with the model? In other words, can a fit be performed that is consistent with general relativity, or PPN modifications thereof? Note that orbits are never constructed in one frame and transformed into another frame: orbits are only constructed once, during the world-line fit to the transformed time coordinates.

As mentioned before, the gravitomagnetic term in the equations of motion produces $\sim 6$ m amplitude effects at $\cos D$ and $\cos 2D$. The transformation steps detailed above do not insert these phenomenological signatures directly, so the claim that the LLR sensitivity to gravitomagnetism is trivially constructed is not borne out by this analysis.

### 3.6 Experimental Conflict Scenario

The crux of the mater is: if some other experiment claimed a departure of gravitomagnetism from general relativity at, say, the 1% level, we would be left asking the question: what part of the LLR analysis did we not understand? And fundamentally, it comes down to basic time transformation. This is another way to understand the statement that *gravitomagnetism is fundamentally tied to transformations*. Gravitomagnetism is the piece that lets gravity be put on a covariant footing, via coordinate transformations. While arbitrarily complex mass currents may mask this fact due to emergent (e.g., "spin") phenomenology, the only reason that gravitomagnetism exists at all is to permit frame transformation (see Schutz 2003). Some have tried to split gravitomagnetism into "intrinsic" and "gauge" forms, but there is no physical basis for this distinction. We do not do the same for magnetic fields in electromagnetism, despite identical frame-dependence features. Similarly, we do not say that the Coriolis "force" is fundamentally different when manifested in a swirling, draining tub (spin system) than when seen in the deflection of a projectile (orbit system), despite seemingly disparate phenomenologies.

Perhaps a fair way to put it is: the mere fact that we can successfully fit the lunar orbit in the SSB frame is striking confirmation that gravitomagnetism plays its expected role. One may wish to call the LLR connection to gravitomagnetism "gauge-dependent," and this is fair enough. But such is gravitomagnetism as a whole. The fact that a frame exists in which the gravitomagnetic field from the moving earth is zero does not diminish the physical reality of that field in shaping orbits in the SSB frame.

If another experiment were to claim a 1% violation of gravitomagnetic phenomenology, the LLR result must be confronted. How can we modify the strength of this term—this physics—to suit one experiment without irreparably damaging the LLR fit? One might investigate which other terms in the equations of motion might also be modified to cover the sins of the delinquent gravitomagnetic term. But each term brings its own unique phenomenological signature. In the case of gravitomagnetism, it is −6.1 m in $\cos D$ and −6.5 m in $\cos 2D$. Other terms will not be able to mask this without unraveling some other periodic signatures. Soffel et al. (2008) recently explored these possibilities with a full model fit to LLR data, and found that indeed one has no freedom to change the scaling of the gravitomagnetic term in an ad-hoc (the most versatile) way by more than 0.15% without destroying the fit to LLR data—even when other terms were permitted to take up the slack.

But modifying the gravitomagnetic strength in isolation—or arbitrarily in relation to the strengths of other terms—in this ad-hoc way lacks physical justification. A more meaningful test is to stay within the framework of the PPN formalism, which amounts to changing $\gamma$ or $\alpha_1$ to influence the strength of the term in (1). In this case, other terms in the equations of motion change in a way that preserves covariance automatically, with the caveat that $\alpha_1$ does describe a preferred frame in space. Here, Soffel et al. (2008) found that LLR provides a similarly restrictive limit on alteration of the gravitomagnetic term, at the 0.2% level.

### 3.7 Summary Statement

The point from this discussion is that the LLR sensitivity to gravitomagnetism represents a real limit in physics. It cannot be dismissed as merely an artifact of frame transformation: a "signal" injected and recovered. There are no $\cos D$ or $\cos 2D$ terms inserted in the transformations of (2) or (3) at the six meter level (few millimeter level, yes). In fact, one can say that gravitomagnetism is intimately connected to transformation properties. Like the magnetic field in electromagnetism, it serves to provide a covariant basis for gravity. We know we can transform magnetic fields away with boosts, but this does not lead us to conclude that such magnetic fields are any less "real" when measured by a compass needle. Likewise, the gravitomagnetic influence on the lunar orbit is real and physical in the SSB. The fact that the LLR clock acts consistently with gravitomagnetism while moving through the SSB frame only serves to highlight the fundamental connection: gravitomagnetism *is by its very nature* a frame-dependent piece of physics. Interesting arrangements of mass currents do not change that simple fact.

## 4 APOLLO

For the last decade or more, LLR data quality has settled to the impressive level of $\sim 2$ cm range uncertainty. Though the range capability has been roughly constant over this time, the science gains represented in Sect. 2 have seen steady improvement over the years as more data accumulates. Because LLR has provided such an extensive array of cutting-edge tests of gravity, it is worth pushing the technique further in order to achieve the best tests possible of the fundamental nature of gravity.

The Apache Point Observatory Lunar Laser-ranging Operation (APOLLO: Murphy et al. 2008) was established in pursuit of this goal. By implementing LLR on a large (3.5 m) astronomical telescope with modern technology, it is possible to greatly exceed the photon rate experienced by other LLR stations, translating to reduced random uncertainty via sheer statistics. APOLLO routinely achieves random uncertainties in the 1–2 mm range (Battat et al. 2009), and has seen record photon yields surpassing previous records by almost two orders-of-magnitude. It will take more time and effort to establish whether APOLLO has systematic errors under control at the millimeter level, but so far nothing obvious has emerged. Few-millimeter precision LLR measurements will permit order-of-magnitude improvements in the various tests of the fundamental nature of gravity detailed in Sect. 2. For more information on the details of APOLLO's construction and performance, please consult the references in this paragraph and information online (Murphy 2009). APOLLO's normal points are made public via this website.

**Acknowledgements** I thank Ken Nordtvedt, Sergei Kopeikin, Eric Michelsen, Jürgen Müller, Slava Turyshev, and Cliff Will for meaningful interactions that have contributed to the articulation of gravitomagnetism herein. I also acknowledge the APOLLO collaboration for their substantial role in establishing the APOLLO experiment, and Jim Williams and Dale Boggs for providing model fits to APOLLO data and associated feedback. APOLLO is funded by the National Science Foundation and by NASA.

**Open Access** This article is distributed under the terms of the Creative Commons Attribution Noncommercial License which permits any noncommercial use, distribution, and reproduction in any medium, provided the original author(s) and source are credited.

## References

J.B.R. Battat, T.W. Murphy, E.G. Adelberger, B. Gillespie, C.D. Hoyle, R.L. McMillan, E.L. Michelsen, K. Nordtvedt, A.E. Orin, C.W. Stubbs, H.E. Swanson, Publ. Astron. Soc. Pac. **121**, 29 (2009)

P.L. Bender, D.G. Currie, R.H. Dicke, D.H. Eckhardt, J.E. Faller, W.M. Kaula, J.D. Mullholland, H.H. Plotkin, S.K. Poultney, E.C. Silverberg, D.T. Wilkinson, J.G. Williams, C.O. Alley, Science **182**, 229 (1973)

J.O. Dickey, P.L. Bender, J.E. Faller, X.X. Newhall, R.L. Ricklefs, J.G. Ries, P.J. Shelus, C. Veillet, A.L. Whipple, J.R. Wiant, J.G. Williams, C.F. Yoder, Science **265**, 482 (1994)

G. Dvali, A. Gruzinov, M. Zaldarriaga, Phys. Rev. D **68**, 024012 (2003)

A. Lue, G. Starkman, Phys. Rev. D **67**, 064002 (2003)

S.M. Kopeikin, Phys. Rev. Lett. **98**, 229001 (2007)

T.D. Moyer, Celest. Mech. **23**, 33 (1981)

J. Müller, K. Nordtvedt, D. Vokrouhlicky, Phys. Rev. D **54**, R5927 (1996)

T. Murphy, APOLLO Website, http://www.physics.ucsd.edu/~tmurphy/apollo. Accessed 19 January 2009

T.W. Murphy Jr., K. Nordtvedt, S.G. Turyshev, Phys. Rev. Lett. **98**, 071102 (2007)

T.W. Murphy, E.G. Adelberger, J.B.R. Battat, L.N. Carey, C.D. Hoyle, P. Leblanc, E.L. Michelsen, K. Nordtvedt, A.E. Orin, J.D. Strasburg, C.W. Stubbs, H.E. Swanson, E. Williams, Publ. Astron. Soc. Pac. **120**, 20 (2008)

K. Nordtvedt, Phys. Rev. **169**, 1014 (1968)

K. Nordtvedt, Astrophys. J. **320**, 871 (1987)

K. Nordtvedt, Class. Quant. Grav. **18**, L133 (2001)

B. Schutz, *Gravity from the Ground Up* (Cambridge University Press, Cambridge, 2003), pp. 245–253

M. Soffel, S. Klioner, J. Müller, L. Biskupek, Phys. Rev. D **78**, 024033 (2008)

C.M. Will, K. Nordtvedt, Astrophys. J. **177**, 757 (1972)

J.G. Williams, X.X. Newhall, J.O. Dickey, Phys. Rev. D **53**, 6730 (1996)

J.G. Williams, S.G. Turyshev, D.H. Boggs, Phys. Rev. Lett. **93**, 261101 (2004)

Space Sci Rev (2009) 148: 225–232
DOI 10.1007/s11214-009-9566-x

# Atom-Based Test of the Equivalence Principle

**Sebastian Fray · Martin Weitz**

Received: 3 April 2009 / Accepted: 7 July 2009 / Published online: 25 July 2009
© The Author(s) 2009. This article is published with open access at Springerlink.com

**Abstract** We describe a test of the equivalence principle with quantum probe particles based on atom interferometry. For the measurement, a light pulse atom interferometer based on the diffraction of atoms from effective absorption gratings of light has been developed. A differential measurement of the Earth's gravitational acceleration g for the two rubidium isotopes $^{85}$Rb and $^{87}$Rb has been performed, yielding a difference $\Delta g/g = (1.2 \pm 1.7) \times 10^{-7}$. In addition, the dependence of the free fall on the relative orientation of the electron to the nuclear spin was studied by using atoms in two different hyperfine states. The determined difference in the gravitational acceleration is $\Delta g/g = (0.4 \pm 1.2) \times 10^{-7}$. Within their experimental accuracy, both measurements are consistent with a free atomic fall that is independent from internal composition and spin orientation.

**Keywords** Quantum mechanics · Gravity · Equivalence principle · Atom interferometer

A unification of gravitational and quantum theory is a still unsolved problem in today's physics. Much of the current research efforts in theoretical approaches like the string theory are motivated by this quest (Schutz and Damour 2009). To verify corresponding theoretical models, critical experiments have to be performed which combine both domains, quantum mechanics and gravity. One class of experiments, which give a deep foundation to gravity, represents tests of Einstein's equivalence principle. This fundamental principle states that all bodies, regardless of their internal composition, are affected by gravity in a universal way, i.e. they in the absence of other forces fall with the same acceleration. For macroscopic objects, tests of the equivalence principle have been performed since the early days of modern

S. Fray (✉)
Max-Planck-Institut für Quantenoptik, Hans-Kopfermann-Str. 1, 85748 Garching, Germany
e-mail: sebf@gmx.net

M. Weitz
Institut für Angewandte Physik, Wegeler Str. 8, 53115 Bonn, Germany

*Present address:*
S. Fray
Qimonda AG, Am Campeon 1-12, 85579 Neubiberg, Germany

physics (Su et al. 1994). Atomic based approaches to the equivalence principle have been proposed (Audretsch et al. 1993; Viola and Onofrio 1997), motivated by the challenge to provide new tests of theories which merge quantum mechanics and relativity.

Before proceeding, let us point out that the first atom based tests of general relativity are the Pound-Rebka experiments that investigate the gravitational redshift. At present, the achieved relative accuracy of this small shift between atomic clocks in different reference frames is $7 \times 10^{-5}$ (Vessot et al. 1980). For a comparison between different tests of the equivalence principle, see (Nordtvedt 2002). We also remark that questions whether a spin-mass coupling exists have been considered theoretically (Lämmerzahl 1998). A measurement on spin-mass interaction was performed by comparing the weight of oppositely spin polarized bulk matter (Hsieh et al. 1989).

In recent years matter-wave interferometric techniques have proven to allow for measurements of inertial effects (Berman 1997). In particular, atom interferometers were used to carry out precision measurements of the rotation (Gustavson et al. 1997, 2000) and the Earth's gravitational acceleration (Peters et al. 1999, 2001).

Here we describe an experiment testing the equivalence principle on an atomic basis. The experiment represents a proof of principle demonstration extending the free fall equivalence principle test to atoms as quantum probe particles, see Fray et al. (2004) for an earlier, detailed account of our measurement. With an atom interferometer, the Earth's gravitational acceleration of the two different isotopes $^{85}$Rb and $^{87}$Rb of the rubidium atom was compared to a relative accuracy of $1.7 \times 10^{-7}$. We also studied the free fall acceleration as a function of relative orientation of nuclear to electron spin to an accuracy of $1.2 \times 10^{-7}$ by comparing interference patterns measured with $^{85}$Rb atoms prepared in two different hyperfine ground states. Within the experimental accuracies, the obtained results of this 2004 measurement are consistent with no violation of the equivalence principle. We expect that technical improvements will in the future allow for extremely critical test of the equivalence principle using atoms as probe particles.

For the measurement, a light pulse atom interferometer based on the diffraction of atoms from standing light waves acting as effective absorption grating has been developed. In each of the matter wave beam splitters, the atomic de Broglie wave is diffracted into a coherent superposition of eigenstates, which differ only in momentum and not in their internal atomic states. We note that this is in contrast to several earlier experiments investigating inertial effects with atom interferometry, where the interfering paths were in different internal atomic states so that the difference in ac-Stark shift can contribute to the systematic error budget (Berman 1997). In our experiment (Fray et al. 2004) the different paths are in the same internal state, so that this source of systematic uncertainties is reduced. We note that several recently developed atom interferometers with large spatial separation between paths and correspondingly high sensitivity also use same internal states of interferometer paths (Müller et al. 2009; Cladé et al. 2009).

We use three standing wave light pulses as atomic beam splitters, which are applied to an ensemble of cold rubidium atoms launched on a vertical ballistic trajectory in an atomic fountain setup. The scheme of the atom interferometer is shown in Fig. 1. The frequency of the atomic beam splitter light is tuned resonantly to an open transition from a ground state $|g_D\rangle$ to a spontaneously decaying excited state $|e\rangle$. The spatially dependent intensity distribution of the standing light wave will then lead to a pumping of the atom that pass through the intensity antinodes into magnetic and hyperfine sublevels that are not any more detected. On the other hand, atoms close to the nodes will remain in the ground state $|g_D\rangle$. After the pulses, only the population in this state is detected. Hence the standing light wave forms an effective absorption grating, where due to the spatial periodicity of $\lambda/2$ an incident atomic

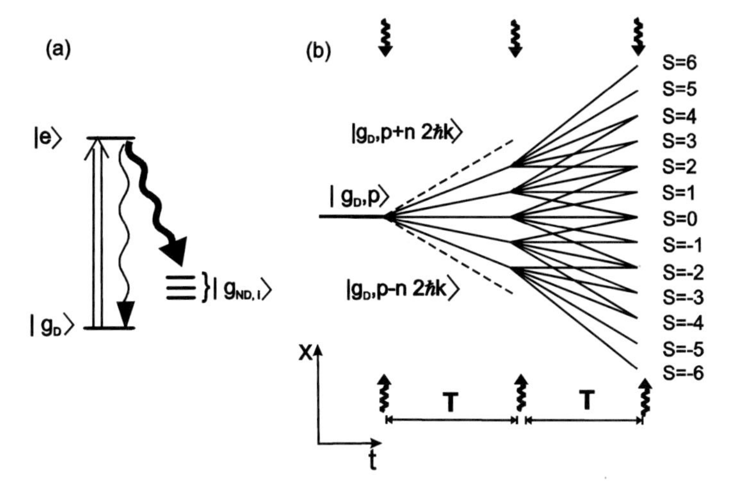


**Fig. 1** (**a**) Simplified level scheme, where $|g_D\rangle$ denotes the ground state detected at the interferometer output, $|g_{ND,i}\rangle$ (ground) states that are not detected, and $|e\rangle$ the electronically excited state. (**b**) Outline of an atomic interferometer realized with three effective absorption gratings of light. The different families of interfering wave packets are numbered by the index $s$

wave function with the initial momentum $p$ is split into a series of eigenstates differing in momentum by an integer multiple of $2\hbar k$. The number of paths increases with higher pulse area, however the transmission efficiency decreases.

In the beginning of the atomic free fall at time $t = 0$ the first optical pulse in our interferometric scheme splits an atomic wave packet into several distinct paths. Later on at a time $t = T$ a second pulse is applied, which again leads to a diffraction and coherently splits up the paths. At the end of the atomic free fall at time $t = 2T$, several paths spatially overlap in a series of families of wave packets and we expect that the wave nature leads to a spatial atomic interference structure. To read out this periodic fringe pattern, we again apply a resonant optical pulse tuned to the open transition and thus overlay the light's intensity distribution with the atomic density. The periodic pattern can now be read out by scanning the position of this third grating and monitoring the number of transmitted atoms in the ground state. For technical reasons, we actually leave the position of the optical grating constant and instead vary the pulse spacing $T$, which due to Earth's gravitational acceleration changes the position of the atomic interference pattern relative to the standing light wave and this way allows for an observation of the fringe pattern.

The expected fringe pattern is calculated in detail elsewhere (Fray et al. 2004). Here we only show the results of the calculation. Considering the spatial density of atoms in the ground state $\rho_D(x) = \int dp\, g(p)\langle x|\psi\rangle\langle\psi|x\rangle$ the following expression can be derived for $\rho(x)$ at the position of the third interferometer pulse:

$$\rho(x) = \sum_{s=-\infty}^{\infty} ||\Psi_s(x)\rangle|^2$$

with

$$|\Psi_s\rangle = \sum_{n=-\infty}^{\infty} I_n\left(\frac{\Omega^2}{\Gamma}\tau_1\right) I_{s-2n}\left(\frac{\Omega^2}{\Gamma}\tau_2\right)$$
$$\times \exp\left\{-(\Omega^2/\Gamma)(\tau_1+\tau_2)\right\} \exp\left\{i(sn-n^2)2\omega_r T\right\} \exp\{-in2kx\},$$

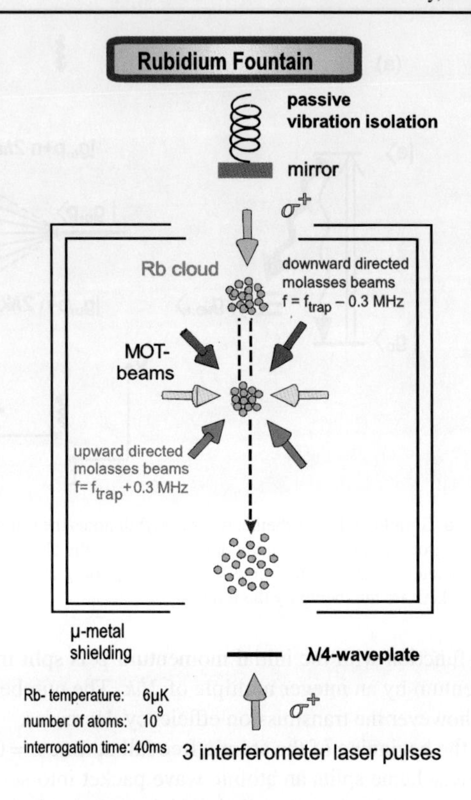


**Fig. 2** Schematic view of the atomic fountain setup. About $10^9$ rubidium atoms are captured in a magneto-optical trap, cooled to a temperature of 6 μK and launched onto a vertical ballistic trajectory by laser forces. A temporal sequence of three resonant optical standing wave pulses is used as beam splitters for the atomic matter waves, realizing an atom interferometer for the two rubidium isotopes respectively

where $\Omega$ denotes the Rabi frequency, $\Gamma$ the spontaneous decay rate of the exited state, $I_n(x)$ the modified Bessel function and $\tau_{1,2}$ the pulse length of the first and second pulse respectively. Further, $\omega_r = 2\hbar k^2/m$ denotes the recoil energy of a two photon transition in frequency units. Introducing the Earth's gravitational acceleration in the above calculation, the exponential factor has to be extended by a phase term $(-inkgT^2)$. While the expected density modulation is near sinusoidal in the unsaturated case [i.e. $(\Omega^2/\Gamma)\tau \leq 1$], the fringe pattern sharpens to an Airy-function-like multiple-beam pattern for larger pulse energies (Weitz et al. 1996; Hinderthür et al. 1997).

Our experimental apparatus is illustrated in Fig. 2. It is based on an ultrahigh vacuum chamber in which a magneto-optic trap captures one billion cold rubidium atoms from the thermal rubidium metal vapor. The atoms are then launched onto a vertical upwards moving ballistic trajectory by acoustooptically shifting the molasses beams' frequencies to a moving molasses traveling upwards. The temperature of the atomic ensemble at the beginning of the launch is 6 μK. On their ballistic flight the atoms reach the upper turning point and fall down. During the parabolic flight of the atoms the three optical beam splitting pulses are applied to realize the atom interferometer. To prepare the atomic ensemble in a magnetic field insensitive ground state sublevel a suitable sequence of microwave pulses and resonant optical pulses is applied. This state selection procedure needs to be repeated also after the atomic beam splitting sequence for state analysis. The remaining atomic population is detected by a FM-fluorescence spectroscopic technique, to allow for a distinction of the cold atoms from the Doppler-broadened background rubidium vapor.

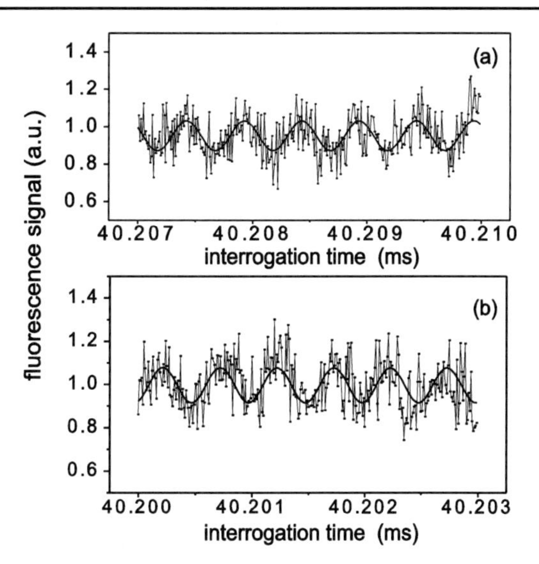


**Fig. 3** Typical atomic interference signals as a function of pulse spacing $T$ varied in narrow regions near $T = 40$ ms. The signals were recorded (**a**) for $^{85}$Rb atoms with the optical field tuned to the $F = 2 \rightarrow F' = 3$ component and (**b**) for $^{87}$Rb atoms using the $F = 1 \rightarrow F' = 2$ component of the rubidium D1-line respectively. The visible phase shift between the two interference patterns is caused by a slightly different transition frequency for the two isotopes

The atomic beam splitter light is generated by an injection locked master-slave diode laser system that is locked to the rubidium D1-line near 795 nm. The emitted radiation passes a series of two acoustooptical modulators used for switching, a single mode optical fiber and is then expanded to a 3 cm Gaussian beam diameter and directed into the vacuum chamber. After passing the chamber, the beam is retroreflected with a mirror. The generated pulsed optical standing wave has a duration of 200 ns and an intensity of 3 mW/cm$^2$ per direction at the position of the atomic cloud, which is sufficient for an effective pumping of atoms near the antinodes of the standing light wave. Since the retroreflecting mirror defines the position of the standing light wave, it is crucial to suppress vibrational coupling onto this mirror. For this reason we use a floating table for the setup and a retroreflecting mirror that is passively isolated from external vibrations. The mirror is suspended on a 2.5 m long ribbon string fixed at the ceiling. Frictionless motion was provided by mounting the mirror on an air bearing.

Measured interference patterns are shown in Fig. 3a and 3b as a function of the spacing $T$ between the beam splitting pulses obtained using $^{85}$Rb and $^{87}$Rb atoms respectively. The data points shown in both plots can be well fitted sinusoidal functions (solid line). We attributed the sinusoidal form of the fringe pattern to the issue that the main contribution to the signal is due to two-path interference. From the fit, the relative phase of the fringe pattern can be determined with an accuracy of 0.02 of a period. The observed fringe contrast is 20 per cent for small interrogation times T and reduces to some 9 per cent for the shown fringe patterns with high gravitational sensitivity recorded near $T = 40$ ms. This loss of fringe contrast for longer coherence times is attributed to residual mechanical vibrations of the interferometer beams retroreflecting mirror. On the other hand, even at small interrogation times the fringe contrast is below the theoretical value of 90%. We attribute this mainly to residual reflection of the vacuum windows, which cause an intensity imbalance between the forward and the backward propagation wave of approximately 2%. The residual reflection is mainly caused by the deposition of rubidium vapor on the vacuum windows of our apparatus.

To determine the total gravitational phase experienced by the atoms at an interrogation time of 40 ms, we record fringe patterns also at the smaller interrogation times of 2, 6,

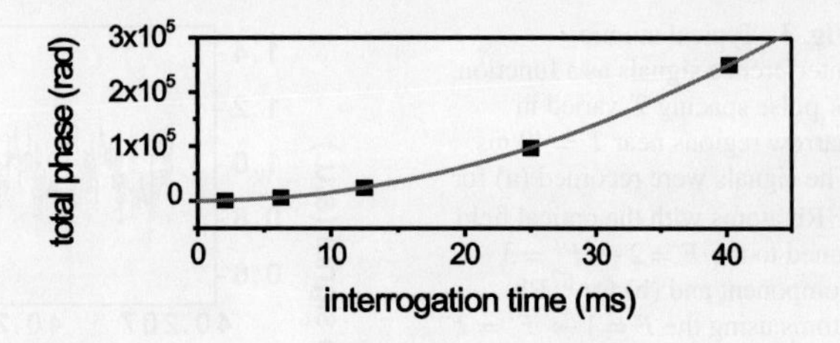


**Fig. 4** Total atom phase between adjacent paths for the $^{85}$Rb interferometer as a function of the interrogation time $T$ between the beam splitting pulses. The shown solid line is a parabolic fit

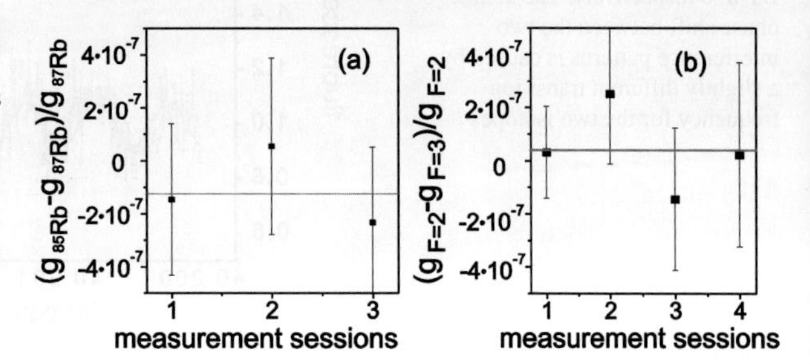


**Fig. 5** Measured difference of the gravitational acceleration in individual measurement sessions (*dots*) and total average (*vertical line*) for (**a**) the two different isotopes $^{85}$Rb and $^{87}$Rb and (**b**) for atoms ($^{85}$Rb) in the different hyperfine ground states $F = 2$ and $F = 3$

12, 18 ms respectively and monitor the accumulation of the gravitational phase. In this way, phase ambiguities due to the periodic character of the fringe pattern can be avoided. Figure 4 shows typical recorded data (dots) along with a fit to the expected parabolic curve (solid), while the gravitational acceleration g was left as a free parameter. The precision of gravitational acceleration measurement is determined by ratio of the phase accuracy and the total phase $\Delta g/g = \Delta\phi/\phi$. At 40 ms the interference pattern starts at roughly the 40000th fringe, and accounting for the determination of the fringe position in a single scan this results in a typical relative accuracy of the total phase: $\Delta g/g = \Delta\phi/\phi \cong 0.02/40000 = 5 \times 10^{-7}$. By recording multiple of such fringe patterns, the earth's gravitational acceleration can in 6 hours of data acquisition be determined to a statistical accuracy of $8 \times 10^{-8}$. We expect that considerable longer interrogation times and thus considerable further increased sensitivities should be possible with active vibration isolation techniques (Hensley et al. 1999).

In our measurement we have compared the Earth's gravitational acceleration of $^{85}$Rb atoms in the $F = 2$, $m_F = 0$ hyperfine ground state with that experienced by $^{87}$Rb atoms in $F = 1$, $m_F = 0$. The data has been recorded in three measurement sessions, each corresponding to the data recorded during one day. In every session fringe patterns at interrogation times $T \approx 40$ ms were recorded for both of the isotopes. A switching between atomic species could be performed very easily by locking the frequency of the cooling and atomic interferometer lasers respectively to the appropriate atomic transition. We typically alternated between measurements investigating the isotopes $^{85}$Rb and $^{87}$Rb respectively in temporal periods of 40 minutes each. As the switching rate between isotopes is faster than the tidal period, and due to the averaging effect we do not expect tidal changes of the Earth's gravitational acceleration (a few $10^{-7}$ in one day, Peters et al. 1999, 2001) to affect the accuracy of our differential measurement at the present level of precision. In future, we expect that atoms based tests of the equivalence principle will at some level of precision require independent laser systems for each of the isotopes, so that the measurements for the two isotopes can be performed (quasi-)simultaneously.

Figure 5a shows the results for the relative difference of the Earth's gravitational acceleration measured in the three sessions along with a horizontal line that represents the total average. This final value for the difference of the gravitational acceleration is $(g_{85\text{Rb}} - g_{87\text{Rb}})/g_{85\text{Rb}} = (1.2 \pm 1.7) \times 10^{-7}$. The quoted error here corresponds to the estimated total uncertainty. In this differential measurement, systematic errors due to misalignment of the beams, wave front curvature, and Coriolis forces largely cancel, and at the present level of accuracy clearly can be neglected. Moreover, as all paths of our atom interferometer are in the same internal state, systematic effects due to the second order Zeeman shift occur only in the presence of a magnetic field gradient. The estimated systematic uncertainty due to field inhomogeneities (8 mG/cm) is for our present apparatus estimated to be $5 \times 10^{-11}$, which is also clearly negligible. Within the quoted uncertainties, our above given value agrees well with the expected result of an identical gravitational acceleration of the two rubidium isotopes in the Earth's field.

In the measurements, the gravitational acceleration experienced by atoms in two different hyperfine ground states has been compared. For this measurement, interference fringes were recorded using $^{85}$Rb atoms prepared selectively in the atomic hyperfine ground states $F = 2$, $m_F = 0$ and $F = 3$, $m_F = 0$ respectively. Results for the difference in the measured Earth's gravitational acceleration for atoms prepared in the corresponding hyperfine ground states are shown in Fig. 5b. The obtained average value is $(g_{F=3} - g_{F=2})/g_{F=2} = (0.4 \pm 1.2) \times 10^{-7}$. Within the quoted experimental uncertainty, no difference in the Earth's gravitational acceleration for atoms in two different hyperfine ground states is observed.

In conclusion, an atom interferometer based on pulsed effective absorption gratings of light has been developed. The atom interferometer has been used to carry out the to our knowledge first atom based test of the equivalence principle.

In the future, technical improvements like an active vibration isolation allowing for longer interrogation times and improvements in wave front quality that can yield multiple beam interference signals with sharp principle maxima, where in contrast to earlier schemes the path number is not limited by the internal structure (Weitz et al. 1996; Hinderthür et al. 1997). Further, the use of dilute Bose-Einstein condensed samples can due to their small velocity spread of atoms increase the obtainable accuracy. We expect that atom interferometers can allow for extremely critical tests of the equivalence principle based on quantum probe particles. Experimental efforts along this goal are described by the contribution of M. Kasevich in this volume. Ultimately space based atom interferometers are expected to yield a still further increased precision (Ertmer et al. 2009).

**Open Access** This article is distributed under the terms of the Creative Commons Attribution Noncommercial License which permits any noncommercial use, distribution, and reproduction in any medium, provided the original author(s) and source are credited.

## References

J. Audretsch, U. Bleyer, C. Lämmerzahl, Phys. Rev. A **47**, 4632 (1993)
See, e.g., P. Berman, *Atom Interferometry* (Academic Press, San Diego, 1997).
P. Cladé, S. Guellati-Khélifa, F. Nez, F. Biraben, Phys. Rev. Lett. **102**, 240402 (2009)
See, e.g.: W. Ertmer et al., Exp. Astron. **23**, 611 (2009)
S. Fray, C. Alvarez Dies, T.W. Hänsch, M. Weitz, Phys. Rev. Lett. **93**, 240404 (2004)
T.L. Gustavson, P. Bouyer, M.A. Kasevich, Phys. Rev. Lett. **78**, 2046 (1997)
T.L. Gustavson, A. Landragin, M.A. Kasevich, Class. Quantum Gravity **17**, 2385 (2000)
J.M. Hensley et al., J. Sci. Instrum. **70**, 2735 (1999)
H. Hinderthür et al., Phys. Rev. A **56**, 2085 (1997)
C.H. Hsieh et al., Mod. Phys. Lett. A **4**, 1597 (1989)
C. Lämmerzahl, in *Proceedings of the International School of Cosmology and Gravitation, Course XV*, ed. by P.G. Bergmann et al. (World Scientific, Singapore, 1998) and references therein

H. Müller, S. Chiow, S. Herrmann, S. Chu, Phys. Rev. Lett. **102**, 240403 (2009)
K. Nordtvedt, gr-qc/0212044 (2002)
A. Peters, C. Keng Yeow, S. Chu, Nature (London) **400**, 849 (1999)
A. Peters, K.Y. Chung, S. Chu, Meterologia **38**, 25 (2001)
See, e.g. the contributions by B. Schutz T. Damour, Space Sci. Rev. (2009, this issue)
Y. Su et al., Phys. Rev. D **50**, 3614 (1994) and references therein
R.F.C. Vessot et al., Phys. Rev. Lett. **45**, 2081 (1980)
L. Viola, R. Onofrio, Phys. Rev. D **55**, 455 (1997)
M. Weitz, T. Heupel, T.W. Hänsch, Phys. Rev. Lett. **77**, 2356 (1996)

Space Sci Rev (2009) 148: 233–247
DOI 10.1007/s11214-009-9539-0

# Testing General Relativity with Atomic Clocks

**S. Reynaud · C. Salomon · P. Wolf**

Received: 5 March 2009 / Accepted: 14 May 2009 / Published online: 28 May 2009
© Springer Science+Business Media B.V. 2009

**Abstract** We discuss perspectives for new tests of general relativity which are based on recent technological developments as well as new ideas. We focus our attention on tests performed with atomic clocks and do not repeat arguments present in the other contributions to the present issue (Space Sci. Rev. 2009, This Issue). In particular, we present the scientific motivations of the space projects ACES (Salomon et al. in CR Acad. Sci. IV-2:1313, 2001) and SAGAS (Wolf et al. in Exp. Astron. 23:651, 2009).

**Keywords** Tests of general relativity · Atomic clocks · Fundamental physics in space

## 1 Introduction

Tests of gravity performed in the solar system show a good agreement with general relativity. The latter is however challenged by observations at larger galactic and cosmic scales which are presently taken care through the introduction of "dark matter" or "dark energy". As long as these components are neither detected through non gravitational means, nor explained as resulting from new physical phenomena, it remains of the uttermost importance to test general relativity at all experimentally accessible scales.

In this paper, we recall the basic elements of general relativity and briefly review the experimental evidences supporting it. We then discuss perspectives for new tests which are based on recent technological developments as well as new ideas. We focus our attention on tests performed with atomic clocks and do not repeat arguments present in the other contributions to the present issue (This Issue 2009).

S. Reynaud (✉)
Laboratoire Kastler Brossel, ENS, UPMC, CNRS, Case 74, Campus Jussieu, 75252 Paris, France
e-mail: serge.reynaud@upmc.fr

C. Salomon
Laboratoire Kastler Brossel, ENS, UPMC, CNRS, Département de Physique de l'Ecole Normale Supérieure, 75231 Paris, France

P. Wolf
LNE-SYRTE, Observatoire de Paris, CNRS, UPMC, 61, Avenue de l'Observatoire, 75014 Paris, France

## 2 Tests of General Relativity

General relativity (GR) is built up on two basic ideas which have to be distinguished. The first one is the metrical (geometrical) interpretation of gravitation, which was proposed by Einstein soon after his pioneering work on special relativity (Einstein 1907, 1911). This identification of gravitation field with the space-time metric is the very core of GR, but it is not sufficient to fix the latter theory. In order to select GR out of the variety of metric theories of gravitation, it is necessary to fix the relation between the geometry of space-time and its matter content. In GR, this relation is given by the Einstein-Hilbert equation which was written only in 1915 (Einstein 1915, 1916; Hilbert 1915).

The metrical interpretation of gravity, often coined under the generic name of the "equivalence principle", is one of the most accurately verified properties of nature (Will 2001). Freely falling test masses follow the geodesics of the Riemannian space-time, that is also the curves which extremize the integral

$$\Delta s \equiv \int \mathrm{d}s, \quad \mathrm{d}s^2 \equiv g_{\mu\nu}\mathrm{d}x^\mu \mathrm{d}x^\nu \tag{1}$$

$g_{\mu\nu}$ is the metric tensor characterizing the space-time and $\mathrm{d}x^\mu$ the displacements in this space-time. As free motions obey a geometrical definition, they are independent of the compositions of the test masses. The potential violations of this "universality of free fall" property are parametrized by a relative difference in the accelerations undergone by two test bodies in free fall from the same location with the same velocity. Modern experiments constrain this parameter to stay below $10^{-12}$, this accuracy being attained in laboratory experiments (Adelberger et al. 2003; Schlamminger et al. 2008) as well as in space tests using lunar laser ranging (Williams et al. 1996, 2004) or planetary probe tracking (Anderson et al. 1996). These results do not preclude the possibility of small violations of the equivalence principle and such violations are indeed predicted by unification models (Damour et al. 2002). Large improvements of this accuracy are expected in the future thanks to the existence of dedicated space projects MICROSCOPE (Touboul et al. 2001) and, on a longer term, STEP (Mester et al. 2001).

As another consequence of the geometrical interpretation of gravity, ideal atomic clocks operating on different quantum transitions measure the same time, because it is also a geometrical quantity, namely the proper time $\Delta s$ integrated along the trajectory (see (1)). This "universality of clock rates" property has also been verified with an extreme accuracy. Its potential variations are measured as a constancy of relative frequency ratios between different clocks at a level of the order of $10^{-16}$ per year, recently a few $10^{-17}$ per year (Marion et al. 2003; Bize et al. 2003; Fischer et al. 2004; Peik et al. 2004; Fortier et al. 2007; Ashby et al. 2007; Rosenband et al. 2008). These results can also be interpreted in terms of a potential "variation of fundamental constants" (Flambaum and Tedesco 2006), thus opening a window on the "new physics" expected to lie "beyond the standard model". In this domain also, large improvements of the accuracy can be expected in the future with the space project ACES (Salomon et al. 2001) and, on a longer term, with projects using optical clocks (Wolf et al. 2009; Schiller et al. 2009).

As already stated, GR is selected out of the large family of metric theories of gravity by the Einstein-Hilbert equation which fixes the coupling between curvature on one hand, matter on the other one, by setting the Einstein curvature tensor $G_{\mu\nu}$ to be proportional to the stress tensor $T_{\mu\nu}$

$$G_{\mu\nu} \equiv R_{\mu\nu} - \frac{1}{2} g_{\mu\nu} R = \frac{8\pi G}{c^4} T_{\mu\nu} \tag{2}$$

The coefficient is fixed by the Newtonian limit and determined by the Newton constant $G$ and the velocity of light $c$. At this point, it is worth emphasizing that it is not possible to deduce the relation (2) only from the geometrical interpretation of gravity. In other words, GR is one member of the family of metric theories of gravity which has to be selected out of this family by comparing its predictions to the results of observations or experiments. The tests performed in the solar system effectively show a good agreement with GR, as shown in particular by the so-called "parametrized post-Newtonian" (PPN) approach (Will 2001).

The main idea of the PPN approach can be described by writing down the solution of GR with a simple model of the solar system where the gravity sources are reduced to the Sun treated as a point-like motion-less mass $M$ (for the real solar system, see for example Petit and Wolf 2005). Using a specific gauge convention where spatial coordinates are isotropic, the metric element thus takes the following form (coordinates are $x^0 \equiv ct$, the radius $r$, the colatitude and azimuth angles $\theta$ and $\varphi$)

$$\mathrm{d}s^2 = g_{00} c^2 \mathrm{d}t^2 + g_{rr}\big(\mathrm{d}r^2 + r^2(\mathrm{d}\theta^2 + \sin^2\theta \mathrm{d}\varphi^2)\big) \tag{3}$$

With this simple model, an exact solution can be found for the metric. It is convenient to write it as a power series expansion in the reduced Newton potential $\phi$ (the length scale $GM/c^2$ has a value of the order of 1.5 km, with $M$ the mass of the Sun, so that $\phi$ is much smaller than unity everywhere in the solar system)

$$g_{00} = 1 + 2\phi + 2\phi^2 + \cdots, \qquad g_{rr} = -1 + 2\phi + \cdots, \qquad \phi \equiv -\frac{GM}{rc^2} \tag{4}$$

The family of PPN metrics can then be introduced by inserting a constant $\beta$ in front of $\phi^2$ in $g_{00}$ and a constant $\gamma$ in front of $\phi$ in $g_{rr}$ (with $\beta = \gamma = 1$ in GR). The values of these PPN parameters affect the predicted motions, and can therefore be confronted to the observations. Experiments on the propagation of light have led to more and more stringent bounds on $|\gamma - 1|$, with the best current results corresponding to deflection measurements using VLBI astrometry (Shapiro et al. 2004) and Doppler tracking of the Cassini probe during its 2002 solar occultation (Bertotti et al. 2003). Meanwhile, analysis of lunar laser ranging data (Williams et al. 2004) have led to bounds on linear superpositions of $\beta$ and $\gamma$, and then in constraints on $|\beta - 1|$. The current status of these tests clearly favors GR as the best description of gravity in the solar system.

Let us emphasize that the common presentation of this status under the form "general relativity is confirmed by the tests" is a bit too loose. The tests discussed so far do not answer by a final "yes" answer to a mere "yes/no" question. They rather select a vicinity of GR as the best current description of gravity within the family of PPN metrics. This warning is not a mere precaution, it is rather a pointer to possible future progress in the domain. There indeed exist theoretical models (see for example Damour et al. 2002) which deviate from GR while staying within the current observational bounds. Furthermore, as discussed below, extensions of GR do not necessarily belong to the PPN family.

This last point is in particular emphasized by the so-called "fifth force" tests which are focused on a possible scale-dependent deviation from the gravity force law (Fischbach and Talmadge 1998). Their main idea is to check the $r$-dependence of the gravity potential, that is also of the component $g_{00}$. Hypothetical modifications of its standard expression, predicted by unification models, are often parametrized in terms of a Yukawa potential added to the standard $g_{00}$. This potential depends on two parameters, an amplitude $\alpha$ measured with respect to Newton potential and a range $\lambda$ related through a Yukawa-like relation to the mass scale of the hypothetical new particle which would mediate the "fifth force". A recent update of the status of such a fifth force is shown on Fig. 1 in Jaekel and Reynaud (2005), reproduced thanks to a courtesy of J. Coy, E. Fischbach, R. Hellings, C. Talmadge and E.M. Standish. It shows that the Yukawa term is excluded with a high accuracy at ranges tested in lunar laser ranging (Williams et al. 1996) and tracking of martian probes (Anderson et al. 1996). At the same time, it also makes it clear that windows remain open for large corrections ($\alpha > 1$) at short ranges as well as long ranges.

The short range window is being actively explored, with laboratory experiments reaching an impressive accuracy at smaller and smaller distances (Kapner et al. 2007). At even shorter ranges, tests of the gravity force law are pursued as careful comparisons between theoretical predictions and experimental measurements of the Casimir force, which becomes dominant at micrometric ranges (Onofrio 2006; Lambrecht et al. 2006). In the long range window, a test of the gravity force was initiated by NASA as the extension of Pioneer 10/11 missions after their primary planetary objectives had been met. This led to the largest scaled test of gravity ever carried out, with the striking output of a signal that failed to confirm the known laws of gravity (Anderson et al. 1998, 2002).

This so-called "Pioneer anomaly" was recorded on Doppler tracking data of the Pioneer 10 & 11 probes by the NASA deep space network. It is thus a result of radionavigation techniques, which are based on the performances of the accurate reference clocks located at reception and emission stations (Asmar et al. 2005). The Doppler observable can equivalently be interpreted as a relative velocity of the probe with respect to the station, which contains not only the effect of motion but also relativistic and gravitational effects. The anomaly has been registered on the two deep space probes showing the best navigation accuracy. This is not an impressive statistics when compared to the large number of tests confirming GR. In particular, when the possibility of an artefact onboard the probe is considered, this artefact could be the same on the two probes.

A number of mechanisms have been considered as attempts of explanations of the anomaly as a systematic effect generated by the spacecraft itself or its environment (see the references in Nieto et al. 2005). In particular, it has been suspected that the anomaly could be due to the thermal radiation force accompanying dissipation of energy from the on-board energy source. Detailed calculations are presently devoted to a re-evaluation of these thermal effects (Bertolami et al. 2008; Lammerzahl 2009; Turyshev 2009). They are expected to produce soon information of interest for the discussion of the anomaly.

The Pioneer anomaly constitutes an intriguing piece of information in a context where the status of gravity theory is challenged by the puzzles of dark matter and dark energy. If confirmed, this signal might reveal an anomalous behaviour of gravity at scales of the order of the size of the solar system and thus have a strong impact on fundamental physics, solar system physics, astrophysics and cosmology. It is therefore important to use as many investigation techniques as might be available for gaining new information.

The Pioneer data which have shown an anomaly have been re-analyzed by several independent groups which have confirmed the presence of the anomaly (Markwardt 2002; Olsen 2007; Levy et al. 2009). The presence of modulated anomalies, superposed to the secular acceleration anomaly, has also been reported (Levy et al. 2009). As data covering the whole period of Pioneer 10 & 11 missions from launch to the last data point have been saved, it is worth investigating not only the Doppler tracking data, but also the telemetry data (Turyshev et al. 2006a, 2006b). These efforts should lead to an improved control of systematics and produce new information of importance on several properties of the force, for example its direction, long-term variation as well as annual or diurnal modulations, spin dependence, a question of particular interest being that of the onset of the anomaly.

If there exist gravity theories where a Pioneer signal can take a natural place, they must be considered with great care. If on the contrary one can prove that there exist no such theories, this result is also important since it allows the range of validity of GR to be extended to the size of the solar system. Anyway, the relevance of the Pioneer anomaly for space navigation is already sufficient to deserve a close scrutiny. This means that the outputs of the data analysis have to be compared with the predictions of possible theoretical explanations of the Pioneer anomaly (see for example Jaekel and Reynaud 2006a, 2006b; Brownstein and Moffat 2006; Bertolami et al. 2007; Bruneton and Esposito-Farèse 2007a, 2007b; Reynaud and Jaekel 2009 and references therein).

In the meantime, new missions have been designed to study the anomaly and understand its origin. In particular, two missions have been proposed to the Cosmic Vision process at ESA, the first one as a medium size mission (Christophe et al. 2009) using accelerometer and radioscience instruments upgraded from existing technology, the second one as a large size mission using new quantum sensors to map the two components of the gravity field with high accuracy (Wolf et al. 2009). These projects are based on new technologies of importance for future fundamental physics and deep space exploration. In the following, we will focus our attention on atomic clocks, as used in the projects ACES (Salomon et al. 2001) and SAGAS (Wolf et al. 2009). We will also discuss their applications, not only for tests in fundamental physics, but also for other important purposes such as earth sciences, solar system physics or navigation.

## 3 Atomic Clock Ensemble in Space (ACES)

The measurement of time intervals has experienced spectacular progress over the last centuries. In the middle of the 20th century, the invention of the quartz oscillator and the first atomic clocks opened a new era for time keeping, with inacuracies improving from 10 microseconds per day in 1967 to 100 picoseconds per day for the primary cesium atomic clocks using laser cooled atomic fountains. The best atomic fountains approach 10 picosecond error per day, i.e. a frequency stability of 1 part in $10^{16}$ (Bize et al. 2005) while the most recent atomic clocks, operating in the optical domain, reach 2 picoseconds per day (Rosenband et al. 2008; Ludlow et al. 2008) and improve at a fast pace.

Because time intervals and frequencies can be measured so precisely, applications of atomic clocks are numerous and diverse. Most precision measurements and units of the SI system can be traced back to frequencies. With the redefinition of the meter 25 years ago and the choice of a conventional value for the speed of light in vacuum, distance measurements have been simply translated into time interval measurements. In other words, the development of quantum technologies has led to large progress in the investigation of our spatio-temporal environment. This is illustrated by the impressive improvement of the measurement of the Einstein redshift effect (Vessot et al. 1980; Pound 1999), which will be pushed further by the ACES project (see the discussion below). This is also made clear by the measurements of distances in the solar system, with the astronomical unit now connected to atomic units and consequently known with a much better accuracy than previously (Shapiro 1999).

The most visible consequence of these ideas is given by the Global Navigation Satellite Systems (GNSS), which are a spectacular and initially unexpected application of the resources of quantum physics (the precision of the system comes from the atomic clocks in the satellites) and general relativity (Ashby 2003). The US Global Positioning System (GPS) enables today any user with a small receiver or modern cell phone to locate its position on the globe with meter accuracy. In a few years from now, the GALILEO system will be the European contribution to GNSS and its vast uses for various applications, in particular in the scientific domain in geodesy, Earth monitoring, and time metrology.

In the following, we review some of the properties of space clocks in the context of the ACES mission (Atomic Clock Ensemble in Space). ACES is an ESA mission in fundamental physics (Salomon et al. 2001, 2007; Cacciapuoti et al. 2007). ACES aims at flying a new generation of atomic clocks onboard the International Space Station (ISS) and comparing them to a network of ultra-stable clocks on the ground. The ISS is orbiting at a mean elevation of 400 km with 90 min. of rotation period and an inclination angle of 51.6°. ACES will be transported on orbit by the Japanese transfer vehicle HTV, and installed at the external payload facility of the Columbus module using the ISS robotic arm.

The ACES payload accommodates two atomic clocks: PHARAO, a primary frequency standard based on samples of laser cooled cesium atoms, and the active hydrogen maser SHM. The performances of the two clocks are combined to generate an on-board timescale with the short-term stability of SHM and the long-term stability and accuracy of the cesium clock PHARAO. The on-board comparison of PHARAO and SHM and the distribution of the ACES clock signal are ensured by the Frequency Comparison and Distribution Package (FCDP), while all data handling processes are controlled by the eXternal PayLoad Computer (XPLC). A GNSS receiver installed on the ACES payload and connected to the on-board time scale will provide precise orbit determination of the ACES clocks. The ACES clock signal will be transferred on ground by a time and frequency transfer link in the microwave domain (MWL). MWL compares the ACES frequency reference to a set of ground clocks, enabling fundamental physics tests and applications in different areas of research.

The planned mission duration is 18 months with a possible extension to 3 years. During the first two months, the functionality of the clocks and of MWL will be tested. Then, a period of 4 months will be devoted to the performance evaluation of the clocks. During this phase, a signal with frequency inaccuracy in the $10^{-15}$ range will be available to ground users. In microgravity, the linewidth of the atomic resonance of the PHARAO clock will be tuned by two orders of magnitude, down to sub-Hertz values (from 11 Hz to 110 mHz), 5 times narrower than in Earth based atomic fountains. After clock optimization, performances in the $10^{-16}$ range are expected both for frequency instability and inaccuracy. In the second part of the mission (12 to 30 months), the on-board clocks will be compared to a network of ground atomic clocks operating both in the microwave and optical domain.

The scientific objectives of the ACES mission cover a wide spectrum. The frequency comparisons between the space clocks and ground clocks will be used to test general relativity to high accuracy. As recalled above identical clocks located at different positions in gravitational fields experience a frequency shift that depends directly on the component $g_{00}$ of the metric, i.e to the Newtonian potential at the clock position. The comparison between the ACES onboard clocks and ground-based atomic clocks will measure the frequency variation due to the gravitational red-shift with a 70-fold improvement on the best previous experiment (Vessot et al. 1980), testing the Einstein prediction at the 2 ppm uncertainty level. Time variations of fundamental constants will be measured by comparing clocks based on different transitions or different atomic species (Fortier et al. 2007). Any transition energy in an atom or a molecule can be expressed in terms of the fine structure constant $\alpha$ and the two dimensionless constants $m_q/\Lambda_{QCD}$ and $m_e/\Lambda_{QCD}$, involving the quark mass $m_q$, the electron mass $m_e$ and the QCD mass scale $\Lambda_{QCD}$ (Flambaum et al. 2004; Flambaum and Tedesco 2006). ACES will perform crossed comparisons of ground clocks both in the microwave and in the optical domain with a resolution of $10^{-17}$ after a few days of integration time. These comparisons will impose strong and unambiguous constraints on time variations of fundamental constants reaching an uncertainty of $10^{-17}$/year in case of a 1-year mission duration and $3 \times 10^{-18}$/year after three years. ACES will also perform tests of the Local Lorentz Invariance (LLI), according to which the outcome of any local test experiment is independent of the velocity of the freely falling apparatus. In 1997, LLI tests based on the measurement of the round-trip speed of light have been performed by comparing clocks on-board GPS satellites to ground hydrogen masers (Wolf and Petit 1997). In such experiments, LLI violations would appear as variations of the speed of light $c$ with the direction and the relative velocity of the clocks. ACES will perform a similar experiment by measuring relative variations of the light velocity at the $10^{-10}$ uncertainty level.

Developed by CNES, the cold atom clock PHARAO will combine laser cooling techniques and microgravity conditions to significantly increase the interaction time and consequently reduce the linewidth of the clock transition. Improved stability and better control of systematic effects will be demonstrated in the space environment. PHARAO will reach a fractional frequency instability of $1 \times 10^{-13} \times \tau^{-1/2}$, where $\tau$ is the integration time expressed in seconds, and an inaccuracy of a few parts in $10^{16}$. The engineering model of the PHARAO clock has been completed and is presently under test at CNES premises in Toulouse. Design and first results are presented in Laurent et al. (2006).

Developed by SPECTRATIME under ESA coordination, SHM provides ACES with a stable fly-wheel oscillator. The main challenge of SHM is represented by the low mass and volume figures required by the space clock with respect to ground H-masers. SHM will provide a clock signal with fractional frequency instability down to $1.5 \times 10^{-15}$ after only $10^4$ s of integration time. Two servo-loops will lock together the clock signals of PHARAO and SHM generating an on-board time scale combining the short-term stability of the H-maser with the long-term stability and accuracy of the cesium clock. The Frequency Comparison and Distribution Package (FCDP) is the central node of the ACES payload. Developed by ASTRIUM and TIMETECH under ESA coordination, FCDP is the on-board hardware which compares the signals delivered by the two space clocks, measures and optimizes the performances of the ACES frequency reference, and finally distributes it to the ACES microwave link MWL.

With the microwave link MWL, frequency transfer with time deviation better than 0.3 ps at 300 s, 7 ps at 1 day, and 23 ps at 10 days of integration time will be demonstrated. The relativistic treatment of space to ground time transfer must be done up to third order in $v/c$ in order to achieve sub $10^{-16}$ accuracy in clock comparisons (Blanchet et al. 2001). The gravitational shift measurement also requires precise orbit determination and knowledge of the Earth gravitational potential as described in Duchayne et al. (2007). MWL is developed by ASTRIUM, KAYSER-THREDE and TIMETECH under ESA coordination. The proposed MWL concept is an upgraded version of the Vessot two-way technique used for the GP-A experiment (Vessot et al. 1980) and the PRARE geodesy instrument.

The frequency resolution that the ACES microwave link should reach in operational conditions is surpassing by one to two orders of magnitude the existing satellite time transfer comparison methods based on the GPS system and TWSTFT (Two Way Satellite Time and Frequency Transfer) (Bauch et al. 2006). It may be compared with T2L2 (Time Transfer by Laser Link) (Exertier et al. 2008) which is now flying onboard the JASON-2 satellite and taking science data since June 2008. The assessment of its time transfer capability is presently ongoing with the interesting prospect to reach a time resolution of $\simeq 10$ ps after one day of averaging for clock comparisons in common view. For the ACES mission, due to the low orbit of the ISS and the limited duration of each pass (300–400 s) over a given ground station, the common view technique will be suitable for comparing ground clocks over continental distances, for instance within Europe or USA or Japan. In a common view comparison, the frequency noise of the onboard clock cancels out to a large degree so that only the link instability remains. ACES mission will also demonstrate the capability to compare ground clocks in non common view with a resolution better than $10^{-13}\Delta t^{1/2}$ for $\Delta t > 1000$ s, that is 3 ps and 10 ps for space-ground comparisons separated by 1000 s and 10000 s respectively. This science objective takes full benefit of the excellent onboard time scale realized by the combination of SHM and PHARAO.

These performances will enable common view and non-common view comparisons of ground clocks with $10^{-17}$ frequency resolution after few days of integration time. The recent development of optical frequency combs (Holzwarth et al. 2000; Diddams et al. 2000) awarded with the 2005 Nobel Prize in Physics, significantly simplifies the link between optical and microwave frequencies. From this point of view, ACES will take full advantage of the recent progress of optical clocks (Rosenband et al. 2008; Ludlow et al. 2008), reaching today instability and inaccuracy levels of 2 parts in $10^{17}$. Multiple frequency comparisons between a variety of advanced ground clocks will be possible among the 35 institutes which have manifested their interest to participate to the ACES mission. This is important for the tests of the variability of fundamental physical constants, and international comparisons of time scales.

Other applications of the ACES clock signal are currently being developed. ACES will demonstrate a new "relativistic geodesy" based on a differential measurement of the Einstein's gravitational red-shift between distant ground clocks. It will take advantage of the accuracy of ground-based optical clocks to resolve differences in the Earth gravitational potential at the $\simeq 10$ cm level. A 10 cm change in elevation on the ground amounts to a frequency shift of 1 part in $10^{17}$, compared to today's optical clock accuracy of 2 parts in $10^{17}$. The GNSS receiver on-board the ACES payload will allow to monitor the GNSS networks and develop interesting applications in Earth remote sensing. This includes studies of the oceans surface via GNSS reflectometry measurements and the analysis of the Earth atmosphere with GNSS radio-occultation experiments.

In addition, ACES will deliver a global atomic time scale with $10^{-16}$ accuracy, it will allow clocks synchronization at an uncertainty level of 100 ps, and contribute to international atomic time scales. At the $10^{-18}$ level, ground clocks will be limited by the fluctuations of the Earth potential suggesting that future high precision time references will have to be placed in the space environment where these fluctuations are significantly reduced. From this point of view, ACES is the pioneer of new concepts for global time keeping and positioning based on a reduced set of ultra-stable space clocks in orbit.

## 4 SAGAS

The SAGAS mission will study all aspects of large scale gravitational phenomena in the Solar System using quantum technology, with science objectives in fundamental physics and Solar System exploration. The large spectrum of science objectives makes SAGAS a unique combination of exploration and science, with a strong basis in both programs. The involved large distances (up to 53 AU) and corresponding large variations of gravitational potential combined with the high sensitivity of SAGAS instruments serve both purposes equally well. For this reason, SAGAS brings together traditionally distant scientific communities ranging from atomic physics through experimental gravitation to planetology and Solar System science.

The payload will include an optical atomic clock optimised for long term performance, an absolute accelerometer based on atom interferometry and a laser link for ranging, frequency comparison and communication. The complementary instruments will allow highly sensitive measurements of all aspects of gravitation via the different effects of gravity on clocks, light, and the free fall of test bodies, thus effectively providing a detailed gravitational map of the outer Solar System whilst testing all aspects of gravitation theory to unprecedented levels.

The SAGAS accelerometer is based on cold Cs atom technology derived to a large extent from the PHARAO space clock built for the ACES mission discussed in the preceding section. The PHARAO engineering model has recently been tested with success, demonstrating the expected performance and robustness of the technology. The accelerometer will only require parts of PHARAO (cooling and trapping region) thereby significantly reducing mass and power requirements. The expected sensitivity of the accelerometer is $1.3 \times 10^{-9}$ m/s$^2$/$\sqrt{\text{Hz}}$ with an absolute accuracy (bias determination) of $5 \times 10^{-12}$ m/s$^2$.

The SAGAS clock will be an optical clock based on trapped and laser cooled single ion technology as pioneered in numerous laboratories around the world. In the present proposal it will be based on a $Sr^+$ ion with a clock wavelength of 674 nm. The expected stability of the SAGAS clock is $1 \times 10^{-14}/\sqrt{\tau}$ (with $\tau$ the integration time), with an accuracy in realising the unperturbed ion frequency of $1 \times 10^{-17}$. The best optical single ion ground clocks presently show stabilities below $4 \times 10^{-15}/\sqrt{\tau}$, slightly better than the one assumed for the SAGAS clock, and only slightly worse accuracies $2 \times 10^{-17}$. So the technology challenges facing SAGAS are not so much the required performance, but the development of reliable and space qualified systems, with reduced mass and power consumption.

The optical link is using a high power (1 W) laser locked to the narrow and stable frequency provided by the optical clock, with coherent heterodyne detection on the ground and on board the spacecraft. It serves the multiple purposes of comparing the SAGAS clock to ground clocks, providing highly sensitive Doppler measurements for navigation and science, and allowing data transmission together with timing and coarse ranging. It is based on a 40 cm space telescope and 1.5 m ground telescopes (similar to lunar laser ranging stations). The short term performance of the link in terms of frequency comparison of distant clocks will be limited by atmospheric turbulence at a few $10^{-13}/\tau$, thus reaching the clock performance after about 1000 s integration time, which is amply sufficient for the SAGAS science objectives. The main challenges of the link will be the required pointing accuracy (0.3″) and the availability of space qualified, robust 1 W laser sources at 674 nm. Quite generally, laser availability and reliability will be the key to achieving the required technological performances, for the clock as well as the optical link.

For this reason a number of different options have been considered for the clock/link laser wavelength, with several other ions that could be equally good candidates (e.g. $Yb^+$ at 435 nm and $Ca^+$ at 729 nm). Given present laser technology, $Sr^+$ was preferred, but this choice could be revised depending on laser developments over the next years. We also acknowledge the possibility that femtosecond laser combs might be developed for space applications in the near future, which would open up the option of using either ion with existing space qualified 1064 nm Nd:YAG lasers for the link.

More generally, SAGAS technology takes advantage of the important heritage from cold atom technology used in PHARAO and laser link technology designed for LISA (Laser Interferometric Space Antenna) (Danzmann et al. 2003). It will provide an excellent opportunity to develop those technologies for general use, including development of the ground segment (Deep Space Network telescopes and optical clocks), that will allow such technologies to be used in many other mission configurations for precise timing, navigation and broadband data transfer throughout the Solar System.

SAGAS will carry out a large number of tests of fundamental physics, and gravitation in particular, at scales only attainable in a deep space experiment. The unique combination of onboard instruments will allow 2 to 5 orders of magnitude improvement on many tests of special and general relativity, as well as a detailed exploration of a possible anomalous scale dependence of gravitation. It will also provide detailed information on the Kuiper belt mass distribution and determine the mass of Kuiper belt objects and possibly discover new ones. During the transits, the mass and mass distribution of the Jupiter system will be measured with unprecedented accuracy. The science objectives are discussed in the following, based on estimated measurement uncertainties of the different observables.

SAGAS will provide three fundamental measurements: the accelerometer readout and the two frequency differences (measured on ground and on board the satellite) between the incoming laser signal and the local optical clock. Auxiliary measurements are the timing of emitted/received signals on board and on the ground, which are used for ranging and time tagging of data. The high precision science observables will be deduced from the fundamental measurements by combining the measurements to obtain information on either the frequency difference between the clocks or the Doppler shift of the transmitted signals. The latter gives access to the relative satellite-ground velocity, from which the gravitational trajectory of the satellite can be deduced by correcting non-gravitational accelerations using the accelerometer readings.

In the following, we assume that Earth station motion and its local gravitational potential can be known and corrected to uncertainty levels below $10^{-17}$ in relative frequency (10 cm on geocentric distance), which, although challenging, are within present capabilities. For the Solar System parameters this requires $10^{-9}$ relative uncertainty for the ground clock parameters ($GM$ and $r$ of Earth), also achieved at present (Groten 1999), and less stringent requirements for the satellite.

For long term integration and the determination of an acceleration bias, the limiting factor will then be the accelerometer noise and absolute uncertainty (bias determination). More generally, modelling of non-gravitational accelerations will certainly allow some improvement on the long term limits imposed by the accelerometer noise and absolute uncertainty, but is not taken into account in our preliminary evaluation (Wolf et al. 2009). Also, depending on the science goal and corresponding signal, the ground and on-board data can be combined in a way to optimise the S/N ratio (see Reynaud et al. 2008 for details). For example, in the Doppler observable, the contribution of one of the clocks (ground or space) can be made negligible by combining signals such that one has coincidence of the "up" and "down" signals on board or on the ground.

We will use a mission profile with a nominal mission lifetime of 15 years and the possibility of an extended mission to 20 years if instrument performance and operation allow this. In that time frame, the trajectory allows the satellite to reach a heliocentric distance of 39 AU in nominal mission and 53 AU with extended duration.

In General Relativity (GR), the frequency difference of two ideal clocks is proportional to (see (1) to first order in the weak field approximation, with $g_{00} \simeq 1 - 2w/c^2$ and $g_{rr} \simeq -1$; $w$ is the Newtonian gravitational potential and $v$ the coordinate velocity)

$$\frac{\mathrm{d}s_\mathrm{g}}{c\mathrm{d}t} - \frac{\mathrm{d}s_\mathrm{s}}{c\mathrm{d}t} \simeq \frac{w_\mathrm{g} - w_\mathrm{s}}{c^2} + \frac{v_\mathrm{g}^2 - v_\mathrm{s}^2}{2c^2} \tag{5}$$

In theories different from GR this relation is modified, leading to different time and space dependence of the frequency difference. This can be tested by comparing two clocks at distant locations (different values of $w$ and $v$) via exchange of an electromagnetic signal. The SAGAS trajectory (large potential difference) and low uncertainty on the observable (5) allows a relative uncertainty on the redshift determination given by the $10^{-17}$ clock bias divided by the maximum value of $\Delta w/c^2$. For a distance of 50 AU this corresponds to a test with a relative uncertainty of $1 \times 10^{-9}$, an improvement by almost 5 orders of magnitude on the uncertainty obtained by the most sensitive experiment at present (Vessot et al. 1980).

Additionally, the mission also provides the possibility of testing the velocity term in (5), which amounts to a test of Special Relativity (Ives-Stilwell test), and thus of Lorentz invariance. Towards the end of the nominal mission, this term is about $4 \times 10^{-9}$, and can therefore be measured by SAGAS with $3 \times 10^{-9}$ relative uncertainty. The best present limit on this type of test is $2 \times 10^{-7}$ (Saathoff et al. 2003), so SAGAS will allow an improvement by a factor $\sim 70$. Considering a particular preferred frame, usually taken as the frame in which the 3 K cosmic background radiation is isotropic, one can set an even more stringent limit. In that case a putative effect will be proportional to $(\mathbf{v}_s - \mathbf{v}_g).\mathbf{v}_{Sun}/c^2$ (see Saathoff et al. 2003), where $\mathbf{v}_s$ and $\mathbf{v}_g$ are the velocity vectors of the satellite and ground while $\mathbf{v}_{Sun}$ is the velocity of the Sun through the CMB frame ($\sim$350 km/s). Then SAGAS will allow a measurement with about $5 \times 10^{-11}$ relative uncertainty, which corresponds to more than 3 orders of magnitude improvement on the present limit. Note that Ives-Stilwell experiments also provide the best present limit on a particularly elusive parameter ($\varkappa_{tr}$) of the Lorentz violating Standard Model Extension (SME) photon sector (Hohensee et al. 2007), so that SAGAS also allows for the same factor 70 to 1000 improvement on that parameter.

Spatial and/or temporal variations of fundamental constants constitute another violation of Local Position Invariance and thus of GR. Over the past few years, there has been great interest in that possibility (see e.g. Uzan 2003 for a review), spurred on the one hand by models for unification theories of the fundamental interactions where such variations appear quite naturally, and on the other hand by recent observational claims of a variation of different constants over cosmological timescales (Murphy et al. 2003; Reinhold et al. 2006). Such variations can be searched for with atomic clocks, as the involved transition frequencies depend on combinations of fundamental constants and in particular, for the optical transition of the SAGAS clock, on the fine structure constant $\alpha$. Such tests take two forms: searches for a drift in time of fundamental constants, or for a variation of fundamental constants with ambient gravitational field. The latter tests for a non-universal coupling between ambient gravity and non-gravitational interactions (clearly excluded by GR) and is well measured by SAGAS, because of the large change in gravitational potential during the mission.

For example, some well-known string theory based models associates a scalar field such as the dilaton to the standard model (Damour and Nordtvedt 1993a, 1993b; Damour and Polyakov 1994; Flambaum and Shuryak 2007). Such scalar fields would couple to ordinary matter and thus their non-zero value would introduce a variation of fundamental constants, in particular $\alpha$ of interest here. The non-zero value of such scalar fields could be of cosmological origin, leading to a constant drift in time of fundamental constants, and/or of local origin, i.e. taking ordinary matter as its source (Flambaum and Shuryak 2007). In the latter case one would observe a variation of fundamental constants with the change in local gravitational potential $w$, which can be simply written as $\delta\alpha/\alpha = k_\alpha \delta w/c^2$. The difference in gravitational potential between the Earth and the SAGAS satellite at the end of nominal mission is about $\delta w/c^2 \simeq 10^{-8}$, which is 30 times more than the variation attainable on Earth. With a $Sr^+$ optical transition used in the SAGAS clock and a ground clock with $10^{-17}$ uncertainty, this yields a limit of $k_\alpha < 2.4 \times 10^{-9}$, a factor 250 improvement over the best present limit (Flambaum and Shuryak 2007).

SAGAS also offers a possibility to improve the test of the PPN family of metric theories of gravitation (Will 2001). The two most common parameters of the PPN framework are the Eddington parameters $\beta$ and $\gamma$. Present limits on $\gamma$ are obtained from measurements on light propagation such as light deflection, Shapiro delay and Doppler velocimetry of the Cassini probe during the 2002 solar occultation (Bertotti et al. 2003). SAGAS will carry out similar measurements during solar conjunctions, with improved sensitivity and at optical rather than radio frequencies, which significantly minimizes errors due to the solar corona and the Earth's ionosphere. When combining the on board and ground measurements such that the "up" and "down" signals coincide at the satellite (classical Doppler type measurement), the noise from the on-board clock cancels to a large extent and one is left with noise from the accelerometer, the ground clock, and the atmosphere. Details on these noise sources, and on the effects of atmospheric turbulence, variations in temperature, pressure and humidity are presented in Wolf et al. (2009). The resulting improvement on the estimation of $\gamma$ is of the order of 100, with some potential for further improvement if several occultations may be analysed.

As already discussed, experimental tests of gravity have shown a good overall agreement with GR, but most theoretical models aimed at inserting GR within the quantum framework predict observable modifications at large scales. The anomalies observed on the Pioneer probes, as well as the phenomena commonly ascribed to "dark matter" and "dark energy", suggest that it is extremely important to test the laws of gravity at interplanetary distances (Reynaud and Jaekel 2009). This situation has motivated an interest in flying new probes to the distances where the anomaly was first discovered, that is beyond the Saturn orbit, and studying gravity with the largely improved spatial techniques which are available nowadays.

SAGAS has the capability to improve our knowledge of the law of gravity at the scale of the solar system, and to confirm or infirm the presence of a Pioneer anomaly (PA). With one year of integration, all SAGAS observables allow a measurement of any effect of the size of the PA with a relative uncertainty of better than 1%. This will allow a "mapping" of any anomalous scale dependence over the mission duration and corresponding distances. Furthermore, the complementary observables available on SAGAS allow a good discrimination between different hypotheses thereby not only measuring a putative effect but also allowing an identification of its origin. SAGAS thus offers the possibility to constrain a significant number of theoretical approaches to scale dependent modifications of GR. Given the complementary observables available on SAGAS the obtained measurements will provide a rich testing ground for such theories with the potential for major discoveries.

Let us emphasize at this point that this gravity law test is important not only for fundamental physics but also for solar system science. The Newton law of gravity is indeed a key ingredient of the models devoted to a better understanding of the origin of the solar system. The Kuiper belt (KB) is a collection of masses, remnant of the circumsolar disk where giant planets of the solar system formed 4.6 billion years ago. Precise measurements of its mass distribution would significantly improve our understanding of planet formation not only in the solar system but also in recently discovered planetary systems. The exceptional sensitivity and versatility of SAGAS for measuring gravity can be used to study the sources of gravitational fields in the outer solar system, and in particular the class of Trans Neptunian Objects (TNOs), of which those situated in the KB have been the subject of intense interest and study over the last years (Morbidelli 2007). Observation of KB objects (KBOs) from the Earth is difficult due to their relatively small size and large distance, and estimates of their masses and distribution are accordingly inaccurate. Estimates of the total KB mass from the discovered objects ($\sim$ 1000 KBOs) range from 0.01 to 0.1 Earth masses, whereas in-situ formation of the observed KBOs would require three orders of magnitude more solid material in a dynamically cold disk.

A dedicated probe like SAGAS will help discriminating different models of the spatial distribution of the KB and for determining its total mass. The relative frequency shift between the ground and space clock due to the KB gravitational potential is indeed a sensitive probe of the spatial distribution of this mass (see Bertolami and Vieira 2006; Bertolami and Páramos 2007; Wolf et al. 2009 for details). The SAGAS frequency observable is well suited to study the large, diffuse, statistical mass distribution of KBOs essentially due to its sensitivity directly to the gravitational potential ($1/r$ dependence), rather than the acceleration ($1/r^2$ dependence). The large diffuse signal masks any signal from individual KBOs. When closely approaching one of the objects, the crossover between acceleration sensitivity (given by the $5 \times 10^{-12}$ $\mathrm{m/s^2}$ uncertainty on non gravitational acceleration) and the frequency sensitivity ($10^{-17}$ uncertainty on $w/c^2$) for an individual object is situated at about 1.2 AU. Below that distance the acceleration measurement is more sensitive than the frequency measurement. This suggests a procedure to study individual objects using the SAGAS observables: use the satellite trajectory (corrected for the non gravitational acceleration) to study the gravity from a close object and subtract the diffuse background from all other KBOs using the frequency measurement. Investigating several known KBOs within the reach of SAGAS (Bernstein et al. 2004) shows that their masses can be determined at the % level when approaching any of them to 0.2 AU or less. Of course, this also opens the way towards the discovery of new such objects, too small to be visible from the Earth. Similarly, during a planetary flyby the trajectory determination (corrected for the non gravitational acceleration) will allow the determination of the gravitational potential of the planetary system. The planned Jupiter flyby with a closest approach of $\sim 600000$ km will improve present knowledge of Jovian gravity by more than two orders of magnitude.

Doppler ranging to deep space missions provides the best upper limits available at present on gravitational waves (GW) with frequencies of order $c/L$ where $L$ is the spacecraft to ground distance i.e. in the 0.01 to 1 mHz range (Armstrong et al. 1987, 2003), and even down to 1 μHz, albeit with lower sensitivity (Anderson and Mashoon 1985; Armstrong et al. 2003). The corresponding limits on GW are determined by the noise PSD of the Doppler ranging to the spacecraft for stochastic GW backgrounds, filtered by the bandwidth of the observations when looking for GW with known signatures. In the case of SAGAS data, this yields a strain sensitivity of $10^{-14}/\sqrt{\mathrm{Hz}}$ for stochastic sources in the frequency range of 0.06 to 1 mHz with a $f^{-1}$ increase at low frequency due to the accelerometer noise. When searching for GW with particular signatures in this frequency region, optimal filtering using a corresponding GW template will allow reaching strain sensitivities as low as $10^{-18}$ with one year of data. This will improve on best present upper limits on GW in this frequency range by about four orders of magnitude. Even if GW with sufficiently large amplitudes are not found, still the results might serve as upper bounds for astrophysical models of known GW sources (Reynaud et al. 2008).

**Acknowledgements** We are indebted to a large number of colleagues with whom we have discussed the topics of this paper over the years. Special thanks are due to CNES for its long term support.

## References

E.G. Adelberger, B.R. Heckel, A.E. Nelson, Ann. Rev. Nucl. Part. Sci. **53**, 77 (2003)
J.D. Anderson, B. Mashoon, Astrophys. J. **290**, 445 (1985)
J.D. Anderson et al., Astrophys. J. **459**, 365 (1996)
J.D. Anderson et al., Phys. Rev. Lett. **81**, 2858 (1998)
J.D. Anderson et al., Phys. Rev. D **65**, 082004 (2002)
J.W. Armstrong et al., Astrophys. J. **318**, 536 (1987)
J.W. Armstrong et al., Astrophys. J. **599**, 806 (2003)
N. Ashby, Living Rev. Rel. 1 (2003). http://www.livingreviews.org/lrr-2003-1
N. Ashby et al., Phys. Rev. Lett. **98**, 070802 (2007)

S.W. Asmar et al., Radio Sci. **40**, RS2001 (2005)
A. Bauch et al., Metrologia **43**, 109120 (2006)
G. Bernstein et al., Astrophys. J. **128**, 1364 (2004)
O. Bertolami, J. Páramos, Int. J. Mod. Phys. D **16**, 1611 (2007)
O. Bertolami, P. Vieira, Class. Quantum Gravity **23**, 4625 (2006)
O. Bertolami et al., Phys. Rev. D **75**, 104016 (2007)
O. Bertolami et al., Phys. Rev. D **78**, 103001 (2008)
B. Bertotti, L. Iess, P. Tortora, Nature **425**, 374 (2003)
S. Bize et al., Phys. Rev. Lett. **90**, 150802 (2003)
S. Bize et al., J. Phys. B **38**, S449 (2005)
L. Blanchet et al., Astron. Astrophys. **370**, 320 (2001)
J.R. Brownstein, J.W. Moffat, Class. Quantum Gravity **23**, 3427 (2006)
J.P. Bruneton, G. Esposito-Farèse, Phys. Rev. D **76**, 124012 (2007a)
J.P. Bruneton, G. Esposito-Farèse, Phys. Rev. D **76**, 129902 (2007b)
L. Cacciapuoti et al., Nucl. Phys. B **166**, 303 (2007)
B. Christophe et al., Exp. Astron. **23**, 529 (2009)
T. Damour, K. Nordtvedt, Phys. Rev. Lett. **70**, 2217 (1993a)
T. Damour, K. Nordtvedt, Phys. Rev. D **48**, 3436 (1993b)
T. Damour, A.M. Polyakov, Nucl. Phys. B **423**, 532 (1994)
T. Damour, F. Piazza, G. Veneziano, Phys. Rev. D **66**, 046007 (2002)
K. Danzmann et al., Adv. Space Res. **32**, 1233 (2003)
S.A. Diddams et al., Phys. Rev. Lett. **84**, 5102 (2000)
L. Duchayne, F. Mercier, P. Wolf, arXiv:0708.2387 (2007)
A. Einstein, Jahrbuch Radioaktivität Elektronik **4**, 411 (1907)
A. Einstein, Ann. Phys. **35**, 898 (1911)
A. Einstein, Sitz. Preuss. Akad. Wissenschaften Berlin 844 (1915)
A. Einstein, Ann. Phys. **49**, 769 (1916)
P. Exertier et al., IDS Workshop, Nice, November (2008), http://ids.cls.fr/documents/report/idsworkshop2008/IDS08s8ExertierT2L2.pdf
E. Fischbach, C. Talmadge, *The Search for Non Newtonian Gravity* (Springer, Berlin, 1998)
M. Fischer et al., Phys. Rev. Lett. **92**, 230802 (2004)
V.V. Flambaum, A.F. Tedesco, Phys. Rev. C **73**, 055501 (2006)
V.V. Flambaum, E.V. Shuryak, arXiv:physics/0701220 (2007)
V.V. Flambaum et al., Phys. Rev. D **69**, 115006 (2004)
T.M. Fortier et al., Phys. Rev. Lett. **98**, 070801 (2007)
E. Groten, Report of the IAG. Special Commission SC3, Fundamental Constants, XXII IAG General Assembly (1999)
D. Hilbert, Nachr. von der Gesellshaft der Wissenschaften zu Göttingen 395 (1915)
M. Hohensee et al., Phys. Rev. D **75**, 049902 (2007)
R. Holzwarth et al., Phys. Rev. Lett. **85**, 2264 (2000)
M.-T. Jaekel, S. Reynaud, Int. J. Mod. Phys. A **20**, 2294 (2005)
M.-T. Jaekel, S. Reynaud, Class. Quantum Gravity **23**, 777 (2006a)
M.-T. Jaekel, S. Reynaud, Class. Quantum Gravity **23**, 7561 (2006b)
D.J. Kapner et al., Phys. Rev. Lett. **98**, 021101 (2007)
A. Lambrecht, P.A. Maia Neto, S. Reynaud, New J. Phys. **8**, 243 (2006)
C. Lammerzahl, Private communication (2009)
P. Laurent et al., Appl. Phys. B **84**, 683 (2006)
A. Levy et al., Adv. Space Res. **43**, 1538 (2009)
A.D. Ludlow et al., Science **319**, 1805 (2008)
H. Marion et al., Phys. Rev. Lett. **90**, 150801 (2003)
C. Markwardt, arXiv:gr-qc/0208046 (2002); http://lheawww.gsfc.nasa.gov/users/craigm/atdf/
J. Mester et al., Class. Quantum Gravity **18**, 2475 (2001)
A. Morbidelli, in *The Kuiper Belt*, ed. by A. Barruci et al. (Univ. of Arizona Press, Tucson, 2007)
M.T. Murphy et al., Mon. Not. R. Astron. Soc. **345**, 609 (2003)
M.M. Nieto, S.G. Turyshev, J.D. Anderson, Phys. Lett. B **613**, 11 (2005)
O. Olsen, Astron. Astrophys. **463**, 393 (2007)
R. Onofrio, New J. Phys. **8**, 237 (2006)
E. Peik et al., Phys. Rev. Lett. **93**, 170801 (2004)
G. Petit, P. Wolf, Metrologia **42**, S138 (2005)
R.V. Pound, Rev. Mod. Phys. **71**, S54 (1999)
E. Reinhold et al., Phys. Rev. Lett. **96**, 151101 (2006)

S. Reynaud et al., Phys. Rev. D **77**, 122003 (2008)
S. Reynaud, M.-T. Jaekel, in *Atom Optics and Space Physics, International School of Physics Enrico Fermi*, vol. 168, ed. by E. Arimondo, W. Ertmer, E.M. Rasel, W.P. Schleich (IOS Press, Amsterdam, 2009)
T. Rosenband et al., Science **319**, 1808 (2008)
G. Saathoff et al., Phys. Rev. Lett. **91**, 190403 (2003)
C. Salomon et al., CR Acad. Sci. **IV-2**, 1313 (2001)
C. Salomon, L. Cacciapuoti, N. Dimarcq, Int. J. Mod. Phys. D **16**, 2511 (2007)
S. Schlamminger et al., Phys. Rev. Lett. **100**, 041101 (2008)
I.I. Shapiro, Rev. Mod. Phys. **71**, S41 (1999)
S.S. Shapiro et al., Phys. Rev. Lett. **92**, 121101 (2004)
S. Schiller et al., Exp. Astron. **23**, 573 (2009)
This Issue, Space Sci. Rev. (2009)
P. Touboul et al., CR Acad. Sci. **IV-2**, 1271 (2001)
S.G. Turyshev, Private communication (2009)
S.G. Turyshev, M.M. Nieto, J.D. Anderson, EAS Publ. Ser. **20**, 243 (2006a)
S.G. Turyshev, V.T. Toth, L.R. Kellogg et al., Int. J. Mod. Phys. D **15**, 1 (2006b)
J.-P. Uzan, Rev. Mod. Phys. **75**, 403 (2003)
R.F.C. Vessot et al., Phys. Rev. Lett. **45**, 2081 (1980)
C.M. Will, Living Rev. Rel. 4 (2001) http://www.livingreviews.org/lrr-2001-4
J.G. Williams, X.X. Newhall, J.O. Dickey, Phys. Rev. D **53**, 6730 (1996)
J.G. Williams, S.G. Turyshev, D.H. Boggs, Phys. Rev. Lett. **93**, 261101 (2004)
P. Wolf et al., Exp. Astron. **23**, 651 (2009)
P. Wolf, G. Petit, Phys. Rev. A **56**, 4405 (1997)

Space Sci Rev (2009) 148: 249–265
DOI 10.1007/s11214-009-9503-z

# Fundamental Constants and Tests of General Relativity—Theoretical and Cosmological Considerations

**Jean-Philippe Uzan**

Received: 5 December 2008 / Accepted: 26 March 2009 / Published online: 23 April 2009
© Springer Science+Business Media B.V. 2009

**Abstract** The tests of the constancy of the fundamental constants are tests of the local position invariance and thus of the equivalence principle. We summarize the various constraints that have been obtained and then describe the connection between varying constants and extensions of general relativity. To finish, we discuss the link with cosmology, and more particularly with the acceleration of the Universe. We take the opportunity to summarize various possibilities to test general relativity (but also the Copernican principle) on cosmological scales.

**Keywords** Fundamental constant · Gravitation · Cosmology

## 1 Introduction

Physical theories usually introduce constants, i.e. numbers that are not, and by construction can not be, determined by the theory in which they appear. They are contingent to the theory and can only be experimentally determined and measured.

These numbers have to be assumed constant for two reasons. First, from a theoretical point of view, we have no evolution equation for them (since otherwise they would be fields) and they cannot be expressed in terms of other more fundamental quantities. Second, from an experimental point of view, in the regimes in which the theories in which they appear have been validated, they should be constant at the accuracy of the experiments, to ensure the reproducibility of experiments. This means that testing for the constancy of these parameters is a test of the theories in which they appear and allow to extend the knowledge of their domain of validity.

Indeed, when introducing new, more unified or more fundamental, theories the number of constants may change so that the list of what we call fundamental constants is a time-dependent concept and reflects both our knowledge and ignorance (Weinberg 1983). Today,

J.-P. Uzan (✉)
Institut d'Astrophysique de Paris, UMR-7095 du CNRS, Université Pierre et Marie Curie,
98 bis bd Arago, 75014 Paris, France
e-mail: uzan@iap.fr

gravitation is described by general relativity, and the three other interactions and whole fundamental fields are described by the standard model of particle physics. In such a framework, one has 22 unknown constants (the Newton constant, 6 Yukawa couplings for the quarks and 3 for the leptons, the mass and vacuum expectation value of the Higgs field, 4 parameters for the Cabibbo-Kobayashi-Maskawa matrix, 3 coupling constants, a UV cut-off to which one must add the speed of light and the Planck constant; see e.g. Hogan 2000).

Since any physical measurement reduces to the comparison of two physical systems, one of them often used to realize a system of units, it only gives access to dimensionless numbers. This implies that only the variation of dimensionless combinations of the fundamental constants can be measured and would actually also correspond to a modification of the physical laws (see e.g. Uzan 2003; Ellis and Uzan 2005). Changing the value of some constants while letting all dimensionless numbers unchanged would correspond to a change of units. It follows that from the 22 constants of our reference model, we can pick 3 of them to define a system of units (such as e.g. $c$, $G$ and $h$ to define the Planck units) so that we are left with 19 unexplained dimensionless parameters, characterizing the mass hierarchy, the relative magnitude of the various interactions etc.

Indeed, this number can change with time. For instance, we know today that neutrinos have to be somewhat massive. This implies that the standard model of particle physics has to be extended and that it will involve at least 7 more parameters (3 Yukawa couplings and 4 CKM parameters). On the other hand, this number can decrease, e.g. if the non-gravitational interactions are unified. In such a case, the coupling constants may be related to a unique coupling constant $\alpha_U$ and a mass scale of unification $M_U$ through

$$\alpha_i^{-1}(E) = \alpha_U^{-1} + \frac{b_i}{2\pi} \ln \frac{M_U}{E},$$

where the $b_i$ are numbers which depends on the explicit model of unification. This would also imply that the variations, if any, of various constants will be correlated.

The tests of the constancy of fundamental constants take all their importance in the realm of the tests of the equivalence principle (Will 1993). This principle, which states the universality of free fall, the local position invariance and the local Lorentz invariance, is at the basis of all metric theories of gravity and implies that all matter fields are universally coupled to a unique metric $g_{\mu\nu}$ which we shall call the physical metric,

$$S_{\text{matter}}(\psi, g_{\mu\nu}).$$

The dynamics of the gravitational sector is dictated by the Einstein-Hilbert action

$$S_{\text{grav}} = \frac{c^3}{16\pi G} \int \sqrt{-g_*} R_* \mathrm{d}^4 x.$$

General relativity assumes that both metrics coincide, $g_{\mu\nu} = g^*_{\mu\nu}$.

The test of the constancy of constants is a test of the local position invariance hypothesis and thus of the equivalence principle. Let us also emphasize that it is deeply related to the universality of free fall since if any constant $c_i$ is a space-time dependent quantity so will the mass of any test particle so that it will experience an anomalous acceleration

$$\delta\vec{a} = \frac{\partial \ln m}{\partial c_i} \dot{c}_i \vec{v},$$

which is composition dependent (see Uzan 2003 for a review).

In particular, this allows to extend tests of the equivalence, and thus tests of general relativity, on astrophysical scales. Such tests are central in cosmology in which the existence of a dark sector (dark energy and dark matter) is required to explain the observations.

Necessity of theoretical physics in our understanding of fundamental constants and on deriving bounds on their variation is, at least, threefold:

1. it is necessary to understand and to model the physical systems used to set the constraints. In particular one needs to determine the effective parameters that can be observationally constrained to a set of fundamental constants;
2. it is necessary to relate and compare different constraints that are obtained at different space-time positions. This often requires a space-time dynamics and thus to specify a model;
3. it is necessary to relate the variation of different fundamental constants through e.g. unification.

This text summarizes these three aspects by first focusing, in Sect. 2, on the various physical systems that have been used, in Sect. 3, on the theories describing varying constants (focusing on unification and on the link with the universality of free fall). Section 4 summarizes the links with cosmology where our current understanding of the dynamics of the cosmic expansion calls for the introduction of another constant: the cosmological constant, if not of a new sector of physics, the "dark energy".

## 2 Physical Systems and Constraints

### 2.1 Physical Systems

The various physical systems that have been considered can be classified in many ways.

First, we can classify them according to their look-back time and more precisely their space-time position relative to our actual position. This is summarized on Fig. 1 which represents our past-light cone, the location of the various systems (in terms of their redshift $z$) and the typical level at which they constrain the time variation of the fine structure constant. This systems include atomic clocks comparisons ($z=0$), the Oklo phenomenon ($z\sim 0.14$), meteorite dating ($z\sim 0.43$), both having a space-time position along the world line of our system and not on our past-light cone, quasar absorption spectra ($z=0.2$–$4$), cosmic microwave background (CMB) anisotropy ($z\sim 10^3$) and primordial nucleosynthesis (BBN, $z\sim 10^8$). Indeed higher redshift systems offer the possibility to set constraints on an larger time scale, but at the prize of usually involving other parameters such as the cosmological parameters. This is particularly the case of the cosmic microwave background and primordial nucleosynthesis, the interpretation of which require a cosmological model (see Uzan 2003 for a review and Uzan and Leclercq 2005 for a non-technical introduction).

The systems can also be classified in terms of the physics that they involve in order to be interpreted (see Table 1). For instance, atomic clocks, quasar absorption spectra and the cosmic microwave background require only to use quantum electrodynamics to draw the primary constraints, so that these constraints will only involve the fine structure constant $\alpha$, the ratio between the proton-to-electron mass ratio $\mu$ and the various gyromagnetic factors $g_I$. On the other hand, the Oklo phenomenon, meteorite dating and nucleosynthesis require nuclear physics and quantum chromodynamics to be interpreted (see below).

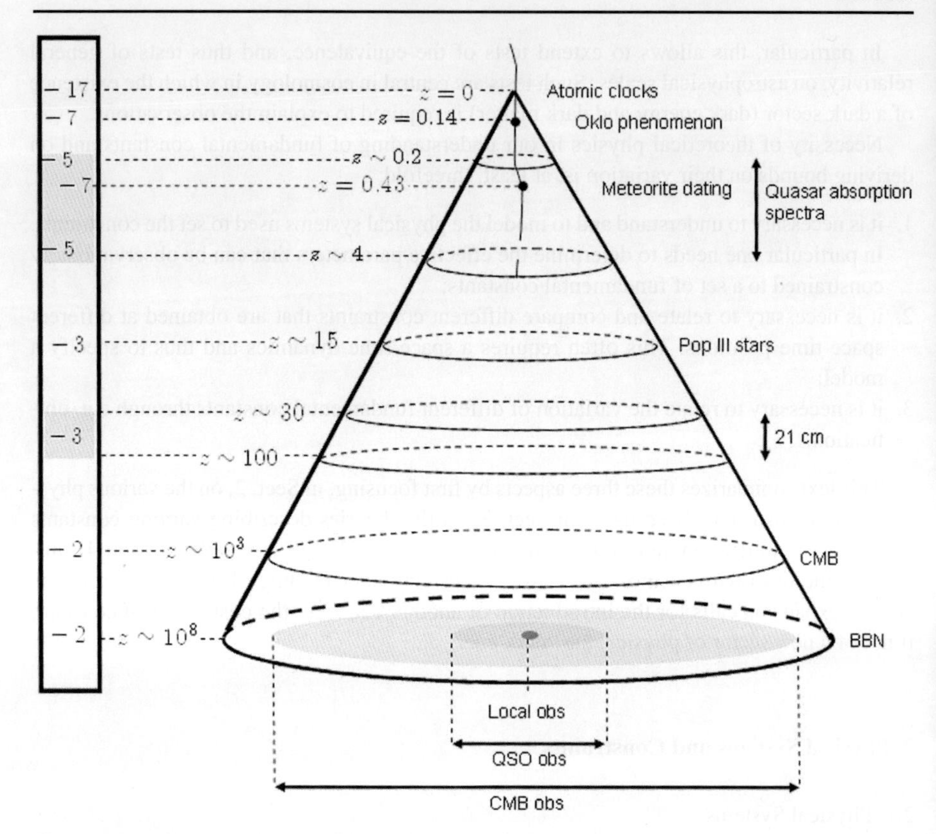


**Fig. 1** Summary of the systems that have been used to probe the constancy of the fundamental constants and their position in a space-time diagram in which the cone represents our past light cone. The *shaded areas* represent the comoving space probed by different tests with respect to the largest scales probed by primordial nucleosynthesis

## 2.2 Setting Constraints

For any system, setting constraints goes through several steps that we sketch here.

First, any system allows us to derive an observational or experimental constraint on an observable quantity $O(G_k, X)$ which depends on a set of primary physical parameters $G_k$ and a set of external parameters $X$, that usually are physical parameters that need to be measured or constrained (e.g. temperature, ...). These external parameters are related to our knowledge of the physical system and the lack of their knowledge is usually referred to as systematic uncertainty.

From a physical model of the system, one can deduce the sensitivities of the observables to an independent variation of the primary physical parameters

$$\kappa_{G_k} = \frac{\partial \ln O}{\partial \ln G_k} \tag{1}$$

As an example, the ratio between various atomic transitions can be computed from quantum electrodynamics to deduce that the ratio of two hyperfine-structure transition depends only on $g_I$ and $\alpha$ while the comparison of fine-structure and hyperfine-structure transitions

**Table 1** Summary of the systems considered to set constraints on the variation of the fundamental constants. We summarize the observable quantities (see text for details), the primary constants used to interpret the data and the other hypotheses required for this interpretation. ($\alpha$: fine structure constant; $\mu$: electron-to-proton mass ratio; $g_i$: gyromagnetic factor; $E_r$: resonance energy of the samarium-149; $\lambda$: lifetime; $B_D$: deuterium binding energy; $Q_{np}$: neutron-proton mass difference; $\tau$: neutron lifetime; $m_e$: mass of the electron; $m_N$: mass of the nucleon)

| System | Observable | Primary constraints | Other hypothesis |
|---|---|---|---|
| Atomic clock | $\delta \ln \nu$ | $g_i, \alpha, \mu$ | – |
| Oklo phenomenon | Isotopic ratio | $E_r$ | Geophysical model |
| Meteorite dating | Isotopic ratio | $\lambda$ | – |
| Quasar spectra | Atomic spectra | $g_p, \mu, \alpha$ | Cloud properties |
| 21 cm | $T_b$ | $g_p, \mu, \alpha$ | Cosmological model |
| CMB | $T$ | $\mu, \alpha$ | Cosmological model |
| BBN | Light element abundances | $Q_{np}, \tau, m_e, m_N, \alpha, B_D$ | Cosmological model |

depend on $g_I$, $\alpha$ and $\mu$. For instance (Dzuba et al. 1999; Karshenboim 2005)

$$\frac{\nu_{\rm Cs}}{\nu_{\rm Rb}} \propto \frac{g_{\rm Cs}}{g_{\rm Rb}}\alpha^{0.49}, \qquad \frac{\nu_{\rm Cs}}{\nu_{\rm H}} \propto g_{\rm Cs}\mu\alpha^{2.83}.$$

The primary parameters are usually not fundamental constants (e.g. the resonance energy of the samarium $E_r$ for the Oklo phenomenon, the deuterium binding energy $B_D$ for nucleosynthesis etc.). The second step is thus to relate the primary parameters to (a choice of) fundamental constants $c_i$. This would give a series of relations (see e.g. Müller et al. 2004)

$$\Delta \ln G_k = \sum_i d_{ki} \Delta \ln c_i. \tag{2}$$

The determination of the parameters $d_{ki}$ requires first to choose the set of constants $c_i$ (do we stop at the masses of the proton and neutron, or do we try to determine the dependencies on the quark masses, or on the Yukawa couplings and Higgs vacuum expectation value, etc.; see e.g. Dent et al. 2008 for various choices) and also requires to deal with nuclear physics and the intricate structure of QCD. In particular, the energy scales of QCD, $\Lambda_{\rm QCD}$, is so dominant that at lowest order all parameters scales as $\Lambda^n_{\rm QCD}$ so that the variation of the strong interaction would not affect dimensionless parameters and one has to take the effect of the quark masses.

As an example, the Oklo phenomenon allows to draw a constraint on the value of the energy of the resonance. The observable $O$ is a set of isotopic ratios that allow to reconstruct the average cross-sections for the nuclear network that involves the various isotopes of the samarium and gadolinium (this involves assumptions about the geometry of the reactor, its temperature that falls into $X$). It was argued (Damour and Dyson 1996) on the basis of a model of the samarium nuclei that the energy of the resonance is mainly sensitive to $\alpha$ so that the only relevant parameter is $d_\alpha \sim -1.1 \times 10^7$. The level of the constraint $-0.9 \times 10^{-7} < \Delta\alpha/\alpha < 1.2 \times 10^{-7}$ that is inferred from the observation is thus related to the sensitivity $d_\alpha$.

## 2.3 Constraints on the Fine Structure Constant

An extended discussion of the different constraints on the time variation of the fine structure constant can be found in Uzan (2003, 2004). We just summarize the state of the art in Fig. 2

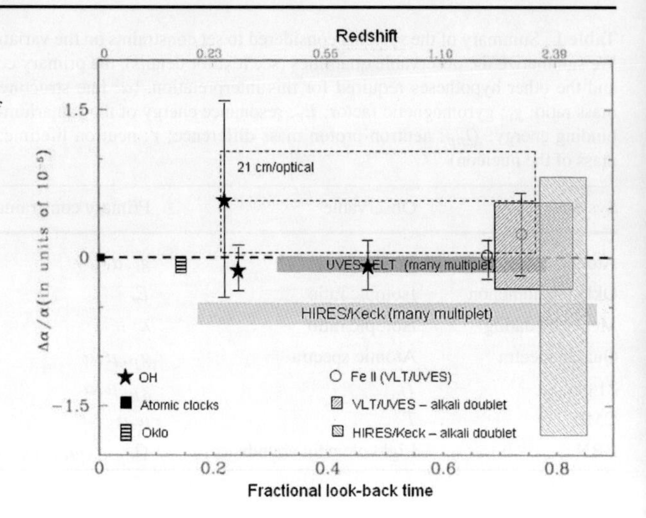


**Fig. 2** Constraints on the variation of the fine structure constant obtained for the various physical systems as a function of redshift or look-back time, assuming the standard ΛCDM model

which depicts the constraints on $\Delta\alpha/\alpha$ for different redshift bands. In the cases where the constraints involve other constants, we have assumed for simplicity that only $\alpha$ was allowed to vary. This is the case in particular for quasar absorption spectra (see the contribution by P. Petitjean in this issue).

The typical constraints on cosmological timescales (of order of 10 Gyr) is $\Delta\alpha/\alpha < 10^{-6}$ which would correspond to a constraint $\Delta\alpha/\alpha < 10^{-16}$ yr$^{-1}$ if one assumes (but there a priori no reason to support it) that the rate of change is constant over time (i.e. $\alpha(t)$ is a linear function of the cosmic time). For comparison, under the same hypothesis, Oklo would give the constraint $\Delta\alpha/\alpha < 5\times10^{-17}$ yr$^{-1}$. But indeed these constraints are complementary since they concern different spacetime positions.

## 2.4 Unification and Correlated Variations

In the context of the unification of the fundamental interactions, it is expected that the variations of the various constants are not independent. By understanding these correlations we can set stronger constraints at the expense of being more model-dependent. We only illustrate this on the example of BBN, along the lines of Coc et al. (2007).

The BBN theory predicts the production of the light elements in the early universe: the abundances synthesized rely on the balance between the expansion of the universe and the weak interaction rates which control the neutron to proton ratio at the onset of BBN (see Peter and Uzan 2005 for a textbook introduction). Basically, the abundance of helium-4 is well approximated by

$$Y_p = 2\frac{(n/p)_\mathrm{f}\exp(-t_\mathrm{N}/\tau)}{1+(n/p)_\mathrm{f}\exp(-t_\mathrm{N}/\tau)} \tag{3}$$

where $(n/p)_\mathrm{f} = \exp(-Q_{np}/kT_\mathrm{f})$ is the neutron to proton ratio at the freeze-out time determined (roughly) by $G_F^2(kT_\mathrm{f})^5 = \sqrt{GN}(kT_\mathrm{f})^2$, $N$ being the number of relativistic degrees of freedom; $Q_{np} = m_n - m_p$, $\tau$ is the neutron lifetime, $G_F$ the Fermi constant and $t_\mathrm{N}$ the time after which the photon density becomes low enough for the photo-dissociation of the deuterium to be negligible. As a conclusion, the predictions of BBN involve a large number of fundamental constants. In particular, $t_\mathrm{N}$ depends on the deuterium binding energy and on the photon-to-baryon ratio, $\eta$. Besides, one needs to include the effect of the fine structure constant in the Coulomb barriers (Bergström et al. 1999). For a different analysis of the effect of varying fundamental constants on BBN predictions see e.g. Müller et al. (2004), Landau et al. (2006), Coc et al. (2007), Dent et al. (2007).

It follows that the predictions of the BBN are mainly dependent of the effective parameters $G_k = (G, \alpha, m_e, \tau, Q_{np}, B_D, \sigma_i)$ while the external parameters are mainly the cosmological parameters $X = (\eta, h, N_\nu, \Omega_i)$. The dependence of the predictions on the independent variation of each of these parameters can be determined and it was found (Flambaum and Shuryak 2002; Coc et al. 2007: Fig. 3) that the most sensitive parameter is the deuterium binding energy, $B_D$ (see also Dent et al. 2007 for a similar analysis including more parameters). The helium and deuterium data allow to set the constraints

$$-7.5 \times 10^{-2} < \frac{\Delta B_D}{B_D} < 6.5 \times 10^{-2}, \qquad -8.2 \times 10^{-2} < \frac{\tau}{\tau} < 6 \times 10^{-2},$$

$$-4 \times 10^{-2} < \frac{\Delta Q_{np}}{Q_{np}} < 2.7 \times 10^{-2}$$

on the independent variations of these parameter. More interestingly (and more speculative!) a variation of $B_D$ in the range $(-7.5, -4) \times 10^{-2}$ would be compatible with the helium and deuterium constraint while reconciling the spectroscopically determined lithium-7 abundance (Coc et al. 2007) with its expected value from WMAP.

In a second step, the parameters $G_k$ can be related to a smaller set of fundamental constants, namely the fine structure constant $\alpha$, the Higgs VEV $v$, the Yukawa couplings $h_i$ and the QCD scale $\Lambda_{\mathrm{QCD}}$ since $Q_{np} = m_n - m_p = \alpha a \Lambda_{\mathrm{QCD}} + (h_d - h_u)v$, $m_e = h_e v$, $\tau_n = G_F^2 m_e^5 f(Q/m_e)$ and $G_F = 1/\sqrt{2}v$. The deuterium binding energy can be expressed in terms of $h_s$, $v$ and $\Lambda_{\mathrm{QCD}}$ (Flambaum and Shuryak 2003) using a sigma nuclear model or in terms of the pion mass (Epelbaum et al. 2003). Assuming that all Yukawa couplings vary similarly, the set of parameters $G_k$ reduces to the set of constants $\{\alpha, v, h, \Lambda_{\mathrm{QCD}}\}$ (again in units of the Planck mass).

Several relations between these constants can however be found. For instance, in grand-unified models the low-energy expression of $\Lambda_{\mathrm{QCD}}$,

$$\Lambda_{\mathrm{QCD}} = \mu \left( \frac{m_c m_b m_t}{\mu^3} \right)^{2/27} \exp\left[ -\frac{2\pi}{9\alpha_3(\mu)} \right]$$

for $\mu > m_t$ yields a relation between $\{\alpha, v, h, \Lambda_{\mathrm{QCD}}\}$ so that one actually has only 3 independent constants. Then, in all models in which the weak scale is determined by dimensional transmutation, changes in the Yukawa coupling $h_t$ will trigger changes in $v$ (Ibánez and Ross 1982). In such cases, the Higgs VEV can be written as

$$v = M_p \exp\left[ -\frac{8\pi^2 c}{h_t^2} \right],$$

where $c$ is a constant of order unity. I follows that we are left with only 2 independent constants. This number can even be reduced to 1 in the case where one assumes that the variation of the constants is trigger by an evolving dilaton (Damour and Polyakov 1994; Campbell and Olive 1995). At each stage, one reduces the number of constants, and thus the level of the constraints, at the expense of some model dependence. In the latter case, it was shown that BBN can set constraints of the order of $|\Delta\alpha/\alpha| < 4 \times 10^{-5}$ (Coc et al. 2007).

## 3 Theories with "Varying Constants"

### 3.1 Making a Constant Dynamical

The question of whether the constants of nature may be dynamical goes back to Dirac (1937) who expressed, in his "Large Number hypothesis", the opinion that very large (or small) dimensionless universal constants cannot be pure mathematical numbers and must not occur in the basic laws of physics. In particular, he stressed that the ratio between the gravitational and electromagnetic forces between a proton and an electron, $Gm_em_p/e^2 \sim 10^{-40}$ is of the same order as the inverse of the age of the universe in atomic units, $e^2H_0/m_ec^3$. He stated that these were not pure numerical coincidences but instead that these big numbers were not pure constants but reflected the state of our universe. This led him to postulate that $G$ varies[1] as the inverse of the cosmic time. Diracs' hypothesis is indeed not a theory and it was shown later that a varying constant can be included in a Lagrangian formulation as a new dynamical degree of freedom so that one gets both a new dynamical equation of evolution for this degree of freedom and a modification of the other field equations with respect to their form derived under the hypothesis it was constant.

Let us illustrate this on the case of scalar-tensor theories, in which gravity is mediated not only by a massless spin-2 graviton but also by a spin-0 scalar field that couples universally to matter fields (this ensures the universality of free fall). In the Jordan frame, the action of the theory takes the form

$$S = \int \frac{\mathrm{d}^4 x}{16\pi G_*}\sqrt{-g}\left[F(\varphi)R - g^{\mu\nu}Z(\varphi)\varphi_{,\mu}\varphi_{,\nu} - 2U(\varphi)\right] + S_{\mathrm{matter}}[\psi; g_{\mu\nu}] \qquad (4)$$

where $G_*$ is the bare gravitational constant. This action involves three arbitrary functions ($F$, $Z$ and $U$) but only two are physical since there is still the possibility to redefine the scalar field. $F$ needs to be positive to ensure that the graviton carries positive energy. $S_{\mathrm{matter}}$ is the action of the matter fields that are coupled minimally to the metric $g_{\mu\nu}$. In the Jordan frame, the matter is universally coupled to the metric so that the length and time as measured by laboratory apparatus are defined in this frame.

It is useful to define an Einstein frame action through a conformal transformation of the metric $g^*_{\mu\nu} = F(\varphi)g_{\mu\nu}$. In the following all quantities labelled by a star (*) will refer to Einstein frame. Defining the field $\varphi_*$ and the two functions $A(\varphi_*)$ and $V(\varphi_*)$ (see e.g. Esposito-Farèse and Polarski 2001) by

$$\left(\frac{\mathrm{d}\varphi_*}{\mathrm{d}\varphi}\right)^2 = \frac{3}{4}\left(\frac{\mathrm{d}\ln F(\varphi)}{\mathrm{d}\varphi}\right)^2 + \frac{1}{2F(\varphi)}, \qquad A(\varphi_*) = F^{-1/2}(\varphi),$$
$$2V(\varphi_*) = U(\varphi)F^{-2}(\varphi),$$

the action (4) reads as

$$S = \frac{1}{16\pi G_*}\int \mathrm{d}^4 x\sqrt{-g_*}\left[R_* - 2g_*^{\mu\nu}\partial_\mu\varphi_*\partial_\nu\varphi_* - 4V\right] + S_{\mathrm{matter}}[A^2 g^*_{\mu\nu}; \psi]. \qquad (5)$$

[1] Dirac hypothesis can also be achieved by assuming that $e$ varies as $t^{1/2}$. Indeed this reflects a choice of units, either atomic or Planck units. There is however a difference: assuming that only $G$ varies violates the strong equivalence principle while assuming a varying $e$ results in a theory violating the Einstein equivalence principle. It does not mean we are detecting the variation of a dimensionful constant but simply that either $e^2/\hbar c$ or $Gm_e^2/\hbar c$ is varying.

The kinetic terms have been diagonalized so that the spin-2 and spin-0 degrees of freedom of the theory are perturbations of $g^*_{\mu\nu}$ and $\varphi_*$ respectively.

The action (4) defines an effective gravitational constant $G_{\rm eff} = G_*/F = G_* A^2$. This constant does not correspond to the gravitational constant effectively measured in a Cavendish experiment. The Newton constant measured in this experiment is $G_{\rm cav} = G_* A_0^2(1 + \alpha_0^2)$ where the first term, $G_* A_0^2$ corresponds to the exchange of a graviton while the second term $G_* A_0^2 \alpha_0^2$ is related to the long range scalar force. The gravitational constant depends on the scalar field and is thus dynamical.

The post-Newtonian parameters can be expressed in terms of the values of $\alpha$ and $\beta$ today as

$$\gamma^{\rm PPN} - 1 = -\frac{2\alpha_0^2}{1+\alpha_0^2}, \qquad \beta^{\rm PPN} - 1 = \frac{1}{2}\frac{\beta_0\alpha_0^2}{(1+\alpha_0^2)^2}. \tag{6}$$

The Solar system constraints imply $\alpha_0$ to be very small, typically $\alpha_0^2 < 10^{-5}$ while $\beta_0$ can still be large. Binary pulsar observations (Esposito-Farèse 2005) impose that $\beta_0 > -4.5$ and that $\dot{G}/G < 10^{-12}\ \mathrm{yr}^{-1}$.

### 3.2 General Dangers

Given the previous discussion, it seems a priori simple to cook up a theory that will describe a varying fine structure constant by coupling a scalar field to the electromagnetic Faraday tensor as

$$S = \int \left[\frac{R}{16\pi G} - 2(\partial_\mu\phi)^2 - \frac{1}{4}B(\phi)F_{\mu\nu}^2\right]\sqrt{-g}\,\mathrm{d}^4x \tag{7}$$

so that the fine structure will evolve according to $\alpha = B^{-1}$.

Such an simple implementation may however have dramatic implications. In particular, the contribution of the electromagnetic binding energy to the mass of any nucleus can be estimated by the Bethe-Weizäcker formula so that

$$m_{(A,Z)}(\phi) \supset 98.25\,\alpha(\phi)\,\frac{Z(Z-1)}{A^{1/3}}\ \mathrm{MeV}.$$

This implies that the sensitivity of the mass to a variation of the scalar field is expected to be of the order of

$$f_{(A,Z)} = \partial_\phi m_{(A,Z)}(\phi) \sim 10^{-2}\frac{Z(Z-1)}{A^{4/3}}\alpha'(\phi). \tag{8}$$

It follows that the level of the violation of the universality of free fall is expected to be of the level of $\eta_{12} \sim 10^{-9} X(A_1, Z_1; A_2, Z_2)(\partial_\phi \ln B)_0^2$. Since the factor $X(A_1, Z_1; A_2, Z_2)$ typically ranges as $\mathcal{O}(0.1\text{–}10)$, we deduce that $(\partial_\phi \ln B)_0$ has to be very small for the Solar system constraints to be satisfied. It follows that today the scalar field has to be very close to the minimum of the coupling function $\ln B$.

Let us mention that such coupling terms naturally appear when compactifying a higher-dimensional theory. As an example, let us recall the compactification of a 5-dimensional Einstein-Hilbert action (Peter and Uzan 2005, Chap. 13)

$$S = \frac{1}{12\pi^2 G_5}\int \bar{R}\sqrt{-\bar{g}}\,\mathrm{d}^5x.$$

Decomposing the 5-dimensional metric $\bar{g}_{AB}$ as

$$\bar{g}_{AB} = \begin{pmatrix} g_{\mu\nu} + \frac{A_\mu A_\nu}{M^2}\phi^2 & \frac{A_\mu}{M}\phi^2 \\ \frac{A_\nu}{M}\phi^2 & \phi^2 \end{pmatrix},$$

where $M$ is a mass scale, we obtain

$$S = \frac{1}{16\pi G}\int \left(R - \frac{\phi^2}{4M^2}F^2\right)\phi\sqrt{-g}\mathrm{d}^4x, \tag{9}$$

where the 4-dimensional gravitational constant is $G = 3\pi G_5/4\int \mathrm{d}y$. The scalar field couples explicitly to the kinetic term of the vector field and cannot be eliminated by a redefinition of the metric: this is the well-known conformal invariance of electromagnetism in four dimensions. Such a term induces a variation of the fine structure constant as well as a violation of the universality of free-fall. Such dependencies of the masses and couplings are generic for higher-dimensional theories and in particular string theory. It is actually one of the definitive predictions for string theory that there exists a dilaton, that couples directly to matter (Taylor and Veneziano 1988) and whose vacuum expectation value determines the string coupling constants (Witten 1984).

For example, in type I superstring theory, the 10-dimensional dilaton couples differently to the gravitational and Yang-Mills sectors because the graviton is an excitation of closed strings while the Yang-Mills fields are excitations of open strings. For small values of the volume of the extra-dimensions, a T-duality makes the theory equivalent to a 10-dimensional theory with Yang-Mills fields localized on a D3-brane. When compactified on an orbifold, the gauge fields couple to fields $M_i$ living only at these orbifold points with couplings $c_i$ which are not universal. Typically, one gets that $M_4^2 = \mathrm{e}^{-2\Phi}V_6 M_I^8$ while $g_{YM}^{-2} = \mathrm{e}^{-2\Phi}V_6 M_I^6 + c_i M_i$. Unfortunately, the 4-dimensional effective couplings depend on the version of the string theory, on the compactification scheme and on the dilaton.

### 3.3 Ways Out

While the tree-level predictions of string theory seem to be in contradiction with experimental constraints, many mechanisms can reconcile it with experiment. In particular, it has been claimed that quantum loop corrections to the tree-level action may modify the coupling function in such a way that it has a minimum (Damour and Polyakov 1994). As explained in the former paragraph, the dilaton needs to be close to the minimum of the coupling function in order for the theory to be compatible with the universality of free fall. In the case of scalar-tensor theories, it was shown that when the coupling function enjoys such a minimum, the theory is naturally attracted toward general relativity (Damour and Nordtvedt 1993). The same mechanism will apply if all the coupling functions have the same minimum (see the contribution by T. Damour in this issue for more details). In that particular model the mass of any nuclei will typically be of the form

$$m_i(\phi) = \Lambda_{\mathrm{QCD}}(\phi) \times \left(1 + a^q \frac{m_q}{\Lambda_{\mathrm{QCD}}} + a^e \alpha\right),$$

where $a^q$ and are $a^e$ are sensitivities. It follows that composition independent effects (i.e. $|\gamma^{\mathrm{PPN}} - 1|$, $|\beta^{\mathrm{PPN}} - 1|$, $\dot{G}/G$) and composition dependent effects ($\eta$, $\dot{\alpha}$, $\dot{\mu}$) will be of the same order of magnitude, dictated by the difference of the value of the dilaton today compared to its value at the minimum of the coupling function.

Another possibility is to invoke an environmental dependence, as can be implemented in scalar-tensor theories by the chameleon mechanism (Khoury and Veltman 2004) which invokes a potential with a minimum that does not coincide with the one of the coupling function.

## 4 Links with Cosmology

The comparison of various constraints requires a full cosmological model. In particular, one cannot assume that the time variation of the constant is linear either with time or redshift (as used to compare the typical magnitudes of the constraints in Sect. 2.3). Besides, cosmology tends to indicate that a new constant, the cosmological constant, has to be included in our description. We briefly address these two points by focusing on the tests that can be performed on cosmological scales and on the status of our cosmological model.

### 4.1 Cosmological Evolution

The cosmological dynamics is central to apply the least coupling principle. Since the dilaton is attracted towards the minimum of the coupling function during its cosmological evolution, deviations from general relativity are expected to be larger in the early universe. In particular BBN can set bounds on the deviation from general relativity that are stronger than those obtained from Solar system experiments (Damour and Pichon 1999; Coc et al. 2006).

### 4.2 About Dark Energy

The construction of any cosmological model relies on 4 main hypotheses: (H1) a theory of gravity; (H2) a description of the matter contained in the Universe and their non-gravitational interactions; (H3) symmetry hypothesis; (H4) an hypothesis on the global structure, i.e. the topology of the Universe.

These hypotheses are not on the same footing since H1 and H2 refer to the physical theories. These two hypotheses are however not sufficient to solve the field equations and we must make an assumption on the symmetries (H3) of the solutions describing our Universe on large scales while H4 is an assumption on some global properties of these cosmological solutions, with same local geometry.

Our reference cosmological model is the $\Lambda$CDM model. It assumes that gravity is described by general relativity (H1), that the Universe contains the fields of the standard model of particle physics plus some dark matter and a cosmological constant, the latter two having no physical explanation at the moment. It also deeply involves the Copernican principle as a symmetry hypothesis (H3), without which the Einstein equations usually can not been solved, and usually assumes that the spatial sections are simply connected (H4). H2 and H3 imply that the description of the standard matter reduces to a mixture of a pressureless and a radiation perfect fluids. This model is compatible with all astronomical data which roughly indicates that $\Omega_{\Lambda 0} \simeq 0.73$, $\Omega_{\mathrm{mat}0} \simeq 0.27$, and $\Omega_{K0} \simeq 0$. Cosmology thus roughly imposes that $|\Lambda_0| \le H_0^2$, that is $\ell_\Lambda \le H_0^{-1} \sim 10^{26}\,\mathrm{m} \sim 10^{41}\,\mathrm{GeV}^{-1}$. Notice that it is disproportionately large compared to the natural scale fixed by the Planck length

$$\rho_{\Lambda_0} < 10^{-120} M_{\mathrm{Pl}}^4 \sim 10^{-47}\,\mathrm{GeV}^4 \tag{10}$$

at the heart of the cosmological constant problem.

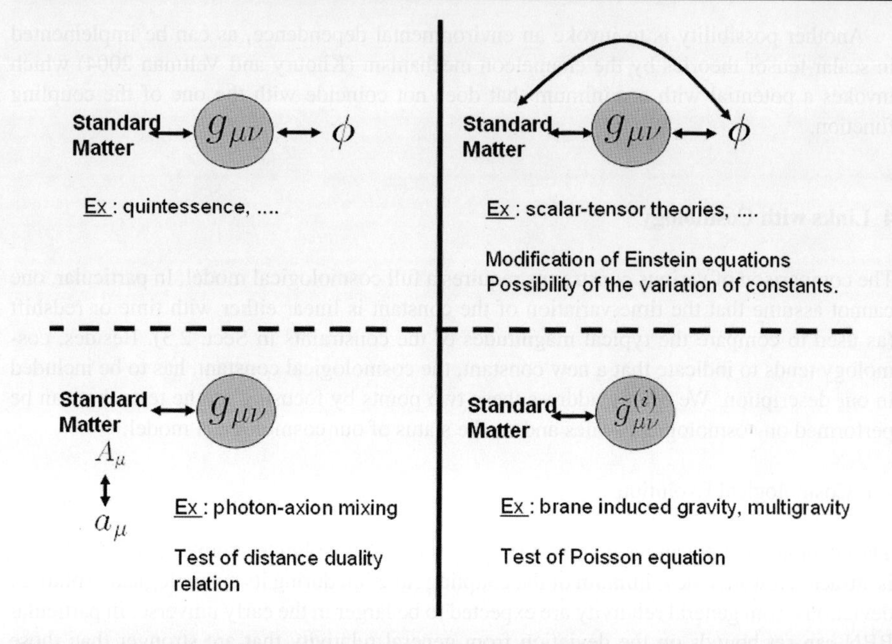


**Fig. 3** Summary of the different classes of physical dark energy models. As discussed in the text, various tests can be designed to distinguish between them. The classes differ according to the nature of the new degrees of freedom and their couplings. *Left column* accounts for models where gravitation is described by general relativity while *right column* models describe a modification of general relativity. In the *upper line* classes, the new fields dominate the matter content of the universe at low redshift. *Upper-left* models (class A) consist of models in which a new kind of gravitating matter is introduced. In the *upper-right* models (class C), a light field induces a long-range force so that gravity is not described by a massless spin-2 graviton only. In this class, Einstein equations are modified and there may be a variation of the fundamental constants. The *lower-right* models (class D) correspond to models in which there may exist an infinite number of new degrees of freedom, such as in some class of braneworld scenarios. These models predict a modification of the Poisson equation on large scales. In the last class (*lower-left*, class B), the distance duality relation may be violated. From Uzan (2007)

The assumption that the Copernican principle holds, and the fact that it is so central in drawing our conclusion on the acceleration of the expansion, splits the investigation into two avenues. Either we assume that the Copernican principle holds and we have to modify the laws of fundamental physics or we abandon the Copernican principle, hoping to explain dark energy without any new physics but at the expense of living in a particular place in the Universe. In the former case, the models can be classified in terms of 4 universality classes summarized on Fig. 3.

The goal in this section is to summarize some attempts to test both the Copernican principle and general relativity on cosmological scales.

## 4.3 Beyond the Copernican Principle

The Copernican principle implies that the spacetime metric reduces to a single function, the scale factor $a(t)$ that can be Taylor expanded as $a(t) = a_0 + H_0(t - t_0) - \frac{1}{2}q_0 H_0^2(t - t_0)^2 + \cdots$. It follows that the conclusions that the cosmic expansion is accelerating ($q_0 < 0$) does not involve any hypothesis about the theory of gravity (other than the one that the spacetime geometry can be described by a metric) or the matter content, as long as this principle holds.

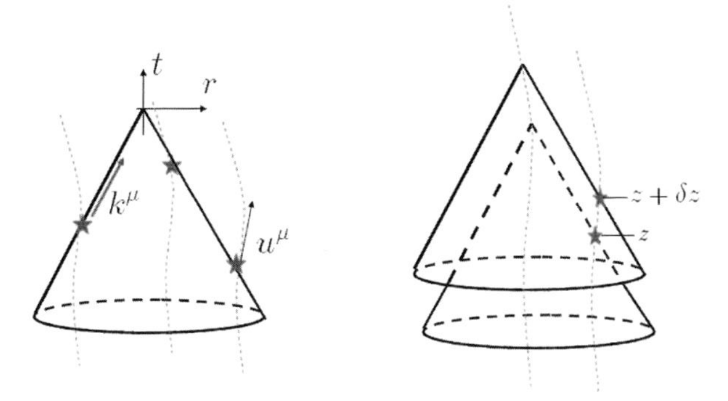


**Fig. 4** *Left*: Most low-redshift data are localized on our past light-cone. In a non-homogeneous spacetime there is no direct relation between the redshift that is observed and the cosmic time, needed to reconstruct the expansion history. *Right*: The time drift of the redshift allows to extract information about two infinitely close past light-cones. $\delta z$ depends on the proper motions of the observer and the sources as well as the spacetime geometry

While isotropy around us seems well established observationally, homogeneity is more difficult to test. The possibility, that we may be living close to the center of a large underdense region has sparked considerable interest, because such models can successfully match the magnitude-redshift relation of type Ia supernovae without dark energy (see the contribution by S. Sarkar in this issue).

The main difficulty in testing the Copernican principle lies in the fact that most observations are located on our past light-cone. Recently, it was realized that cosmological observations may however provide such a test (Uzan et al. 2008a, 2008b). It exploits the time drift of the redshift that occurs in any expanding spacetime, as first pointed out in the particular case of Robertson-Walker spacetimes for which it takes the form $\dot{z} = (1+z)H_0 - H(z)$. Such an observation would give informations on the dynamics outside the past light-cone since it compares the redshift of a given source at two times and thus on two infinitely close past light-cones (see Fig. 4-right). It follows that it contains informations about the spacetime structure along the worldlines of the observed sources that must be compatible with the one derived from the data along the past light-cone.

For instance, in a spherically symmetric spacetime, the expression depends on the shear, $\sigma(z)$, of the congruence of the wordlines of the comoving observers evaluated along our past light-cone (Uzan et al. 2008a),

$$\dot{z} = (1+z)H_0 - H(z) - \frac{1}{\sqrt{3}}\sigma(z).$$

It follows that, when combined with other distance data, it allows to determine the shear on our past light-cone and one can check whether it is compatible with zero, as expected for any Robertson-Walker spacetime.

In a RW universe, one can go further and determine a consistency relation between several observables since $H^{-1}(z) = D'(z)[1 + \Omega_{K0}H_0^2 D^2(z)]^{-1/2}$, where a prime stands for $\partial_z$ and $D(z) = D_L(z)/(1+z)$; this relation is independent of the Friedmann equations. It follows that in any Robertson-Walker spacetime the *consistency relation*,

$$1 + \Omega_{K0}H_0^2\left(\frac{D_L(z)}{1+z}\right)^2 - [H_0(1+z) - \dot{z}(z)]^2\left[\frac{\mathrm{d}}{\mathrm{d}z}\left(\frac{D_L(z)}{1+z}\right)\right]^2 = 0,$$

between observables must hold whatever the matter content and the field equations, since it derives from pure kinematical relations that do not rely on the dynamics.

$\dot{z}(z)$ has a typical amplitude of order $\delta z \sim -5 \times 10^{-10}$ on a time scale of $\delta t = 10$ yr, for a source at redshift $z = 4$. This measurement is challenging, and impossible with present-day facilities. However, it was recently revisited in the context of Extremely Large Telescopes (ELT), arguing they could measure velocity shifts of order $\delta v \sim 1$–$10$ cm/s over a 10 years period from the observation of the Lyman-$\alpha$ forest. It is one of the science drivers in design of the CODEX spectrograph (Pasquini et al. 2005) for the future European ELT. Indeed, many effects, such as proper motion of the sources, local gravitational potential, or acceleration of the Sun may contribute to the time drift of the redshift. It was shown (Uzan et al. 2008a), however, that these contributions can be brought to a 0.1% level so that the cosmological redshift is actually measured.

Let us also stress that another idea was also proposed recently (Goodman 1995; Caldwell and Stebbins 2008; Clarkson et al. 2008, see the contribution by R. Caldwell in this issue). This idea is based on the distortion of the Planck spectrum of the cosmic microwave background.

## 4.4 Testing General Relativity on Astrophysical Scales

Extracting constraints on deviations from GR on cosmological scales is difficult because large scale structures entangle the properties of matter and gravity. On sub-Hubble scales, one can, however, construct tests reproducing those in the Solar system. For instance, light deflection is a test of GR because we can measure independently the deflection angle and the mass of the Sun.

On sub-Hubble scales, relevant for the study of the large-scale structure, the Einstein equations reduce to the Poisson equation

$$\Delta\Psi = 4\pi G\rho_{\rm mat}a^2\delta_{\rm mat} = \frac{3}{2}\Omega_{\rm mat}H^2a^2\delta_{\rm mat}, \tag{11}$$

relating the gravitational potential and the matter density contrast.

As first pointed out by Uzan and Bernardeau (2001), this relation can be tested on astrophysical scales, since the gravitational potential and the matter density perturbation can be measured independently from the use of cosmic shear measurements and galaxy catalogs. The test was recently implemented with the CFHTLS-weak lensing data and the SDSS data to conclude that the Poisson equation holds observationally to about 10 Mpc (Doré et al. 2007). The main limitation in the applicability of this test is due to the biasing mechanisms (i.e. the fact that galaxies do not necessarily trace faithfully the matter field) even if it is thought to have no significant scale dependence at such scales.

### *4.4.1 Toward a Post-ΛCDM Formalism*

The former test of the Poisson equation exploits one rigidity of the field equations on sub-Hubble scales. It can be improved by considering the full set of equations.

Assuming that the metric of spacetime takes the form

$$\mathrm{d}s^2 = -(1+2\Phi)\mathrm{d}t^2 + (1-2\Psi)a^2\gamma_{ij}\mathrm{d}x^i\mathrm{d}x^j \tag{12}$$

on sub-Hubble scales, the equation of evolution reduces to the continuity equation $\delta'_{\rm mat} + \theta_{\rm mat} = 0$, where $\theta$ is the divergence of the velocity perturbation and a prime denotes a derivative with respect to the conformal time, the Euler equation $\theta'_{\rm mat} + \mathcal{H}\theta_{\rm mat} = -\Delta\Phi$, where $\mathcal{H}$ is the comoving Hubble parameter, the Poisson equation (11) and $\Phi = \Psi$.

These equations imply many relations between the cosmological observables. For instance, decomposing $\delta_{\rm mat}$ as $D(t)\epsilon(x)$ where $\epsilon$ encodes the initial conditions, the growth rate $D(t)$ evolves as $\ddot{D}+2H\dot{D}-4\pi G\rho_{\rm mat}D=0$. This equation can be rewritten in terms of $p=\ln a$ as time variable (Peter and Uzan 2005) and considered not as a second order equation for $D(t)$ but as a first order equation for $H^2(a)$

$$(H^2)'+2\left(\frac{3}{a}+\frac{D''}{D'}\right)H^2=3\frac{\Omega_{\rm mat0}H_0^2D}{a^2D'}$$

where a prime denotes a derivative with respect to $p$. It can be integrated as (Chiba and Nakamura 2007)

$$\frac{H^2(z)}{H_0^2}=3\Omega_{\rm mat0}\left(\frac{1+z}{D'(z)}\right)^2\int\frac{D}{1+z}(-D')\mathrm{d}z. \tag{13}$$

This exhibits a rigidity between the growth function and the Hubble parameter. In particular the Hubble parameter determined from background data and from perturbation data using (13) must agree. This was used in the analysis of Wang et al. (2007).

Another relation exists between $\theta_{\rm mat}$ and $\delta_{\rm mat}$. The Euler equation implies that $\theta_{\rm mat}=-\beta(\Omega_{\rm mat0},\Omega_{\Lambda 0})\delta_{\rm mat}$, with $\beta(\Omega_{\rm mat0},\Omega_{\Lambda 0})\equiv \mathrm{d}\ln D(a)/\mathrm{d}\ln a$.

We conclude that the perturbation variables are not independent and the relation between them are inherited from some assumptions on the dark energy. Phenomenologically, we can generalize the sub-Hubble equations to

$$\delta'_{\rm mat}+\theta_{\rm mat}=0,\qquad \theta'_{\rm mat}+\mathcal{H}\theta_{\rm mat}=-\Delta\Phi+S_{\rm de}, \tag{14}$$

$$-k^2\Phi=4\pi GF(k,H)\delta_{\rm mat}+\Delta_{\rm de},\qquad \Delta(\Phi-\Psi)=\pi_{\rm de}. \tag{15}$$

We assume that there is no production of baryonic matter so that the continuity equation is left unchanged. $S_{\rm de}$ describes the interaction between dark energy and standard matter. $\Delta_{\rm de}$ characterizes the clustering of dark energy, $F$ accounts for a scale dependence of the gravitational interaction and $\pi_{\rm de}$ is an effective anisotropic stress. It is clear that the $\Lambda$CDM corresponds to $(F,\pi_{\rm de},\Delta_{\rm de},S_{\rm de})=(1,0,0,0)$. The expression of $(F,\pi_{\rm de},\Delta_{\rm de},S_{\rm de})$ for quintessence, scalar-tensor, $f(R)$ and DGP models and more generally for models of the classes A–D can be found in Uzan (2007).

From an observational point of view, weak lensing survey gives access to $\Phi+\Psi$, galaxy maps allow to reconstruct $\delta_g=b\delta_{\rm mat}$ where $b$ is the bias, velocity fields give access to $\theta$. In a $\Lambda$CDM, the correlations between these observables are not independent since, for instance $\langle\delta_g\delta_g\rangle=b^2\langle\delta^2_{\rm mat}\rangle$, $\langle\delta_g\theta_m\rangle=-b\beta\langle\delta^2_{\rm mat}\rangle$ and $\langle\delta_g\kappa\rangle=8\pi G\rho_{\rm mat}a^2b\langle\delta^2_{\rm mat}\rangle$. Various ways of combining these observables have been proposed, construction of efficient estimators and forecast for possible future space missions designed to make these tests as well as the possible limitations (arising e.g. from non-linear bias, the effect of massive neutrinos or the dependence on the initial conditions) are now being extensively studied (Zhang et al. 2007; Jain and Zhang 2007; Amendola et al. 2008; Song and Koyama 2008).

## 5 Conclusion

The study of fundamental constants provides tests of general relativity that can be extended on astrophysical scales. These constrains can be useful in our understanding of the origin of the acceleration of the cosmic expansion. They can be combined with tests of general relativity based on the large scale structure of the universe.

In the coming future, we can hope to obtain new constraints from population III stars ($z \sim 15$) and 21 cm absorption ($z \sim 30$–100) and improved constraints from QSO by about a factor 10 (and probably more with ELT).

Interpreting the coupled variations of various constants is also challenging since it usually implies to deal with nuclear physics and QCD. It also offers a window on unification mechanisms and on string theory.

All these aspects make the study of fundamental constants a lively and promising topic.

## References

L. Amendola, M. Kunz, D. Sapone, J. Cosmol. Astropart. Phys. **04**, 013 (2008)
L. Bergström, S. Iguri, H. Rubinstein, Phys. Rev. D **60**, 045005 (1999)
R. Caldwell, A. Stebbins, Phys. Rev. Lett. **100**, 191302 (2008)
B.A. Campbell, K.A. Olive, Phys. Lett. B **345**, 429 (1995)
T. Chiba, T. Nakamura, Prog. Theor. Phys. **118**, 815 (2007)
C. Clarkson, B. Basset, T. Lu, Phys. Rev. Lett. **101**, 011301 (2008)
A. Coc et al., Phys. Rev. D **73**, 083525 (2006)
A. Coc et al., Phys. Rev. D **76**, 023511 (2007)
T. Damour, F.J. Dyson, Nucl. Phys. B **480**, 37 (1996)
T. Damour, K. Nordtvedt, Phys. Rev. Lett. **70**, 2217 (1993)
T. Damour, B. Pichon, Phys. Rev. D **59**, 123502 (1999)
T. Damour, A.M. Polyakov, Nucl. Phys. B **423**, 532 (1994)
T. Dent, S. Stern, C. Wetterich, Phys. Rev. D **76**, 063513 (2007)
T. Dent, S. Stern, C. Wetterich, Phys. Rev. D **78**, 103518 (2008)
P.A.M. Dirac, Nature (London) **139**, 323 (1937)
O. Doré et al., arXiv:0712.1599 (2007)
V.A. Dzuba, V.V. Flambaum, J.K. Webb, Phys. Rev. A **59**, 230 (1999)
G.F.R. Ellis, J.-P. Uzan, Am. J. Phys. **73**, 240 (2005)
E. Epelbaum, U.G. Meissner, W. Gloeckle, Nucl. Phys. A **714**, 535 (2003)
G. Esposito-Farèse, eConf C0507252 SLAC-R-819, T025 (2005)
G. Esposito-Farèse, D. Polarski, Phys. Rev. D **63**, 063504 (2001)
V.V. Flambaum, E.V. Shuryak, Phys. Rev. D **65**, 103503 (2002)
V.V. Flambaum, E.V. Shuryak, Phys. Rev. D **67**, 083507 (2003)
J. Goodman, Phys. Rev. D **52**, 1821 (1995)
C.J. Hogan, Rev. Mod. Phys. **72**, 1149 (2000)
L.E. Ibánez, G.G. Ross, Phys. Lett. B **100**, 215 (1982)
B. Jain, P. Zhang, arXiv:0709.2375 (2007)
S.G. Karshenboim, http://arXiv.org/abs/hep-ph/0509010 (2005)
J. Khoury, A. Veltman, Phys. Rev. Lett. **92**, 171104 (2004)
S.J. Landau, M.E. Mosquera, H. Vucetich, Astrophys. J. **637**, 38 (2006)
C.M. Müller, G. Schäfer, C. Wetterich, Phys. Rev. D **70**, 083504 (2004)
L. Pasquini et al., Messenger **122**, 10 (2005)
P. Peter, J.-P. Uzan, *Cosmologie Primordiale* (Belin, Paris, 2005). Translated as Primordial Cosmology (Oxford University Press, 2009)
Y.-S. Song, K. Koyama, arXiv:0802.3897 (2008)
T.R. Taylor, G. Veneziano, Phys. Lett. B **213**, 450 (1988)
J.-P. Uzan, Rev. Mod. Phys. **75**, 403 (2003)
J.-P. Uzan, AIP Conf. Proc. **736**, 3 (2004)
J.-P. Uzan, Gen. Relativ. Gravit. **39**, 307 (2007)
J.-P. Uzan, F. Bernardeau, Phys. Rev. D **64**, 083004 (2001)
J.-P. Uzan, B. Leclercq, *De l'importance d'être une constante* (Dunod, Paris, 2005). Translated as The Natural Laws of the Universe—Understanding Fundamental Constants (Praxis, 2008)
J.-P. Uzan, C. Clarkson, G.F.R. Ellis, Phys. Rev. Lett. **100**, 191303 (2008a)

J.-P. Uzan, F. Bernardeau, Y. Mellier, Phys. Rev. D **70**, 021301 (2008b)
Y. Wang et al., Phys. Rev. D **76**, 063503 (2007)
E. Witten, Phys. Lett. B **149**, 351 (1984)
S. Weinberg, Phil. Trans. R. Soc. Lond. A **310**, 249 (1983)
C.M. Will, *Theory and Experiment in Gravitational Physics* (Cambridge University Press, Cambridge, 1993)
P. Zhang et al., Phys. Rev. Lett. **99**, 141302 (2007)

Space Sci Rev (2009) 148: 267–288
DOI 10.1007/s11214-009-9564-z

# Testing the Stability of the Fine Structure Constant in the Laboratory

**N. Kolachevsky · A. Matveev · J. Alnis · C.G. Parthey · T. Steinmetz · T. Wilken · R. Holzwarth · T. Udem · T.W. Hänsch**

Received: 10 April 2009 / Accepted: 8 June 2009 / Published online: 5 August 2009
© The Author(s) 2009. This article is published with open access at Springerlink.com

**Abstract** In this review we discuss the progress of the past decade in testing for a possible temporal variation of the fine structure constant $\alpha$. Advances in atomic sample preparation, laser spectroscopy and optical frequency measurements led to rapid reduction of measurement uncertainties. Eventually laboratory tests became the most sensitive tool to detect a possible variation of $\alpha$ at the present epoch. We explain the methods and technologies that helped to make this possible.

**Keywords** Drift of the fine structure constant · Frequency comb · Laser stabilization

**PACS** 06.20.Jr · 06.30.Ft · 32.30.Jc

## 1 Introduction

The fine structure constant $\alpha = e^2/\hbar c$ is a dimensionless parameter that measures the strength of all electromagnetic interactions. As such it appears in a large variety of phenomena such as forces between charged objects that in turn determine the structure of atoms and molecules. Further examples are the propagation of electromagnetic waves, chemical reactions or even macroscopic phenomena like friction. The value of the fine structure constant can be thought of as the electromagnetic force between two electrons at a distance of

N. Kolachevsky · A. Matveev · J. Alnis · C.G. Parthey · T. Steinmetz · T. Wilken · T. Udem · T.W. Hänsch
Max-Planck Institut für Quantenoptik, Hans-Kopfermann-Str 1, 85748 Garching, Germany

N. Kolachevsky (✉) · A. Matveev
P.N. Lebedev Physical Institute, Leninsky prosp. 53, 119991 Moscow, Russia
e-mail: kolik@lebedev.ru

T. Steinmetz · R. Holzwarth
MenloSystems GmbH, Am Klopferspitz 19, 82152 Martinsried, Germany

T.W. Hänsch
Ludwig-Maximilians-Universität, Munich, Germany

one meter measured in units where the speed of light $c$ and Planck's constant $\hbar$ are set to unity. The fine structure constant is used as an expansion parameter in the quantum theory of electromagnetic interactions, Quantum Electrodynamics (QED) which is one of the most successful theories in physics.

Unfortunately, neither this theory nor any other known theory makes any prediction on the value of the fine structure constant which is determined experimentally to $\alpha \approx 1/137$. Unlike many other dimensionless numbers that we find in nature, such as the number of particles in the Universe, the fine structure constant represents a *small* number. This fact has led P.A.M. Dirac to formulate his "large number hypothesis" (Dirac 1937), where he constructed small dimensionless numbers from known physical constants assuming these are the fundamental parameters. One of these small numbers is the age of the Universe in atomic units divided by the electromagnetic force between an electron and a proton measured in units of their gravitational force and was believed to be $\approx 3$ in 1937. Following this hypothesis, the gravitational constant $G$ or any other constant that appears in the construction of these small numbers should vary in time as the Universe expands. This was the first alternative theory assuming time-dependent coupling constants after Einstein's General Relativity. Indeed, there is no theory yet that predicts the value of $\alpha$ to be stable or drifting so that there is no reason to expect one or the other behavior. Even though Dirac's estimated drift rate of $\alpha$ has been ruled out by repeated measurements, the general possibility of "variable constants" remains open.

Modern theories that go beyond the large number hypothesis which allow for the drift of fundamental constants rely on coupling between gravitation and other fundamental interactions. Attempts to unify gravity with electromagnetic, weak and strong interactions encounter severe difficulties. To build such a "theory of everything" it seems that one has to extend the number of dimensions of our usual space-time world. String theories may allow for temporal and spatial variation of the coupling constants that could be associated with cosmic dynamics. Some possible mechanisms that lead to a drift or spatial variations of the fundamental constants are discussed in Taylor and Veneziano (1988), Damour et al. (2002), Flambaum (2007) and Flambaum and Shuryak (2008). As of now there is not sufficient theoretical evidence to make any well-grounded prediction of the size of such variations. The effect, if existing at all, should be extremely small since the gravitational interaction seems to be almost decoupled at the low-energy limit. For this reason experimental research is the appropriate way to probe this type of new physics that goes beyond the standard model.

Concerning the time interval $\Delta t$ separating two measurements, there are two extreme classes of experiments: (i) astronomical or geological observations and (ii) high precision laboratory measurements. The investigation of absorption or emission lines of distant galaxies back illuminated by the white light of quasars at even larger distances takes advantage of the extremely long look back time of up to $10^{10}$ years while the relative sensitivity to the $\alpha$ variation reaches $10^{-6}$. In contrast to that, laboratory frequency comparisons are restricted to short time intervals of a few years but can be as sensitive if precision measurements with uncertainties of better than $10^{-15}$ are performed. Currently such a low uncertainty can only be realized by frequency measurements in the radio or optical domain. For this reason frequency comparisons of atomic, molecular or ionic transitions are used. The important advantages of laboratory experiments are: The variety of different systems that may be tested, the possibility to change parameters of the experiments in order to control systematic effects and the straightforward determination of the drift rates from the measured values. Modern precision frequency measurements deliver information about the stability of the to-date values of the constants, which can only be tested with laboratory measurements. At the same time only non-laboratory methods are sensitive to processes that occurred in the early Universe, which may be much larger than at present times. As both classes of experiments

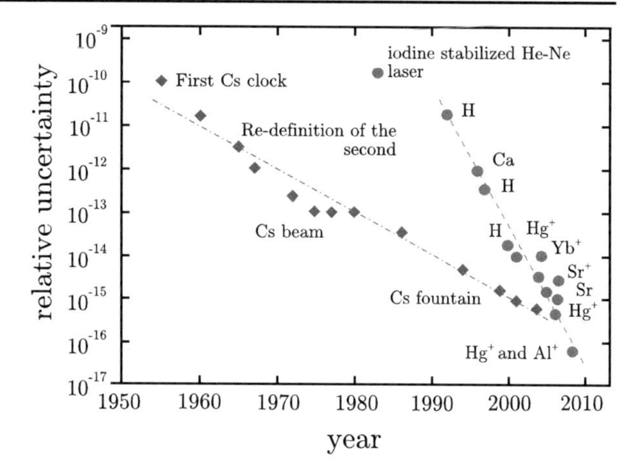


**Fig. 1** Uncertainties of the microwave Cs ground state hyperfine splitting (*diamonds*) and optical frequency standards based on a variety of atoms and ions (*circles*)

(i) and belong to different epochs, they supplement each other to get a more detailed view of the possible time variation of fundamental constants.

In 2000 J.K. Webb and co-workers introduced the many-multiplet method (Webb et al. 2001), which is an extension of the alkali-doublet method (Savedoff 1956; Minkowski and Wilson 1956), to extract the value of the fine structure constant from quasar absorption spectra. As the detected absorption lines emerged billions of light years away they conserve the value of $\alpha$ over that period of time. Application of the many-multiplet method to KECK/HIRES QSO data indicated that $\alpha$ was smaller by $\Delta\alpha/\alpha = (5.4 \pm 1.2) \times 10^{-6}$ about $10^{10}$ years ago (Webb et al. 2001). This $5\sigma$ deviation from a non drifting value stimulated further investigation in the field. In 2003–2004 another set of astrophysical data obtained by the Very Large Telescope was analyzed independently using the same approach (Chand et al. 2004; Quast et al. 2004). The conclusion was that $\alpha$ was stable within $|\Delta\alpha/\alpha| < 10^{-6}$ in the past. Meanwhile M.T. Murphy and co-workers pointed out some possible flaws in the data evaluation (Murphy et al. 2008) so that astrophysical data remain contradictory (see also Varshalovich et al. 2003). Unlike laboratory measurements astrophysical data analysis strongly relies on cosmological evolution, i.e. expansion, isotopic abundances, magnetic field distribution, etc. which are also debated.

Laboratory frequency measurements have become competitive very recently in terms of sensitivity to a possible variation of $\alpha$ in the present epoch. Figure 1 summarizes the progress achieved during the last decades in the field of optical frequency measurements (only a few are selected for the plot). For comparison, the progress in microwave frequency standards used for realization of the SI second is shown in the same plot. Improvements of the last years have been due to new ultra-cold atomic samples, laser stabilization techniques as well as breakthroughs in optical frequency measurements so that relative uncertainties in the optical domain are approaching $10^{-17}$.

With the introduction of frequency combs (see Sect. 2.2) high-precision optical frequency measurements became a routine procedure, readily available for a broad scientific community (Holzwarth et al. 2000; Udem et al. 2002). Repeated frequency measurements of some atomic transitions allowed to tighten the upper limit for the variation of frequency ratios. The latter can be used in a variety of fundamental tests, including the search for the variation of $\alpha$. Optical frequency measurements from 2000 and 2003 in ytterbium (Peik et al. 2004) and mercury (Bize et al. 2003) ions as well as in atomic hydrogen (Fischer et al. 2004) allowed to impose a model-independent restriction of $\dot{\alpha}/\alpha = (-0.9 \pm 2.9) \times 10^{-15}$ yr$^{-1}$

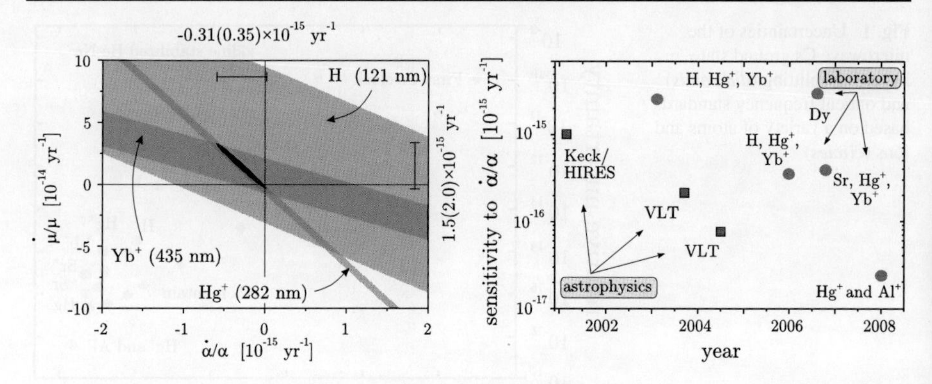


**Fig. 2** (*Left*)—Illustration of the laboratory method for detecting possible variations of the fine structure constant $\alpha$. The method is based on combinations of absolute frequency measurements in different atomic systems (Peik et al. 2004; Fischer et al. 2004; Fortier et al. 2007). Here $\mu$ is the reduced magnetic moment of the $^{133}$Cs nucleus. (*Right*)—Comparison of sensitivities of astrophysical and laboratory methods for a presumed linear drift of $\alpha$

(see Fig. 2, left). The sensitivity of this test was already competitive to the sensitivity of the KECK/HIRES data analysis (Webb et al. 2001) (Fig. 2, right) which was the most sensitive analysis from astrophysical observations at that time.

The laboratory method is based on optical frequency measurements and allowed to increase the sensitivity of probing the possible $\alpha$ variation by an order of magnitude already in 2005 (Peik et al. 2006; Blatt et al. 2008). Thus far the lowest limit on the present drift rate of $\alpha$ has been obtained by T. Rosenband and co-workers at NIST (USA) by direct comparison of optical clock transitions in mercury and aluminum ions via a frequency comb (Rosenband et al. 2008; Lorini et al. 2008). Their result reads $\dot{\alpha}/\alpha = (-1.6 \pm 2.3) \times 10^{-17}\ \mathrm{yr}^{-1}$ and is an order of magnitude more accurate than astrophysical observations albeit at a different epoch.

In what follows we present modern techniques used for spectroscopy and frequency measurement of narrow optical transitions (Sect. 2), discuss the model-independent laboratory method for restricting the variation of $\alpha$ (Sect. 3) and point out some perspectives opened by optical frequency metrology for astrophysics (Sect. 4).

## 2 Precision Optical Spectroscopy and Optical Frequency Measurements

The principle of modern optical frequency measurement is presented in Fig. 3. A laser is tuned to the wavelength of a narrow metrological transition (usually referred to as a "clock transition") in an atomic, ionic or molecular sample. Most commonly, the laser frequency is stabilized by active feedback to a transmission peak of a well isolated optical cavity ("reference cavity") which allows to achieve sub-hertz spectral line width of the interrogating laser. Some recent advances in laser stabilization techniques will be described in Sect. 2.1. The laser frequency is then scanned across the transition which allows to find the line center $\omega_0$ using an appropriate line shape model. The measured transition quality factor can reach $10^{15}$ which provides extremely high resolution. To obtain the transition frequency the beat note $\omega_{\text{beat}}$ between the laser and one of the modes of a stabilized frequency comb is measured with the help of a frequency counter. Details of this type of measurement are presented in Sect. 2.2. If the comb is stabilized to a primary frequency reference (i.e. a Cs atomic clock),

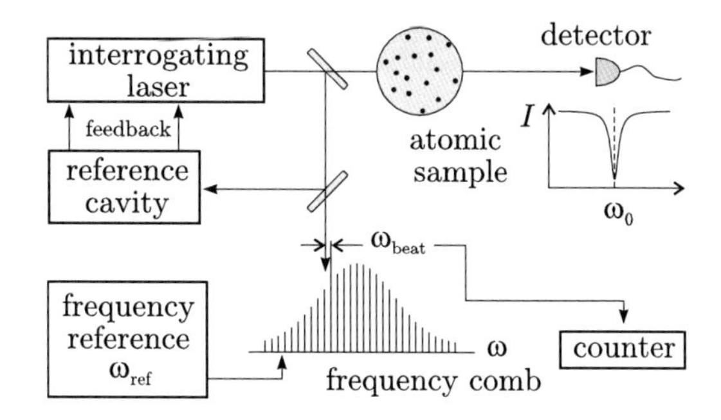


**Fig. 3** Setup for the measurement an optical transition frequency in an atomic sample with the help of an optical frequency comb

the measurement presented in Fig. 3 will yield the *absolute* frequency of the optical transition. Absolute frequency measurements allow a comparison of different results obtained at laboratories all over the world. An example of such a comparison is given in Sect. 3.1. On the other hand, if the comb is stabilized with the help of some other reference, which can be e.g. another optical frequency, the measurement will yield the ratio $\omega_0/\omega_{\text{ref}}$. One can thus compare transition frequencies in different atomic samples avoiding time-consuming absolute frequency measurements and avoiding the additional uncertainty introduced by primary frequency standards (Fig. 1).

## 2.1 Laser Stabilization

Interrogation of a clock transition in an atomic sample requires a narrow-band laser source. Due to vibrations, fluctuations of the pump intensity as well as some intrinsic noise sources (e.g. the excessive phase noise in semiconductor lasers, Petermann 1979; Henry 1982), the typical laser line width turns out to be many orders of magnitude broader than the Schawlow-Townes limit (Schawlow and Townes 1958). A passive isolation of the laser resonator itself is not sufficient to suppress these noise sources so that active stabilization to an external reference cavity is implemented.

The reference cavity should be well isolated from the environment by placing it in a separate vibrationally and thermally stabilized vacuum chamber. If the laser frequency is stabilized to the transmission peak of such a cavity, e.g. by means of the Pound-Drever-Hall technique (Drever et al. 1983), the laser frequency fluctuations $\delta\nu$ will be directly coupled to the fluctuations of the cavity length $\delta l$ according to $\delta\nu/\nu = \delta l/l$, where $\nu$ is the laser frequency, and $l$ is the cavity length. If one desires to achieve $\delta\nu = 1$ Hz using a cavity of $l = 10$ cm, the distance between the mirrors should remain constant at the level of $10^{-16}$ m, which is a fraction of the proton radius!

The first demonstration of a sub-Hz laser line width by B.C. Young and co-workers in 1999 made use of a heavy optical bench suspended with rubber tubes from the lab ceiling (Young et al. 1999) for vibration isolation. In the meantime cavity designs and mountings emerged where deformations due to vibrations do not change the critical length that separates the mirrors (Notcutt et al. 2005; Nazarova et al. 2006). This has not only led to much more compact setups but also to a number of laser sources successfully stabilized to sub-hertz level (Stoehr et al. 2006; Ludlow et al. 2007). In our laboratory we use vertically mounted Fabry-Pérot (FP) cavities, with a spacer design from A.D. Ludlow and co-workers (2007) and reach 40 dB suppression of vertical vibration sensitivity (Alnis et al. 2008).

The principle of such a cavity mounting is shown in the left hand side of Fig. 4. The cavity spacer is suspended at its midplane such that the influence of vibration induced ver-

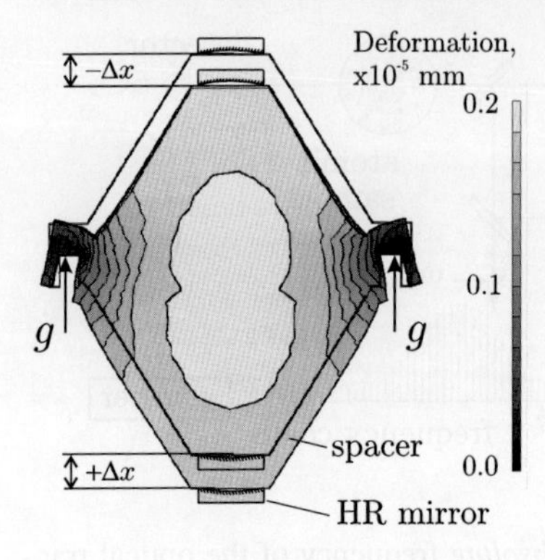


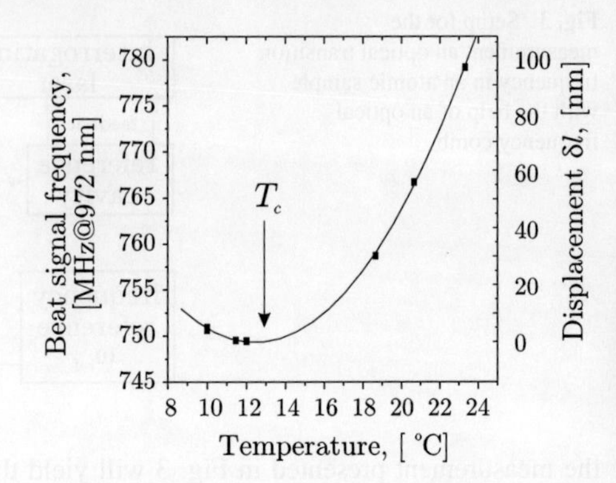


**Fig. 4** (*Left*)—Deformations of a vertically mounted Fabry-Pérot (FP) cavity under the influence of vertical acceleration of $1g$ determined with finite elements analysis. The mounting is such that the compression of upper part $-\Delta x$ is compensated by the stretch of the lower part $+\Delta x$ maintaining the critical distance of the mirrors. (*Right*)—Determination of the zero-expansion temperature $T_C$ by measuring the beat note frequency between the FP cavity and the laser stabilized to the second FP cavity maintained at a constant temperature. At $T_C$ the cavity length $l$ reaches the minimum

tical acceleration to the mirror separation is significantly suppressed. This makes the setup virtually immune for vertical vibrations that are more difficult to suppress than horizontal ones. If such a cavity is placed on a vibration-isolated platform, the acoustic and seismic vibrations from the environment have no detectable influence on the cavity frequency.

Another issue that affects the stability for averaging times larger than several seconds is the dimensional stability due to temperature variations. Certain glass ceramics can be made with very low thermal expansion and the one made by Corning is called ultra low expansion glass (ULE). ULE is a Titania-doped silicate glass that has a specified thermal expansion minimum at some temperature $T_c$ around room temperature according to

$$\delta l/l \sim 10^{-9}(T - T_c)^2, \tag{1}$$

where $T$ is the cavity temperature. To reduce the quadratic dependence the temperature of the material should be stabilized as close as possible to $T_c$. Unfortunately, the measured composite thermal stationary point of the cavity resonance frequency usually ends up below room temperature (Fox 2008) (Fig. 4, right). This poses a problem because cooling of the vacuum chamber is more difficult than heating as water condensation on the windows prevents coupling of laser light into the chamber. The problem was solved by cooling the FP cavity directly in the vacuum chamber by Peltier elements (Alnis et al. 2008).

This type of vibration- and thermal compensation allows to set up extremely stable and compact laser sources. For example, the diode lasers operating at 972 nm designed for two-photon spectroscopy of the $1S$–$2S$ clock transition in atomic hydrogen are characterized as shown in Fig. 5. Two nearly identical systems have been built in our lab with one FP cavity maintained near $T_c$ of its resonance frequency, while the other stabilized to a temperature 25°C above that point. Both lasers demonstrate excellent short-time stability (up to 10 s) approaching the thermal noise limit of $10^{-15}$ which is set by the Brownian particle motion on the mirror surfaces. Concerning the long-term stability, the FP cavity at $T_c$ demonstrates a much better performance since it is much less influenced by ambient temperature fluctuations. It possesses a linear drift of about +50 mHz/s mainly caused by ULE aging, while its resonance frequency deviates from that linear drift by only ±10 Hz on a time scale of 10 hrs.

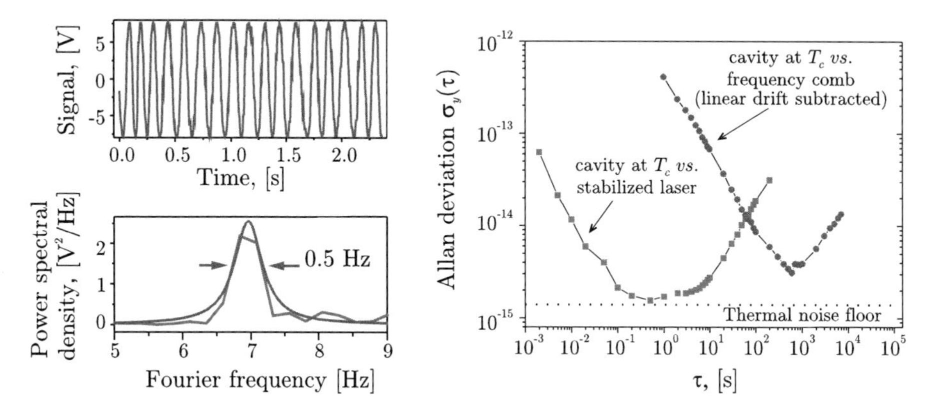


**Fig. 5** (*Left*)—Oscillogram of the beat note signal between two diode lasers locked to two independent vertically mounted FP cavities, its Fourier transformation and Lorentzian fit. (*Right*)—Allan deviation of the beat note signal from the left part of the figure (*squares*). The cavity kept at $T_C$ is extremely stable as can be seen from its absolute frequency measurement performed with an optical frequency comb (see next section) which was referenced to a hydrogen maser (*circles*)

Excellent spectral characteristics and small size of the setup allow to use such lasers in the most demanding applications of frequency metrology. Possible routes to overcome the thermal noise limit are currently being discussed and may result in further improvements of the laser spectral purity.

## 2.2 Ultra-Short Pulse Lasers and Frequency Combs

Frequency can be measured with by far the highest precision of all physical quantities. In the radio frequency domain (say up to 100 GHz), frequency counters have existed for a long time. Almost any of the most precise measurements in physics have been performed with such a counter that uses an atomic clock as a time base. To extend this accurate technique to higher frequencies, so called harmonic frequency chains have been constructed since the late 1960ies (Hocker et al. 1967; Evenson et al. 1973). Because of the large number of steps necessary to build a long harmonic frequency chain, it was not before 1995 when visible laser light was first referenced phase coherently to a cesium atomic clock using this method (Schnatz et al. 1996).

The disadvantage of these harmonic frequency chains was not only that they could easily fill several large laser laboratories at once, but that they could be used to measure a single optical frequency only. Even though mode locked lasers for optical frequency measurements have been used in rudimentary form in the late 1970ies (Eckstein et al. 1978), this method became only practical with the advent of femtosecond (fs) mode locked lasers. Such a laser necessarily emits a very broad spectrum, comparable in width to the optical carrier frequency.

In the frequency domain a train of short pulses from a femtosecond mode locked laser is the result of a phase coherent superposition of many continuous wave (cw) longitudinal cavity modes. These modes at $\omega_n$ form a series of frequency spikes that is called a frequency comb. As has been shown, the modes are remarkably uniform, i.e. the separation between adjacent modes is constant across the frequency comb (Holzwarth et al. 2000; Udem et al. 1999; Diddams et al. 2002; Ma et al. 2004). This strictly regular arrangement is the most important feature used for optical frequency measurement and may be expressed as:

$$\omega_n = n\omega_r + \omega_{CE}. \tag{2}$$

Here the mode number $n$ of some $10^5$ may be enumerated such that the frequency offset $\omega_{CE}$ lies in between 0 and $\omega_r = 2\pi/T$. The mode spacing is thereby identified with the pulse repetition rate, i.e. the inverse pulse repetition time $T$. With the help of that equation two radio frequencies $\omega_r$ and $\omega_{CE}$ are linked to the optical frequencies $\omega_n$ of the laser. For this reason mode locked lasers are capable to replace the harmonic frequency chains of the past.

To derive the frequency comb properties (Reichert et al. 1999) as detailed by (2), it is useful to consider the electric field $E(t)$ of the emitted pulse train. We assume that the electric field $E(t)$, measured for example at the lasers output coupling mirror, can be written as the product of a periodic envelope function $A(t)$ and a carrier wave $C(t)$:

$$E(t) = A(t)C(t) + c.c. \tag{3}$$

The envelope function defines the pulse repetition time $T = 2\pi/\omega_r$ by demanding $A(t) = A(t-T)$. The only thing about dispersion that should be added for this description, is that there might be a difference between the group velocity and the phase velocity inside the laser cavity. This will shift the carrier with respect to the envelope by a certain amount after each round trip. The electric field is therefore in general not periodic with $T$. To obtain the spectrum of $E(t)$ the Fourier integral has to be calculated:

$$\tilde{E}(\omega) = \int_{-\infty}^{+\infty} E(t)e^{i\omega t}dt. \tag{4}$$

Separate Fourier transforms of $A(t)$ and $C(t)$ are given by:

$$\tilde{A}(\omega) = \sum_{n=-\infty}^{+\infty} \delta(\omega - n\omega_r)\tilde{A}_n \quad \text{and} \quad \tilde{C}(\omega) = \int_{-\infty}^{+\infty} C(t)e^{i\omega t}dt. \tag{5}$$

A periodic frequency chirp imposed on the pulses is accounted for by allowing a complex envelope function $A(t)$. Thus the "carrier" $C(t)$ is defined to be whatever part of the electric field that is non-periodic with $T$. The convolution theorem allows us to calculate the Fourier transform of $E(t)$ from $\tilde{A}(\omega)$ and $\tilde{C}(\omega)$:

$$\tilde{E}(\omega) = \frac{1}{2\pi}\int_{-\infty}^{+\infty} \tilde{A}(\omega')\tilde{C}(\omega - \omega')d\omega' + c.c. = \frac{1}{2\pi}\sum_{n=-\infty}^{+\infty} \tilde{A}_n\tilde{C}(\omega - n\omega_r) + c.c. \tag{6}$$

The sum represents a periodic spectrum in frequency space. If the spectral width of the carrier wave $\Delta\omega_c$ is much smaller than the mode separation $\omega_r$, it represents a regularly spaced comb of laser modes just like (2), with identical spectral line shapes. If $\tilde{C}(\omega)$ is centered at say $\omega_c$, then the comb is shifted by $\omega_c$ from containing only exact harmonics of $\omega_r$. The frequencies of the mode members are calculated from the mode number $n$ (Udem et al. 2002; Eckstein et al. 1978; Reichert et al. 1999):

$$\omega_n = n\omega_r + \omega_c. \tag{7}$$

The measurement of the $\omega_c$ as described below (see also Holzwarth et al. 2000; Reichert et al. 1999; Diddams et al. 2000; Udem et al. 1999, 2002) usually yields a value modulo $\omega_r$,

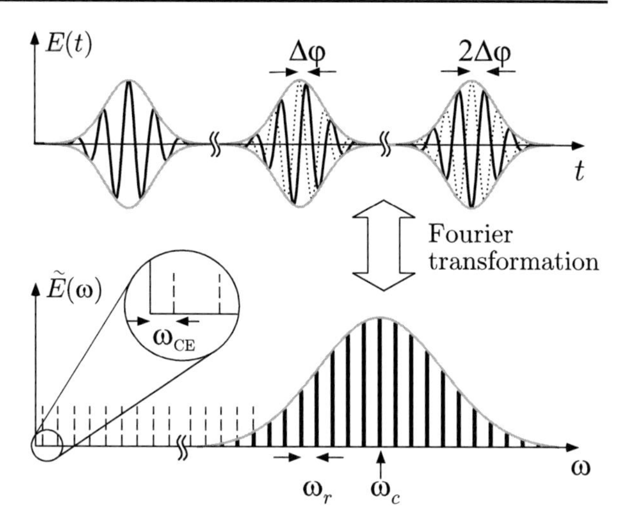


**Fig. 6** Consecutive un-chirped pulses ($A(t)$ is real) with carrier frequency $\omega_c$ and the corresponding spectrum (not to scale). Because the carrier propagates with a different velocity within the laser cavity than the envelope (with phase- and group velocity respectively), the electric field does not repeat itself after one round trip. A pulse-to-pulse phase shift $\Delta\varphi$ results in an offset frequency of $\omega_{CE} = \Delta\varphi/T$. The mode spacing is given by the repetition rate $\omega_r$. The width of the spectral envelope is given by the inverse pulse duration up to a factor of order unity that depends on the pulse shape

so that renumbering the modes will restrict the offset frequency to smaller values than the repetition frequency and (2) and (7) are identical.

If the carrier wave is monochromatic $C(t) = e^{-i\omega_c t - i\varphi}$, its spectrum will be $\delta$-shaped and centered at the carrier frequency $\omega_c$. The individual modes are also $\delta$-functions $\tilde{C}(\omega) = \delta(\omega - \omega_c)e^{-i\varphi}$. The frequency offset (7) is identified with the carrier frequency. According to (3) each round trip will shift the carrier wave with respect to the envelope by $\Delta\varphi = \arg(C(t-T)) - \arg(C(t)) = \omega_c T$ so that the frequency offset may also be identified by $\omega_{CE} = \Delta\varphi/T$ (Udem et al. 2002; Eckstein et al. 1978; Reichert et al. 1999). In a typical laser cavity this pulse-to-pulse carrier-envelope phase shift is much larger than $2\pi$, but measurements usually yield a value modulo $2\pi$. The restriction $0 \leq \Delta\varphi \leq 2\pi$ is synonymous with the restriction $0 \leq \omega_{CE} \leq \omega_r$ introduced above. Figure 6 sketches this situation in the time domain for a chirp free pulse train.

### *2.2.1 Extending the Frequency Comb*

The spectral width of a pulse train emitted by a fs laser can be significantly broadened in a single mode fiber (Agrawal 2001) by self phase modulation. Assuming a single mode carrier wave, a pulse that has propagated the length $L$ acquires a self induced phase shift of

$$\Phi_{NL}(t) = -n_2 I(t)\omega_c L/c, \tag{8}$$

where the pulse intensity is given by $I(t) = \frac{1}{2}c\varepsilon_0|A(t)|^2$. For fused silica the non-linear Kerr coefficient $n_2$ is comparatively small but almost instantaneous even on the time scale of fs pulses. This means that different parts of the pulse travel at different speed. The result is a frequency chirp across the pulse without affecting its duration. The pulse is no longer at the Fourier limit so that the spectrum is much broader than the inverse pulse duration where the extra frequencies are determined by the time derivative of the self induced phase shift $\dot{\Phi}_{NL}(t)$. Therefore pure self-phase modulation would modify the envelope function in (3) according to

$$A(t) \longrightarrow A(t)e^{i\Phi_{NL}(t)}. \tag{9}$$

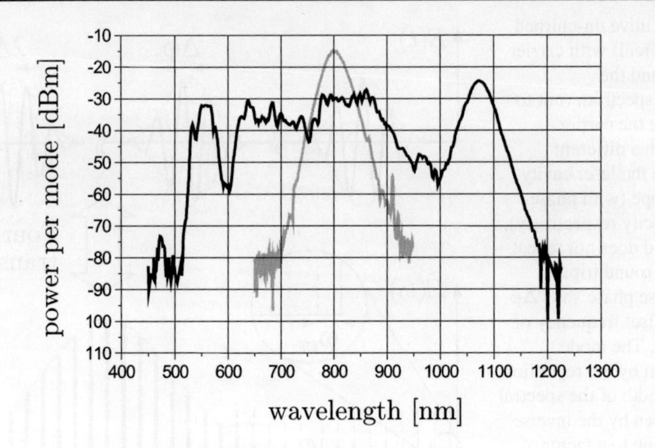


**Fig. 7** Power per mode of the frequency comb on a logarithmic scale (0 dBm = 1 mW). The lighter 30 nm (14 THz at −3 dB) wide spectrum displays the laser intensity and the darker octave spanning spectrum (532 nm through 1064 nm) is observed after spectral broadening in a 30 cm microstructured fiber (Knight et al. 1996). The laser was operated at $\omega_r = 2\pi \times 750$ MHz (modes not resolved) with 25 fs pulse duration. An average power of 180 mW was coupled through the microstructure fiber (Holzwarth et al. 2001)

Because $\Phi_{NL}(t)$ has the same periodicity as $A(t)$ the comb structure of the spectrum is maintained and the derivations (6) remain valid because periodicity of $A(t)$ was the only assumption made. An optical fiber is most appropriate for this process because it can maintain the necessary small focus area over a virtually unlimited length. In practice, however, other pulse reshaping mechanism, both linear and non-linear, are present so that the above explanation might be too simple.

A microstructured fiber uses an array of submicron-sized air holes that surround the fiber core and run the length of a silica fiber to obtain a desired effective dispersion (Knight et al. 1996). This can be used to maintain the high peak power over an extended propagation length and to significantly increase the spectral broadening. With these fibers it became possible to broaden low peak power, high repetition rate lasers to beyond one optical octave as shown in Fig. 7.

Another class of frequency combs that can stay in lock for longer times are fs fiber lasers (Nelson et al. 1997). The most common type is the erbium doped fiber laser that emits within the telecom band around 1550 nm. For this reason advanced and cheap optical components are available to build such a laser. The mode locking mechanism is similar to the Kerr lens method, except that non-linear polarization rotation is used to favor the pulsed high peak intensity operation. Up to a short free space section that can be build very stable, these lasers have no adjustable parts. Continuous stabilized operation for many hours (Kubina et al. 2005; Adler et al. 2004) has been reported. The Max-Planck Institut für Quantenoptik in Garching (Germany) operates a fiber based self referenced frequency comb that stays locked without interruption for months.

### 2.2.2 *Self-Referencing*

The measurement of $\omega_{CE}$ fixes the position of the whole frequency comb and is called self-referencing. The method relies on measuring the frequency gap between *different* harmonics derived from the *same* laser or frequency comb. The simplest approach is to fix the absolute position of the frequency comb by measuring the gap between $\omega_n$ and $\omega_{2n}$ of modes taken

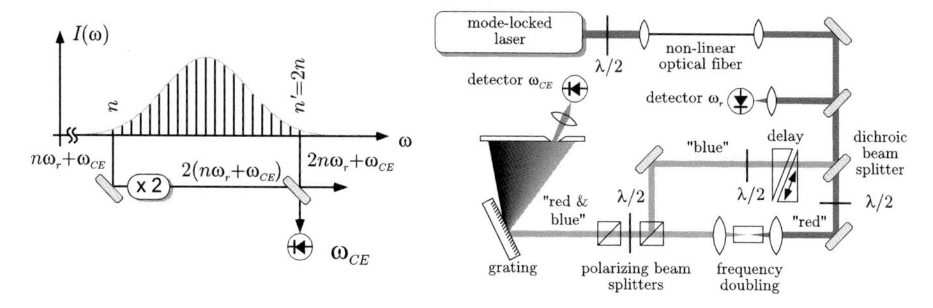


**Fig. 8** (*Left*)—The principle of the $f-2f$ self referencing relies on detecting a beat note at $\omega_{CE}$ between the frequency doubled "*red*" wing $2(n\omega_r+\omega_{CE})$ of the frequency comb and the "*blue*" modes at $2n\omega_r+\omega_{CE}$. (*Right*)—More detailed layout of the self referencing scheme. See text for details

directly from the frequency comb (Holzwarth et al. 2000; Udem et al. 2002; Diddams et al. 2000; Reichert et al. 2000). In this case the carrier-envelope offset frequency $\omega_{CE}$ is directly produced by beating the frequency doubled[1] red wing of the comb $2\omega_n$ with the blue side of the comb at $\omega_{2n}$: $2\omega_n-\omega_{n'}=(2n-n')\omega_r+\omega_{CE}=\omega_{CE}$ where again the mode numbers $n$ and $n'$ are chosen such that $(2n-n')=0$. This approach requires an octave spanning comb, i.e. a bandwidth of 375 THz if centered at the titanium-sapphire gain maximum at 800 nm.

Figure 8 sketches the $f-2f$ self referencing method. The spectrum of a mode locked laser is first broadened to more than one optical octave with an optical fiber. A broad band $\lambda/2$ wave plate allows to choose the polarization with the most efficient spectral broadening. After the fiber a dichroic mirror separates the infrared ("red") part from the green ("blue"). The former is frequency doubled in a non-linear crystal and reunited with the green part to create a wealth of beat notes, all at $\omega_{CE}$. These beat notes emerge as frequency difference between $2\omega_n-\omega_{2n}$ according to (2) for various values of $n$. The number of contributing modes is given by the phase matching bandwidth $\Delta\nu_{pm}$ of the doubling crystal and can easily exceed 1 THz.

As described, both degrees of freedom $\omega_r$ and $\omega_{CE}$ of the frequency comb can be measured up to a sign in $\omega_{CE}$ that will be discussed below. For stabilization of these frequencies, say relative to a radio frequency reference, it is also necessary to control them. Again the repetition rate turns out to be simpler. Mounting one of the laser's cavity mirrors on a piezo electric transducer allows to control the pulse round trip time. Controlling the carrier envelope frequency requires some effort. Any laser parameter that has a different influence on the cavity round trip phase delay and the cavity round trip group delay may be used to change $\omega_{CE}$ (Haus and Ippen 2001). Experimentally it turned out that the energy of the pulse stored inside the mode locked laser cavity has a strong influence on $\omega_{CE}$. To phase lock the carrier envelope offset frequency $\omega_{CE}$, one can therefore control the laser power through its energy source (pump laser).

### 2.2.3 Frequency Conversions

Given the above we conclude that the frequency comb may serve as a frequency converter between the optical and radio frequency domains allowing to perform the following phase coherent operations:

[1] It should be noted that this does not simply mean the doubling of each individual mode, but the general sum frequencies generation of all modes. Otherwise the mode spacing, and therefore the repetition rate, would be doubled as well.

- convert a radio frequency into an optical frequency. In this case both $\omega_r$ and $\omega_{CE}$ from (2) are directly locked to the radio frequency source.
- convert an optical frequency into a radio frequency. In this case the frequency of one of the comb modes $\omega_n$ is locked to a clock laser while the carrier envelope frequency $\omega_{CE}$ is phase locked to $\omega_r$. The repetition rate will then be used as the countable clock output.
- convert an optical frequency to another optical frequency, i.e. measuring optical frequency ratios. In this case the comb is stabilized to one of the lasers as described in the second case, but instead of measuring $\omega_r$ one measures the beat note frequency between another laser and its closest comb mode $\omega_n'$.

### 2.3 Frequency Measurement of the $1S$–$2S$ Transition in Atomic Hydrogen

One of the first optical frequency measurements performed with an optical frequency comb was the $1S$–$2S$ transition frequency in atomic hydrogen in our lab. During the last decades precision spectroscopic experiments on hydrogen and deuterium atoms have yielded new accurate values for the Rydberg constant (Biraben et al. 2001), the ground-state Lamb shift (Weitz et al. 1994), the deuteron structure radius (Huber et al. 1998), and the $2S$ hyperfine structure (Kolachevsky et al. 2004a, 2004b). Accurate optical frequency measurements allow for sensitive tests of quantum electrodynamics (QED), which are based on comparisons between experimental values and results from corresponding QED calculations (for review see Eides et al. 2001; Karshenboim and Ivanov 2002a, 2002b).

To measure the frequency $\omega_L$ of the continuous wave (cw) interrogation laser (486 nm) that drives the $1S$–$2S$ transition, a beat note $\omega_b$ with a stabilized frequency comb is generated (see Fig. 8). For this purpose the beam of the cw laser is spatially overlapped with the beam that contains the frequency comb and guided to a photo detector. The frequency of the interrogation laser is then given by

$$\omega_L = n\omega_r \pm \omega_{CE} \pm \omega_b. \tag{10}$$

The signs can be determined by introducing small changes to $\omega_r$ and $\omega_{CE}$ and observing the corresponding shift in $\omega_b$. This uniquely fixes both signs if $\omega_L$ is held fixed during this test. The mode number $n$ may be determined by a coarse measurement say with a high-resolution wave meter, by re-measuring with different $\omega_r$ or by comparison with previous results of lower accuracy.

In 1999 the absolute frequency measurement resulted in a relative uncertainty of 1.8 parts in $10^{14}$ (Niering et al. 2000). In 2003 this measurement has been repeated and the results of both campaigns are shown in Fig. 9. In both cases a transportable Cs atomic fountain clock (FOM) from LNM-SYRTE Paris (Santarelli et al. 1999) has been transported to our lab at Garching. Its accuracy has been evaluated to $8 \times 10^{-16}$, but during the experiments a verification at the level of $2 \times 10^{-15}$ only has been performed which is still one order of magnitude better than required for the $1S$–$2S$ transition. Unfortunately, the uncertainty of the 2003 measurement remained nearly the same due to an excessive day-to-day scatter. A further improvement of the accuracy is expected by $1S$–$2S$ spectroscopy on hydrogen atoms decelerated magnetically from a supersonic beam (Vanhaecke et al. 2007; Narevicius et al. 2008).

The measurements allowed not only to determine the absolute frequency of the transition, but also to set an upper limit of the difference of $(-29 \pm 57)$ Hz between the measurements that are 44 months apart. This is equivalent to a fractional time variation of the ratio $f_{\mathrm{H}}/f_{\mathrm{Cs}}$ equal to $(-3.2 \pm 6.3) \times 10^{-15}\ \mathrm{yr}^{-1}$, where the ground state hyperfine splitting of Cs-133,

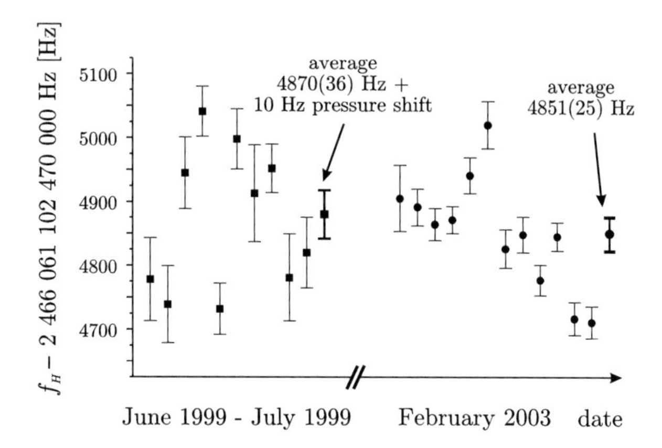


**Fig. 9** Absolute frequency measurements of the $1S$–$2S$ transition in atomic hydrogen

which is used as a reference in these measurements, is given by $f_{Cs}$. This limit on the temporal variation of the absolute optical frequency opens the possibility to derive an upper limit of the variation of $\alpha$ as detailed in the next section.

## 3 High-Precision Laboratory Measurements and Variation of the Fine Structure Constant

So far all laboratory measurements of the drift rates of fundamental constants are based on comparisons of electromagnetic transitions that depend in a different way on these constants. The non-relativistic scalings of gross-, fine- and hyperfine transitions in atoms, ions and molecules are summarized in Table 1. A first order theory is sufficient here, as none of the drifts have been detected yet with small relative uncertainty. To evaluate the possible drift of $\alpha$ one measures a frequency ratio of two transitions. Pioneering astrophysical measurements (Savedoff 1956; Minkowski and Wilson 1956) used that method which is now called the "alkali-doublet method".

In a real situation the values in that table need to be multiplied with a relativistic correction $F_{\text{rel}}(Z\alpha)$ that depends only on the fine structure constant and may be determined from relativistic Hartree-Fock calculations. For hyperfine transitions in alkali atoms there exists an approximate expression for the relativistic correction called the Casimir correction (Casimir 1963) which reads as

$$F_{\text{rel}}(Z\alpha) = \frac{3}{\lambda(4\lambda^2 - 1)}, \quad \text{where } \lambda \equiv \sqrt{1 - (Z\alpha)^2}. \tag{11}$$

For heavy atomic systems the correction to the hyperfine levels $F_{\text{rel}}(Z\alpha)$ differs significantly from 1 (e.g. $F_{\text{rel}} = 1.39$ for Cs) so that the sensitivity for $\alpha$ variations may be expressed as:

$$L_\alpha^{(\text{HFS})} \equiv \alpha \frac{\partial}{\partial \alpha} \ln[F_{\text{rel}}(Z\alpha)] = (Z\alpha)^2 \frac{12\lambda^2 - 1}{\lambda^2(4\lambda^2 - 1)}. \tag{12}$$

For the Cs ground state hyperfine splitting this equals $L_\alpha^{(\text{HFS})}(\text{Cs}) \approx 0.8$. As only ratios of frequencies can be determined in a real measurement, the sensitivity of any experiment will be given by the ratio of the sensitivity functions of the involved frequencies. Therefore the sensitivity function defined above is meaningless until it is referenced to another one.

**Table 1** Scaling factors for different atomic systems in the non-relativistic approximation. Here $Ry$ is the Rydberg constant in hertz, $g_{\text{nucl}}$ is the nuclear $g$-factor, $\mu_N$ and $\mu_B$—nuclear and Bohr magnetons respectively, $m_e$ and $m_p$—electron and proton mass respectively. In the relativistic case, i.e. for heavier atoms, it is necessary to multiply the scalings with the relativistic correction $F_{\text{rel}}(Z\alpha)$ that depends only on the fine structure constant and may be determined from relativistic Hartree-Fock calculations

| Sample | Transition | Scaling factor |
|---|---|---|
| Atom, ion | gross structure | $Ry$ |
| | fine structure | $\alpha^2 Ry$ |
| | hyperfine structure | $g_{\text{nucl}}(\mu_N/\mu_B)\alpha^2 Ry$ |
| Molecule | gross structure | $Ry$ |
| | vibration structure | $(m_e/m_p)^{1/2} Ry$ |
| | rotational structure | $(m_e/m_p) Ry$ |

For optical transition frequencies $f^{(\text{opt})}$ no approximation such as the Casimir correction exists that would be useful for deriving the leading order dependence on the fine structure constant. For this reason relativistic Hartree-Fock calculations have been used. V.A. Dzuba and co-workers have expressed the results of their calculation in terms of the two parameters $q_1$ and $q_2$ according to:

$$f^{(\text{opt})} = f_0^{(\text{opt})} + q_1\left[\left(\frac{\alpha}{\alpha_0}\right)^2 - 1\right] + q_2\left[\left(\frac{\alpha}{\alpha_0}\right)^4 - 1\right]. \tag{13}$$

Here $f_0^{(\text{opt})}$ and $\alpha_0$ are the present day (or laboratory) values of the optical transition frequency and the fine structure constant respectively. This equation was used to describe quasar absorption spectra but may be used for laboratory measurements. In this case $f_0^{(\text{opt})}$ and $\alpha_0$ are also laboratory values but at different times. Results for the parameters $q_1$ and $q_2$ for various atoms and ions including some important optical clock transitions are published in Dzuba et al. (1999a, 1999b), Dzuba and Flambaum (2000). Only even powers in $\alpha$ enter the expansion (13) because the relativistic correction is proportional to $\sqrt{m_e^2 + p^2}$, which contains even powers of electron momentum $p \sim Z\alpha$. Re-expressing Dzuba's notation in terms of the relativistic correction introduced above yields:

$$L_\alpha^{(\text{opt})} \equiv \alpha \frac{\partial}{\partial \alpha} \ln F_{\text{rel}}(Z\alpha) = \frac{2q_1 + 4q_2}{f_0^{(\text{opt})}}. \tag{14}$$

Table 2 lists a few values of this quantity adapted from Dzuba et al. (1999a, 1999b), Dzuba and Flambaum (2000) that are relevant for metrological transitions. Note, that for these calculations the value of $Ry$ has been assumed to be fixed which imposes a constrain on the value of the product $m_e c^2 \alpha^2 / h$. Another way of interpreting this is by picking $Ry$ as the unit of frequency. Using the same unit for all frequencies it will eventually drop out of all measurable quantities as only frequency ratios can be determined in practice. This will become more clear from the further analysis.

Comparing optical transitions with different relativistic corrections became a powerful instrument to set upper limits to the drift of fundamental constants. This method is widely used in astrophysical observations ("many-multiplet" method Webb et al. 2001) and in laboratory comparisons. An elegant realization of this method has been used by A. Cingöz et al. (2007), by utilizing different relativistic corrections of two nearly degenerate levels of opposite parity in neutral dysprosium. Monitoring the radio frequency transitions at 3.1 MHz for

**Table 2** Sensitivity of relativistic corrections $F_{\text{rel}}(Z\alpha)$ to $\alpha$ for some atomic transitions according to Dzuba et al. (1999a, 1999b), Dzuba and Flambaum (2000)

| $Z$ | Atom | Transition | $\lambda$ [nm] | $L_\alpha^{(\text{opt})}$ |
|---|---|---|---|---|
| 1 | H | $1\text{s}\ \text{S}_{1/2}(F=1, m_F=\pm1) \to 2\text{s}\ \text{S}_{1/2}(F'=1, m'_F=\pm1)$ | 121 | 0 |
| 20 | Ca | ${}^1\text{S}_0(m_J=0) \to {}^3\text{P}_1(m_J=0)$ | 657 | 0.03 |
| 49 | $\text{In}^+$ | $5\text{s}^2\ {}^1\text{S}_0 \to 5\text{s}5\text{p}\ {}^3\text{P}_0$ | 237 | 0.21 |
| 70 | $\text{Yb}^+$ | $6\text{s}\ {}^2\text{S}_{1/2}(F=0) \to 5\text{d}\ {}^2\text{D}_{3/2}(F=2)$ | 435 | 0.9 |
| 80 | $\text{Hg}^+$ | $5\text{d}^{10}6\text{s}\ {}^2\text{S}_{1/2}(F=0) \to 5\text{d}^9 6\text{s}^2\ {}^2\text{D}_{5/2}(F'=2, m'_F=0)$ | 282 | −3.2 |

${}^{163}$Dy and 325 MHz for ${}^{162}$Dy during 8 months only, the authors set a stringent limit on the drift of the fine-structure constant of $\partial\ln(\alpha)/\partial t = (-2.7 \pm 2.6) \times 10^{-15}\ \text{yr}^{-1}$ without any assumptions about the drift of other constants. In the next section we will describe how one can deduce a model-independent restriction to $\dot{\alpha}$ from different absolute optical frequency measurements.

## 3.1 Upper Limit for the Drift of the Fine Structure Constant from Optical Frequency Measurements

### *3.1.1 Absolute Frequency Measurements*

Thanks to the optical frequency comb the determination of absolute optical transition frequencies became a simple task, where the attribute "absolute" means that the frequency is measured in hertz, i.e. in terms of the Cs ground state hyperfine splitting. For this a ratio like $f^{(\text{opt})}/f_{\text{Cs}}^{(\text{HFS})}$ is determined. According to Table 1 such a ratio depends on two fundamental constants, $\alpha$ and the Cs nuclear magnetic moment measured in Bohr magneton's $\mu_{\text{Cs}}/\mu_B$. One may argue that the latter is not a fundamental quantity, but one has to keep in mind that the nuclear moment is mostly determined by the strong interaction. In that sense it measures the strong interaction in some units. The only difference to the electromagnetic interaction measured by $\alpha$ is that, lacking a precise model for the Cs nucleus we are not sure what those units are. For this reason there are two parameters that need to be determined from an absolute optical frequency measurement and it is impossible to disentangle the contribution from the drift rate of just one absolute optical frequency. At the other hand, the task is solvable if one has more than one absolute frequency measurement at hand under the condition, that the values $L_\alpha^{\text{opt}}$ are different for these transitions.

For the general case let's assume that there are $N$ repeated absolute frequency measurements of corresponding transitions $\text{T}_i$. For each of the transitions one can derive the relative drift of its absolute frequency $b_i$ $(i = 1 \ldots N)$ as well as the corresponding uncertainty $\sigma_i$ (one standard deviation)

$$\frac{\partial}{\partial t}\ln\frac{f_{\text{Cs}}^{(\text{HFS})}}{f_{\text{T}_i}^{(\text{opt})}} = b_i \pm \sigma_i. \tag{15}$$

One can rewrite (15) using the results from Table 2 and (14)

$$\frac{\partial}{\partial t}\ln\frac{f_{\mathrm{Cs}}^{(\mathrm{HFS})}}{f_{\mathrm{T}_i}^{(\mathrm{opt})}} = \frac{\partial}{\partial t}\left[\ln\left(\frac{\mu_{\mathrm{Cs}}}{\mu_B}\right) + (2 + L_\alpha^{(\mathrm{HFS})}(\mathrm{Cs}) - L_\alpha^{(\mathrm{opt})}(\mathrm{T}_i))\ln\alpha\right]$$
$$= y + A_i x, \tag{16}$$

where we introduced the definitions $y \equiv \partial \ln(\mu_{\mathrm{Cs}}/\mu_B)/\partial t$ and $x \equiv \partial \ln(\alpha)/\partial t$ respectively. The coefficient $A_i$ incorporates sensitivities of the corresponding relativistic corrections $L_\alpha$ as well as the $\alpha^2$ scaling for hyperfine transitions. Thus, experiments relate $x$ and $y$ to measured values $b_i$ with uncertainties $\sigma_i$ through:

$$y = A_i x + b_i \pm \sigma_i. \tag{17}$$

Let's assume further that the measured data follows a Gaussian distribution $P(x, y)$:

$$P(x, y) \propto e^{-\frac{1}{2}R^2(x,y)}, \quad \text{where} \quad R^2 = \sum_i \frac{1}{\sigma_i^2}(y - A_i x - b_i)^2. \tag{18}$$

The expectation values for the relative drift rates $x$ and $y$ are determined by the maximum likelihood method corresponding to the minimum of $R^2(x, y)$:

$$\begin{aligned}\frac{\partial R^2}{\partial x} &= -2\sum \frac{1}{\sigma_i^2}(y - A_i x - b_i)A_i = 0\\ \frac{\partial R^2}{\partial y} &= -2\sum \frac{1}{\sigma_i^2}(y - A_i x - b_i) = 0.\end{aligned} \tag{19}$$

With the definitions $B_1 \equiv \sum 1/\sigma_i^2$, $B_2 \equiv \sum A_i^2/\sigma_i^2$, $B_3 \equiv \sum b_i^2/\sigma_i^2$, $B_4 \equiv \sum A_i/\sigma_i^2$, $B_5 \equiv \sum b_i/\sigma_i^2$, $B_6 \equiv \sum A_i b_i/\sigma_i^2$ we can solve system (19) for $x$ and $y$ and obtain expressions for the expectation values:

$$\langle x\rangle = \frac{B_4 B_5 - B_1 B_6}{B_1 B_2 - B_4^2}, \quad \langle y\rangle = \frac{B_2 B_5 - B_4 B_6}{B_1 B_2 - B_4^2}. \tag{20}$$

The standard deviation for $\langle x\rangle$ can be calculated from the integral:

$$\int_{-\infty}^{+\infty} e^{-\frac{1}{2}R^2(x,y)}\,dy \propto \exp\left[\frac{(B_5 + x B_4)^2 - B_1(B_3 + x(x B_2 + 2B_6))}{2B_1}\right] \tag{21}$$

Rewriting the exponent

$$\exp\left[-\frac{(x - \langle x\rangle)^2}{2\sigma_x^2} + const_x\right], \tag{22}$$

one gets the standard deviation for $x$

$$\sigma_x = \sqrt{\frac{B_1}{B_1 B_2 - B_4^2}} \tag{23}$$

and, similarly, for $y$

$$\sigma_y = \sqrt{\frac{B_2}{B_1 B_2 - B_4^2}}. \tag{24}$$

The evaluation may be represented graphically as on the left hand side of Fig. 2.

As an example consider the results of the absolute frequency measurements of the $1S$–$2S$ transition in atomic hydrogen (data taken during 2001–2003 at our lab Fischer et al. 2004):

$$-\frac{\partial}{\partial t}\ln\frac{f_{\mathrm{H}}^{(\mathrm{opt})}}{f_{\mathrm{Cs}}^{(\mathrm{HFS})}} = \frac{\partial}{\partial t}\left[\ln\left(\frac{\mu_{\mathrm{Cs}}}{\mu_B}\right) + (2+0.8)\ln\alpha\right]$$
$$= y + 2.8x = (3.2 \pm 6.4) \times 10^{-15}\ \mathrm{yr}^{-1}, \tag{25}$$

the frequency measurement of the electric quadrupole transition $5\mathrm{d}^{10}6\mathrm{s}\ ^2\mathrm{S}_{1/2}\ (F=0)$–$5\mathrm{d}^9 6\mathrm{s}^2\ ^2\mathrm{D}_{5/2}\ (F'=2, m'_F=0)$ at $\lambda = 282$ nm in a single laser cooled $^{199}\mathrm{Hg}^+$ ion (data taken during 2000–2006 at NIST (Fortier et al. 2007)):

$$-\frac{\partial}{\partial t}\ln\frac{f_{\mathrm{Hg}}^{(\mathrm{opt})}}{f_{\mathrm{Cs}}^{(\mathrm{HFS})}} = \frac{\partial}{\partial t}\left[\ln\left(\frac{\mu_{\mathrm{Cs}}}{\mu_B}\right) + (2+0.8+3.2)\ln\alpha\right]$$
$$= y + 6x = (-0.37 \pm 0.39) \times 10^{-15}\ \mathrm{yr}^{-1}, \tag{26}$$

and the frequency measurement of the $6\mathrm{s}\ ^2\mathrm{S}_{1/2}(F=0)$–$6\mathrm{s}\ ^2\mathrm{D}_{3/2}(F=3)$ electric quadrupole transition at $\lambda = 436$ nm of a single trapped and laser cooled $^{171}\mathrm{Yb}^+$ ion (data taken during 2000–2006 at PTB (Peik et al. 2004, 2006):

$$-\frac{\partial}{\partial t}\ln\frac{f_{\mathrm{Yb}}^{(\mathrm{opt})}}{f_{\mathrm{Cs}}^{(\mathrm{HFS})}} = \frac{\partial}{\partial t}\left[\ln\left(\frac{\mu_{\mathrm{Cs}}}{\mu_B}\right) + (2+0.8-0.9)\ln\alpha\right]$$
$$= y + 1.9x = (0.78 \pm 1.4) \times 10^{-15}\ \mathrm{yr}^{-1}. \tag{27}$$

Using the experimental data from (25), (26), (27) and expressions (20), (23), (24) stringent restrictions for fractional variations of the fundamental constants can be derived (Fortier et al. 2007):

$$x = \frac{\partial}{\partial t}\ln\alpha = (-0.31 \pm 0.35) \times 10^{-15}\mathrm{yr}^{-1}, \tag{28}$$

$$y = \frac{\partial}{\partial t}\ln\frac{\mu_{\mathrm{Cs}}}{\mu_B} = (1.5 \pm 2.0) \times 10^{-15}\mathrm{yr}^{-1}. \tag{29}$$

This result does not use any assumption on correlation of fundamental constants. It is an important advantage of the method opening the possibility to test some extensions of the grand unification theories where the strong, weak and electromagnetic coupling constants are expected to merge for higher energies. The drifts (if existing) of corresponding coupling constants should be correlated; one can even derive a relation of the drift rates of hadron masses and nuclear $g$-factors that are determined by the strong interaction, and the relative drift rate of the fine structure constant: $\Delta m_p/m_p \approx \Delta g_{\mathrm{nucl}}/g_{\mathrm{nucl}} \approx \pm 35\Delta\alpha/\alpha$ (Calmet and Fritzsch 2002a, 2002b). Of course, the theory can be tested only if a non zero drift rate is detected as the relation holds even if none of the constants is actually drifting.

### *3.1.2 Coupling to Gravity*

Besides setting an upper limit to a variation of $\alpha$, repeated absolute frequency measurements deliver important information about the coupling between gravity and other fundamental interactions. Since 2005 the $^1S_0$–$^3P_0$ clock transition frequency in $^{87}$Sr has been measured relative to the Cs standard at three laboratories in Paris, Boulder and Tokyo with gradually improving accuracy (Blatt et al. 2008). In these experiments Sr atoms are placed at the minima of a periodic optical potential (an "optical lattice") tuned to a selected wavelength (Katori et al. 2003) which prevents the clock transition to be shifted by the optical potential. This generates extremely narrow (down to 2 Hz at 698 nm) optical resonances in a large ensemble of atoms. Results agree at a level of $10^{-15}$ so that this type of optical clocks is among the most accurate.

Besides improvement of the null-result $\dot{\alpha}/\alpha = (-3.3 \pm 3.0) \times 10^{-16}\ \mathrm{yr}^{-1}$, an upper limit for coupling between gravity and other fundamental interactions was set. The Earth moves on an elliptic orbit in a varying solar gravitational potential with a fractional variation of up to $3.3 \times 10^{-10}$. If the coupling between $\alpha$ and the variation of the gravitational potential $\Delta U(t)$ is assumed to be of the form

$$\frac{\delta\alpha}{\alpha} = k_\alpha \frac{\Delta U(t)}{c^2}, \tag{30}$$

where $k_\alpha$ is the coupling constant, Blatt et al. (2008) have set a restriction of

$$k_\alpha = (2.5 \pm 3.1) \times 10^{-6}. \tag{31}$$

The same coupling constant has previously been limited with transitions in dysprosium (see also the beginning of Sect. 3) with approximately half the sensitivity (Ferrell et al. 2007).

Of course, the sensitivity of this type of measurement depends on the fractional variation of the gravitational potential which is rather small for the Sun-Earth system. The idea of performing atomic clock frequency comparisons at larger values of $\Delta U/c^2$ was considered previously within the *SpaceTime* satellite mission (Lämmerzahl and Dittus 2002) in which a fly-by maneuver near Jupiter could increase $\Delta U/c^2$ to $5 \times 10^{-7}$.

### 3.1.3 Direct Comparison of Optical Frequencies

Improving the accuracy of absolute optical frequency measurements one eventually encounters the limit set by the stability and accuracy of the best Cs clocks (see Fig. 1). The accuracy of the currently best state-of-the art fountain clocks is around $5 \times 10^{-16}$ (Bize et al. 2005). As the frequency combs are not limiting at this level (see e.g. Zimmermann et al. 2004) direct comparison of two optical clocks (Sect. 2.2.3) can provide improved data if both of these clocks are more accurate than the Cs clocks would be.

Indeed at NIST (Boulder, USA) two of these clocks are available that are based on optical clock transition frequencies in $Hg^+$ and $Al^+$ (Rosenband et al. 2008; Lorini et al. 2008). The frequency ratio measured with the help of a frequency comb has a relative uncertainty of only $5.2 \times 10^{-17}$ which is an order of magnitude smaller than for any absolute frequency measurement. This breakthrough became possible after implementation of new concepts in probing of clock transitions in cold ions, development of narrow-band lasers and progress in optical frequency transfer. Comparison of the highly relativistic system of $Hg^+$ (the sensitivity to $\alpha$ variation is $L_\alpha^{\mathrm{opt}}(\mathrm{Hg}^+) = -3.2$) with the nearly non-relativistic system of $Al^+$ ($L_\alpha^{\mathrm{opt}}(\mathrm{Al}^+) = +0.008$)) over a time interval of only one year allowed to derive a restriction to the variation of $\alpha$ of

$$\dot{\alpha}/\alpha = (-1.6 \pm 2.3) \times 10^{-17}\ \mathrm{yr}^{-1}. \tag{32}$$

This is the lowest limit obtained yet of any type of measurement and is consistent with zero. On top of that the result is model-independent and opens the possibility to draw conclusions about variations of other fundamental constants when combined with other types of experiments.

## 4 Frequency Combs for Astrophysics

The laser frequency comb turned out to be not only an indispensable element in laboratory optical frequency measurements, but is also useful tool for astronomical observations. It has been demonstrated recently that a frequency comb with a resolvable large mode spacing can be used as an accurate calibration tool for high-resolution spectrographs. This is particular interesting for astrophysical applications (Steinmetz et al. 2008). An accurate frequency axis for spectrometers is required for a number of sensitive fundamental measurements like testing the drift of redshifts of different astrophysical objects (Sandage 1962), the search for extrasolar planets by the reflex Doppler motion (Mayor and Queloz 1995; Marcy and Butler 1996; Lovis et al. 2006) as well as the search for cosmological variations of fundamental constants (Bahcall and Salpeter 1965; Thompson 1975; Webb et al. 1999).

If Doppler shifts on the order of 1 cm $s^{-1}$ ($3 \times 10^{-11}c$) could be measured the presumed acceleration of the cosmic expansion could be verified in real time in a largely model independent way, i.e. without assuming the validity of general relativity (Steinmetz et al. 2008). For a typical high-resolution spectrometer used in astrophysics such a resolution corresponds to a physical size of one silicon atom on a CCD substrate. This indicates that only with the statistics of a very large number of calibration lines the sensitivity can be achieved at the condition that the systematics are under control at the same level.

In 2008 a first implementation of the laser frequency comb as a calibration tool for the German Vacuum Tower Telescope (VTT) (Schröter et al. 1985) has been demonstrated (Steinmetz et al. 2008). The very high resolution of the VTT (0.8 GHz) is still too low to resolve individual modes of the erbium fiber laser frequency comb with $\omega_r = 2\pi \times 250$ MHz used for its calibration. Filtering of the desirable frequency comb modes by an external Fabry-Pérot cavity was suggested as one possible solution (e.g. Li et al. 2008). The Fabry-Pérot cavity used by Steinmetz et al. (2008) has a free spectral range of $m\omega_r$ where the integer $m$ can be set between 4 and 60. This cavity interferometrically suppresses all modes generated by the laser except every $m$th. The resulting well resolved comb is used to illuminate the spectrograph slit.

A CCD image of the fragment of the solar spectrum with atmospheric absorption lines is presented in Fig. 10. The filtered frequency comb radiation was overlaid with the output of the telescope and sent to the VTT spectrometer. The comb modes were stabilized to a Rb atomic clock. One can recognize the resolved filtered comb modes used as frequency markers separated by 15 GHz intervals ($m = 60$) super imposed on the solar spectrum. Even for

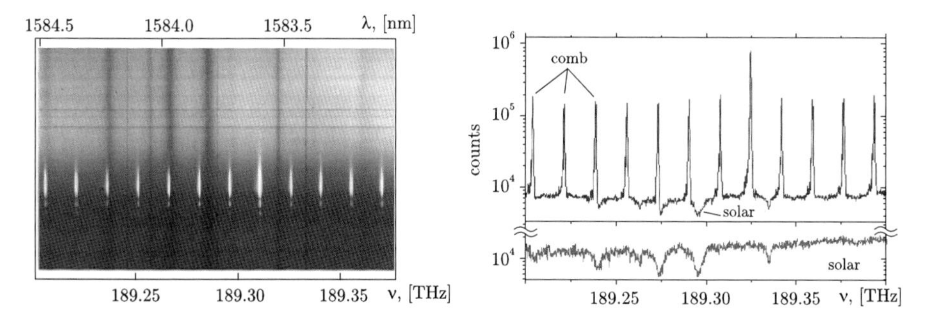


**Fig. 10** (*Left*)—a CCD image of a fragment of the solar photosphere spectrum (*dark Fraunhofer lines*) overlaid with a frequency comb with 15 GHz mode spacing (*bright regular lines*). (*Right*)—horizontal cuts through the CCD image that contain the frequency comb and solar spectrum (*top*) and the solar spectrum only (*bottom*)

**Table 3** Model-independent restrictions of the variation of the fine structure constant $\alpha$ from laboratory measurements. Results of Fischer et al. (2004); Peik et al. (2004, 2006), Blatt et al. (2008) were obtained by absolute optical frequency measurements, the result of Rosenband et al. (2008) was obtained by direct comparison of two optical frequencies with the help of a frequency comb, while in Cingöz et al. (2007) a radio-frequency transition between highly-excited nearly degenerative levels has been measured.

| Year | Atomic samples | $\dot{\alpha}/\alpha$, yr$^{-1}$ | Method | Ref. |
|---|---|---|---|---|
| 2004 | H, $Yb^+$, $Hg^+$ | $(-0.9 \pm 2.9) \times 10^{-15}$ | absolute frequency | Fischer et al. (2004) |
| 2004 | H, $Yb^+$, $Hg^+$ | $(-0.3 \pm 2.0) \times 10^{-15}$ | absolute frequency | Peik et al. (2004) |
| 2006 | $Yb^+$, $Hg^+$ | $(-2.6 \pm 3.9) \times 10^{-16}$ | absolute frequency | Peik et al. (2006) |
| 2007 | Dy | $(-2.7 \pm 2.6) \times 10^{-15}$ | rf transition | Cingöz et al. (2007) |
| 2008 | Sr, H, $Yb^+$, $Hg^+$ | $(-3.3 \pm 3.0) \times 10^{-16}$ | absolute frequency | Blatt et al. (2008) |
| 2008 | $Hg^+$, $Al^+$ | $(-1.6 \pm 2.3) \times 10^{-17}$ | direct comparison | Rosenband et al. (2008) |

this very first demonstration a calibration uncertainty of only 9 m s$^{-1}$ (root mean square) in cosmic velocity units was achieved which compares very well with the uncertainties of traditional calibration techniques. By increasing the number of modes (up to $10^4$) by a spectrally broader frequency comb, it is feasible to reduce the statistical uncertainty to the desirable 1 cm s$^{-1}$ level. The approach also opens possibilities to analyze the systematic uncertainties of the spectrograph and remove their contribution. Implementation of frequency combs for astrophysics allows to observe small variations of spectral lines on a large time scale referenced directly to the SI unit of hertz.

## 5 Conclusions

Table 3 summarizes the results of recent laboratory measurements aiming at the search for a time varying $\alpha$ (see also Fig. 2, right). Combinations with other precision laboratory measurements like comparisons of fountain clocks (see e.g. Marion et al. 2003) or precision molecular spectroscopy (Amy-Klein et al. 2005) deliver important information on the variation of reduced magnetic moments and the electron-to-proton mass ratio. The field is rapidly evolving and we included only a few results most relevant to the topic reviewed.

As seen from the table, repeated frequency measurements in cold atoms, ions and molecules allow to set stringent restrictions on the variation of fundamental constants. At the moment the sensitivity of these methods resides at a level of $\dot{\alpha}/\alpha \sim 10^{-17}$ yr$^{-1}$ which is the lowest model-independent restriction at the present epoch. A further increase in sensitivity is expected due to improvements of frequency measurements, the increasing observation time interval, and the increase of the number of atomic samples under study. On the other hand, frequency combs open perspectives for improving the accuracy of astrophysical observations and potentially push forward the sensitivity of astrophysical tests for a variation of $\alpha$ billions of years ago.

**Acknowledgements** J.A. is supported by EU Marie Curie fellowship. N.K. acknowledges support from MPG and RF Presidential grant MD-887.2008.2 and the Russian Science Support Foundation.

**Open Access** This article is distributed under the terms of the Creative Commons Attribution Noncommercial License which permits any noncommercial use, distribution, and reproduction in any medium, provided the original author(s) and source are credited.

## References

F. Adler et al., Opt. Express **12**, 5872 (2004)
G.P. Agrawal, *Nonlinear Fiber Optics* (Academic Press, New York, 2001)
J. Alnis, A. Matveev, N. Kolachevsky, Th. Udem, T.W. Hänsch, Phys. Rev. A **77**, 1 (2008)
A. Amy-Klein et al., Opt. Lett. **30**, 3320 (2005)
J.N. Bahcall, E.E. Salpeter, Astrophys. J. **142**, 1677 (1965)
F. Biraben et al., in *The Hydrogen Atom. Precision Physics of Simple Atomic Systems*, ed. by S.G. Karshenboim, F.S. Pavone, G.F. Bassani, M. Inguscio, T.W. Hänsch (Springer, Berlin, 2001), p. 18
S. Bize et al., Phys. Rev. Lett. **90**, 150802 (2003)
S. Bize et al., J. Phys. B: At. Mol. Opt. Phys. **38**, S449-68 (2005)
S. Blatt et al., Phys. Rev. Lett. **100**, 140801 (2008)
X. Calmet, H. Fritzsch, Eur. Phys. J. C **24**, 639 (2002b)
X. Calmet, H. Fritzsch, Phys. Lett. B **540**, 173 (2002a)
H.B.G. Casimir, *On the Interaction Between Atomic Nuclei and Electrons* (Freeman, San Francisco, 1963), p. 54
H. Chand, R. Srianand, P. Petitjean, B. Aracil, Astron. Astrophys. **417**, 853 (2004)
A. Cingöz et al., Phys. Rev. Lett. **98**, 040801 (2007)
T. Damour, F. Piazza, G. Veneziano, Phys. Rev. Lett. **89**, 081601 (2002)
S.A. Diddams, L. Hollberg, L.S. Ma, L. Robertsson, Opt. Lett. **27**, 58 (2002)
S.A. Diddams et al., Phys. Rev. Lett. **84**, 5102 (2000)
P.A.M. Dirac, Nature (London) **139**, 323 (1937)
R.W.P. Drever et al., Appl. Phys. **31**, 97 (1983)
V.A. Dzuba, V.V. Flambaum, Phys. Rev. A **61**, 034502 (2000)
V.A. Dzuba, V.V. Flambaum, J.K. Webb, Phys. Rev. A **59**, 230 (1999a)
V.A. Dzuba, V.V. Flambaum, J.K. Webb, Phys. Rev. Lett. **82**, 888 (1999b)
J.N. Eckstein, A.I. Ferguson, T.W. Hänsch, Phys. Rev. Lett. **40**, 847 (1978)
M.I. Eides, H. Grotch, V.A. Shelyuto, Phys. Rep. **342**, 63 (2001)
K.M. Evenson, J.S. Wells, F.R. Petersen, B.L. Danielson, G.W. Day, Appl. Phys. Lett. **22**, 192 (1973)
S.J. Ferrell et al., Phys. Rev. A **76**, 062104 (2007)
M. Fischer et al., Phys. Rev. Lett. **92**, 230802 (2004)
V.V. Flambaum, Int. J. Mod. Phys. A **22**, 4937 (2007)
V.V. Flambaum, E.V. Shuryak, AIP Conf. Proc. **995**, 1 (2008)
T.M. Fortier et al., Phys. Rev. Lett. **98**, 070801 (2007)
R.W. Fox, Proc. SPIE **7099**, 70991R (2008)
H.A. Haus, E.P. İppen, Opt. Lett. **26**, 1654 (2001)
C. Henry, IEEE J. Quantum Electron. **18**, 259 (1982)
L.O. Hocker, A. Javan, D.R. Rao, L. Frenkel, T. Sullivan, Appl. Phys. Lett. **10**, 147 (1967)
R. Holzwarth et al., Phys. Rev. Lett. **85**, 2264 (2000)
R. Holzwarth et al., Laser Phys. **11**, 1100 (2001)
A. Huber, Th. Udem, B. Gross, J. Reichert, M. Kourogi, K. Pachucki, M. Weitz, T.W. Hänsch, Phys. Rev. Lett. **80**, 468 (1998)
S.G. Karshenboim, V.G. Ivanov, Euro. Phys. J. D **19**, 13 (2002b)
S.G. Karshenboim, V.G. Ivanov, Phys. Lett. B **524**, 259 (2002a)
H. Katori, M. Takamoto, V.G. Pal'chikov, V.D. Ovsiannikov, Phys. Rev. Lett. **91**, 173005 (2003)
J.C. Knight et al., Opt. Lett. **21**, 1547 (1996)
N. Kolachevsky, M. Fischer, S.G. Karshenboim, T.W. Hänsch, Phys. Rev. Lett. **92**, 033003 (2004a)
N. Kolachevsky, P. Fendel, S.G. Karshenboim, T.W. Hänsch, Phys. Rev. A **70**, 063503 (2004b)
P. Kubina et al., Opt. Express **13**, 909 (2005)
C. Lämmerzahl, H. Dittus, Ann. Phys. (Leipzig) **11**, 95 (2002)
C.-H. Li et al., Nature **452**, 610 (2008)
L. Lorini et al., Eur. Phys. J. Special Topics **163**, 19 (2008)
C. Lovis et al., Nature **441**, 305 (2006)
A.D. Ludlow et al., Opt. Lett. **32**, 641 (2007)
L.S. Ma et al., Science **303**(5665), 1843 (2004)
G.W. Marcy, R.P. Butler, Astrophys. J. **464**, L147 (1996)
H. Marion et al., Phys. Rev. Lett. **90**, 150801 (2003)
M. Mayor, D. Queloz, Nature **378**, 355 (1995)
R. Minkowski, O.C. Wilson, Astrophys. J. **123**, 373 (1956)
M.T. Murphy, J.K. Webb, V.V. Flambaum, Mon. Not. R. Astron. Soc. **384**, 1053 (2008)

E. Narevicius, A. Libson, C.G. Parthey, I. Chavez, J. Narevicius, U. Even, M.G. Raizen, Phys. Rev. Lett. **100**, 093003 (2008)
T. Nazarova, F. Riehle, U. Sterr, Appl. Phys. B **83**, 531 (2006)
L.E. Nelson et al., Appl. Phys. B **65**, 277 (1997)
M. Niering et al., Phys. Rev. Lett. **84**, 5496 (2000)
M. Notcutt, L.-S. Ma, J. Ye, J.L. Hall, Opt. Lett. **30**, 1815 (2005)
E. Peik et al., Phys. Rev. Lett. **93**, 170801 (2004)
E. Peik et al., in *Proceedings of the 11th Marcel Grossmann Meeting*, Berlin (2006). arXiv:physics/0611088
K. Petermann, IEEE J. Quantum Electron. **15**, 566 (1979)
R. Quast, D. Reimers, S.A. Levshakov, Astron. Astrophys. **415**, 27 (2004)
J. Reichert, R. Holzwarth, Th. Udem, T.W. Hänsch, Opt. Commun. **172**, 59 (1999)
J. Reichert et al., Phys. Rev. Lett. **84**, 3232 (2000)
T. Rosenband et al., Science **319**, 1808 (2008)
A. Sandage, Astrophys. J. **136**, 319 (1962)
G. Santarelli et al., Phys. Rev. Lett. **82**, 4619 (1999)
M.P. Savedoff, Nature **178**, 689 (1956)
A.L. Schawlow, C.H. Townes, Phys. Rev. **112**, 1940 (1958)
H. Schnatz, B. Lipphardt, J. Helmcke, F. Riehle, G. Zinner, Phys. Rev. Lett. **76**, 18 (1996)
E.H. Schröter, D. Soltau, E. Wiehr, Vistas Astron. **28**, 519 (1985)
T. Steinmetz et al., Science **321**, 1335 (2008)
H. Stoehr, F. Mensing, J. Helmcke, U. Sterr, Opt. Lett. **31**, 736 (2006)
T.R. Taylor, G. Veneziano, Phys. Lett. B **213**, 450 (1988)
R.I. Thompson, Astron. Lett. **16**, 3 (1975)
Th. Udem, R. Holzwarth, T.W. Hänsch, Nature **416**, 233 (2002)
Th. Udem, J. Reichert, R. Holzwarth, T.W. Hänsch, Opt. Lett. **24**, 881 (1999)
N. Vanhaecke, U. Meier, M. Andrist, B.H. Meier, F. Merkt, Phys. Rev. A **75**, 031402(R) (2007)
D.A. Varshalovich, A.V. Ivanchik, A.V. Orlov, A.Y. Potekhin, P. Petitjean, Current status of the problem of cosmological variability of fundamental physical constants, in *Lecture Notes in Physics: Precision Physics of Simple Atomic Systems*, ed. by S.G. Karshenboim, V.B. Smirnov (Springer, Berlin, 2003), pp. 199
J.K. Webb, V.V. Flambaum, C.W. Churchill, M.J. Drinkwater, J.D. Barrow, Phys. Rev. Lett. **82**, 884 (1999)
J.K. Webb et al., Phys. Rev. Lett. **87**, 091301 (2001)
M. Weitz, A. Huber, F. Schmidt-Kaler, D. Leibfried, T.W. Hänsch, Phys. Rev. Lett. **72**, 328 (1994)
B.C. Young, F.C. Cruz, W.M. Itano, J.C. Bergquist, Phys. Rev. A **82**, 3799 (1999)
M. Zimmermann, C. Gohle, R. Holzwarth, Th. Udem, T.W. Hänsch, Opt. Lett. **29**, 310 (2004)

Space Sci Rev (2009) 148: 289–300
DOI 10.1007/s11214-009-9520-y

# Constraining Fundamental Constants of Physics with Quasar Absorption Line Systems

**Patrick Petitjean · Raghunathan Srianand · Hum Chand · Alexander Ivanchik · Pasquier Noterdaeme · Neeraj Gupta**

Received: 20 February 2009 / Accepted: 23 April 2009 / Published online: 14 May 2009
© Springer Science+Business Media B.V. 2009

**Abstract** We summarize the attempts by our group and others to derive constraints on variations of fundamental constants over cosmic time using quasar absorption lines. Most upper limits reside in the range 0.5–1.5 $\times 10^{-5}$ at the $3\sigma$ level over a redshift range of approximately 0.5–2.5 for the fine-structure constant, $\alpha$, the proton-to-electron mass ratio, $\mu$ and a combination of the proton gyromagnetic factor and the two previous constants, $g_p(\alpha^2/\mu)^\nu$, for only one claimed variation of $\alpha$. It is therefore very important to perform new measurements to improve the sensitivity of the numerous methods to at least $<0.1 \times 10^{-5}$ which should be possible in the next few years. Future instrumentations on ELTs in the optical and/or ALMA, EVLA and SKA pathfinders in the radio will undoutedly boost this field by allowing to reach much better signal-to-noise ratios at higher spectral resolution and to perform measurements on molecules in the ISM of high redshift galaxies.

**Keywords** Quasars: absorption lines · Physics: Fundamental constants

---

P. Petitjean (✉)
UPMC Paris 06, Institut d'Astrophysique de Paris, UMR7095 CNRS, 98bis Boulevard Arago, 75014, Paris, France
e-mail: petitjean@iap.fr

R. Srianand · P. Noterdaeme
IUCAA, Post Bag 4, Ganesh Khind, Pune 411 007, India

R. Srianand
e-mail: anand@iucaa.ernet.in

H. Chand
ARIES, Manora Peak, Nainital 263129 (Uttarakhand), India
e-mail: hum@aries.ernet.in

A. Ivanchik
Ioffe Physical Technical Institute, St. Petersburg 194021, Russia
e-mail: iav@astro.ioffe.ru

N. Gupta
Australia Telescope National Facility, CSIRO, Epping, NSW 1710, Australia
e-mail: neeraj.gupta@atnf.csiro.au

## 1 Introduction

As most of the successful physical theories rely on the constancy of few fundamental quantities (the speed of light, $c$, the fine-structure constant, $\alpha$, the proton-to-electron mass ratio, $\mu$, etc.), constraining the possible time variations of these fundamental physical quantities is an important step toward understanding the rules of nature. Current laboratory constraints exclude any significant time variation of the dimensionless constants in the low-energy regime. It is not excluded however that they could have varied over cosmological time-scales. Savedoff (1956) first pointed out the possibility of using redshifted atomic lines from distant objects to test the evolution of dimensionless physical constants. The idea is to compare the wavelengths of the same transitions measured in the laboratory on earth and in the remote universe. This basic principle has been first applied to QSO absorption lines by Bahcall et al. (1967). The field has been given tremendous interest recently with the advent of 10 m class telescopes.

For comparison with constraints obtained from laboratory experiments see other papers in this volume and reviews by others e.g. Uzan (2003) or Flambaum (2008).

## 2 The Method

The idea is simply to compare wavelengths of the same transition measured in the remote universe and in the laboratory. As we live in an expanding universe we need at least two transitions that have different sensitivities to the changes in fundamental constants. Ideally, transitions with no sensitivity to a variation of constants are used to measure the redshift and transitions with a large sensitivity to a variation of constants are used to constrain this variation once the redshift is known.

### 2.1 Atomic Data

Modern spectrographs mounted on 10 m class telescopes provide high signal-to-noise ratio data on faint remote quasars at very high spectral resolution (typically $R \sim 50{,}000$). Since for a given spectral resolution, rest-frame measurements at high redshift have a better precision by a factor of 3 to 5 than laboratory determinations (see e.g. Petitjean and Aracil 2004), the method often requires dramatic improvements in laboratory measurements (see references in Murphy et al. 2003 for $\alpha$, Ubachs and Reinhold 2004; Philip et al. 2004; Ivanov et al. 2008 for $\mu$).

Atomic calculations are also needed to determine the sensitivity coefficients that characterize the change in rest wavelength due to a change in a given constant. This has been done for $\mu$ from $H_2$ lines (see Varshalovich and Potekhin 1995) and from molecular lines (e.g. Flambaum and Kozlov 2007) and for $\alpha$ (e.g. Dzuba et al. 2002 and references therein; Kozlov et al. 2008a, 2008b).

### 2.2 Quasar Absorption Line Systems

The measurements are performed using absorption systems seen in the spectra of remote quasars. Figure 1 shows a quasar spectrum obtained after a typical observation of 10 hours with UVES at the European Very Large Telescope. The quasar, at a redshift of $z_{\rm em} = 2.58$, is quite bright and can be observed at high spectral resolution. Its spectrum is characterized by emission lines from the Lyman series of neutral hydrogen (Lyman-$\alpha$, normally at 1215 Å is

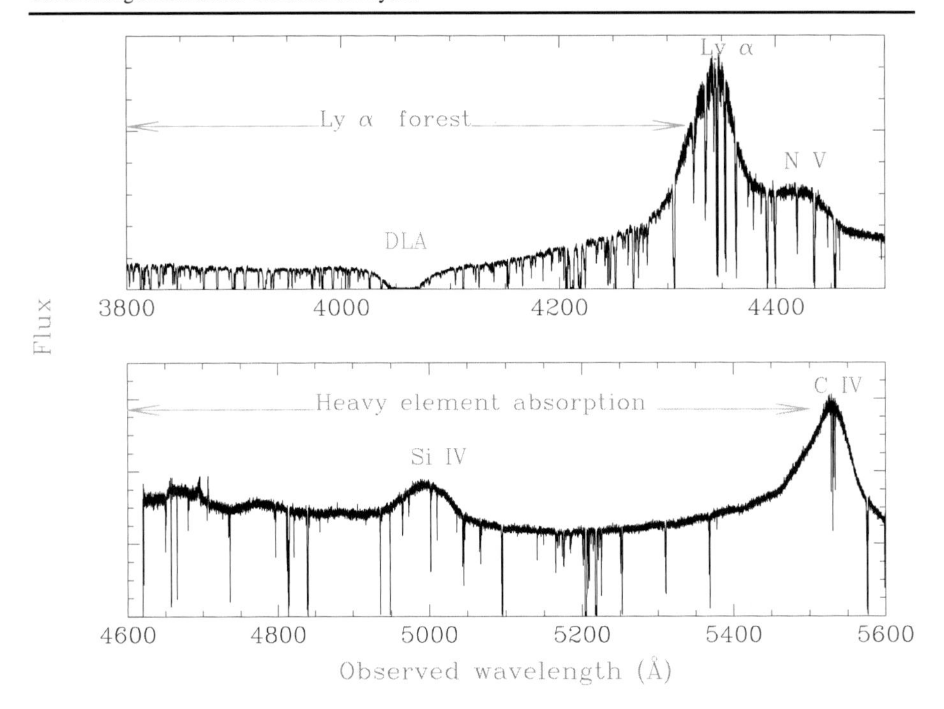


**Fig. 1** UVES-VLT spectrum of a quasar with emission redshift $z_{\rm em} = 2.58$. The quasar is characterised by broad emission (H I Lyman-$\alpha$ at $\lambda \sim 4340$ Å, or C IV$\lambda$1550 at $\lambda \sim 4340$ Å). Below 4350 Å, numerous H I Lyman-$\alpha$ absorption lines are seen that are produced by intergalactic clouds (*narrow lines*) or galactic disks (the so-called "Damped Lyman-$\alpha$ systems or DLAs") located by chance at smaller redshift along the line of sight to the quasar. A DLA system is present along this line of sight at $z_{\rm abs} = 2.33$ (Lyman-$\alpha$ absorption at ~4050 Å). *Metal lines* are seen above 4350 Å

redshifted by a factor $1 + z_{\rm em} = 3.58$ at ~4340 Å) or the resonance transitions from $C^{3+}$ (or C IV) normally at ~1550 Å and seen here at ~5550 Å. It can be seen that numerous absorption lines are superimposed on top of the continuum from the quasar. These absorptions arise when the line of sight crosses by chance a gaseous cloud. At wavelengths smaller than the Lyman-$\alpha$ emission line from the quasar, any tiny amount of neutral hydrogen (corresponding to intergalactic clouds) will produce a narrow Lyman-$\alpha$ absorption line. This region is called the Lyman-$\alpha$ forest. Measurements try to avoid this region of the spectrum as any absorption can be blended with an intervening H I Lyman-$\alpha$ line. This is not possible however for $H_2$ which UV lines are always redshifted in the Lyman-$\alpha$ forest. When the line of sight passes through the halo or disk of a galaxy, a strong Lyman-$\alpha$ absorption is seen together with metal lines. The strongest Lyman-$\alpha$ lines (with column densities $\log N(\text{H I}) > 20.3$) correspond to the so-called damped Lyman-$\alpha$ systems (see Fig. 1).

Small variations in the constants induce small positive or negative shifts in the wavelengths of atomic or molecular species. It must be realized that these shifts are quite small. For a relative variation of $\sim 10^{-5}$ in the fine-structure constant $\alpha$, the typical shift of transitions easily observable is ~0.5 km/s (the situation is slightly better in the radio, see below). This means about 20 mÅ for a redshift of about $z \sim 2$. This corresponds to about a third of a pixel at the spectral resolution of $R \sim 40000$ achieved with UVES-VLT or HIRES-Keck. This kind of measurement is not easy because (i) the number of independent absorption lines used in an individual measurement is not large, typically five or six, and (ii) several sources of uncertainties hamper the measurement.

The dependence of rest wavelengths to the variation of $\alpha$ is parameterized using the fitting function given by Dzuba et al. (1999), $\omega = \omega_0 + q_1 x + q_2 y$. Here $\omega_0$ and $\omega$ are, respectively, the vacuum wave number (in units of $\mathrm{cm}^{-1}$) measured in the laboratory and in the absorption system at redshift $z$. $x$ and $y$ are dimensionless numbers defined as $x = (\alpha_z/\alpha_0)^2 - 1$ and $y = (\alpha_z/\alpha_0)^4 - 1$. The sensitivity coefficients $q_1$ and $q_2$ are obtained using many-body relativistic calculations (see Dzuba et al. 1999).

### 2.3 Source of Errors

The absorption lines have complex profiles because they are the result of the QSO photons travelling through the highly inhomogeneous medium that is associated with the potential wells of cosmological halos. These profiles are usually fitted using a combination of Voigt-profiles. For each component, the exact redshift, the column density and the width of the line (Doppler parameter) are fit parameters to be determined in addition to the shift from a possible variation of constants. These parameters are constrained assuming that the profiles are the same for all transitions. This is obviously true for transitions from the same species (as in Fig. 2) but is not necessarily true in case transitions from different species are used for the same measurement. Indeed, for testing the variations of $\alpha$, transitions from Mg II, Si II and/or Fe II are commonly used. To avoid this problem, one could use transitions from one species only (Quast et al. 2004; Chand et al. 2005; Levshakov et al. 2006) but suitable systems are rare and the sensitivity of the method is reduced.

Another difficulty is that the fit is usually not unique. This is not a severe problem if the lines are not strongly saturated however (as in the bottom panel of Fig. 2) because in that case the positions of the components are well defined. Simulations have shown (Chand et al. 2004) that the presence of strongly saturated lines can increase errors on the determination of $\Delta\alpha/\alpha$ by a factor of two in the case of a simple profile.

Spectra in the optical are taken with high resolution echelle spectrographs that have usually a red and a blue arm. In each arm the spectrum is split into a large number of orders and recorded on different CCDs. The wavelength calibration solution is calculated for each CCD. This can introduce *local* deviations from the correct solution that may be difficult to control. Wavelength calibration is of high importance here and should be carefully controlled (see Thompson et al. 2009 in the case of UVES). A way to overcome the difficulty is to use completely independent instruments to observe the same object. Chand et al. (2005) used the very stable spectrograph HARPS mounted on the 3.6 m telescope at the La Silla observatory to observe the bright quasar HE 0515–4414 that was previously observed with UVES. They find that although the dispersion of the wavelength solution is much smaller for HARPS than for UVES, errors are well constrained except for *local* artifacts. This means that eventhough wavelength calibration is not a concern at the sensitivity level we can expect to achieve, local problems can spoil the measurement for a few systems. Also for some methods, two absorptions lines in very different wavelength bands (e.g. the radio and the UV) are used (see Sect. 5.1) and the intercalibration has to be checked carefully.

Note that temperature must be controlled precisely or registered carefully so that adequate air-vacuum correction can be applied. Flexures in the instrument are dealt with by recording a calibration lamp spectrum before and after the science exposure which is usually 1 hour long. The signal-to-noise ratio of the data is a crucial issue. Simulations by Chand et al. (2004) showed that errors are inversely correlated with SNR. A minimum of SNR $\sim 50$ at the position of the absorption lines is required to achieve measurements at the level considered here.

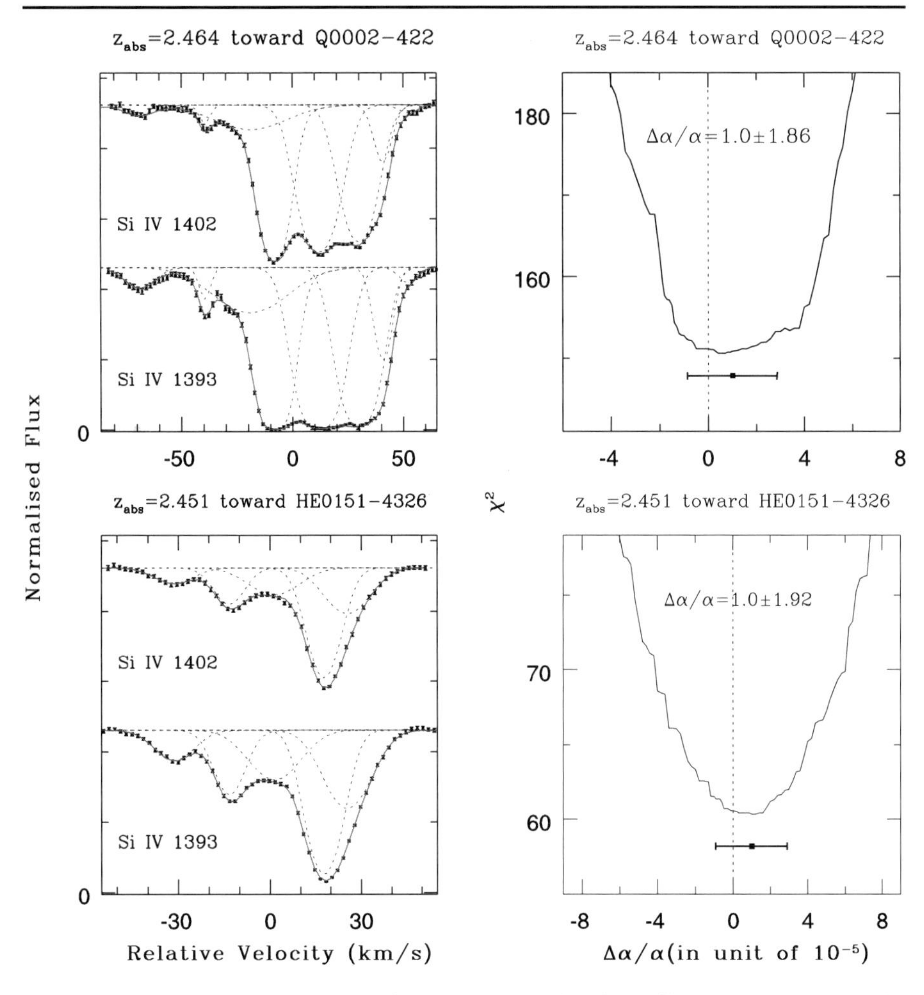


**Fig. 2** Fits of Si IV doublets. Absorption profiles are decomposed in Voigt-profile components (shown as *dotted lines*). Errors are increased in case the profile is strongly saturated. $\chi^2$ curves are shown in the *right-hand panels*. Note that the SNR is particularly good in the present cases

## 3 Variation of the Fine-Structure Constant $\alpha$

### 3.1 The Many-Multiplet Method

The power of the Many-Multiplet Method (MMM) is to use a large number of transitions to constrain the variation of $\alpha$. At least five transitions are used, usually from different species. The transitions are choosen so that their sensitivities to a change in $\alpha$ are different. For example rest wavelengths of Mg II doublets and Mg I are fairly insensitive to small changes in $\alpha$ thereby providing good anchors for measuring the systemic redshift. Whereas the rest wavelengths of Fe II multiplets are very sensitive to small variations in $\alpha$. The accuracy depends on how well the absorption line profiles are modeled. The recent application of the many-multiplet method (Dzuba et al. 1999; Webb et al. 1999) has improved by an order of magnitude the accuracy of the $\Delta\alpha/\alpha$ measurements based on QSO absorption lines (Webb et al. 2001). Analysis of HIRES/Keck data has resulted in the claim of a variation in $\alpha$, $\Delta\alpha/\alpha = (-0.54 \pm 0.12) \times 10^{-5}$, over a redshift range $0.2 < z < 3.7$ (Murphy et al. 2003).

In order to check this result independently, we have applied the MM method to very high quality (SNR $\sim 60-80$, $R \geq 44{,}000$) UVES/VLT data. In view of the numerous systematic errors involved in the MM method, we have carried out detailed simulations to define proper selection criteria to choose suitable absorption systems in order to perform the best analysis (see Chand et al. 2004 for details). Application of these selection criteria to the full sample of 50 Mg II/Fe II systems lead us to restrict the study to 23 Mg II/Fe II systems over a redshift range $0.4 \leq z \leq 2.3$. The weighted mean of the individual measurements from this analysis is a non detecton with a $3\sigma$ upper limit of $\Delta\alpha/\alpha < 0.20 \times 10^{-5}$ (Srianand et al. 2004; Chand et al. 2004).

All further analysis performed with UVES spectra fail to confirm any variation in $\alpha$ (Quast et al. 2004; Levshakov et al. 2005). In particular, Chand et al. (2006) analyse spectra of the bright quasar HE 0515–4414 taken with two different instruments, UVES at the VLT and HARPS at the 3.6 m telescope in La Silla. They show that the results of a non-evolving $\alpha$ reported in the literature based on UVES/VLT data should not be heavily influenced by problems related to wavelength calibration uncertainties and multiple component Voigt profile decomposition. Considering that different procedures can be used, a robust $3\sigma$ limit on the variation of $\alpha$ at $z \sim 1.5$ obtained with UVES data is $\Delta\alpha/\alpha < 0.30 \times 10^{-5}$.

Note that several authors have used the five transitions of Fe II to obtained a limit from absorption lines of only one species in order to be certain that the profile structures are identical (Quast et al. 2004; Chand et al. 2006; Levshakov et al. 2007). Useful atomic data are given by Porsev et al. (2007). The number of systems suitable for such measurements is unfortunately very small and limits achieved are of the same order of magnitude.

It is now of high importance to improve the procedure and to increase the number of measurements in order to decrease this limit to below $10^{-6}$ which is a reasonable goal for present day instrumentation.

### 3.2 The Alkali Doublet (AD) Method

Alkali doublets are conspicuous in astrophysical spectra both in emission (for example the [O III]$\lambda\lambda$4969,5007 doublet) and in absorption (for example the Si IV$\lambda\lambda$1393,1402 or the C IV$\lambda\lambda$1548,1550 doublets). In the later case however atomic data are not well known (e.g. Petitjean and Aracil 2004). The method, although less sensitive than the MM method, has the advantage to use only one species and to be applicable to higher redshifts. Bahcall et al. (1967) were the first of a long list to apply this technique to QSO spectra.

More recently, Murphy et al. (2001) analysed a KECK/HIRES sample of 21 Si IV doublets observed along 8 QSO sight lines and derived $\Delta\alpha/\alpha < 3.9 \times 10^{-5}$. The analysis of 15 Si IV doublets selected from a ESO-UVES sample yielded the strongest constraint obtained with this method: $\Delta\alpha/\alpha < 1.3 \times 10^{-5}$ over the redshift range $1.59 \leq z \leq 2.92$ (Chand et al. 2005).

The AD method can be applied to emission as well as absorption lines. However emission lines are usually broad as compared to absorption lines. Errors are therefore larger on individual measurements and must be beaten by large statistics. As a result, the constraints obtained from emission lines are not as strong as those derived from absorption lines. Bahcall et al. (2004) have recently found $\Delta\alpha/\alpha < 4.2 \times 10^{-4}$ using O III emission lines from SDSS QSOs.

## 4 Variation of the Proton-to-Electron Mass Ratio $\mu$

In the framework of unified theories (e.g. SUSY GUT) with a common origin of the gauge fields, variations of the gauge coupling $\alpha_{\rm GUT}$ at the unified scale ($\sim 10^{16}$ GeV) will induce variations of all the gauge couplings in the low energy limit, $\alpha_{\rm i} = f_{\rm i}(\alpha_{\rm GUT}, E)$, and provide a relation $\Delta\mu/\mu \simeq R\Delta\alpha/\alpha$, where $R$ is a model dependent parameter and $|R| \leq 50$ (e.g. Dine et al. 2003, and references therein). Thus, independent estimates of $\Delta\alpha/\alpha$ and $\Delta\mu/\mu$ could constrain the mass formation mechanisms in the context of unified theories.

On earth, the proton-to-electron mass ratio has been measured with a relative accuracy of $2 \times 10^{-9}$ and equals $\mu_0 = 1836.15267261(85)$. Laboratory metrological measurements rule out considerable variation of $\mu$ on a short time scale but do not exclude its changes over the cosmological scale, $\sim 10^{10}$ years. Moreover, one can not reject the possibility that $\mu$ (as well as other constants) could be different in widely separated regions of the Universe.

### 4.1 $H_2$

The method using $H_2$ transitions to constrain the possible variations of $\mu$ was proposed by Varshalovich and Levshakov (1993). It is based on the fact that wavelengths of electron-vibro-rotational lines depend on the reduced mass of the molecule, with the dependence being different for different transitions. It enables us to distinguish the cosmological redshift of a line from the shift caused by a possible variation of $\mu$.

Thus, the measured wavelength $\lambda_{\rm i}$ of a line formed in the absorption system at the redshift $z_{\rm abs}$ can be written as, $\lambda_{\rm i} = \lambda_{\rm i}^0(1 + z_{\rm abs})(1 + K_{\rm i}\Delta\mu/\mu)$, where $\lambda_{\rm i}^0$ is the laboratory (vacuum) wavelength of the transition, and $K_{\rm i} = {\rm d}\ln\lambda_i^0/{\rm d}\ln\mu$ is the sensitivity coefficient for the Lyman and Werner bands of molecular hydrogen (Varshalovich and Potekhin 1995). This expression can be represented in terms of the individual line redshift $z_{\rm i} \equiv \lambda_{\rm i}/\lambda_{\rm i}^0 - 1$ as, $z_{\rm i} = z_{\rm abs} + bK_{\rm i}$, where $b = (1 + z_{\rm abs})\Delta\mu/\mu$.

In reality, $z_{\rm i}$ is measured with some uncertainty which is caused by statistical errors of the astronomical measurements, $\lambda_{\rm i}$, and by errors of the laboratory measurements of $\lambda_{\rm i}^0$. Nevertheless, if $\Delta\mu/\mu$ is nonzero, there must be a correlation between $z_{\rm i}$ and $K_{\rm i}$ values. Thus, a linear regression analysis of these quantities yields $z_{\rm abs}$ and $b$ (as well as its statistical significance), consequently an estimate of $\Delta\mu/\mu$.

Several studies have yielded tight upper limits on $\mu$-variations, $|\Delta\mu/\mu| < 7 \times 10^{-4}$ (Cowie and Songaila 1995), $|\Delta\mu/\mu| < 2 \times 10^{-4}$ (Potekhin et al. 1998), $|\Delta\mu/\mu| < 5.7 \times 10^{-5}$ (Levshakov et al. 2002) and $\Delta\mu/\mu < 7 \times 10^{-5}$ (Ivanchik et al. 2003). Recently, a new limit was estimated, $\Delta\mu/\mu < 2.2 \times 10^{-5}$ at the $3\sigma$ level, by measuring wavelengths of 76 $H_2$ lines of Lyman and Werner bands from two absorption systems at $z_{\rm abs} = 2.5947$ and 3.0249 in the spectra of quasars Q 0405–443 and Q 0347–383, respectively. Data were of the highest spectral resolution ($R = 53000$) and $S/N$ ratio (30–70) for these kind of studies (Ivanchik et al. 2005).

This result is subject to important systematic errors of two kinds: (i) using different sets of laboratory wavelengths yield different results; (ii) the molecular lines are located in the Lyman-$\alpha$ forest where they can be strongly blended with intervening H I Lyman-$\alpha$ absorption lines. The first type of systematics are addressed by new laboratory measurements (Philip et al. 2004; Reinhold et al. 2006). The second type of systematics needs careful fitting of the Lyman-$\alpha$ forest. This has been performed recently by King et al. (2008). These authors use principally the same set of data as above and derive $\Delta\mu/\mu < 1.2. \times 10^{-5}$ at the $3\sigma$ level.

### 4.2 HD

The detection of several HD transitions makes it possible to test the possible time variation of the proton-to-electron mass ratio, in the same way as with $H_2$ but in a completely independent way. As these measurements may involve various unknown systematics, it is important to use different sets of lines and different techniques. Sensitivity coefficients and accurate wavelengths for HD transitions have been published very recently (Ivanov et al. 2008). It must be noted however that till now $H_2$ is detected in absorption in only 14 Damped Lyman-$\alpha$ systems whereas HD is detected in only two places in the whole universe.

Deuterated molecular hydrogen was detected very recently together with carbon monoxide (CO; Srianand et al. 2008) and $H_2$ in a Damped Lyman-$\alpha$ cloud at $z_{\rm abs} = 2.418$ toward the quasar SDSS1439 + 11. Five lines of HD in three components were detected together with more than a hundred $H_2$ transitions in seven components. Although each HD component is associated to one of the $H_2$ components, the strong blending of the latter, especially in low rotational levels, does not allow for the exact determination of the relative positions of the HD and $H_2$ components. In passing, the column densities integrated over the whole profile for both HD and $H_2$ yield $N(\rm HD)/2N(\rm H_2) = 1.5^{+0.6}_{-0.4} \times 10^{-5}$ (Noterdaeme et al. 2008). Five HD absorption lines (L3-0 R0, L5-0 R0, L7-0 R0, L8-0 R0 and W0-0 R0) are clearly detected and were fitted simultaneously. The $3\sigma$ limit reached here is $\Delta\mu/\mu < 9 \times 10^{-5}$.

Although the number of available lines and the signal-to-noise ratio do not allow to reach the level of accuracy achieved with $H_2$, it is important to pursue in this direction and to measure $\Delta\mu/\mu$ also from HD lines. This is very important given the scarcity of possible independent measurements.

### 4.3 $NH_3$

Recently, Flambaum and Kozlov (2007) showed that the high sensitivity of the $NH_3$ inversion transitions to a change in $\mu$ could be used to constrain the variations of this constant. The only intermediate redshift system where this molecule is detected is the $z = 0.685$ lens toward B 0218 + 357 (Combes and Wiklind 1995; Henkel et al. 2005). They obtain a $3\sigma$ upper limit on the variation of $\mu$ at this redshift of $6 \times 10^{-6}$. Murphy et al. (2008) refined this limit to $2 \times 10^{-6}$. This technique has been applied by Levshakov et al. (2008) to $NH_3$ and other nitrogen rich molecules observed in our Galaxy with a sensitivity reaching $10^{-7}$–$10^{-8}$.

## 5 Combinations of Constants

### 5.1 The 21 cm Absorbers

As the energy of the hyperfine H I 21-cm transition is proportional to the combination of three fundamental constants, $x = \alpha^2 g_{\rm p}/\mu$, high resolution optical and 21-cm spectra can be used together to probe the combined cosmological variation of these constants (Tubbs and Wolfe 1980). In the definition of $x$, $\alpha$ is the fine-structure constant, $\mu$ is the proton-to-electron mass ratio and $g_{\rm p}$ is the gyromagnetic factor (dimensionless) of the proton (see e.g. Tzanavaris et al. 2005).

To apply this technique, the redshift of the 21 cm line must be compared to that of UV lines of Si II, Fe II and/or Mg II. Two difficulties arise: (i) the radio and optical sources must coincide: as QSOs in the optical can be considered as pointlike sources, it must be checked on VLBI maps that the corresponding radio source is also pointlike which is not true for all quasars; (ii) the gas at the origin of the 21 cm and UV absorptions must be co-spatial: it is likely to be the case if the lines are narrow. Therefore systems in which the measurement can be performed must be selected carefully. Since the overall number of suitable systems is very small, they must be searched for.

For this reason we have embarked on a large survey to search for 21 cm absorbers at intermediate and high redshifts. For this we first selected strong Mg II systems ($W_r > 1$ Å) from the Sloan Digital Sky Survey in the redshift range suitable for a follow-up with the Giant Meterwave Telescope (GMRT), $1.10 < z_{abs} < 1.45$. We then cross-correlated the ~3000 SDSS systems we found with the FIRST radio survey to select the background sources having at least a $S_{1.4\ GHz} > 50$ mJy bright component coincident with the optical QSO. There are only 63 sources fulfilling these criteria out of which we observed 35 over ~400 hours of GMRT observing time.

We detected 9 new 21 cm absorption systems. This is by far the largest number of 21-cm detections from any single survey. Prior to our survey no intervening 21-cm system was known in the above redshift range and only one system was known in the redshift range $0.7 \leq z \leq 1.5$. Our GMRT survey thus provides systems in a narrow redshift range where variations of $x$ can be constrained. For this, high resolution and high signal-to-noise ratio observations of the absorbers must be performed to detect the UV absorption lines that will provide the anchor to fix the exact redshift, the variations of $x$ being then constrained by the position of the 21 cm absorption line.

These UV observations exist for one of the system at $z_{abs} = 1.3608$ (see Fig. 3). Although the UV data could be even better, a preliminary constraint was obtained: $\Delta x/x < 10^{-5}$ at the $3\sigma$ level.

### 5.2 Other Molecules: CO, OH, $NH_3$, $HCO^+$

Other molecules can be used to derive strong constraints on fundamental constants. Wiklind and Combes (1999) noticed that a potential application of the observation of radio molecular absorption lines at high redshift is to check the invariance of constants. Radio lines are well suited for this purpose because spectral resolution better than 1 km s$^{-1}$ can be achieved in the radio wavelength range. By comparing the redshift of a molecular transition to that of the 21 cm hyperfine H I line, it is possible to constrain a combination $\alpha^2 g_p(M_{red}/m_p)$ of the proton gyromagnetic ratio $g_p$, $\alpha$ and the ratio of the reduced mass of the molecule to the proton mass. They already put a $3\sigma$ limit of $10^{-5}$ on the variations of the above coefficient at $z_{abs} = 0.25$ and 0.68 using radio transitions from CO and $HCO^+$. A strong limitation of this technique is that different absorption lines may probe different volumes along the line of sight. This is also true when comparing several CO lines as the opacity depends on the excitation conditions at each point of the cloud. Note that this is also true for all techniques using absorption lines from different species. In addition, the systems where the test can be performed are again, at the moment, very rare.

Similarly, beautiful observations of conjugate absorption and emission OH lines have been performed recently. Kanekar et al. (2005) have detected the four 18 cm OH lines from the $z_{abs} = 0.765$ gravitational lens toward PMN J0134–0931 with the 1612 and 1720 MHz lines in conjugate absorption and emission (see also Kanekar and Chengalur 2004). They compare the H I and OH absorption redshifts of the different components in this system and also in the absorber arising from the $z = 0.685$ lens toward B 0218 + 357 to place stringent constraints on changes in $F = g_p(\alpha^2/\mu)^{1.57}$. They obtain $\Delta F/F < 4 \times 10^{-5}$.

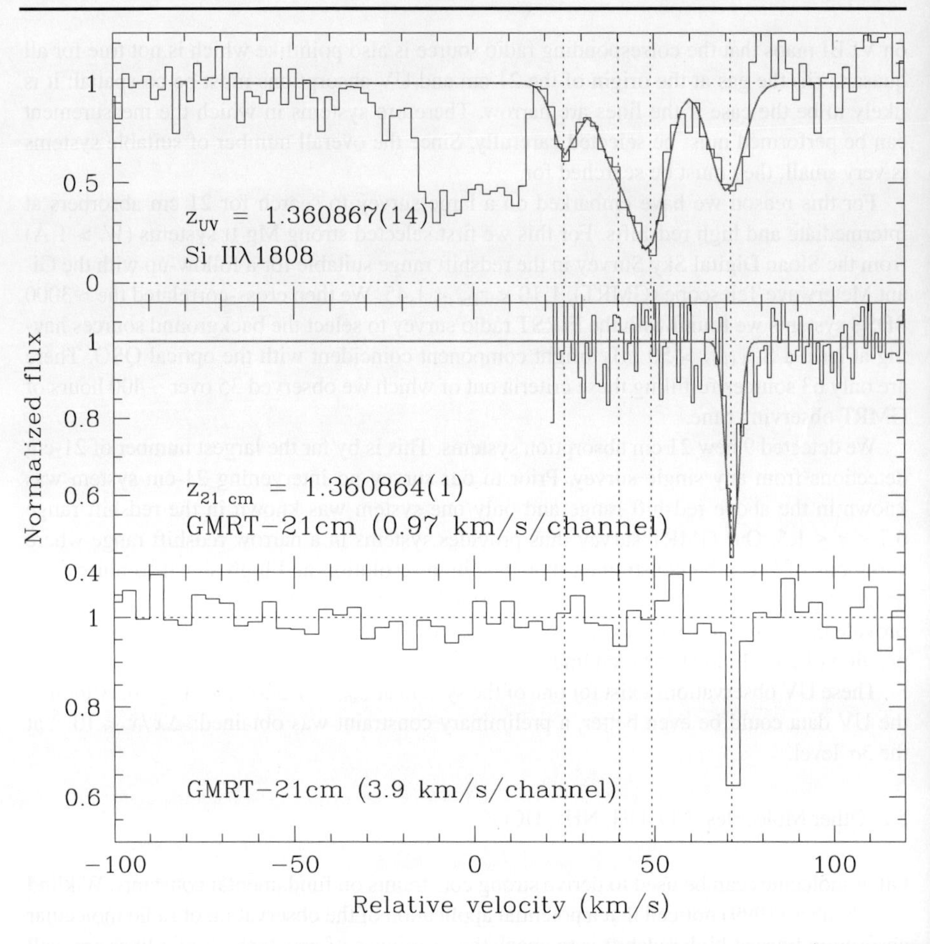


**Fig. 3** Detection of 21 cm absorption in a cloud at $z_{abs} = 1.3608$. The 21 cm component is very narrow and is associated with a UV component well detached from the bulk of the Si II absorption profile. A comparison of the positions of these two transitions sets strong constraints on the parameter $\alpha^2 g_p/\mu$. It is apparent that the errors will come from the determination of the position of the UV Si II absorption line

## 6 Conclusion

Results are summarized in Table 1. Column #1 gives the constant under study (see definitions in the Text above), #2 is for the method used, #3 indicates the redshift at or the redshift range over which the measurement is performed, #4 gives the constraints or measurement obtained and #5 gives partial references (other references are given in the Text above). It is apparent that constraints reside in the range ∼0.3–1.5 × $10^{-5}$ at the 3$\sigma$ level over a redshift range of approximately 0.5–2.5. Note that Reinhold et al. (2006) do not claim detection. They are cautious enough to state that systematics dominate the measurements. Indeed, data and observed wavelength determinations are the same as those in Ivanchik et al. (2005). The only claimed detection of varying $\alpha$ is from Murphy et al. (2003). It is therefore very important to increase the number of measurements and to improve the measurements themselves (e.g. Thompson et al. 2009) to reach the sensitivity level of at least <0.1 × $10^{-5}$ which should be possible in the next few years. Future instrumentation on ELTs will undoutedly

**Table 1** Constraints on the cosmological variations of fundamental constants

| | Method | Redshift | Constraint ($10^{-5}$)* | References** |
|---|---|---|---|---|
| $\alpha$ | MMM | 0.2–3.7 | $-0.54 \pm 0.12$ | Murphy et al. (2003) |
| | | 0.5–2.5 | <0.30 | Srianand et al. (2004) |
| | FeII | 1.515 | <0.45 | Quast et al. (2004), Levshakov et al. (2005) |
| | AD (SiIV) | 1.6–2.8 | <1.3 | Chand et al. (2005) |
| | AD ([OIII]) | 0.15–0.8 | <42 | Bahcall et al. (2004) |
| $\mu$ | $H_2$ | 2.595, 3.025 | <2.1 | Ivanchik et al. (2005), Reinhold et al. (2006) |
| | | 2.595, 3.025 | <1.2 | King et al. (2008) |
| | HD | 2.418 | <9 | Noterdaeme et al. (2008) |
| | $NH_3$ | 0.685 | <0.27 | Flambaum and Kozlov (2007), Murphy et al. (2008) |
| $g_p(\alpha^2/\mu)^\nu$ | 21 cm | 1.361 | <1.0 | Srianand et al. (2009) |
| | OH | 0.685 | <4.0 | Kanekar et al. (2005) |
| | $NH_3$ | 0.685 | <5.0 | Flambaum and Kozlov (2007) |
| | CO, $HCO^+$ | 0.25, 0.685 | <1.0 | Wiklind and Combes (1999) |

*$3\sigma$ for upper limits

**See Text for other references

boost this field by allowing to reach much better signal-to-noise ratios at higher spectral resolution (e.g. Liske et al. 2008). As discussed above, the strongest constraints may come from radio observations of molecules in the ISM of high redshift galaxies. A new area will be opened in this field by the upcoming facilities such as ALMA and EVLA for high-redshift molecular studies and SKA pathfinders for 21 cm and OH surveys.

**Acknowledgements** This work is based on observations collected during several observing programmes at the European Southern Observatory with the Ultra-violet and Visible Echelle Spectrograph mounted on the 8.2 m KUEYEN telescope operated at the Paranal Observatory, Chile and at the Giant Meter-wave Radio Telescope in India. We gratefully acknowledge support from the Indo-French Centre for the Promotion of Advanced Research (Centre Franco-Indien pour la Promotion de la Recherche Avancée). P.N. was supported by an ESO Ph.D. fellowship.

## References

J.N. Bahcall, W.L.W. Sargent, M. Schmidt, Astrophys. J. **149**, L11 (1967)
J.N. Bahcall, C.L. Steinhardt, D. Schlegel, Astrophys. J. **600**, 520 (2004)
H. Chand, R. Srianand, P. Petitjean et al., Astron. Astrophys. **417**, 853 (2004)
H. Chand, P. Petitjean, R. Srianand et al., Astron. Astrophys. **430**, 47 (2005)
H. Chand, R. Srianand, P. Petitjean et al., Astron. Astrophys. **451**, 45 (2006)
F. Combes, T. Wiklind, Astron. Astrophys. **303**, L61 (1995)
L.L. Cowie, A. Songaila, Astrophys. J. **453**, 596 (1995)
M. Dine, Y. Nir, G. Raz, T. Volansky, Phys. Rev. D **67**, 015009 (2003)
V.A. Dzuba, V.V. Flambaum, J.K. Webb, Phys. Rev. A **59**, 230 (1999)
V.A. Dzuba, V.V. Flambaum, M.G. Kozlov, F. Marchenko, Phys. Rev. A **66**, 022501 (2002)
V.V. Flambaum, Eur. Phys. J. **163**, 159 (2008)
V.V. Flambaum, M.G. Kozlov, Phys. Rev. Lett. **98**, 240801 (2007)

C. Henkel, N. Jethava, A. Kraus et al., Astron. Astrophys. **440**, 893 (2005)
A. Ivanchik, P. Petitjean, E. Rodriguez, D. Varshalovich, Astrophys. Space Sci. **283**, 583 (2003)
A. Ivanchik, P. Petitjean, D. Varshalovich et al., Astron. Astrophys. **440**, 45 (2005)
T.L. Ivanov, M. Roudjane, M.O. Vieitez et al., Phys. Rev. Lett. **100**, 093007 (2008)
N. Kanekar, J.N. Chengalur, Mon. Not. R. Astron. Soc. **350**, L17 (2004)
N. Kanekar, C.L. Carilli, G.I. Langston et al., Phys. Rev. Lett. **95**, 261301 (2005)
J.A. King, J.K. Webb, M.T. Murphy, R.F. Carswell, Phys. Rev. Lett. **101**, 251304 (2008)
M.G. Kozlov, I.I. Tupitsyn, D. Reimers (2008a). arXiv:0812.3210
M.G. Kozlov, S.G. Porsev, S.A. Levshakov, D. Reimers, P. Molaro, Phys. Rev. A **77**, 032119 (2008b)
S. Levshakov, M. Dessauges-Zavadsky, S. D'Odorico, P. Molaro, Mon. Not. R. Astron. Soc. **333**, 373 (2002)
S.A. Levshakov, M. Centurión, P. Molaro et al., Astron. Astrophys. **434**, 827 (2005)
S.A. Levshakov, M. Centurión, P. Molaro et al., Astron. Astrophys. **449**, 879 (2006)
S.A. Levshakov, P. Molaro, S. Lopez et al., Astron. Astrophys. **466**, 1077 (2007)
S.A. Levshakov, D. Reimers, M.G. Kozlov, S.G. Porsev, P. Molaro, Astron. Astrophys. **479**, 719 (2008)
J. Liske, L. Pasquini, P. Bonifacio et al. (2008). arXiv:0802.1926
M.T. Murphy, J.K. Webb, V.V. Flambaum et al., Mon. Not. R. Astron. Soc. **327**, 1237 (2001)
M.T. Murphy, J.K. Webb, V.V. Flambaum, Mon. Not. R. Astron. Soc. **345**, 609 (2003)
M.T. Murphy, V.V. Flambaum, S. Muller, C. Henkel, Science **320**, 1611 (2008)
P. Noterdaeme, P. Petitjean, C. Ledoux et al., Astron. Astrophys. **491**, 397 (2008)
P. Petitjean, B. Aracil, Astron. Astrophys. **422**, 523 (2004)
J. Philip, J.P. Sprengers, Th. Pielage et al., Can. J. Chem. **82**, 713 (2004)
S.G. Porsev, K.V. Koshelev, I.I. Tupitsyn et al., Phys. Rev. A **76**, 052507 (2007)
A. Potekhin, A. Ivanchik, D. Varshalovich et al., Astrophys. J. **505**, 523 (1998)
R. Quast, D. Reimers, S.A. Levshakov, Astron. Astrophys. **415**, L7 (2004)
E. Reinhold, R. Buning, U. Hollenstein, A. Ivanchik, P. Petitjean, W. Ubachs, Phys. Rev. Lett. **96**, 151101 (2006)
S. Savedoff, Nature **178**, 688 (1956)
R. Srianand, H. Chand, P. Petitjean, B. Aracil, Phys. Rev. Lett. **92**, 121302 (2004)
R. Srianand, P. Noterdaeme, C. Ledoux, P. Petitjean, Astron. Astrophys. **482**, L39 (2008)
R. Srianand et al. (2009, in preparation)
R.I. Thompson, J. Bechtold, J.H. Black, C.J.A.P. Martins, New Ast. **14**, 379 (2009)
A.D. Tubbs, A.M. Wolfe, Astrophys. J. **236**, L105 (1980)
P. Tzanavaris, J.K. Webb, M.T. Murphy, V.V. Flambaum, S.J. Curran, Phys. Rev. Lett. **95**, 041301 (2005)
W. Ubachs, E. Reinhold, Phys. Rev. Lett. **92**, 101302 (2004)
J.P. Uzan, Rev. Mod. Phys. **75**, 403 (2003)
D. Varshalovich, S. Levshakov, JETP Lett. **58**, 231 (1993)
D. Varshalovich, A. Potekhin, Space Sci. Rev. **74**, 259 (1995)
J.K. Webb, V.V. Flambaum, C.W. Churchill et al., Phys. Rev. Lett. **87**, 884 (1999)
J.K. Webb, M.T. Murphy, V.V. Flambaum et al., Phys. Rev. Lett. **87**, 091301 (2001)
T. Wiklind, F. Combes, in *IAU Colloquium 183*. Kyoto, August 18–22, 1997, p. 167 (1999)

Space Sci Rev (2009) 148: 301–313
DOI 10.1007/s11214-009-9500-2

# The Role of Dark Matter and Dark Energy in Cosmological Models: Theoretical Overview

**A.F. Zakharov · S. Capozziello · F. De Paolis · G. Ingrosso · A.A. Nucita**

Received: 17 February 2009 / Accepted: 24 March 2009 / Published online: 7 April 2009
© Springer Science+Business Media B.V. 2009

**Abstract** Concepts of dark matter (DM) and dark energy (DE) are introduced. As for other anomalies we describe two ways to solve DM and DE problems, namely a conservative way when we have to find substances with DM and DE properties or we have to change a fundamental gravity law. We discuss constraints on DM concentration near the Galactic Center from apocenter shift data. We note that Solar system data put stringent constraints on alternative theories of gravity.

**Keywords** General relativity and gravitation · Cosmology · Observational cosmology · Dark matter · Λ-term · Dark energy · Cosmological tests

A.F. Zakharov
National Astronomical Observatories of Chinese Academy of Sciences, Beijing 100012, China

A.F. Zakharov (✉)
Institute of Theoretical and Experimental Physics, B. Cheremushkinskaya 25, 117259 Moscow, Russia
e-mail: zakharov@itep.ru

S. Capozziello
Dipartimento di Scienze Fisiche, Universitá "Federico II" di Napoli, 80126 Naples, Italy
e-mail: capozziello@na.infn.it

S. Capozziello
Istituto Nazionale di Fisica Nucleare, Sezione di Napoli, 80126 Naples, Italy

F. De Paolis · G. Ingrosso
Dipartimento di Fisica, Università del Salento and INFN Sezione di Lecce, CP 193, 73100 Lecce, Italy

F. De Paolis
e-mail: depaolis@le.infn.it

G. Ingrosso
e-mail: ingrosso@le.infn.it

A.A. Nucita
XMM-Newton Science Operations Centre, ESAC, ESA, PO Box 50727, 28080 Madrid, Spain
e-mail: nucita@le.infn.it

## 1 Introduction: Urbain Jean Joseph Le Verrier: Neptune Discovery (1846) & Mercury's Anomaly (1859) (an Invisible ("Dark") Object or a Violation of the Newtonian Gravity Law)

A connection between cosmology and the Le Verrier's discoveries of anomalies was noted out by Juszkiewicz (2008). In 1846, analyzing trajectories of known objects (planets) and reconstructing potentials and masses and trajectories all objects in the game, Le Verrier (1846) predicted the existence of an extra (initially unknown (dark)) planet, Neptune and soon afterwards the planet was detected by the German astronomer J.G. Galle (1846).[1]

Le Verrier (1846) discovered the Mercury pericenter anomaly and explained 93% of the observed value, but a supplementary advance of 38 arcseconds/century was without explanation (later on, the value was corrected to 43 arcseconds/century).

Le Verrier (1859) had analyzed the following options in order to explain the anomaly.

- the gravitational field of an invisible matter (planet, asteroids near Sun), implying the introduction of a new object;
- the deviation from the Newtonian law, implying the change of a law of Nature;
- or the lack in precision, implying that the model has to be clarified.

Le Verrier (1876) analyzed the information from 25 transits of Vulcan (according to his opinion 19 transits were reliable) and predicted a further transit in March 1877 (the planet was not observed).[2]

Similarly as in todays cosmological DM and DE problems, different options were considered such as the existence of an extra planet between the Sun and Mercury (the Vulcan prediction), the modification mass of Venus by more than 10% and modification of Newton gravity law (for example, such as Newcomb's (1895) modification of the Newton's laws such as $1/r^n$ ($n = 2.0000001574$ for $d\tilde{\omega}/century = 42.34''$, earlier Hall (1894) used $n = 2.00000016$ for $d\tilde{\omega}/century = 43''$). A solution about a modification of the Newton gravity law to explain anomalies such as the Mercury anomaly was given by I. Newton (1686) in his Principia, where he considered a generalization of the gravitational force

$$F = \frac{br^p + cr^m}{r^3}, \tag{1}$$

and he obtained

$$d\tilde{\omega} = 2\pi\sqrt{\left|\frac{b-c}{mb-pc}\right|}. \tag{2}$$

For $p = 1$ we have $F = (br + cr^m)/r^3$, so, if $m \geq 2$, we have generalizations of the Newtonian force with an extra term. Thus, for the specific case $c = 0$, $F = br^{m-3}$ we obtain $d\tilde{\omega} = 2\pi\sqrt{|\frac{b}{mb}|} = 2\pi\sqrt{\frac{1}{m}}$, for $m = 3 - n$ we have $d\tilde{\omega} = 2\pi\sqrt{\frac{1}{3-n}}$ and if $n = 2 + \delta$, then $d\tilde{\omega} = 2\pi\sqrt{\frac{1}{1-\delta}} \approx 2\pi(1 + \delta/2)$.

Therefore, following Le Verrier's way and analyzing carefully trajectories of celestial bodies we can reconstruct the gravitational potentials and mass distributions governing the motions of celestial bodies as it will be shown for the Galactic Center.

[1] Oort (1932) and Zwicky (1933, 1937) used the same scheme leading to the introduction of dark matter (DM) concept, when he discovered high velocity dispersions he concluded about an existence of dark matter rather than a violation a fundamental gravity law at large distances.

[2] Le Verrier died on 23 September 1877.

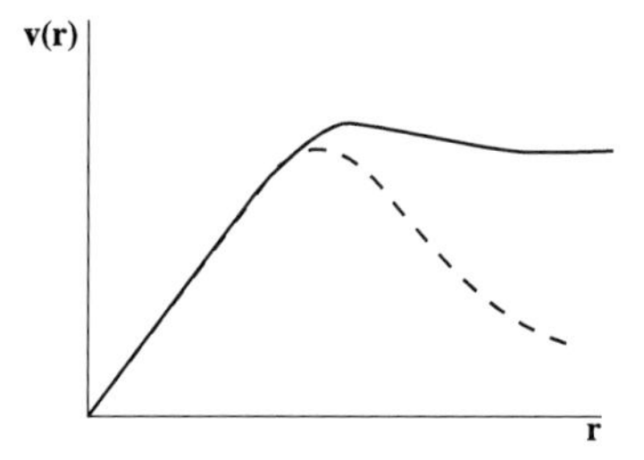


**Fig. 1** The sketch of a typical rotation curve for a spiral galaxy. *Solid line* is an observable rotation curve while *dashed line* corresponds to a theoretical rotation curve if one takes into account only a visible matter

## 2 Dark Matter in the Galactic Clusters and in Spiral Galaxies

Oort (1932) (for the Galaxy) and Zwicky (1933, 1937) (for the Coma cluster) found that an evaluated gravitational potential is different from a potential reconstructed from visible matter. Like Le Verrier for the Mercury anomaly, Oort (1932) used a conservative approach assuming that a gravity law has to be Newtonian one and one has to introduce an invisible substance which was started to be called after Oort (1932) Dark Matter (DM). This claim has been supported by studies of rotation curves for spiral galaxies (Rubin and Ford 1970; Rubin et al. 1977, 1978, 1980) (see Fig. 1). From an inspection of the figure, one concludes that the velocities grow linearly with radius, after that the velocities are roughly constant (or in other words, rotation curves are flat). From the virial theorem for a steady state configuration and the Newtonian potential we have (Goldstein 1980)

$$2T + V = 0, \tag{3}$$

where $T$ is the average kinetic energy, while $V$ is the average potential energy or

$$mV^2 = \frac{GmM(R)}{R}. \tag{4}$$

If we wish to fit rotation curves with spherical symmetrical distributions, assuming $\rho = const$ (thus, we have $M(R) \propto R^3$ and $V \propto R$), for the flat part of rotation curves we suggest $\rho = const/R^2$ (therefore, we have $M(R) \propto R$ and $V = const$).

Again like for the Mercury anomaly we have the dilemma that we have to introduce a new substance (DM) (a conservative solution) or to change the fundamental gravity law (a more radical solution). Initially, people prefer the conservative solution and they try to find DM in different ways but at the moment we can not exclude the possibility that a modification of the law of gravity is needed such as MOND, MOG (Milgrom 1983; Bekenstein 2004; Brownstein and Moffat 2006; Blanchet and Le Tiece 2008) or $f(R)$ gravity (see below).

### 2.1 Different Components of Matter

Counting the luminous sources yields for the luminous matter a fraction $\Omega_{\text{lum}} < 1\%$, while Big Bang nucleosynthesis and CMB data give an estimate for the baryonic matter of $\Omega_b \approx 5\%$, and for non-baryonic dark matter $\Omega_{nb} \approx 20$–$25\%$ (Bahcall and Fan 1998; Roos 2003).

One could expect the existence of baryonic DM in faint astronomical objects such as brown dwarfs, Jupiter or Earth-like planets, black holes as a result of stellar evolution, snow balls etc, while for non-baryonic dark matter we have to introduce new objects such as primodial black holes or new particles and ones of the most popular from these are neutralinos which are Weakly Interacting Massive Particle (WIMPs). There are different ways of DM searches, namely, direct and indirect ones (Smith and Lewin 1990;

Klapdor-Kleingrothaus and Zuber 2000). Basically, in direct searches people measure interactions (recoils) of nuclei with DM particles, while in indirect searches observational data ($\gamma$-radiation, for instance) are used to fit them with annihilation of DM particles.

There are claims about discoveries of neutralino existence with both direct and indirect searches. The DAMA collaboration has claimed that they discovered a modulation which is related with neutralino features (Bernabei et al. 2008), but other collaborations do not confirm the DAMA result, moreover, they insist that the DAMA result has to be ruled out with their own data. However, an independent analysis shows that DAMA and other experiments data may be not contradictory, but at the moment it is very hard to neutralino mass from these data (Savage et al. 2008). So, we have to wait for the independent confirmations (or disproofs) of the DAMA conclusions.

### 2.2 Indirect Searches of DM

In the last years intensive searches for dark matter (DM), especially its non-baryonic component, both in galactic halos and at galaxy centers have been undertaken (see for example Bertone et al. 2005; Bertone and Merritt 2005a for recent results). It is generally accepted that the most promising candidate for the non-baryonic DM component is the neutralino. In this case, the $\gamma$-flux from galactic halos (and from our Galactic halo in particular) could be explained by neutralino annihilation (Gurevich and Zybin 1997; Bergström et al. 1998; Tasitsiomi and Olinto 2002; Stoehr et al. 2003; Prada et al. 2004; Profumo 2005; Mambrini et al. 2005; de Boer et al. 2005). Since $\gamma$-rays are detected not only from high galactic latitude, but also from the Galactic Center, there is a wide spread of hypotheses (see Evans et al. 2004 for a discussion) that a DM concentration might be present at the Galactic Center. In this case the Galactic Center could be a strong source of $\gamma$-rays and neutrinos (Bertone et al. 2004, 2005; Bergström et al. 1998, 2005; Bouquet et al. 1989; Stecker 1988; Berezinsky et al. 1994; Gnedin and Primack 2004; Horns 2005; Bertone and Merritt 2005b) due to DM annihilation. Since it is also expected that DM forms spikes at galaxy centers (Gondolo and Silk 1999; Ullio et al. 2001; Merritt 2003) the $\gamma$-ray flux from the Galactic Center should increase significantly in that case.

## 3 Dark Matter in the Galactic Center

For the black hole in the Galactic Center, Hall and Gondolo (2006) used estimates of the enclosed mass obtained in various ways and tabulated by Ghez et al. (2003, 2005). The black hole, stellar cluster and DM could contribute to the mass inside stellar orbits. Moreover, if a DM cusp existed around the Galactic Center it could modify the trajectories of stars moving around it in a sensible way depending on the DM mass distribution. Progress in monitoring bright stars near the Galactic Center has been reached recently (Ghez et al. 2003, 2005; Genzel et al. 2003). The astrometric limit for bright stellar sources near the Galactic Center with 10 meter telescopes is today $\delta\theta_{10} \sim 1$ mas and the Next Generation Large Telescope (NGLT) will be able to improve this number at least down to $\delta\theta_{30} \sim 0.5$ mas (Weinberg et al. 2005a, 2005b) or even to $\delta\theta_{30} \sim 0.1$ mas (Ames et al. 2002; Weinberg et al. 2005a, 2005b) in the K-band. Therefore, it will be possible to measure the proper motion for about $\sim 100$ stars with astrometric errors several times smaller than errors in current observations. Recently it was shown (Zakharov et al. 2007a, 2008; Ghez et al. 2008) that it is possible to constrain the parameters of the DM possibly present distribution around the Galactic Center by considering the induced apoastron shift due to the presence of this DM sphere and either available data obtained with the present generation of telescopes (the so called *conservative* limit) and also expectations from future NGLT observations or with other advanced observational facilities. Recent advancements in infrared astronomy allow testing the scale of the mass profile at the center of our galaxy down to tens of AU. With the Keck 10 m telescope, the proper motion of several stars orbiting the Galactic Center black hole have been monitored and almost entire orbits, as for example that of the S2 star, have been measured allowing for an unprecedent description of the Galactic Center region. Measurements of the amount of mass $M(<r)$ contained within a distance $r$ from the Galactic Center are continuously improved as more precise data are collected. Recent observations (Ghez et al. 2003) extend down to the periastron distance ($\simeq 3 \times 10^{-4}$ pc) of the S16 star and correspond to a value of the enclosed mass within $\simeq 3 \times 10^{-4}$ pc of $\simeq 3.67 \times 10^{6}\ M_{\odot}$. Here and in the following, we use the three component model for the central region of our galaxy based on estimates of the enclosed mass given by Ghez et al. (2003, 2005) which have recently been proposed (Hall and Gondolo 2006). This model is constituted by the central black hole, the central stellar cluster and the DM sphere (made of WIMPs), i.e.

$$M(<r) = M_{BH} + M_{*}(<r) + M_{DM}(<r), \tag{5}$$

where $M_{BH}$ is the mass of the central black hole Sagittarius $A^{*}$.

In Fig. 2, for example, assuming that the test particle orbiting the Galactic Center region is the S2 star, we show the Post-Newtonian orbits obtained by the black hole only and the black hole plus the stellar cluster plus the contribution of DM mass density with $R_{DM} = 10^{-3}$ pc. As one can see, for selected parameters for DM and stellar cluster masses and radii the effect of the stellar cluster is almost negligible while the effect of the DM distribution is crucial since it enormously overcomes the shift due to the black hole (for $R_{DM} = 10^{-3}$ pc). Moreover, as expected, its contribution is opposite in sign with respect to that of the black hole (Nucita et al. 2007). We note that the expected apoastron (or, equivalently, periastron) shifts (mas/revolution), $\Delta\Phi$ (as seen from the center) and the corresponding values $\Delta\phi_{E}^{\pm}$ as seen from Earth (at the distance $R_0 \simeq 8$ kpc from the GC) are related by

$$\Delta\phi_{E}^{\pm} = \frac{d(1 \pm e)}{R_0}\Delta\Phi, \tag{6}$$

where the sign $\pm$ indicates the shift angles of the apoastron (+) and periastron (−), respectively. The S2 star semi-major axis and eccentricity are $d = 919$ AU and $e = 0.87$ (Ghez et al. 2005). In Fig. 3, the S2 apoastron shift as a function of the DM distribution size $R_{DM}$ is given for $\alpha = 0$ and $M_{DM} \simeq 2 \times 10^{5}\ M_{\odot}$. Taking into account that the present day precision for the apoastron shift measurements is of about 10 mas, one can say that the S2 apoastron shift cannot be larger than 10 mas. Therefore, any DM configuration that gives a total S2 apoastron shift larger than 10 mas (in the opposite direction due to the DM sphere) is excluded. The same analysis is done for two different values of the DM mass distribution slope, i.e. $\alpha = 1$ and $\alpha = 2$. In any case, we have calculated the apoastron shift for the S2 star orbit assuming a total DM mass $M_{DM} \simeq 2 \times 10^{5}\ M_{\odot}$. As one can see, the upper limit of about 10 mas on the S2 apoastron shift may allow to conclude that DM radii in the range of about $10^{-3}$–$10^{-2}$ pc are excluded by present observations for DM mass distribution slopes. We notice that the results of the present analysis allows to further constrain the results of Hall and Gondolo (2006), who have concluded that if the DM sphere radius is in the range $10^{-3}$–1 pc, configurations with DM mass up to $M_{DM} = 2 \times 10^{5}\ M_{\odot}$ are acceptable. The present analysis shows that DM configurations of the same mass are acceptable only for $R_{DM}$ from out the range between $10^{-3}$–$10^{-2}$ pc, almost irrespectively of the $\alpha$ value.

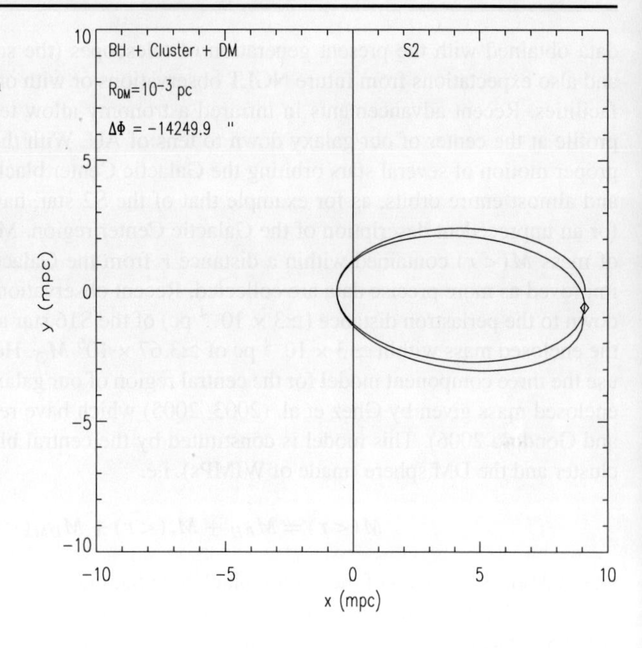


**Fig. 2** PN-orbits for different mass configurations at the Galactic Center. Deviations from elliptical orbit are clearly seen (here we assume that DM mass $M_{DM} \simeq 2 \times 10^5\ M_\odot$ and $R_{DM} = 10^{-3}$ pc)

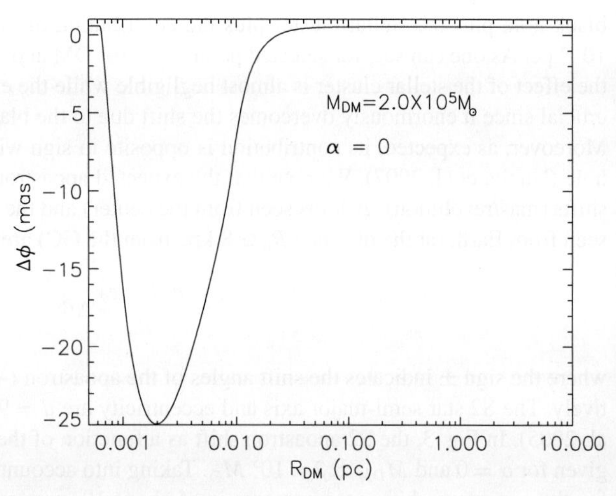


**Fig. 3** Apoastron shift as a function of the DM radius $R_{DM}$ for $\alpha = 0$ and $M_{DM} \simeq 2 \times 10^5\ M_\odot$. Taking into account present day precision for the apoastron shift measurements (about 10 mas) one can say that DM radii $R_{DM}$ in the range $8 \times 10^{-4}$–$10^{-2}$ pc are not acceptable

We have considered the constraints that the upper limit (presently of about 10 mas) of the S2 apoastron shift may put on the DM configurations at the galactic center considered by Hall and Gondolo (2006). When (in about 10–15 years, even without considering improvements in observational facilities) the precision of the S2 apoastron shift will be about 1 mas (that is equal to the present accuracy in the S2 orbit reconstruction) our analysis will allow to further constrain the DM distribution parameters. In particular, the asymmetric shape of the curves in Fig. 3 imply that any improvement in the apoastron shift measurements will allow to extend the forbidden region especially for the upper limit for $R_{DM}$. Quantitatively, we have a similar behavior for other choices of slope parameters $\alpha$ for DM concentrations. In this context, future facilities for astrometric measurements at a level 10 μ as of faint infrared stars will be extremely useful (Eisenhauer et al. 2005) and they give an opportunity to put even more severe constraints on the DM distribution. In addition, it is also expected to detect faint infrared stars or even hot spots (Genzel and Karas 2007) orbiting the Galactic Center. In this case, consideration of higher order relativistic corrections for an adequate analysis of the stellar orbital motion have to be taken into account. Due to the great progress in the precision of measurements, one could not exclude the possibility that matter density will be so low that alternative scenarios (to DM annihilation model) will be needed to explain the $\gamma$-flux from the Galactic Center. Electromagnetic processes in plasma in a presence of a strong gravitational field near the Galactic Center may be important components of such alternative scenarios for the detected $\gamma$-flux. In our considerations we adopted simple analytical expressions and reliable values for the $R_{DM}$ and $M_{DM}$ parameters following (Hall and Gondolo 2006) just to illustrate the relevance of the apoastron shift phenomenon in constraining the DM mass distribution at the Galactic Center. If other models for the DM distributions are considered (see, for instance Merritt et al. 2007 and references therein) the qualitative aspects of the problem are preserved although, of course, quantitative results on apoastron shifts may be different.

## 4 Λ-Term, Dark Energy or Alternative Theories of Gravity

All cosmological models are based on solutions of Einstein equations (Einstein 1915; Weinberg 1972)

$$R_{ij} - \frac{1}{2} R g_{ij} = -\kappa T_{ij}, \tag{7}$$

where $R_{ij}$ is the Ricci tensor, $g_{ij}$ is the metric tensor, $R$ is the scalar curvature, $T_{ij}$ is the stress-energy tensor, $\kappa = \frac{8\pi G}{c^2}$, $G$ is the Newton constant, $c$ is the speed of light. We note that initially, (7) was written in the equivalent form

$$R_{ij} = -\kappa \left( T_{ij} - \frac{1}{2} T_l^l g_{ij} \right). \tag{8}$$

To find a static cosmological solution Einstein (1917) introduced the basic cosmological principles such as assumptions about isotropy and homogeneity of the Universe and Λ-term to balance the gravitational attraction. He wrote the gravitational field equations in the following form

$$R_{ij} - \frac{1}{2} R g_{ij} - \Lambda g_{ij} = -\kappa T_{ij}. \tag{9}$$

### 4.1 Λ-Term Revisited

Friedman (1922) found solutions of the Einstein equations with a substitution of metric in the following form

$$ds^2 = c^2 dt^2 - a^2(t) ds_3^2, \tag{10}$$

where $a$ is the scale factor, $ds_3^2$ describes a metric in 3-dimensional space. The Friedmann equations could be written in the following form (Weinberg 1972)

$$\left(\frac{\dot{a}}{a}\right)^2 = \frac{8\pi G}{3}\rho - \frac{k}{a^2} + \frac{\Lambda}{3}, \quad \frac{\ddot{a}}{a} = -\frac{4\pi G}{3}\left(\rho + \frac{3p}{c^2}\right) + \frac{\Lambda}{3}, \tag{11}$$

where $k$ is the spatial curvature. Denoting the Hubble constant $H_0 = \dot{a}/a$, we obtain for the flat Universe case $k = 0$, $\rho_c(t_0) = 1.88h^2 \times 10^{-29}$ g/cm$^3$, where $h$ is the dimensionless Hubble constant ($H_0 = 100h$ km s$^{-1}$ Mpc$^{-1}$). Typically densities of different components of matter are expressed in term of critical density units with corresponding indices $\Omega = \rho/\rho_c$. Introducing the notation $\rho_\Lambda = \Lambda/(3H_0^2)$, one can re-write the first Friedmann equation in the form

$$\left(\frac{\dot{a}}{a}\right)^2 = \frac{8\pi G}{3}(\rho + \rho_\Lambda) - \frac{k}{a^2}, \tag{12}$$

therefore, for a flat Universe ($k = 0$), we have $\Omega + \Omega_\Lambda = 1$. In spite of theoretical predictions of inflationary cosmological models on the flatness of the Universe (Linde 1990), there were no observational data indicating a non-vanishing $\Lambda$-term until 1998 and usually the $\Lambda$-term was assumed to be zero. Moreover, from the elementary particle theory point of view it is very hard to create so small a $\Lambda$-term satisfying the observational astronomical constraints (Weinberg 1989). On the other hand, before 1998 it was not precisely known how large $\Omega_m$ would be, but in 1998 the evaluation of the mass in distant clusters showed that $\Omega_m = 0.2^{+0.3}_{-0.1}$ and high density cosmological models $\Omega_m = 1$ have a very small probability ($p \leq 10^{-6}$) (Bahcall and Fan 1998). Meanwhile, assuming that supernovae type Ia (SNe Ia) are standard cosmological candles in 1998 it was found that $\Omega_\Lambda = 0.7$ and $\Omega_m = 0.3$ and the Universe is flat (Riess et al. 1998; Perlmutter et al. 1999) and this claim was supported with Large Scale Structure (LSS) and Cosmic Microwave Background (CMB) radiation data (Bahcall et al. 1999). These conclusions were supported further by LSS data and CMB experiments such as Boomerang, MAXIMA, WMAP (Komatsu et al. 2008).

### 4.2 $\Lambda$-Term or Dark Energy?

The $\Lambda$-term may be in the LHS of (9) and in this case it is interpreted as a part of space-time geometry (in the gravitational field equations according to Einstein's idea a geometrical quantity (tensor) describing space–time manifold has to be equal to a quantity (stress-energy tensor) describing a matter). However, we could transfer the $\Lambda$-term into the RHS of (9), where the stress-energy tensor is and in this case, the term is interpreted as a matter with equation of state $p = -\rho c^2$. Generalizing the relation and assuming that $\Lambda$ is not precisely a constant we introduce the equation of state for dark energy such as $p = w\rho c^2$, where $w = -1$ corresponds to the $\Lambda$-term, $-1 \leq w \leq 1$ corresponds to quintessence, $w \leq -1$ corresponds to phantom energy, if $w$ crosses $-1$, it could be the quintom model (Feng et al. 2006). At the moment the simplest and the most economic model ($w = -1$) is still satisfying observational data, but one can not exclude that $w$ may vary in space and time and in this case future studies of $w$ (Albrecht et al. 2006), especially with space-borne facilities like SNAP,[3] are extremely important.

### 4.3 Standard Cosmology vs. $f(R)$ Gravity

The standard cosmological model has problems to understand the origin of Dark Matter (DM) and dark energy (DE). Another approach was proposed and it was based on an assumption that gravity is different from standard general relativity it can be described by a modified Lagrangian (Carroll et al. 2004) (however, there are tensions with Solar system constraints (Shao et al. 2006)). The model was successful to explain an acceleration of

[3] http://snap.lbl.gov/.

the Universe, but it has problems to fit Solar system data. Conformal theories are also not excluded (Behnke et al. 2002; Barbashov et al. 2006a, 2006b). Recently, a generalization of the approach has been proposed in the framework of higher order theories of gravity—also referred to as $f(R)$ (or fourth order) theories—a modification of the gravity action of the form (Capozziello 2007; Capozziello and Garattini 2007; Capozziello and Francaviglia 2007; Capozziello et al. 2005, 2006a, 2006b, 2006d, 2006e, 2006f, 2006g, 2007b, 2007c, 2007d; Capozziello and Troisi 2005; Carloni et al. 2005; Borowiec et al. 2006, 2007)

$$\mathcal{A}=\int d^4x\sqrt{-g}[f(R)+\mathcal{L}_m], \tag{13}$$

where $f(R)$ is a generic function of the Ricci scalar curvature and $\mathcal{L}_m$ is the standard matter Lagrangian. For example, if $f(R)=R+2\Lambda$ the theory coincides with General Relativity (GR) with $\Lambda$ term. In particular, a power law function $f(R)$ theories of the form $f(R)=f_0R^n$ was considered. As a result, in the weak field limit, the gravitational potential is found to be

$$\Phi(r)=-\frac{Gm}{2r}\left[1+\left(\frac{r}{r_c}\right)^{\beta}\right], \tag{14}$$

where

$$\beta=\frac{12n^2-7n-1-\sqrt{36n^4+12n^3-83n^2+50n+1}}{6n^2-4n+2}. \tag{15}$$

The dependence of the $\beta$ parameter on the $n$ power is shown in Zakharov et al. (2006). Of course, for $n\to\infty$ it follows $\beta\to 1$, while for $n=1$ the parameter $\beta$ reduces to zero and the Newtonian gravitational field is recovered. On the other hand, while $\beta$ is a universal parameter, $r_c$ in principle is an arbitrary parameter, depending on the considered system and its typical scale. Consider for example the Sun as the source of the gravitational field and the Earth as the test particle. Since Earth's velocity is $\simeq$30 km s$^{-1}$, it has been found that the parameter $r_c$ varies in the range $\simeq$1–$10^4$ AU. Once $r_c$ and $\beta$ has been fixed, authors of the paper (Capozziello et al. 2006c) used them to study deviations from the standard Paczynski light curve for gravitational microlensing (Paczynski 1986) and claimed that the implied deviation can be measured. It is clear that for gravitational microlensing one could detect observational differences between GR and an alternative theory (the fourth order gravity in particular), so that one should have different potentials at the scale $R_E$ (the Einstein radius) of the gravitational microlensing. For the Galactic microlensing case $R_E$ is about 1 AU. This is a reason why the authors (Capozziello et al. 2006c) have selected $r_c$ at a level of astronomical units to obtain observable signatures for non-vanishing $\beta$. This considerations can be extended considering the Newtonian limit of *any* analytic $f(R)$-gravity model (Capozziello et al. 2007b). By imposing the spherical symmetry and developing the metric up to the second order, we have

$$ds^2=g_{00}(x^0,r)dx^{0^2}+g_{rr}(x^0,r)dr^2-r^2d\Omega, \tag{16}$$

where $x^0=ct$ and $d\Omega$ is the angular element. To develop the post-Newtonian limit of the theory, one can consider a perturbed metric with respect to a Minkowski background $g_{\mu\nu}=\eta_{\mu\nu}+h_{\mu\nu}$. The metric coefficients can be developed as:

$$\begin{aligned} &g_{tt}(t,r)\simeq 1+g_{tt}^{(2)}(t,r)+g_{tt}^{(4)}(t,r), \qquad g_{rr}(t,r)\simeq -1+g_{rr}^{(2)}(t,r),\\ &g_{\theta\theta}(t,r)=-r^2, \qquad g_{\phi\phi}(t,r)=-r^2\sin^2\theta, \end{aligned} \tag{17}$$

where we put, for the sake of simplicity, $c = 1$, $x^0 = ct \to t$. We want to obtain the most general result without imposing particular forms for the $f(R)$-Lagrangian. We only consider analytic Taylor expandable functions

$$f(R) \simeq f_0 + f_1 R + f_2 R^2 + f_3 R^3 + \cdots. \tag{18}$$

To obtain the post-Newtonian approximation of $f(R)$-gravity, one has to plug the expansions (17) and (18) into the field equations and then expand the system up to the orders $O(0)$, $O(2)$ and $O(4)$. This approach provides general results and specific (analytic) Lagrangians are selected by the coefficients $f_i$ in (18). If we now consider the $O(2)$-order of approximation in the field equations, the general solution is:

$$\begin{cases} g_{tt}^{(2)} = \delta_0 - \frac{2GM}{f_1 r} - \frac{\delta_1(t)e^{-r\sqrt{-\xi}}}{3\xi r} + \frac{\delta_2(t)e^{r\sqrt{-\xi}}}{6(-\xi)^{3/2} r}, \\ g_{rr}^{(2)} = -\frac{2GM}{f_1 r} + \frac{\delta_1(t)[r\sqrt{-\xi}+1]e^{-r\sqrt{-\xi}}}{3\xi r} - \frac{\delta_2(t)[\xi r + \sqrt{-\xi}]e^{r\sqrt{-\xi}}}{6\xi^2 r}, \\ R^{(2)} = \frac{\delta_1(t)e^{-r\sqrt{-\xi}}}{r} - \frac{\delta_2(t)\sqrt{-\xi}e^{r\sqrt{-\xi}}}{2\xi r}, \end{cases} \tag{19}$$

where $\xi \doteq \frac{f_1}{6f_2}$, $f_1$ and $f_2$ are the expansion coefficients obtained by the $f(R)$-Taylor series. In the limit $f \to R$, for a point-like source of mass $M$ we recover the standard Schwarzschild solution. Let us notice that the integration constant $\delta_0$ is dimensionless, while the two arbitrary time-functions $\delta_1(t)$ and $\delta_2(t)$ have respectively the dimensions of *lenght*$^{-1}$ and *lenght*$^{-2}$; $\xi$ has the dimension *lenght*$^{-2}$. The functions $\delta_i(t)$ $(i = 1, 2)$ are completely arbitrary since the field equations, in spherical symmetry, depend only on spatial derivatives. Besides, the integration constant $\delta_0$ can be set to zero, as in the standard theory of potential, since it represents an unessential additive quantity. In order to obtain the physical prescription of the asymptotic flatness at infinity, we can discard the Yukawa growing mode in (19) and then the metric is:

$$ds^2 = \left[1 - \frac{2GM}{f_1 r} - \frac{\delta_1(t)e^{-r\sqrt{-\xi}}}{3\xi r}\right]dt^2$$

$$\left[1 + \frac{2GM}{f_1 r} - \frac{\delta_1(t)(r\sqrt{-\xi}+1)e^{-r\sqrt{-\xi}}}{3\xi r}\right]dr^2 - r^2 d\Omega. \tag{20}$$

The Ricci scalar curvature is

$$R = \frac{\delta_1(t)e^{-r\sqrt{-\xi}}}{r}. \tag{21}$$

The solution can be given also in terms of the gravitational potential. In particular, we have an explicit Newtonian-like term into the definition. The first expression of (19) provides the second order solution in terms of the metric expansion (see definition (17)). In particular, it is $g_{tt} = 1 + 2\phi_{\text{grav}} = 1 + g_{tt}^{(2)}$ and then the gravitational potential of an analytic $f(R)$-theory is

$$\phi_{\text{grav}} = -\frac{GM}{f_1 r} - \frac{\delta_1(t)e^{-r\sqrt{-\xi}}}{6\xi r}, \tag{22}$$

where an *interaction length* can be defined. Such a length is a function of the series coefficients, $f_1$ and $f_2$. This generalize the above positions to any analytic function $f(R)$. Besides these results, fourth order gravity theories were very successful to explain standard cosmological data such as SNe Ia fits, an acceleration of the Universe (Capozziello and Garattini 2007; Capozziello et al. 2005, 2006a, 2006b, 2006d, 2006e, 2006f, 2006g, 2007b, 2007c, 2007d; Capozziello and Troisi 2005; Carloni et al. 2005; Capozziello and Francaviglia 2007; Capozziello 2007; Borowiec et al. 2006, 2007) rotation curves for galaxies (Capozziello et al. 2007a; Martins and Salucci 2007) and it was suggested that the standard general relativity plus DM and DE may be distinguished from $R^n$ approaches with gravitational microlensing (Capozziello et al. 2006c), but Solar system data (planetary orbital periods, in particular) put severe constraints on parameters of the theories (Zakharov et al. 2006, 2007b).

GR and Newtonian theory (as its weak field limit) were verified by a very precise way at different scales. There are observational data which constrain parameters of alternative theories as well. As a result, the parameter $\beta$ of fourth order gravity should be very close to zero (it means that the gravitational theory should be very close to GR). In particular, the $\beta$ parameter values considered for microlensing (Capozziello et al. 2006c), for rotation curves (Capozziello and Garattini 2007) and cosmological SN type Ia (Borowiec et al. 2006) are ruled out by solar system data. No doubt that one could also derive further constraints on the fourth order gravity theory by analyzing other physical phenomena such as Shapiro time delay, frequency shift of radio photons (Bertotti et al. 2003), laser ranging for distant objects in the solar system, deviations of trajectories of celestial bodies from ellipses, parabolas and hyperbolas and so on. But our aim was only to show that only $\beta \simeq 0$ values are not in contradiction with solar system data in spite of the fact that there are a lot of speculations to fit observational data with $\beta$ values significantly different from zero.

## 5 Conclusions

DM and DE problems are the most challenging in modern physics. In spite of a great progress in theory and observational facilities, this puzzle is far from its final understanding, especially in the case of DE ($\Lambda$-term) problem. Probably, further observational data given with the joint NASA–DOE space mission SNAP could help to put new constraints on models, to select the reliable theoretical approach, to do a significant step in understanding these problems. Concerning the DM problem there are great expectations with the start of Large Hadron Collider, because an evaluated energy range for DM particles may be investigated with the accelerator.

**Acknowledgements** AFZ acknowledges the National Natural Science Foundation of China (Grant # 10233050) and National Basic Research Program of China (2006CB806300) for partial financial support.

## References

A. Albrecht et al., (2006). arXiv:astro-ph/0609591
G. Ames et al., (2002). http://tmt.ucolick.org/reports_and_notes/reports/Web_final_Greenbook.pdf
N. Bahcall, X. Fan, Astrophys. J. **504**, 1 (1998)
N. Bahcall et al., Science **284**, 1481 (1999)
B.M. Barbashov et al., Int. J. Mod. Phys. A **21**, 5957 (2006a)
B.M. Barbashov et al., Phys. Lett. B **633**, 458 (2006b)
D. Behnke et al., Phys. Lett. B **530**, 20 (2002)
J.D. Bekenstein, Phys. Rev. D **70**, 083509 (2004); erratum-ibid. D **71**, 069901 (2005)
V. Berezinsky, A. Bottino, G. Mignola, Phys. Lett. B **325**, 136 (1994)
L. Bergström, P. Ullio, J.H. Buckley, Astropart. Phys. **94**, 131301 (1998)
L. Bergström et al., Phys. Rev. Lett. **9**, 138 (2005)

R. Bernabei et al., (2008). arXiv:0804.2741 [astro-ph]
G. Bertone, D. Merritt, Modern Phys. Lett. A **20**, 1021 (2005a)
G. Bertone, D. Merritt, Phys. Rev. D **72**, 103502 (2005b)
G. Bertone et al., Phys. Rev. D **70**, 063503 (2004)
G. Bertone, D. Hooper, J. Silk, Phys. Reports **405**, 279 (2005)
B. Bertotti, L. Iess, P. Tortora, Nature **425**, 374 (2003)
L. Blanchet, A. Le Tiece, (2008). arXiv:0804.3518v2 [astro-ph]
A. Borowiec, W. Godlowski, M. Szydlowski, Phys. Rev. D **74**, 043502 (2006)
A. Borowiec, W. Godlowski, M. Szydlowski, Int. J. Geom. Meth. Mod. Phys. **4**, 183 (2007)
A. Bouquet, P. Salati, J. Silk, Phys. Rev. D **40**, 3168 (1989)
J.R. Brownstein, J.W. Moffat, Astrophys. J. **636**, 721 (2006)
S. Capozziello, Intern. J. Geom. Meth. Mod. Phys. **4**, 53 (2007)
S. Capozziello, M. Francaviglia, (2007). arXiv:0706.1146 [astro-ph]
S. Capozziello, R. Garattini, Class. Quant. Grav. **24**, 1627 (2007)
S. Capozziello, A. Troisi, Phys. Rev. D **72**, 044022 (2005)
S. Capozziello, V.F. Cardone, A. Troisi, Phys. Rev. D **71**, 043503 (2005)
S. Capozziello, V.F. Cardone, M. Francaviglia, Gen. Rel. Grav. **38**, 711 (2006a)
S. Capozziello, V.F. Cardone, A. Troisi, J. Cosm. Astropart. **8**, 1 (2006b)
S. Capozziello, V.F. Cardone, A. Troisi, Phys. Rev. D **73**, 104019 (2006c)
S. Capozziello, S. Nojiri, S.D. Odintsov, Phys. Lett. B **634**, 93 (2006d)
S. Capozziello et al., Phys. Lett. B **639**, 135 (2006e)
S. Capozziello et al., Phys. Rev. D **73**, 043512 (2006f)
S. Capozziello et al., Class. Quant. Grav. **23**, 1205 (2006g)
S. Capozziello, V.F. Cardone, A. Troisi, Mon. Not. R. Astron. Soc. **375**, 1423 (2007a)
S. Capozziello, A. Stabile, A. Troisi, Phys. Rev. D (2007b), 104019
S. Capozziello, A. Stabile, A. Troisi, Class. Quant. Grav. **24**, 2153 (2007c)
S. Capozziello, A. Troisi, V.F. Cardone, New Astron. Rev. **5**, 341 (2007d)
S. Carloni et al., Class. Quant. Grav. **22**, 4839 (2005)
S.M. Carroll, V. Duvvuri, M. Trodden, M.S. Turner, Phys. Rev. D **70**, 043528 (2004)
W. de Boer et al., Phys. Rev. Lett. **95**, 209001 (2005)
A. Einstein, Sitzungsber. d. Berl. Koniglich Preussische Akademie der Wissenschaften **48**(2), 844 (1915)
A. Einstein, Sitzungsber. d. Berl. Koniglich Preussische Akademie der Wissenschaften **1**, 142 (1917)
F. Eisenhauer, G. Perrin, S. Rabien, Astron. Nachr. **326**, 561 (2005)
N.W. Evans, F. Ferrer, S. Sarkar, Phys. Rev. D **69**, 123501 (2004)
B. Feng, M. Li, Y.-S. Piao, X. Zhang, Phys. Lett. B **634**, 101 (2006)
A. Friedman, Z. Phys. **10**, 377 (1922) (English translation in: Gen. Rel. Grav. **31**, 1991 (1999))
J.G. Galle, Month. Not. R. Astron. Soc. **7**, 153 (1846)
R. Genzel, V. Karas, (2007). Preprint arXiv:0704.1281v1 [astro-ph]
R. Genzel et al., Astrophys J. **594**, 812 (2003)
A.M. Ghez et al., Astron. Nachr. **324**, 527 (2003)
A.M. Ghez et al., Astrophys. J. **620**, 744 (2005)
A.M. Ghez et al., (2008). Preprint arXiv:0808.2870v1 [astro-ph]
O.Y. Gnedin, J.R. Primack, Phys. Rev. Lett. **93**, 061302 (2004)
H. Goldstein, *Classical Mechanics*, 2nd edn. (Addison-Wesley, Reading, 1980)
P. Gondolo, J. Silk, Phys. Rev. Lett. **83**, 1719 (1999)
A.V. Gurevich, K.P. Zybin, Phys. Lett. A **225**, 217 (1997)
A. Hall, Astron. J. **14**, 49 (1894)
J. Hall, P. Gondolo, Phys. Rev. D **74**, 063511 (2006)
D. Horns, Phys. Lett. B **607**, 225 (2005)
R. Juszkiewicz, (2008). Private communication
H.-V. Klapdor-Kleingrothaus, K. Zuber, *Astroparticle Physics* (IOP Publ. House, 2000)
E. Komatsu et al., (2008). arXiv:0803.0547v2 [astro-ph]
U.J.J. Le Verrier, (1846). Compte Rendu 31(Aug)
U.J.J. Le Verrier, R. Acad. Sci. Paris **59**, 379 (1859)
U.J.J. Le Verrier, R. Acad. Sci. Paris **83**, 583 (1876)
A.D. Linde, *Particle Physics and Inflationary Cosmology* (Harwood Acad., Chur, 1990)
Y. Mambrini et al., (2005). Preprint hep-ph/0509300
C.F. Martins, P. Salucci, (2007). arXiv:astro-ph/0703243
D. Merritt, (2003). arXiv:astro-ph/0301365
D. Merritt, S. Harfst, G. Bertone, Phys. Rev. D **75**, 043517 (2007)
M. Milgrom, Astrophys. J. **270**, 365 (1983)

S. Newcomb, The elements of the four inner planets and the fundamental constants of astronomy, in *Suppl. Am. Ephem. Naut. Aim.* (U.S. Govt. Printing Office, Washington, 1895)
I. Newton, *Philosophia Naturalis Principia Mathematica* (Impramatur, S. Pepys, London, 1686), p. 141
A.A. Nucita et al., Publ. Astron. Soc. Pacific **119**, 349 (2007)
J.H. Oort, Bull. Astron. Inst. Netherlands **6**, 249 (1932)
B. Paczynski, Astrophys. J. **506**, 1 (1986)
S. Perlmutter et al., Astrophys. J. **517**, 565 (1999)
F. Prada et al., Phys. Rev. Lett. **93**, 241301 (2004)
S. Profumo, Phys. Rev. D **72**, 103521 (2005)
A.G. Riess et al., Astron. J. **116**, 1009 (1998)
M. Roos, *Introduction to Cosmology* (Wiley, New York, 2003)
V. Rubin, W.K. Ford, Astrophys. J. **159**, 379 (1970)
V. Rubin, N. Thonnard, W.K. Ford, Astrophys. J. **217**, L1 (1977)
V. Rubin, W.K. Ford, N. Thonnard, Astrophys. J. **225**, L107 (1978)
V. Rubin, W.K. Ford, N. Thonnard, Astrophys. J. **238**, 401 (1980)
C. Savage et al., (2008). arXiv:0808.3607v2 [astro-ph]
C.-G. Shao, R.-G. Cai, B. Wang, R.-K. Su, Phys. Lett. B **633**, 164 (2006)
P.F. Smith, J.D. Lewin, Phys. Rep. **187**, 203 (1990)
F.W. Stecker, Phys. Lett. B **201**, 529 (1988)
F. Stoehr et al., Mon. Not. R. Astron. Soc. **345**, 1313 (2003)
A. Tasitsiomi, A.V. Olinto, Phys. Rev. D **66**, 023502 (2002)
P. Ullio, H.-S. Zhao, M. Kamionkowski, Phys. Rev. D **64**, 043504 (2001)
S. Weinberg, *Gravitation and Cosmology* (Wiley, New York, 1972)
S. Weinberg, Rev. Mod. Phys. **61**, 1 (1989)
N. Weinberg, M. Miloslavljević, A.M. Ghez, (2005a). Preprint arXiv:astro-ph/0512621
N. Weinberg, M. Miloslavljević, A.M. Ghez, Astrophys J. **622**, 878 (2005b)
A.F. Zakharov, A.A. Nucita, F. De Paolis, G. Ingrosso, Phys. Rev. D **74**, 107101 (2006)
A.F. Zakharov, A.A. Nucita, F. De Paolis, G. Ingrosso, Phys. Rev. D **76**, 62001 (2007a)
A.F. Zakharov et al., AIP Conf. Proc. **966**, 173 (2007b)
A.F. Zakharov, A.A. Nucita, F. De Paolis, G. Ingrosso, J. Phys.: Conf. Ser. **133**, 012032 (2008)
F. Zwicky, Helv. Phys. Acta **6**, 110 (1933)
F. Zwicky, Astrophys, J. **86**, 217 (1937)

Space Sci Rev (2009) 148: 315–328
DOI 10.1007/s11214-009-9509-6

# Some Uncomfortable Thoughts on the Nature of Gravity, Cosmology, and the Early Universe

**L.P. Grishchuk**

Received: 31 March 2009 / Accepted: 9 April 2009 / Published online: 5 May 2009
© Springer Science+Business Media B.V. 2009

**Abstract** A specific theoretical framework is important for designing and conducting an experiment, and for interpretation of its results. The field of gravitational physics is expanding, and more clarity is needed. It appears that some popular notions, such as 'inflation' and 'gravity is geometry', have become more like liabilities than assets. A critical analysis is presented and the ways out of the difficulties are proposed.

**Keywords** Gravitation · Cosmology · Theory · Experiment

**PACS** 98.70.Vc · 04.30.-w · 04.20.Fy

## 1 Introduction

The proximity of the site of this gravitational conference to CERN and its recently completed Large Hadron Collider (LHC) reminds us of the affinity of our ultimate goals in the study of micro- and macro-worlds. Although the LHC will be investigating the minute particles—hadrons, the outreach page of the LHC web-site explains to the wide public that the "aim of the exercise is to smash protons ... into each other and so recreate conditions a fraction of a second after the big bang" (LHC web-site). One can also see clarifications in other sources of information, according to which the "LHC experiments ... will probe matter as it existed at the very beginning of time", and that this is a "new era of understanding about the origins and evolution of the universe".

These cosmological, rather than particle-physical, explanations can perhaps be justified, at least in part, by the alleged difficulties that our colleagues encountered in communication

Contribution to the 'Nature of Gravity' conference at the International Space Science Institute, October 2008, Bern, Switzerland.

L.P. Grishchuk (✉)
School of Physics and Astronomy, Cardiff University, Cardiff CF234AA, UK
e-mail: grishchuk@astro.cf.ac.uk

L.P. Grishchuk
Sternberg Astronomical Institute, Moscow State University, Moscow 119899, Russia

with general public. According to a circulating rumor, at the time when the opening of the LHC was widely covered by TV media, the BBC received numerous messages from angry parents who complained that "you keep talking about all these hardons while children may be watching". Surely, the descriptive cosmology is safer and easier to convey to the public than the notions of high-energy physics. But in the long run, it is indeed true that the fundamental question of the birth of the Universe is always in the background of our research intentions. This problem fascinates both scientific communities, as well as some part of the rest of population.

Undoubtedly, new important discoveries will be made at LHC and they will bring us closer to answers to very deep issues in physics. However, the record-breaking energies of $10^4$ GeV at our accelerators are still very far away from energies we need to understand in order to tackle the problem of the origins of the Universe. The ability of LHC to answer the questions on the physics of the very early Universe at $10^{15}$–$10^{19}$ GeV can be compared with the ability of a telescope hardly resolving a planetary system to answer the questions on the structure of a human hair. In these matters, our best hopes are associated with cosmic rather than laboratory studies. It is now recognized that there is no better way of trying to unveil the origins of the Universe than by measuring the relic gravitational waves that were generated by a strong variable gravitational field of the emerging Universe. We may be close to discovering relic gravitational waves, and thus answering some of the most fundamental questions. These efforts, as well as some principal issues on the nature of gravity, will be discussed below.

## 2 Spontaneous Birth of the Universe

The question of origins arises inevitably given the nonstationary character of the world that we observe at large scales. The broad-brush picture of the expanding homogeneous and isotropic distribution of matter and fields is pretty accurate as a zero-order approximation. We believe that this approximation was even better in the past, because the existing deviations from homogeneity and isotropy seem to have been growing in the course of time. It is known from observations that the present size $l_p$ of our patch of approximate large-scale homogeneity and isotropy is at least as big as the present-day Hubble radius $l_H = c/H_0 \approx 10^{28}$ cm. Combining available observations with plausible statistical assumptions one can conclude that the size of this patch is significantly bigger than $l_H$. It was evaluated to be about $500 l_H$ (Grishchuk 1992) and maybe larger. In the present discussion, we will limit ourselves with $l_p = 10^3 l_H$. It is also known from observations that the present averaged energy density $\rho_p c^2$ of all sorts of matter in our patch is close to $\rho_p = 3H_0^2/8\pi G \approx 10^{-29}$ g/cm$^3$.

Given the observed expansion, one can extrapolate $\rho(t)$ and $l(t)$ back in time by the known laws of physics. Temporarily leaving aside the possible recent interval of accelerated evolution governed by "dark energy" (see Sect. 7), we will first have $\rho(t) \propto 1/l^3(t)$ at the matter-dominated stage, up to $\rho_{eq} \approx 10^{-20}$g/cm$^3$, and then $\rho(t) \propto 1/l^4(t)$ at the radiation-dominated stage. Eventually one encounters a cosmological singularity characterized by the infinitely large energy density. It is reasonable to think that there must exist a smarter answer to the question of initial state of the observed Universe than singularity. Singularity is likely to be only a sign of breakdown of the currently available theories.

It is tempting to begin from the limits of applicability of current theories, that is, from the Planck density $\rho_{Pl} = c^5/G^2\hbar \approx 10^{94}$ g/cm$^3$ and the Planck size $l_{Pl} = (G\hbar/c^3)^{1/2} \approx 10^{-33}$ cm, and imagine that the embryo Universe was somehow created by a quantum-gravity (or by a 'theory-of-everything') process in this state and then started to expand (see

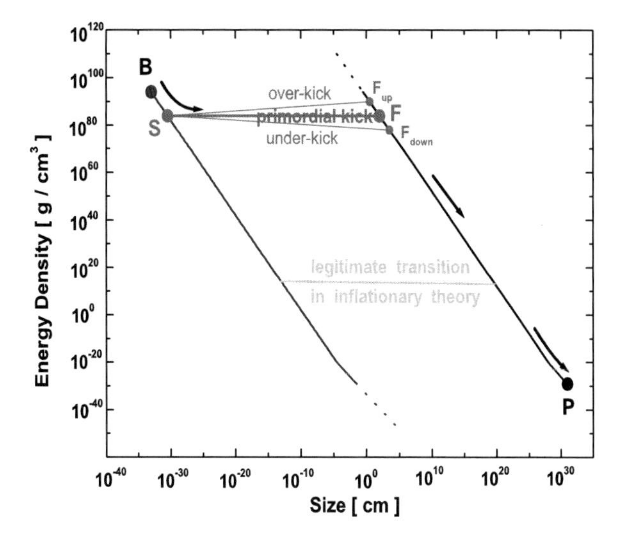


**Fig. 1** (Color online) A primordial kick (*red lines*) is required in order to reach the present state of the Universe P from the birth event B

Zel'dovich 1986 and references there). Conceptually, it is easier to imagine that the created Universe was spatially closed, which means that its total energy, including gravity, was zero and remains zero. Also, the concept of a closed universe helps avoid the question of environment. But a closed universe is not strictly necessary for our line of argument, so one can think of a small configuration in a possibly larger system. It is also plausible to think that the classical evolution of our patch of the Universe could have started with a natal size somewhat larger than $l_{Pl}$ and a natal energy density, of whatever matter that was there, somewhat lower than $\rho_{Pl}$.

The trouble is, however, that the very natural hypothesis of spontaneous birth of the observed Universe will not bring us anywhere near our present state characterized by $\rho_p$ and $l_p$, unless we make additional assumptions (Zel'dovich 1986).

In Fig. 1, the present state of the accessible Universe is marked by the point P. The hypothesized birth of the Universe is marked by the point B. If we go in the past from the point P according to the laws of matter-dominated and radiation-dominated evolutions (black curve), we totally miss the desired point B. Indeed, the energy density reaches the Planck value $\rho_{Pl}$ when the size of the model universe is 0.3 cm instead of the required $10^{-33}$ cm. On the other hand, if we descend from the point B according to the laws of radiation-dominated and then matter-dominated evolutions (blue curve), we totally miss the point P. Indeed, the present energy density $\rho_p$ is reached while the size of the model universe is 0.03 cm instead of the required $10^{31}$ cm. And the present size $10^{31}$ cm is reached when the density drops to $10^{-126}$ g/cm$^3$ instead of the required $\rho_p$. If the size $l_p$ of the homogeneous isotropic patch is larger than the assumed $l_p = 10^3 l_H$, the point P and the black curve in Fig. 1 move to the right, and the mis-match of the curves is only exacerbated.

The only way to reach P from B is to assume that the newly born Universe has experienced a 'primordial kick' allowing the point of evolution to jump over from the blue curve to the black curve. During the kick, the size of the Universe (or, better to say, the size of the patch of homogeneity and isotropy) should increase by about 33 orders of magnitude, but the energy density may not change too much or may stay constant. In the zero-order approximation, when the homogeneity and isotropy of the patch are maintained, the level of the energy density and the specific points of the start S and finish F of the kick transition are not very important, as we can reach P from B by many ways. However, the actual route becomes extremely important in the first-order approximation, when we have to take into account the quantum-mechanical generation of primordial cosmological perturbations.

If we are on the right track with this whole picture, the observations indicate (see below) that the actual kick trajectory took place at the energy densities around $10^{-10}\rho_{Pl}$ when the Hubble parameter $H$ was around $H_i = 10^{-5}H_{Pl} = 10^{-5}/t_{Pl}$ (horizontal red line in Fig. 1), with some possibility of a slight 'over-kick' to point $F_{up}$ or 'under-kick' to point $F_{down}$, as shown by tilted red lines in Fig. 1.

The general relativity allows the required kick trajectories, but only if the properties of the primeval matter were not like those that we deal with in laboratories (even at LHC). The energy density $\rho c^2$ and the effective pressure $p$ should satisfy the condition $\rho c^2 + p = 0$ for the kick to proceed with constant energy density, and $\rho c^2 + p < 0$ for over-kick with increasing energy density or $\rho c^2 + p > 0$ for under-kick with decreasing energy density. The kick Hubble parameter $H_i(t)$ remains constant ($\dot{H} = 0$) for the $\rho c^2 + p = 0$ case, and it is slightly increasing ($\dot{H} > 0$) or decreasing ($\dot{H} < 0$) for over-kick and under-kick trajectories, respectively. If the existing indications (see below) that the initial kick did indeed happen are further supported by observations, the detailed inquiry in the properties of the substance that might have driven the kick will be paramount.

Solutions with $\rho c^2 + p \geq 0$ and $p \approx -\rho c^2$ are allowed by some versions of the scalar field within general relativity. They are usually associated with the notion of inflation (see, for example, Weinberg 2008). The case with $p = -\rho c^2$ is known as the standard (de Sitter) inflation. The scalar field plus gravity model may not be what we are searching for, but at least it was shown (Belinsky et al. 1985a; Belinsky et al. 1985b) that the required solutions with $p \approx -\rho c^2$ are attractors in the space of all solutions of this dynamical system, so these are typical solutions which can bring us from the quantum boundary S (Belinsky et al. 1985a; Belinsky et al. 1985b) to the end point $F_{down}$. (Scalar field models do not admit the over-kick trajectories.)

It is often stated that inflation was invented by particle physicists in order to solve outstanding cosmological problems. The length of the list of solved problems depends on the enthusiasm of the writing inflationary author. To me personally, most of these problems and solutions smack of the Soviet-style management, when the government creates a problem and then demands the credit for heroically solving it. In any case, inflation was always assumed to have lasted sufficiently long; otherwise the point of evolution would have fell short of reaching the point $F_{down}$ on the black curve in Fig. 1. Whatever the problems the inflationary hypothesis solves, they are automatically solved by the hypothesis of primordial kick, whose only rationale is to serve as an 'umbilical cord' facilitating the joining of the birth event B with the present state of the Universe P.

While in the approximation of homogeneity and isotropy the hypothesis of inflation and the hypothesis of initial kick are equal, in the sense that they both solve problems by making plausible assumptions, at the level of cosmological perturbations the inflationary theory got everything wrong, as we shall see below, in Sect. 4.

## 3 Primordial Cosmological Perturbations

The essentially classical and highly symmetric, i.e. homogeneous and isotropic, solution describing the initial kick should be augmented by quantum fluctuations of the fields that were present there. It is natural to assume that these quantized fields were initially in their ground (vacuum) states. If the field is properly, superadiabatically, coupled to the strong gravitational field of the kick solution, the quantum-mechanical Schrödinger evolution transforms the initial vacuum state of the field into a multiparticle (strongly squeezed vacuum) state. This is called the superadiabatic, or parametric, amplification; for a recent review of the subject, see (Grishchuk to be published). Not all fields couple properly to the highly symmetric solutions under discussion. For example, electromagnetic fields do not. But the gravitational field perturbations do couple superadiabatically to the gravitational field of the kick, and therefore they can be amplified. Apparently, this is how the present-time complexity arises from the past-time simplicity.

The chief gravitational field perturbations are two transverse-traceless degrees of freedom describing gravitational waves and two degrees of freedom, scalar and longitudinal–longitudinal one, which in general relativity exist only if they are accompanied by perturbations of matter. These latter two degrees of freedom describe density perturbations, and usually only one of them is independent. For models of matter such as scalar fields, the equations for density perturbations are almost identical to the equations for gravitational waves. The arising multiparticle states at each frequency are conveniently combined in the power spectra of these quantum-mechanically generated perturbations. The crucial quantity, both for gravitational waves and density perturbations, is the primordial power spectrum of the gravitational field (metric) perturbations. This is because the gravitational field perturbations survive numerous transformations of the matter content of the Universe at the end of the kick and later, and therefore they provide the unambiguous input values for physics and calculations at the radiation-dominated and matter-dominated stages.

The values of the Hubble parameter along the kick trajectory and the shape of the kick trajectory itself are very important, because they define the numerical levels and spectral slopes of the primordial metric spectra. The Hubble parameter $H_i$ determines the amplitude of metric perturbations, whereas the tilt of the red lines in Fig. 1 determines the tilt of the spectrum. The primordial metric power spectrum $P(n)$ is a function of the time-independent wavenumber $n$, where the wavenumber $n_H = 4\pi$ corresponds to the wavelength which will be equal to $l_H$ today.

Very similar forms of equations for gravitational waves (gw) and density perturbations (dp), and identically the same physics of their superadiabatic amplification, translate into very similar power spectra $P(n)$. In a good approximation, sufficient for our purposes, the generated gw and dp power spectra can be written in the power-law forms

$$P(n)\,(gw) = \left(\frac{H_i}{H_{Pl}}\right)^2 \left(\frac{n}{n_H}\right)^{n_t}, \qquad P(n)\,(dp) = \left(\frac{H_i}{H_{Pl}}\right)^2 \left(\frac{n}{n_H}\right)^{n_s-1}, \tag{1}$$

where the spectral indices $n_t$, $n_s$ are constants, and $n_s - 1 \approx n_t$. For flat kick evolutions, i.e. for horizontal lines like the red line in Fig. 1, the generated spectra have flat (Harrison-Zeldovich-Peebles) shape, i.e. $n_t = n_s - 1 = 0$. The over-kick or under-kick evolutions tilt the spectrum toward the 'blue' or 'red' shapes, respectively. Of course, it is simplistic to expect that the red transitions in Fig. 1 should be strictly straight lines and the spectral indices $n_t$, $n_s$ strict numbers independent of the wavenumber $n$. In simple models driven by scalar fields, the spectral indices are actually slowly decreasing functions of $n$.

The derivation of primordial spectra is based only on general relativity and quantum mechanics. The comparison of spectra and their consequences with observations is our best chance to learn whether the kick did indeed take place and how it looked like. In particular, assuming that the observed anisotropies of the cosmic microwave background radiation (CMB) are indeed caused by cosmological perturbations of quantum-mechanical origin, we conclude that the kick Hubble parameter $H_i$ was close to $10^{-5} H_{Pl}$, because this kick has generated the amplitudes of metric perturbations at the level of $10^{-5}$, which in their turn produced the observed large-scale CMB anisotropy at the level of $10^{-5}$.

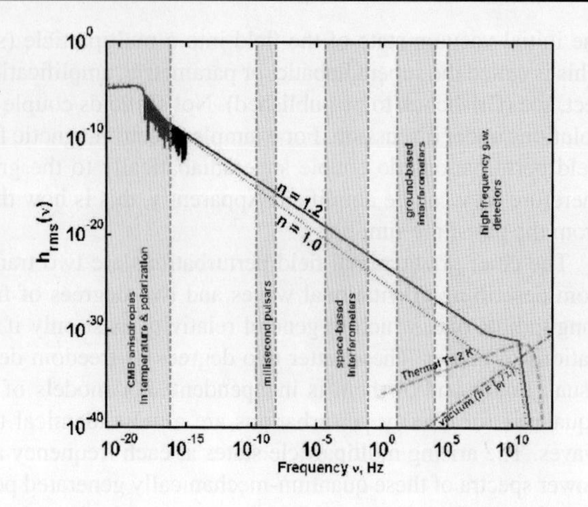


**Fig. 2** The present-day spectrum of the root-mean-square amplitude $h_{rms}(\nu)$ of relic gravitational waves. The *solid line* corresponds to the primordial spectral index $n_t = 0.2$ $(n = 1.2)$, while the *dashed line* is for $n_t = 0$ $(n = 1.0)$

The perturbations with wavelengths much shorter than the radius $l_p$ of our patch were processed in the course of evolution, and therefore the form of the spectrum in the recombination era (and today) differs from primordial. The spectrum is no longer a smooth function of the wavenumber (or frequency) but contains maxima and minima. This is a consequence of the standing-wave pattern of the primordial metric perturbations—the inevitable feature of the underlying quantum-mechanical squeezing. The power-law slope of the envelope of the oscillations also changes.

As an example, in Fig. 2, taken from (Grishchuk to be published), we show today's power spectra as functions of frequency $\nu$ for relic gravitational waves normalized to the observed CMB anisotropies. Only a few first cycles of oscillations at lowest frequencies are shown. Two possibilities are depicted in this figure: the dashed line is the resulting spectrum which originated in the superadiabatic amplification during the horizontal red line transition of Fig. 1, whereas the solid line is for the resulting spectrum originated in a slightly over-kick transition. At the highest relevant frequencies, before the spectrum sharply goes down, one can see a noticeable increase of power in $h_{rms}$. This is the result of the after-kick piece of evolution governed by matter with a stiff (Zeldovich) equation of state $p = \rho c^2$. This piece of evolution was assumed in this calculation, but it is not guaranteed. We may be not so lucky to have to deal in the future high-frequency experiments with this increased gravitational-wave power.

## 4 Cosmological Perturbations in Inflationary Theory

Now we turn to what is broadly addressed as inflation, and specifically to predictions of inflationary theory on cosmological perturbations. Predictions of inflationary theory are dramatically different from what was described above, in Sect. 3. In attempt of deriving density perturbations, inflationary authors invariably begin with vacuum fluctuations of the scalar field (inflation) in de Sitter space–time. Then, after some jumping between variables and gauges, they conclude that the amplitudes of scalar metric perturbations (often denoted by letters $\zeta$ or $\mathcal{R}$) should be infinitely large, in the same physical system and right from the very beginning. Instead of (1), the inflation-predicted primordial metric power spectrum for density perturbations reads:

$$P(n)\ (dp)|_{\text{inflation}} = \frac{1}{\epsilon}\left(\frac{H_i}{H_{Pl}}\right)^2\left(\frac{n}{n_H}\right)^{n_s-1}, \tag{2}$$

where the parameter $\epsilon$ is $\epsilon \equiv -\frac{\dot{H}}{H^2}$. This parameter is almost a constant in cases under discussion. The standard (de Sitter) inflation is characterized by $\epsilon = 0$ ($p = -\rho c^2$), and therefore the inflation-predicted standard spectrum blows up to infinity at every wavelength and for any non-zero value of $H_i$.

Formula (2) is the main contribution of inflationary theory to the subject of cosmological perturbations. Having arrived at the incorrect formula with the arbitrarily small factor $\epsilon$ in the denominator of this formula, inflationary theory elevates the question of the "energy scale of inflation" to the status of a major scientific problem, which inflationists will be happily solving for decades to come. Indeed, one is now free to start with energy densities $\rho_i$, say, 80 orders of magnitude smaller than $\rho_{Pl}$ and the Hubble parameter $H_i$ 40 orders of magnitude smaller than $H_{Pl}$ (grey line in Fig. 1). One can still claim that (s)he has built a successful inflationary model, because the required level $10^{-5}$ of primordial metric amplitudes for density perturbations can now be achieved by simply making the grey line sufficiently horizontal (but not exactly horizontal) and thus making the denominator $\epsilon$ in (2) sufficiently small (but not exactly zero). Astrophysics does not permit one to take $H_i$ smaller than the value of $H$ in the era of nucleosynthesis, i.e. below the grey line in Fig. 1. Otherwise, one could have started with the "energy scale of inflation" equal to the energy density of water and still build a successful inflationary cosmology.

Inflationary theory substitutes the predicted divergency of density perturbations at $\epsilon = 0$, (2), by the claim that it is the amount of relic gravitational waves that should be small. This is meant to be represented by the "tensor-to-scalar" ratio $r$, which is the ratio of the gw power spectrum from (1) to the dp power spectrum from (2). In inflationary theory, it is written as

$$r = 16\epsilon = -8n_t. \tag{3}$$

The WMAP (Wilkinson Microwave Anisotropy Probe) team reports the limits on the parameter $r$ that were found from the observations (Komatsu et al. 2008). One can see these limits in the form of the likelihood function for $r$ in the left panel of Fig. 3 in Komatsu et al. (2008). The maximum of the likelihood function is at $r = 0$, which means that the most likely value of the parameter $r$ is $r = 0$. Then, according to (3), the most likely value of $\epsilon$ is $\epsilon = 0$. Since $\epsilon$ is in the denominator of the inflation-predicted power spectrum (2), one concludes that if the inflationary predictions are correct, then the most likely values of density perturbations responsible for the data observed and analyzed by the WMAP team are infinitely large. Nevertheless, this situation is often qualified as the "striking success of inflation" (Baumann et al. 2008) demonstrated by the WMAP findings.

For the parameter $r$, the inflationary theory predicts practically everything what one can possibly imagine, including something like $r \leq 10^{-24}$ as the inflationary outcome, based on (2), of the most sophisticated string-inspired models (for a review, see Baumann et al. 2008 and references there). The big problem is not that the seemingly most advanced theories predict $r \leq 10^{-24}$, thus making the search for relic gravitational waves look ridiculous. The big problem is that when the relic gravitational waves are discovered, the proposers, being misguided by their own derivations, will reject microphysical theories which in fact may be perfectly viable, and will accept theories which in fact may be totally wrong.

## 5 Discovering Relic Gravitational Waves in the Cosmic Microwave Background Radiation

There are several reasons to believe that the observed CMB anisotropies are caused by cosmological perturbations of quantum-mechanical origin. Probably the major one is the observed oscillations in the CMB power spectra, which are likely to be a reflection of the quantum-mechanical squeezing and the associated standing-wave pattern of the primordial metric perturbations. If cosmological perturbations do indeed have quantum-mechanical origin, (1) implies that the gw and dp contributions to the lower-order CMB multipoles $\ell$ should be of the same order of magnitude. Relic gravitational waves have not been scattered or absorbed in any significant amount since the time of their quantum-mechanical generation. The explicit identification of relic gravitational waves in the data would be a monumental step in the study of the primordial kick and the origins of the Universe.

Density perturbations and gravitational waves affect temperature and polarization of the CMB. The polarization is usually characterized by the two components—$E$ and $B$. The $B$ component is not generated by density perturbations, and therefore it is often stated that the aim of the exercise is to detect B-modes. This is not so. The aim of the exercise is to detect relic gravitational waves, not to detect B-modes. Gravitational waves are present in all correlation functions of the CMB, and one should be smart enough to distinguish their contribution from competing contributions. The B-mode channel of information has some advantages, but many disadvantages too. For example, the 5-year WMAP data on the BB correlation function are not informative because they are mostly noise. At the same time, the much stronger TE signal is recorded at a large number of data points, which allows one to derive meaningful conclusions about the gw contribution.

The 5-year WMAP TE data were thoroughly analyzed in Zhao et al. (2009). The conclusion of this investigation is such that the lower-$\ell$ TE and TT data do contain a hint of presence of the gravitational wave contribution. In terms of the parameter $R$, which gives the ratio of contributions of gw and dp to the temperature quadrupole, the best-fit model produced $R = 0.24$. This means that 20% of the temperature quadrupole is accounted for by gravitational waves and 80% by density perturbations. The residual WMAP noise is high, so the uncertainty of this determination is large, and it easily includes the hypothesis that there is no gw contribution at all. However, the uncertainty will be much smaller in the forthcoming more sensitive observations, most notably with the *Planck* satellite. It is likely that the *Planck* mission will be capable of strengthening our belief that the primordial kick did take place, and at the near-Planckian values of the Hubble parameter, $H_i \approx 10^{-5} H_{Pl}$.

The future TE and BB data from *Planck* satellite were simulated and analyzed in Zhao et al. (2009). The quality of the future performance of the BB channel is not clear, so the authors discuss the 'optimistic' and 'realistic' options. The result of the analysis is shown in Fig. 3, where the signal to noise parameter $S/N$ is plotted as a function of $R$. It is seen from the graph that if the maximum likelihood value $R = 0.24$ derived from the WMAP5 TE and TT data is taken as a real signal, it will be present at a better than $3\sigma$ level in the *Planck's* TE observational channel, and at a better than $2\sigma$ level in the 'realistic' BB channel. The time of discovery of relic gravitational waves may be nearer than is usually believed.

## 6 Field-Theoretical Formulation of General Relativity

One of important premises in the above conclusions is the assumption that the general relativity remains valid up to energy densities approaching Planckian limit. Although there is no

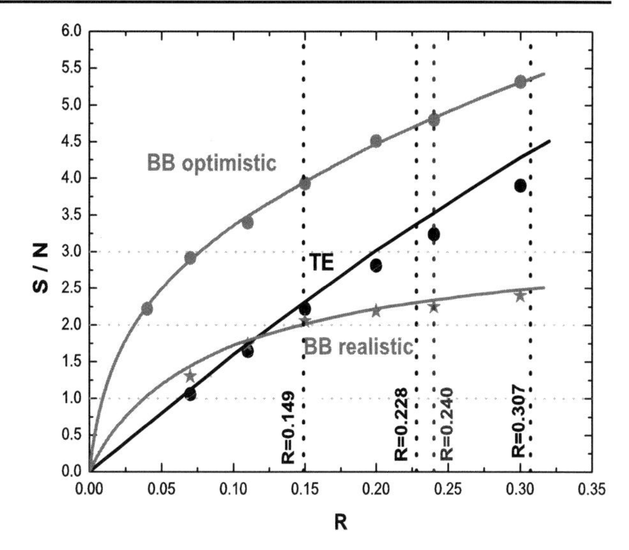


**Fig. 3** (Color online) The signal to noise ratio $S/N$ as a function of $R$ for the $TE$ (*black*) and $BB$ (*red*) observational channels. The *points* show numerical results, whereas the *curves* are analytical approximations

obvious reason to doubt this, any piece of new information on the domain of applicability of a fundamental theory is useful. The tests of gravitational theories in various circumstances and conditions is an important part of current research, including the results and plans discussed at this conference. As usual, the set-up of an experiment and interpretation of its results partially depend on the accepted theoretical framework, which in the case of general relativity seems to be obvious and well established: "gravity is geometry". I personally feel that the emphasis on the geometrical aspect of gravity has exhausted its usefulness and has become an impeding rather than a driving factor in gravitational physics.

There seems to be something odd in the conviction of a large part of our community that the gravitational waves are "oscillations in the fabric of space–time", and that the ongoing and planned gravitational-wave experiments are attempting to measure the "strain in the fabric of space–time". This understanding is usually presented as a consequence of the equivalence principle which forbids, so is believed, such things as rigorously defined local gravitational energy density and flux of energy. At best, this understanding allows surrogates, such as the averaged pseudotensors. The argument seems to be especially appropriate for the current research, wherein the gravitational-wave detectors are usually small and, in a sense, local. The size of the detector is much smaller than the wavelength which the detector is most sensitive to, so the gravitational-wave flux through the detector is supposed to be completely or partially removable by a coordinate transformation. (In fact, the four pseudotensor components describing energy density and flux are removable by four coordinate transformation functions everywhere, not only locally.) Therefore, the argument goes, it is the "squeezing and stretching of space–time itself" that is a proper description of the phenomenon, and not the absorption of the part of the gravitational wave flux by the detector, which makes the detector 'click' and register the event.

The problem appears to be more than simply linguistic. The concept "gravity is geometry" can be misleading in the analysis of the detector's response and in the interpretation of results. More importantly, it obscures the deeper understanding of gravity. Namely, the fact that the Einstein's general relativity is a perfectly consistent non-linear gravitational field theory in flat (Minkowski) space–time, with rigorously defined gravitational energy-momentum tensor, and with no need for geometrical notions of curved space–time.

The field-theoretical approach to general relativity is based on the concept of gravitational field $h^{\mu\nu}(x^\alpha)$ defined in flat space–time with the metric tensor $\gamma_{\mu\nu}(x^\alpha)$. The curvature tensor constructed from $\gamma_{\mu\nu}(x^\alpha)$ is identically zero:

$$\breve{R}_{\alpha\beta\mu\nu}(\gamma_{\rho\sigma}) = 0. \tag{4}$$

The choice of coordinates $x^\alpha$ is in one's hands, so the $\gamma_{\mu\nu}(x^\alpha)$ can always be transformed, if necessary, to the Minkowski matrix $\eta_{\mu\nu}$ (class of Lorentzian coordinates). In arbitrary curvilinear coordinates, the covariant derivatives are denoted by a semicolon ";".

The gravitational Lagrangian $L^g$ is a quadratic function of first derivatives of the field $h^{\mu\nu}$ (see Babak and Grishchuk 1999, 2003 and references there). The variation of $L^g$ with respect to the field variables $h^{\mu\nu}$ leads to the gravitational equations of motion:

$$\frac{1}{2}\left[(\gamma^{\mu\nu} + h^{\mu\nu})(\gamma^{\alpha\beta} + h^{\alpha\beta}) - (\gamma^{\mu\alpha} + h^{\mu\alpha})(\gamma^{\nu\beta} + h^{\nu\beta})\right]_{;\alpha;\beta} = \kappa t^{\mu\nu}. \tag{5}$$

In the right hand side (r.h.s) of these equations stands the gravitational energy-momentum tensor $t^{\mu\nu}$ (and $\kappa \equiv 8\pi G/c^4$). The $t^{\mu\nu}$ is rigorously defined as variational derivative of $L^g$ with respect to variations of the metric tensor $\gamma_{\mu\nu}$, with the constraint (4) properly taken into account. The $t^{\mu\nu}$ contains squares of first derivatives of the field $h^{\mu\nu}$, but not higher derivatives. The gravitational field equations were deliberately rearranged to the form of (5), where $t^{\mu\nu}$ is the manifest source for the generalized wave (d'Alembert) operator standing in the left hand side (l.h.s.) of the equations.

In the presence of matter Lagrangian $L^m$, the right hand side of (5) changes:

$$\frac{1}{2}\left[(\gamma^{\mu\nu} + h^{\mu\nu})(\gamma^{\alpha\beta} + h^{\alpha\beta}) - (\gamma^{\mu\alpha} + h^{\mu\alpha})(\gamma^{\nu\beta} + h^{\nu\beta})\right]_{;\alpha;\beta} = \kappa (t^{\mu\nu}|_m + \tau^{\mu\nu}). \tag{6}$$

The $\tau^{\mu\nu}$ is the matter energy-momentum tensor derived as variational derivative of $L^m$ with respect to the metric tensor $\gamma_{\mu\nu}$, whereas $t^{\mu\nu}|_m$ is the modified version of the gravitational energy momentum-tensor $t^{\mu\nu}$. The $t^{\mu\nu}|_m$ includes the term arising from $L^m$ and describing the interaction of gravity with matter.

The divergence (contracted covariant derivative) of the left hand side of (5), (6) vanishes identically. This means that the equations of motion contain the differential conservation laws

$$t^{\mu\nu}{}_{;\nu} = 0, \qquad (t^{\mu\nu}|_m + \tau^{\mu\nu})_{;\nu} = 0. \tag{7}$$

The differential conservation laws can be converted into conserved integrals for isolated non-radiating systems.

The geometrical description of gravity can be introduced as follows. The special form of $L^g$, which is the field-theoretical analog of the Hilbert-Einstein Lagrangian, and the universal coupling of gravity to matter assumed in $L^m$, which is a realization of the equivalence principle, allow one to 'glue together' the metric tensor $\gamma^{\mu\nu}$ and the field variables $h^{\mu\nu}$ into one tensorial object $g^{\mu\nu}$. The Lagrangians $L^g$ and $L^m$, as well as the gravitational and matter field equations, can now be rewritten in terms of this object and its derivatives alone, plus matter variables.

Specifically, one introduces $g^{\mu\nu}$ according to the rule

$$\sqrt{-g}g^{\mu\nu} = \sqrt{-\gamma}(\gamma^{\mu\nu} + h^{\mu\nu}), \tag{8}$$

and $g_{\mu\nu}$ according to the definition $g_{\mu\alpha}g^{\alpha\nu} = \delta^{\nu}_{\mu}$. Then, the field equations (5), (6) absorb $\gamma^{\mu\nu}$ and $h^{\mu\nu}$ into $g^{\mu\nu}$ and take the form of Einstein's geometrical equations

$$R_{\mu\nu} = 0, \qquad R_{\mu\nu} - \frac{1}{2} g_{\mu\nu} R = \kappa T_{\mu\nu}, \tag{9}$$

where $R_{\mu\nu}$ is the Ricci curvature tensor constructed from $g_{\mu\nu}$ in the usual manner. The $g_{\mu\nu}$ can be interpreted as a metric tensor of some effective curved space–time. The matter energy-momentum tensor $T_{\mu\nu}$ in (9) is variational derivative of $L^m$ with respect to $g_{\mu\nu}$, in contrast to $\tau^{\mu\nu}$ which was variational derivative of $L^m$ with respect to $\gamma_{\mu\nu}$.

Note that the field equations as a whole can be rewritten in terms of $g_{\mu\nu}$, but not individual parts of these equations. The gravitational energy-momentum tensor $t^{\mu\nu}$ cannot be rewritten as a function of the tensor $g_{\mu\nu}$ and its first derivatives alone. This is something to be expected, as there is no tensor that one could build from $g_{\mu\nu}$ and first derivatives of $g_{\mu\nu}$, apart of $g_{\mu\nu}$ itself. Therefore, there does not exist any meaningful gravitational energy-momentum tensor in the geometrical version of general relativity.

The universal character of the gravitational field, which allows one to 'glue together' $\gamma^{\mu\nu}$ and $h^{\mu\nu}$ into a single object $g^{\mu\nu}$ in the total Lagrangian and in the equations of motion, makes the flat space–time 'non-observable' in the presence of gravitational fields. This is not surprising. We would have encountered the same 'non-observability' of flat space–time in classical electrodynamics had we had access only to test particles with one and the same charge to mass ratio $e/m$. In the absence of neutral particles, which are capable of drawing the lines and angles of the Minkowski world in the region occupied by electromagnetic field, one would be given the option to interpret the motion of charged particles as arising due to the 'curvature of space–time itself' rather than due to the external electromagnetic field. This is a possible interpretation, but not particularly illuminating, at least in this case.

The generally covariant theories, including general relativity, admit arbitrary coordinate transformations. A coordinate transformation acting on an object, such for example as the metric tensor or the energy-momentum tensor of the electromagnetic field, can always be rearranged to state that the coordinates are not touched, but the object itself receives an increment in the same point. The rule of changing the numerical values of the object to the new ones involves the technique of Lie derivatives and depends on the transformation properties of the object and a given coordinate transformation. If a dynamical equation is formulated as equality to zero of some tensor composed of the underlying objects, the increments of this tensor vanish when the dynamical equation is satisfied. So, a solution of this dynamical equation translates into a new solution. This procedure of changing the values of objects in the same coordinate frame and generating new solutions can be called a gauge transformation.

The ability to make arbitrary coordinate transformations in the field-theoretical general relativity can be rearranged to state even more. Namely, that the coordinate system $x^{\alpha}$ and the metric tensor $\gamma_{\mu\nu}(x^{\alpha})$ remain untouched, but the set of gravitational field variables $h^{\mu\nu}(x^{\alpha})$, as well as matter variables, change to another set of variables in the same coordinate system $x^{\alpha}$. Under this operation, the gravitational field equations (5), (6), as well as the matter field equations and the conservation laws (7), transform into a combination of themselves, so that a solution of field equations translates into another solution. A coordinate transformation interpreted this way can be called a 'true' gauge transformation, because it looks very much similar to the gauge (gradient) transformation of classical electrodynamics, which changes electromagnetic potentials, but not coordinates and metric. However, the origin of gauge transformations in gravity is different—it is all the same arbitrary coordinate transformations. In general, there is no any other symmetries.

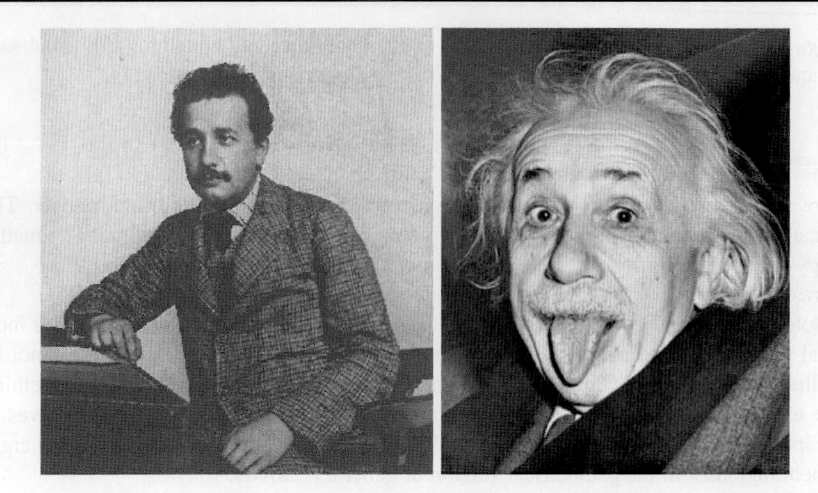

**Fig. 4** Two famous photos of Einstein. *Left*: 1905, Bern. *Right*: 1951, Princeton

While the equations as a whole are gauge-invariant (transform into a combination of themselves) under the action of 'true' gauge transformations, individual parts of the equations are not. For example, in (5) the l.h.s. and r.h.s. receive individual non-zero increments, but such that they precisely cancel each other, leaving (5) gauge-invariant. To demand the gauge-invariance of $t^{\mu\nu}$ on its own, thus attempting to make $t^{\mu\nu}$ "physically significant" and not "devoid of physical meaning" (Butcher et al. 2008), would be equivalent to demanding that the field equations should be violated after a gauge transformation.

The same holds true for conservation laws (7). Even if the energy-momentum tensor $\tau^{\mu\nu}$ represents only a couple of free particles, $\tau^{\mu\nu}$ changes under the action of 'true' gauge transformations. So, the associated change of $t^{\mu\nu}|_m$ must take care of new positions and new dynamics of the particles. The $t^{\mu\nu}|_m$ cannot be gauge-invariant on its own, but the gauge-related solutions are observationally equivalent, at least in the classical domain of the theory, and as long as one ignores initial and boundary conditions.

As far as one can see at present, the geometrical and field-theoretical pictures of gravity are representations of one and the same theory of general relativity, not different theories. Each of the viewpoints has its advantages and disadvantages, and we have to become eloquent in both of them. Feynman once remarked (Feynman 1965): "if the peculiar viewpoint taken is truly experimentally equivalent to the usual in the realm of the known there is always a range of applications and problems in this realm for which the special viewpoint gives one a special power and clarity of thought, which is valuable in itself".

In Fig. 4 one can see two famous photographs of Einstein. He is shown at times of the beginning and the end of his scientific career. In the left photo, the physicist Einstein is here, in Bern, in 1905. He was denouncing the notions of absolute space and time, but was not yet under the influence of geometrical techniques. In the right photo, Einstein is in Princeton, in 1951, when the idea that the space–time is something like a "fabric", which can curve, wrap, expand, oscillate, etc. was accepted. By that time, Einstein has twice changed his opinion about the reality of gravitational waves. This second picture, is it not addressed to those of us who believe too much literally that "gravity is geometry"?

## 7 Generalizing the General Relativity

The special power and clarity of thought mentioned by Feynman may be especially valuable now, when we are facing the situation that some modifications of general relativity may be required. Hopefully, the recent indications on the accelerated expansion of our approximated homogeneous isotropic Universe can be resolved by more accurate understanding of the limits of applicability of this approximation, and this will be so much for the "dark energy". But if not, the prospect of modifications of Einstein's gravity will be looming large. The internal consistency of the candidate modified theory will be a larger hurdle to overcome than agreement with observations.

Without a physical guidance, we are in front of an ocean of possibilities. The geometrical route "gravity is geometry" will definitcly lead to undesirable higher-order differential equations in terms of $g_{\mu\nu}$, as there is no way of modifying the second-order dynamical equations (9) except adding a new fundamental constant—the cosmological $\Lambda$-term. In contrast, the very logic of the field-theoretical approach allows natural and consistent generalizations of general relativity without raising the order of differential field equations (Babak and Grishchuk 1999, 2003). These generalizations are based on the possibility (and maybe necessity) of adding the 'mass'-term

$$\sqrt{-\gamma}\left[k_1 h^{\rho\sigma} h_{\rho\sigma} + k_2 h^2\right] \tag{10}$$

to the gravitational Lagrangian $L^g$. In this expression, the constants $k_1$ and $k_2$ have dimensionality of $[length]^{-2}$, and $h \equiv h^{\mu\nu}\gamma_{\mu\nu}$.

The linearized approximation of this generalized gravity allows one to interpret the introduced constants as masses $m_2$ and $m_0$ of $spin-2$ and $spin-0$ gravitons,

$$\left(\frac{cm_2}{\hbar}\right)^2 = 4k_1, \qquad \left(\frac{cm_0}{\hbar}\right)^2 = -2k_1\frac{k_1+4k_2}{k_1+k_2}. \tag{11}$$

The interpretation in terms of masses implies $m_0 > 0$ and $m_2 > 0$, but the theory allows also $(m_0)^2 < 0$ and $(m_2)^2 < 0$. When both constants $(m_0)^2$ and $(m_2)^2$ are sent to zero, this finite-range gravitational theory smoothly approaches the equations and observational predictions of the massless general relativity—the property not shared by many other proposed modifications.

It is amazing to see how much the class of allowed solutions broadens and changes under the impact of the seemingly innocent modification (10) of general relativity. The modifications affect the polarization states and propagation of gravitational waves, the event horizons of black holes, the early-time and late-time cosmological evolution. In cosmological applications, the signs and values of $(m_0)^2$ and $(m_2)^2$ are crucial for the character of changes in the early-time and late-time evolution. In particular, if one allows $(m_0)^2$ to be negative, the late-time evolution of a homogeneous isotropic universe experiences an accelerated expansion, which we may be required to explain if the observations keep demanding this phenomenon. The graph of the modified scale factor $a$ as a function of time is shown in Fig. 5, taken from the second paper in Babak and Grishchuk (1999), Babak and Grishchuk (2003).

## 8 Conclusions

The quest for better understanding of gravity is far from over. This conference is certainly not the last one on the topic of nature of gravity. Hopefully, important progress will be

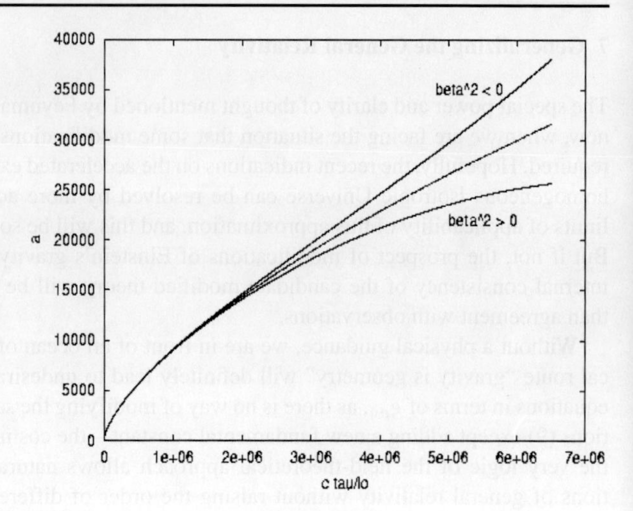


**Fig. 5** The *dashed line* is a Friedman solution of general relativity. The *upper solid line* is a solution of the finite-range gravity for $(\frac{cm_0}{\hbar})^2 \equiv \beta^2 < 0$, while the lower *solid line* is a solution for $\beta^2 > 0$

reached in the near future, particularly as a result of new experiments, such as *Planck* and others. Having discovered relic gravitational waves, we will be more confident about the limits of applicability of general relativity and will better understand how cosmic gravity works in concert with quantum mechanics. More excitement is anticipated.

**Acknowledgements** I am grateful to D. Baskaran and W. Zhao for help in preparation of this manuscript.

## References

S.V. Babak, L.P. Grishchuk, Phys. Rev. D **61**, 024038 (1999)
S.V. Babak, L.P. Grishchuk, Int. J. Modern Phys. D **12**(10), 1905 (2003)
D. Baumann et al., CMBPol mission concept study. Probing inflation with CMB polarization, 2008. arXiv: 0811.3919
V.A. Belinsky, L.P. Grishchuk, I.M. Khalatnikov, Ya.B. Zeldovich, Phys. Lett. B **155**, 232 (1985a)
V.A. Belinsky, L.P. Grishchuk, I.M. Khalatnikov, Ya.B. Zeldovich, Sov. Phys. JETP **62**, 427 (1985b)
L.M. Butcher, A. Lasenby, M. Hobson, Phys. Rev. D **78**, 064034 (2008)
R.P. Feynman, in *Physics Nobel Prize Lectures* (Elsevier, Amsterdam, 1965) p. 155
L.P. Grishchuk, Phys. Rev. D **45**, 4717 (1992)
L.P. Grishchuk, Discovering relic gravitational waves in cosmic microwave background radiation, in Wheeler Book, ed. by I. Ciufolini, R. Matzner (Springer, Berlin, to be published). arXiv:0707.3319
E. Komatsu et al., Five-year WMAP Observations: Cosmological Interpretation, 2008. arXiv:0803.0547
LHC web-site: http://lhc-machine-outreach.web.cern.ch/lhc-machine-outreach/
S. Weinberg, *Cosmology* (Oxford University Press, London, 2008)
Y.B. Zel'dovich, Cosmological field theory for observational astronomers. Sov. Sci. Rev. E Astrophys. Space Phys. **5**, 1–37 (1986). http://nedwww.ipac.caltech.edu/level5/Zeldovich/Zel_contents.html
W. Zhao, D. Baskaran, L.P. Grishchuk, Phys. Rev. D **79**, 023002 (2009)

Space Sci Rev (2009) 148: 329–346
DOI 10.1007/s11214-009-9616-4

# The Cosmic Microwave Background and Fundamental Physics

**Anthony Lasenby**

Received: 9 September 2009 / Accepted: 23 November 2009 / Published online: 8 January 2010
© Springer Science+Business Media B.V. 2009

**Abstract** A brief overview of the links between the Cosmic Microwave Background (CMB) and fundamental physics is given. After a summary of the basics of the CMB, and of current observations, topics are considered including the generation of perturbations, inflation and string theory, evidence for parity violation, variation of the fundamental constants, restrictions on alternative theories of gravity, primordial non-Gaussianity, early- and late-time Bianchi universes and topological defects.

**Keywords** Cosmic microwave background · Early universe · Cosmology · Gravity

## 1 Introduction

This review attempts to give a brief overview of the current state of Cosmic Microwave Background (CMB) observations and theory, and to look at the impact of the CMB as regards the contribution it can make to fundamental physics and to theories of gravitation.

The CMB stands in a primary position amongst current cosmological data in terms of its ability to give us information about inflation, gravity waves, string cosmology, parity violation, universal rotation, topological defects and non-Gaussianity. It is currently *not* as good as regards giving unambiguous constraints on alternative gravity theories and fundamental constant variation. Some of this is due to the relevant calculations not having been fully carried out yet. However, it is also because in the CMB we can currently accommodate many forms of scalar field, for example in inflation or quintessence, and in many features 'alternative gravity' theories look like GR plus a scalar field. However, we will nevertheless attempt to point out many unique things coming from the CMB.

A. Lasenby (✉)
Astrophysics Group, Cavendish Laboratory, J.J. Thomson Avenue, Cambridge, CB3 0HE, UK
e-mail: a.n.lasenby@mrao.cam.ac.uk

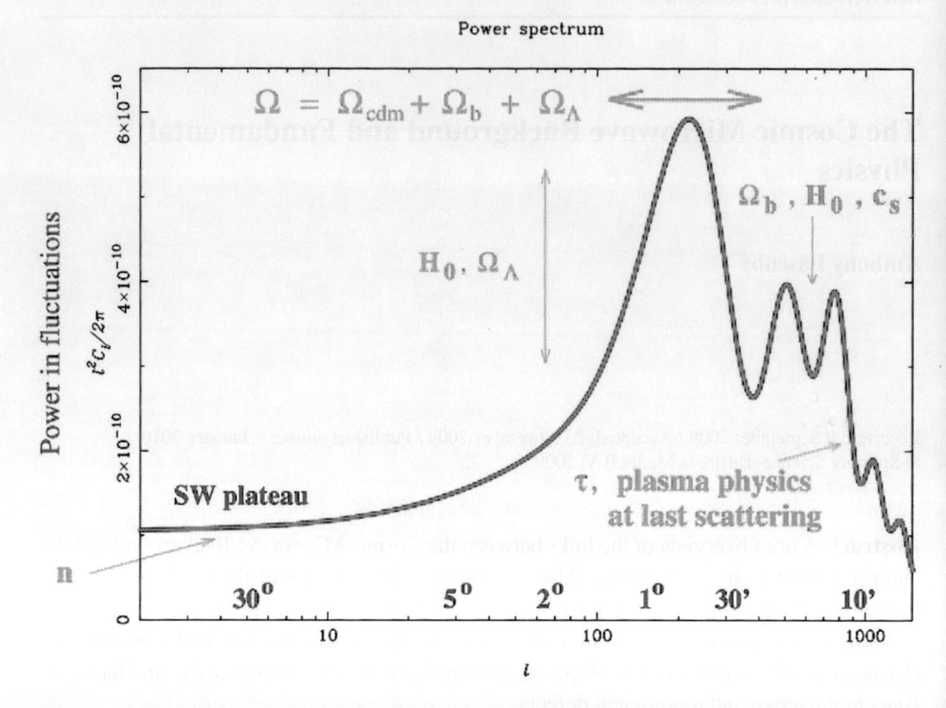


**Fig. 1** Schematic form of expected CMB power spectrum, showing important dependencies

## 2 The CMB

Several fine reviews exist about CMB physics and the latest observational situation—for an up to date and thorough treatment, see Challinor and Peiris (2009). Here we shall simply give a brief overview, since our main focus is implications of the CMB for fundamental physics.

The CMB is a thermal relic of hot 'big bang' with almost perfect blackbody spectrum. Its temperature today of 2.726 K implies a CMB photon number density $4 \times 10^8$ m$^{-3}$. Of these photons, ~90% last-interacted with matter at recombination ($z \sim 1000$); the remaining 10% suffered further Thomson scattering once the Universe reionised around $z \sim 11$–$12$. We can infer this from fits to the optical depth back to recombination, for which accurate figures started arriving with the third year of WMAP data (Page et al. 2007).

Fluctuations in photon phase space density and gravitational potential give rise to small temperature anisotropies ($\sim 10^{-5}$). A key quantity we can construct from observations of these is the CMB temperature power spectrum. In Fig. 1 we show a typical theoretical power spectrum with its dependence on major parameters indicated. The power spectrum is defined in terms of a decomposition of the sky into spherical harmonics:

$$\frac{\delta T}{T}(\theta, \phi) = \sum_{\ell,m} a_{\ell m} Y_{\ell m}(\theta, \phi).$$

The rotationally invariant power spectrum is defined from this by

$$C_\ell = \langle |a_{\ell m}|^2 \rangle.$$

What is plotted in Fig. 1 is $\ell(\ell+1)C_\ell$, which is the power per unit log interval in $\ell$. A rough way of translating between $\ell$ and angular scale is indicated in the figure.

Physical lengths at recombination get translated to an angle on the sky via the angular diameter distance formula. This is mainly a function of $\Omega_{\text{total}}$, so the left/right position of the peaks in the power spectrum as a function of $\ell$ is a sensitive indicator of the total energy density of the universe (baryonic, dark and an effective cosmological constant component). This is shown as the double arrow over the first peak in the spectrum: a universe *above* critical density has its peaks shifted to the *left* relative to this diagram, while a universe below critical density has its peaks shifted to the right. The spacing and height of the secondary peaks in the power spectrum depend on the detailed physics which occurs during the recombination epoch, and this is indicated by the dependence upon the baryon density $\Omega_b$, the Hubble constant $H_0$ and the sound speed at that epoch. There is a roughly exponential cutoff in primordial perturbations expected as one goes to higher $\ell$ (smaller angular scale) dictated by the combined effects of the total optical depth, $\tau$, between recombination and the present, and processes such as Silk damping during recombination.

Two other important parameters concern the initial input power spectrum coming from inflation. The parameter $n_s$, the slope of the primordial power-law power spectrum of scalar perturbations, is defined via

$$\langle|\delta_k|^2\rangle \propto k_s^n$$

where the $\delta_k$s are the initial matter power spectrum. Inflation (in its most common forms) predicts $n_s \lesssim 1$. This is what would give a roughly flat slope in the CMB power spectrum on large scales, and corresponds to the region labelled 'SW plateau' in the diagram, named after the Sachs–Wolfe effect (Sachs and Wolfe 1967), which dominates in this region.

The second is the ratio $r$ of tensor (gravitational wave) to scalar (density) perturbations. Gravitational waves (direct perturbations in the metric) are produced during inflation, and their amplitude is a measure of the energy scale at which inflation occurred, which is so far unknown. A key fact about these waves is that they cause a particular mode of CMB polarisation, called the $B$ mode. The CMB polarisation field on the sky can be divided into two components in a way which is rotationally invariant (unlike the decomposition into the $Q$ and $U$ Stokes parameter combinations, which depends on reference frame). These two components are called $E$ and $B$ modes, after their analogy to divergence and curl components of a vector field. Due to its curl nature, the $B$ mode flips sign under parity inversion. This handedness is a quick way of seeing that the $B$ mode of polarisation cannot be sourced by scalar perturbations. Detection of primordial $B$ modes is therefore a 'smoking gun' for the presence of primordial gravitational waves, and the ratio $r$ of the tensor component of the perturbations to the amplitude of scalar perturbations, at a given scale, is a very important parameter to attempt to measure. From $r$, we can directly constrain the energy scale of inflation, via the relation

$$r \sim 0.008\left(\frac{E_{\text{inf}}}{10^{16}\ \text{GeV}}\right)^4. \tag{1}$$

Figure 2 shows the predicted power spectra arising from scalar and tensor perturbations (left and right panels respectively) for the temperature ($T$), $E$ and $B$ modes, and for the cross-correlation of $T$ and $E$. (Note: parity considerations predict zero $EB$ and $TB$ cross-power spectra.) In the left panel (scalars), there is a non-zero $B$-mode spectrum drawn, but this is the result of lensing of primordial $E$-modes into $B$-mode by intervening matter at late times, and is not primordial itself. The gravity wave $B$-mode spectrum in the right panel is plotted at the maximum value allowed by current data, $r=0.22$, and one can see clearly the huge

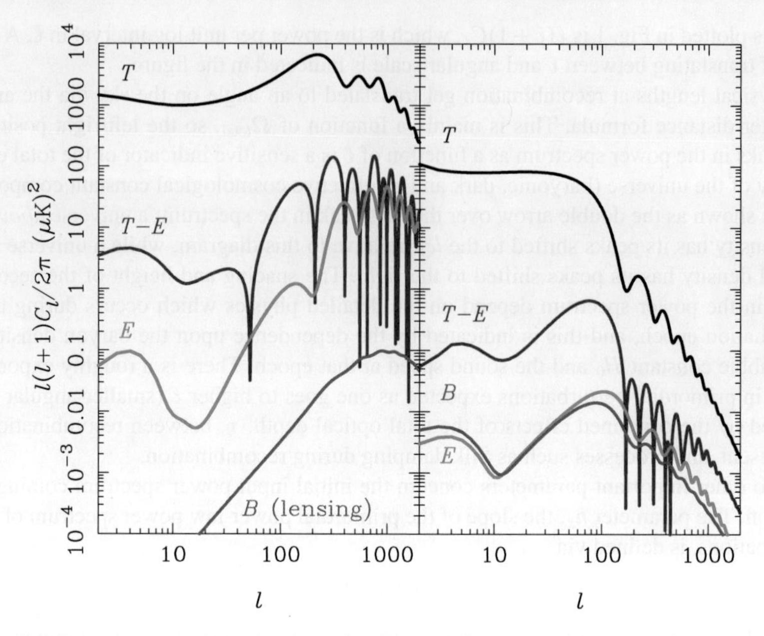


**Fig. 2** Temperature (*black*), $E$-mode (*green*), $B$-mode (*blue*) and $T$-$E$ cross-correlation (*red*) CMB power spectra from scalar perturbations (*left*) and tensor perturbations (gravitational waves; *right*). The amplitude of the tensor perturbations is shown at the maximum amplitude allowed by current data ($r = 0.22$, Komatsu et al. 2009). The $B$-mode spectrum induced by weak gravitational lensing is also shown in the *left-hand panel* (*blue*). (Courtesy A. Challinor.)

increase in sensitivity and accuracy that will be necessary to get down to these levels (and beyond, since from (1) we may well need to reach levels of $r \sim 0.01$, even for GUT-scale inflation), as compared to what was necessary to measure and characterise the temperature anisotropies.

## 3 Current CMB Observations

Figure 3 shows a compilation of temperature anisotropy results, taken from Reichardt et al. (2009) (see this reference for details of the three sets of observations plotted). The results shown here are not the most up to date possible, but do allow a convenient comparison with the form of the predicted power spectrum shown in Fig. 1, at least out to $\ell \sim 1000$. We see that there is extremely good agreement between the observations and the predicted power spectrum for a $\Lambda$-CDM model, except perhaps at the $\ell = 2$ point (quadrupole), where there is a deficit in the WMAP5 value, as compared to the expected value. This is discussed further below.

Rather than give plots of the current $E$-mode and $B$-mode results in comparison with the expected power spectra, we simply note that recent results from the QUAD experiment at the South Pole (Brown et al. 2009) show that the expected peak structure in the $E$-mode at scales between about 200 and 2000 in $\ell$ (see Fig. 2) has been definitely detected, at high significance, and that recent results from BICEP (also at the South Pole), give a direct limit to the $B$ mode level of $r < 0.73$ at 95% confidence (Chiang et al. 2009). We note that this is much larger than the limit of $r < 0.22$ at 95% given by Komatsu et al. (2009), and used

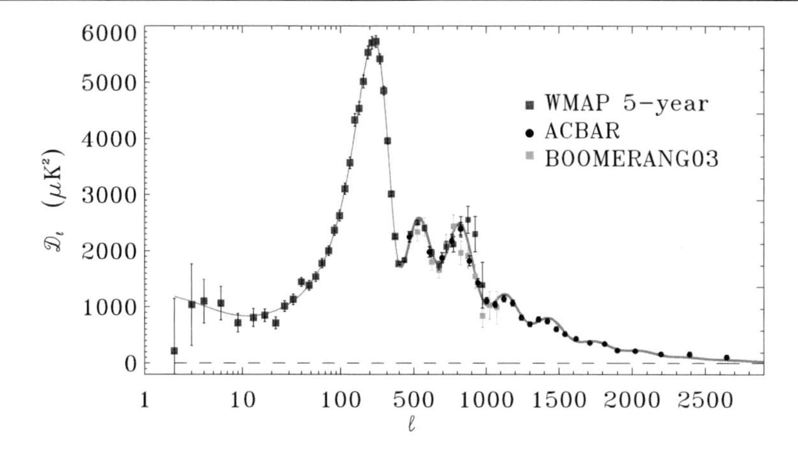


**Fig. 3** Observations of the power spectrum of CMB temperature anisotropies, taken from Reichardt et al. (2009). Note the $D_\ell$ plotted on the $y$ axis is the same as $\ell(\ell+1)C_\ell/2\pi$

in Fig. 2. This is because the latter is not a direct limit, but comes via a combination of constraints from $T$ and $E$ mode CMB, together with large-scale structure data and supernovae. Chiang et al. (2009) show that the direct upper limit on $B$-modes from WMAP data is $r < 6$, i.e. considerably larger again.

The next two years should see a very considerable improvement in CMB measurements on all scales with data starting to come through from the Planck satellite, which was successfully launched on May 14th 2009. Detection of an $r$ value as low as 0.05 should be possible with a Planck mission that includes 4 sky coverages (see Efstathiou and Gratton 2009), as well as a much improved measurement of $n_s$. The current constraints in the $(n_s, r)$ plane are shown in Fig. 4, taken from Komatsu et al. (2009). This shows that simple inflation models with a $\lambda\phi^4$ type potential, and of order 50 to 60 e-folds of inflation, are now ruled out by the data. However, an $m^2\phi^2$ type chaotic potential is still allowed, and if this *is* the correct model, we note it would give an encouragingly large value of $r$, around 0.1, and that this should certainly be measurable by the next generation of terrestrial polarisation experiments or Planck. Irrespective of model, we can also see from the figure that there is currently just over $2\sigma$ evidence that $n_s$ is less than one. Beyond this, there is currently no compelling evidence for any other features in the primordial power spectrum, although a cutoff at low $k$, which could give rise to the low quadrupole value, is a possibility which will be discussed further below. In Bridges et al. (2008a), a Bayesian reconstruction of the primordial spectrum using current data is carried out, and this shows that the Bayesian evidence values peak at a reconstruction involving just 3 nodes, again pointing to the fact that current data quality is insufficient for more detailed analysis.

## 4 The CMB and the Nature of Gravity

Having introduced the basics of the CMB and current results, we now discuss some important areas where there is an interplay between the CMB and fundamental theories of gravity. This is a large field, so the following is essentially a personal selection of what could be discussed under this topic, with some examples to give a flavour of what is involved.

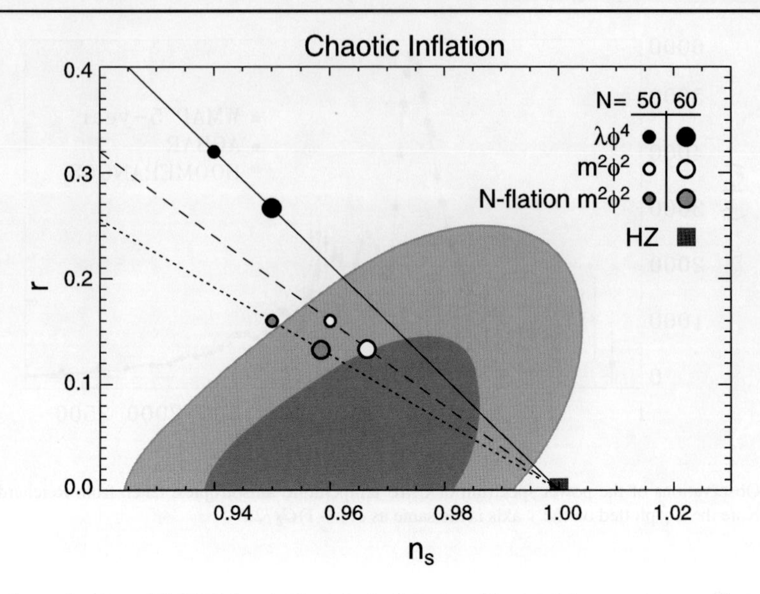


**Fig. 4** Constraints from WMAP5 data in the $(n_s, r)$ plane. 1 and $2\sigma$ confidence regions are shown, along with the position expected on the plot for a chaotic inflation potential of the types $\lambda\phi^4$ and $m^2\phi^2$. The N-flation model consists of many massive axion fields. (Taken from Komatsu et al. (2009).)

## 4.1 The Generation of Perturbations

Where do the seed fluctuations which the CMB power spectrum measures come from? As we know, the basic idea is that they are generated during inflation via the amplification of initial quantum fluctuations. Thus the CMB immediately links us to a very profound and still not fully understood area of gravitational physics—particle production in changing background fields. The standard 'slow roll inflation' answers for the perturbation amplitudes in gravitational waves and scalars are (respectively):

$$\begin{aligned} \mathcal{P}_{\text{grav}}(k) &\sim \left(\frac{H}{2\pi}\right)^2, \\ \mathcal{P}_{\mathcal{R}}(k) &\sim \left(\frac{H}{2\pi}\right)^2 \left(\frac{H}{\dot{\phi}}\right)^2, \end{aligned} \tag{2}$$

where $H$ is the Hubble parameter, and $\dot{\phi}$ is the rate of change of the inflaton field, and both are evaluated at the moment the $k$ mode of interest passes out of the horizon. However, these expressions make an assumption about the nature of the 'initial' vacuum—basically that the perturbations come out of a state where the background geometry is evolving in a de Sitter-like way. In cases where background fields are not evolving like this, for example when there are a small number of e-folds on the largest scales of interest (which happens for closed universe models, or where the initial $\phi$ field is kinetic energy dominated (see e.g. Boyanovsky et al. 2006)) or there exists shear (e.g. in Bianchi models), then there can be modes for which this approach is not possible. Then one can attempt a generalised WKB analysis in which one uses the adiabatic invariance technique developed in the context of curved space field theory in the 1970s (see e.g. Fulling 1989).

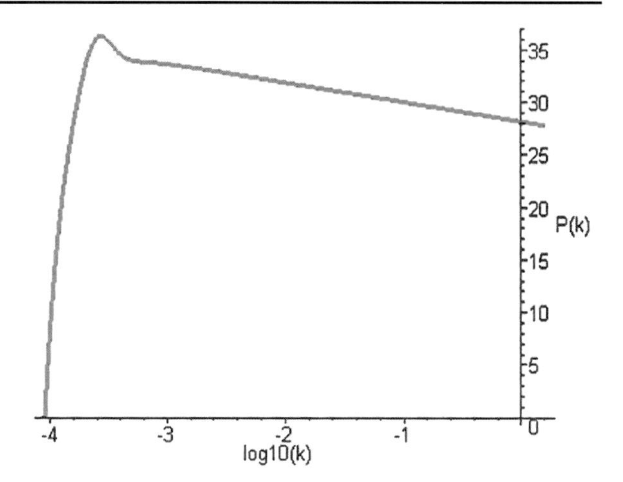


**Fig. 5** Prediction for the primordial scalar power spectrum carried out using a full WKB adiabatic invariance approach for a closed universe model of the type considered in Lasenby and Doran (2005). $k$ is comoving wavenumber measured in units of $\mathrm{Mpc}^{-1}$ and the vertical scale is a measure of amplitude

For example, in Lasenby and Doran (2005), a slightly closed universe model was proposed, based on a novel boundary condition at infinite future time. We used the $H^2/\dot{\phi}$ formula to approximate the perturbation production in order to calculate the primordial spectrum in this model. This gave a spectrum closely matching the normal expectations from slow roll, except for an exponential cutoff at low $k$ which displayed an interesting fit to a phenomenological form suggested by Efstathiou (2003) in response to the low WMAP quadrupole. Later (in work so far unpublished), calculations were carried out more properly, in a full WKB approach, which makes no assumptions about an initial de Sitter phase. A typical result from this is shown in Fig. 5. The spectrum matches very well with the results of using the formula (2), except for an interesting signature 'bump' at low $k$. This translates through to give a small change in the CMB power spectrum at low $\ell$, but given the effects of cosmic variance it is unclear whether this will ever be observable.

In the context of Bianchi I models, Pitrou et al. (2008) have recently shown that there are some modes on large scales for which even a generalised WKB process breaks down. It is so far unclear what to do next! As will be argued later, in Sect. 4.9, our own universe may actually be like this initially (or at least Bianchi IX), so this problem may be something that needs to be faced up to. In any case, undoubtedly the CMB gives us a laboratory in which to study profound questions at the interface of quantum theory and gravity.

### 4.2 Inflation and String Theory

If inflation comes out of string theory, then one can regard the CMB as a key tool giving us information about low-energy effective string theory, i.e. on gravity coupled to particle physics at energies around the GUT scale. However, it goes further than this. String theory does not work with explicit potentials in 4d space–time, but with moduli fields and fluxes in 10 or 11 dimensions. The existence of something like an inflationary era is then by no means guaranteed, since there may not be an effective potential of the right form that emerges out of the string phase. If an inflationary sector of the theory can be found, and several candidates have emerged over recent years, one can ask about its properties in relation to the key CMB observables discussed above. In this respect one is getting direct information about string theory itself, or at least one corner of the 'landscape' of possibilities that it encompasses.

The parameter which has turned out to be most crucial and informative in this respect is the parameter $r$ discussed above, which gives the ratio of tensor to scalar perturbations in the CMB power spectra. (This is part of the reason that measurement of $r$ is currently so crucial.) The current situation is that almost all models of string inflation give small levels of gravitational waves.

We can understand this problem as follows. In canonical single field models, Lyth (1997) showed that

$$r = \frac{8}{M_{\rm Pl}^2}\left(\frac{d\phi}{dN}\right)^2 \tag{3}$$

where $M_{\rm Pl}$ is the Planck mass, and $N$ is the number of e-folds. Thus field evolution over 50–60 e-folds implies a total change in the inflaton field of $\Delta\phi \sim (r/0.002)^{1/2}$ in units of $M_{\rm Pl}$. Thus detectable gravity waves means the inflaton evolved through a *super-Planckian distance*. One problem is that there may be geometrical effects in string theory moduli which make this super-Planckian evolution difficult to achieve, as discussed in e.g. Baumann and McAllister (2007) and Kallosh and Linde (2007).

Additionally, it is now believed that having a smooth potential over $\Delta\phi > M_{\rm Pl}$ is problematic for an effective field theory with a cutoff $\Lambda_c < M_{\rm Pl}$ unless a *shift symmetry* exists, which protects against higher order corrections to the potential. This view has been strongly advocated by Daniel Baumann (see e.g. Baumann 2009 and Baumann and McAllister 2009), and would mean that an experimental confirmation that tensor modes exist at a detectable level, would imply that such a symmetry *must* exist for the potential, thus giving us a direct insight into string theory. The first 'stringy' models incorporating this type of symmetry, with axion-like potentials, are now starting to appear, for example the Axion Monodromy model of Flauger et al. (2009). This leads to a broad $\phi^2$ type potential, but with superposed oscillations, and thus potentially unique observable effects in the CMB.

### 4.3 Further Links with Fundamental Theory—Parity Violation?

The Komatsu et al. (2009) WMAP 5-year paper considers the possibility of parity-violating interactions between photons and dark matter.

The Lagrangian would be of the form

$$\mathcal{L} = -\frac{1}{2} p_\alpha A_\beta \overline{F}^{\alpha\beta} \tag{4}$$

where $A_\alpha$ is the standard EM vector potential, $\overline{F}^{\alpha\beta}$ is the dual of the standard Faraday tensor, and $p_\alpha$ is taken as the gradient of a light scalar field (standing in for some aspect of dark matter).

This interaction (of Chern–Simons form) violates parity, and makes the two polarisation states of a photon propagate with different group velocities, and hence causes the polarisation plane to rotate. This is called cosmological birefringence, and was discussed along with proposals for how to observe it by Carroll (1998).

The rotation of the polarisation plane makes CMB $E$-modes transform into $B$-modes, like the effects of gravitational lensing. However, in this case there is a parity violation (*unlike* lensing), which means $TB$ and $EB$ cross-power spectra are produced, which is a distinctive signature of such effects. Thus these spectra (now distributed with the WMAP likelihood code) are included in the analysis for this purpose, and a likelihood analysis carried out to determine if there is evidence for rotation of the plane of polarisation. The results from this are shown in Fig. 6, and expressed as a confidence interval the result is

$$\Delta\alpha = -1.7^\circ \pm 2.1^\circ \quad (95\%\ \text{conf.}) \tag{5}$$

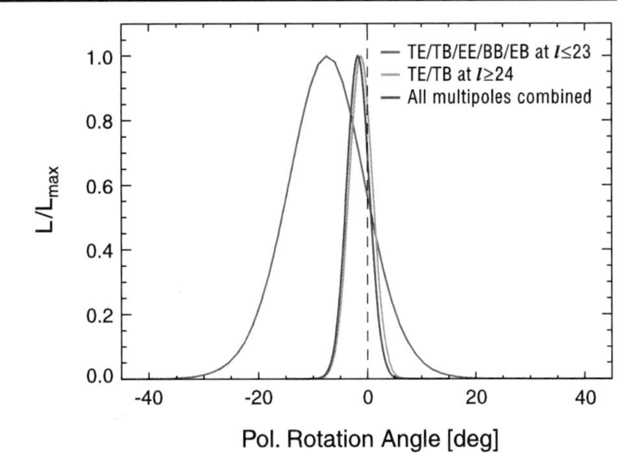


**Fig. 6** Constraint on the polarisation rotation angle from WMAP 5-year data (from Komatsu et al. 2009). The *red curve* shows the results from all multipoles combined

when all $\ell$ ranges combined. (We note lower $\ell$ ranges measure rotation from reionisation to us, higher from recombination to us.)

An analysis of this kind has also recently been carried out using the QUAD data. This yields the somewhat tighter constraints (Brown et al. 2009):

$$\Delta\alpha = 0.64^\circ \pm 0.50^\circ \text{ (random)} \pm 0.50^\circ \text{ (systematic)}. \quad (6)$$

### 4.4 Constraints on $\alpha$ Variation

The issue of variation with time of the fundamental constants is one where the CMB can potentially help a great deal. Provided the constant involved is dimensionless, then questions about the effects of its variation can be physically well-posed, although even so one must be very careful to try to think through all the changes that might be relevant, since uses of particular values may be deeply embedded within the physics with which one is familiar. Variation in the fine structure constant

$$\alpha = \frac{e^2}{4\pi\epsilon_0 \hbar c} \quad (7)$$

affects the CMB primarily via the changes it induces in the Thomson scattering cross section, and various derived constants important during the recombination era. Summaries of these effects are given in Kaplinghat et al. (1999) and Stefanescu (2007). The type of constraints one can obtain from the CMB (see e.g. Rocha et al. 2003 for results using WMAP first-year data) are many orders of magnitude less restrictive than those using e.g. quasar absorption line data, discussed elsewhere in this volume. However, the CMB results constitute constraints at a redshift of $z \sim 1100$, and thus can nevertheless represent useful input to theories in which $\alpha$ has a particular form of variation with time.

Figure 7 shows some results from work using WMAP3 data to constrain $\alpha$, and illustrates the difference in CMB power spectra for $\alpha$ at its best-fit value when it is allowed to vary, versus when it is fixed (to the present day value). The resulting constraints on $\alpha$ variation are

$$-0.039 < \frac{(\alpha_{\rm rec} - \alpha_0)}{\alpha_0} < 0.010. \quad (8)$$

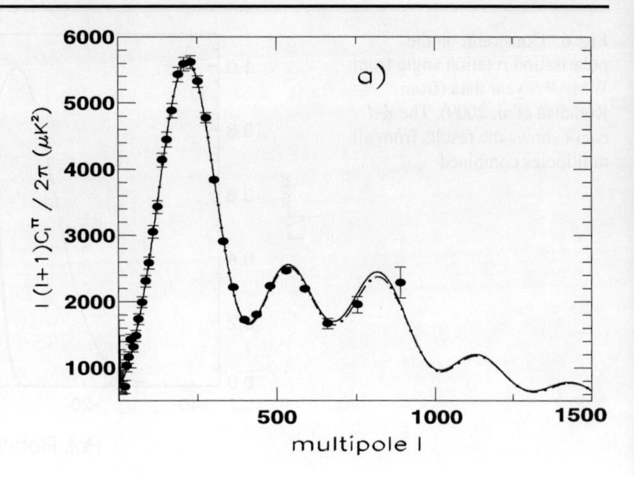


**Fig. 7** The CMB temperature power spectrum CTT for a best-fit $\Lambda$-CMD model with varying $\alpha$ (*continuous lines*) and for the best-fit $\Lambda$-CDM model with constant $\alpha$ (*dot-dash lines*). The data shown are WMAP3. (Taken from Stefanescu 2007.)

This is at 65% confidence. The best-fit value is negative and would correspond to

$$\alpha^{-1}\frac{d\alpha}{dt} = 4.65 \times 10^{-13}\ \text{yr}^{-1}. \tag{9}$$

We note that there is definitely room for a repeated analysis of this problem, using the latest CMB data.

## 4.5 Constraints on MOND and Modified Gravity

Modified Newtonian Dynamics (MOND) theories attempt to explain the effects attributed to dark matter, and in particular galactic rotation curves, by a modification of the Newtonian law of gravitation at low accelerations. The theories were, as the name suggests, initially non-relativistic (Milgrom 1983), but a relativistic version was proposed by Bekenstein (2004), known as TeVeS (for tensor, vector, scalar), and this provides an alternative to GR which, although somewhat complicated, can be tested in the setting of relativistic cosmology.

The generic approach to trying to test theories of this kind is to note that the essence of MOND and TeVeS is that $\Omega_{\text{cdm}} = 0$, i.e., there is no dark matter, only baryonic!

It was found quite early that one can match the amplitude and position of the 3rd and subsequent peaks in the CMB power spectrum by adding in neutrinos in some combination. However, one then has to get the scale right with respect to the 1st and 2nd peaks. Again, generically, cold dark matter means the 3rd peak will be higher than the 2nd, whereas baryon domination gives strictly decreasing peaks.

With 2005 data, Slosar et al. (2005) showed that cold dark matter was favoured against a purely baryonic composition, using just CMB data alone, at Bayesian odds of 200:1. A comparison between such models and the data is shown in Fig. 8. This seems fairly definitive, but again there is definitely room for another go at this with the latest data, since the QUAD, ACBAR and CBI experiments have brought strong improvements to the accuracy to which the 3rd and subsequent peaks are delineated.

A further very important point to note is that comparisons such as these are not taking proper account of what the theories themselves are potentially predicting for the evolution of perturbations and effects on the CMB—all that is used from the theory is the idea that $\Omega_{\text{cdm}} = 0$, and analyses are thereafter carried out in a conventional context.

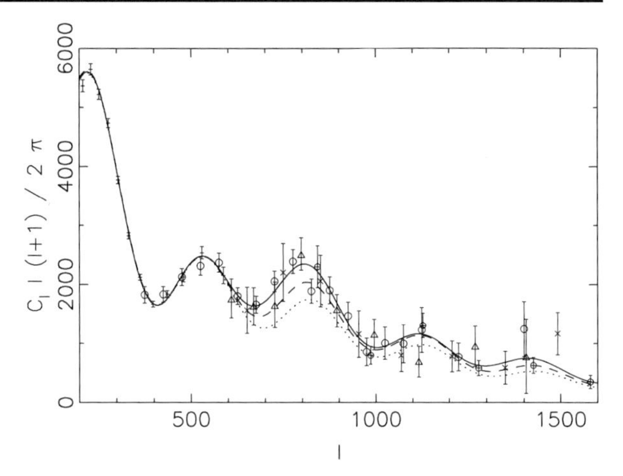


**Fig. 8** Comparison between experimental data and predicted spectra to test theories in which the dark matter component is very small. The *solid line* is for a standard $\Lambda$-CDM model, and the *dashed* and *dotted lines* are for the best-fitting models with $\Omega_{\rm cdm} < 0.01$. Taken from (Slosar et al. 2005), which gives details of the data used

This is obviously a big omission, but it is starting to be repaired in more recent attempts at using 'modified gravity' to explain dark matter and dark energy. For example, Zuntz et al. (2008) gives a summary of comparison between data and theory for the 'Einstein-aether' theory, which contains a Lorentz-invariance violating vector field. Here computations are properly carried out within the context of the theory itself. Interestingly, they find a best-fit aether model which is mildly superior to the standard $\Lambda$-CDM cosmology, at about $2\sigma$, at a cost of two extra parameters. Thus there seems still a great deal of room for theories of this kind to survive in the context of current CMB data.

## 4.6 Primordial Non-Gaussianity

Over the past two years it has become clear to the wider CMB community that primordial non-Gaussianity is a powerful discriminant among competing theories of inflation. The current focus is on the quantity $f_{\rm NL}$ defined for a curvature perturbations $\Phi(\boldsymbol{r})$ by

$$\Phi(\boldsymbol{r}) = \Phi_{\rm L}(\boldsymbol{r}) + f_{\rm NL}\big(\Phi_{\rm L}^2(\boldsymbol{r}) - \langle\Phi_{\rm L}^2(\boldsymbol{r})\rangle\big). \tag{10}$$

This gives the 'non-linear' correction to an underlying linear field. In single-field, slow roll inflation, one expects $f_{\rm NL} \sim 1$. This is actually a very low degree of non-Gaussianity (remember the basic $\Phi(\boldsymbol{r})$ value is $\sim 10^{-5}$) and corresponds to the expectation that the initial quantum states are like the ground state of a harmonic oscillator, i.e. Gaussian!

The claims that have caught attention are from Yadav and Wandelt (2008), who analysed WMAP3 data and found

$$27 < f_{\rm NL} < 147 \quad (95\%\ \text{confidence}). \tag{11}$$

This would exclude the hypothesis of Gaussian fluctuations at 99.5% confidence. If real, this would be extremely significant, and favour non-standard forms of inflation, such as DBI (Dirac–Born–Infeld, Alishahiha et al. 2004) for which one expects $f_{\rm NL} \sim 50$, or ekpyrotic scenarios (Lehners and Steinhardt 2008).

However, galactic foregrounds, which are themselves highly non-Gaussian, are a big problem! The most authoritative current analysis is probably from the WMAP 5-year results of Komatsu et al. (2009). They obtain

$$-9 < f_{\rm NL} < 111 \quad (95\%\ \text{confidence}) \tag{12}$$

and thus do *not* exclude Gaussianity, although the skew towards positive values is similar to that found by Yadav and Wandelt (2008). The jury is definitely still out on this, but the claims have certainly increased awareness that non-Gaussianity is a crucial indicator of the type of inflation. The Planck satellite is going to offer substantial improvements in this area. The increased number of $\ell$-modes (due to higher resolution), better foreground frequency coverage and better polarisation sensitivity mean that it is likely to be able to measure $f_{NL}$ to an accuracy better than 5, which would be sufficient to resolve theories such as single-field chaotic inflation, and DBI inflation, with a signal to noise ratio of 10.

## 4.7 A Bianchi Model Universe?

The sky revealed by WMAP appears to have some interesting possible departures from what would be expected for a perfect statistically homogeneous and isotropic Gaussian field. Several authors have commented on a significant North/South asymmetry in the WMAP data (e.g. Eriksen et al. 2004 and Hoftuft et al. 2009). Furthermore, a strange alignment has been reported between low multipoles of the CMB distribution on the sky (see e.g. de Oliveira-Costa et al. (2004) who discuss the alignment of the quadrupole and octupole.) Finally, a significantly non-Gaussian 'cold spot' was found in Vielva et al. (2004) and drawn further attention to in Cruz et al. (2005).

Partially as a response to this, Jaffe et al. (2005) returned to a question which has arisen at several different points in the history of CMB research: Is there evidence for universal shear and rotation imprinted in the CMB? This issue deserves a place in a discussion of the role of the CMB in fundamental physics, since it could in principle be bound up with questions such as Mach's hypothesis, and worries as to whether the notion of the universe *as a whole* rotating make philosophical sense. In fact, both vorticity and shear, which was what the CMB is sensitive to, can be defined in terms of locally measurable quantities within a fluid, so probably worries such as these can be avoided here, although the questions are intriguing.

The role of the CMB in constraining universal rotation was first addressed in Hawking (1969), and a thorough theoretical study carried out in Barrow et al. (1985). Using results from this latter paper, Jaffe et al. (2005) fitted a particular form of Bianchi template to WMAP sky. The model used was Bianchi $\text{VII}_h$, which is a generalisation of an open FRW universe to the anisotropic case. They found a best fit with $\Omega_0 = 0.5$, and the coldest part of template corresponded with the non-Gaussian spot drawn attention to in Cruz et al. (2005). As an example of the sensitivity of the CMB to universal rotation, one can note that the magnitude of the angular velocity fitted by Jaffe et al. (2005) is about $6 \times 10^{-10}$ radians per Hubble time. The relation between the WMAP sky, the cold spot, and the Bianchi template is illustrated in Fig. 9. A very intriguing feature of this fit is that removal of the Bianchi template at the level found simultaneously removes the non-Gaussianity effects associated with the hot-spot, the anomalous nature of the low order alignments and the major component of North–South asymmetry alluded to above. However, the best-fit value of $\Omega_0 = 0.5$ is of course in dramatic conflict with most other astrophysical indicators, and does not result in a viable cosmological model. There was thus interest in extending the Barrow et al. (1985) results to the case including a cosmological constant, which was not considered in the original work. Specifically there was hope that a low matter content (necessary to give the right sort of 'pitch' to the Bianchi spiral pattern) could be accompanied by a cosmological constant sufficient to give an almost flat universe, consonant with the rest of the observations. This extension was first carried out by Jaffe et al. (2006), and then extended to a full Markov Chain Monte Carlo (MCMC) analysis of cosmological models by Bridges et al. (2007). This

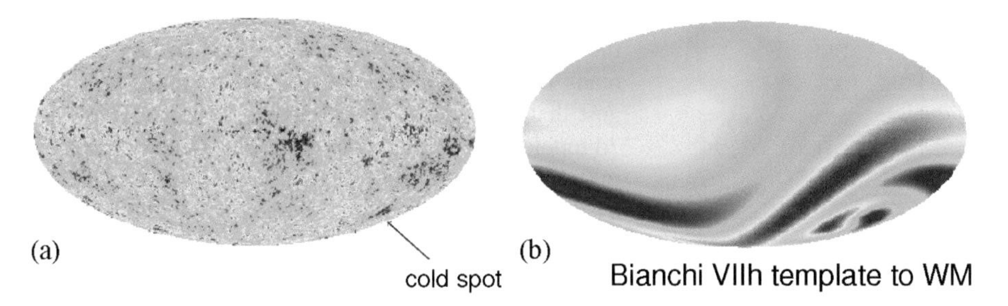


**Fig. 9** WMAP3 map with the cold spot position indicated. (**b**) the Bianchi $\text{VII}_h$ template fitted by Jaffe et al. (2005)

study found that even including a $\Lambda$, it was not possible to find a region in parameter space jointly compatible with a Bianchi template and with the cosmological data (specifically the WMAP 3-year power spectrum, and large-scale structure data).

### *4.7.1 Bianchi Polarisation*

An interesting further area with respect to the issue of a 'rotating universe', comes from the question of *polarisation* in Bianchi models.

Polarisation in Bianchi $\text{VII}_h$ models has recently been considered by Pontzen and Challinor (2007). They extended the calculation of the type carried out in Collins and Hawking (1973) and Barrow et al. (1985) to a full radiative transfer calculation including polarisation. They find a polarisation peak at low $\ell$ due to rapid decay of shear, with roughly equal $E$ and $B$ modes' power. An interesting feature of this is that the effects of the Bianchi model do quite well at 'simulating' the reionisation bump in the $E$-mode power spectrum, i.e. the raised section for $\ell \lesssim 10$. The big surprise, however, is that the predicted $B$-mode peak is big enough, so that already existing limits on $B$-mode anisotropy already rule out the possibility of a Bianchi component as large as that in the Jaffe et al. model! Thus these models of late-time Bianchi anisotropy, which have been interesting in simultaneously explaining more than one anomaly in the CMB data, appear definitely not to be viable.

There is one more possibility in this area, which arises from the fact that only a particular mode of shear and vorticity was explored in the analyses described above. In fact, the Bianchi $\text{VII}_h$ cosmology supports several more such modes, and it would be worthwhile to explore the parameter space belonging to these, as well as other Bianchi universes, such as the closed Bianchi IX models, as well. (See Pontzen (2009) for a review of CMB effects across a wide range of Bianchi models.)

## 4.8 Could the Cold Spot Be a Cosmic Texture?

The 'cold spot' referred to above, has raised interest in another way in which the CMB is relevant to fundamental physics. Currently, although it is believed very likely that symmetry-breaking phase transitions in the early universe would have led to topological defects which persisted to later times, there is no direct evidence for such defects. Detection of one would be extremely important for high-energy physics in reflecting, somewhat like inflation, energy scales and regimes beyond the reach of Earth-bound accelerators. The most likely way such a defect could be observed is via either its effects imprinted onto the CMB, or via gravitational lensing phenomena induced by it.

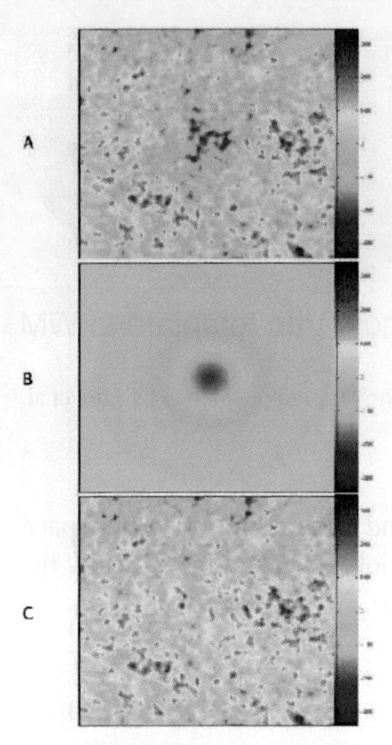


From (Cruz et al. 2007).

The discovery of the 'cold spot' mentioned above raises the interesting issue of whether we may in fact have already discovered such a defect. Cruz et al. (2007) suggest that this cold spot is due to a 'texture'. These are 3d topological defects, coming from symmetry breaking in (typically) the SU(2) gauge group. The type modelled by them is a global texture: these unwind gradually and can have late time effects on the CMB. The particular model considered by them has the texture at $z \sim 6$, and so in principle effects might also be visible via lensing of background galaxies. In the figure at the left, *A* shows piece of actual CMB sky, *B* the texture model and *C* the CMB after model subtraction. The authors consider other possible explanations, including the possibility of a giant void at this position, but find greater statistical evidence for the texture explanation as compared to the alternative explanations. A fit to the amplitude of the effect gives a symmetry-breaking scale of $8.7 \times 10^{15}$ GeV, highlighting again that, as with cosmic strings, detection would certainly be very important for our understanding of particle theory.

We note that Bridges et al. (2008b) re-examined previous conclusions about fitting Bianchi models to the WMAP data in the light of the possibility that the cold spot is not the focus of a Bianchi spiral, but due to another separate effect (e.g. the texture model). Using WMAP data 'corrected' for the texture fit they found that both right- and left-handed spiral Bianchi models were left almost entirely unconstrained by the data and consequently exhibited significantly reduced Bayesian evidences. Both Bianchi models were in fact significantly disfavoured by the data. This result reinforces the suggestion that it was the cold spot that was driving the previous apparent Bianchi $\text{VII}_h$ detections.

## 4.9 Effects of Anisotropy During Inflation?

Section 4.7 was concerned with what we can call late-time Bianchi models. This is where our current universe is taken to have shear and rotation, and the effects are laid down during recombination, and during propagation to us from then.

An alternative, which has recently begun to be explored, is the investigation of effects of anisotropy during inflation itself. In particular, could some of the large-scale anomalies in the CMB be laid down during inflation, during isotropisation from some earlier phase? Some recent works which explore this include Emir Gümrükçüoglu et al. (2007), Pereira et al. (2007) and Pitrou et al. (2008). Using a rather different approach, Dechant et al. (2009) have been looking at closed Bianchi models (Bianchi IX) containing a scalar field. In particular, they look at a biaxially symmetric model in which as one approaches the 'big bang', one radius of curvature obeys $R_1 \propto t$ while the remaining two satisfy $R_2 = R_3 \mapsto$ const. (as does the scalar field $\phi$).

The surprise is that this model gets smoothly through the 'big bang', with no singularities in any physical quantities, to a symmetric (though presumably parity-inverted) universe beforehand.

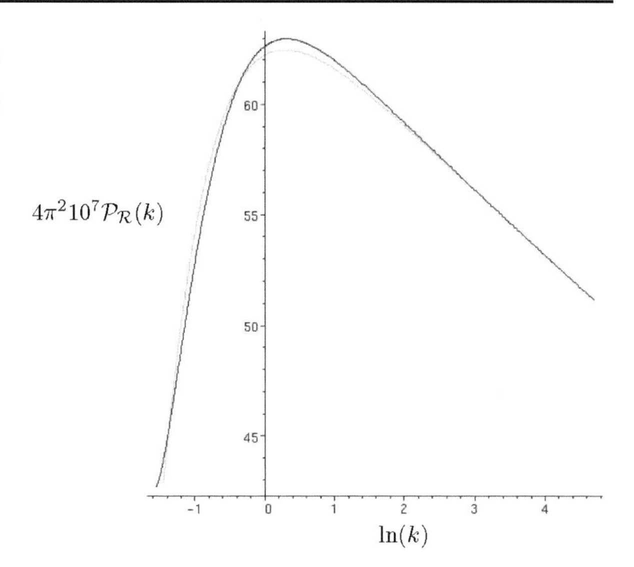


**Fig. 10** Predicted perturbation power spectrum in the early-time Bianchi IX model of Dechant et al. (2009). The *grey line* is a fit to the spectrum with an exponential cutoff proposed by Efstathiou (2003). (Note that the $k$ values are offset by a constant compared to those in Fig. 5.)

The period of inflation due to the scalar field makes sure the universe is spherically symmetric by the end of inflation, and also that $\Omega$ ends up close to 1. However, on the largest scales the universe is just oblate ($\sim$ 0.2% level) at the point where the perturbations are being laid down. This suggests one will be able to get phase correlations and some large-scale asymmetries present due to this, which is very interesting given the claims for such effects mentioned in Sect. 4.7. Calculating these effects will take getting to grips fully with the issues of principle mentioned in Sect. 4.1 in connection with perturbation calculations in anisotropic models. However, ignoring this, and performing a crude power spectrum calculation, displays a very nice fit to the phenomenological form suggested by Efstathiou (2003) in response to the low WMAP quadrupole; see Fig. 10. Also, the tensor mode looks acceptable at $r \sim 0.17$, although we note that the joint constraint from WMAP5 and large-scale structure mentioned above ($r < 0.22$ at 95% confidence), is starting to approach this level.

Investigating the stability of the biaxial setup, Dechant et al. (2009) find that the full triaxial case with $R_3(t) = R_2(t)(1+\delta(t))$, and $\delta$ small, oscillates until inflation kicks in, and then 'freezes out' in a fashion very similar to linear perturbation modes at horizon crossing in FRW, as shown in Fig. 11.

## 5 Conclusions

In conclusion, we hope to have shown that the CMB is a very rich field with many applications and implications for fundamental physics. In terms of experimental quantities we wish to measure, non-Gaussianity and polarisation (which quantifies gravity waves and the energy scale of inflation) are two key things for the future, and there are good prospects of measuring these quite soon at levels which are highly significant in constraining theories of the early universe. We also note that there are still many interesting theoretical calculations remaining for the CMB, including calculation of the spectrum of perturbations in non-standard cases of current interest (e.g. early shear), and effects coming out of a string cosmology era.

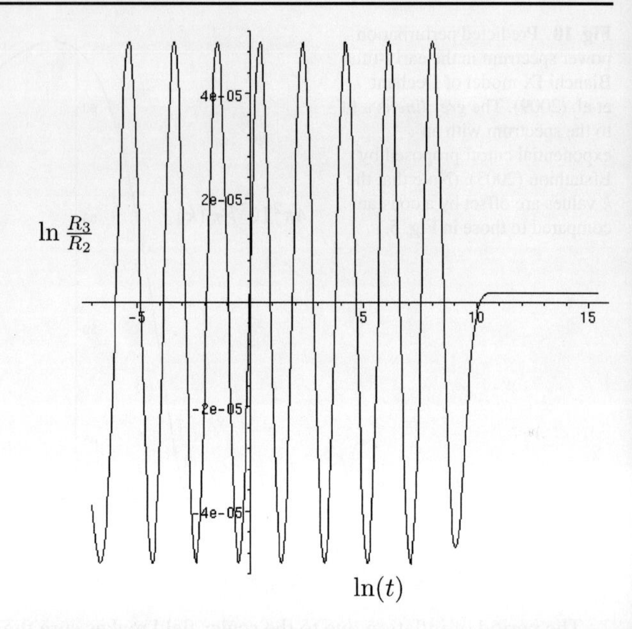


**Fig. 11** Dynamics of a full triaxial Bianchi IX model: The natural logarithm of the ratio $R3/R2$ of the nearly degenerate radii is plotted as a function of log time (in Planck units). (From Dechant et al. 2009.)

**Acknowledgements** I would like to thank Anthony Challinor and Hiranya Peiris for help in preparation of some of the material and for useful discussions. I would also like to thank Martin Huber and Rudolf Treumann for organising a very interesting meeting, and being patient editors.

## References

M. Alishahiha, E. Silverstein, D. Tong, DBI in the sky: Non-Gaussianity from inflation with a speed limit. Phys. Rev. D **70**(12), 123505 (2004). doi:10.1103/PhysRevD.70.123505

J.D. Barrow, R. Juszkiewicz, D.H. Sonoda, Universal rotation – How large can it be? Mon. Not. R. Astron. Soc. **213**, 917–943 (1985)

D. Baumann, TASI lectures on inflation. ArXiv e-prints (2009). arXiv:0907.5424

D. Baumann, L. McAllister, A microscopic limit on gravitational waves from D-brane inflation. Phys. Rev. D **75**(12), 123508 (2007). doi:10.1103/PhysRevD.75.123508

D. Baumann, L. McAllister, Advances in inflation in string theory. ArXiv e-prints (2009). arXiv:0901.0265

J.D. Bekenstein, Relativistic gravitation theory for the modified Newtonian dynamics paradigm. Phys. Rev. D **70**(8), 083509 (2004). doi:10.1103/PhysRevD.70.083509

D. Boyanovsky, H.J. de Vega, N.G. Sanchez, CMB quadrupole suppression. II. The early fast roll stage. Phys. Rev. D **74**(12), 123007 (2006). doi:10.1103/PhysRevD.74.123007

M. Bridges, J.D. McEwen, A.N. Lasenby, M.P. Hobson, Markov chain Monte Carlo analysis of Bianchi $\mathrm{VII}_h$ models. Mon. Not. R. Astron. Soc. **377**, 1473–1480 (2007). doi:10.1111/j.1365-2966.2007.11616.x

M. Bridges, F. Feroz, M.P. Hobson, A.N. Lasenby, Bayesian optimal reconstruction of the primordial power spectrum. ArXiv e-prints (2008a). arXiv:0812.3541v1

M. Bridges, J.D. McEwen, M. Cruz, M.P. Hobson, A.N. Lasenby, P. Vielva, E. Martínez-González, Bianchi $\mathrm{VII}_h$ models and the cold spot texture. Mon. Not. R. Astron. Soc. **390**, 1372–1376 (2008b). doi:10.1111/j.1365-2966.2008.13835.x

M.L. Brown, P. Ade, J. Bock, M. Bowden, G. Cahill, P.G. Castro, S. Church, T. Culverhouse, R.B. Friedman, K. Ganga, W.K. Gear, S. Gupta, J. Hinderks, J. Kovac, A.E. Lange, E. Leitch, S.J. Melhuish, Y. Memari, J.A. Murphy, A. Orlando, C. O'Sullivan, L. Piccirillo, C. Pryke, N. Rajguru, B. Rusholme, R. Schwarz, A.N. Taylor, K.L. Thompson, A.H. Turner, E.Y.S. Wu, M. Zemcov, Improved measurements of the temperature and polarization of the CMB from QUaD. ArXiv e-prints (2009). arXiv:0906.1003

S.M. Carroll, Quintessence and the rest of the world: suppressing long-range interactions. Phys. Rev. Lett. **81**, 3067–3070 (1998). doi:10.1103/PhysRevLett.81.3067

A. Challinor, H. Peiris, Lecture notes on the physics of cosmic microwave background anisotropies, in *American Institute of Physics Conference Series*, ed. by M. Novello, S. Perez. American Institute of Physics Conference Series, vol. 1132, 2009, pp. 86–140. doi:10.1063/1.3151849
H.C. Chiang, P.A.R. Ade, D. Barkats, J.O. Battle, E.M. Bierman, J.J. Bock, C.D. Dowell, L. Duband, E.F. Hivon, W.L. Holzapfel, V.V. Hristov, W.C. Jones, B.G. Keating, J.M. Kovac, C.L. Kuo, A.E. Lange, E.M. Leitch, P.V. Mason, T. Matsumura, H.T. Nguyen, N. Ponthieu, C. Pryke, S. Richter, G. Rocha, C. Sheehy, Y.D. Takahashi, J.E. Tolan, K.W. Yoon, Measurement of CMB polarization power spectra from two years of BICEP data. ArXiv e-prints (2009). arXiv:0906.1181
C.B. Collins, S.W. Hawking, The rotation and distortion of the universe. Mon. Not. R. Astron. Soc. **162**, 307 (1973)
M. Cruz, E. Martínez-González, P. Vielva, L. Cayón, Detection of a non-Gaussian spot in WMAP. Mon. Not. R. Astron. Soc. **356**, 29–40 (2005). doi:10.1111/j.1365-2966.2004.08419.x
M. Cruz, N. Turok, P. Vielva, E. Martínez-González, M. Hobson, A cosmic microwave background feature consistent with a cosmic texture. Science **318**, 1612 (2007). doi:10.1126/science.1148694
A. de Oliveira-Costa, M. Tegmark, M. Zaldarriaga, A. Hamilton, Significance of the largest scale CMB fluctuations in WMAP. Phys. Rev. D **69**(6), 063516 (2004). doi:10.1103/PhysRevD.69.063516
P.P. Dechant, A.N. Lasenby, M.P. Hobson, Anisotropic, nonsingular early universe model leading to a realistic cosmology. Phys. Rev. D **79**(4), 043524 (2009). doi:10.1103/PhysRevD.79.043524
G. Efstathiou, Is the low cosmic microwave background quadrupole a signature of spatial curvature? Mon. Not. R. Astron. Soc. **343**, 95–98 (2003). doi:10.1046/j.1365-8711.2003.06940.x
G. Efstathiou, S. Gratton, B-mode detection with an extended Planck mission. J. Cosmol. Astropart. Phys. **6**, 11 (2009). doi:10.1088/1475-7516/2009/06/011
A. Emir Gümrükçüoglu, C.R. Contaldi, M. Peloso, Inflationary perturbations in anisotropic backgrounds and their imprint on the cosmic microwave background. J. Cosmol. Astropart. Phys. **11**, 5 (2007). doi:10.1088/1475-7516/2007/11/005
H.K. Eriksen, F.K. Hansen, A.J. Banday, K.M. Górski, P.B. Lilje, Asymmetries in the cosmic microwave background anisotropy field. Astrophys. J. **605**, 14–20 (2004). doi:10.1086/382267
R. Flauger, L. McAllister, E. Pajer, A. Westphal, G. Xu, Oscillations in the CMB from axion monodromy inflation. ArXiv e-prints (2009). arXiv:0907.2916
S.A. Fulling, *Aspects of Quantum Field Theory in Curved Space-Time* (Cambridge University Press, Cambridge, 1989)
S. Hawking, On the rotation of the universe. Mon. Not. R. Astron. Soc. **142**, 129 (1969)
J. Hoftuft, H.K. Eriksen, A.J. Banday, K.M. Górski, F.K. Hansen, P.B. Lilje, Increasing evidence for hemispherical power asymmetry in the five-year WMAP data. Astrophys. J. **699**, 985–989 (2009). doi:10.1088/0004-637X/699/2/985
T.R. Jaffe, A.J. Banday, H.K. Eriksen, K.M. Górski, F.K. Hansen, Evidence of vorticity and shear at large angular scales in the WMAP data: a violation of cosmological isotropy? Astrophys. J. **629**, 1–4 (2005). doi:10.1086/444454
T.R. Jaffe, S. Hervik, A.J. Banday, K.M. Górski, On the viability of Bianchi type $\mathrm{VII}_h$ models with dark energy. Astrophys. J. **644**, 701–708 (2006). doi:10.1086/503893
R. Kallosh, A. Linde, Testing string theory with cosmic microwave background. J. Cosmol. Astropart. Phys. **4**, 17 (2007). doi:10.1088/1475-7516/2007/04/017
M. Kaplinghat, R.J. Scherrer, M.S. Turner, Constraining variations in the fine-structure constant with the cosmic microwave background. Phys. Rev. D **60**(2), 023516 (1999). doi:10.1103/PhysRevD.60.023516
E. Komatsu, J. Dunkley, M.R. Nolta, C.L. Bennett, B. Gold, G. Hinshaw, N. Jarosik, D. Larson, M. Limon, L. Page, D.N. Spergel, M. Halpern, R.S. Hill, A. Kogut, S.S. Meyer, G.S. Tucker, J.L. Weiland, E. Wollack, E.L. Wright, Five-year Wilkinson microwave anisotropy probe observations: Cosmological Interpretation. Astrophys. J. Suppl. Ser. **180**, 330–376 (2009). doi:10.1088/0067-0049/180/2/330
A. Lasenby, C. Doran, Closed universes, de Sitter space, and inflation. Phys. Rev. D **71**(6), 063502 (2005). doi:10.1103/PhysRevD.71.063502
J.L. Lehners, P.J. Steinhardt, Intuitive understanding of non-Gaussianity in ekpyrotic and cyclic models. Phys. Rev. D **78**(2), 023506 (2008). doi:10.1103/PhysRevD.78.023506
D.H. Lyth, What would we learn by detecting a gravitational wave signal in the cosmic microwave background anisotropy? Phys. Rev. Lett. **78**, 1861–1863 (1997). doi:10.1103/PhysRevLett.78.1861
M. Milgrom, A modification of the Newtonian dynamics as a possible alternative to the hidden mass hypothesis. Astrophys. J. **270**, 365–370 (1983). doi:10.1086/161130
L. Page, G. Hinshaw, E. Komatsu, M.R. Nolta, D.N. Spergel, C.L. Bennett, C. Barnes, R. Bean, O. Doré, J. Dunkley, M. Halpern, R.S. Hill, N. Jarosik, A. Kogut, M. Limon, S.S. Meyer, N. Odegard, H.V. Peiris, G.S. Tucker, L. Verde, J.L. Weiland, E. Wollack, E.L. Wright, Three-year Wilkinson microwave anisotropy probe (WMAP) observations: polarization analysis. Astrophys. J. Suppl. Ser. **170**, 335–376 (2007). doi:10.1086/513699

T.S. Pereira, C. Pitrou, J.P. Uzan, Theory of cosmological perturbations in an anisotropic universe. J. Cosmol. Astropart. Phys. **9**, 6 (2007). doi:10.1088/1475-7516/2007/09/006
C. Pitrou, T.S. Pereira, J.P. Uzan, Predictions from an anisotropic inflationary era. J. Cosmol. Astropart. Phys. **4**, 4 (2008). doi:10.1088/1475-7516/2008/04/004
A. Pontzen, Rogues' gallery: The full freedom of the Bianchi CMB anomalies. Phys. Rev. D **79**(10), 103518 (2009). doi:10.1103/PhysRevD.79.103518
A. Pontzen, A. Challinor, Bianchi model CMB polarization and its implications for CMB anomalies. Mon. Not. R. Astron. Soc. **380**, 1387–1398 (2007). doi:10.1111/j.1365-2966.2007.12221.x
C.L. Reichardt, P.A.R. Ade, J.J. Bock, J.R. Bond, J.A. Brevik, C.R. Contaldi, M.D. Daub, J.T. Dempsey, J.H. Goldstein, W.L. Holzapfel, C.L. Kuo, A.E. Lange, M. Lueker, M. Newcomb, J.B. Peterson, J. Ruhl, M.C. Runyan, Z. Staniszewski, High-resolution CMB power spectrum from the complete ACBAR data set. Astrophys. J. **694**, 1200–1219 (2009). doi:10.1088/0004-637X/694/2/1200
G. Rocha, R. Trotta, C.J.A.P. Martins, A. Melchiorri, P.P. Avelino, P.T.P. Viana, New constraints on varying $\alpha$. New Astron. Rev. **47**, 863–869 (2003). doi:10.1016/S1387-6473(03)00153-2
R.K. Sachs, A.M. Wolfe, Perturbations of a cosmological model and angular variations of the microwave background. Astrophys. J. **147**, 73 (1967)
A. Slosar, A. Melchiorri, J.I. Silk, Test of modified Newtonian dynamics with recent Boomerang data. Phys. Rev. D **72**(10), 101301 (2005). doi:10.1103/PhysRevD.72.101301
P. Stefanescu, Constraints on time variation of fine structure constant from WMAP-3yr data. New Astron. **12**, 635–640 (2007). doi:10.1016/j.newast.2007.06.004
P. Vielva, E. Martínez-González, R.B. Barreiro, J.L. Sanz, L. Cayón, Detection of non-Gaussianity in the Wilkinson microwave anisotropy probe first-year data using spherical wavelets. Astrophys. J. **609**, 22–34 (2004). doi:10.1086/421007
A.P.S. Yadav, B.D. Wandelt, Evidence of primordial non-Gaussianity ($f_{NL}$) in the Wilkinson microwave anisotropy probe 3-year data at $2.8\sigma$. Phys. Rev. Lett. **100**(18), 181301 (2008). doi:10.1103/PhysRevLett.100.181301
J.A. Zuntz, P.G. Ferreira, T.G. Zlosnik, Constraining Lorentz violation with cosmology. Phys. Rev. Lett. **101**(26), 261102 (2008). doi:10.1103/PhysRevLett.101.261102

Space Sci Rev (2009) 148: 347–362
DOI 10.1007/s11214-009-9552-3

# Perspectives on Dark Energy

**R.R. Caldwell**

Received: 5 June 2009 / Accepted: 9 June 2009 / Published online: 9 July 2009
© Springer Science+Business Media B.V. 2009

**Abstract** The problem of dark energy is reviewed. The basic description of dark energy properties and parameterization is presented. Two alternative scenarios are examined, one invoking new gravitational phenomena, and another introducing a more fundamental revision of the standard cosmological model.

**Keywords** Cosmology · Dark energy · Gravitation

## 1 Introduction

Observations and experiments are consistent with the hypothesis that the majority of the energy of the Universe is in the form of a heretofore undiscovered substance, referred to simply as "dark energy", that is causing the cosmic expansion to accelerate. The physical origin of dark energy is unknown [see Durrer and Maartens 2008; Frieman et al. 2008; Caldwell and Kamionkowski 2009 for reviews]. The leading theories propose that dark energy is a static, cosmological constant consisting of quantum zero-point energy, or a low-mass, slowly-rolling scalar field. Alternatively, it has been proposed that the phenomena attributed to dark energy are due to a change in the form of gravitation on cosmological length scales. Or possibly due to a violation statistical homogeneity, a fundamental tenet of cosmology. Confirmation of any one of these ideas would have a profound impact on physics. Determining the nature of dark energy is widely regarded as one of the most important problems in physics and astronomy (Albrecht et al. 2006).

The leading hypothesis is that the dark energy is due to a cosmological constant. The modern quest to characterize or measure the cosmological constant may have been initiated by Zeldovich (1968), but the first estimates of its size were made by deSitter. On 14 April 1917, shortly after the publication of Einstein's first paper on cosmology, Einstein wrote to deSitter (Einstein 1998):

Presented at "The Nature of Gravity" meeting at the ISSI, Bern, Switzerland on October 6–10, 2008.

R.R. Caldwell (✉)
Dartmouth College, Department of Physics & Astronomy, Hanover, NH 03755, USA
e-mail: robert.r.caldwell@dartmouth.edu

> In any case, one thing stands. The general theory of relativity allows the addition of the [cosmological constant] in the field equations. One day, our actual knowledge of the composition of the fixed-star sky, the apparent motions of the fixed stars, and the position of spectral lines as a function of distance, will probably have come far enough for us to be able to decide empirically the question of whether or not $\lambda$ vanishes.

On 18 April 1917, the reply came from deSitter:

> "The main point in our difference in creed is that you have a specific belief and I am a skeptic... Observations will never be able to prove that $\lambda$ vanishes, only that $\lambda$ is smaller than a given value. Today I would say that $\lambda$ is certainly smaller than $10^{-45}$ cm$^{-2}$ and is probably smaller than $10^{-50}$. Maybe one day observations will also provide a specific value for $\lambda$, but up to now I have no knowledge of anything pointing to this...

We may consider this bound, $\lambda < 10^{-45}$ cm$^{-2}$, indicating a radius of curvature roughly equal to the size of the galaxy, to be the first limit on the cosmological term.

## 2 Cosmological Constant

The evidence for dark energy comes along three main lines. First, observations of large-scale structure have long indicated that the gravitating matter density is far less than the critical value: $\Omega_M < 1$. Second, the largest geometry experiment in the Universe, carried out with the cosmic microwave background (CMB), has shown that spatial curvature is negligible on cosmological scales: $|\Omega_k| \ll 1$. Third, the type 1a supernova Hubble diagram has been used to show that the cosmic expansion is accelerating: the deceleration parameter is negative, $q < 0$. Interpreting these three lines of evidence within the context of Einstein's general relativity (GR), the simple conclusion is reached that some sort of dark energy is responsible, very possibly a cosmological constant. Indeed, most analyses of observational data find that the cosmological constant provides an adequate description. The big question, however, is to determine the physical nature of such a constant. In the absence of any such theoretical explanation, we must regard the cosmological constant as a placeholder until we can get a better understanding of dark energy.

### 2.1 Cosmological Constant Problems

No realistic attempt at a physical interpretation of the cosmological term was made by Einstein or deSitter. We know that the quantum vacuum was identified as a source of the cosmological term by Pauli (Straumann 2008) and then Zeldovich (1968). Both must have recognized the absurdity of the cutoff necessary in the momenta of the particle physics vacuum that would be necessary to obtain an energy density not in conflict with observations. The cosmological term is just as enigmatic today—Weinberg's seminal review (Weinberg 1989) is arguably as good a summary as any of what we do, or rather do not, know about the cosmological constant (Nobbenhuis 2006).

It is worthwhile to point out that there are two cosmological constant problems. The first, original problem is to explain why the vacuum energy density is not $\sim M_P^4$ (with $M_P$ the Planck mass); an argument that requires $\Lambda$ to vanish would suffice, although it would be difficult to envision testing such an explanation. The second, newer problem is to explain why the vacuum energy density is approximately $\sim(10^{-3}\ \text{eV})^4$. Again, the momentum cut-off in the particle physics vacuum is inadequate to address either problem, as it conflicts with experience—vacuum excitations above the milli-eV scale do exist. A novel reaction to these problems is to speculate that our description of the gravitational coupling is too broad, and in fact the graviton is unable to convey the full energy of the quantum vacuum. Such ideas have been recently proposed in the guise of "fat gravitons" whereby a putative finite size or composite nature of the graviton prevents it from coupling to high-frequency modes of Standard Model fields (Sundrum 2004), or in the form of a sort of low-pass filter, to borrow from electronics, operating on Einstein's equations that prohibits high frequency or high energy modes of stress-energy from gravitating (Dvali et al. 2007). The consequences of a specific realization of this idea have been explored (Caldwell and Grin 2008), as we now describe.

## 2.2 Regulating the Gravitation of the Quantum Vacuum

The energy scale above which the gravitational coupling is suppressed is estimated by matching the predicted quantum vacuum energy density with the energy density of a cosmological constant, $\Lambda$, necessary to explain the accelerated cosmic expansion. Using $N$ equivalent, massless scalar particles as a proxy for the gravitating energy density of the particle physics vacuum, then

$$\rho_\Lambda = \frac{N}{2}\int \frac{d^3k}{(2\pi\hbar)^3} kcf(k). \tag{1}$$

We introduce the function $f(k) = \mathrm{e}^{-k/\mu}$ to regulate the momentum at the vertex where vacuum bubbles connect to gravitons, in order to limit the gravitating energy density. We refer to $\mu$ as a "cutoff" scale in the sense that the standard gravitational interactions are severely weakened above this scale. We match $\rho_\Lambda = \Omega_\Lambda \rho_{crit}$ and obtain as the desired cutoff scale $\mu = 0.0048(\Omega_\Lambda h^2/N)^{1/4}$ eV/c. Current measurements give $\Omega_\Lambda h^2 = 0.35 \pm 0.04(1\sigma)$ (Komatsu et al. 2009) so that $\mu = 0.0037(1 \pm 0.03)/N^{1/4}$ eV/c. We now examine the consequences of this cutoff.

Short-distance gravitational phenomena below the length $\ell_0 = \hbar/\mu \sim 0.05$ mm are affected by such a cutoff, which we impose on the graviton four-momentum $q^\mu$ so that $q^2 \equiv q^\mu q_\mu < \mu^2$. For real gravitons, $q^\mu q_\mu = 0$ and so the constraint is trivially satisfied. For virtual gravitons, the cutoff may be imposed by suppressing the graviton propagator in the ultraviolet: $1/q^2 \to \mathcal{G}(q^2/\mu^2)/q^2$, where $\mathcal{G}$ is a function of the graviton momentum. For example, our exponential cutoff follows if $\mathcal{G}(x) = e^{-\sqrt{x}}$. Such a modified propagator follows naturally from modified gravitational Lagrangians.

An exponential cutoff to the momentum-space integral for the virtual gravitons exchanged between two static masses, $m_1$ and $m_2$, changes the Newtonian potential to

$$\begin{aligned} V &= -8\pi G m_1 m_2 \int \frac{d^3q}{(2\pi)^3\hbar}\frac{1}{2q^2} e^{\frac{i}{\hbar}\vec{q}\cdot(\vec{x}_1-\vec{x}_2)} \times f(q) \\ &= -\frac{G m_1 m_2}{r} \times \frac{2}{\pi}\arctan\frac{r}{\ell_0}. \end{aligned} \tag{2}$$

Relativistic corrections to the potential are similarly modified (Corinaldesi 1956; Barker et al. 1966). The above expression asymptotes to the standard result for $r \gg \ell_0$ but reaches a finite minimum as $r/\ell_0 \to 0$. Hence, static masses become free of gravitation at short distances.

The possibility of new gravitational phenomena at submillimeter distances has motivated laboratory tests of the Newtonian force law (Hoyle et al. 2001; Chiaverini et al. 2003; Long

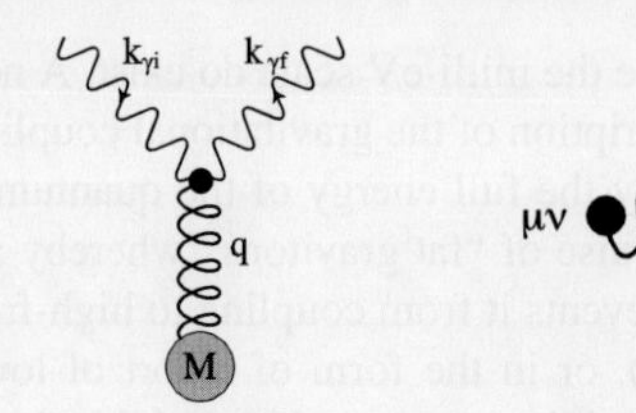


**Fig. 1** (*Left*) The Feynman diagram for the gravitational deflection of light is shown. (*Right*) The modification to the graviton propagator, $D^{\mu\nu\sigma\tau}$, in the form of a cutoff function $f(q)$ that suppresses graviton models with momentum $q$ above some critical value, is illustrated

et al. 2003; Hoyle et al. 2004; Smullin et al. 2005; Kapner et al. 2007). These experiments look for departures from the Newtonian force law, which are interpreted as bounds on a Yukawa-type modification of the potential, $V = -\frac{Gm_1m_2}{r} \times (1 + \alpha e^{-r/\lambda})$. The potential (2) roughly corresponds to $\alpha \simeq -1$ and $\lambda \simeq \ell_0$. Recent measurements show that the Newtonian force law holds down to 56 μm for $|\alpha| = 1$ so that $\mu > 0.0035$ eV/c at the 95% confidence level (Kapner et al. 2007). These efforts are at the threshold of the scale inferred from $\Lambda$.

Long-distance gravitational phenomena are also sensitive to such modifications and provide a tighter bound on $\mu$, the scale of new physics. The key is the limited range of graviton momenta mediating the gravitational force exerted by a massive body on a test particle. Considering the deflection of light as an elastic, quantum mechanical scattering process, as illustrated in Fig. 1, the photon energy is conserved but its momentum is redirected. A maximum graviton momentum implies a maximum deflection angle, and so $|\vec{k}_{\gamma i} - \vec{k}_{\gamma f}| \approx 2k_\gamma\theta < \mu$, where $k_\gamma$ is the photon momentum. Defining $\theta_{GR} \equiv 4GM/(bc^2)$ for the standard light-deflection result without the cutoff, then a tree-level calculation of the light-deflection yields $\theta/\theta_{GR} = F(2\theta k_\gamma/\mu)$, where $F(x) = \sqrt{(1-x)e^{-x} - x^2 Ei(-x)}$ and $Ei(x) \equiv -\int_{-x}^{\infty} e^{-t} dt/t$ is the exponential-integral function. We note that the static, frequency-independent metric potential is insufficient to describe the photon's path past the lensing source when $\theta_{GR} \gtrsim \mu/2k_\gamma$. It would be necessary to introduce an effective force into the geodesic equation, based on the modified graviton propagator. We thus find that the deflection is half the standard prediction when $2\theta k_\gamma/\mu \sim 1$. In the limit $\theta \ll \mu/2k_\gamma$, $F \to 1$, but for $\theta \gtrsim \mu/2k_\gamma$ the deflection angle is suppressed. Hence, we would expect a dearth of gravitationally lensed images of high-frequency light if there were a cutoff in graviton momentum.

Numerous gravitational lens systems have been observed from radio to X-ray frequencies. The tightest constraint to $\mu$ comes from X-ray observations of the gravitationally lensed system Q0957+561 (Chartas et al. 2002). For this lens system, image A due to the quasar at $z = 1.4$ appears $5.2''$ away from the primary lensing galaxy at $z = 0.36$ (www.cfa.harvard.edu/castles). Using the angular-diameter distances to the source and from lens to the source, $D_S$, $D_{LS}$, to reconstruct the lensing geometry, we estimate a deflection angle of $\theta = 5.2'' \times D_S/D_{LS} = 7.8''$. The lens image locations are unchanged for $E_\gamma < 5$ keV (Chartas 2005), which yields the lower bound $\mu > 0.38$ eV/c. This result pushes the threshold for departures from the Newtonian force law down to 0.5 μm. This lower limit is nearly two orders of magnitude higher than, and therefore rules out, the cutoff $f(q)$ motivated by the cosmological constant. (See Caldwell and Grin 2008 for further details.)

## 3 Cosmological Parameters

The dark energy is characterized by its abundance, $\Omega_{DE}$, as a fraction of critical density and its equation of state, $w$ defined to be the ratio of the homogeneous pressure to the energy density. The current state of the art on these parameters is obtained primarily from the joint analysis of the type 1a supernova Hubble diagram and the cosmic microwave background anisotropy pattern. Quoting results from the Wilkinson Microwave Anisotropy Probe (WMAP) 5 year data analysis (Komatsu et al. 2009), under the assumption that the dark energy is a cosmological constant with equation of state $w = -1$, then the abundance is $\Omega_\Lambda = 0.721 \pm 0.015(1\sigma)$. Allowing for time-varying dark energy with a constant equation of state, then $w = -0.984^{+0.065}_{-0.064}(1\sigma)$ and $\Omega_m h^2 = 0.1369 \pm 0.0037(1\sigma)$. For a full explanation of the conditions and caveats for these results, we refer to the original references. Comparable results have been obtained by the Supernova Cosmology Project (Kowalski et al. 2008).

A constant equation of state has been widely adopted as the principle parameter to describe the time evolution and degree of "gravitational repulsion" for a dynamical dark energy. As of yet, there has been no need to model the time-evolution in greater detail. A recent study (Sullivan et al. 2007) in which the dark energy is attributed a constant equation of state over several redshift intervals has found that the current data support $w \sim -1$ for all intervals. An update from 2008 yields $w = -0.983^{+0.083}_{-0.050}$ in the redshift range $0 < z < 0.2$, $w = -0.933^{+0.083}_{-0.084}$ in the range $0.2 < z < 0.5$, and $w = -0.767^{+0.117}_{-0.166}$ in the range $0.5 < z < 1.8$ (Sarkar 2008), consistent with no evidence for a sharp change in the equation of state.

As an aside, it is quite remarkable that the dark energy equation of state is constrained for even two redshift bins. A crude but intuitive interpretation of cosmological parameters would attribute one parameter to every identifiable feature in the data-curve. As examples, consider the matter power spectrum obtained from galaxy redshift surveys. The power spectrum has two basic features—it has an approximate power-law slope and an overall amplitude—from which one can attempt to infer a shape parameter and a spectral tilt. (Additional features, such as a turnover in the spectrum and slight wiggles, can help determine the comoving scale of radiation-matter equality, and the baryon abundance.) A similar identification of features and parameters can be made with the cosmic microwave background anisotropy spectrum. But the supernova Hubble diagram, a plot of magnitude versus redshift, is alarmingly featureless. It seems to be a very nearly straight line, the slope of which tells us the Hubble constant, with a slight bit of curvature at increasing redshift. It is from this slight bit of curvature—which Robertson (1955) first elucidated, Humason et al. (1956) first attempted to measure, and Hoyle and Sandage (1956) first highlighted as an indicator of world-models—that cosmologists attempt to tease out dark energy parameters. Yet, a futuristic, satellite-based dark energy experiment to measure the supernova Hubble diagram with greater precision may realistically aim to constrain more than two equation of state parameters (Sullivan et al. 2008).

The equation of state parameter allows us to easily model the effects of a dynamical dark energy without the need to supply a theory of dark energy. This is immensely useful because it allows us to test the hypothesis that the dark energy is static—and any preference in the data for $w \neq -1$ would be ample stimulus to explore new models. Until then, the cosmological constant is a placeholder for new physics.

## 4 The Shape of New Physics

What could the new physics look like? A reasonable guess is that the dark energy, if it is dynamical, can be described in the language of our fundamental theories as a scalar field. Because a scalar field can carry negative pressure when the field evolution is dominated by its potential energy, it can satisfy the requirement that the dark energy is responsible for the cosmic acceleration. A cosmic scalar field is also the primary mechanism for most theories of cosmic inflation, and elsewhere it has been employed as dark matter. To distinguish from other scenarios, we may refer to the cosmic scalar as "quintessence" (Caldwell et al. 1998). To begin to discuss a particular field theory, let us write the action

$$S = \int d^4x\sqrt{-g}\left(\frac{R}{16\pi G} - \frac{\Lambda}{8\pi G} + \mathcal{L}_{SM} + \mathcal{L}_Q\right) \tag{3}$$

where $\mathcal{L}_{SM}$ is the Lagrangian for Standard Model particles and the quintessence Lagrangian is $\mathcal{L}_Q = -\frac{1}{2}\nabla_\mu\phi\nabla^\mu\phi - V(\phi)$. We use metric signature $(-+++)$ and adopt standard curvature conventions (Will 1993). We do not attempt to solve the cosmological constant problem, and so trivially set $\Lambda = 0$ and ignore the vacuum stress-energy. In order that the scalar field potential energy dominates today, we require $V \simeq (1-\Omega_m)\times 3H^2/8\pi G$. And in order that the scalar field does not fluctuate, lest we confuse it with dark matter, then $|V''| \lesssim H^2$. Together, these constraints imply a mass scale $10^{-42}$ GeV and a Planck-scale field strength. It is hard to conceive of a scalar field that is realistically embedded in a model of physics beyond the Standard Model with such extreme numbers. However, there is one outstanding model—a cosmic axion or pseudo-Nambu Goldstone boson is arguably the only natural way to have an extremely low mass scalar (Frieman et al. 1995), and keep it "dark" or non-interacting with the Standard Model (Carroll 1998). Here, the potential is $V = m^4(1+\cos\phi/f)$, where $m \simeq 0.002$ eV and $f \sim 10^{18}$ GeV (Coble et al. 1997; Dutta and Sorbo 2007). The field has been frozen by Hubble friction through most of cosmic history; it is currently relaxing to its ground state; in the future, the field will oscillate rapidly in the bottom of the potential, redshifting like nonrelativistic matter. This model offers no explanation for the coincidence problem, as to why the dark energy has come to dominate today, or the fine-tuning problem, since the early-time values of $\phi, \dot\phi$ directly determine the present-day properties of the dark energy.

The fine-tuning problem can be solved by a tracker-quintessence model. In this case the potential has the form $V = M^4(\phi/M_P)^{-n}$ with $n > 0$, motivated by the order parameter in models of dynamical breaking of supersymmetry (Affleck et al. 1984). This scalar field potential was first explored in the cosmological context in Ratra and Peebles (1988), but the tracking behavior was identified in Zlatev et al. (1999), Steinhardt et al. (1999). What the tracker means is that, for a broad range of initial conditions, the evolution of the field approaches, and then locks onto, a universal track whereby the scalar field equation of state is $w \approx (nw_B - 2)/(n+2)$ and $w_B$ is the equation of state of the dominant or background component. In the matter-dominated era, $w_B = 0$ so that $w < 0$, so that the dark energy must gain on the matter density. Thus, the field has been rolling down the potential for most of cosmic history; due to the curvature of the potential, the rolling slows as if the field begins to freeze in its tracks, and the equation of state becomes more negative; as the field comes to dominate, the equation of state asymptotes to $w \to -1$ as the rolling slows; since $V \to 0$ only at $\phi \to \infty$, it will never fully relax. Note that the universal track is uniquely determined by the mass $M$ and the index $n$, so that there is a one-to-one relationship between the dark energy density, $\Omega_Q$ and the equation of state as a function of time. The broad insensitivity of the late-time behavior to the initial conditions is appealing—this model solves the above-mentioned fine-tuning problem. It does not entirely address the coincidence problem—the equation of state indeed becomes more negative as the radiation era gives way to the matter era, but there is no explanation as to why the acceleration is happening now, as opposed to any other time.

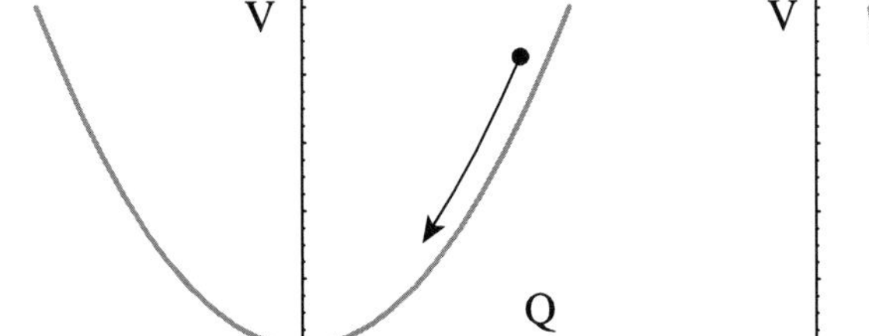


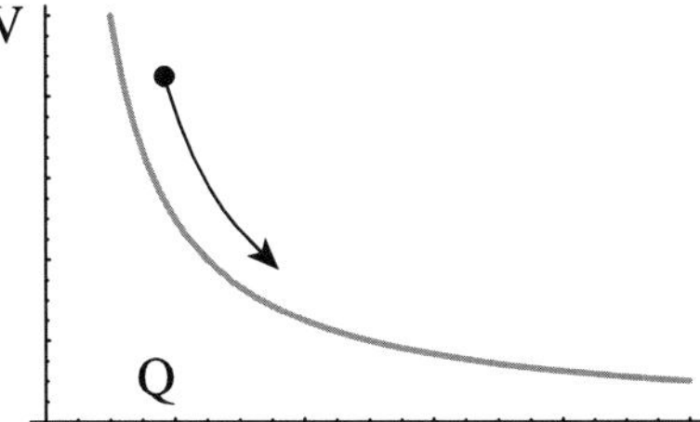


**Fig. 2** Two examples of potentials for the quintessence field. In the first, representative of a conventional massive scalar or PNGB, the field is relaxing towards the local minimum as the equation of state runs from $w = -1$ to $w = 0$. In the second, representative of vacuumless potentials such as the tracker, the field is evolving towards the global minimum, as the equation of state asymptotes to $w = -1$

Many other cosmic scalar field models have been proposed, with more complicated potentials, complex scalar fields, and non-canonical kinetic terms known as k-essence (Armendariz-Picon et al. 2001). The two models described above are among the simplest, and are representative of a wide range of models thought to be realistic from the perspective of fundamental physics.

## 5 Parameterizing the Equation of State

The equation of state of dynamical dark energy is unlikely to be a constant if the above discussion of theoretical models is any guidance. Using the cosmic axion and the tracker field as examples, we may identify two classes of quintessence models, thawing and freezing. Thawing models have a potential with a $V = 0$ minimum accessible within a finite range of $\phi$; the field started up the potential, frozen by Hubble friction; as the Hubble constant decayed, the field began to thaw and roll down. In these cases, the equation of state starts at $w = -1$ and evolves towards $w = 0$. In contrast, freezing models are vacuumless, as the minimum is not accessible within a finite range of $\phi$, despite the absence barriers. In such cases the equation of state evolves towards $w \to -1$.

The trajectories of thawing and freezing models occupy rather well-defined regions of the $w$ vs. $dw/d\ln a$ parameter plane. The boundaries of these regions have been identified for a range of simple thawing and freezing models (Caldwell and Linder 2005). Of course, there are plenty of models not contained in these regions (Huterer and Peiris 2007; Crittenden et al. 2007), but the exceptional models tend to be more complicated or contain metastable vacua, e.g. an effective cosmological constant. The evolutionary trajectories can be used as a guide for modeling the time-dependence of the equation of state, and the size of these regions can be used as a guide for the sensitivity of experiments that test for dynamical dark energy. For the time evolution, the trajectories suggest $dw/d\ln a = A(1+w)$ where $1 < A < 3$ for thawing fields and $dw/d\ln a = Aw(1+w)$ where $0 < A < 3$ for freezing fields. It remains to attempt to place constraints on such a parameter $A$ and an integration constant, although there have been related studies (Li et al. 2007). Rather, a simpler relation $w = w_0 + (1-a)w_a$ has been widely adopted. Current constraints on these parameters are shown in Fig. 3. These constraint regions are much larger than the expected range of thawing and freezing models: the current constraints are not prohibitive.

The preceding discussion has ignored the possibility that the equation of state might be more negative than $-1$. There are good reasons to avoid such a possibility—a fluid component with $w < -1$ is pathological in a variety of ways (Caldwell 2002), and if unchecked

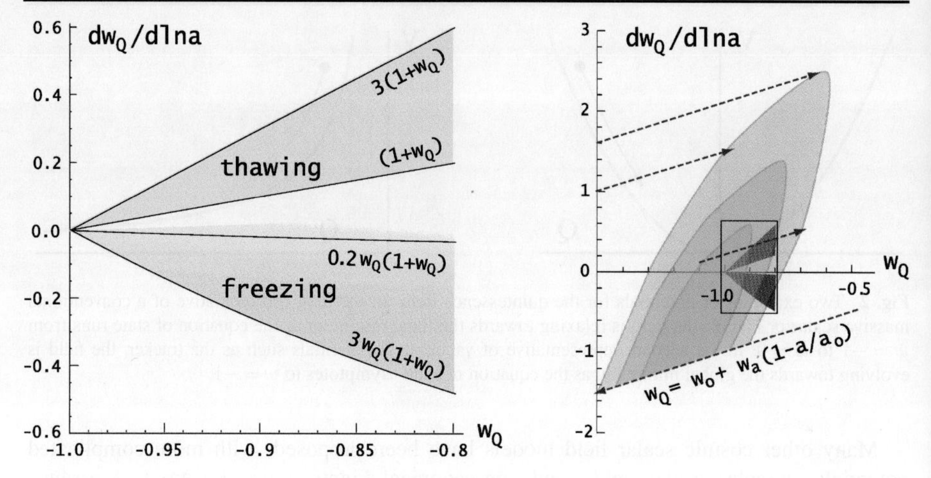


**Fig. 3** The $w_Q$ vs. $dw_Q/d\ln a$ parameter plane of dynamical dark-energy models is shown. *On the left*, the likely range of thawing and freezing models is shown. *On the right*, the current 68.3%, 95.4%, and 99.7% confidence-level constraints on the dark energy parameterization $w_Q = w_0 + w_a(1 - a/a_0)$ (Kowalski et al. 2008) have been converted into the present-day values of $w_Q, dw_Q/d\ln a$. The *dashed lines* show the direction of evolution of models located at particular points on the 99.7% confidence-level boundary

then it can lead to singular behavior in the future (Caldwell et al. 2003). There are less pathological mechanisms that might lead to the inference $w < -1$, through modifications of gravity, of dark matter, of the distance-redshift relation. Before proposing new physics, however, it is important to recognize that distance measurements are intrinsically biased to indicate an equation of state that is more negative than the true value.

The familiar likelihood contours in the $w$–$\Omega_m$ plane are more banana-like than elliptical—this is a sign of the non-linearity between the observables and the parameters. If there was a linear relationship between the observables, such as the magnitude $\mu$, and parameters $w$ and $\Omega_m$, then Gaussian errors on $\mu$ would translate into Gaussian errors on a linear combination of $w$ and $\Omega_m$, and the likelihood contour would resemble an ellipse. But we know that the magnitude and the luminosity distance are nonlinear functions of $\Omega_m$ and $w$. Consider the following: take a fiducial $\Lambda$CDM model with $\Omega_m = 0.3$ and calculate the luminosity distance to $z = 1$. Now calculate the distance for $w = -1 - 0.2$ and for $w = -1 + 0.2$: the change in distance is smaller for $w = -1.2$ than for $w = -0.8$. So if we have a measurement of the distance to $z = 1$ with symmetric, Gaussian errors, then the uncertainty in $w$ will not be symmetric, Gaussian; moreover, the probability distribution function will be skewed towards more negative values of $w$. If $w = -1$ gives the best fit, then there will still be a long tail at $w < -1$. This can be seen in the likelihood contours, illustrated in Fig. 4. By taking a slice at constant $\Omega_m$ (horizontal line) the probability distribution function can be seen to cut off quickly at $w > -1$ but extends far for $w < -1$. Marginalizing over $\Omega_m$, it is clear that the skewness will cause $\langle w \rangle$ to be biased towards values below $-1$. However, the marginalization does something else tricky. Look at slices through the contours at constant $w$ (vertical lines); the amount of red that the curve passes through determines the relative contribution to the marginalized distribution for $w$. Because the banana turns downwards as $w$ increases past $-1$, the amount of red also increases, with the result that the peak of the marginalized distribution will occur at $w > -1$. Both of these effects, the skewness towards more negative values of $w$, and the shift in the peak towards more positive values of $w$, are seen in simulations. Thus, claims that $w < -1$ may be pre-

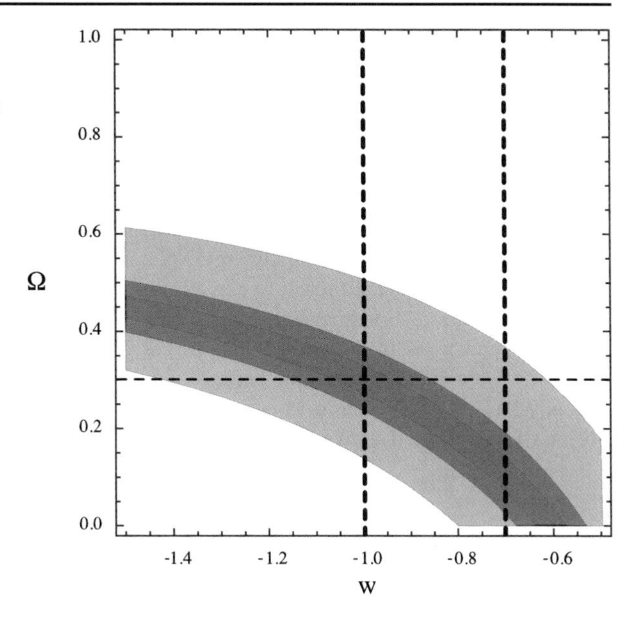


**Fig. 4** Likelihood contours for a model universe with matter density $\Omega_m$ and constant dark energy EOS, $w$ based on fictional measurements of the luminosity distance to redshift $z$

mature. Because distance measurements at different redshift rotate the degenerate likelihood contours slightly in the $w$–$\Omega_m$ plane, if enough measurements can be assembled, each of which rotated the likelihood contours, then the overlap may begin to approach an ellipse or perhaps a circle, thereby removing the bias in $w$.

## 6 New Gravitational Phenomena

The inference that some form of dark energy is necessary to explain cosmological measurements relies on the validity of general relativity. Yet, our most basic notion about how gravity works is obviously violated—the cosmic acceleration seems to imply that distant galaxies repel rather than attract each other. Hence, it seems entirely possible that the balance between stress-energy and metrical curvature is guided by a different theory of gravitation on cosmological scales. Unfortunately, there does not seem to be any savings in terms of economy of description. No alternative theory—scalar-tensor gravity, Dvali-Gabadadze-Porrati (DGP) (Dvali et al. 2000) or brane-world motivated scenarios, massive gravity and the like—provides as simple a description of gravitational phenomena as GR with a cosmological constant. Be that as it may, it is still interesting and necessary to test gravitation on the largest scales.

Cosmic acceleration can be caused by new fluids, new theories of gravity, or some admixture of both. This uncertainty places a premium on descriptions of the so-called "dark physics" which remain useful across different models and in spite of varying assumptions. In the case of new fluids (dark energy), the literature chooses to speak in terms of the equation of state $w$ and its derivative. In the case of new gravitational physics, the model-independent *lingua franca* is the relationship between the Newtonian ($\psi$) and longitudinal ($\phi$) gravitational potentials. The potentials, implicitly defined through the perturbed Robertson-Walker metric

$$ds^2 = a^2[-(1+2\psi)d\tau^2 + (1-2\phi)d\vec{x}^2], \tag{4}$$

are most familiar for their roles in Newton's equation, $\ddot{\vec{x}} = -\vec{\nabla}\psi$, and the Poisson equation, $\nabla^2\phi = 4\pi G a^2 \delta\rho$. The gravitational potentials are equal in the presence of non-relativistic stress-energy under GR. Alternate theories of gravity make no such guarantee. Scalar-tensor and $f(R)$ theories, braneworld scenarios such as DGP gravity, and massive gravity all predict a systematic difference or "slip", so that $\phi \neq \psi$ in the presence of non-relativistic stress-energy.

We use this common behavior as a launching point to describe an as-yet unnamed theory of gravity that dispenses with dark energy. We naively expect that the non-relativistic dark- and baryonic-matter are sufficient to explain the evolution of the scale factor $a(t)$, through a modification to the Friedmann equation whereby the amount of spacetime curvature generated per unit mass is changed relative to GR by a factor of roughly $1 + \Omega_\Lambda/\Omega_m$. For simplicity, we assume the resulting background evolution is indistinguishable from that of a spatially-flat, ΛCDM scenario. However, there is a similar shift in the ratio of gravitational potentials, which we (Caldwell et al. 2007) parameterize as

$$\psi = [1 + \varpi(z)]\phi, \tag{5}$$

$$\varpi(z) = \varpi_0(1+z)^{-3}. \tag{6}$$

Our naive expectation is that $\varpi \simeq \Omega_\Lambda/\Omega_m$ by today. The departure from GR kicks in only when the cosmic expansion begins to accelerate.

A phenomenological description of gravity in the cosmological context, and the implications for observations and experiment, have been widely pursued in the literature. The parameter $\varpi$ is not completely free—it must satisfy a constraint, valid under general assumptions (Bertschinger 2006). A parameterized post-Friedmannian scheme, appropriate for $f(R)$ theories (in which the Ricci scalar in the Einstein-Hilbert action is replaced by a general function of the Ricci scalar, as in Carroll et al. 2004) that successfully drive cosmic acceleration whilst satisfying Solar System tests of gravity, has been proposed (Hu and Sawicki 2007). Other approaches have introduced a second function to modify the Poisson equation (Jain and Zhang 2008; Amendola et al. 2008; Song and Koyama 2009; Zhao et al. 2009).

Full details of the model employed in Caldwell et al. (2007), and comparison with observations are found in subsequent publications: Daniel et al. (2008) discuss the compatibility with other parameterizations (Hu and Sawicki 2007; Bertschinger and Zukin 2008); current constraints on $\varpi_0$, primarily due to the cosmic microwave background, large scale structure, and weak lensing, are evaluated in Daniel et al. (2009); and Serra et al. (2009) looks ahead to the use of weak lensing of the cosmic microwave background.

The results of our multiparameter investigation (Daniel et al. 2009) are shown in Fig. 5. The left panel shows the 68% and 95% contours in $(\Omega_m, \varpi_0)$ space marginalized over all other parameters. The blue contour shows the constraint resulting from the WMAP 5 year data set only (Dunkley et al. 2009); the red contour includes the Union type 1a supernova data (Kowalski et al. 2008), galaxy weak lensing data from the Canada–France–Hawaii Telescope Legacy Survey (Fu et al. 2008), and cross-correlations between galaxy surveys and the cosmic microwave background (Ho et al. 2008). Based on these current results, we arrive at the following conclusions. First, present cosmological data constrains gravity to agree with GR, $\varpi_0 = 0.09^{+0.74}_{-0.59}(2\sigma)$, assuming a background evolution consistent with ΛCDM. Second, very negative values of $\varpi_0$ are ruled out. This should not be surprising, since a sign difference between the longitudinal and Newtonian gravitational potentials would mean that test particles are repelled by overdense regions. Third, large-scale structure data (in our case, weak lensing and the galaxy–CMB correlation) are critical to constraining $\varpi_0$. To estimate

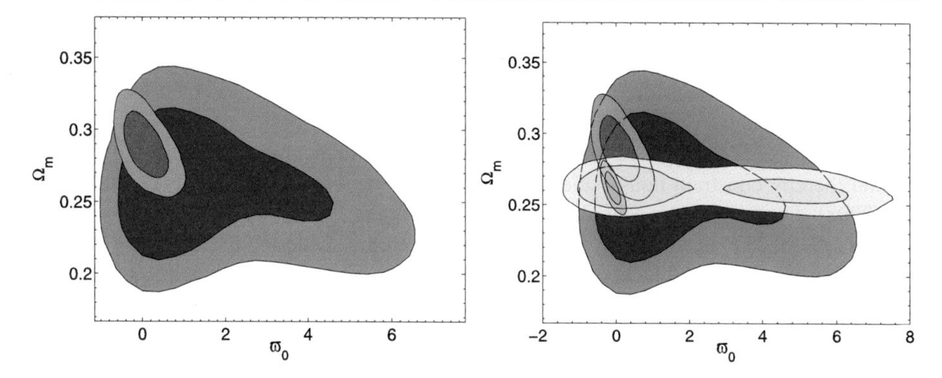


**Fig. 5** (*Left*) The current 68% and 95% likelihood contours in the $\varpi_0$–$\Omega_m$ parameter space are shown. The *blue contours* are based on CMB data alone. The *red contours* add weak lensing, type 1a supernovae, and galaxy-CMB cross-correlation data. (*Right*) The projected 68% and 95% likelihood contours based on mock Planck data are shown in *yellow*; the *green contours* add mock DUNE/Euclid weak lensing data. Marginalizing over all other parameters, the mock datasets (assuming $\varpi_0 = 0$) give the constraint $\varpi_0 = -0.07^{+0.13}_{-0.16}(2\sigma)$, a factor of $\sim 4$ improvement over the current constraint. (Figs. 6 and 8 of Daniel et al. 2009)

how much better the constraints can be under future experiments, mock data sets were generated, simulating the results of the upcoming Planck CMB experiment, and a future weak lensing survey modeled after the proposed ESA experiment DUNE/Euclid. The right panel of Fig. 5 shows the results. Assuming the underlying model is $\Lambda$CDM in GR, the bound $|\varpi_0| \lesssim 0.2$ may be achieved, representing a test of GR at the 20% level, well below our naive expectations for the magnitude of a departure from GR that explains the cosmic acceleration without dark energy.

## 7 A Fundamental Change in Our View of the Universe

The cosmic acceleration may indicate that one of the basic tenets of the standard model of cosmology is invalid, rather than indicate the existence of some new, exotic dark energy. One such tenet is the Cosmological Principle, the assumption of approximate homogeneity and isotropy of matter and radiation throughout the universe. The Cosmological Principle is known to be partly satisfied. The Universe is observed to be very nearly isotropic on our celestial sphere, on the basis of the near-isotropy of the CMB temperature pattern (Jaffe et al. 2005; Hajian and Souradeep 2006). The Universe is observed to be approximately homogeneous across the distances probed by large-scale structures (Hogg et al. 2005). Yet, radial homogeneity on scales $\gtrsim 1$ Gpc, a cosmic version of the Copernican Principle, remains to be proven. If the assumption of radial homogeneity is relaxed, and if we observe from a preferred vantage point, then it may be possible to explain the apparent cosmic acceleration in terms of a peculiar distribution of matter centered upon our location (Dabrowski and Hendry 1998; Celerier 2000; Barrett and Clarkson 2000; Tomita 2001; Iguchi et al. 2002; Moffat 2006; Alnes et al. 2006; Vanderveld et al. 2006; Garfinkle 2006; Chung and Romano 2006; Biswas et al. 2007; Enqvist and Mattsson 2007; Chuang et al. 2008). In fact, models of the universe consisting of a spherically-symmetric distribution of matter, described by a Lemaitre-Tolman-Bondi spacetime (Bondi 1947), have been shown to produce a Hubble diagram that is consistent with observations. These models require no cosmological constant or other form of dark energy, and locally resemble a matter-dominated, low-density universe or void. The observed near-isotropy constrains us to occupy a very special location, at or near the center of the void, in violation of the Copernican Principle. Although the Copernican Principle may be widely accepted by *fiat*, it is imperative that such a foundational principle be proven.

Recently, we have proposed to test the Copernican Principle, to verify radial homogeneity and thereby constrain non-accelerating void cosmological models (Caldwell and Stebbins 2008). The test follows a similar argument (Goodman 1995), that constrains distortions of the CMB blackbody spectrum produced by anisotropic scattering (Stebbins 2007). The CMB is initially thermal (blackbody), but small inhomogeneities cause variations in the temperature at different locations and along different lines-of-sight that preserve the blackbody spectrum. However, scattering of this anisotropic radiation into our line-of-sight by ionized gas produces observable spectral distortions. This allows us to indirectly detect large anisotropies in other parts of the universe.

Here we are interested in anisotropies caused by a large, local void. Such a structure causes ionized gas to move outward, in motion relative to the CMB frame which leads to a Doppler anisotropy in the gas frame. The gravitational potential of such a structure also leads to a Sachs-Wolfe (SW) effect for photons which originate inside of the void and scatter back toward us. The geometry of these effects is illustrated in Fig. 6. A large void, or any other non-Copernican structure, will lead to large anisotropies in other places that will be reflected back at us in the form of spectral distortions. Hence, deviations from a blackbody spectrum can indicate a violation of the Copernican Principle. In essence, we use the reionized universe as a mirror to look at ourselves in CMB light. If we see ourselves in the mirror it is because ours is a privileged location. If we see nothing in the mirror, then the Copernican Principle is upheld.

The distortion of the CMB blackbody spectrum due to scattering by anisotropic CMB radiation is (Stebbins 2007)

$$u[\hat{\mathbf{n}}] = \frac{3}{16\pi} \int_0^\infty dz \frac{d\tau}{dz} \int d^2\hat{\mathbf{n}}' \left(1 + (\hat{\mathbf{n}} \cdot \hat{\mathbf{n}}')^2\right) \left( \frac{\Delta T}{T}[\hat{\mathbf{n}}, \hat{\mathbf{n}}, z] - \frac{\Delta T}{T}[\hat{\mathbf{n}}', \hat{\mathbf{n}}, z] \right)^2, \quad (7)$$

where $\Delta T/T[\hat{\mathbf{n}}', \hat{\mathbf{n}}, z]$ is the CMB temperature anisotropy in the direction $\hat{\mathbf{n}}'$, as observed at redshift $z$ in the direction $\hat{\mathbf{n}}$ from the central observer, and $\tau$ is the optical depth. For cosmic voids extending out to redshifts $z \lesssim 1$, reflections back at us may occur up to $z \lesssim 3$. The optical depth to Thomson scattering is small, so that it is appropriate to consider single scattering. Since the mean CMB temperature is not known *a priori*, but rather is fit to the observations, $u$ is observationally degenerate with the Compton $y$-distortion parameter according to the relation $u = 2y$. Thus observational constraints on $2y$ can be treated as constraints on $u$.

The $u$-distortion is evaluated according to the procedure described in Caldwell and Stebbins 2008. However, we have since improved our calculation to model the void as a low density region in a LTB universe with metric

$$ds^2 = -dt^2 + \frac{R'^2}{1+f} dr^2 + R^2 d\Omega^2 \quad (8)$$

where $R = R(t, r)$ and $f = f(r)$. We use primes to indicate partial derivatives with respect to radius, and dots to indicate time derivatives. In the case of constant bang-time models there is one free function, $f(r)$. To make a connection with the familiar notation of Robertson-Walker spacetimes, we define $a(t, r) \equiv R(t, r)/r$ and $k(r) \equiv f(r)/r^2$.

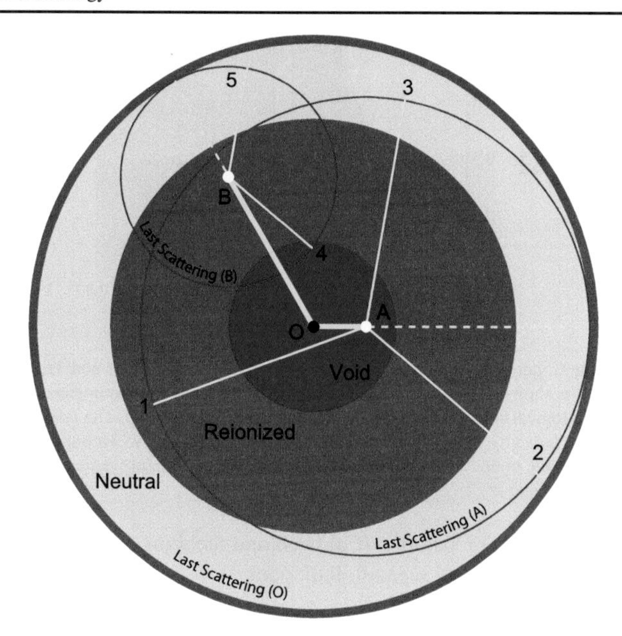


**Fig. 6** Illustrated is a cross-section through a model universe with the observer (O) at the center of a void, in violation of the Copernican Principle. CMB photons traveling in any direction may Thomson scatter off reionized gas toward the observer. The final spectrum of the observed light will be a mixture of blackbody spectra with different (anisotropic) temperatures, producing a distorted blackbody. The *yellow lines* represent: incoming beams of unscattered, primary CMB photons (*dashed*); incoming beams of scattered photons (*thin*), and the observed beams (*thick*) for representative scattering centers with last scattering surfaces represented by the dark circles. A is in the *Doppler zone*: Beams 1–3 experience the same SW temperature shift, introducing no anisotropy. However, gradients in the void gravitational potential cause the gas to move with respect to the CMB frame, so A sees a differential Doppler anisotropy, resulting in spectral distortions. B is in the *reflection zone*: B is at rest with respect to the CMB frame and sees no Doppler anisotropy. However, some of the incoming photons, e.g. beam *4*, originate inside the void so there will be an anisotropic SW temperature shift, leading to spectral distortions. (From Caldwell and Stebbins 2008)

For a specific model, we set $k(r) = (1 + (cr)^n)^{-1}$ where $c$ is a constant set by the redshift to the edge of the void, and $n$ controls the thickness of the wall. More details will be provided in a forthcoming publication. The constraints on $u$ in the case $n = 2$ are shown in Fig. 7. The blue contours show the $1, 2, 3\sigma$ bounds on LTB cosmological parameters, evaluated following a similar procedure to Garcia-Bellido and Haugboelle (2008), required for the LTB model to satisfy the type 1a supernova Hubble diagram (Kowalski et al. 2008), the angular-diameter distance of baryon acoustic oscillations (Eisenstein et al. 2005; Percival et al. 2007), and the acoustic peak of the CMB (Dunkley et al. 2009), and the HST measurement of the Hubble constant (Freedman et al. 2001). The best current bound on $u$ is due to FIRAS (Mather et al. 1994; Boggess et al. 1992; Fixsen et al. 1996) which constrains $y < 15 \times 10^{-6}$ or $u < 3 \times 10^{-5}$ at the 95% confidence level. The corresponding constraint on Hubble bubble parameters are shown in Fig. 7. Also shown are constraints for projected bounds $y < 15 \times 10^{-7}$. The limits are expected to improve (Fixsen and Mather 2002; Kogut et al. 2006), but a $y$-distortion from the IGM would likely mask the signal discussed here if $u \lesssim 10^{-6}$ (Zhang et al. 2004). The bounds on $u$ show that a sliver of parameter space is compatible with current observations. In the case $n = 4$, the void wall is thinner, sharper but the blue contours shift far to the right, towards higher $z_v$, and there is no overlap with the red contours. And when $n < 2$ the void wall is wider and the blue contours

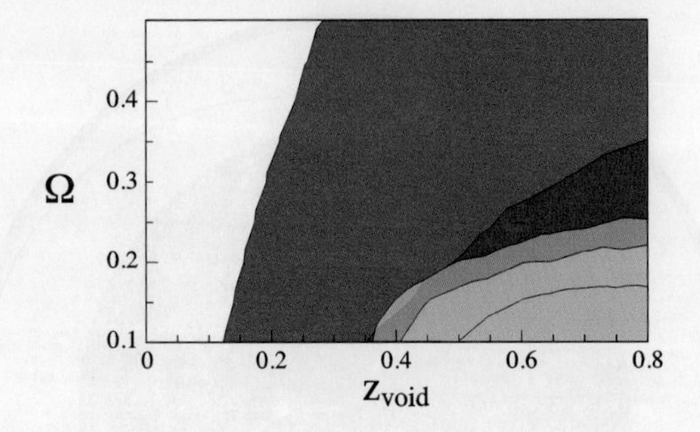


**Fig. 7** The test of the Copernican Principle, in terms of constraints on the size and depth of a local, spherically-symmetric void, is shown. The *blue shaded regions* show the range of parameters *required* for the LTB model to satisfy cosmological constraints to the distance–redshift relationship. The *dark red region* shows the range of parameters *excluded* by the $u$-distortion test, based on the FIRAS 95% confidence level bound. The *lighter red region* shows the bound that would result with a factor of 10 improvement over the FIRAS bound

shrink. A full analysis is in preparation (Caldwell and Stebbins 2009), but the prospects for an anti-Copernican cosmological scenario look poor.

## 8 Summary

The physical origin of the cosmic acceleration is currently a mystery. A vast catalog of observations and experiments can be put in order if the existence of a new exotic fluid—dark energy, with properties similar to that of a cosmological constant—is presumed to dominate the Universe. Other explanations, perhaps lacking in economy but no less valid, have been proposed that also fit the data. These alternatives include a dynamical form of dark energy, new theories of gravity that depart from general relativity on cosmological scales, and gigaparsec voids that dispense with dark energy and cosmic acceleration altogether. Despite the variety of possible solutions, the discussion of the preceding sections shows that we are very close to ruling out or tightly constraining many of these alternative proposals. One can expect that, by the centenary anniversary of Einstein's introduction of the cosmological constant in 2017, we will know if he was ultimately justified.

**Acknowledgements** I would like to thank the organizers for a provocative meeting, bringing together diverse perspectives on gravitational physics, in the birthplace of special relativity, Einstein's 1905 home. Thanks also to my collaborators Asantha Cooray, Scott Daniel, Daniel Grin, Marc Kamionkowski, Alessandro Melchiorri, Paolo Serra, and Albert Stebbins for permission to share these results. This work was supported in part by NSF AST-0349213.

## References

I. Affleck, M. Dine, N. Seiberg, Nucl. Phys. B **241**, 493 (1984)
A. Albrecht et al., arXiv:astro-ph/0609591 (2006)
H. Alnes, M. Amarzguioui, O. Gron, Phys. Rev. D **73**, 083519 (2006)
L. Amendola, M. Kunz, D. Sapone, J. Cosmol. Astropart. Phys. **0804**, 013 (2008)
C. Armendariz-Picon, V.F. Mukhanov, P.J. Steinhardt, Phys. Rev. D **63**, 103510 (2001)
B. Barker, S. Gupta, R. Haracz, Phys. Rev. **149**, 1027 (1966)
R.K. Barrett, C.A. Clarkson, Class. Quant. Grav. **17**, 5047 (2000)
E. Bertschinger, Astrophys. J. **648**, 797 (2006)

E. Bertschinger, P. Zukin, Phys. Rev. D **78**, 024015 (2008)
T. Biswas, R. Mansouri, A. Notari, J. Cosmol. Astropart. Phys. **0712**, 017 (2007)
N.W. Boggess, J.C. Mather, R. Weiss et al., Astrophys. J. **397**, 420 (1992)
H. Bondi, Mon. Not. R. Astron. Soc. **107**, 410 (1947)
R.R. Caldwell, Phys. Lett. B **545**, 23 (2002)
R.R. Caldwell, D. Grin, Phys. Rev. Lett. **100**, 031301 (2008)
R.R. Caldwell, M. Kamionkowski, arXiv:0903.0866 [astro-ph.CO] (2009)
R.R. Caldwell, E.V. Linder, Phys. Rev. Lett. **95**, 141301 (2005)
R.R. Caldwell, A. Stebbins, Phys. Rev. Lett. **100**, 191302 (2008)
R. Caldwell, A. Stebbins (2009, in preparation)
R.R. Caldwell, R. Dave, P.J. Steinhardt, Phys. Rev. Lett. **80**, 1582 (1998)
R.R. Caldwell, M. Kamionkowski, N.N. Weinberg, Phys. Rev. Lett. **91**, 071301 (2003)
R. Caldwell, A. Cooray, A. Melchiorri, Phys. Rev. D **76**, 023507 (2007)
S.M. Carroll, Phys. Rev. Lett. **81**, 3067 (1998)
S.M. Carroll, V. Duvvuri, M. Trodden, M.S. Turner, Phys. Rev. D **70**, 043528 (2004)
M.N. Celerier, Astron. Astrophys. **353**, 63 (2000)
G. Chartas et al., Astrophys. J. **565**, 96 (2002)
G. Chartas, private communication (2005)
J. Chiaverini et al., Phys. Rev. Lett. **90**, 151101 (2003)
C.H. Chuang, J.A. Gu, W.Y. Hwang, Class. Quant. Grav. **25**, 175001 (2008)
D.J.H. Chung, A.E. Romano, Phys. Rev. D **74**, 103507 (2006)
K. Coble, S. Dodelson, J.A. Frieman, Phys. Rev. D **55**, 1851 (1997)
E. Corinaldesi, Proc. Phys. Soc. (Lond.) **A69**, 189 (1956)
R. Crittenden, E. Majerotto, F. Piazza, Phys. Rev. Lett. **98**, 251301 (2007)
M.P. Dabrowski, M.A. Hendry, Astrophys. J. **498**, 67 (1998)
S.F. Daniel, R.R. Caldwell, A. Cooray, A. Melchiorri, Phys. Rev. D **77**, 103513 (2008)
S.F. Daniel, R.R. Caldwell, A. Cooray, P. Serra, A. Melchiorri, arXiv:0901.0919 [astro-ph.CO] (2009)
J. Dunkley et al. (WMAP Collaboration), Astrophys. J. Suppl. **180**, 306 (2009)
R. Durrer, R. Maartens, Gen. Relativ. Gravit. **40**, 301 (2008)
K. Dutta, L. Sorbo, Phys. Rev. D **75**, 063514 (2007)
G.R. Dvali, G. Gabadadze, M. Porrati, Phys. Lett. B **485**, 208 (2000)
G. Dvali, S. Hofmann, J. Khoury, Phys. Rev. D **76**, 084006 (2007)
A. Einstein, The collected papers of Albert Einstein, vol. 8: The Berlin Years: Correspondence, 1914–1918. (English supplement translation.) translated by Ann M. Hentschel, Klaus Hentschel, consultant (Princeton University Press, Princeton, 1998)
D.J. Eisenstein et al. (SDSS Collaboration), Astrophys. J. **633**, 560 (2005)
K. Enqvist, T. Mattsson, J. Cosmol. Astropart. Phys. **0702**, 019 (2007)
D. Fixsen, E. Cheng et al., Astrophys. J. **473**, 576 (1996)
D.J. Fixsen, J.C. Mather, Astrophys. J. **581**, 817 (2002)
W.L. Freedman et al. (HST Collaboration), Astrophys. J. **553**, 47 (2001)
J.A. Frieman, C.T. Hill, A. Stebbins, I. Waga, Phys. Rev. Lett. **75**, 2077 (1995)
J. Frieman, M. Turner, D. Huterer, Ann. Rev. Astron. Astrophys. **46**, 385 (2008)
L. Fu et al., Astron. Astrophys. **479**, 9 (2008)
J. Garcia-Bellido, T. Haugboelle, J. Cosmol. Astropart. Phys. **0804**, 003 (2008)
D. Garfinkle, Class. Quant. Grav. **23**, 4811 (2006)
J. Goodman, Phys. Rev. D **52**, 1821 (1995)
A. Hajian, T. Souradeep, Phys. Rev. D **74**, 123521 (2006)
S. Ho, C. Hirata, N. Padmanabhan, U. Seljak, N. Bahcall, Phys. Rev. D **78**, 043519 (2008)
D.W. Hogg et al., Astrophys. J. **624**, 54 (2005)
F. Hoyle, A. Sandage, Publ. Astron. Soc. Pac. **68**, 301 (1956)
C.D. Hoyle et al., Phys. Rev. Lett. **86**, 1418 (2001)
C.D. Hoyle et al., Phys. Rev. D **70**, 042004 (2004)
W. Hu, I. Sawicki, Phys. Rev. D **76**, 104043 (2007)
M.L. Humason, N.U. Mayall, A.R. Sandage, Astron. J. **61**, 97 (1956)
D. Huterer, H.V. Peiris, Phys. Rev. D **75**, 083503 (2007)
H. Iguchi, T. Nakamura, K.i. Nakao, Prog. Theor. Phys. **108**, 809 (2002)
T.R. Jaffe, A.J. Banday, H.K. Eriksen, K.M. Gorski, F.K. Hansen, Astrophys. J. **629**, L1 (2005)
B. Jain, P. Zhang, Phys. Rev. D **78**, 063503 (2008)
D.J. Kapner et al., Phys. Rev. Lett. **98**, 021101 (2007)
A. Kogut et al., New Astron. Rev. **50**, 925 (2006)
E. Komatsu et al. (WMAP Collaboration), Astrophys. J. Suppl. **180**, 330 (2009)

M. Kowalski et al., Astrophys. J. **686**, 749 (2008)
C. Li, D.E. Holz, A. Cooray, Phys. Rev. D **75**, 103503 (2007)
J.C. Long et al., Nature **421**, 922 (2003)
J.C. Mather et al., Astrophys. J. **420**, 439 (1994)
J.W. Moffat, J. Cosmol. Astropart. Phys. **0605**, 001 (2006)
S. Nobbenhuis, Found. Phys. **36**, 613 (2006)
W.J. Percival, S. Cole, D.J. Eisenstein, R.C. Nichol, J.A. Peacock, A.C. Pope, A.S. Szalay, Mon. Not. R. Astron. Soc. **381**, 1053 (2007)
B. Ratra, P.J.E. Peebles, Phys. Rev. D **37**, 3406 (1988)
H.P. Robertson, Publ. Astron. Soc. Pac. **67**, 82 (1955)
D. Sarkar, private communication (2008)
P. Serra, A. Cooray, S.F. Daniel, R. Caldwell, A. Melchiorri, arXiv:0901.0917 [astro-ph.CO] (2009)
S.J. Smullin et al., Phys. Rev. D **72**, 122001 (2005) [Erratum-ibid. D **72**, 129901 (2005)]
Y.S. Song, K. Koyama, J. Cosmol. Astropart. Phys. **0901**, 048 (2009)
A. Stebbins, arXiv:astro-ph/0703541 (2007)
P.J. Steinhardt, L.M. Wang, I. Zlatev, Phys. Rev. D **59**, 123504 (1999)
N. Straumann, arXiv:0810.2213 [physics.hist-ph] (2008)
S. Sullivan, A. Cooray, D.E. Holz, J. Cosmol. Astropart. Phys. **0709**, 004 (2007)
S. Sullivan, D. Sarkar, S. Joudaki, A. Amblard, D.E. Holz, A. Cooray, Phys. Rev. Lett. **100**, 241302 (2008)
R. Sundrum, Phys. Rev. D **69**, 044014 (2004)
K. Tomita, Mon. Not. R. Astron. Soc. **326**, 287 (2001)
R.A. Vanderveld, E.E. Flanagan, I. Wasserman, Phys. Rev. D **74**, 023506 (2006)
S. Weinberg, Rev. Mod. Phys. **61**, 1 (1989)
C.M. Will, Cambridge, UK: Univ. Pr. (1993) 380 p.
Y.B. Zel'dovich, Sov. Phys. Usp. **11**, 381 (1968)
P. Zhang, U. Pen, H. Trac, Mon. Not. R. Astron. Soc. **355**, 451 (2004)
G.B. Zhao, L. Pogosian, A. Silvestri, J. Zylberberg, arXiv:0905.1326 [astro-ph.CO] (2009)
I. Zlatev, L.M. Wang, P.J. Steinhardt, Phys. Rev. Lett. **82**, 896 (1999)

Space Sci Rev (2009) 148: 363–381
DOI 10.1007/s11214-008-9478-1

# An Assessment of the Systematic Uncertainty in Present and Future Tests of the Lense-Thirring Effect with Satellite Laser Ranging

**Lorenzo Iorio**

Received: 23 November 2008 / Accepted: 5 December 2008 / Published online: 25 December 2008
© Springer Science+Business Media B.V. 2008

**Abstract** We deal with the attempts to measure the Lense-Thirring effect with the Satellite Laser Ranging (SLR) technique applied to the existing LAGEOS and LAGEOS II terrestrial satellites and to the recently approved LARES spacecraft. According to general relativity, a central spinning body of mass $M$ and angular momentum $\boldsymbol{S}$ like the Earth generates a gravitomagnetic field which induces small secular precessions of the orbit of a test particle geodesically moving around it. Extracting this signature from the data is a demanding task because of many classical orbital perturbations having the same pattern as the gravitomagnetic one, like those due to the centrifugal oblateness of the Earth which represents a major source of systematic bias. The first issue addressed here is: are the so far published evaluations of the systematic uncertainty induced by the bad knowledge of the even zonal harmonic coefficients $J_\ell$ of the multipolar expansion of the Earth's geopotential reliable and realistic? Our answer is negative. Indeed, if the differences $\Delta J_\ell$ among the even zonals estimated in different Earth's gravity field global solutions from the dedicated GRACE mission are assumed for the uncertainties $\delta J_\ell$ instead of using their covariance sigmas $\sigma_{J_\ell}$, it turns out that the systematic uncertainty $\delta\mu$ in the Lense-Thirring test with the nodes $\Omega$ of LAGEOS and LAGEOS II may be up to 3 to 4 times larger than in the evaluations so far published (5–10%) based on the use of the sigmas of one model at a time separately. The second issue consists of the possibility of using a different approach in extracting the relativistic signature of interest from the LAGEOS-type data. The third issue is the possibility of reaching a realistic total accuracy of 1% with LAGEOS, LAGEOS II and LARES, which should be launched in November 2009 with a VEGA rocket. While LAGEOS and LAGEOS II fly at altitudes of about 6000 km, LARES will be likely placed at an altitude of 1450 km. Thus, it will be sensitive to much more even zonals than LAGEOS and LAGEOS II. Their corrupting impact has been evaluated with the standard Kaula's approach up to degree $\ell = 60$ by using $\Delta J_\ell$ and $\sigma_{J_\ell}$; it turns out that it may be as large as some tens percent. The different orbit

L. Iorio (✉)
INFN-Sezione di Pisa, Pisa, Italy
e-mail: lorenzo.iorio@libero.it

*Present address:*
L. Iorio
Viale Unità di Italia 68, 70125 Bari, BA, Italy

of LARES may also have some consequences on the non-gravitational orbital perturbations affecting it which might further degrade the obtainable accuracy in the Lense-Thirring test.

**Keywords** Experimental tests of gravitational theories · Satellite orbits · Harmonics of the gravity potential field

**PACS** 04.80.Cc · 91.10.Sp · 91.10.Qm

## 1 Introduction

In the weak-field and slow motion approximation, the Einstein field equations of general relativity get linearized resembling to the Maxwellian equations of electromagnetism. As a consequence, a gravitomagnetic field $\boldsymbol{B}_{\mathrm{g}}$, induced by the off-diagonal components $g_{0i}, i = 1, 2, 3$ of the space-time metric tensor related to the mass-energy currents of the source of the gravitational field, arises (Mashhoon 2007). The gravitomagnetic field affects orbiting test particles, precessing gyroscopes, moving clocks and atoms and propagating electromagnetic waves (Ruggiero and Tartaglia 2002; Schäfer 2004). Perhaps, the most famous gravitomagnetic effects are the precession of the axis of a gyroscope (Pugh 1959; Schiff 1960) and the Lense-Thirring[1] precessions (Lense and Thirring 1918) of the orbit of a test particle, both occurring in the field of a central slowly rotating mass like, e.g., our planet. Direct, undisputable measurements of such fundamental predictions of general relativity are not yet available.

The measurement of the gyroscope precession in the Earth's gravitational field has been the goal of the dedicated space-based[2] GP-B mission (Everitt 1974; Everitt et al. 2001) launched in 2004 and carrying onboard four superconducting gyroscopes; its data analysis is still ongoing. The target accuracy was originally 1%, but it is still unclear if the GP-B team will succeed in reaching such a goal because of some unmodelled effects affecting the gyroscopes: (1) a time variation in the polhode motion of the gyroscopes and (2) very large classical misalignment torques on the gyroscopes.

In this Chapter we will focus on the attempts to measure of the Lense-Thirring effect in the gravitational field of the Earth; for Mars and the Sun see Iorio (2006a, 2007a, 2008a) and Krogh (2007), respectively. Far from a localized rotating body with angular momentum $\boldsymbol{S}$ the gravitomagnetic field can be written as

$$\boldsymbol{B}_{\mathrm{g}} = \frac{G}{cr^3}[\boldsymbol{S} - 3(\boldsymbol{S}\cdot\hat{\boldsymbol{r}})\hat{\boldsymbol{r}}], \tag{1}$$

where $G$ is the Newtonian gravitational constant and $c$ is the speed of light in vacuum. It acts on a test particle orbiting with a velocity $\boldsymbol{v}$ with the non-central acceleration (Soffel 1989)

$$\boldsymbol{A}_{\mathrm{LT}} = -\frac{2}{c}\boldsymbol{v} \times \boldsymbol{B}_{\mathrm{g}} \tag{2}$$

[1] According to an interesting historical analysis recently performed by Pfister (2007), it would be more correct to speak about an Einstein-Thirring-Lense effect.

[2] See on the WEB http://einstein.stanford.edu/.

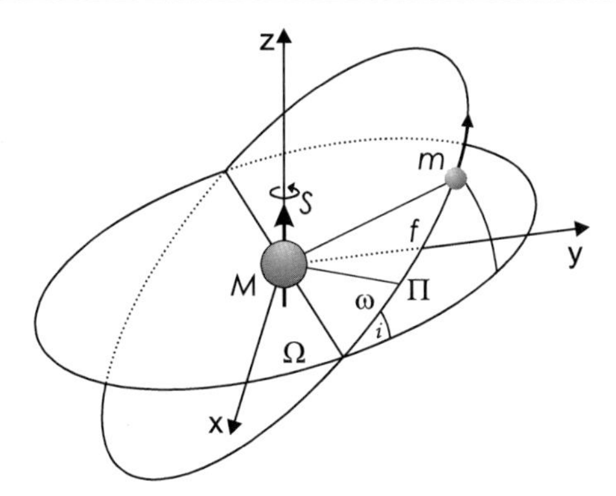


**Fig. 1** Keplerian orbit. The longitude of the ascending node $\Omega$ is counted from a reference X direction in the equator of the central body, assumed as reference plane {X, Y}, to the line of the nodes which is the intersection of the orbital plane with the equatorial plane of the central body. It has mass $M$ and proper angular momentum $\boldsymbol{S}$. The argument of pericentre $\omega$ is an angle in the orbital plane counted from the line of the nodes to the location of the pericentre, here marked with $\Pi$. The time-dependent position of the moving test particle of mass $m$ is given by the true anomaly $f$, counted anticlockwise from the pericentre's position. The inclination between the orbital and the equatorial planes is $i$. Thus, $\Omega, \omega, i$ can be viewed as the three (constant) Euler angles fixing the configuration of a rigid body, i.e. the orbit which in the unperturbed Keplerian case does change neither its shape nor its size, in the inertial {X, Y, Z} space. Courtesy by H.I.M. Lichtenegger, IWF, Graz

which induces secular precessions of the longitude of the ascending node $\Omega$

$$\dot{\Omega}_{\rm LT} = \frac{2GS}{c^2a^3(1-e^2)^{3/2}}, \tag{3}$$

and the argument of pericentre $\omega$

$$\dot{\omega}_{\rm LT} = -\frac{6GS\cos i}{c^2a^3(1-e^2)^{3/2}}, \tag{4}$$

of the orbit of a test particle. In (3) and (4) $a$ and $e$ are the semimajor axis and the eccentricity, respectively, of the test particle's orbit and $i$ is its inclination to the central body's equator. The semimajor axis $a$ fixes the size of the ellipse, while its shape is determined by the eccentricity $0 \leq e < 1$; an orbit with $e = 0$ is a circle. The angles $\Omega$ and $\omega$ establish the orientation of the orbit in the inertial space and in the orbital plane, respectively. $\Omega$, $\omega$ and $i$ can be viewed as the three Euler angles which determine the orientation of a rigid body with respect to an inertial frame. In Fig. 1 we illustrate the geometry of a Keplerian orbit.

In this Chapter we will critically discuss the following topics

- Section 2. The realistic evaluation of the total accuracy in the test performed in recent years with the existing Earth's artificial satellites LAGEOS and LAGEOS II (Ciufolini and Pavlis 2004; Ciufolini et al. 2006; Ries et al. 2008). LAGEOS was put into orbit in 1976, followed by its twin LAGEOS II in 1992; they are passive, spherical spacecraft entirely covered by retroreflectors which allow for their accurate tracking through laser pulses sent from Earth-based ground stations according to the Satellite Laser Ranging (SLR) technique (Degnan 1985). They orbit at altitudes of about 6000 km ($a_{\rm LAGEOS} = 12\,270$ km, $a_{\rm LAGEOS\ II} = 12\,163$ km) in nearly circular paths ($e_{\rm LAGEOS} = 0.0045$, $e_{\rm LAGEOS\ II} = 0.014$) inclined by 110 deg and 52.65 deg, respectively, to the Earth's equator. The Lense-Thirring effect for their nodes amounts to about 30 milliarcseconds per year (mas yr$^{-1}$)

which correspond to about 1.7 $\mathrm{m\,yr^{-1}}$ in the cross-track direction[3] at the LAGEOS altitudes.

The idea of measuring the Lense-Thirring node rate with the just launched LAGEOS satellite, along with the other SLR targets orbiting at that time, was put forth by Cugusi and Proverbio (1978). Tests have started to be effectively performed later by using the LAGEOS and LAGEOS II satellites (Ciufolini et al. 1996), according to a strategy by Ciufolini (1996) involving the use of a suitable linear combination of the nodes $\Omega$ of both satellites and the perigee $\omega$ of LAGEOS II. This was done to reduce the impact of the most relevant source of systematic bias, i.e. the mismodelling in the even ($\ell = 2, 4, 6\ldots$) zonal ($m = 0$) harmonic coefficients $J_\ell$ of the multipolar expansion of the Newtonian part of the terrestrial gravitational potential due to the diurnal rotation (they induce secular precessions on the node and perigee of a terrestrial satellite much larger than the gravitomagnetic ones. The $J_\ell$ coefficients cannot be theoretically computed but must be estimated by suitably processing long data sets from the dedicated satellites like CHAMP and GRACE; see Sect. 2): the three-elements combination used allowed for removing the uncertainties in $J_2$ and $J_4$. In Ciufolini et al. (1998a) a $\approx 20\%$ test was reported by using the[4] EGM96 Earth gravity model (Lemoine et al. 1998); subsequent detailed analyses showed that such an evaluation of the total error budget was overly optimistic in view of the likely unreliable computation of the total bias due to the even zonals (Iorio 2003; Ries et al. 2003a, 2003b). An analogous, huge underestimation turned out to hold also for the effect of the non-gravitational perturbations (Milani et al. 1987) like the direct solar radiation pressure, the Earth's albedo, various subtle thermal effects depending on the physical properties of the satellites' surfaces and their rotational state (Inversi and Vespe 1994; Vespe 1999; Lucchesi 2001, 2002, 2003, 2004; Lucchesi et al. 2004; Ries et al. 2003a), which the perigees of LAGEOS-like satellites are particularly sensitive to. As a result, the realistic total error budget in the test reported in Ciufolini et al. (1998a) might be as large as 60–90% or (by considering EGM96 only) even more.

The observable used by Ciufolini and Pavlis (2004) with the GRACE-only EIGEN-GRACE02S model (Reigber et al. 2005) and by Ries et al. (2008) with other more recent Earth gravity models was the following linear combination[5] of the nodes of LAGEOS and LAGEOS II, explicitly computed by Iorio and Morea (2004) following the approach put forth by Ciufolini (1996)

$$f = \dot{\Omega}^{\text{LAGEOS}} + c_1 \dot{\Omega}^{\text{LAGEOS II}}, \tag{5}$$

where

[3] A perturbing acceleration like $\boldsymbol{A}_{\text{LT}}$ is customarily projected onto the radial $\hat{\boldsymbol{r}}$, transverse $\hat{\boldsymbol{\tau}}$ and cross-track $\hat{\boldsymbol{\nu}}$ directions of an orthogonal frame comoving with the satellite (Soffel 1989); it turns out that the Lense-Thirring node precession affects the cross-track component of the orbit according to $\Delta\nu_{\text{LT}} \approx a \sin i \, \Delta\Omega_{\text{LT}}$ ((A65), p. 6233 in Christodoulidis et al. 1988).

[4] Contrary to the subsequent models based on the dedicated satellites CHAMP (http://www-app2.gfz-potsdam.de/pb1/op/champ/index_CHAMP.html) and GRACE (http://www-app2.gfz-potsdam.de/pb1/op/grace/index_GRACE.html), EGM96 relies upon multidecadal tracking of SLR data of a constellation of geodetic satellites including LAGEOS and LAGEOS II as well; thus the possibility of a sort of *a priori* 'imprinting' of the Lense-Thirring effect itself, not solved-for in EGM96, cannot be neglected.

[5] See also Pavlis (2002), Ries et al. (2003a, 2003b).

$$c_1 \equiv -\frac{\dot{\Omega}_{.2}^{\text{LAGEOS}}}{\dot{\Omega}_{.2}^{\text{LAGEOS II}}}$$

$$= -\frac{\cos i_{\text{LAGEOS}}}{\cos i_{\text{LAGEOS II}}}\left(\frac{1-e_{\text{LAGEOS II}}^2}{1-e_{\text{LAGEOS}}^2}\right)^2\left(\frac{a_{\text{LAGEOS II}}}{a_{\text{LAGEOS}}}\right)^{7/2}. \quad (6)$$

The coefficients $\dot{\Omega}_{.\ell}$ of the aliasing classical node precessions (Kaula 1966) $\dot{\Omega}_{\text{class}} = \sum_{\ell} \dot{\Omega}_{.\ell} J_{\ell}$ induced by the even zonals have been analytically worked out up to $\ell = 20$ in, e.g. Iorio (2003); they yield $c_1 = 0.544$. The Lense-Thirring signature of (5) amounts to 47.8 mas $\text{yr}^{-1}$. The combination (5) allows, by construction, to remove the aliasing effects due to the static and time-varying parts of the first even zonal $J_2$. The nominal (i.e. computed with the estimated values of $J_{\ell}$, $\ell = 4, 6\ldots$) bias due to the remaining higher degree even zonals would amount to about $10^5$ mas $\text{yr}^{-1}$; the need of a careful and reliable modeling of such an important source of systematic bias is, thus, quite apparent. Conversely, the nodes of the LAGEOS-type spacecraft are directly affected by the non-gravitational accelerations at a $\approx 1\%$ level of the Lense-Thirring effect (Lucchesi 2001, 2002, 2003, 2004; Lucchesi et al. 2004). For a comprehensive, up-to-date overview of the numerous and subtle issues concerning the measurement of the Lense-Thirring effect see, e.g., Iorio (2007b).

- Section 3. Another approach which could be followed in extracting the Lense-Thirring effect from the data of the LAGEOS-type satellites.
- Section 4. The possibility that the LARES mission, recently approved by the Italian Space Agency (ASI), will be able to measure the Lense-Thirring node precession with an accuracy of the order of 1%.

In Van Patten and Everitt (1976a, 1976b) it was proposed to measure the Lense-Thirring precession of the nodes $\Omega$ of a pair of counter-orbiting spacecraft to be launched in terrestrial polar orbits and endowed with drag-free apparatus. A somewhat equivalent, cheaper version of such an idea was put forth in 1986 by Ciufolini (1986) who proposed to launch a passive, geodetic satellite in an orbit identical to that of LAGEOS apart from the orbital planes which should have been displaced by 180 deg apart. The measurable quantity was, in the case of the proposal by Ciufolini (1986), the sum of the nodes of LAGEOS and of the new spacecraft, later named LAGEOS III, LARES, WEBER-SAT, in order to cancel to a high level of accuracy the corrupting effect of the multipoles of the Newtonian part of the terrestrial gravitational potential which represent the major source of systematic error (see Sect. 2). Although extensively studied by various groups (Ries et al. 1989; Ciufolini et al. 1998b), such an idea was not implemented for many years. In Iorio et al. (2002) it was proposed to include also the data from LAGEOS II by using a different observable. Such an approach was proven in Iorio (2005a) to be potentially useful in making the constraints on the orbital configuration of the new SLR satellite less stringent than it was originally required in view of the recent improvements in our knowledge of the classical part of the terrestrial gravitational potential due to the dedicated CHAMP and, especially, GRACE missions.

Since reaching high altitudes and minimizing the unavoidable orbital injection errors is expensive, it was explored the possibility of discarding LAGEOS and LAGEOS II using a low-altitude, nearly polar orbit for LARES (Lucchesi and Paolozzi 2001; Ciufolini 2006), but in Iorio (2002, 2007c) it was proven that such alternative approaches are not feasible. It was also suggested that LARES would be able to probe alternative theories of gravity (Ciufolini 2005), but also in this case it turned out to be impossible (Iorio 2005b, 2007d).

The stalemate came to an end when ASI recently made the following official announcement (http://www.asi.it/SiteEN/MotorSearchFullText.aspx?keyw=LARES): "On

February 8, the ASI board approved funding for the LARES mission, that will be launched with VEGA's maiden flight before the end of 2008. LARES is a passive satellite with laser mirrors, and will be used to measure the Lense-Thirring effect." The Italian version of the announcement yields some more information specifying that LARES, designed in collaboration with National Institute of Nuclear Physics (INFN), is currently under construction by Carlo Gavazzi Space SpA; its Principal Investigator (PI) is I. Ciufolini and its scientific goal is to measure at a 1% level the Lense-Thirring effect in the gravitational field of the Earth. Concerning the orbital configuration of LARES, In one of the latest communication to INFN, Rome, 30 January 2008, (Ciufolini 2008a) writes that LARES will be launched with a semimajor axis of approximately 7600 km and an inclination between 60 and 80 deg. More precise information can be retrieved in Sect. 5.1, page 9 of the document Educational Payload on the Vega Maiden Flight Call For CubeSat Proposals, European Space Agency, Issue 1 11 February 2008, downloadable at http://esamultimedia.esa.int/docs/LEX-EC/CubeSat%20CFP%20issue%201.pdf. It is written there that LARES will be launched into a circular orbit with altitude $h = 1200$ km, corresponding to a semimajor axis $a_{\mathrm{LARES}} = 7578$ km, and inclination $i = 71$ deg to the Earth's equator. Latest information[6] point towards a launch at the end of 2009 with a VEGA rocket in a circular orbit inclined by 71 deg to the Earth's equator at an altitude of[7] 1450 km corresponding to a semimajor axis of $a_{\mathrm{LR}} = 7828$ km. More or less the same has been reported by Ciufolini (2008b) to the INFN in Villa Mondragone, 3 October 2008.

## 2 The Systematic Error of Gravitational Origin in the LAGEOS-LAGEOS II Test

The realistic evaluation of the total error budget of the LAGEOS-LAGEOS II node test (Ciufolini and Pavlis 2004) raised a lively debate (Ciufolini and Pavlis 2005; Ciufolini et al. 2006; Iorio 2005a, 2006b, 2006c, 2007e; Lucchesi 2005), mainly focussed on the impact of the static and time-varying parts of the Newtonian component of the Earth's gravitational potential through the secular precessions induced on a satellite's node.

In the real world the path of a probe is not only affected by the relativistic gravitomagnetic field but also by a huge number of other competing classical orbital perturbations of gravitational and non-gravitational origin. The most insidious disturbances are those induced by the static part of the Newtonian component of the multipolar expansion in spherical harmonics[8] $J_\ell, \ell = 2, 4, 6, \ldots$ of the gravitational potential of the central rotating mass (Kaula 1966): they affect the node with effects having the same signature of the relativistic signal of interest, i.e. linear trends which are orders of magnitude larger and cannot be removed from the time series of data without affecting the Lense-Thirring pattern itself as well. The only thing that can be done is to model such a corrupting effect as most accurately as possible and assessing the impact of the residual mismodelling on the measurement of the frame-dragging effect. The secular precessions induced by the even zonals of the geopotential can be written as

$$\dot{\Omega}^{\mathrm{geopot}} = \sum_{\ell=2} \dot{\Omega}_{.\ell} J_\ell, \tag{7}$$

[6] See on the WEB http://www.esa.int/esapub/bulletin/bulletin135/bul135f_bianchi.pdf.

[7] I thank Dr. D. Barbagallo (ESRIN) for having kindly provided me with the latest details of the orbital configuration of LARES.

[8] The relation among the even zonals $J_\ell$ and the normalized gravity coefficients $\overline{C}_{\ell 0}$ which are customarily determined in the Earth's gravity models, is $J_\ell = -\sqrt{2\ell + 1}\,\overline{C}_{\ell 0}$.

where the coefficients $\dot{\Omega}_{.\ell}, \ell = 2, 4, 6, \ldots$ depend on the parameters of the Earth ($GM$ and the equatorial radius $R$) and on the semimajor axis $a$, the inclination $i$ and the eccentricity $e$ of the satellite. For example, for $\ell = 2$ we have

$$\dot{\Omega}_{.2} = -\frac{3}{2} n \left(\frac{R}{a}\right)^2 \frac{\cos i}{(1-e^2)^2}; \tag{8}$$

$n = \sqrt{GM/a^3}$ is the Keplerian mean motion. They have been analytically computed up to $\ell = 20$ in, e.g., Iorio (2003). Their mismodelling can be written as

$$\delta\dot{\Omega}^{\text{geopot}} \leq \sum_{\ell=2} |\dot{\Omega}_{.\ell}|\,\delta J_{\ell}, \tag{9}$$

where $\delta J_{\ell}$ represents our uncertainty in the knowledge of the even zonals $J_{\ell}$

A common feature of all the competing evaluations so far published is that the systematic bias due to the static component of the geopotential was always calculated by using the released (more or less accurately calibrated) sigmas $\sigma_{J_\ell}$ of one Earth gravity model solution at a time for the uncertainties $\delta J_{\ell}$. Thus, it was said that the model X yields a $x\%$ error, the model Y yields a $y\%$ error, and so on.

Since a trustable calibration of the formal, statistical uncertainties in the estimated zonals of the covariance matrix of a global solution is always a difficult task to be implemented in a reliable way, a much more realistic and conservative approach consists, instead, of taking the difference[9]

$$\Delta J_{\ell} = |J_{\ell}(\text{X}) - J_{\ell}(\text{Y})|, \quad \ell = 2, 4, 6, \ldots \tag{10}$$

of the estimated even zonals for different pairs of Earth gravity field solutions as representative of the real uncertainty $\delta J_{\ell}$ in the zonals (Lerch et al. 1994). In Tables 1, 2, 3, 4, 5, 6, 7, 8, and 9 we present our results for the most recent GRACE-based models released so far by different institutions and retrievable on the Internet at[10] http://icgem.gfz-potsdam.de/ICGEM/ICGEM.html. The models used are EIGEN-GRACE02S (Reigber et al. 2005) from GFZ (Potsdam, Germany), GGM02S (Tapley et al. 2005) and GGM03S (Tapley et al. 2007) from CSR (Austin, Texas), ITG-Grace02s (Mayer-Gürr et al. 2006) and ITG-Grace03s (Mayer-Gürr 2007) from IGG (Bonn, Germany), JEM01-RL03B from JPL (NASA, USA) and AIUB-GRACE01S (Jäggi et al. 2008) from AIUB (Switzerland). Note that this approach was explicitly followed also by Ciufolini (1996) with the JGM3 and GEMT-2 models. In Tables 1–9 we quote both the sum $\sum_{\ell=4}^{20} f_{\ell}$ of the absolute values of the individual mismodelled terms

$$f_{\ell} = |\dot{\Omega}^{\text{LAGEOS}}_{.\ell} + c_1 \dot{\Omega}^{\text{LAGEOS II}}_{.\ell}|\,\Delta J_{\ell}, \quad \ell = 4, 6, 8, \ldots \tag{11}$$

(SAV), and the square root of the sum of their squares $\sqrt{\sum_{\ell=4}^{20} f_{\ell}^2}$ (RSS); in both cases we normalized them to the combined Lense-Thirring total precession of 47.8 mas yr$^{-1}$.

The systematic bias evaluated with a more realistic approach is about 3 to 4 times larger than one can obtain by only using this or that particular model. The scatter is still quite large

[9]See Fig. 5 of Lucchesi (2007) for a comparison of the estimated $\overline{C}_{40}$ in different models.

[10]I thank M. Watkins (JPL) for having provided me with the even zonals and their sigmas of the JEM01-RL03B model.

**Table 1** Impact of the mismodelling in the even zonal harmonics on $f_\ell = |\dot{\Omega}_\ell^{\text{LAGEOS}} + c_1 \dot{\Omega}_\ell^{\text{LAGEOS II}}|\Delta J_\ell, \ell = 4, \ldots, 20$, in mas yr$^{-1}$. Recall that $J_\ell = -\sqrt{2\ell+1}\,\overline{C}_{\ell 0}$; for the uncertainty in the even zonals we have taken here the difference $\Delta\overline{C}_{\ell 0} = |\overline{C}_{\ell 0}^{(\text{X})} - \overline{C}_{\ell 0}^{(\text{Y})}|$ between the model X = GGM02S (Tapley et al. 2005) and the model Y = ITG-Grace02s (Mayer-Gürr et al. 2006). GGM02S is based on 363 days of GRACE-only data (GPS and intersatellite tracking, neither constraints nor regularization applied) spread between April 4, 2002 and Dec 31, 2003. The $\sigma$ are formal for both models. $\Delta\overline{C}_{\ell 0}$ are always larger than the linearly added sigmas, apart from $\ell = 12$ and $\ell = 18$. Values of $f_\ell$ smaller than 0.1 mas yr$^{-1}$ have not been quoted. The Lense-Thirring precession of the combination of (5) amounts to 47.8 mas yr$^{-1}$. The percent bias $\delta\mu$ have been computed by normalizing the linear sum of $f_\ell, \ell = 4, \ldots, 20$ (SAV) and the square root of the sum of $f_\ell^2, \ell = 4, \ldots, 20$ to the Lense-Thirring combined precessions

| $\ell$ | $\Delta\overline{C}_{\ell 0}$ (GGM02S-ITG-Grace02s) | $\sigma_{\text{X}} + \sigma_{\text{Y}}$ | $f_\ell$ (mas yr$^{-1}$) |
|---|---|---|---|
| 4 | $1.9 \times 10^{-11}$ | $8.7 \times 10^{-12}$ | 7.2 |
| 6 | $2.1 \times 10^{-11}$ | $4.6 \times 10^{-12}$ | 4.6 |
| 8 | $5.7 \times 10^{-12}$ | $2.8 \times 10^{-12}$ | 0.2 |
| 10 | $4.5 \times 10^{-12}$ | $2.0 \times 10^{-12}$ | – |
| 12 | $1.5 \times 10^{-12}$ | $1.8 \times 10^{-12}$ | – |
| 14 | $6.6 \times 10^{-12}$ | $1.6 \times 10^{-12}$ | – |
| 16 | $2.9 \times 10^{-12}$ | $1.6 \times 10^{-12}$ | – |
| 18 | $1.4 \times 10^{-12}$ | $1.6 \times 10^{-12}$ | – |
| 20 | $2.0 \times 10^{-12}$ | $1.6 \times 10^{-12}$ | – |
| | $\delta\mu = 25\%$ (SAV) | $\delta\mu = 18\%$ (RSS) | |

**Table 2** Bias due to the mismodelling in the even zonals of the models X = ITG-Grace03s (Mayer-Gürr 2007), based on GRACE-only accumulated normal equations from data out of September 2002–April 2007 (neither a priori information nor regularization used), and Y = GGM02S (Tapley et al. 2005). The $\sigma$ for both models are formal. $\Delta\overline{C}_{\ell 0}$ are always larger than the linearly added sigmas, apart from $\ell = 12$ and $\ell = 18$

| $\ell$ | $\Delta\overline{C}_{\ell 0}$ (ITG-Grace03s-GGM02S) | $\sigma_{\text{X}} + \sigma_{\text{Y}}$ | $f_\ell$ (mas yr$^{-1}$) |
|---|---|---|---|
| 4 | $2.58 \times 10^{-11}$ | $8.6 \times 10^{-12}$ | 9.6 |
| 6 | $1.39 \times 10^{-11}$ | $4.7 \times 10^{-12}$ | 3.1 |
| 8 | $5.6 \times 10^{-12}$ | $2.9 \times 10^{-12}$ | 0.2 |
| 10 | $1.03 \times 10^{-11}$ | $2 \times 10^{-12}$ | – |
| 12 | $7 \times 10^{-13}$ | $1.8 \times 10^{-12}$ | – |
| 14 | $7.3 \times 10^{-12}$ | $1.6 \times 10^{-12}$ | – |
| 16 | $2.6 \times 10^{-12}$ | $1.6 \times 10^{-12}$ | – |
| 18 | $8 \times 10^{-13}$ | $1.6 \times 10^{-12}$ | – |
| 20 | $2.4 \times 10^{-12}$ | $1.6 \times 10^{-12}$ | – |
| | $\delta\mu = 27\%$ (SAV) | $\delta\mu = 21\%$ (RSS) | |

and far from the 5–10% claimed in Ciufolini and Pavlis (2004). In particular, it appears that $J_4$, $J_6$, and to a lesser extent $J_8$, which are just the most relevant zonals for us because of their impact on the combination of (5), are the most uncertain ones, with discrepancies $\Delta J_\ell$ between different models, in general, larger than the sum of their sigmas $\sigma_{J_\ell}$, calibrated or not.

**Table 3** Bias due to the mismodelling in the even zonals of the models X = GGM02S (Tapley et al. 2005) and Y = GGM03S (Tapley et al. 2007) retrieved from data spanning January 2003 to December 2006. The $\sigma$ for GGM03S are calibrated. $\Delta\overline{C}_{\ell 0}$ are larger than the linearly added sigmas for $\ell = 4, 6$. (The other zonals are of no concern)

| $\ell$ | $\Delta\overline{C}_{\ell 0}$ (GGM02S-GGM03S) | $\sigma_X + \sigma_Y$ | $f_\ell$ (mas yr$^{-1}$) |
|---|---|---|---|
| 4 | $1.87 \times 10^{-11}$ | $1.25 \times 10^{-11}$ | 6.9 |
| 6 | $1.96 \times 10^{-11}$ | $6.7 \times 10^{-12}$ | 4.2 |
| 8 | $3.8 \times 10^{-12}$ | $4.3 \times 10^{-12}$ | 0.1 |
| 10 | $8.9 \times 10^{-12}$ | $2.8 \times 10^{-12}$ | 0.1 |
| 12 | $6 \times 10^{-13}$ | $2.4 \times 10^{-12}$ | – |
| 14 | $6.6 \times 10^{-12}$ | $2.1 \times 10^{-12}$ | – |
| 16 | $2.1 \times 10^{-12}$ | $2.0 \times 10^{-12}$ | – |
| 18 | $1.8 \times 10^{-12}$ | $2.0 \times 10^{-12}$ | – |
| 20 | $2.2 \times 10^{-12}$ | $1.9 \times 10^{-12}$ | – |
| | $\delta\mu = 24\%$ (SAV) | $\delta\mu = 17\%$ (RSS) | |

**Table 4** Bias due to the mismodelling in the even zonals of the models X = EIGEN-GRACE02S (Reigber et al. 2005) and Y = GGM03S (Tapley et al. 2007). The $\sigma$ for both models are calibrated. $\Delta\overline{C}_{\ell 0}$ are always larger than the linearly added sigmas apart from $\ell = 14, 18$

| $\ell$ | $\Delta\overline{C}_{\ell 0}$ (EIGEN-GRACE02S-GGM03S) | $\sigma_X + \sigma_Y$ | $f_\ell$ (mas yr$^{-1}$) |
|---|---|---|---|
| 4 | $2.00 \times 10^{-11}$ | $8.1 \times 10^{-12}$ | 7.4 |
| 6 | $2.92 \times 10^{-11}$ | $4.3 \times 10^{-12}$ | 6.3 |
| 8 | $1.05 \times 10^{-11}$ | $3.0 \times 10^{-12}$ | 0.4 |
| 10 | $7.8 \times 10^{-12}$ | $2.9 \times 10^{-12}$ | 0.1 |
| 12 | $3.9 \times 10^{-12}$ | $1.8 \times 10^{-12}$ | – |
| 14 | $5 \times 10^{-13}$ | $1.7 \times 10^{-12}$ | – |
| 16 | $1.7 \times 10^{-12}$ | $1.4 \times 10^{-12}$ | – |
| 18 | $2 \times 10^{-13}$ | $1.4 \times 10^{-12}$ | – |
| 20 | $2.5 \times 10^{-12}$ | $1.4 \times 10^{-12}$ | – |
| | $\delta\mu = 30\%$ (SAV) | $\delta\mu = 20\%$ (RSS) | |

Another way to evaluate the uncertainty in the LAGEOS-LAGEOS II node test may consist of computing the nominal values of the total combined precessions for different models and comparing them, i.e. by taking

$$\left| \sum_{\ell=4} (\dot{\Omega}_{.\ell}^{\mathrm{LAGEOS}} + c_1 \dot{\Omega}_{.\ell}^{\mathrm{LAGEOS\ II}})[J_\ell(\mathrm{X}) - J_\ell(\mathrm{Y})] \right|. \quad (12)$$

The results are shown in Table 10.

A different approach that could be followed to take into account the scatter among the various solutions consists in computing mean and standard deviation of the entire set of values of the even zonals for the models considered so far, degree by degree, and taking the standard deviations as representative of the uncertainties $\delta J_\ell, \ell = 4, 6, 8, \ldots$. It yields $\delta\mu = 15\%$, in agreement with Ries et al. (2008).

**Table 5** Bias due to the mismodelling in the even zonals of the models X = JEM01-RL03B, based on 49 months of GRACE-only data, and Y = GGM03S (Tapley et al. 2007). The $\sigma$ for GGM03S are calibrated. $\Delta\overline{C}_{\ell 0}$ are always larger than the linearly added sigmas apart from $\ell = 16$

| $\ell$ | $\Delta\overline{C}_{\ell 0}$ (JEM01-RL03B-GGM03S) | $\sigma_X + \sigma_Y$ | $f_\ell$ (mas yr$^{-1}$) |
|---|---|---|---|
| 4 | $1.97 \times 10^{-11}$ | $4.3 \times 10^{-12}$ | 7.3 |
| 6 | $2.7 \times 10^{-12}$ | $2.3 \times 10^{-12}$ | 0.6 |
| 8 | $1.7 \times 10^{-12}$ | $1.6 \times 10^{-12}$ | – |
| 10 | $2.3 \times 10^{-12}$ | $8 \times 10^{-13}$ | – |
| 12 | $7 \times 10^{-13}$ | $7 \times 10^{-13}$ | – |
| 14 | $1.0 \times 10^{-12}$ | $6 \times 10^{-13}$ | – |
| 16 | $2 \times 10^{-13}$ | $5 \times 10^{-13}$ | – |
| 18 | $7 \times 10^{-13}$ | $5 \times 10^{-13}$ | – |
| 20 | $5 \times 10^{-13}$ | $4 \times 10^{-13}$ | – |
| | $\delta\mu = 17\%$ (SAV) | $\delta\mu = 15\%$ (RSS) | |

**Table 6** Bias due to the mismodelling in the even zonals of the models X = JEM01-RL03B and Y = ITG-Grace03s (Mayer-Gürr 2007). The $\sigma$ for ITG-Grace03s are formal. $\Delta\overline{C}_{\ell 0}$ are always larger than the linearly added sigmas

| $\ell$ | $\Delta\overline{C}_{\ell 0}$ (JEM01-RL03B-ITG-Grace03s) | $\sigma_X + \sigma_Y$ | $f_\ell$ (mas yr$^{-1}$) |
|---|---|---|---|
| 4 | $2.68 \times 10^{-11}$ | $4 \times 10^{-13}$ | 9.9 |
| 6 | $3.0 \times 10^{-12}$ | $2 \times 10^{-13}$ | 0.6 |
| 8 | $3.4 \times 10^{-12}$ | $1 \times 10^{-13}$ | 0.1 |
| 10 | $3.6 \times 10^{-12}$ | $1 \times 10^{-13}$ | – |
| 12 | $6 \times 10^{-13}$ | $9 \times 10^{-14}$ | – |
| 14 | $1.7 \times 10^{-12}$ | $9 \times 10^{-14}$ | – |
| 16 | $4 \times 10^{-13}$ | $8 \times 10^{-14}$ | – |
| 18 | $4 \times 10^{-13}$ | $8 \times 10^{-14}$ | – |
| 20 | $7 \times 10^{-13}$ | $8 \times 10^{-14}$ | – |
| | $\delta\mu = 22\%$ (SAV) | $\delta\mu = 10\%$ (RSS) | |

It must be recalled that also the further bias due to the cross-coupling between $J_2$ and the orbit inclination, evaluated to be about 9% in Iorio (2007e), must be added.

## 3 A New Approach to Extract the Lense-Thirring Signature from the Data

The technique adopted so far in Ciufolini and Pavlis (2004) and Ries et al. (2008) to extract the gravitomagnetic signal from the LAGEOS and LAGEOS II data is described in detail in Lucchesi and Balmino (2006) and Lucchesi (2007). In both the approaches the Lense-Thirring force is not included in the dynamical force models used to fit the satellites' data. In the data reduction process no dedicated gravitomagnetic parameter is estimated, contrary to, e.g., station coordinates, state vector, satellites' drag and radiation coefficients $C_D$ and $C_R$, respectively, etc.; its effect is retrieved with a sort of post-post-fit analysis in which

**Table 7** Aliasing effect of the mismodelling in the even zonal harmonics estimated in the X = ITG-Grace03s (Mayer-Gürr 2007) and the Y = EIGEN-GRACE02S (Reigber et al. 2005) models. The covariance matrix $\sigma$ for ITG-Grace03s are formal, while the ones of EIGEN-GRACE02S are calibrated. $\Delta\overline{C}_{\ell 0}$ are larger than the linearly added sigmas for $\ell = 4, \ldots, 20$, apart from $\ell = 18$

| $\ell$ | $\Delta\overline{C}_{\ell 0}$ (ITG-Grace03s-EIGEN-GRACE02S) | $\sigma_X + \sigma_Y$ | $f_\ell$ (mas yr$^{-1}$) |
|---|---|---|---|
| 4 | $2.72 \times 10^{-11}$ | $3.9 \times 10^{-12}$ | 10.1 |
| 6 | $2.35 \times 10^{-11}$ | $2.0 \times 10^{-12}$ | 5.1 |
| 8 | $1.23 \times 10^{-11}$ | $1.5 \times 10^{-12}$ | 0.4 |
| 10 | $9.2 \times 10^{-12}$ | $2.1 \times 10^{-12}$ | 0.1 |
| 12 | $4.1 \times 10^{-12}$ | $1.2 \times 10^{-12}$ | – |
| 14 | $5.8 \times 10^{-12}$ | $1.2 \times 10^{-12}$ | – |
| 16 | $3.4 \times 10^{-12}$ | $9 \times 10^{-13}$ | – |
| 18 | $5 \times 10^{-13}$ | $1.0 \times 10^{-12}$ | – |
| 20 | $1.8 \times 10^{-12}$ | $1.1 \times 10^{-12}$ | – |
| | $\delta\mu = 37\%$ (SAV) | $\delta\mu = 24\%$ (RSS) | |

**Table 8** Bias due to the mismodelling in the even zonals of the models X = JEM01-RL03B, based on 49 months of GRACE-only data, and Y = AIUB-GRACE01S (Jäggi et al. 2008). The latter one was obtained from GPS satellite-to-satellite tracking data and K-band range-rate data out of the period January 2003 to December 2003 using the Celestial Mechanics Approach. No accelerometer data, no de-aliasing products, and no regularisation was applied. The $\sigma$ for AIUB-GRACE01S are formal. $\Delta\overline{C}_{\ell 0}$ are always larger than the linearly added sigmas

| $\ell$ | $\Delta\overline{C}_{\ell 0}$ (JEM01-RL03B–AIUB-GRACE01S) | $\sigma_X + \sigma_Y$ | $f_\ell$ (mas yr$^{-1}$) |
|---|---|---|---|
| 4 | $2.95 \times 10^{-11}$ | $2.1 \times 10^{-12}$ | 11 |
| 6 | $3.5 \times 10^{-12}$ | $1.3 \times 10^{-12}$ | 0.8 |
| 8 | $2.14 \times 10^{-11}$ | $5 \times 10^{-13}$ | 0.7 |
| 10 | $4.8 \times 10^{-12}$ | $5 \times 10^{-13}$ | – |
| 12 | $4.2 \times 10^{-12}$ | $5 \times 10^{-13}$ | – |
| 14 | $3.6 \times 10^{-12}$ | $5 \times 10^{-13}$ | – |
| 16 | $8 \times 10^{-13}$ | $5 \times 10^{-13}$ | – |
| 18 | $7 \times 10^{-13}$ | $5 \times 10^{-13}$ | – |
| 20 | $1.0 \times 10^{-12}$ | $5 \times 10^{-13}$ | – |
| | $\delta\mu = 26\%$ (SAV) | $\delta\mu = 23\%$ (RSS) | |

the time series of the computed[11] "residuals" of the nodes with the difference between the orbital elements of consecutive arcs, combined with (5), is fitted with a straight line.

In order to enforce the reliability of the ongoing test it would be desirable to proceed following other approaches as well. For instance, the gravitomagnetic force could be explicitly modelled in terms of a dedicated solve-for parameter (not necessarily the usual PPN $\gamma$ one) to be estimated in the least-square sense along with all the other parameters usually determined in fitting the LAGEOS-type satellites data, and the resulting correlations

[11]The expression "residuals of the nodes" is used, strictly speaking, in an improper sense because the Keplerian orbital elements are not directly measured.

**Table 9** Bias due to the mismodelling in the even zonals of the models X = EIGEN-GRACE02S (Reigber et al. 2005) and Y = AIUB-GRACE01S (Jäggi et al. 2008). The $\sigma$ for AIUB-GRACE01S are formal, while those of EIGEN-GRACE02S are calibrated. $\Delta\overline{C}_{\ell 0}$ are larger than the linearly added sigmas for $\ell = 4, 6, 8, 16$

| $\ell$ | $\Delta\overline{C}_{\ell 0}$ (EIGEN-GRACE02S−AIUB-GRACE01S) | $\sigma_X + \sigma_Y$ | $f_\ell$ (mas yr$^{-1}$) |
|---|---|---|---|
| 4 | $2.98 \times 10^{-11}$ | $6.0 \times 10^{-12}$ | 11.1 |
| 6 | $2.29 \times 10^{-11}$ | $3.3 \times 10^{-12}$ | 5.0 |
| 8 | $1.26 \times 10^{-11}$ | $1.9 \times 10^{-12}$ | 0.4 |
| 10 | $6 \times 10^{-13}$ | $2.5 \times 10^{-12}$ | – |
| 12 | $5 \times 10^{-13}$ | $1.6 \times 10^{-12}$ | – |
| 14 | $5 \times 10^{-13}$ | $1.6 \times 10^{-12}$ | – |
| 16 | $2.9 \times 10^{-12}$ | $1.4 \times 10^{-12}$ | – |
| 18 | $6 \times 10^{-13}$ | $1.4 \times 10^{-12}$ | – |
| 20 | $2 \times 10^{-13}$ | $1.5 \times 10^{-12}$ | |
| | $\delta\mu = 34\%$ (SAV) | $\delta\mu = 25\%$ (RSS) | |

**Table 10** Systematic uncertainty $\delta\mu$ in the LAGEOS-LAGEOS II test evaluated by taking the absolute value of the difference between the nominal values of the total combined node precessions due to the even zonals for different models X and Y, i.e. $|\dot{\Omega}^{\rm geopot}(\mathrm{X}) - \dot{\Omega}^{\rm geopot}(\mathrm{Y})|$

| Models compared | $\delta\mu$ |
|---|---|
| AIUB-GRACE01S−JEM01-RL03B | 20% |
| AIUB-GRACE01S−GGM02S | 27% |
| AIUB-GRACE01S−GGM03S | 3% |
| AIUB-GRACE01S−ITG-Grace02 | 2% |
| AIUB-GRACE01S−ITG-Grace03 | 0.1% |
| AIUB-GRACE01S−EIGEN-GRACE02S | 33% |
| JEM01-RL03B−GGM02S | 7% |
| JEM01-RL03B−GGM03S | 17% |
| JEM01-RL03B−ITG-Grace02 | 18% |
| JEM01-RL03B−ITG-Grace03s | 20% |
| JEM01-RL03B−EIGEN-GRACE02S | 13% |
| GGM02S−GGM03S | 24% |
| GGM02S−ITG-Grace02 | 25% |
| GGM02S−ITG-Grace03s | 27% |
| GGM02S−EIGEN-GRACE02S | 6% |
| GGM03S−ITG-Grace02 | 1% |
| GGM03S−ITG-Grace03s | 3% |
| GGM03S−EIGEN-GRACE02S | 30% |
| ITG-Grace02−ITG-Grace03s | 2% |
| ITG-Grace02−EIGEN-GRACE02S | 31% |
| ITG-Grace03s−EIGEN-GRACE02S | 33% |

among them could be inspected. Moreover, one could also look at the changes in the values of the complete set of the estimated parameters with and without the Lense-Thirring effect modelled.

A first, tentative step towards the implementation of a similar strategy with the LAGEOS satellites in terms of the PPN parameter $\gamma$ has been recently taken by Combrinck (2008).

## 4 On the LARES Mission

The combination that should be used for measuring the Lense-Thirring effect with LAGEOS, LAGEOS II and LARES is (Iorio 2005a)

$$\dot{\Omega}^{\text{LAGEOS}} + k_1 \dot{\Omega}^{\text{LAGEOS II}} + k_2 \dot{\Omega}^{\text{LARES}}. \tag{13}$$

The coefficients $k_1$ and $k_2$ entering (13) are defined as

$$\begin{aligned} k_1 &= \frac{\dot{\Omega}_{.2}^{\text{LARES}} \dot{\Omega}_{.4}^{\text{LAGEOS}} - \dot{\Omega}_{.2}^{\text{LAGEOS}} \dot{\Omega}_{.4}^{\text{LARES}}}{\dot{\Omega}_{.2}^{\text{LAGEOS II}} \dot{\Omega}_{.4}^{\text{LARES}} - \dot{\Omega}_{.2}^{\text{LARES}} \dot{\Omega}_{.4}^{\text{LAGEOS II}}} = 0.3586, \\ k_2 &= \frac{\dot{\Omega}_{.2}^{\text{LAGEOS}} \dot{\Omega}_{.4}^{\text{LAGEOS II}} - \dot{\Omega}_{.2}^{\text{LAGEOS II}} \dot{\Omega}_{.4}^{\text{LAGEOS}}}{\dot{\Omega}_{.2}^{\text{LAGEOS II}} \dot{\Omega}_{.4}^{\text{LARES}} - \dot{\Omega}_{.2}^{\text{LARES}} \dot{\Omega}_{.4}^{\text{LAGEOS II}}} = 0.0751. \end{aligned} \tag{14}$$

The combination (13) cancels out, by construction, the impact of the first two even zonals; we have used $a_{\text{LR}} = 7828$ km, $i_{\text{LR}} = 71.5$ deg. The total Lense-Thirring effect, according to (13) and (14), amounts to 50.8 mas yr$^{-1}$.

### 4.1 A Conservative Evaluation of the Impact of the Geopotential on the LARES Mission

The systematic error due to the uncancelled even zonals $J_6, J_8, \dots$ can be conservatively evaluated as

$$\delta\mu \leq \sum_{\ell=6} \left| \dot{\Omega}_{.\ell}^{\text{LAGEOS}} + k_1 \dot{\Omega}_{.\ell}^{\text{LAGEOS II}} + k_2 \dot{\Omega}_{.\ell}^{\text{LARES}} \right| \delta J_\ell \tag{15}$$

Of crucial importance is how to assess $\delta J_\ell$. By proceeding as in Sect. 2 and by using the same models up to degree $\ell = 60$ because of the lower altitude of LARES with respect to LAGEOS and LAGEOS II which brings into play more even zonals, we have the results presented in Table 11. They have been obtained with the standard and widely used Kaula approach (Kaula 1966) in the following way. We, first, calibrated our numerical calculation with the analytical ones performed with the explicit expressions for $\dot{\Omega}_{.\ell}$ worked out up to $\ell = 20$ in Iorio (2003); then, after having obtained identical results, we confidently extended our numerical calculation to higher degrees by means of two different softwares.

It must be stressed that they may be still optimistic: indeed, computations for $\ell > 60$ become unreliable because of numerical instability of the results.

In Table 12 we repeat the calculation by using for $\delta J_\ell$ the covariance matrix sigmas $\sigma_{J_\ell}$; also in this case we use the approach by Kaula (1966) up to degree $\ell = 60$.

If, instead, one assumes $\delta J_\ell = s_\ell,\ \ell = 2, 4, 6, \dots$ i.e., the standard deviations of the sets of all the best estimates of $J_\ell$ for the models considered here the systematic bias, up to $\ell = 60$, amounts to 12% (SAV) and 6% (RSS). Again, also this result may turn out to be optimistic for the same reasons as before.

It must be pointed out that the evaluations presented here rely upon calculations of the coefficients $\dot{\Omega}_{.\ell}$ performed with the well known standard approach by Kaula (1966); it would be important to try to follow also different computational strategies in order to test them.

**Table 11** Systematic percent uncertainty $\delta\mu$ in the combined Lense-Thirring effect with LAGEOS, LAGEOS II and LARES according to (15) and $\delta J_\ell = \Delta J_\ell$ up to degree $\ell = 60$ for the global Earth's gravity solutions considered here; the approach by Kaula (1966) has been followed. For LARES we adopted $a_{\rm LR} = 7828$ km, $i_{\rm LR} = 71.5$ deg, $e_{\rm LR} = 0.0$

| Models compared ($\delta J_\ell = \Delta J_\ell$) | $\delta\mu$ (SAV) | $\delta\mu$ (RSS) |
|---|---|---|
| AIUB-GRACE01S–JEM01-RL03B | 23% | 16% |
| AIUB-GRACE01S–GGM02S | 16% | 8% |
| AIUB-GRACE01S–GGM03S | 22% | 13% |
| AIUB-GRACE01S–ITG-Grace02 | 24% | 15% |
| AIUB-GRACE01S–ITG-Grace03 | 22% | 14% |
| AIUB-GRACE01S–EIGEN-GRACE02S | 14% | 7% |
| JEM01-RL03B–GGM02S | 14% | 9% |
| JEM01-RL03B–GGM03S | 5% | 3% |
| JEM01-RL03B–ITG-Grace02 | 4% | 2% |
| JEM01-RL03B–ITG-Grace03s | 5% | 2% |
| JEM01-RL03B–EIGEN-GRACE02S | 26% | 15% |
| GGM02S–GGM03S | 13% | 7% |
| GGM02S–ITG-Grace02 | 16% | 8% |
| GGM02S–ITG-Grace03s | 14% | 7% |
| GGM02S–EIGEN-GRACE02S | 14% | 7% |
| GGM03S–ITG-Grace02 | 3% | 2% |
| GGM03S–ITG-Grace03s | 2% | 0.5% |
| GGM03S–EIGEN-GRACE02S | 24% | 13% |
| ITG-Grace02–ITG-Grace03s | 3% | 2% |
| ITG-Grace02–EIGEN-GRACE02S | 25% | 14% |
| ITG-Grace03s–EIGEN-GRACE02S | 24% | 13% |

**Table 12** Systematic percent uncertainty $\delta\mu$ in the combined Lense-Thirring effect with LAGEOS, LAGEOS II and LARES according to (15) and $\delta J_\ell = \sigma_{J_\ell}$ up to degree $\ell = 60$ for the global Earth's gravity solutions considered here; the approach by Kaula (1966) has been followed. For LARES we adopted $a_{\rm LR} = 7828$ km, $i_{\rm LR} = 71.5$ deg, $e_{\rm LR} = 0.0$

| Model ($\delta J_\ell = \sigma_\ell$) | $\delta\mu$ (SAV) | $\delta\mu$ (RSS) |
|---|---|---|
| AIUB-GRACE01S (formal) | 11% | 9% |
| JEM01-RL03B | 1% | 0.9% |
| GGM03S (calibrated) | 5% | 4% |
| GGM02S (formal) | 20% | 15% |
| ITG-Grace03s (formal) | 0.3% | 0.2% |
| ITG-Grace02s (formal) | 0.4% | 0.2% |
| EIGEN-GRACE02S (calibrated) | 21% | 17% |

## 4.2 The Impact of Some Non-Gravitational Perturbations

It is worthwhile noting that also the impact of the subtle non-gravitational perturbations will be different with respect to the original proposal because LARES will fly in a lower orbit and its thermal behavior will probably be different with respect to LAGEOS and LAGEOS II.

The reduction of the impact of the thermal accelerations, like the Yarkovsky-Schach effects, should have been reached with two concentric spheres. However, as explained by Andrés (2007), this solution will increase the floating potential of LARES because of the much higher electrical resistivity and, thus, the perturbative effects produced by the charged particle drag. Moreover, the atmospheric drag will increase also because of the lower orbit of the satellite, both in its neutral and charged components. Indeed, although it does not affect directly the node $\Omega$, it induces a secular decrease of the inclination $i$ of a LAGEOS-like satellite (Milani et al. 1987) which translates into a further bias for the node itself according to

$$\delta\dot{\Omega}_{\mathrm{drag}} = \frac{3}{2}n\left(\frac{R}{a}\right)^2 \frac{\sin i\ J_2}{(1-e^2)^2}\delta i, \tag{16}$$

in which $\delta i$ accounts not only for the measurement errors in the inclination, but also for any unmodelled/mismodelled dynamical effect on it. According to Iorio (2008b), the secular decrease for LARES would amount to

$$\left\langle \frac{di}{dt} \right\rangle_{\mathrm{LR}} \approx -0.6\ \mathrm{mas\,yr^{-1}} \tag{17}$$

yielding a systematic uncertainty in the Lense-Thirring signal of (13) of about 3–9% $\mathrm{yr}^{-1}$. An analogous indirect node effect via the inclination could be induced by the thermal Yarkovski-Rubincam force as well (Iorio 2008b). Also the Earth's albedo, with its anisotropic components, may have a non-negligible effect.

Let us point out the following issue as well. At present, it is not yet clear how the data of LAGEOS, LAGEOS II and LARES will be finally used by the proponent team in order to try to detect the Lense-Thirring effect. This could turn out to be a non-trivial matter because of the non-gravitational perturbations. Indeed, if, for instance, a combination[12]

$$\dot{\Omega}^{\mathrm{LARES}} + h_1\dot{\Omega}^{\mathrm{LAGEOS}} + h_2\dot{\Omega}^{\mathrm{LAGEOS\ II}} \tag{18}$$

was adopted instead of that of (13), the coefficients of the nodes of LAGEOS and LAGEOS II, in view of the lower altitude of LARES, would be

$$\begin{aligned} h_1 &= \frac{\dot{\Omega}^{\mathrm{LAGEOS\ II}}_{.2}\dot{\Omega}^{\mathrm{LARES}}_{.4} - \dot{\Omega}^{\mathrm{LARES}}_{.2}\dot{\Omega}^{\mathrm{LAGEOS\ II}}_{.4}}{\dot{\Omega}^{\mathrm{LARES}}_{.2}\dot{\Omega}^{\mathrm{LAGEOS\ II}}_{.4} - \dot{\Omega}^{\mathrm{LAGEOS\ II}}_{.2}\dot{\Omega}^{\mathrm{LAGEOS}}_{.4}} = 13.3215, \\ h_2 &= \frac{\dot{\Omega}^{\mathrm{LARES}}_{.2}\dot{\Omega}^{\mathrm{LAGEOS}}_{.4} - \dot{\Omega}^{\mathrm{LAGEOS}}_{.2}\dot{\Omega}^{\mathrm{LARES}}_{.4}}{\dot{\Omega}^{\mathrm{LAGEOS}}_{.2}\dot{\Omega}^{\mathrm{LAGEOS\ II}}_{.4} - \dot{\Omega}^{\mathrm{LAGEOS\ II}}_{.2}\dot{\Omega}^{\mathrm{LAGEOS}}_{.4}} = 4.7744. \end{aligned} \tag{19}$$

and the combined Lense-Thirring signal would amount to 676.8 $\mathrm{mas\,yr^{-1}}$. As a consequence, the direct and indirect effects of the non-gravitational[13] perturbations on the nodes of LAGEOS and LAGEOS II would be enhanced by such larger coefficients and this may yield a degradation of the total obtainable accuracy.

[12] The impact of the geopotential is, by construction, unaffected with respect to the combination of (13).

[13] The same may hold also for time-dependent gravitational perturbations affecting the nodes of LAGEOS and LAGEOS II, like the tides.

## 5 Conclusions

The so far published evaluations of the total systematic uncertainty induced by the even zonal harmonics of the geopotential in the Lense-Thirring test with the combined nodes of the SLR LAGEOS and LAGEOS II satellites are likely optimistic. Indeed, they are all based on the use of the covariance sigmas, more or less reliably calibrated, of the covariance matrices of various Earth gravity model solutions used one at a time separately in such a way that the model X yields an error of $x\%$, the model Y yields an error $y\%$, etc. Instead, comparing the estimated values of the even zonals for different pairs of models allows for a more conservative evaluation of the real uncertainties in our knowledge of the static part of the geopotential. As a consequence, the uncertainty in the Lense-Thirring signal is about 3–4 times larger than the figures so far claimed (5–10%), amounting to various tens percent (37% for the pair EIGEN-GRACE02S and ITG-GRACE03s, about 25–30% for the other most recent GRACE-based solutions).

Concerning the extraction of the Lense-Thirring signal from the data of the LAGEOS-type satellites, different approaches with respect to the one followed so far should be implemented in order to do something really new. For instance, the gravitomagnetic force should be explicitly included in the dynamical force models of the LAGEOS satellites and an ad-hoc parameter should be estimated in the least-square sense in addition to those determined so far without modelling the Lense-Thirring effect. Moreover, also the variation of the values of all the other estimated parameters with and without the gravitomagnetic force modelled should be inspected along with their mutual correlations.

Applying the strategy of the difference of the estimated even zonals to the ongoing LARES mission shows that reaching a 1% measurement of the Lense-Thirring effect with LAGEOS, LAGEOS II and LARES might be difficult. Indeed, since LARES will orbit at a lower altitude with respect to the LAGEOS satellites, more even zonal harmonics are to be taken into account. Assessing realistically their impact is neither easy nor unambiguous. Straightforward calculations up to degree $\ell = 60$ with the standard Kaula's approach yield errors as large as some tens percent; the same holds if the sigmas of the covariance matrices of several global Earth's gravity models are used. Such an important point certainly deserves great attention. Another issue which may potentially degrade the expected accuracy is the impact of some non-gravitational perturbations which would have a non-negligible effect on LARES because of its lower orbit. In particular, the secular decrease of the inclination of the new spacecraft due to the neutral and charged atmospheric drag induces an indirect bias in the node precessions by the even zonals which, in the case of LARES, should be of the order of $\approx$3–9% yr$^{-1}$.

**Acknowledgements** I thank M.C.E. Huber, R.A. Treumann and the entire staff of ISSI for the organization of the exquisite workshop which I had the pleasure and the honor to attend. I am grateful to D. Barbagallo (ESA-ESRIN) for the information on the LARES orbital configuration. I acknowledge the financial support of INFN-Sezione di Pisa and ISSI.

## References

J.I. Andrés, *Enhanced Modelling of LAGEOS Non-Gravitational Perturbations. PhD Thesis Book* (Ed. Sieca Repro Turbineweg, Delft, 2007)

D.C. Christodoulidis, D.E. Smith, R.G. Williams, S.M. Klosko, Observed tidal braking in the Earth/Moon/Sun system. J. Geophys. Res. **93**(B6), 6216–6236 (1988)

I. Ciufolini, Measurement of the Lense-Thirring drag on high-altitude, laser-ranged artificial satellites. Phys. Rev. Lett. **56**(4), 278–281 (1986)

I. Ciufolini, On a new method to measure the gravitomagnetic field using two orbiting satellites. Nuovo Cim. A **109**(12), 1709–1720 (1996)

I. Ciufolini, LARES/WEBER-SAT, frame-dragging and fundamental physics. http://arxiv.org/abs/gr-qc/0412001. Accessed 3 January 2005

I. Ciufolini, On the orbit of the LARES satellite. http://arxiv.org/abs/gr-qc/0609081. Accessed 20 September 2006

I. Ciufolini, http://www.infn.it/indexen.php. Astroparticle Physics. Calendario riunioni. Roma, 30 gennaio 2008. 14:30 Aggiornamento LARES (20'). lares_dellagnello.pdf (2008a), p. 17

I. Ciufolini, http://www.infn.it/indexen.php. Astroparticle Physics. Calendario riunioni. Villa Mondragone, 30 sett.–4 ott. Friday 03 October 2008. 10:20 LARES (20') (2008b)

I. Ciufolini, E.C. Pavlis, A confirmation of the general relativistic prediction of the Lense–Thirring effect. Nature **431**, 958–960 (2004)

I. Ciufolini, E.C. Pavlis, On the measurement of the Lense-Thirring effect using the nodes of the LAGEOS satellites, in reply to "On the reliability of the so far performed tests for measuring the Lense-Thirring effect with the LAGEOS satellites" by L. Iorio. New Astron. **10**(8), 636–651 (2005)

I. Ciufolini, D.M. Lucchesi, F. Vespe, A. Mandiello, Measurement of dragging of inertial frames and gravitomagnetic field using laser-ranged satellites. Nuovo Cim. A **109**(5), 575–590 (1996)

I. Ciufolini, E.C. Pavlis, F. Chieppa, E. Fernandes-Vieira, J. Pérez-Mercader, Test of general relativity and measurement of the Lense-Thirring effect with two Earth satellites. Science **279**(5359), 2100–2103 (1998a)

I. Ciufolini et al., *LARES Phase A* (University La Sapienza, Rome, 1998b)

I. Ciufolini, E.C. Pavlis, R. Peron, Determination of frame-dragging using Earth gravity models from CHAMP and GRACE. New Astron. **11**(8), 527–550 (2006)

L. Combrinck, Evaluation of PPN parameter Gamma as a test of General Relativity using SLR data, in *16th Int. Laser Ranging Workshop*, Poznań (PL), 13–17 October 2008

L. Cugusi, E. Proverbio, Relativistic effects on the motion of Earth's artificial satellites. Astron. Astrophys. **69**, 321–325 (1978)

J.J. Degnan, Satellite laser ranging: current status and future prospects. IEEE Trans. Geosci. Remote Sens. **GE-23**(4), 398–413 (1985)

C.W.F. Everitt, The gyroscope experiment I. General description and analysis of gyroscope performance, in *Proc. Int. School Phys. "Enrico Fermi" Course LVI*, ed. by B. Bertotti (New Academic Press, New York, 1974), pp. 331–360

C.W.F. Everitt et al., Gravity Probe B: Countdown to launch, in *Gyros, Clocks, Interferometers...: Testing Relativistic Gravity in Space*, ed. by C. Lämmerzahl, C.W.F. Everitt, F.W. Hehl (Springer, Berlin, 2001), pp. 52–82

P. Inversi, F. Vespe, Direct and indirect solar radiation effects acting on LAGEOS satellite: Some refinements. Adv. Space Res. **14**(5), 73–77 (1994)

L. Iorio, Letter to the editor: A critical approach to the concept of a polar, low-altitude LARES satellite. Class. Quantum Gravity **19**(17), L175–L183 (2002)

L. Iorio, The impact of the static part of the Earth's gravity field on some tests of General Relativity with satellite laser ranging. Celest. Mech. Dyn. Astron. **86**(3), 277–294 (2003)

L. Iorio, The impact of the new Earth gravity models on the measurement of the Lense-Thirring effect with a new satellite. New Astron. **10**(8), 616–635 (2005a)

L. Iorio, On the possibility of testing the Dvali Gabadadze Porrati brane-world scenario with orbital motions in the Solar system. J. Cosmol. Astropart. Phys. **7**, 8 (2005b)

L. Iorio, Comments, replies and notes: A note on the evidence of the gravitomagnetic field of Mars. Class. Quantum Gravity **23**(17), 5451–5454 (2006a)

L. Iorio, A critical analysis of a recent test of the Lense-Thirring effect with the LAGEOS satellites. J. Geod. **80**(3), 128–136 (2006b)

L. Iorio, The impact of the new Earth gravity model EIGEN-CG03C on the measurement of the Lense-Thirring effect with some existing Earth satellites. Gen. Relativ. Gravit. **38**(3), 523–527 (2006c)

L. Iorio, Reply to "Comment on 'Evidence of the gravitomagnetic field of Mars' ", by Kris Krogh. J. Gravit. Phys. (2007a, in press). http://arxiv.org/abs/gr-qc/0701146

L. Iorio (ed.), *The Measurement of Gravitomagnetism: A Challenging Enterprise* (NOVA, Hauppauge, 2007b)

L. Iorio, A comment on the paper "On the orbit of the LARES satellite", by I. Ciufolini. Planet. Space Sci. **55**(10), 1198–1200 (2007c)

L. Iorio, LARES/WEBER-SAT and the equivalence principle. Europhys. Lett. **80**(4), 40007 (2007d)

L. Iorio, An assessment of the measurement of the Lense-Thirring effect in the Earth gravity field, in reply to: "On the measurement of the Lense-Thirring effect using the nodes of the LAGEOS satellites, in reply to "On the reliability of the so far performed tests for measuring the Lense-Thirring effect with the LAGEOS satellites" by L. Iorio", by I. Ciufolini and E. Pavlis. Planet. Space Sci. **55**(4), 503–511 (2007e)

L. Iorio, Advances in the measurement of the Lense-Thirring effect with planetary motions in the field of the Sun. Sch. Res. Exch. **2008**, 105235 (2008a)

L. Iorio, On the impact of the atmospheric drag on the LARES mission (2008b). http://arxiv.org/abs/gr-qc/0809.3564. Accessed 8 October 2008

L. Iorio, A. Morea, The impact of the new Earth gravity models on the measurement of the Lense-Thirring effect. Gen. Relativ. Gravit. **36**(6), 1321–1333 (2004)

L. Iorio, D.M. Lucchesi, I. Ciufolini, The LARES mission revisited: an alternative scenario. Class. Quantum Gravity **19**(16), 4311–4325 (2002)

A. Jäggi, G. Beutler, L. Mervart, GRACE gravity field determination using the celestial mechanics approach—first results. Presented at the IAG Symposium on "Gravity, Geoid, and Earth Observation 2008", Chania, GR, 23–27 June 2008

W.M. Kaula, *Theory of Satellite Geodesy* (Blaisdell, Waltham, 1966)

K. Krogh, Comments, replies and notes: Comment on 'Evidence of the gravitomagnetic field of Mars'. Class. Quantum Gravity **24**(22), 5709–5715 (2007)

F.G. Lemoine, S.C. Kenyon, J.K. Factor, R.G. Trimmer, N.K. Pavlis, D.S. Chinn, C.M. Cox, S.M. Klosko, S.B. Luthcke, M.H. Torrence, Y.M. Wang, R.G. Williamson, E.C. Pavlis, R.H. Rapp, T.R. Olson, The Development of the Joint NASA GSFC and the National Imagery Mapping Agency (NIMA) Geopotential Model EGM96. NASA/TP-1998-206861, 1998

J. Lense, H. Thirring, Über den Einfluss der Eigenrotation der Zentralkörper auf die Bewegung der Planeten und Monde nach der Einsteinschen Gravitationstheorie. Phys. Z. **19**, 156–163 (1918)

F.J. Lerch, R.S. Nerem, B.H. Putney, T.L. Felsentreger, B.V. Sanchez, J.A. Marshall, S.M. Klosko, G.B. Patel, R.G. Williamson, D.S. Chinn, A geopotential model from satellite tracking, altimeter, and surface gravity data: GEM-T3. J. Geophys. Res. **99**(B2), 2815–2839 (1994)

D.M. Lucchesi, Reassessment of the error modelling of non-gravitational perturbations on LAGEOS II and their impact in the Lense-Thirring determination. Part I. Planet. Space Sci. **49**(5), 447–463 (2001)

D.M. Lucchesi, Reassessment of the error modelling of non-gravitational perturbations on LAGEOS II and their impact in the Lense-Thirring determination. Part II. Planet. Space Sci. **50**(10–11), 1067–1100 (2002)

D.M. Lucchesi, The asymmetric reflectivity effect on the LAGEOS satellites and the germanium retroreflectors. Geophys. Res. Lett. **30**(18), 1957 (2003)

D.M. Lucchesi, LAGEOS satellites germanium cube-corner-retroreflectors and the asymmetric reflectivity effect. Celest. Mech. Dyn. Astron. **88**(3), 269–291 (2004)

D.M. Lucchesi, The impact of the even zonal harmonics secular variations on the Lense-Thirring effect measurement with the two Lageos satellites. Int. J. Mod. Phys. D **14**(12), 1989–2023 (2005)

D.M. Lucchesi, The LAGEOS satellites orbital residuals determination and the way to extract gravitational and non-gravitational unmodeled perturbing effects. Adv. Space Res. **39**(10), 1559–1575 (2007)

D.M. Lucchesi, G. Balmino, The LAGEOS satellites orbital residuals determination and the Lense Thirring effect measurement. Planet. Space Sci. **54**(6), 581–593 (2006)

D.M. Lucchesi, A. Paolozzi, A cost effective approach for LARES satellite, in *XVI Congresso Nazionale AIDAA*, Palermo, IT, 24–28 September 2001

D.M. Lucchesi, I. Ciufolini, J.I. Andrés, E.C. Pavlis, R. Peron, R. Noomen, D.G. Currie, LAGEOS II perigee rate and eccentricity vector excitations residuals and the Yarkovsky-Schach effect. Planet. Space Sci. **52**(8), 699–710 (2004)

B. Mashhoon, Gravitoelectromagnetism: a brief review, in *The Measurement of Gravitomagnetism: A Challenging Enterprise*, ed. by L. Iorio (NOVA, Hauppauge, 2007), pp. 29–39

T. Mayer-Gürr, A. Eicker, K.-H. Ilk, ITG-GRACE02s: a GRACE gravity field derived from short arcs of the satellite's orbit, in *1st Int. Symp. of the International Gravity Field Service "Gravity Field of the Earth"*, Istanbul, TR, 28 August–1 September 2006

T. Mayer-Gürr, ITG-Grace03s: The latest GRACE gravity field solution computed in Bonn, in *Joint Int. GSTM and DFG SPP Symp.*, Potsdam, 15–17 October 2007. http://www.geod.uni-bonn.de/itg-grace03.html

A. Milani, A.M. Nobili, P. Farinella, *Non-Gravitational Perturbations and Satellite Geodesy* (Adam Hilger, Bristol, 1987)

E.C. Pavlis, Geodetic contributions to gravitational experiments in space, in *Recent Developments in General Relativity: Proc. 14th SIGRAV Conf. on General Relativity and Gravitational Physics*, ed. by R. Cianci, R. Collina, M. Francaviglia, P. Fré. Genova, IT, 18–22 September 2000 (Springer, Milan, 2002), pp. 217–233

H. Pfister, On the history of the so-called Lense-Thirring effect. Gen. Relativ. Gravit. **39**(11), 1735–1748 (2007)

G.E. Pugh, WSEG Research Memorandum No. 11, 1959

Ch. Reigber, R. Schmidt, F. Flechtner, R. König, U. Meyer, K.-H. Neumayer, P. Schwintzer, S.Y. Zhu, An Earth gravity field model complete to degree and order 150 from GRACE: EIGEN-GRACE02S. J. Geodyn. **39**(1), 1–10 (2005)

J.C. Ries, R.J. Eanes, M.M. Watkins, B.D. Tapley, Joint NASA/ASI Study on Measuring the Lense-Thirring Precession Using a Second LAGEOS Satellite CSR-89-3 Center for Space Research, Austin, 1989

J.C. Ries, R.J. Eanes, B.D. Tapley, Lense-Thirring precession determination from laser ranging to artificial satellites, in *Nonlinear Gravitodynamics. The Lense–Thirring Effect*, ed. by R.J. Ruffini, C. Sigismondi (World Scientific, Singapore, 2003a), pp. 201–211

J.C. Ries, R.J. Eanes, B.D. Tapley, G.E. Peterson, Prospects for an improved Lense-Thirring test with SLR and the GRACE gravity mission, in *Proc. 13th Int. Laser Ranging Workshop, NASA CP (2003-212248)*, ed. by R. Noomen, S. Klosko, C. Noll, M. Pearlman (NASA Goddard, Greenbelt, 2003b). http://cddisa.gsfc.nasa.gov/lw13/lw_proceedings.html#science

J.C. Ries, R.J. Eanes, M.M. Watkins, Confirming the frame-dragging effect with satellite laser ranging, in *16th Int. Laser Ranging Workshop*, Poznań (PL), 13–17 October 2008

M.L. Ruggiero, A. Tartaglia, Gravitomagnetic effects. Nuovo Cim. B **117**(7), 743–768 (2002)

G. Schäfer, Gravitomagnetic effects. Gen. Relativ. Gravit. **36**(10), 2223–2235 (2004)

L. Schiff, Possible new experimental test of general relativity theory. Phys. Rev. Lett. **4**(5), 215–217 (1960)

M. Soffel, *Relativity in Astrometry, Celestial Mechanics and Geodesy* (Springer, Berlin, 1989)

B.D. Tapley, J.C. Ries, S. Bettadpur, D. Chambers, M. Cheng, F. Condi, B. Gunter, Z. Kang, P. Nagel, R. Pastor, T. Pekker, S. Poole, F. Wang, GGM02-An improved Earth gravity field model from GRACE. J. Geod. **79**(8), 467–478 (2005)

B.D. Tapley, J.C. Ries, S. Bettadpur, D. Chambers, M. Cheng, F. Condi, S. Poole, American Geophysical Union, Fall Meeting 2007, Abstract #G42A-03, 2007

R.A. Van Patten, C.W.F. Everitt, Possible experiment with two counter-orbiting drag-free satellites to obtain a new test of Einstein's general theory of relativity and improved measurements in geodesy. Phys. Rev. Lett. **36**(12), 629–632 (1976a)

R.A. Van Patten, C.W.F. Everitt, A possible experiment with two counter-rotating drag-free satellites to obtain a new test of Einstein's general theory of relativity and improved measurements in geodesy. Celest. Mech. Dyn. Astron. **13**(4), 429–447 (1976b)

F. Vespe, The perturbations of Earth penumbra on LAGEOS II perigee and the measurement of Lense-Thirring gravitomagnetic effect. Adv. Space Res. **23**(4), 699–703 (1999)

Space Sci Rev (2009) 148: 383–395
DOI 10.1007/s11214-009-9516-7

# Misalignment and Resonance Torques and Their Treatment in the GP-B Data Analysis

**G.M. Keiser · J. Kolodziejczak · A.S. Silbergleit**

Received: 7 April 2009 / Accepted: 16 April 2009 / Published online: 12 May 2009
© Springer Science+Business Media B.V. 2009

**Abstract** Classical torques acting on the GP-B gyroscopes decrease the accuracy in the measurement of the relativistic drift rate. Based on measurements made during the year-long science data collection, tests done following the science data collection, and a theoretical analysis of potential torques, there are two dominant classical torques acting on the gyroscopes. The first torque, known as the misalignment torque, has a magnitude proportional to the misalignment between the gyroscope spin axis and the satellite roll axis and is aligned perpendicular to the plane containing these two vectors. The second torque, known as the resonance torque, mainly produces a permanent offset in the orientation of the gyroscope spin axis when a harmonic of the gyroscope polhode frequency is in the vicinity of the satellite roll frequency. These two torques have the same physical origin: an electrostatic interaction between the patch effect fields on the surfaces of the rotor and the housing. In the post-mission data analysis, the change in the gyroscope orientation due to both of these torques can be clearly separated from the relativistic drift rate.

**Keywords** Gravity probe B · Experimental tests of gravitational theories · Patch effect · Gyroscope torques

**PACS** 04.80.Cc · 07.05.Kf · 45.40.Cc

## 1 Introduction

As discussed in the accompanying papers (Everitt et al. 2009; Heifetz et al. 2009; Muhlfelder et al. 2009; Silbergleit et al. 2009), the Gravity Probe B (GP-B) satellite was designed to accurately measure the geodetic and frame-dragging precessions predicted by the general

G.M. Keiser (✉) · A.S. Silbergleit
Gravity Probe B, HEPL, Stanford University, Stanford, CA 94305-4085, USA
e-mail: mac@relgyro.stanford.edu

J. Kolodziejczak
NASA/Marshall Space Flight Center, Huntsville, AL 35812, USA

theory of relativity (de Sitter 1916; Lense and Thirring 1918). Averaged over the orbital motion of the satellite (Adler and Silbergleit 2000), both of these predicted effects produce a linear drift in the orientation of the gyroscope spin axis with time. A classical torque acting on a gyroscope can produce an error in an accurate measurement of the relativistic drift rate.

This paper describes the two classical torques acting on the GP-B gyroscopes that had a measurable effect on the orientation of the gyroscope spin axis. The misalignment torque is described in Sect. 2 of this paper, while the resonance torque is described in Sect. 3 of this paper. In each of these sections, we present the observational evidence for these torques, discuss the physical origin of each torque, and demonstrate post-mission data analysis methods of separating the gyroscope drift rate caused by each torque from the relativistic drift rate. Both torques are caused by an interaction of the patch effect potential on the surface of the gyroscope rotor with a patch effect potential on the surface of the gyroscope housing. Additional observations that can be explained by the patch effect will be described in a forthcoming paper (Buchman et al. 2009). Because the misalignment torque acts in a specific direction and because the resonance torque produces a unique time signature in the orientation of the gyroscope spin axis, the effects of these torques on the gyroscope spin axis orientation can clearly be separated from the relativistic drift rate. The final section of this paper discusses the impact of the torques on the measurement of the relativistic drift rate.

## 2 Misalignment Torques

### 2.1 Observations

The Gravity Probe B mission timeline was divided into three phases. The initialization phase, which began after launch on April 20, 2004, through the spin-up and final alignment of the gyroscopes on August 29, 2004, included the initial calibration of the instrument, tests of each of the four gyroscopes at slow spin speeds, adjustment of the spacecraft's attitude and translation control system, spin-up of each of the four gyroscopes, and a final alignment of the spin axes. The main science data collection phase began on August 29, 2004, and continued until August 15, 2005. The final phase of the mission was called the calibration phase, where potential disturbances were deliberately increased to determine their impact on the measurement of the spin axis orientation with respect to the guide star and to measure other possible classical torques acting on the gyroscope.

During the calibration phase, 17 spacecraft operations were performed to study the effect of increasing the angle between the satellite roll axis and the four gyroscope spin axes. Throughout the science data collection phase, while the guide star, IM Pegasi (HR 8703), was not occulted by the earth and was being used to measure the satellite's attitude, the satellite's attitude control system maintained the axis of the science telescope and the satellite roll axis, which coincided with the science telescope's optical axis, within 200 mas (milliarcsec) of the apparent position of the guide star. Because of the 20 arcsec annual optical aberration and the relativistic linear drift in the gyroscopes' spin axis orientation, the angle between the roll axis and the spin axis slowly varied over the course of the year but was maintained within 30 arc seconds during normal operations.

Each of the 17 spacecraft operations included (1) measurement of the gyroscope orientation for 24 hours, (2) maneuver of the satellite roll axis to a nearby star or virtual star, (3) maintenance of the gyroscope spin axis in this orientation for a period of 12 or 24 hours,

(4) another spacecraft maneuver back to the guide star, IM Pegasi, and, finally, (5) another period of 24 hours to measure the gyroscope spin axis orientation after the spacecraft maneuvers. The gyroscope drift rate at the increased misalignment could then be found from the change in each gyroscope spin axis orientation divided by the time at the increased misalignment. The majority of these spacecraft maneuvers were to stars or virtual stars within one degree of the IM Pegasi (HR Pegasi is 0.4 deg to the East of IM Pegasi, and HD 216635 is 1 deg to the North of IM Pegasi), but maneuvers were also made to one star as far as 7 deg from IM Pegasi.

In addition to increasing the misalignment between the satellite roll axis and the gyroscope spin axis, during some of the these operations the spacecraft was deliberately accelerated as much as $10^{-7}$ m/s$^2$ using the helium thrusters to measure the combined effect of increasing the misalignment and the spacecraft acceleration. Furthermore, at enhanced misalignment angles, tests were done to measure the effect of increasing the 20 Hz control voltage of the electrostatic suspension system from its nominal 200 mV level. The operating mode of the electrostatic suspension system was also changed so that the mean control voltages on the pair of electrodes on one axis were held at a constant dc value rather than being modulated at 20 Hz.

The average gyroscope drift rate for gyroscope 3 and the orientation of the spacecraft relative to IM Pegasi during the maneuver are shown in Fig. 1 for those operations where the satellite roll axis lay within 1 degree of the guide star, and where the control voltages applied by the electrostatic suspension system were modulated at 20 Hz. In the polar plot, each vector represents the magnitude and direction of the average drift rate during the maneuver, and the base of the vector is at the location of the misalignment between the roll axis and spin axes during the maneuver. In this figure, note that the gyroscope drift rate lies in a direction perpendicular to the misalignment and is approximately proportional to the magnitude of the misalignment. The right hand panel of Fig. 1 shows the measured magnitude of the drift rate compared to the magnitude of the misalignment for gyroscope 3. The drift rate is proportional to the misalignment for misalignments less than 0.4 degrees but begins to show evidence of nonlinearity at a misalignment 1 degree. At angles larger than 1 degree, the relation between the misalignment and the drift rate was clearly nonlinear al-

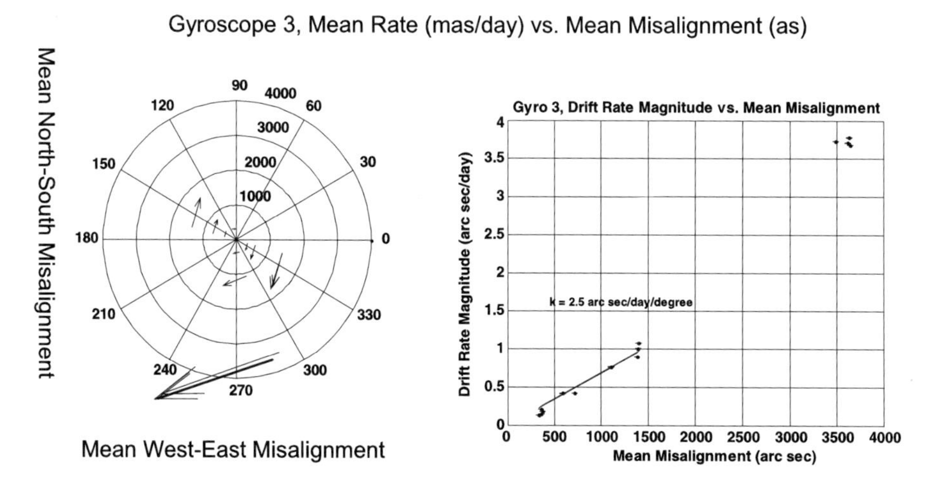


**Fig. 1** Mean drift rate vs. misalignment for gyroscope 3 during calibration maneuvers

though the direction of the drift rate of the gyroscope spin axis continues to be in a direction perpendicular to the misalignment.

Changing the operating mode of the electrostatic suspension system provided clear evidence that the source of these torques is electrostatic. While the voltages applied to the electrodes were modulated at 20 Hz, the proportionality factor between the misalignment angle and the gyroscope drift rate remained the same regardless of the magnitude of the modulated 20 Hz control voltage applied to the electrodes and regardless of the acceleration of the spacecraft. However, when steady voltages were instead applied to the electrodes, there was a significant change in the magnitude of the drift rate for a given misalignment angle, although the direction of the gyroscope spin axis drift rate continued to lie in a direction perpendicular to the misalignment. In addition, the magnitude and direction of the drift rate change depended on the magnitudes of the steady voltages applied to the six electrodes.

## 2.2 Explanation and Calculation of Torque

The electrostatic patch effect (Darling 1989) produces a nonuniform potential on the surface of a metal. Charges in a metal with a finite conductivity move until the electrostatic potential is uniform. However, if there is a nonuniform dipole layer on the surface of the metal due to the crystalline structure or impurities on the surface of the metal, then the electrostatic potential on the surface is no longer uniform. In this case, the electric field and the force are no longer necessarily perpendicular to the surface of the metal. For an isolated metal surface, the net force and torque on the body are zero, but with two metallic surfaces in close proximity, the net force or torque on each surfaces will, in general, be nonzero. For two plane parallel surfaces, the net force due to patch effect fields has been calculated by Speake (1996).

To investigate the effect of a nonuniform potential on the surface of the rotor and the housing, the electrostatic potential on the surface of the rotor and the housing was expanded in terms of spherical harmonics:

$$V_R(\theta,\phi) = \sum_{l=0}^{\infty}\sum_{m=-l}^{l} R_{lm} Y_{lm}(\theta,\phi),$$
$$V_H(\theta,\phi) = \sum_{l=0}^{\infty}\sum_{m=-l}^{l} H_{lm} Y_{lm}(\theta,\phi), \tag{1}$$

where $R_{lm}$ and $H_{lm}$ are coefficients in the expansion of the potentials in terms of the orthonormal spherical harmonics, $Y_{lm}(\theta,\phi)$. With these boundary conditions, Laplace's equation may be solved for the potential, $\Phi$, in the region between the two surfaces. With this solution and the approximation that the gap, $\Delta$, between the rotor and the housing is much smaller than the radius, $a$, of the rotor, the gradient in the potential at the surface of the rotor and the housing is given by

$$\left.\frac{\partial\Phi}{\partial r}\right|_{r=a} = \left.\frac{\partial\Phi}{\partial r}\right|_{r=b} \approx \frac{1}{\Delta}\sum_{l=0}^{\infty}\sum_{m=-l}^{l} (H_{lm} - R_{lm})\, Y_{lm}(\theta,\phi). \tag{2}$$

The energy stored in the electric field may be calculated from the potential and the gradient in the potential at the two surfaces:

$$W = \frac{\varepsilon_0}{2} \int_{gap} (\nabla \Phi)^2 \, dV = \frac{\varepsilon_0}{2} \left( \int_{r=b} \Phi \frac{\partial \Phi}{\partial r} dA - \int_{r=a} \Phi \frac{\partial \Phi}{\partial r} dA \right)$$
$$\approx \frac{\varepsilon_0 a^2}{2} \int d\Omega \, (\Phi_b - \Phi_a) \left. \frac{\partial \Phi}{\partial r} \right|_{r=a} = \frac{\varepsilon_0 a^2}{2\Delta} \sum_{l=0}^{\infty} (H_{lm} - R_{lm})(H_{lm} - R_{lm})^* . \quad (3)$$

In this expression, $\varepsilon_0$ is the permittivity of free space. Here and in the equations which follow, the asterisk denotes the complex conjugate of the coefficient. The orthonormal properties of the spherical harmonics have been used here: it is important to note that the result (3) is valid only if the spherical harmonic expansions for both surfaces are written in the same reference frame.

The expansion of both surface potentials are originally done in the frame fixed to the corresponding surface. Their coefficients do not change as long as the patch potential remains constant. To calculate the energy stored in the electrostatic field and the gyroscope torques, rotation matrices may be used to find the expansion of the surface potential in a common reference frame. In the principal axis reference frame, with the $z'$-axis chosen as the principal axis corresponding to the maximum moment of inertia, the rotor potential may be expanded in terms of spherical harmonics in that reference frame $(r, \theta', \phi')$:

$$V_R(\theta', \phi') = \sum_{l=0}^{\infty} \sum_{m=-l}^{l} R'_{lm} Y_{lm}(\theta', \phi'). \quad (4)$$

Since the rotor potential is real, $R'_{lm} = (-1)^m {R'_{l,-m}}^*$. We chose a common reference frame to be the guide star reference frame, where the $z$-axis lies in the apparent direction of the guide star and the $x$-axis lies in the plane of the $z$-axis and the Earth's rotation axis. The spherical harmonics in the principal axis reference frame may be rotated to the guide star reference frame through two successive sets of Euler angle rotations. The first is a transformation from the principal axis reference frame to a reference frame with the $z$-axis aligned with the instantaneous spin axis. This 3-2-3 transformation uses the three Euler angles $\phi_P$, $\gamma$, and $\phi_S$ which correspond to a rotation about the principal axis with the largest moment of inertia, a rotation about the new $y$-axis by an angle $\gamma$ which is the angle between the principal axis and the spin axis, and finally a rotation about the spin axis by the spin phase, $\phi_S$. The second set of Euler angle transformations is a 2-3 transformation: the first angle rotates about the $y$-axis of the coordinate system by an angle $-\beta$ so the $z$-axis is aligned with the apparent direction to the guide star, while the second angle rotates about the direction to the guide star by an angle $-\alpha$. These two angles, $\alpha$ and $\beta$, determine the orientation of the gyroscope spin axis in the guide star reference frame. With these two sets of rotations, the rotor potential in the guide star reference frame becomes

$$V_R(\theta, \phi) = \sum_{l=0}^{\infty} \sum_{m=-l}^{l} R'_{lm} \sum_{q=-l}^{l} D^l_{qp}(\alpha, \beta, 0) \sum_{p=-l}^{l} D^l_{pm}(-\phi_S, -\gamma, -\phi_P) Y_{lq}(\theta, \phi)$$
$$= \sum_{l=0}^{\infty} \sum_{m=-l}^{l} R'_{lm} \sum_{q=-l}^{l} e^{-iq\alpha} d^l_{qp}(\beta) \sum_{p=-l}^{l} e^{ip\phi_S} d^l_{pm}(-\gamma) e^{im\phi_P} Y_{lq}(\theta, \phi). \quad (5)$$

The elements of the rotation matrices, $D$ and $d$, are defined in Rose (1995). If the housing potential is initially expanded in a reference frame where the $z$-axis coincides with both the satellite roll axis and the apparent direction to the guide star, then transforming to the inertially fixed reference frame involves only a rotation about the $z$-axis by the satellite roll phase, $\phi_R$. In the guide star reference frame, the housing potential may be written as

$$V_H(\theta, \phi) = \sum_{k=0}^{\infty} \sum_{j=-k}^{k} H'_{kj} e^{ij\phi_R} Y_{kj}(\theta, \phi). \tag{6}$$

With these two expressions for the rotor and housing potentials in a common reference frame, the variable part of the energy in the electrostatic field averaged over the spin of the gyroscope is

$$\bar{W} = -\frac{\varepsilon_0 a^2}{\Delta} \sum_{l=0}^{\infty} \sum_{m=-l}^{l} \sum_{q=-l}^{l} R'_{lm} H'^{*}_{lq} d^l_{q0}(\beta) d^l_{0m}(-\gamma) e^{im\phi_P} e^{-iq(\alpha+\phi_R)}. \tag{7}$$

In addition to averaging over the spin of the gyroscope, the energy in the electrostatic field may be averaged over the roll of the housing and the polhode period. In this case, the averaged energy is

$$\begin{aligned} \bar{\bar{W}} &= -\frac{\varepsilon_0 a^2}{\Delta} \sum_{l=0}^{\infty} R'_{l0} H'_{l0} d^l_{00}(\beta) d^l_{00}(-\gamma) \\ &= -\frac{\varepsilon_0 a^2}{\Delta} \sum_{l=0}^{\infty} R'_{l0} H'_{l0} P_l(\beta) P_l(-\gamma). \end{aligned} \tag{8}$$

Here $P_l$ is the Legendre polynomial of order $l$.

The torque acting on the gyroscope may be found from the derivative of the average energy with respect to the angles, $\alpha$ and $\beta$, which define the orientation of the gyroscope spin axis in the guide star reference frame. Since the average energy given above only depends on the angle $\beta$, the torque is given by

$$\vec{\tau} = -\frac{\partial \bar{\bar{W}}}{\partial \beta} \hat{e}_\beta = \frac{\varepsilon_0 a^2}{\Delta} \sum_{l=1}^{\infty} R_{l0} H_{l0} P_l(-\gamma) \frac{\partial P_l(\beta)}{\partial \beta} \hat{e}_\beta. \tag{9}$$

The torque is the negative derivative of the energy with respect to the angle. The unit vector, $e_\beta$, lies in a direction perpendicular to the plane of the misalignment angle, $\beta$, so this torque will always be perpendicular to the misalignment. From this result, the torque is clearly a nonlinear function of the misalignment angle, but for small angles this expression becomes

$$\vec{\tau} = \frac{\varepsilon_0 a^2}{\Delta} \beta \hat{e}_\beta \sum_{l=1}^{\infty} l(l+1) R'_{l0} H'_{l0} P_l(-\gamma), \quad \beta \ll 1. \tag{10}$$

This result agrees with the observations. The torque always lies in a direction perpendicular to the misalignment, and, for small angles, the torque is proportional to the misalignment. In addition, averaged over the polhode period, the magnitude of the torque for a given misalignment will slowly change if the angle, $\gamma$, between the rotor's principal axis and the spin axis slowly changes even though the patch effect potentials on the surface of the rotor and the housing remain fixed. Since the absolute value of the Legendre polynomial is always less than or equal to unity, for a given value of $l$, the polhode motion reduces the average torque compared to the torque in the absence of polhode motion ($\gamma = 0$). The magnitude of the torque depends on the magnitude and distribution of the patch effect potentials on the surface of the rotor and the housing, but it is important to note that this torque is due to the interaction of the patch effect on the rotor with the patch effect on the housing. Neither surface alone will produce a torque.

In addition, this result explains the change in the magnitude of the torque when the operating mode of the electrostatic suspension system was changed. Since the torque is linearly proportional to the potential on the surface of the housing, a modulated voltage applied to the electrodes will, on the average, produce no net torque regardless of its magnitude. However, when a steady voltage is applied to the electrodes, the electrode potentials are the sum of the electrode patch effect potential and the voltage applied. This change in the electrode potential produces a corresponding change in the magnitude of the misalignment torque. The magnitude of the torque depends on the magnitude of the applied steady voltage as well as the orientation of each electrode with respect to the gyroscope spin axis.

### 2.3 Data Analysis in the Presence of Misalignment Torques

In an inertial reference frame (Silbergleit et al. 2001), the equations for the components of the gyroscope drift rate in the North–South and West–East directions in the presence of a misalignment torque may be written as

$$\begin{aligned}\frac{ds_{NS}}{dt} &= r_{NS} + k\mu_{WE},\\ \frac{ds_{WE}}{dt} &= r_{WE} - k\mu_{NS}.\end{aligned} \tag{11}$$

Where $s_{NS}$ and $s_{WE}$ are the components of the gyroscope spin axis in the North–South and West–East directions, $r_{NS}$ and $r_{WE}$ are the uniform components of the gyroscope drift rate, $\mu_{NS}$ and $\mu_{WE}$ are the components of the misalignment, $\beta$, between the spin and roll axes in the North–South and West–East directions. From (10), this coefficient, $k$, is given by

$$k = \frac{1}{I\omega_s}\frac{\varepsilon_0 a^2}{\Delta}\sum_{l=1}^{\infty} l(l+1)R'_{l0}H'_{l0}P_l(-\gamma). \tag{12}$$

This coefficient may be time dependent because the polhode angle, $\gamma$, is changing due to polhode damping even though the patch effect potentials are fixed on the surface of the rotor and the housing.

There are several methods of analyzing the data to determine simultaneously the uniform components of the drift rate and the misalignment torque coefficient based on the time history of the gyroscope spin axis orientation. One method is to estimate the uniform components of the drift rate as well as the misalignment torque coefficient, using an appropriate model of the time variation of the latter. This method along with preliminary results is described in the accompanying paper by Heifetz et al. (2009). A second, complementary method takes advantage of the linear combination of the two drift rate equations (11) which eliminates the torque coefficient. Defining the misalignment phase as

$$\theta = \tan^{-1}\frac{\mu_{WE}}{\mu_{NS}}, \tag{13}$$

the linear combination of the drift rate equations

$$\frac{ds_R}{dt} = \cos\theta \frac{ds_{NS}}{dt} + \sin\theta \frac{ds_{WE}}{dt} = r_{NS}\cos\theta + r_{WE}\sin\theta \tag{14}$$

is independent of the misalignment torque coefficient. For any given misalignment phase, it is possible to determine this radial component of the drift rate without any knowledge of the torque coefficient. Furthermore, with sufficient variation in the misalignment phase, it is possible to determine independently the two drift rate components, $r_{NS}$ and $r_{WE}$.

Equation (14) describes the rate of change of the gyroscope spin axis orientation in the presence of a continuously changing, but known, misalignment. In the analysis of the data, the rate of change of the spin axis orientation must be determined from the measured orientation. This rate may be estimated by dividing the data into batches of 4 or 5 days and finding the rate of change of the orientation over this interval. Then, the radial rate is a sinusoidal function of the average misalignment phase. The amplitude and phase of this sinusoidal curve determine the uniform components of the gyroscope drift rate, $r_{NS}$ and $r_{WE}$. Instead of dividing the data into 4 or 5 day batches, a better method is to find the gyroscope spin axis drift rate on an orbit-by-orbit basis. In this case, the gyroscope drift rates in the North–South and West–East directions may be determined from the first differences in the spin axis orientation,

$$r_i = \left.\frac{ds}{dt}\right|_i = \frac{s_{i+1} - s_i}{t_{i+1} - t_i}. \tag{15}$$

The noise in these gyroscope drift rates is sequentially correlated, and it is essential to explicitly include this sequential correlation in the measurement noise matrix. [See, for example, Reasenberg (1972).] The first differences in the orientation estimates may then be used to construct a series of derived measurements of the gyroscope drift rate. These drift rate measurements are a function only of the uniform components and the misalignment phase. If the misalignment phase is known, then a least squares fit may be used to determine the uniform components of the drift rate as long as the sequential correlation and the dependence on the misalignment phase are explicitly included.

## 3 Resonance Torques

### 3.1 Observations

In the course of the data analysis, we found that significant changes in the orientation of a gyroscope spin axis occurred during those intervals when a harmonic of the gyroscope polhode frequency was nearly equal to the satellite roll frequency. Although measurable changes did not always occur at those times when this resonance condition held, each measurable change occurred at one of these resonance conditions.

Figure 2 shows the magnitude of the components of the gyroscope 2 readout signal at the cosine and sine of the roll frequency for each orbit during one of these resonances. The cosine component is proportional to the angle between the gyroscope spin axis and the apparent direction to the guide star in the plane of the orbit, while the sine component is proportional to the same angle in the perpendicular direction. Figure 3 is a plot of these components versus one another.

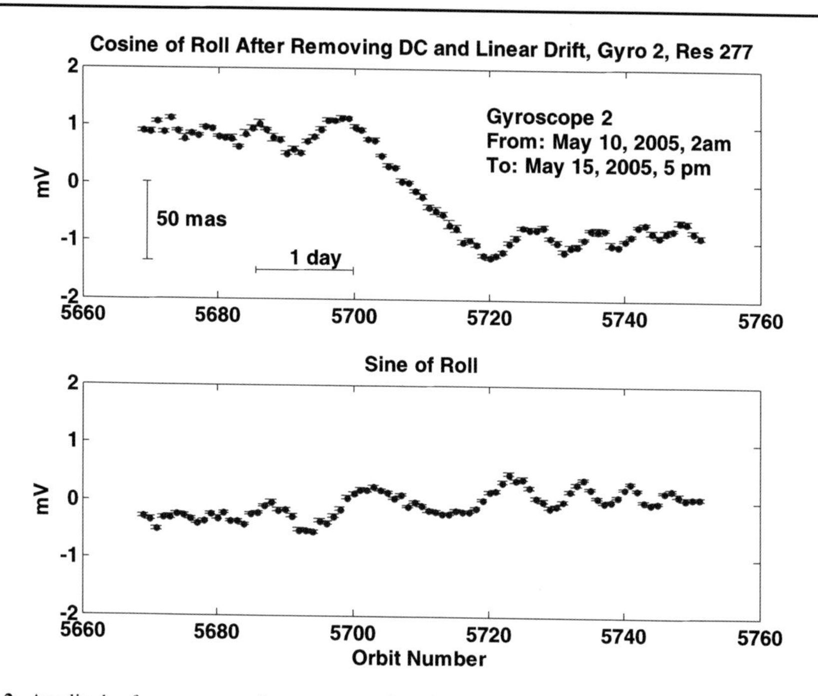


Fig. 2 Amplitude of components of gyroscope readout signal at cosine and sine of satellite roll frequency

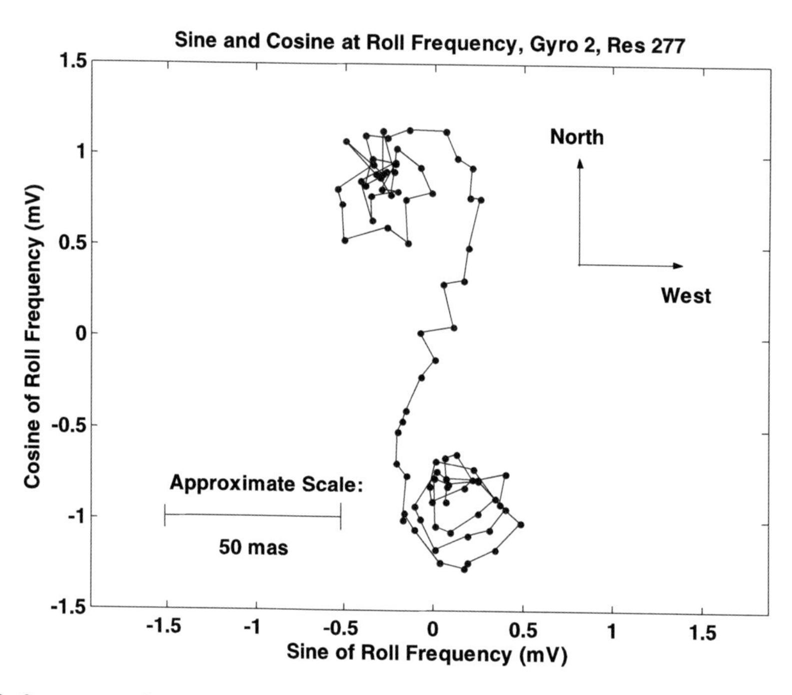


Fig. 3 Components of gyroscope readout signal at satellite roll frequency. Each *point* represents the estimated average orientation for one orbit of the satellite

### 3.2 Explanation and Calculation of Torque

From (7), the spin averaged torque may be calculated during those intervals where a harmonic of the gyroscope polhode frequency is equal to the satellite roll frequency. In this case, there are two components of the torque because the spin averaged energy in the electrostatic field depends on both the angles, $\alpha$ and $\beta$, which define the orientation of the gyroscope spin axis with respect to the guide star reference frame:

$$\tau_\alpha = -\frac{\partial \bar{W}}{\partial \alpha} = -\frac{\varepsilon_0 a^2}{\Delta}\sum_{l=0}^{\infty}\sum_{m=-l}^{l}\sum_{q=-l}^{l} \mathrm{Re}(R'_{lm} H'^{*}_{lq} iq e^{-iq\alpha} d^l_{q0}(\beta) d^l_{0m}(-\gamma) e^{im\phi_P} e^{-iq\phi_R}),$$
$$\tau_\beta = -\frac{\partial \bar{W}}{\partial \beta} = \frac{\varepsilon_0 a^2}{\Delta}\sum_{l=0}^{\infty}\sum_{m=-l}^{l}\sum_{q=-l}^{l} \mathrm{Re}\left( R'_{lm} H'^{*}_{lq} e^{-iq\alpha} \frac{\partial d^l_{q0}(\beta)}{\partial \beta} d^l_{0m}(-\gamma) e^{im\phi_P} e^{-iq\phi_R}\right). \tag{16}$$

A third component of the spin-averaged torque, which lies along the instantaneous spin axis, is zero. These components of the torque are not orthogonal. The torque in the guide reference frame in the two perpendicular directions which are also perpendicular to the direction to the guide star may be found from the relations

$$\tau_x = -\tau_\alpha \cos\alpha \cot\beta - \tau_\beta \sin\alpha,$$
$$\tau_y = -\tau_\alpha \sin\alpha \cot\beta + \tau_\beta \cos\alpha, \tag{17}$$
$$\tau_z = \tau_\alpha.$$

Combining these last two equations and transforming to the inertial reference frame (Silbergleit et al. 2001), the components of the spin axis drift rate in the North–South and West–East directions are

$$\frac{ds_{NS}}{dt} = a_m \cos\Delta\phi - b_m \sin\Delta\phi,$$
$$\frac{ds_{WE}}{dt} = a_m \sin\Delta\phi + b_m \cos\Delta\phi, \tag{18}$$

where $\Delta\phi$ is the phase difference between the $m^{\mathrm{th}}$ harmonic of the polhode phase and the satellite roll phase. For small misalignment angles, $\beta \ll 1$, the coefficients $a_m$ and $b_m$ are given by

$$a_m = \frac{\varepsilon_0 a^2}{\Delta}\sum_{l=0}^{\infty}\sqrt{l(l+1)}\, d^l_{0m}(\gamma)\,\mathrm{Re}\{R_{lm} H_{l,-1}\},$$
$$b_m = \frac{\varepsilon_0 a^2}{\Delta}\sum_{l=0}^{\infty}\sqrt{l(l+1)}\, d^l_{0m}(\gamma)\,\mathrm{Im}\{R_{lm} H_{l,-1}\}. \tag{19}$$

Assuming the polhode frequency is changing approximately linearly with time at a rate $r$ close to the resonance, the phase difference changes quadratically with time, the time history of the orientation of the gyroscope spin axis may be found by integrating the drift

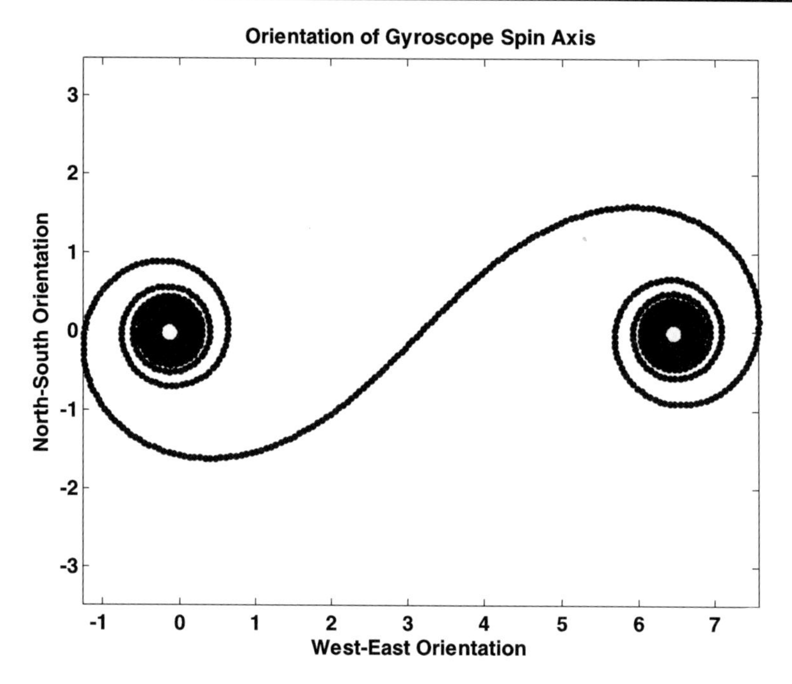


**Fig. 4** Expected change in orientation due to resonance torques. The units in this figure are arbitrary but are equal in the North–South and West–East directions

rate equations

$$\begin{aligned} s_{NS}(t) &= s_{NS0} + a_m \int_{-\infty}^{t} \cos\frac{rt^2}{2} + b_m \int_{-\infty}^{t} \sin\frac{rt^2}{2}, \\ s_{WE}(t) &= s_{WE0} - a_m \int_{-\infty}^{t} \sin\frac{rt^2}{2} + b_m \int_{-\infty}^{t} \cos\frac{rt^2}{2}. \end{aligned} \tag{20}$$

The integrals in these equation are Fresnel's integrals, and the path traced out by the components of the gyroscope spin axis is the Cornu spiral as shown in Fig. 4. This curve agrees well with the measurements shown in Fig. 3.

## 3.3 Data Analysis in the Presence of Resonance Torques

As long as the polhode phase is known well enough, the times at which the resonances occur and the interval over which the phase difference is less than $2\pi$ radians may be accurately determined. Outside of this interval, although these resonance torques have a constant amplitude but time dependent frequency, the deviation of the gyroscope's spin axis orientation is less a small fraction of the magnitude of the total offset produced by the resonance torque. The most straightforward approach to the analysis of the data in the presence of these resonance torques is simply to exclude the data during these intervals and to assume that the torques produce an unknown change in the orientation during this time. With such a data analysis approach, it is, of course, necessary to determine the sensitivity of the overall result to the width of the interval over which the data is excluded.

An alternative approach would be to use the data during each of the resonances to determine the two coefficients, $a_m$ and $b_m$, for each resonance. Although this approach increases the number of coefficients which need to be determined from the data analysis, the advantage is that additional, unsegmented data may be used. Covariance analyses have shown that there is a net increase in the precision of the results when data during the resonance is included and the two coefficients are determined along with the other parameters. The caveat here is that it is necessary to clearly demonstrate that the model used for the orientation or rate during each of these resonances is accurate.

## 4 Summary and Conclusion

The misalignment and resonance torques produced significant disturbances to the orientation of the gyroscope spin axes. These torques may be explained by the same physical effect—an interaction of patch effect fields on the surface of the rotor with patch effect fields on the surface of the housing. The misalignment torque is proportional to the misalignment between the gyroscope spin axis and the satellite roll axis. All four gyroscopes exhibited a drift rate due to the misalignment torque, but the magnitude of the torque varied considerably from one gyroscope to another and over the duration of the mission. The resonance torques, on the other hand, are independent of the misalignment to lowest order, but their frequency depends on the difference between a harmonic of the polhode frequency and the satellite roll frequency. When this frequency difference is small, they can produce a permanent offset of the gyroscope spin axis. Because of the significantly different polhode frequencies and rates of change of the polhode frequencies, the frequency of the offsets due to the resonance torques was very different for the different gyroscopes. As the polhode motion was damped out, the frequency of these resonance conditions decreased.

In the analysis of the telemetry data from the satellite, the effects of these torques may be clearly separated from the uniform drift rate. In the case of the misalignment torques, the drift rate due to these torques may be separated from the uniform drift rate because of the known direction of the misalignment drift rate which is perpendicular to the misalignment. Two alternative data analysis methods can be used: one only uses the component of the drift rate in a direction parallel to the misalignment, the other uses the drift rate information in both directions but includes the misalignment torque coefficient as one of the parameters in the data analysis. The effects of the resonance torques may also be separated from the uniform drift rate either by not using the data during those times when the effects of the resonance torques are most pronounced or by explicitly including two additional parameters for each resonance in the data analysis.

**Acknowledgements** The authors would like to acknowledge the remarkable contributions of numerous people at Stanford University, Lockheed-Martin, and NASA to the construction and operation of the GP-B Satellite. We would also like to acknowledge the contributions of S. Buchman, B. Clarke, J. Conklin, F. Everitt, M. Heifetz, D. Hipkins, J. Li, B. Muhlfelder, Y. Ohshima, J. Turneaure, and P. Worden for their insights and contributions to this paper.

## References

R.J. Adler, A.S. Silbergleit, General treatment of orbiting gyroscope precession. Int. J. Theor. Phys. **39**(5), 1291–1316 (2000)

S. Buchman, J.P. Turneaure et al. (2009, to be published)

T.W. Darling, Electric fields on metal surfaces at low temperatures, in *School of Physics* (University of Melbourne, Parkville, 1989), p. 88

W. de Sitter, On Einstein's theory of gravitation and its astronomical consequences. Mon. Not. R. Astron. Soc. **77**, 155–184 (1916)

C.W.F. Everitt et al., Gravity Probe B data analysis – status and potential for improved acuracy of scientific results. Space Sci. Rev. (2009, this issue)

M. Heifetz et al., The Gravity Probe B data analysis filtering approach. Space Sci. Rev. (2009)

J. Lense, H. Thirring, On the influence of the proper rotations of central bodies on the motions of planets and moons according to Einstein's theory of gravitation. Zeitschr. Phys. **19**, 156 (1918). English Translation: B. Mashhoon, F.W. Hehl et al., On the gravitational effects of rotating masses: the lense-thirring papers. Gen. Relativ. Gravitat. 16, 711–750 (1984)

B. Muhlfelder et al., GP-B systematic error. Space Sci. Rev. (2009, this issue)

R. Reasenberg, Filtering with perfectly correlated measurement noise. AIAA J. **10**(7), 942–943 (1972)

M.E. Rose, *Elementary Theory of Angular Momentum* (Dover, New York, 1995)

A.S. Silbergleit, M.I. Heifetz, A.S. Krechetov, Model of starlight deflection and parallax for the GP-B data reduction. Gravity Probe B Document S0393, Stanford University, Stanford, CA, 2001

A.S. Silbergleit et al., Polhode motion, trapped flux, and the GP-B science data analysis. Space Sci. Rev. (2009, this issue)

C.C. Speake, Forces and force gradients due to patch fields and contact-potential differences. Class. Quantum Gravity **13**, A291–A297 (1996)

Space Sci Rev (2009) 148: 397–409
DOI 10.1007/s11214-009-9548-z

# Polhode Motion, Trapped Flux, and the GP-B Science Data Analysis

**A. Silbergleit · J. Conklin · D. DeBra · M. Dolphin · G. Keiser · J. Kozaczuk · D. Santiago · M. Salomon · P. Worden**

Received: 2 January 2009 / Accepted: 9 June 2009 / Published online: 17 June 2009
© Springer Science+Business Media B.V. 2009

**Abstract** Magnetic field trapped in the Gravity Probe B (GP-B) gyroscope rotors contributes to the scale factor of the science readout signal. This contribution is modulated by the rotor's polhode motion. In orbit, polhode period was observed to change due to a small energy dissipation, which significantly complicates data analysis. We present precise values of spin phase, spin down rate, polhode phase and angle, and scale factor variations obtained from the data by Trapped Flux Mapping. This method finds the (unique) trapped field distribution and rotor motion by fitting a theoretical model to the harmonics of high (gyroscope spin) frequency signal. The results are crucial for accurately determining the gyroscope relativistic drift rate from the science signal.

**Keywords** Gravity Probe B · Polhode motion · Dissipation · Trapped Flux Mapping

**PACS** 04 · 04.20.-q · 45.40.Cc · 74.25.-q · 02.70.-c

## 1 Introduction

The GP-B relativity science mission to measure the geodetic and frame–dragging effects on the four spinning superconducting gyroscopes (see Adler and Silbergleit 2000 and the references therein) is described in Everitt et al. (1980, 2009); Turneaure et al. (2003). It uses the low frequency (LF) London Moment magnetic readout provided by a high-precision low-noise SQUID. In addition, a high frequency (HF) signal from residual magnetic field trapped in the type II superconductor (niobium) is present, as a superposition of multiple harmonics of the spin frequency. This paper deals with the analysis of the HF signal, and presents the analysis results, which are crucial for obtaining the desired accuracy of the measurements.

A. Silbergleit (✉) · J. Conklin · D. DeBra · M. Dolphin · G. Keiser · J. Kozaczuk · D. Santiago · M. Salomon · P. Worden
Gravity Probe B, HEPL, Stanford University, Stanford, CA 94305-4085, USA
e-mail: gleit@stanford.edu

In Sect. 2 the relation between the gyro polhode motion, trapped field, and science readout is explained, and the sources of HF signal are described. The on-orbit discovery of the time variation of the gyro polhode period, and the explanation of this effect as kinetic energy dissipation, are given in Sect. 3, along with some results obtained from the analysis of the measured polhode period time history using a general dissipation model. Section 4 explains the concept of Trapped Flux Mapping (TFM; determination of trapped magnetic field and characteristics of gyro motion from the HF signal analysis), its relevance and importance for the final result of GP-B experiment, and the key points of the procedure. The actual approach to TFM in the form of a three level computer analysis scheme is detailed in Sect. 5, while the results of TFM are given in Sect. 6. Section 7 contains some conclusions.

## 2 Gyroscope Polhode Motion, Trapped Flux, and GP-B Readout

Polhode motion is important in the context of GP-B data analysis: the trapped magnetic flux couples to the SQUID pick-up loop, which contributes to the overall readout scale factor. Below we describe the model of this phenomenon.

### 2.1 Free Gyroscope Motion: Polhoding

Let us choose the principal inertia axes of a GP-B gyroscope, $\hat{I}_1$, $\hat{I}_2$, $\hat{I}_3$, as a Cartesian body-fixed frame, and number the moments of inertia in their non-decreasing order, $0 < I_1 \le I_2 \le I_3$. The GP-B flight rotors are the best spheres ever made, as certified by the Guinness World Records,[1] with $I_i - I_j/I_k \sim 10^{-6}$. Still, the value of the inertial *asymmetry parameter*, defined as

$$0 \le Q \equiv \frac{I_2 - I_1}{I_3 - I_1} \le 1, \tag{1}$$

can be anywhere between zero and one. The case $Q = 0$ corresponds to a rotor symmetric about the maximum inertia axis, $\hat{I}_3$.

The instanteneous position of the spin axis, $\boldsymbol{\omega}_s$, in the rotor body can be described by the angle, $\gamma_p$, between it and the axis $\hat{I}_3$, and the second angle, $\phi_p$, between the projection of $\boldsymbol{\omega}_s$ on the plane $\{\hat{I}_1, \hat{I}_2\}$ and the axis $\hat{I}_1$ (see Fig. 1). The motion of the spin axis in the

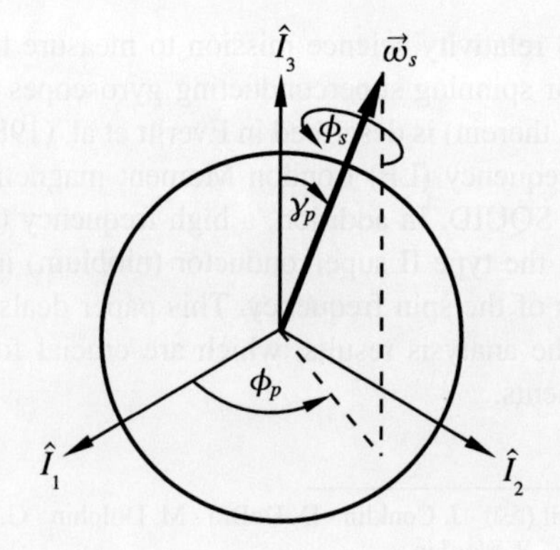


**Fig. 1** Body-fixed frame and the instantaneous position of the spin axis

[1] See http://einstein.stanford.edu/MISSION/mission6.html.

**Table 1** Rotor spin frequency, $f_s$, and spin-down time, $\tau_{sd}$

| Parameter | Gyro 1 | Gyro 2 | Gyro 3 | Gyro 4 |
|---|---|---|---|---|
| $f_s$ (Hz) | 79.40 | 61.81 | 82.11 | 64.84 |
| $\tau_{sd}$ (years) | 15,800 | 13,400 | 7,000 | 25,700 |

body, $\phi_p(t)$, $\gamma_p(t)$, obeys the Euler equations. If the torques are negligible and the motion is free, it is described by the exact Euler solution (e.g. Landau and Lifshitz 1959; MacMillan 1960) corresponding to the precession of the spin axis $\boldsymbol{\omega}_s$ about the inertia axis ($\hat{I}_3$, in the case of GP-B gyros). This precession is called the polhode motion, or polhoding; we also call $\gamma_p$ and $\phi_p$ the polhode angle and phase, respectively. The period of the polhode motion, $T_p = const$, as well as the path of the vector $\boldsymbol{\omega}_s$ in the body, is determined by the values of the conserved angular momentum, $L = const$, and energy, $E = const$. The polhode angular rate,

$$\Omega_p \equiv 2\pi/T_p \propto \left[\sqrt{(I_3 - I_1)(I_3 - I_2)}/I_3\right]\omega_s \approx (1-4) \times 10^{-6}\omega_s \quad \text{(or smaller)},$$

for GP-B gyros. It is worthwhile noting that in the case of a symmetric rotor ($Q = 0$), the polhode path is a circular cone and the motion is uniform, i.e., with $\gamma_p = const$ and $\dot{\phi}_p = \Omega_p = const$, so the polhode phase is a linear function of time. In contrast, for an asymmetric rotor ($Q > 0$) the polhode cone is *not* circular, and the motion of $\boldsymbol{\omega}_s$ in the body is no longer uniform. In particular, $\gamma_p(t)$ oscillates with the period $T_p(t)/2$, $\phi_p(t)$ is not a linear function of time (modulated with the same period), and the angular velocity $\dot{\phi}_p(t) \neq const$ does not coincide identically with $\Omega_p(t)$.

Before turning to the GP-B readout, we give the values of the spin frequencies and spin-down times of GP-B gyros determined by TFM as described in Sect. 5.2. They are found in Table 1; the spin-down times, of 7 to 25 thousand years, are remarkably large.

One more question related to gyro motion is worthwhile of mentioning here, namely, the effect of the elastic deformation of the rotor (centrifugal bulge) on the polhode motion. The role of this effect in the relativity experiment was first considered by Barker and O'Connell (1975). Later T. Duhamel, following the general approach of Egarmin (1980), gave an accurate and complete analysis of the effect as applied to GP-B in his thesis (Duhamel 1984). This analysis clearly shows that the motion of a near homogeneous and near spherical elastic rotor is the same as the motion of a rigid body with the principal moments of inertia slightly changed by the centrifugal bulge. The additions to the moments, $\Delta I_b$, are proportional, of course, to the square of the spin frequency, and their relative values for all the GP-B rotors are $\Delta I_b/I < 10^{-6}$. Since the change in the spin frequency over the whole mission due to the spin-down is very small ($\Delta f_s/f_s \sim 10^{-4}$ or smaller, see Table 1), the time variation of the moments of inertia (one part in $10^{10}$ or less) is completely negligible. Moreover, to lowest order in the inertial asymmetry ($\sim 10^{-6}$) of undeformed GP-B rotors, the bulge additions on all the axes are *identical*. Thus the parameter $Q$, defined in (1) as the ratio of *differences* of moments of inertia, is entirely independent of time, not to mention within the needed accuracy given in Table 3.

## 2.2 GP-B Readout: London Moment and Trapped Flux

The main source of the GP-B magnetic readout is the dipole field of the London Moment (LM) aligned with the gyroscope spin axis (London 1961, Sect. 12). However, this is not the only source: when a type II superconductor is cooled below the critical temperature, residual magnetic field (even under the GP-B conditions!) is trapped inside the rotor. Magnetic quanta $\pm\Phi_0$ form (macroscopically) point sources on the rotor surface (fluxons). The multipole field of these point sources contributes to the total flux through the SQUID pick-up loop, which rolls with the spacecraft in inertial space with the roll period $T_r \approx 77.5$ s.

The SQUID signal is proportional to the total magnetic flux through the pick-up loop, which is the sum of the LM and trapped flux (TF) contributions,

$$\Phi(t) = \Phi^{LM}(t) + \Phi^{TF}(t). \tag{2}$$

It is straightforward to see that the LM flux is proportional to the small angle $\beta$ ($\sim 10^{-4}$ or smaller) between the LM and pick-up loop plane, $\Phi^{LM}(t) = C_g^{LM}\beta(t)$. The LM scale factor $C_g^{LM}$, proportional to the spin speed, is constant, up to a very small spin-down rate of GP-B gyros (spin-down time $\sim 10^4$ years, see Everitt et al. 2009). The angle $\beta(t)$ carries the *relativity signal* at the low roll frequency $f_r \approx 0.01$ Hz.

The time signature of the TF signal emerges from the following argument. Fluxons are frozen in the rotor surface and spin with it; the function that converts a fluxon's position into its magnetic flux through the pick-up loop is strongly nonlinear (almost a step–function, due to a relatively small gap between the rotor and the loop). Therefore the TF signal consists of multiple harmonics of spin, and, since the pick-up loop is rolling, those are actually the harmonics of spin $\pm$ roll frequency. Finally, since the spin axis moves in the rotor body (polhoding), the spin harmonics are modulated by the polhode frequency, so that

$$\begin{aligned}\Phi^{TF}(t) &= \sum_n H_n(t)e^{in(\phi_s \pm \phi_r)} \\ &= \sum_{|n|=odd} H_n(t)e^{in(\phi_s \pm \phi_r)} + \beta(t)\sum_{|n|=even} h_n(t)e^{in(\phi_s \pm \phi_r)},\end{aligned} \tag{3}$$

where $H_n(t) = H_n(\phi_p(t), \gamma_p(t)), h_n(t) = h_n(\phi_p(t), \gamma_p(t))$. The accurate derivation of the formula (3) is given in Keiser and Silbergleit (1991); Nemenmann and Silbergleit (1999), and for the case of non-rolling spacecraft outlined in Keiser and Cabrera (1983), where an alternative approach was used (see Sect. 5.1).

The fact that even harmonics of spin are multiplied by the (same as above) small angle $\beta$ is of purely geometric nature: the spin axis is set to reside almost in the pick-up loop plane in the GP-B experiment.

According to (2), the LM flux and d.c. part of the TF ($n = 0$ in the expression (3)) combine to provide the GP-B *low frequency* (LF) *science readout* (sampled after the additional lowpass filter with a 4 Hz cutoff) given by:

$$\Phi_{LF}(t) = \Phi^{LM}(t) + \Phi^{TF}_{DC}(t) = C_g^{LM}\beta(t) + C_g^{TF}(t)\beta(t), \quad C_g^{TF}(t) \equiv h_0(t). \tag{4}$$

Thus, polhode motion combined with the TF creates polhode variations of the readout scale factor. These variations are not larger than 5% of the (constant) LM scale factor for all the GP-B gyros.

### 2.3 GP-B High Frequency Data

There were two telemetry sources of the high frequency (HF) SQUID signal: (1) FFT of first six harmonics of spin, and (2) *snapshots*, i.e., $\sim 2$ s stretches of original SQUID signal sampled at 2200 Hz. Both were received only during the part of an orbit when the Guide Star was occulted by the earth. At time when the snapshots were available, a snapshot would come about every 40 s, but the gaps between the snapshot arrays were up to 2 days.

HF FFT harmonics were analyzed during the mission as soon as the telemetry was in place. Snapshots, being the most accurate source of HF information, were thoroughly treated after the mission. All 976,478 snapshots available from the mission science period were processed after the flight (see Sect. 4).

## 3 Changing Polhode Period: On-Orbit Discovery and Its Explanation

One of the two on-orbit discoveries which affected GP-B data analysis most strongly was the changing polhode period. The full time history of the polhode period for each of the GP-B gyroscopes is shown in Fig. 2, with very good agreement of the results obtained using the two HF signals. Moreover, an entirely different gyro position signal (polhode modulation of its the spin component) gives the same result (see Dolphin 2007). Notably, the $T_p(t)$ plot for gyros 1, 2 has a peak where formally $T_p = \infty$. Nevertheless, throughout the science period (starting Sept. 13, 2004), polhode periods of all the four gyros decrease monotonically and tend to specific asymptotic values $T_{pa}$.

Such behavior can only be explained by *energy dissipation*, which reduces the kinetic energy of the rotor without changing its angular momentum. This phenomenon was observed, in particular, in rolling satellites starting with the 'Explorer' launched in 1958 (Modi 1973). Dissipation moves the spin axis in the body to the *maximum* inertia axis where the energy is easily seen to have a *minimum*, under the conserved angular momentum constraint. Thus the polhode path, controlled by the parameter $L^2/2E$, also changes (note that the spin-down torque changes both quantities $L$ and $E$, but *not* this ratio, hence *not the polhode path*). The two types of behavior seen in the plots of Fig. 1 are also explained: by pure chance, gyros 1, 2 start their evolution with the spin vector precessing about $\hat{I}_1$, so $T_p$ tends to infinity when $\boldsymbol{\omega}_s$ crosses the separatrix (Landau and Lifshitz 1959; MacMillan 1960). In contrast with this, the evolution of gyros 3, 4 starts when $\boldsymbol{\omega}_s$ is already precessing about $\hat{I}_3$, so $T_p(t)$ just decreases monotonically.

Writing $\omega_s = \omega_{1,3}$ for the case when $\boldsymbol{\omega}_s$ is parallel to $\hat{I}_{1,3}$, respectively, and using the angular momentum conservation $[I_1\omega_1 = I_3\omega_3, \omega_1 = (I_3/I_1)\omega_3]$ one can readily determine the relative energy loss (and spin speed reduction) from the minimum, $\hat{I}_1$, to the maximum,

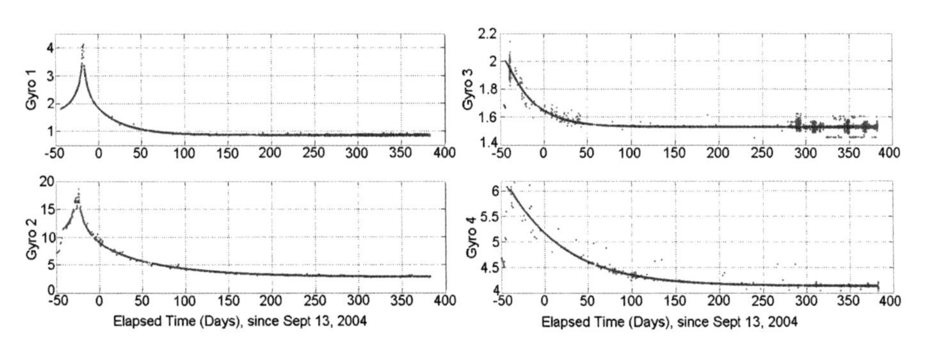


**Fig. 2** Polhode period history of GP-B gyros: *red*—from HF FFT, *blue*—from snapshots

**Table 2** Asymptotic polhode period, $T_{pa}$, and dissipation time, $\tau_{dis}$

| Parameter | Gyro 1 | Gyro 2 | Gyro 3 | Gyro 4 |
|---|---|---|---|---|
| $T_{pa}$ (hours) | 0.867 | 2.851 | 1.529 | 4.137 |
| $\tau_{dis}$ (days) | 31.9 | 74.6 | 30.7 | 61.2 |

$\hat{I}_3$, inertia axis:

$$\frac{E_1 - E_3}{E_1} = \frac{I_1\omega_1^2 - I_3\omega_3^2}{I_1\omega_1^2} = \frac{\omega_1 - \omega_3}{\omega_1} = \frac{I_3 - I_1}{I_3} < 4 \times 10^{-6} \tag{5}$$

for GP-B gyros. The total energy loss needed to move the spin axis all the way from min to max inertia axis is thus less than 4 μJ ($E \sim 1$ J); in one year, the average dissipated power required for this is just $10^{-13}$ W!

The physical origin of this energy dissipation is not completely clear. It might be related to inelastic deformations of the rotor, but probably is dominated by dissipative patch effect torques (see Buchman and Gill 2007). Independent of the origin, we have found a *general dissipation model* in the form of an additional term in the Euler equations unique up to a scalar factor. The detailed description of these new equations along with their solution will be published separately (a brief derivation of the model can be found in Salomon 2008). Fitting the polhode period time history obtained from the model to the measurements allowed us to determine the rotor asymmetry parameter (also found, in some cases, from the gyro position signal), the asymptotic period $T_{pa} \sim 1{-}2$ hours, and the characteristic time of dissipation $\tau_{dis} \sim 1{-}2$ months, depending on the gyro (see Table 2). Thus the dissipation is *slow* ($T_p \ll \tau_{dis}$), so that the polhode motion of the GP-B gyros is *quasi-adiabatic*.

## 4 Trapped Flux Mapping: Concept and Importance

### 4.1 Trapped Flux Mapping

*Trapped Flux Mapping* (TFM) is the procedure of finding the distribution of trapped magnetic field and characteristics of gyro motion from *odd* harmonics of HF SQUID signal, by fitting them to their theoretical model. The latter is based on the following solution for the scalar magnetostatic potential outside the spherical rotor with $K$ pairs of fluxons and anti-fluxons on its surface (in the body-fixed frame described in Sect. 2.1):

$$\Psi(r,\theta,\phi) = \frac{\Phi_0}{2r_g}\sum_{l=1}^{\infty}\left(\frac{r_g}{r}\right)^{l+1}\frac{1}{l+1}\sum_{m=-l}^{l} A_{lm}Y_{lm}(\theta,\phi),$$

$$A_{lm} = \sum_{k=1}^{K}\left[Y_{lm}^*(\theta_k^+,\phi_k^+) - Y_{lm}^*(\theta_k^-,\phi_k^-)\right]. \tag{6}$$

Here $\Phi_0$ is the magnetic flux quantum, and $r_g$ is the rotor radius (more on this model in Sect. 5.1). *If* the fluxon number, $K$, and positions, $\theta_k^{\pm}, \phi_k^{\pm}$, were known, *then* the coefficients $A_{lm}$ would be found uniquely (and easily!) by the above formula. However, neither are known, in reality, so the coefficients $A_{lm}$ *are to be estimated* by the TFM procedure, along with the main functions describing the gyro motion, such as the spin phase, the polhode phase, etc. As shown below, TFM provides the asymmetry parameter $Q$ and all coefficients

$A_{lm}$ up to $l_{max} = 21, 25, 21, 21$ (gyros 1–4 respectively) for each rotor. Likewise, for the entire duration of the mission, TFM gives the spin speeds to 10 nHz; the spin down rates to 1 pHz/s; the spin phases to 0.05 rad; the polhode phases to 0.02 rad, and the polhode angles to 0.01–0.1 rad. With these results we compute the complete scale factor variations $C_g^{TF}(t)$ for the whole mission using formula (9).

The TFM products play a crucial role in the GP-B science analysis, which simply cannot be accomplished without them with the needed accuracy. First of all, accurate polhode phase and angle values at any time of the mission are required for modeling both the scale factor variations and patch effect torques (Heifetz et al. 2009; Keiser et al. 2009), because all the torque coefficients are modulated by the polhode harmonics in the same fashion as the LF SQUID scale factor. In addition, TFM brings in an independent determination of those scale factor variations, which, first, allows for a separate determination of the LM scale factor and slowly varying D.C. part of the TF scale factor. Second, $C_g^{TF}(t)$ found by TFM can be directly used in the LF science analysis thus dramatically reducing the number of estimated parameters, and even turning the (originally non-linear) estimation problem into a linear one.

## 4.2 Key Points of TFM

To carry out TFM, one first of all needs to process the measured HF SQUID signal, $z^{HF}(t) \propto \Phi^{HF}(t)$ (those almost one million snapshots mentioned in Sect. 2.3), to extract the multiple harmonics of spin, $H_n(t)$, according to the formula (3):

$$\text{measured} \quad z^{HF}(t) = \sum_n H_n(t) e^{in(\phi_s \pm \phi_r)}. \tag{7}$$

Then, starting with expression (6) for the magnetic potential in the rotor-fixed frame, one transforms it to the frame of the pick-up loop using the rotation matrices for spherical harmonics (Rose 1995), and computes the magnetic field. The flux through the loop can then be found by integrating the normal field component (for details of this derivation see Kozaczuk 2007; Conklin 2009). As in (3), the result is a sum of the spin harmonics, but with an explicit expression for each of them. In particular, the odd harmonics relevant to TFM are given by

$$H_n(t) = \frac{\Phi_0}{2} \sum_{\substack{l=|n| \\ l \text{ odd}}}^{\infty} \left(\frac{r_g}{b}\right)^l \sum_{m=-l}^{l} A_{lm} d^l_{0n}(\pi/2) d^l_{mn}(\gamma_p) e^{im\phi_p} I_l, \quad n \text{ odd}. \tag{8}$$

Here $b$ is the pick-up loop radius, $r_g/b \sim 1$, $\phi_p(t)$ and $\gamma_p(t)$ are the polhode phase and angle defined in Sect. 2.1 (and roughly known from the polhode period modelling, Sect. 3), $d^l_{mn}(\alpha)$ is the rotation matrix, essentially, a hypergeometric polynomial of the argument $\tan^2(\alpha/2)$ (see Rose 1995, formula (4.14)), and $I_l$ is the known coefficient.

Formula (8) is exactly the expression to fit to the measured harmonics $H_n(t)$. The Euler angles $\phi_p(t)$, $\gamma_p(t)$, and $\phi_s(t)$, which enter it in a *nonlinear* way, are originally derived from measurements, but their accuracy is insufficient for properly determining $A_{lm}$. Expressions for these functions based on the dissipation model and Euler solution are used in a *nonlinear* fitting process described in Sect. 5, which was developed to find them. Then a *linear* fit is carried out to estimate the coefficients $A_{lm}$ and to complete TFM.

Concluding this section we note that the same derivation that yields formula (8) for odd harmonics provides similar expressions for even ones as well. The latter are proportional to the spin-to-pick-up loop misalignment, $\beta$, to lowest order (see (3)). According to (4), the

D.C. harmonics ($n = 0$) gives the scale factor of the trapped flux contribution to the science signal, namely:

$$C_g^{TF}(t) = h_0(t) = \frac{\Phi_0}{2} \sum_{\substack{l=1 \\ l \text{ odd}}}^{\infty} \left(\frac{r_g}{b}\right)^l \sum_{m=-l}^{l} A_{lm} d_{m0}^l(\gamma_p) e^{im\phi_p} I_l l P_{l-1}(0). \tag{9}$$

Therefore, as soon as accurate values of $\phi_p(t)$, $\gamma_p(t)$ and $A_{lm}$ are determined from TFM, the scale factor variations can be immediately computed.

## 5 Trapped Flux Mapping: Three Analysis Levels

The TFM estimation process is broken up into three parts called *levels*, in which various groups of parameters are estimated somewhat independently. Level A and B data processing leads to the estimates of the polhode phase and angle, $\phi_p(t)$ and $\gamma(t)$, the spin phase, $\phi_s(t)$, and the asymmetry parameter, $Q$. Once we know the rotor orientation (these three Euler angles) and its asymmetry, in Level C we perform a linear fit to find the coefficients $A_{lm}$.

### 5.1 Level A: Asymmetry Parameter, Polhode Phase and Angle

The polhode phase, $\phi_p(t)$, can be written as the sum of the *symmetric polhode phase*, $\phi_p(t, Q = 0)$, and the polhode modulation, $\Delta\phi_p(t, Q)$, which has a period $T_p/2$, and is due to the inertial asymmetry described by the parameter $Q$ from (1). The symmetric polhode phase is the integral of the polhode rate (we choose $t_0$ to be any moment of time when $\omega_s$ lies in the $\{\hat{I}_1, \hat{I}_3\}$ plane),

$$\phi_p(t, Q = 0) = \int_{t_0}^{t} \Omega_p(t')dt' = \int_{t_0}^{t} \frac{2\pi}{T_p(t')} dt', \tag{10}$$

and it is just a straight line if no dissipation is present, $T_p = const$, $\Omega_p = const$. An expression for the deviation from the straight line (a sum of exponents with the time scale $\tau_{dis}$) is implied by the dissipation model, so the total model for the polhode phase becomes:

$$\phi_p(t) = \phi_{p0} + \Omega_{pa}(t - t_{\text{ref}}) - \sum_{m=1}^{M} D_m e^{-n\frac{t-t_{\text{ref}}}{\tau_{dis}}} + \Delta\phi_p(t, Q). \tag{11}$$

Here $\phi_{p0}$ is the value of the symmetric polhode phase at time $t_{\text{ref}}$ with no dissipation present, $\Omega_{pa} = 2\pi/T_{pa}$ is the asymptotic polhode rate, and $D_m$ are constant coefficients. Note that the total number of the parameters in the formula (11), $\{\phi_{p0}, \Omega_{pa}, Q, D_1, D_2, \ldots, D_M\}$, that need to be estimated is $3 + M$, with $M < 12$ sufficient for a good enough accuracy for all the gyros. The polhode angle, $\gamma_p$, is computed from the Euler solution given the polhode rate, the spin rate $\omega_s(t)$ (discussed in the next section), and $Q$.

Estimating $Q$ is the most computationally intensive part of the TFM data processing. This is because $Q$ estimation is sensitive to all fit parameters, even though $Q$ is entirely independent physically. As a result, only three separate batches of data spanning 1–2 days each are used to estimate $Q$ for every gyroscope. The batches were chosen to be relatively long and early in the mission, when the polhode variations are largest and the influence of $Q$ is strongest. The Level A code is run many times on a single batch of data while incrementing $Q$ from 0 to 1 by 0.02 at the each next run. The value of $Q$ for which the RMS of

**Table 3** Asymmetry parameter, $Q$

| Gyro 1 | Gyro 2 | Gyro 3 | Gyro 4 |
|---|---|---|---|
| $0.303 \pm 0.069$ | $0.143 \pm 0.029$ | $0.127 \pm 0.072$ | $0.190 \pm 0.048$ |

the post-fit residuals is minimum is taken as the best estimate of $Q$, with its associated error. Table 3 shows the resulting values of $Q$ from Level A processing. The accuracy of these results is $\sim 20\%$, enough for estimation of the scale factor variations due to trapped flux to $\sim 1\%$. This is due to the relative insensitivity of the signal to the asymmetry parameter.

Once $Q$ is estimated, individual fits are performed on one batch of data at a time depending on how the data are grouped. Spectral analysis of the data provides initial estimates of the polhode rate $\Omega_p(t)$ accurate to $\sim$ 100 nHz (Salomon 2008). Integrating this preliminary value of $\Omega_p(t)$ is sufficient for the Level A fitting process, but the initial polhode phase, $\phi_{p0}$, is unknown. Therefore the initial polhode phase is estimated for each batch of data using a standard Nelder-Mead simplex method (see Lagarias et al. 1998) implemented by the function *fminsearch.m* in the MATLAB software. This gives the estimates of $\phi_{p0}$ at roughly 100 different times throughout the science mission.

Each estimate of the initial polhode phase has some error associated with it. To improve the overall accuracy of the polhode phase determination, a single model, (11), is fit to all the preliminary polhode phase estimates. Not only does this smooth the estimates, reducing the overall error, but it also produces a polhode phase time-history that is continuous and self-consistent over the entire mission. The unknown parameters $\phi_{p0}$, $\Omega_{pa}$ and $D_m$ all appear linearly in this model, allowing for a simple least squares fit. An estimate of the polhode phase during each batch of data is constructed from the estimates of $\phi_{p0}$ for the same batch, and the integration of the estimated polhode rate. To construct a self-consistent polhode phase, the estimated polhode rate accurate to $\sim$ 600 nrad/sec is adequate for resolving $\pi$ ambiguities between batches of data that are typically not more that 4 days apart. The RMS of the polhode phase post-fit residuals for each gyroscope is on the order of 0.1 rad.

### 5.2 Level B: Spin Phase

The model used to describe the spin phase is based on the observed behavior of the measured spin frequency, which is accurately described over short stretches by a straight line plus a small contribution due to the changing polhode frequency. This results in a spin phase model consisting of a quadratic polynomial plus two corrections,

$$\phi_s(t) = \phi_{si} + \omega_{si}(t - t_i) + \frac{1}{2}d\omega_s(t - t_i)^2 + \phi_p(t, Q = 0) + \Delta\phi_s(t, Q). \tag{12}$$

The first correction, the symmetric polhode phase $\phi_p(t, Q = 0)$, is added because the measured HF harmonics prove to be of the spin plus polhode frequency rather than the spin frequency. The last term, $\Delta\phi_s(t, Q)$, is a small modulation of the spin phase caused by the precession of the angular velocity, $\boldsymbol{\omega}_s$, about the angular momentum vector in inertial space. This modulation term can be explicitly computed from the solution to Euler's equations transformed into the inertial frame (see Landau and Lifshitz 1959; Salomon 2008), so it requires no additional parameters to estimate. The model (12) is independently fit to every batch of data about 1 day duration. Since at this step the parameters in the symmetric polhode phase model are considered known, the three parameters to estimate for each batch are $\omega_{si}$, $d\omega_s$, and $\phi_{si}$.

The nonlinear search routine for these three parameters is a modified version of the Nelder-Mead simplex method developed by Kozaczuk (2007) and implemented in the properly modified version of the MATLAB function *fminsearch.m*. The nonlinear search for the initial spin rate $\omega_{si}$ and the decay rate $d\omega_s$ is performed separately from the search for the initial spin phase $\phi_{s0}$, in order to simplify the nonlinear estimation process. The Level B code iterates between a search for $(\omega_{si}, d\omega_s)$ and a search for $\phi_{s0}$. A total of four iterations is a good compromise between accuracy and computation time for all the four GP-B gyros.

## 5.3 Level C: Trapped Field Distribution and Polhode Phase Refining

With the parameters estimated in the Level A and Level B data processing, the time histories of the three Euler angles, $\phi_p(t)$, $\theta_p(t)$ and $\phi_s(t)$, are computed for the whole science period. Then a linear least squares fit of the model (6) to the measured odd harmonics of spin, $H_{2k+1}(t)$, is performed to estimate the coefficients $A_{lm}$ using the data from the entire mission. Figure 3 shows the best-fit values of the real and imaginary parts of $A_{lm}$ with the various maximum number of odd harmonics, $N_{max}$, and coefficients, $L_{max}$, used, $N_{max} = L_{max} = 19, 23, 29$. A good agreement in the common values from different fits is seen, with a rather regular distribution, which turns out to be normal with zero mean.

After this step, estimates exist for all necessary parameters in the model (8). An iterative approach is used to refine these estimates further.

One of the most fundamental quantities in the determination of the trapped magnetic potential and the prediction of the scale factor variations $C_g^{TF}(t)$ is the polhode phase, $\phi_p(t)$. The estimation of $\phi_p(t)$ from the Level A data processing is typically accurate to 50–100 mrad, based on the post-fit residuals. A more accurate technique for determining the polhode phase error comes in at Level C and involves the absolute value of the measured first harmonics, $|H_1(t)|$. Using this quantity has two significant advantages for polhode phase determination. The first is that $H_1$ is the largest signal present in the HF SQUID output, providing confidence in its accuracy. The second, more important advantage is that the absolute value of any measured harmonics, $|H_n(t)|$, is totally independent of errors in the spin phase,

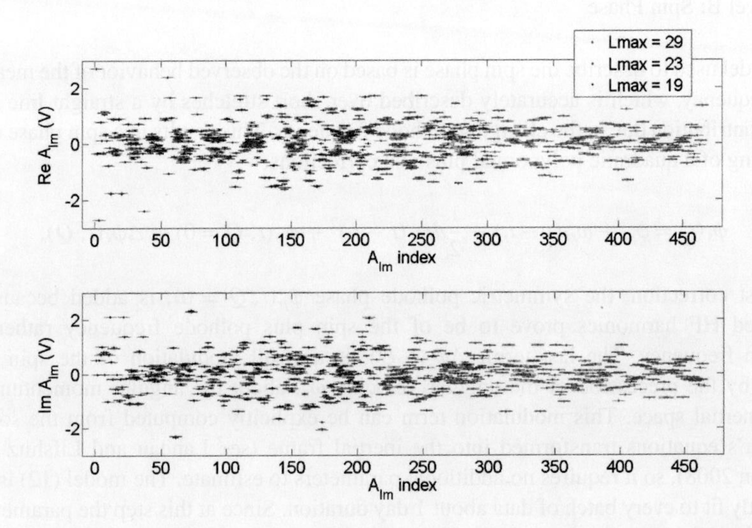


**Fig. 3** Best-fit $\mathrm{Re}(A_{lm}), \mathrm{Im}(A_{lm})$, with error bars, for gyroscope 1 using up to $L_{max}$ odd harmonics, $L_{max} = 19, 23, 29$—in *black*, *blue*, and *red*, respectively

**Table 4** Trapped flux mapping results of Level C fits

| Gyro No. | Relative Residuals | Number of Harmonics | Relative Size of Variations | Formal Error in $C_g^{TF}$ Relative to $C_g$ |
|---|---|---|---|---|
| 1 | 1.1% | 21 | 0.2%–3.0% | $1.5 \times 10^{-4}$–$7.0 \times 10^{-5}$ |
| 2 | 1.5% | 25 | 0.5%–1.5% | $6.0 \times 10^{-5}$–$3.0 \times 10^{-5}$ |
| 3 | 2.6% | 21 | 0.01%–1.0% | $2.0 \times 10^{-4}$–$1.6 \times 10^{-4}$ |
| 4 | 2.8% | 21 | 0.1%–0.3% | $8.5 \times 10^{-5}$–$6.5 \times 10^{-5}$ |

which is apparent from (7). In fact, $|H_1(t)|$ effectively depends only on the polhode phase and $A_{lm}$, since the polhode angle is simply computed from the measured parameters.

Upon completion of Levels A, B and C, a best-fit $|H_1(t)|$ can be constructed from the estimated parameters and compared with the measured value. This comparison gives a correction, $\delta\phi_p$, to the best-fit polhode phase from Level A. The exponential polhode phase model, (11), is then re-fit to the corrected polhode phase. With the new polhode phase, Level B processing is carried out again to refine the spin phase parameters, and then Level C is rerun to refine the coefficients $A_{lm}$. The polhode phase refinement, the Level B processing, and then the Level C processing constitutes a single iteration of the process that converges to the optimal solution. We continue these iterations until the RMS of the Level C post-fit residuals reaches a stable value.

## 6 Trapped Flux Mapping: Results

The accuracy achieved in the Level C fits and trapped flux scale factor, $C_g^{TF}(t)$, is given in Table 4. The post-fit residuals are roughly a few percent for all gyroscopes, which translates into $\sim 10^{-4}$ error in the predicted trapped flux scale factor relative to the total scale factor $C_g = C_g^{LM} + C_g^{TF}$. The highest uncertainty in the estimated $C_g^{TF}$ is for gyroscope 3, even though the post-fit residuals are larger for gyroscope 4. This is because the polhode damps out relatively quickly for gyroscope 3, causing poor observability of the coefficients $A_{lm}$.

The calculated distribution of trapped magnetic potential of the gyro 1 surface is shown in Fig. 4, along with the polhode paths and scale factor variations, $C_g^{TF}$, on different dates separated by more than 4 months. Due to the dissipation, the polhode path shrinks over time, and the magnitude of scale factor oscillations drops accordingly.

Figure 5 shows polhode variations of the scale factor, $C_g(t)^{TF}$, for all four gyroscopes during a few hours of the same day, 10 November 2004. An independent estimate of the total gyroscope readout scale factor, $C_g(t)$, comes from the analysis of the LF SQUID signal. This determination is limited by the constantly changing trapped flux contribution, and it is not possible to separate the London Moment scale factor from the d.c. part of the Trapped Flux. Nevertheless, this determination provides a useful comparison of TFM. The LF results agree with the TFM results to $\sim 10^{-3}$ or better, relative to the total scale factor, $C_g$. The best agreement is for gyroscope 3, which has the highest uncertainty in the $C_g^{TF}$ estimated by TFM: for gyroscope 3 the relative polhode variations, $C_g^{TF}/C_g$, are by far smaller than for any other gyro.

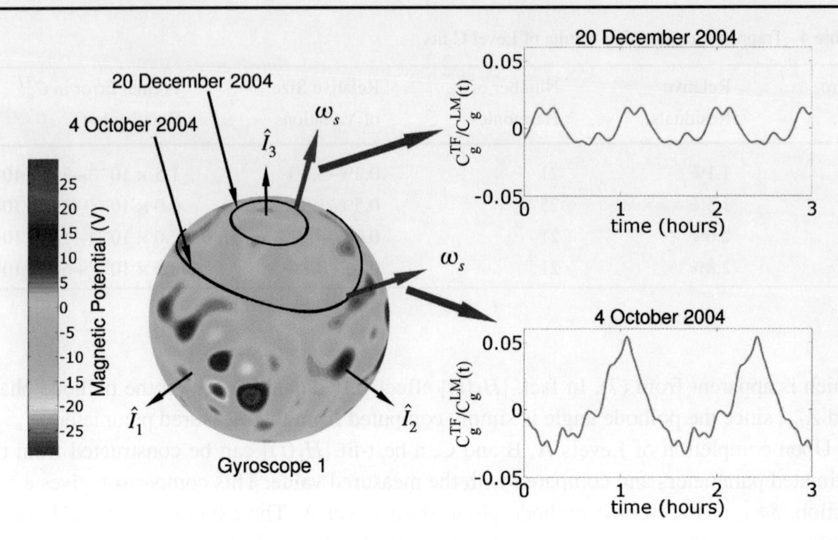


**Fig. 4** Gyroscope 1: magnetic potential distribution, and polhode paths and scale factor variations, $C_g^{TF}(t)$, on Oct. 4, 2004, and Feb. 20, 2005. The path shrinks due to dissipation, and the magnitude of variations goes down accordingly

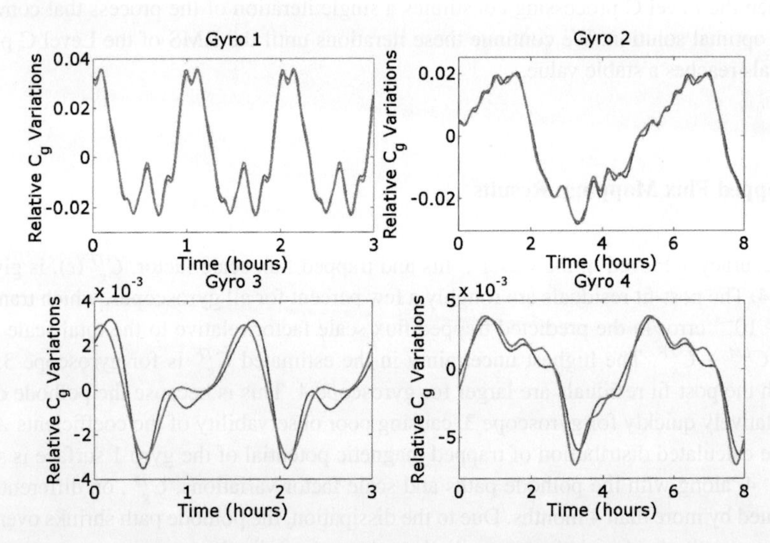


**Fig. 5** Scale factor variations, $C_g^{TF}(t)$, relative to the total scale factor, $C_g(t)$, on Nov. 10, 2004, as estimated by both TFM (*green*) and the LF analysis (*red*)

## 7 Conclusions

The change of polhode period and path discovered on orbit is explained by a slow rotation energy loss while conserving the angular momentum, and is properly analyzed using a new general model of dissipative gyroscope motion. This lays the ground for the developed procedure of mapping magnetic trapped field. Trapped Flux Mapping has been carried out successfully for each of the four GP-B gyroscopes using the odd harmonics of the HF SQUID signal. Its results are necessary to determine the LF scale factor variations and modulation of patch effect torque coefficients, thus they are crucial for the best measurement of relativistic drift rate.

**Acknowledgements** This work was supported by NASA contract NAS 8-39225, by donations from Richard Fairbank and Stanford University, and recently through a collaborative agreement with KACST. The authors are extremely grateful to W. Bencze, J. Berberian, S. Buchman, B. Clarke, F. Everitt, M. Heifetz, T. Holmes, J. Li, B. Muhlfelder, I. Nemenman, V. Solomonik and J. Turneaure for their numerous insights and help in completing this work.

## References

R.J. Adler, A.S. Silbergleit, Int. J. Theor. Phys. **39**(5), 1287 (2000)
B.M. Barker, R.F. O'Connell, Phys. Rev. D **11**(4), 711 (1975)
S. Buchman, D.K. Gill, *Evidence for Patch Effect Forces on the Gravity Probe B Gyroscopes*, APS Meeting, April 2007. Abstract: Bulletin of the APS, **52**(3) (2007)
J.W. Conklin, *Estimation of the Mass Center and Dynamics of a Spherical Test Mass for Gravitational Reference Sensors*. Ph.D. Thesis, Aero/Astro, Stanford University (Stanford, 2009)
M. Dolphin, *Polhode Dynamics and Gyroscope Asymmetry Analysis on Gravity Probe B Using Gyroscope Position Data*. Ph.D. Thesis, Aero/Astro, Stanford University (Stanford, 2007)
T.G. Duhamel, *Contributions to the Error Analysis in the Relativity Experiment*. Ph.D. Thesis, Aero/Astro, Stanford University (Stanford, 1984)
N.A. Egarmin, Mech. Solids **15**, 33 (1980)
C.W.F. Everitt, W.W. Fairbank, D.B. DeBra et al., *Report on a Program to Develop a Gyro Test of General Relativity in a Satellite and Associated Control Technology*. HEPL, Aero/Astro, Stanford University (Stanford, 1980)
C.W.F. Everitt, M. Adams, W. Bencze et al., Space Sci. Rev. (2009, this issue) (Chap. 1 of this report)
M. Heifetz, W. Bencze, T. Holmes et al., Space Sci. Rev. (2009, this issue) (Chap. 4 of this report)
G.M. Keiser, B. Cabrera, in *Proc. of The National Aerospace Meeting* (The Institute of Navigation, Washington, 1983)
G.M. Keiser, J. Kolodziejczak, A.S. Silbergleit, Space Sci. Rev. (2009, this issue) (Chap. 3 of this report)
G.M. Keiser, A.S. Silbergleit, *Pick-up Loop Symmetry and Centering*. Gravity Probe B document S0243 (Stanford University, Stanford, 1991)
J.A. Kozaczuk, *Precise Determination of the Spin Speed and Spin Down Rate of Gravity Probe B Gyroscopes*. Physics Honor Thesis, Stanford University (Stanford, 2007)
J. Lagarias et al., SIAM J. Optim. **9**(1), 112 (1998)
L.D. Landau, E.M. Lifshitz, *Mechanics* (Pergamon, Elmsford, 1959)
F. London, *Superfluids, vol. 1* (Dover, New York, 1961)
W.D. MacMillan, *Dynamics of Rigid Bodies* (Dover, New York, 1960)
V.J. Modi, J. Spacecr. Rockets **11**, 743 (1973)
I.M. Nemenmann, A.S. Silbergleit, J. Appl. Phys. **86**(1), 614 (1999)
M.E. Rose, *Elementary Theory of Angular Momentum* (Dover, New York, 1995)
M. Salomon, *Properties of Gravity Probe B Gyroscopes Obtained from High Frequency SQUID Signal*. Ph.D. Thesis, Aero/Astro, Stanford University (Stanford, 2008)
J.P. Turneaure, C.W.F. Everitt, B.W. Parkinson et al., Adv. Space Res. **32**(7), 1387 (2003)

Space Sci Rev (2009) 148: 411–427
DOI 10.1007/s11214-009-9525-6

# The Gravity Probe B Data Analysis Filtering Approach

**M. Heifetz · W. Bencze · T. Holmes · A. Silbergleit · V. Solomonik**

Received: 7 April 2009 / Accepted: 19 April 2009 / Published online: 4 June 2009
© Springer Science+Business Media B.V. 2009

**Abstract** A simple pre-flight strategy of the Gravity Probe B (GP-B) data analysis has evolved in the elaborate multi-level structure after the discovery of the complex polhode motion, and of the patch effect torques. We describe a cascade of estimators (filters) that reduce the science data (SQUID and telescope signals) to the estimates of the relativistic drift rates. Those estimators, structured in two "floors", are based on the polhode-related models for the readout scale factor and patch effect torque. Results of the 1st Floor processing—gyro orientation profiles—manifest clearly the strong geodetic effect but also the presence of classical torque. Modeling of the patch effect torque at the 2nd Floor provides a successful compensation of the torque contributions, and leads to consistent estimates of the relativistic drift rates.

**Keywords** Gravity Probe B · Data analysis · Torque models · Relativistic drift estimates

**PACS** 04 · 04.80.Cc · 02.70-c · 07.05.Kf

## 1 Introduction

Data analysis in the GP-B experiment (Everitt et al. 2009)—estimation of the relativistic drift rate for each of the four unique GP-B gyroscopes—was supposed to be rather straightforward (Haupt 1996; Heifetz et al. 2003). One should take the low frequency (LF) SQUID readout signal for the duration of the mission, calibrate the scale factor based on orbital and annual aberration known from the GPS orbit data and earth ephemeris; obtain the time-history of the gyroscope inertial orientation in the projections on the directions in the orbital plane and perpendicular to it, and then estimate the slopes of those projections, which were supposed to be almost straight lines. The slope of the line in the orbital plane gives the measurement of the geodetic effect. The slope of the projection perpendicular to the orbital

M. Heifetz (✉) · W. Bencze · T. Holmes · A. Silbergleit · V. Solomonik
Gravity Probe B, Hansen Experimental Physics Laboratory, Stanford University, Stanford, CA, 94305, USA
e-mail: misha@relgyro.stanford.edu

plane gives the measurement of the frame-dragging effect [this is exactly true for the ideal polar orbit with the Guide Star (GS) in the orbit plane].

In reality, an unexpected damped polhode motion of all the GP-B rotors, and some larger than expected classical torques on them, were discovered on orbit. In addition, about a year of science data was cut into 10 segments by various spacecraft (S/C) anomalies. This has turned a "simple" data analysis strategy into a challenging adaptive estimation process involving a multi-level filtering machinery.

There are three cornerstones of the GP-B filtering (estimation) method. 1) SQUID readout signal structure: *measurement models*—these models are based on the underlying physics and engineering of the GP-B instrument. 2) Gyroscope motion: *torque models*—accumulated understanding of the underlying physics. 3) Filter implementation: *numerical techniques*—algorithms from the estimation theory (Kailath et al. 2000; Bierman 2006) specifically adapted to GP-B data.

To successfully estimate the relativity parameters, the filtering approach must meet several major challenges: a) the complicated variation of the readout scale factor (see details in Sect. 3); b) a continuously acting torque due to electrostatic patch effect (see Sect. 4); c) simultaneous estimation of the relativity parameters, multiple torque and scale factor coefficients; d) noisy measurements that depend on unknown parameters in a nonlinear fashion; e) segmented science data (see Sect. 3); f) the large volume of data ($\sim$one Terabyte): four gyroscopes $\times$4605 orbits of science data with the 2-second time step.

In Sect. 2 the main original idea of a "simple" data analysis is explained and the science signal measurement model is introduced. Section 3 presents the current two-floor structure of the filtering approach, gives the description of the model of scale factor variations, and shows the results of the science signal analysis. Those are the gyroscope orientation profiles revealing the presence of Newtonian torques. A model of the patch effect torque is introduced in Sect. 4, and a concept of the "reconstruction" of the gyroscope's relativistic motion by means of modeling, estimation, and elimination of the torque contributions from the orientation profiles is described. A further expansion of the data analysis, the "two-second filter", is described in Sect. 5. Section 6 contains some conclusions.

## 2 "Simple" GP-B Data Analysis: Pre-Launch Concept

### 2.1 Inertial Frame and the Relevant Vectors

The drift experienced by the GP-B gyroscopes is measured in the inertial Cartesian frame $\{\hat{e}_{GS}, \hat{e}_{EW}, \hat{e}_{NS}\}$ introduced by Silbergleit et al. (2001) (see Fig. 1). Its first unit vector $\hat{e}_{GS}$ points from the Earth center to the true position of the GS (IM Pegassi). The second unit vector is $\hat{e}_{EW} = (\hat{e}_{GS} \times \hat{z})/|\hat{e}_{GS} \times \hat{z}|$, where $\hat{z}$ is the unit vector of the inertial frame JE2000 (see Seidelmann 1992; McCarthy 1996). The last axis $\hat{e}_{NS}$ is naturally defined as $\hat{e}_{NS} = \hat{e}_{EW} \times \hat{e}_{GS}$. The $\hat{z}$ axis coincides with the Earth rotation axis at noon, January 1, 2000, and stays very close to it later. Thus the geodetic drift is in the direction of $\hat{e}_{NS}$, and the frame-dragging drift is along $\hat{e}_{EW}$, if the orbit is polar and contains $\hat{e}_{GS}$ in its plane. The real GP-B orbit is close enough to this ideal one, so the frame is a natural choice for GP-B data analysis.

The unit vector $\hat{s}$ represents the gyroscope spin axis, while $\hat{\tau}$ points along the S/C roll axis. The *misalignment angle*, $\psi$, between these vectors is only $\sim 10^{-4}$ or smaller, and they are both at about the same small angles to $\hat{e}_{GS}$. For this reason, the *misalignment vector, $\vec{\mu}$*, defined as $\vec{\mu} = \hat{s} - \hat{\tau}$, lies, to lowest order, in the NS–EW plane, and $\psi \approx \mu \sim 10^{-4}$.

When the GS is visible by the telescope (GS valid, GSV), the S/C attitude and translation control system points the S/C roll axis in the apparent direction to the GS. The latter differs

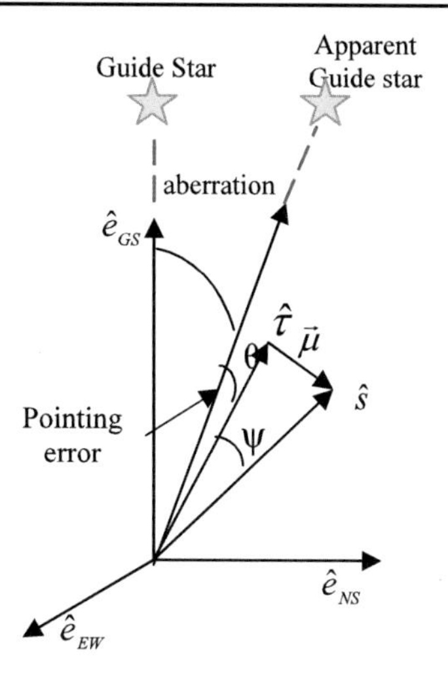


**Fig. 1** Inertial reference frame

from true direction to the GS by the sum of the orbital and annual aberrations (up to 25 arcs), the pointing error, $\theta$, (within ~50 marcs on average), and the contribution of the stellar parallax and starlight bending by the Sun (up to 25 mas). In the pre-launch data reduction concept, only GSV periods were planned to be analyzed.

### 2.2 Measurement Model

The GP-B readout system provides measurements of the time-varying angle between the direction of the gyroscope spin axis aligned with the London moment (Turneaure et al. 2003; Everitt et al. 2008), and the SQUID pick-up loop that rolls with the S/C. The corresponding measurement model is shown, to lowest order, to be (Silbergleit 2006):

$$Z(t) = C_g(t)\{[\tau_{NS}(t) - s_{NS}(t)]\cos(\Phi_r(t) + \delta\varphi) + [\tau_{EW}(t) - s_{EW}(t)]\sin(\Phi_r(t) + \delta\varphi)\} + bias + noise. \quad (1)$$

There are two groups of variables in this equation. The first group consists of the available data: $Z(t)$ is the SQUID signal, $\Phi_r(t)$ is the roll phase signal, and $\tau_{NS}(t)$, $\tau_{EW}(t)$ are known (see the previous section; pointing error is determined by the telescope signal). The second group of the unknown variables that need to be estimated includes the readout scale factor $C_g(t)$, the gyroscope inertial orientation $s_{NS}(t)$, $s_{EW}(t)$, the roll phase offset $\delta\varphi$, and the bias components, such as the SQUID calibration signal.

In the original pre-launch strategy of the data analysis, the idea of a "simple" batch-based estimator was implemented. The science signals were processed in a sequence of short batches; within each batch, the scale factor and the roll phase offset can be calibrated using the known total aberration, and subsequently, the estimates $\tilde{s}_{NS}(t)$ and $\tilde{s}_{EW}(t)$ of the gyro orientation time history could be obtained. The estimated curve of each of the orientations would then be fitted to a line, and the estimated slopes would give the measurement of the relativity parameters.

Two surprising on-orbit findings led to the development of a much more elaborate two-floor structure of the data analysis described below.

## 3 Two-Floor Structure of Data Analysis

### 3.1 Modification of the Pre-launch Scheme

Pre-launch data analysis strategy was built on the assumption that no modeling of classical torques is needed. In this case processing of the SQUID signal could be done in one batch, or, to check the instrument stability, in a sequence of batches. During the mission, after the discovery of the polhode-related behavior of the scale factor $C_g(t)$ (see Silbergleit et al. 2009) and of the so called *misalignment* torque (see Keiser et al. 2009), it became evident that more sophisticated modeling of both the scale factor and torque is necessary. The magnitude of the misalignment torque is proportional to the (small) misalignment, $\mu$, and its direction is perpendicular to the misalignment vector (Sect. 2.1). This torque acted continuously, so it requires continuous modeling for the entire orbit, both its GSV and GSI parts (GSI = GS Invalid, the period when the GS is occulted by the Earth).

To simplify the development of a cumbersome analysis structure, save the computational resources and stabilize filtering algorithms, we decided to carry out data analysis in two steps called *floors*. The 1st Floor is a modification of the pre-launch scheme using batch processing with no torque modeling within each batch. At the 2nd Floor torque and more accurate scale factor modeling occurs, as a way of integrating information from all the batches. For various reasons, we have chosen the batch length equal to one orbit ($\sim$97 min).

The implementation of the two-floor analysis structure faced an additional difficulty in the form of several different characteristic time scales present in the signals. The longest of them is 1 year, the period of the annual aberration. Unfortunately, the year-long duration of the experiment was interrupted by several spacecraft anomalies (e.g. the solar flare on January 20, 1995), so science data came in 10 segments separated by day long gaps (the list of segments is given in Table 1). Naturally, one of the questions to be resolved was how to connect those segments in the data analysis.

The next important time scale is the orbital period, the period of orbital aberration calibrating the scale factor within each batch. A short interval ($\sim$2 min) of the GS reacquisition by the telescope is also a part of the data for each orbit.

The third meaningful time scale is the period of the GP-B satellite roll (77.5 sec): the science signal is at the roll frequency, as seen from (1). Other time scales are related to the gyro polhode motion: it is the polhode period ($\sim$ few hours), which itself is changing slowly

**Table 1** Segmented science data

| Segment | Science data | Duration (days) |
|---|---|---|
| 1 | September 13, 2004 – September 23 | 11 |
| 2 | September 25 – November 10 | 47 |
| 3 | November 12 – December 04 | 23 |
| 4 | December 05 – December 09 | 5 |
| 5 | December 10 – January 20, 2005 | 42 |
| 6 | January 21 – March 04 | 43 |
| 7 | March 07 – March 15 | 9 |
| 8 | March 16 – March 18 | 3 |
| 9 | March 19 – May 27 | 70 |
| 10 | May 31 – July 23, 2005 | 54 |

with the characteristic time of kinetic energy dissipation ($\sim$ 1 or 2 months, depending on the gyro) (Silbergleit et al. 2009).

All these time scales were taken into account in the design of the data analysis software architecture. The orbital period was the main batch length for the 1st Floor of the data reduction (see Sect. 3.2 below). The changing polhode period and the energy dissipation times were the key parameters in the development of the trapped flux mapping (TFM) (Salomon 2008; Conklin 2009; Silbergleit et al. 2009), and the outputs of the TFM, the polhode phase and polhode angle, were key ingredients of the scale factor and torque modeling (see Sect. 3.3 and Sect. 4 below). The annual aberration, which is defined by the annual period, is the main long-term calibrator of the scale factor, and also the main contributor to the misalignment time signature. The latter is essential for the observability and identification of the misalignment torque coefficients (see Sect. 4). Lastly, the roll frequency is, naturally, the fundamental frequency of the SQUID science signal. But in addition, it is also one of the key elements of the roll-resonance torque and the algorithms of its compensation (see Sect. 4 and Sect. 5).

### 3.2 Floor 1: One-Orbit Batch Data Reduction

The idea of the 1st Floor is the compression of science data within each orbit. The inputs are the LF SQUID and telescope signals, roll phase signal, and pre-computed aberration data. The filtering (estimation) is carried out independently for each orbit based on the measurement model (1). The state vector includes the two components of the gyro inertial orientation, the coefficients of the scale factor polhode variations, the roll phase offset, the two telescope scale factors (of two directional channels), and the coefficients in the bias variation model. The $1/f$ nature of the SQUID noise is addressed by applying a band-pass digital filter to the SQUID and telescope signals. The resulting signal spectrum is within the roll $\pm$ orbit frequency band. No torque modeling is performed at the 1st Floor. The output of the 1st Floor data reduction consists of the estimates of the state vector (1 point per orbit), full information (inverse covariance) matrix, and the post-fit residuals. An additional output, the misalignment and pointing signals, is used for the misalignment torque modeling at the 2nd Floor (see Sect. 4).

Among various modules in the 1st Floor structure (Fig. 2) several should be highlighted: 1) scale factor model; 2) pointing error compensation—gyroscope/telescope scale factor matching; 3) data grading (using only high quality data); 4) bias model (including polhode variations, bias jumps, calibration signal components, etc.); 5) nonlinear least-squares estimator. The latter is implemented in the form of the square-root information filter Bierman (2006) that provides numerical stability and computational fidelity.

### 3.3 Floors 1 and 2: Readout Scale Factor Modeling

A mathematical model of the scale factor variations due to trapped magnetic flux and gyro polhoding has been shown to be ($\gamma \mathrm{C_g}$ -model):

$$C_g(t) = C_{g0}\left\{1 + \sum_{m=0}^{M} a_m \cos(m\Phi_0(t)) + b_m \sin(m\Phi_0(t))\right\}, \tag{2a}$$

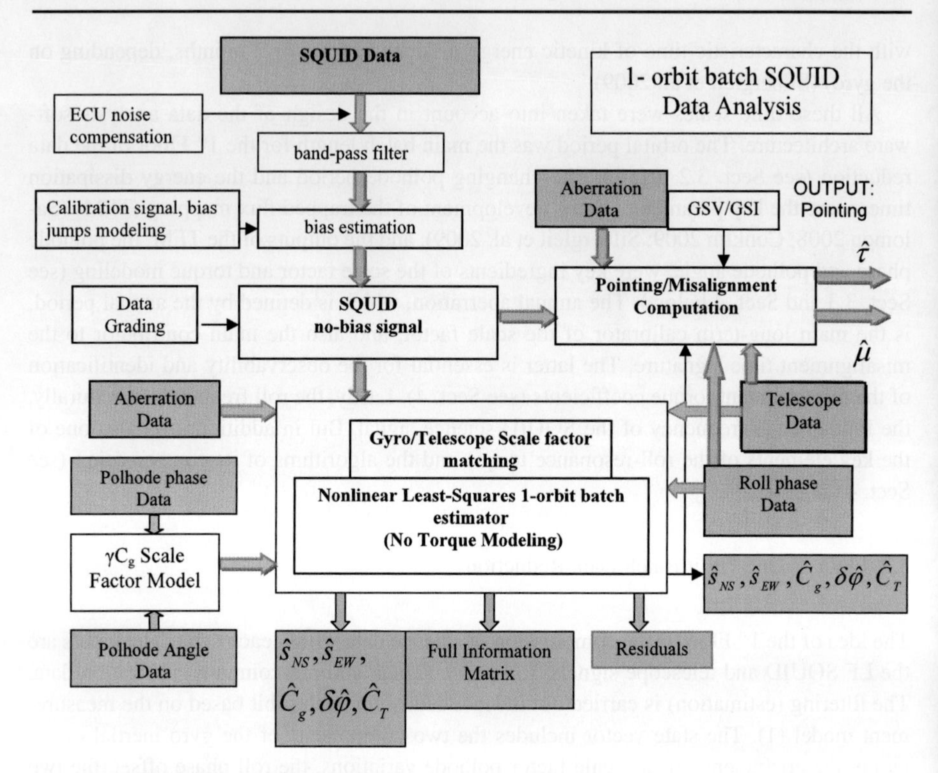


**Fig. 2** First Floor structure block-diagram

$$
\begin{aligned}
&\begin{pmatrix} a_0(\gamma_0) \\ b_0(\gamma_0) \end{pmatrix} = \sum_{n=0}^{N_0} \begin{pmatrix} a_{0n} \\ b_{0n} \end{pmatrix} \varepsilon_0^n(t), \\
&\begin{pmatrix} a_m(\gamma_0) \\ b_m(\gamma_0) \end{pmatrix} = \sum_{n=1}^{N_n} \begin{pmatrix} a_{mn} \\ b_{mn} \end{pmatrix} \varepsilon_0^{n+1}(t), \quad m = 1, 2, \ldots; \\
&\varepsilon_0(t) = \tan(\gamma_0(t)/2).
\end{aligned} \tag{2b}
$$

Here $\Phi_0$ and $\gamma_0$ are the polhode phase and polhode angle calculated for a symmetric gyroscope based on the information from the TFM (see Silbergleit et al. 2009 for all the details). The numbers $M$, $N_0$, $N_n$ are chosen so to make these approximations good enough for our analysis accuracy without including unnecessary terms (in theory, all the expansions are, of course, infinite). The number of terms required for the needed accuracy decreases towards the end of the mission, with the damping of polhode variations.

In this model the scale factor is represented as an expansion in harmonics of the polhode phase, but the coefficients of this expansion turn out to be functions of the polhode angle. The polhode frequency and angle are slowly changing due to energy dissipation, with the characteristic time of this process (1–2 months). At Floor 1 only (2a) is used, and the coefficients $a_m$ and $b_m$ are estimated once per orbit. At the 2nd Floor these estimates are fit to the model (2b), with the unknown parameters $a_{mn}$ and $b_{mn}$ to be estimated. If the fit is successful, then the complete physical model (2a), (2b) is justified. The fit results for real data are plotted in Fig. 3; clearly, the model explains the complicated profiles of the coefficients $a_m$ and $b_m$ very well. An identical model is used at the 2nd Floor for the torque coefficients

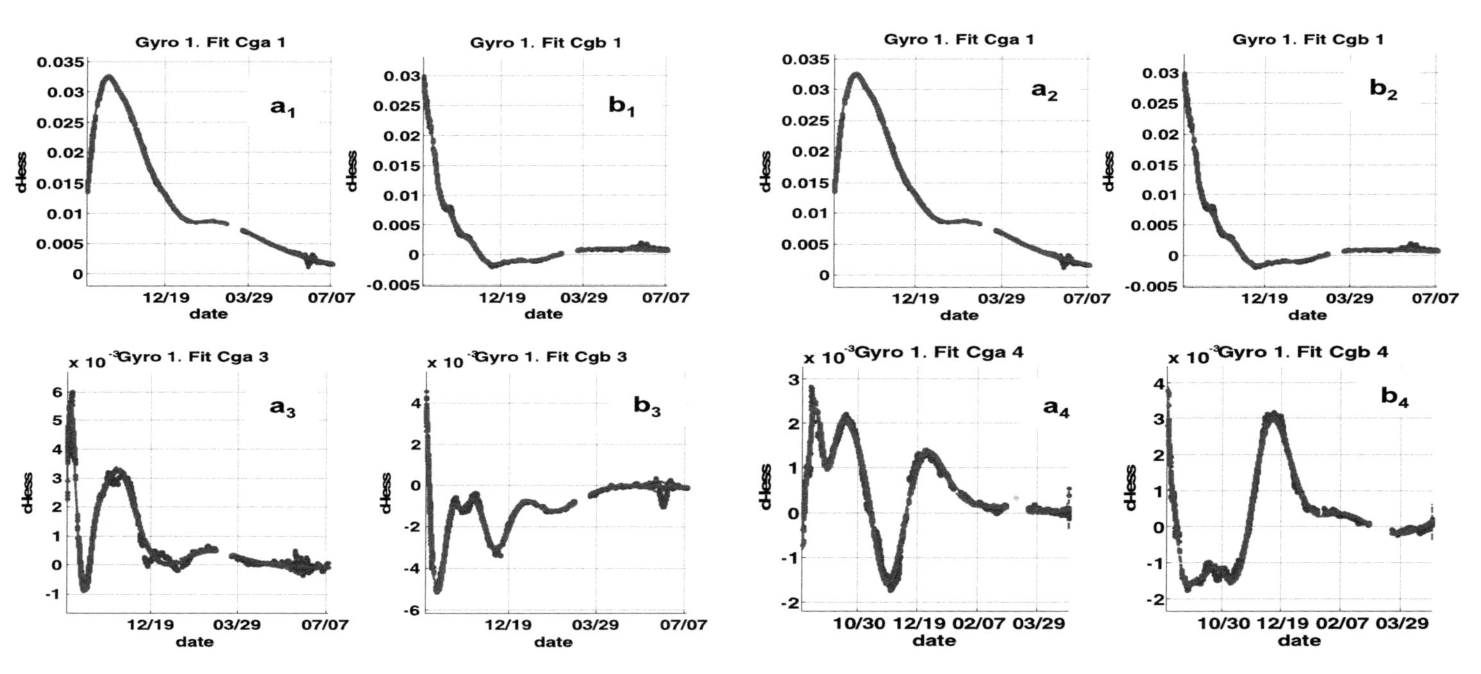


**Fig. 3** Gyro 1 polhode harmonics from batch analysis with fits to $\gamma C_g$-model (dimensionless values). *Blue*—measured coefficients, *red*—fit to $\varepsilon(t)$ model

(Sect. 4). The scale factor model (2a), (2b) will also be used in a new two-second filter described in Sect. 5.

The successful use of the $\gamma C_g$-model to a significant extent is due to TFM that computes a highly accurate polhode phase and angle for each gyroscope for the whole duration of the science mission. It also provides an alternative to the estimation of scale factor variations by the model (2a), (2b). Namely, scale factor variations were computed from the TFM results (again, for each gyro and the whole mission), and could be used as a known function, simplifying significantly the 1st Floor processing.

### 3.4 Gyroscope Orientation Profiles: Moving Window Method

The main inputs of the 2nd Floor are the gyro orientation profiles from the 1st Floor, along with their covariance matrix. However, the 1st Floor profiles are not smooth enough for the torque modeling. Smoothing of the noisy 1-orbit batch orientation estimates is performed by means of a "moving window" concept.

A window of a length of several polhode periods moves along the time axis with one-orbit shifts. Within each window the information obtained by the one-orbit batch processing is integrated: components of the 1st Floor state vector for each orbit are treated as the "measurements" of the window state vector parameters, and the information matrices available for each orbit are summed up. To propagate the gyro orientation within a window, we apply the misalignment torque model (see Sect. 4.1). If there is a jump in one or both orientation components, it is modeled as a "ramp". The output of the "moving window" estimator is a pair of smoothed profiles of gyro orientation (1 point per orbit) in the NS and EW directions, along with their uncertainty at each point (see Fig. 4 and Fig. 5). There is a clear linear trend in the NS plot that should probably represent the geodetic effect, and about the right size, relative to the GR prediction. However, the EW component is by far not a straight line: the small frame-dragging effect is completely concealed by classical torque contributions.

## 4 Floor 2: Misalignment and Roll-Resonance Torque Modeling. Torque Compensation and Relativity Estimation

The "moving window" smoothing allows one to observe the fine structure of the gyro orientation. It was instrumental in understanding and modeling of the resonance phenomena, i.e. jumps in the orientation profiles. Kolodziejczak (2007) identified many significant jumps with the so called *roll-resonance* torques which are not averaged out over the roll period around times, $t_m$, when the constant roll frequency, $\omega_r$, coincides with a harmonic of the time-varying polhode frequency, $\omega_p$, i.e., $\omega_r = m\omega_p(t_m)$ (roll-to-polhode resonance; see Everitt et al. 2008; Keiser et al. 2009). A physical model of both the misalignment and roll-resonance torque and its use in the 2nd Floor analysis is described in this section.

### 4.1 Misalignment and Roll-Resonance Torque Model

One of the assumptions in the early phases of the post-flight GP-B data analysis was that only the roll-averaged patch effect torque should be considered. Patch effect torque is caused by the variation of electrostatic potential on the rotor and housing surfaces and their relative motion. Discovery of the resonance torques and subsequent derivation of the roll-resonance torque model (Keiser et al. 2009) showed that the non-roll-averaged torque should be taken into account, that is, should be modeled and compensated.

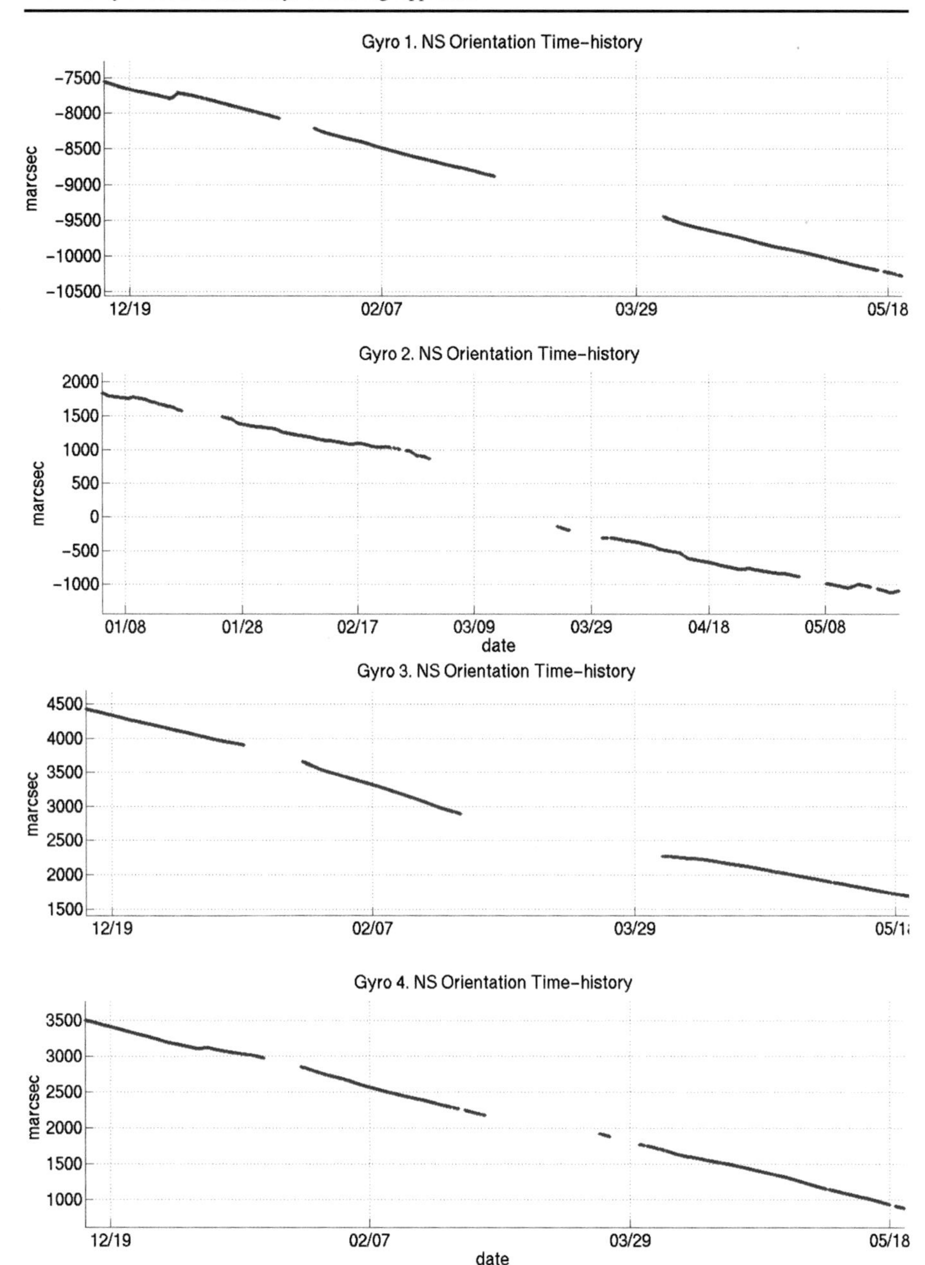


**Fig. 4** Gyroscope orientation time history (NS direction; units of the vertical axis are arc-seconds)

As mentioned, the patch effect produces two torques that have to be taken into account in the data reduction. The misalignment torque is proportional to the misalignment $\mu$, with the coefficient determined by the patch patterns. There is also a contribution of zero order in $\mu$

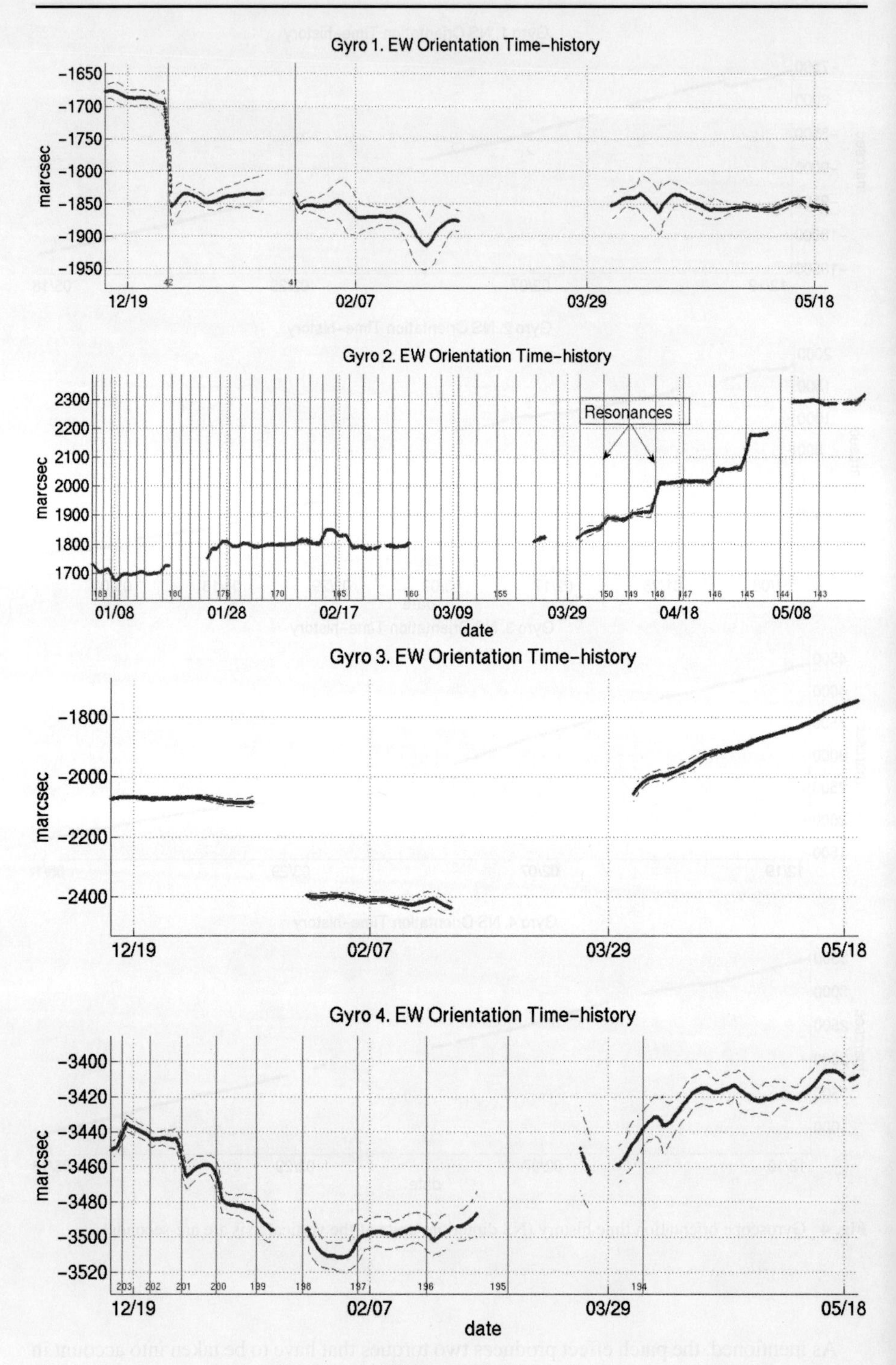


**Fig. 5** Gyroscope orientation time history (EW direction). *Red vertical lines* correspond to the resonances $\omega_r = m\omega_p(t)$ (units of the *vertical axis* are arc-seconds)

(not vanishing when $\mu = 0$), which varies at the harmonics of roll; the coefficients involved in it are also determined by the patches.

Due to the polhode motion of the rotor, all the torque coefficients are superpositions of multiple polhode frequency harmonics (see formula (4a) below). When one of the multiple frequencies coincides with the roll frequency, a non-averaging d.c. contribution appears and affects the long-term gyroscope drift. In addition, there is some evidence that the resonance torque does not completely average to zero over every roll period even between the resonances. Therefore it should also be modeled continuously in the estimation.

Equation of motion of the GP-B gyroscope spin axis ($\hat{s}$) in the presence of the misalignment and roll-resonance torque is:

$$\begin{aligned} \frac{ds_{NS}}{dt} &= r_{NS} + k(t)\mu_{EW} + [c^+(t)\cos\Phi_r + c^-(t)\sin\Phi_r] \\ \frac{ds_{EW}}{dt} &= r_{EW} - k(t)\mu_{NS} + [-c^+(t)\sin\Phi_r + c^-(t)\cos\Phi_r], \end{aligned} \tag{3}$$

where $r_{NS}$ and $r_{EW}$ are the relativistic drift rates, $\mu_{NS}$ and $\mu_{EW}$ are the inertial components of the misalignment, $\vec{\mu}$, (see Fig. 1), and $k(t)$, $c^{\pm}(t)$ are the torque coefficients. Expressions in brackets represent the non-roll-average, roll-resonance torque model, where $\Phi_r$ is the roll phase.

The torque coefficients $k(t)$, $c^{\pm}(t)$ related to the patch distribution can be represented in exactly the same way as the scale factor $C_g$ in (2a), (2b), namely:

$$c(t) = \sum_{m=0}^{\tilde{M}} [c_{1m}(\gamma_0)\cos m\Phi_0(t) + c_{2m}(\gamma_0)\sin m\Phi_0(t)], \tag{4a}$$

$$\begin{aligned} &\begin{pmatrix} c_{10}(\gamma_0) \\ c_{20}(\gamma_0) \end{pmatrix} = \sum_{n=0}^{\tilde{N}_0} \begin{pmatrix} c_{10n} \\ c_{20n} \end{pmatrix} \varepsilon_0^n(t); \\ &\begin{pmatrix} c_{1m}(\gamma_0) \\ c_{2m}(\gamma_0) \end{pmatrix} = \sum_{n=1}^{\tilde{N}_n} \begin{pmatrix} c_{1mn} \\ c_{2mn} \end{pmatrix} \varepsilon_0^{n+1}(t), \quad m = 1, 2, \dots; \\ &\varepsilon_0(t) = \tan(\gamma_0(t)/2). \end{aligned} \tag{4b}$$

This formulas follow (Silbergleit 2009) from the expressions for the roll resonance torque obtained in Keiser et al. (2009). A particular case of (3) in the vicinity of a roll-to-polhode resonance is also given there (see (18) of the Keiser et al. paper).

Here $\Phi_0$ and $\gamma_0$ are the polhode phase and polhode angle calculated for a symmetric gyroscope, and $c(t)$ stands for any of the torque coefficients. Thus $c_{1m}(t) = k^{\pm}_{1m}(t)$, $c^{\pm}_{1m}(t)$, $c_{1mn} = k^{\pm}_{1mn}$, $c^{\pm}_{1mn}$, and $c_{2m}(t) = k^{\pm}_{2m}(t)$, $c^{\pm}_{2m}(t)$, $c_{2mn} = k^{\pm}_{2mn}$, $c^{\pm}_{2mn}$, respectively. The only difference between the formulas (2b) and (4b) is in the meaning of the states to be estimated, $\{a_{mn}, b_{mn}\}$ in (2b), and $\{c_{1mn}, c_{2mn}\}$ in (4b). The values of the first set are determined by the trapped field distribution on the rotor surface, while in the second case they depend on the electrostatic patch patterns and the inertial asymmetry of the rotor.

The torque model (3), (4a), (4b) exploits the combinations of phases $\Phi^{\mp}_m = \Phi_r \mp m\Phi_0$, and produces shifts in the orientations $s_{NS}$ and $s_{EW}$ at the resonance times when the frequency of the phase $\Phi^-_m(t)$ passes through zero, i.e., the above resonance condition holds, $\omega_r = m\omega_p(t_m)$. Naturally, the following conceptual questions arise: can the torque model (3)–(4) explain the observed multiple jumps in the gyroscope orientation profiles? Could all the significant torque contributions be estimated using this model and subtracted from the gyro orientation profiles, providing thus a reconstruction of the gyroscope relativistic motion and measurement of its rate? As shown below, both answers are affirmative.

### 4.2 Elimination of Torque Contributions from the 1st Floor Orientation Profiles: Proof of Concept

The output of the 1st Floor includes, in general, the multidimensional state vector (gyro orientation, scale factor coefficients, roll phase offset, SQUID bias components, etc.) and the full information matrix. To address the above questions, we consider only the orientation profiles and their uncertainties as inputs to the 2nd Floor processing. The Kalman filter is a natural tool for solving this simplified estimation problem: orientation profiles give the measurement of the orientations $s_{NS}$ and $s_{EW}$ at the discrete, 1 point per orbit, known time points. Their behavior, $s_{NS}(t)$ and $s_{EW}(t)$, is described by the dynamic model (3), (4a), (4b). One complication should be immediately addressed, namely, science data segmentation, according to Table 1. The segments are separated by the S/C anomalies. Some of them, such as, for instance, solar flare between segments 5 and 6, could change the patch patterns, and thus the values of the torque coefficients. To overcome this obstacle, the dynamic estimator was designed in the following framework (Segments 5, 6, and 9 were analyzed for gyros 1, 2, 4; the results for gyro 3 are from Segments 5 and 6 only).

*Step 1. Independent Kalman filter/smoother for each segment and each gyroscope*

Introduce the state vector $x = [s_{NS}, s_{EW}, r_{NS}, r_{EW}, \{c^{\pm}_{1mn}, c^{\pm}_{2mn}\}, \{k_{1mn}, k_{2mn}\}]$, where the first two components are dynamically changing states according to the model (3), and all other components are constants. The time-varying roll-resonance torque coefficients $k(t)$, $c^{-}(t)$, $c^{+}(t)$ are represented by the model (4a), (4b); the parameters of this model, $c^{\pm}_{1mn}$, $c^{\pm}_{2mn}$, $k_{1mn}$, $k_{2mn}$, remain constant within a given segment.

A Kalman filter technique was implemented in the form of a smoother with the forward-backward propagation-update structure (Bryson 2002; Bar-Shalom et al. 2001). The observability analysis showed that the roll-resonance torque coefficients in the vicinity of the corresponding resonances are strongly observable, and their estimates are statistically significant. The outputs of the Step 1 estimators are, specifically, the estimates of the torque coefficients and the modeled orientation profiles: the solution of the equations (3) with the estimated torque coefficient values. An example of the modeled orientation profiles, that has the same shape as the input, is presented in Fig. 6.

*Step 2. Reconstruction of "Relativistic Trajectory": Subtracting Torque Contributions*

We can combine the segments by subtracting the torque contributions estimated for each of the segments in Step 1 from the input orientation profiles. The resulting curve is the reconstructed "relativistic trajectory"; it consists of the segment pieces that have the same "relativistic" slope. The reconstructed trajectory can then be fit to the same-slope line over each segment; thus the state vector includes a common slope and individual y-intercepts. The estimate of the slope gives the value of the corresponding relativistic drift rate. The reconstructed trajectory noise is larger than the orientation profile noise by the known amount defined by the uncertainty in the subtracted torque contributions.

The described two-step approach can be used not only for one particular gyroscope, but also for any combination of the gyroscope signals. Step 2 easily allows for fitting the pieces of the reconstructed trajectories of different gyroscopes over different segments, since they

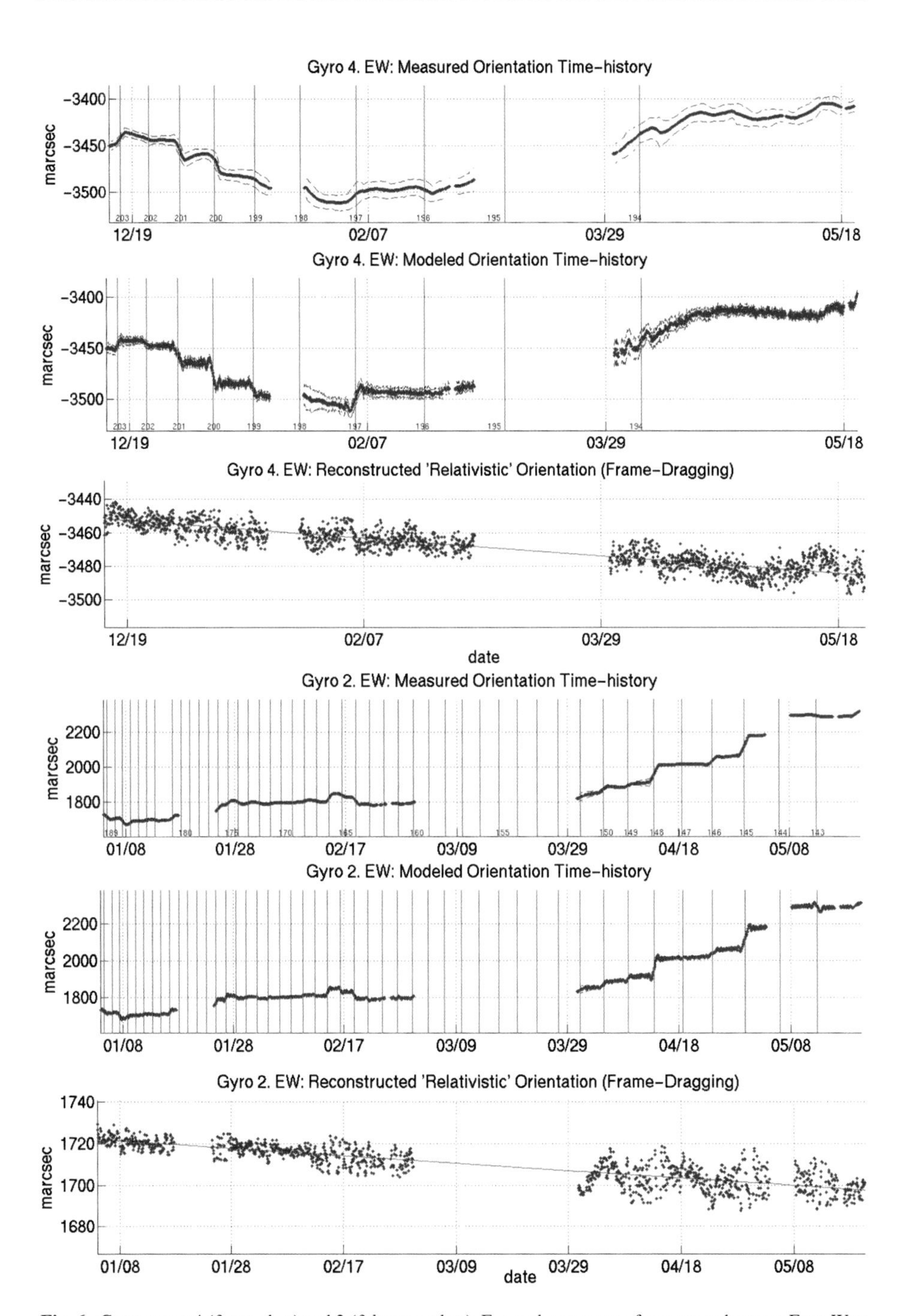


**Fig. 6** Gyroscopes 4 (*3 top plots*) and 2 (*3 bottom plots*). For each gyroscope from top to bottom: East–West orientation profile; modeled orientation profile; reconstructed "relativistic" trajectory

all have the same relativistic drift rate (slope). Reconstructed trajectories (corrected for the segment shifts) of the gyroscopes 4 (10 resonances) and 2 (73 resonances) are presented in Fig. 6. Both the NS and EW reconstructed trajectories now contain a clear linear trend, *the EW curve for the first time ever*.

Estimates of the relativistic drift rates can also be obtained directly from the output of Step 1 estimators. Each individual estimator results in the state vector and information matrix whose dimensions differ from segment to segment owing to the different number of resonances and torque coefficients. Nevertheless, they also have a common part: relativistic drift rates. The integration of the information from all one-segment estimators can be done by creating the augmented state vector and information matrix. The augmented state vector includes the relativistic parameters and all torque coefficients from all segments, and the augmented information matrix consists of blocks of the individual information matrices. The combined covariance matrix is the inverse of the augmented information matrix, and the combined state vector estimate is a linear combination of the individual estimates with weighting coefficients, which depend on the combined covariance matrix and the individual information matrices. These two described methods of relativity parameters estimation are statistically equivalent.

### 4.3 Two-Floor Analysis Results

The estimates of the relativistic drift rates with 50%—probability error ellipses obtained by the second method described above are shown in Fig. 7. The "individual" gyro estimates are consistent with each other (the ellipses overlap). The error bars of the estimates given in Fig. 7 are formal statistical 50% probability errors.

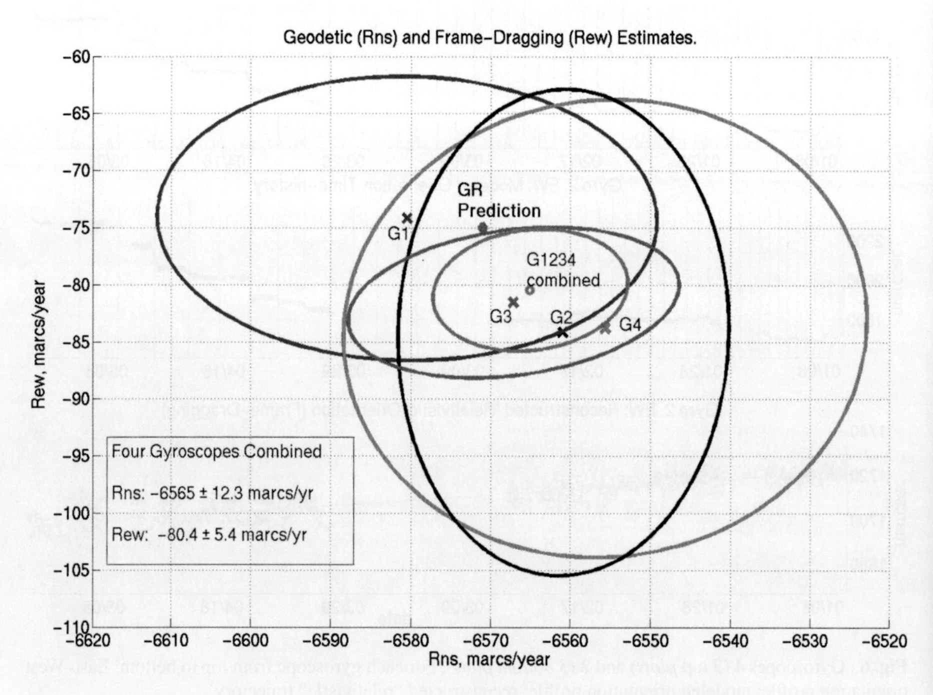


**Fig. 7** Preliminary relativity estimates with 50% uncertainty ellipses. Gyroscopes 1, 2, 4: segments 5, 6, and 9; (gyro 3: segments 5, 6). Systematic error or a model sensitivity analysis not included

Data analysis using the methods outlined above is still preliminary since a number of factors have not been fully investigated: (1) more data segments for gyroscope analysis; (2) variation of the number of parameters used to characterize the torque coefficients; (3) unmodeled error in the measurement of the telescope pointing during GSI; (4) use of the maximum information matrix to determine the error ellipse; (5) inclusion of systematic errors, etc. (The accompanying paper Muhlfelder et al. 2009 discusses approaches to establishing systematic errors bounds.)

Nevertheless, the preliminary results given in Fig. 7 are, of course, subject to the limitations just described. There are two important observations that can be made in this data analysis example: (1) the residual errors are substantially smaller than found with earlier data analysis methods, and (2) the relativity estimates for the individual gyroscopes are statistically consistent with each other. The combined result for all four gyroscopes is in agreement with the GR values. (GR predictions for the real GP-B orbit, taking into account the GS proper motion, the solar geodetic effect, and the EW projection of the Earth geodetic effect, are −6571 marcs/year in the NS direction, and −75 marcs/year in the EW direction).

## 5 Data Analysis Expansion: Two-Second Filter

The accuracy of the results of the above two-floor structure is badly limited by the orientation profile time step, which is one data point per orbit ($1^{st}$ Floor data compression). Effective direct modeling of the roll-resonance torques and their compensation in the gyro orientation profiles, implemented as a proof of concept, emphasized the existence of torques at the roll frequency corresponding to ~1 min roll period, apparently, much smaller than the duration of one orbit (~97 min). Consequently the current two-floor estimation structure does not fully address the roll-resonance torque. To improve the relativity measurements, it is desirable to replace it with a single-floor estimator/filter, with the data sampling rate much higher than the roll frequency, so that continuous dynamical modeling of all the pertinent features takes place.

We are currently working on the design and implementation of the single-floor "two-second filter" where the SQUID and telescope science signals, sampled at the rate of 2 seconds, are processed according to the readout model (1), scale factor model (2a)–(2b) *simultaneously* with the patch effect torque model (3), (4a), (4b). The state vector includes the relativity parameters, scale factor coefficients, and all the parameters currently estimated at the $1^{st}$ Floor, as well as the torque coefficients, which are so far estimated separately at the $2^{nd}$ Floor. For the first time, S/C pointing will be determined consistently with all other data. During a GSI mode it will be estimated and iteratively updated using the available four SQUID signals, and the estimates of all the needed parameters obtained from the analysis of the previous GSV mode. Information about the GSI pointing is needed for the gyro orientation propagation due to continuously acting torques.

This enhanced two-second filter will provide accurate modeling of the underlying physical phenomena implemented in a flexible, module-based structure. It will permit fast incorporation and evaluation of instrument models, and the check of sensitivity to variations in the source data and parameter sets. The architecture of the two-second filter incorporates the "one system" approach: all the GP-B instruments (gyroscopes, telescope, spacecraft, etc.) should be treated as a single GP-B system. It includes the model identification and verification on the sensor-to-system level (gyroscope, SQUID, telescope, spacecraft, etc.). It also encompasses the development, verification and testing of filtering algorithms based on distributed computing and parallel processing. The enhanced direct torque model toolset brings in important benefits to the overall GP-B data analysis picture and increases confidence in final results by testing for consistency during the analysis.

## 6 Conclusion

A "simple" strategy of the GP-B data analysis has evolved in the complex two-floor structure after on-orbit discoveries of the changes in the rotor's polhode period and path, and of patch effect torques. Direct modeling of the readout scale factor at the 1st Floor, and the misalignment and roll-resonance torque modeling at the 2nd Floor, allowed us to separate the relativistic drift from the drift induced by classical torques. A cascade of four interconnected estimators applied to the GP-B science data has demonstrated a consistent determination of the geodetic and frame-dragging effects for all GP-B gyroscopes, as well as fidelity of the physical models used.

There is a clear way to improve the accuracy of GP-B results (Everitt et al. 2009). Existence of a roll component in the patch effect torque suggests the development of a one-level estimator ("2-second filter"), with a sampling data rate much higher than the roll period. This filter will simultaneously exploit all the models currently used in the estimators of the 1st and 2nd Floors. Though the number of estimated parameters in the new filter will be much larger than the one in the reduced 1st and 2nd Floor filters, the new estimator will be free of the shortcomings associated with the current two-floor data analysis structure.

**Acknowledgements** The authors would like to thank D. Bartlett, B. Clarke, J. Conklin, D. DeBra, F. Everitt, G. Keiser, A. Krechetov, J. Li, I. Mandel, B. Muhlfelder, B. Parkinson, R. Reasenberg, J. Ries, P. Saulson, J. Turneaure, and P. Worden for numerous discussions, insights and support in our data analysis journey.

## References

Y. Bar-Shalom, X.-R. Li, T. Kirubarajan, *Estimation with Applications to Tracking and Navigation* (Wiley, New York, 2001)

G.J. Bierman, *Factorization Methods for Discrete Sequential Estimation* (Dover, New York, 2006)

A.E. Bryson, *Applied Linear Optimal Control: Examples and Algorithms* (Cambridge University Press, New York, 2002)

J.W. Conklin, Estimation of the mass center and dynamics of a spherical test mass for gravitational reference sensors. Ph.D. Thesis, Aero/Astro, Stanford University, Stanford, 2009

C.W.F. Everitt, M. Adams, W. Bencze et al., Class. Quantum Gravity **25**, 114002 (2008)

C.W.F. Everitt, M. Adams, W. Bencze et al., Space Sci. Rev. (2009, this issue)

G.T. Haupt, Development and experimental verification of a nonlinear data reduction algorithm for the Gravity Probe B Relativity Mission. Ph.D. Thesis, Aero/Astro, Stanford University, Stanford, 1996

M.I. Heifetz, G.M. Keiser, A.S. Silbergleit, in *Nonlinear Gravitodynamics*, ed. by R.J. Ruffini, C. Sigismondi (World Scientific, New Jersey, 2003)

T. Kailath, A.H. Sayed, B. Hassibi, *Linear Estimation* (Prentice Hall, Upper Saddle River, 2000)

G.M. Keiser, J. Kolodziejczak, A.S. Silbergleit, Space Sci. Rev. (2009, this issue)

J. Kolodziejczak, Correlation between times of observed gyro orientation changes and high polhode harmonic—roll frequency resonances. Science Advisory Committee to GP-B, Meeting 16, Stanford University, March 23–24, 2007

D.D. McCarthy, *IERS Conventions (1996)* (U.S. Naval Observatory, 1996)

B. Muhlfelder, M. Adams, B. Clarke et al., Space Sci. Rev. (2009, this issue)

M. Salomon, Properties of gravity Probe B gyroscopes obtained from high frequency SQUID signal. Ph.D. Thesis, Aero/Astro, Stanford University, Stanford, 2008

P.K. Seidelmann, *Explanatory Supplement to the Astronomical Almanac* (University Science Books, Mill Valley, 1992)

A.S. Silbergleit, On the linearity of the gravity Probe B measurement model. Gravity Probe B document S0632, Rev. C, Stanford University, Stanford, 2006
A.S. Silbergleit, GP-B science signal models and their analysis. Gravity Probe B document S0996, Stanford University, Stanford, 2009
A.S. Silbergleit, M.I. Heifetz, A.S. Krechetov, Model of starlight deflection and parallax for the GP-B data reduction. Gravity Probe B document S0393, Stanford University, Stanford, 2001
A.S. Silbergleit, J. Conklin, D. DeBra et al., Space Sci. Rev. (2009, this issue)
J.P. Turneaure, C.W.F. Everitt, B.W. Parkinson et al., Adv. Space Res. **32**(7), 1387 (2003)

Space Sci Rev (2009) 148: 429–439
DOI 10.1007/s11214-009-9523-8

# GP-B Systematic Error Determination

**B. Muhlfelder · M. Adams · B. Clarke · G.M. Keiser ·
J. Kolodziejczak · J. Li · J.M. Lockhart · P. Worden**

Received: 8 April 2009 / Accepted: 18 April 2009 / Published online: 16 May 2009
© Springer Science+Business Media B.V. 2009

**Abstract** We have evaluated the systematic error in the GP-B experiment using five different approaches and estimated the individual contributions of many error sources. The systematic effects we consider include those due to gyroscope torques, gyroscope readout, telescope readout, and guide star proper motion. Effects with an estimated impact on the experiment error larger than 1 mas/yr are discussed in detail. Examples of analyses that bound other sources to less than 1 mas/yr are included to show the range of techniques employed to perform this work. We describe the remaining tasks to complete the systematic error analysis and estimate the total experiment uncertainty.

**Keywords** Gravity probe B · General relativity · Systematic error

**PACS** 04 · 04.20.-q · 07.05.Tp · 07.87.+v · 45.40.Cc

## 1 Introduction

The bounding of systematic error or uncertainty is an essential part of every high accuracy physics experiment. An early discussion of GP-B systematic uncertainty (Everitt et al. 1980) gave a description of many ground-based tests and analyses, and provided a theoretical framework for the experiment, thus allowing an early quantification of the GP-B systematic uncertainty.

The GP-B experiment was designed to provide an accurate measurement of non-Newtonian (general relativistic) gyroscope precession without a need to model systematic

B. Muhlfelder (✉) · M. Adams · B. Clarke · G.M. Keiser
Gravity Probe B, HEPL, Stanford Univ., Stanford, CA 94305-4085, USA
e-mail: barry@relgyro.stanford.edu

J. Kolodziejczak · J. Li
NASA/Marshall Space Flight Center, Huntsville, AL 35812, USA

J.M. Lockhart · P. Worden
Physics and Astronomy Dept., San Francisco State Univ., San Francisco, CA 94132, USA

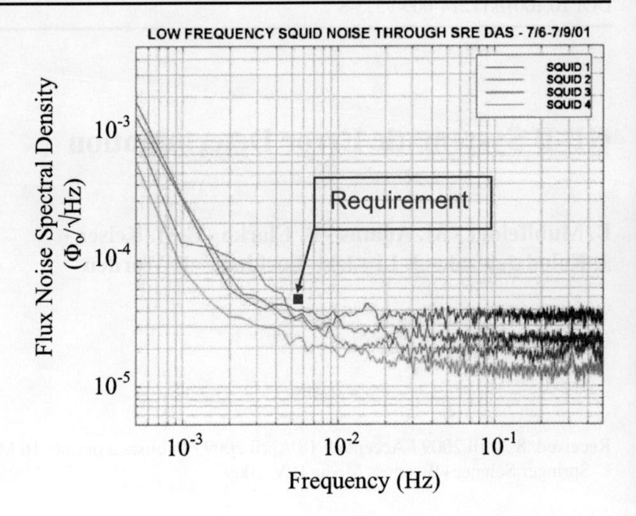


**Fig. 1** Low frequency noise of the four SQUID readout systems. $\Phi_0$ is the magnetic flux quantum

effects. A detailed error tree, containing both stochastic and systematic sources for every gyro, allowed trade studies to be performed for various design parameters. Prior to launch in April 2004, the experiment uncertainty was expected to be limited by stochastic noise in the SQUID-based London Moment gyroscope readout system. Based upon test and analysis techniques, the expectation was that the impact of systematics on the overall experiment accuracy would be small, and that without modeling systematic effects, the total experiment uncertainty would be less than 0.5 mas/yr.

During the first two weeks of on-orbit GP-B operations, we measured the SQUID noise. The results are shown in Fig. 1. Using a systematics-free measurement model, the experiment uncertainty based upon this noise level is ~0.2 mas/yr for each gyroscope, or ~0.1 mas/yr when combining the results from all 4 gyroscopes. Although we expected that there would be little impact from systematic effects, after measuring the SQUID noise, we began an exhaustive set of calibrations designed to test the pre-launch expectation.

The on-orbit mission plan comprised three phases: (1) Initialization, (2) Science, and (3) Calibration. These phases were designed to perform three objectives: configure the hardware for science during Initialization, collect the science data during Science, and calibrate systematic effects during all phases. Keiser (2009) has described the numerous SQUID and telescope readout operations, acquisition of the guide star, and gyroscope suspension and spin up. All science hardware, including each of the four gyroscopes and their readout systems, the telescope and its eight detector channels, and many instrumentation channels functioned successfully. Initialization of the experiment was completed in August 2004. By comparison, few operations were required during the Science phase. The goal of this phase was to collect a year of science data with no changes to the configuration of the science instrument and minimum interference of any kind. More than 1 Terabyte of science data was collected prior to the completion of the science phase on August 15, 2005. The final on-orbit operations were performed during the 6½ week long Calibration phase.

The full set of GP-B calibrations probes more than 100 physical sources including guide star proper motion, telescope readout, and gyroscope readout and torques. The uncertainty in the proper motion is known from published sources to better than ~1 mas/yr. To date, all GP-B analyses have been performed with information available from these sources. Recent measurements (Shapiro et al. 2007) will provide even tighter limits on proper motion uncertainty, however, by agreement, the actual proper motion values have been withheld from the

GP-B data analysis team to allow for future comparison and incorporation. In this paper we focus on readout issues and gyroscope torques discovered in the course of calibrations and data analysis. Both of these effects are much larger in magnitude than the known guide star proper motion uncertainty.

Analyses of data collected on-orbit demonstrate that all but three GP-B systematic effects are small, namely an evolving scale factor due to damped gyroscope polhode motion and two related Newtonian gyroscope torques. In accompanying papers, Silbergleit et al. (2009) describe a technique to model the effect of the gyroscope's changing polhode on the gyroscope scale factor and Keiser et al. (2009) describe the Newtonian torques generated by the interaction of patch potentials on the rotor and housing surfaces. Assuming only gyroscope spin and roll averaging, two distinct manifestations of these patches are found: (1) Misalignment torques that are perpendicular and proportional to the gyroscope misalignment and (2) Resonance torques that occur at the difference frequency of harmonics of polhode frequency and the spacecraft roll frequency. Left unmodeled, these effects would cause experiment error up to 1 arcsec/yr. Contrary to the initial plan, we have had to model these three effects in the data analysis to achieve the best experiment accuracy. These modeling efforts, begun in 2004 and continuing to the present, have resulted in a dramatically reduced systematic error. In the discussion here, we focus on the residual uncertainty associated with imperfect modeling of these effects. Before describing these and other smaller effects, we give the methodology for evaluating the total systematic experiment uncertainty.

## 2 Approaches to Estimating Experiment Uncertainty

The study of individual systematic effects involves three steps: identification, modeling (if required), and bounding. Identification connects a potential physical mechanism disturbing the measurement to the experiment. Modeling is only required for large effects, specifically those associated with patch potentials. Physics-based models governing the patch effect are incorporated into the science data analysis to separate the relativistic drift from that induced by patch effect torques. The residual experiment uncertainty associated with these effects added to the inherent experiment uncertainty associated with small effects ($<1$ mas/yr) bounds the total experiment uncertainty.

Five approaches are used to bound the total GP-B experiment uncertainty:

A. Bottom-up analyses
B. Sensitivity tests
C. Gyro-to-gyro comparisons
D. Data sampling tests
E. Comparison of results of independent analysis methods.

### 2.1 Bottom-up Analyses

These analyses give the uncertainty associated with individual sources and are based upon physical models of the hardware with parameter values determined from ground and/or flight measurement. For instance, a bottom-up analysis gives the gyroscope precession rate resulting from the support force acting on the gyroscope mass unbalance vector. The physical model here is the mass unbalance torque causing a change in the angular momentum of the spinning mass. The parameters for this model are the support force (measured on-orbit), the mass unbalance (measured on the ground and refined on-orbit), the angular velocity (measured on-orbit), and the gyroscope moment of inertia. Using these known parameter values gives a Newtonian drift rate below 0.08 mas/yr.

### 2.2 Sensitivity Tests

The assessment of the impact of an individual systematic effect on experiment uncertainty is determined by using the science analysis algorithms to determine the sensitivity to a particular effect. One sensitivity test method propagates the numerical uncertainty in a system parameter through the modeling software. The resulting variation in the science results gives the experiment uncertainty due to this parameter.

Example: Timing error. The gyroscope orientation evolves in time, and therefore our analysis is susceptible to timing error. To assess the impact of timing error, we find the science result using the time stamp provided by the onboard clock for each data point. The uncertainty in our knowledge of the timing is determined by independent analysis of on-orbit timing data. The time stamp for each data point is modified by this uncertainty and the analysis is rerun to give a new science result. The change in the science result gives the timing systematic uncertainty. This analysis bounds the impact of the timing error on the relativity determination to less than 0.5 mas/yr.

### 2.3 Gyro-to-Gyro Variation

The range of relativistic drift rate estimates for the four gyroscopes gives a measure of the experiment uncertainty due to both stochastic and systematic sources, although common-mode systematic effects are not reflected. The dominant stochastic error results from noise in each gyro readout system. This noise directly leads to gyro-to-gyro differences. Additional gyro-to-gyro differences arise from gyro-specific systematic uncertainty associated with deficiencies in modeling gyro readout and gyro torque.

### 2.4 Data Selection Tests

The science result should be independent of which data are selected for analysis. For example, data analyzed from different times of the year should yield statistically equivalent science results. Variation in the experiment result based on the analysis of various combinations of the total data set provides a measure of the experiment uncertainty.

### 2.5 Two Independent Analysis Methods and Teams

Two GP-B teams use independent modeling approaches to perform the science analysis. The most significant difference in their methodologies is in the treatment of the misalignment torque. One team takes a torque-free projection of the gyroscope motion whereas the other explicitly models this effect. The two teams also model the resonance torque and the evolving gyroscope scale factor with differing techniques described below. Agreement in science results using these two modeling approaches increases our confidence in the science result.

## 3 Initial Models of Scale Factor, Torques, and the Associated Early Science Results

The modeling of the evolving gyroscope scale factor and the patch effect torques has dramatically improved during the past 4½ years. Our progress has resulted from insights into the underlying physics and from key advances in essential supporting analyses.

Prior to launch, we had expected the scale factor to be periodic at polhode frequency and patch effect torques to be negligible. Early in the Science phase, it became clear that the gyroscope polhode path and frequency were changing with time due to polhode damping, thereby resulting in a slowly evolving gyroscope scale factor. Initially, we modeled this evolving scale factor as a Fourier expansion of polhode harmonics. The year-long data set was divided into multi-day-long batches with a set of Fourier coefficients determined for each batch. Although over the past four years we have developed significantly more sophisticated techniques to estimate the gyro scale factor, this first method was used during the Science and Calibration phases of the mission to track the orientation of the gyroscopes. The gyroscope orientation information was essential to the discovery of the patch effect torques. This connection of one modeling advance leading to another has occurred repeatedly in our work and underlies our progress over the past several years.

The first manifestation of patch effect torque was identified during the calibration phase of the mission. The geometrical nature of this misalignment torque was established through a series of tests in which the space vehicle was pointed to neighboring stars, thereby dramatically increasing the misalignment of the gyroscope spin axis relative to the spacecraft roll axis (Keiser et al. 2009). We model this torque either explicitly by including a torque-induced drift term perpendicular to the measured misalignment or implicitly by taking a torque-free projection of the gyroscope drift parallel to the misalignment.

The resonance torque was identified after completing the science phase. It was observed that on resonance, i.e. when a harmonic of polhode frequency coincided with the spacecraft roll frequency, the gyroscope orientation would often shift. In some cases this shift, typically occurring in one or two days, was as large as 0.15 arcsec. The original method to model this resonance behavior was to excise the data near the shift. Although the underlying physical explanation for these shifts was unknown, this approach significantly improved the fit of the remaining data to the model.

In April of 2007, using the above techniques, we reported an experiment uncertainty of 97 mas/yr. This result was obtained by determining the sensitivity of the science results to scale factor and torque modeling deficiencies. Scale factor modeling deficiencies were bounded by varying the number of polhode harmonic terms included in the model and by changing the batch length used to determine these terms. Misalignment torque error due to misalignment uncertainty was derived from two independent determinations of the misalignment profiles. Figure 2 shows the result of these two analyses for gyroscope 1, plotted here as gyroscope orientation (the orientation is given by vehicle pointing minus gyro misalign-

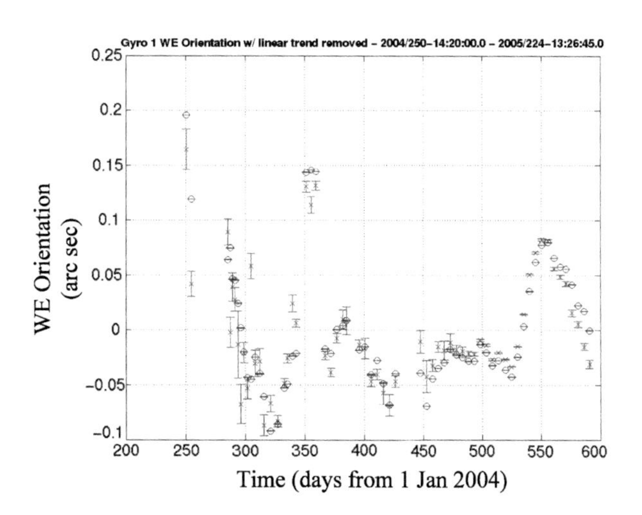


**Fig. 2** Gyroscope 1 WE orientation are shown using two different analyses, performed in early 2007. Differences give a measure of the uncertainty

ment). The difference in results gives a misalignment uncertainty of 30 mas. This uncertainty was propagated through the science analysis yielding only a small change (4 mas/yr) shift to the science result, thereby suggesting some other source for the leading cause of uncertainty.

## 4 Model Improvements

Although we lacked a detailed understanding of the modeling deficiency, we found that the largest contribution to experiment uncertainty in 2007 was caused by resonance torque mis-modeling. In those analyses, data was excised near resonances, governed by the phase difference between a harmonic of the changing polhode and the steady roll signal. We defined this difference to be zero on resonance and excised data near resonance when the phase difference was less than a specified value. Sensitivity tests, performed by varying this value, led to more than 50 mas/yr variation in the science result. A subsequent method using linear ramps to model the NS and WE orientations during resonances demonstrated approximately three times lower sensitivity to the resonance torque than the excising method. A study of flight data however shows subtle oscillations in the orientation profiles before and after each resonance (Keiser et al. 2009). This is contrary an implicit assumption made by each of the methods above, that the resonance torque vanishes before and after each resonance. Our most recent efforts make explicit use of a physics-based patch effect model that predicts the observed subtle gyroscope behavior. The similarity between the data and this patch effect model increases our confidence in our understanding of the underlying physics. In addition, this model improves the numerical fit of the model to the data, thereby reducing the residuals. Initial science results making use of the improved model are encouraging (Heifetz et al. 2009).

Use of the full resonance torque model requires accurate knowledge of the polhode phase, only recently available. Given that the observed resonances occurred for high harmonics of the polhode frequency, and that the uncertainty in the polhode phase scales with harmonic number, the need for accurate phase information is clear. Trapped Flux Mapping (TFM) now gives an accurate determination of polhode phase (Silbergleit et al. 2009) by fitting the gyroscope's body fixed spin signal to a continuous polhode path. The uncertainty in these phase estimates has improved dramatically over the past year with a current uncertainty of $< 0.5°$. Sensitivity tests are in progress to determine the impact of this uncertainty on the overall experiment uncertainty.

## 5 Sensitivity Tests

The resonance torque modeling improvements described above were motivated by large science result sensitivity. These recent improvements suggest that we reexamine the other modeling sensitivities to determine an updated total uncertainty. Table 1 gives a list of the most important tests. Modeling the 25 day orbital period of the GP-B binary guide star, IM Pegasi, caused less than a 5 mas/yr change to the science result. The resulting orbital parameter estimates will be published once the guide star results from the SAO are released. Readout scale factor induced experiment uncertainty has been probed by comparing the science results obtained using the harmonic expansion method described above with the results using two other techniques. The first of these, the $\gamma C_g$ model (Heifetz et al. 2009), uses the low frequency gyroscope signal to determine a body fixed representation of the trapped field contribution to the scale factor. The other, TFM, also determines a body fixed representation

**Table 1** Sensitivity tests and the associated range of parameter values

| Sensitivity parameter | Parameter range |
|---|---|
| Guide star orbital motion | 0 to fitted |
| Resonance torque | Continuous fit vs. fixed width (0.75, 1.5, 2) |
| Scale factor modulation | Harmonic expansion, $\gamma C_g$, TFM |
| Polhode phase | $-0.5°$ to $0.5°$ |
| Misalignment torque averaging window | 15–25 orbits |
| Misalignment torque uncertainty | $-0.1$ to 0.1 radians misalignment phase |

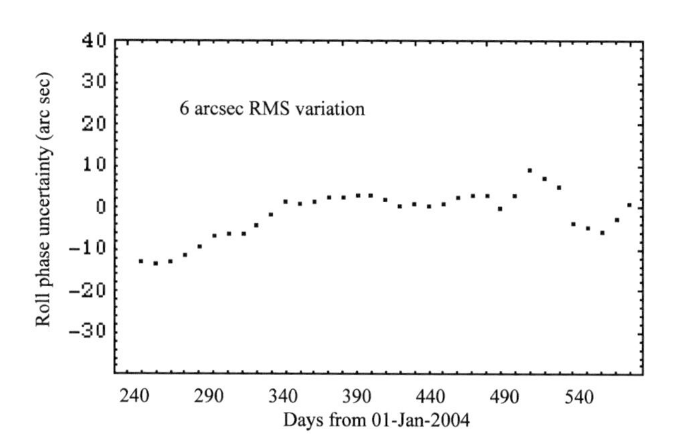


**Fig. 3** Spacecraft roll phase uncertainty vs. time

of the trapped field contribution to the scale factor, but it does so using the high frequency gyroscope signal. Use of the TFM results also simplifies the science analysis from a nonlinear to a linear fit and thus provides a check of sophisticated non-linear estimation techniques. Finally, tests were performed to probe the sensitivity of the science result to misalignment and resonance torque modeling.

A preliminary estimate of the uncertainty in science result for gyroscope 1, based upon all of these sensitivity analyses gives a preliminary uncertainty of 30 mas/yr. Gyro-to-gyro comparisons by Heifetz et al. (2009) suggest a still lower bound.

Following launch, analysis of the flight data has confirmed many potential systematic effects are less than 1 mas/yr. Currently, we do not model these effects, but we may need to in the future as the total experiment uncertainty decreases.

Rather than provide a long list of bounded systematic effects, it is instructive to describe several representative analyses.

## 5.1 Roll Phase Uncertainty Coupling to Science Result

Roll phase uncertainty causes the science signal in the NS direction to be rotated into the WE direction and the science signal in the WE direction to be rotated into the NS direction. A 10 arcsec roll phase uncertainty results in a 1 mas/yr uncertainty in the science result. Figure 3 shows the roll phase uncertainty for the duration of the mission. The roll phase and roll phase uncertainty are determined using a primary and backup CCD-based roll star telescopes. As the spacecraft rolls about the line of sight to the guide star, other stars 30–40 degrees from the roll axis are tracked as they rotate through each of the CCD's field of view.

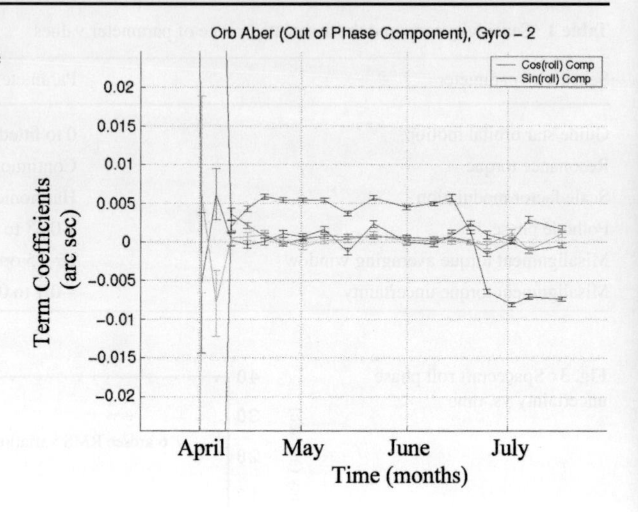


**Fig. 4** Out-of-phase orbital aberration term estimates. Cosine and sine terms of out-of-phase orbital aberration. 2× orbital not modeled: *Black/Red*. 2× orbital modeled: *Blue/Green*

The uncertainty in the roll phase is found by comparing the primary and backup roll phase measurements.

## 5.2 Gyroscope Position Coupling to the Science Result

The gyroscope readout signal will be modified by the gyroscope's position change, if the position couples to the readout signal. We know that a gravity gradient force acts on the gyroscope's position at roll +/− twice orbit frequency, and therefore, if the aforementioned coupling exists, there would be a readout signal at roll +/− twice orbit. This readout signal would mix with and modify the signal at roll +/− orbit. The mixing is due to the use of only the guide-star-visible portion of each aberration cycle to determine the gyroscope scale factor. Figure 4 shows the results of two analyses for gyroscope 2. The signal that is out-of-phase with orbital aberration should be very small. Confirming a small out-of-phase term increases confidence in the scale factor as determined from the in-phase term. When the analysis is done without including the roll +/− twice orbit term, a signal of 5 mas amplitude is observed out-of-phase at roll +/− orbit. The switch of the sign in the middle of June occurred when the drag free sensor was switched from gyroscope 1 to gyroscope 3. When the model includes a roll +/− twice orbit term, the estimated out-of-phase signal nearly vanishes. This analysis allows us to place a 1 mas/yr limit on the experiment uncertainty associated with the coupling of the position of gyroscope 2 to its readout system. Analyses for the other gyroscopes demonstrated even weaker coupling than that found for gyroscope 2.

## 5.3 Electromagnetic Interference (EMI)

The gyroscope readout system is based on the detection of magnetic fields of $10^{-16}$ T near the satellite roll frequency. Extreme care was taken in the design and operation of the science instrument to limit spurious magnetic signals. Although these efforts were largely successful, one extraneous signal source, the Experiment Control Unit (ECU) was identified during science data taking. Figure 5 shows approximately 1 minute of the science signal for gyroscope 1. The raw time series data profile is shown in black; the processed data profile with the ECU signal removed is shown in red. As is the case for more than 90% of the science mission, and as is seen in the figure, the science signal at 12.9 mHz is far removed from

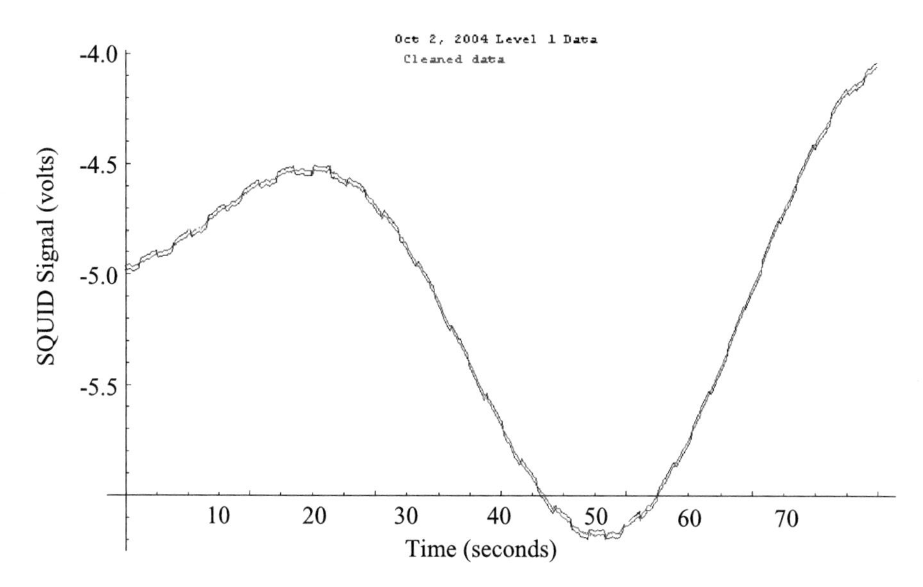


**Fig. 5** Gyroscope 1 low frequency science data

the ECU signal frequency. The readout data at these times are used in the science analysis. When the ECU signal frequency is near the science signal frequency, we excise the data. Sensitivity tests show negligible variation in the science result for varied amounts of excised data. To facilitate these and other tests the science database incorporates a data-grading tool that allows consistent, convenient, unbiased testing.

### 5.4 SQUID Non-linearity

One area of on-going investigation is SQUID readout non-linearity. Thus far, analyses show evidence for such a non-linearity although we do not yet have sufficient understanding to incorporate a specific model into the science analysis. Prior to launch the SQUID was specified to have a non-linearity of $<-76$ dBFS. Although this specification was successfully confirmed, tests on the ground and later on-orbit revealed subtle behavior in the digitizing hardware of the SQUID's electronics. A misconfiguration of the SQUID's A/D converter caused some of the $2^{16}$ digital output levels in the 16 bit A/D converter to be preferred over adjacent levels. This preferential-levels behavior may result in system non-linearity. A recent analysis of on-orbit data highlighting possible non-linear behavior is shown in Fig. 6.

Shifts in the apparent misalignment occur when the readout calibration signal was toggled on/off and they occur in the radial direction relative to zero misalignment. These shifts in gyroscope misalignment are suggestive of offsets in the dc gyroscope scale factor and are likely related to non-linear behavior in the gyroscope readout system.

## 6 Torques Other than Those Due to Patch Effect Potentials

Gyroscope torques cause Newtonian gyroscope drift. There are two approaches to limit the impact of this drift on the experiment error. One approach that has been used with patch

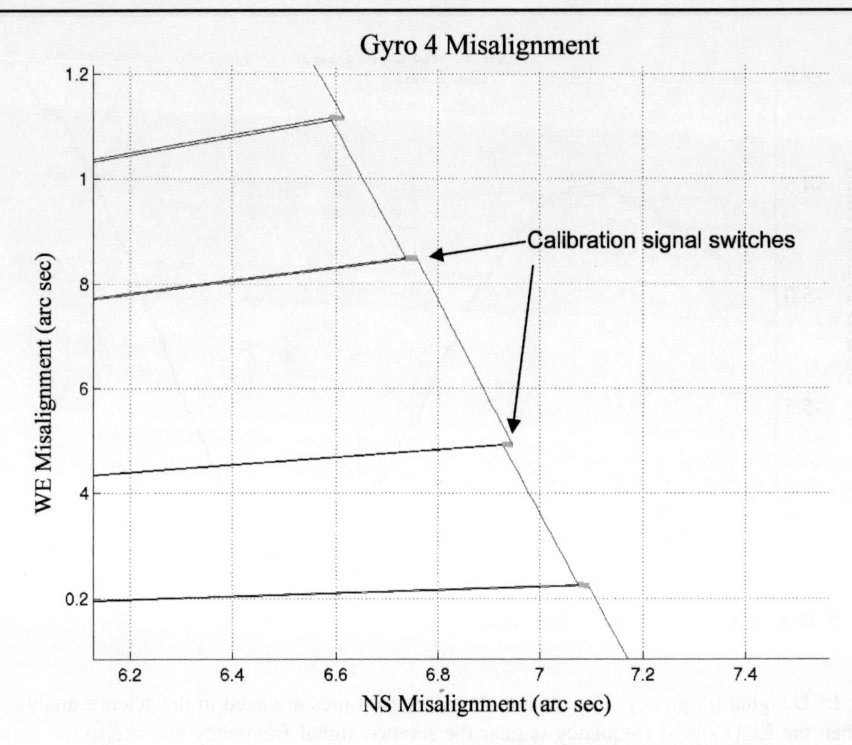


**Fig. 6** Gyroscope 4 misalignment June–July, 2005. Apparent shifts in gyroscope misalignment correspond to times when the gyroscope readout calibration signal was toggled on/off

potential torques is to model the underlying physics. Sensitivity tests and gyro-to-gyro comparisons provide a measure of residual uncertainty. A bottom-up approach has been used for other Newtonian torques including support dependent and support independent effects. Sources for these torques include gyroscope physical contamination, magnetic contamination, differential gas pressure damping, and gyroscope mass unbalance and gyroscope geometry imperfections. We have performed a preliminary analysis and find that the Newtonian drift associated with these torques is less than 1 mas/yr.

## 7 Summary

There has been significant progress in bounding the total GP-B systematic uncertainty. Many effects have been bounded to less than 0.1 mas/yr to 1 mas/yr. We have described five approaches to determining the total uncertainty. The most recent models yield reduced scatter in sensitivity test results and provide very encouraging gyro-to-gyro agreement in the relativity rate estimates: however, more work remains. We need to extend the analysis of support dependent and support independent torques to include the effects of dissipation on the experiment uncertainty (Buchman et al. 2009). Completion of these analyses will provide the underpinnings for a report detailing the contributions to the uncertainty from more than 100 sources. We continue to improve the science analysis, refining the modeling of the physics governing the behavior of the GP-B hardware. A large effort, now in the early stages of implementation, will track the gyroscope dynamics using a 2 second time step rather than the current once per orbit time step. It is likely that reduced uncertainty due to systematic effects will be a vital by-product. Sensitivity tests and gyro-to-gyro comparison of the science results using these refined models will provide the final elements in determining the total experiment uncertainty.

**Acknowledgements** The authors thank S. Buchman, J. Conklin, C.W.F. Everitt, M. Heifetz, D. Hipkins, Y. Ohshima, A. Silbergleit, and J. Turneaure for their useful discussions and insights into this work.

## References

Buchman et al., Rev. Sci. Instrum (2009, to be submitted)
C.W.F. Everitt et al., Report on a program to develop a gyro test of general relativity in a satellite and associated control technology, 1980
M. Heifetz et al., Space Sci. Rev. (2009, this issue)
G.M. Keiser, Gravity Probe B, in *Proceedings of the International School of Physics 'Enrico Fermi', Course CLXVII—Atom Optics and Space Physics* (2009, to be published)
G.M. Keiser et al., Space Sci. Rev. (2009, this issue)
I. Shapiro et al., APS Conference, Jacksonville, FL, 2007
A. Silbergleit et al., Space Sci. Rev. (2009, this issue)

Space Sci Rev (2009) 148: 441–454
DOI 10.1007/s11214-009-9521-x

# Space-Time Metrology for the LISA Gravitational Wave Observatory, and Its Demonstration on LISA Pathfinder

**Stefano Vitale**

Received: 25 April 2009 / Accepted: 30 April 2009 / Published online: 14 May 2009
© Springer Science+Business Media B.V. 2009

**Abstract** The paper describes the key instrumental features of LISA, and discusses its limiting sources of uncertainty. It is shown how the basic element of LISA is a "Doppler link" where the frequency of a laser beam is compared by two far apart, free falling "observers". These observers are two of the six LISA free falling test-masses. Each of the three LISA arms carries two such Doppler links, propagating along opposite directions. It is shown that a link like this acts as a time-delayed, differential accelerometer. Its performance is limited by three main sources of disturbance. Parasitic forces accelerate the test masses relative to their geodetic trajectories. Interferometry measurement noise is interpreted as relative motion of test-masses. Finally, because of intervening optical components fixed to spacecraft, the measurement is also corrupted by the relative acceleration of spacecraft and the test mass. We analyze these sources within a unifying picture, and show that LISA has a large "science margin". Even with a sensitivity substantially poorer of what has been calculated, LISA remains a powerful astrophysics observatory.

**Keywords** Gravitational wave astronomy · Space-time metrology

## 1 Introduction

LISA is a space-borne Gravitational Wave (GW) observatory currently under formulation as a joint effort between ESA and NASA (Bender et al. 1998). Contrary to the ground based observatories developed so far, it is designed to be a signal dominated instrument aiming at a wealth of bright GW sources (Prince et al. 2009). Some of these sources, like binaries made out of compact stars in our galaxy, are entirely known. Their detection, with signal to noise ratio in excess of 100 in one year of integration, is then "guaranteed". For similar sources, the emission of GW has indirectly been observed from the acceleration of their

S. Vitale (✉)
Department of Physics, University of Trento, and Istituto Nazionale di Fisica Nucleare, 38100, Povo, Trento, Italy
e-mail: Stefano.Vitale@unitn.it

orbital period, due to the energy loss by GW emission, while they spiral in (Weisberg and Taylor 2005).

Even more powerful is the emission expected from binaries of super-massive black-holes formed during galaxy mergers. These sources may reach signal to noise ratios >100 at $z = 15$. Their detection will open the possibility both for gravitational cosmology, as they allow precise ranging and location of the source, and for tracing back the merger history of black-holes and galaxies throughout the universe.

Less intense but quite abundant are the sources constituted by the in-spiraling of a star-size compact object into a super-massive black-hole. For more that 10,000 orbits the system emits waves that will give LISA a signal to noise ratio >10. These waves will give in fact a map of the gravitation field at the event horizon. This promises to be a quite unique General Relativity laboratory allowing, for instance, the verification of the so called "no-hair" theorem for black-holes.

Discussing the science return of LISA is outside the scope of this paper. However the discussion above is aimed at making clear that LISA is not a single experiment in gravity, but rather an astrophysical observatory using gravitational radiation as a mean for observing the universe. As in all space-borne observatories, the performance requirements for LISA are not set as a yes or no threshold. On the contrary they contain substantial margin to allow for a reasonable graceful degradation of the science return in case of partial failures or underperformance.

## 2 LISA Measurement Scheme

The LISA instrument has been described in many papers and articles (Bender et al. 1998; Vitale et al. 2002; Jennrich 2009). We summarize here the basic features useful for the following discussion.

LISA consists of a constellation of 3 spacecraft (SC) in heliocentric orbit. Orbits are adjusted such that the three SC keep an equilateral triangle formation with a $5 \times 10^6$ km side without the need of maneuvers. Each SC contains a pair of cubic, metallic, 2 kg, test-masses (TM) nominally in pure geodesic motion. Each TM is the end point of a single arm interferometer, the other end point being in one of the other two SC. Thus the constellation is formed by three interferometer arms arranged as the sides of an equilateral triangle. Each interferometer arm measures then the relative displacement (of the center of mass) of two free-falling TM, $5 \times 10^9$ m apart. As in all gravitational wave interferometric detectors, the gravitational signal appears as a tide-shaped oscillation of these displacements (see next section).

Nominally two adjoining arms constitute one Michelson interferometer, though with a 60° opening between the arms instead of the usual 90°. However there are a few remarkable differences between LISA and a standard Michelson interferometer.

1. An interferometer based on true reflection of the light beam at the end mirrors, would lose all power due to the widening of the beam over $5 \times 10^9$ m. To avoid this, in LISA two independent beams travel each way within one arm (Fig. 1).

   The phase relation between these beams is measured locally and the two beams are then phase-locked. This laser transponder synthesizes in fact one complete interferometric arm.
2. Each TM-to-TM interferometer arm is in reality split in three distinct measurements (Fig. 1). In each of the two SC, a local interferometer measures the relative displacement of the TM relative to the SC, nominally along the direction joining the two SC.

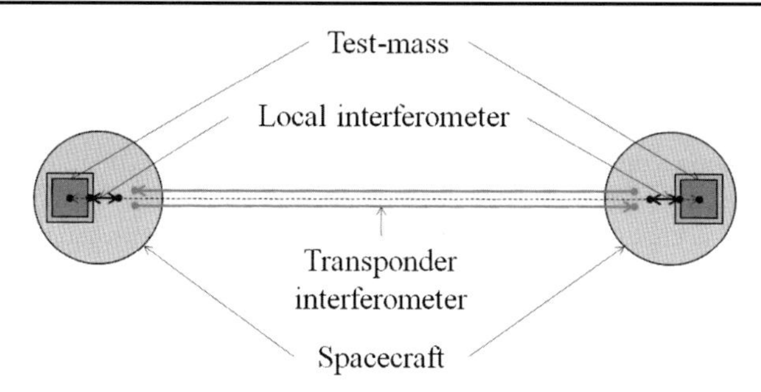


**Fig. 1** The measurement scheme within each of the three LISA arms. Two local interferometers measure the displacement between each TM and the SC in which the TM is hosted. One transponder interferometer, consisting of two phase locked counter propagating laser beams, measures the relative displacement of the two SC. The measurement objective is the displacement between the centers of mass of the TM (*black dots* joined by the *dashed arrow*). The measurement is built up as the sum of the three interferometric measurements above. The figures also shows that each of the various interferometric measurements takes place in between two fiducial points. In the ideal measurement, all these fiducial points need to be aligned along the *dashed line*

3. In addition, the transponder interferometer, mentioned in the previous paragraph, measures the relative displacement between the two SC. The overall TM to TM displacement is then obtained as the sum of these three measurements.

   The four bodies entering in one arm, two SC and two TM, are extended bodies. Thus each interferometer measures the relative displacement between two point-like small areas, of some millimeter width, each located on one of the two bodies, the relative displacement of which the interferometer is supposed to measure.

   The measurement of the relative displacement of the center of mass of the two TM, is only accurate then, to the extent that these six points all lie along the line joining the two centers of mass. Any misalignment folds in the large displacements due to the relative rotations of SC and TM.
4. In an ordinary Michelson interferometer, the frequency noise of the laser is coherent in both arms, and does not affect then the measurement of the difference of phase or frequency between the arms, the physical quantity that carries the gravitational wave signal. This cancellation is only true if the light beams converging onto the interferometer vertex from the two different arms, have been sent out at the same time. For this, the arms must have same length, the noise suppression being proportional to $\delta L/L$, where $\delta L$ is the unbalance of the arm-length $L$. In LISA the arms can only be made equal to within some $10^4$ km. Thus the light beams collected from two arms at one SC, have been generated at times that can be shifted by up to a fraction of a second. The resulting noise suppression is too poor, if any. Thus one requires combining and time-shifting, within data post-processing, the signals coming from different arms, in order to reconstruct the lost simultaneity.

In order to get the TM in pure geodesic motion, but still placed inside the SC at the right location, 12 different dynamical controls minimize the motion of the TM relative to the SC. No force can be exerted on the TM along the directions of their respective gravitational measurement, that is, along the directions of the laser interferometer arm they belong too. Along these two directions, the displacement of the SC relative to the TM is measured by the local interferometer and kept constant by acting on the SC via a set of $\mu N$ force thrusters. Thus effectively, along these directions the SC follows the two TM. If the interferometer arms were at 90°, one could say that the SC follows one TM in one direction and the other along a second direction normal to the first one. With a 60° interferometer, affine transformations guarantee that this two-directions, two-TM scheme is still feasible.

The motion along a direction normal to the first two can still be controlled by steering the SC to follow one of the two TM. For all other 9 degrees of freedom, the motion of each TM relative to the SC is kept nominally at zero by exerting proper electrostatic forces or torques on the relevant TM. This is obtained by applying proper voltages on a set of electrodes surrounding the TM. These electrodes are also used to sense the motion of the TM from the variation of their capacitance to it. Capacitance variation are detected with an ac-voltage drive, capacitive bridge.

The above set of electrodes, and their supporting structure, form a quasi-closed housing for the TM (the electrode housing from now on) that leave gaps from 3 to 4 mm around it, depending on direction.

The TM has no mechanical contact to the enclosing housing. To keep it electrically neutral, despite the steady flow of cosmic rays, a flux of electrons compensates the flux of, mostly, protons generated by cosmic rays. Electrons are obtained by proper UV illumination of the electrode housing and the TM. The system works in closed loop, with the TM charge being continuously measured by applying voltages to a specific set of electrodes, and by measuring the resulting torque on the charged TM. This charge signal is used to regulate the relative illumination of the electrodes and of the TM to keep the charge nominally at zero.

## 3 The Basic Element: The Doppler Link

Neglecting the details, each of the LISA arm consists of two counter-propagating laser beams. Each beam conceptually propagates form one test-mass to the other. This arrangement may be represented with the textbook scheme of a particle, the emitter, emitting an electromagnetic wave, and a second one, the receiver, collecting it. In LISA these two particles are the TM. Both at the emitter and at the receiver, the frequency[1] of the light is measured in their relative rest frame. The difference of these two measurements is given by (Synge 1964; Chauvineau et al. 2005):

$$\Delta\nu \equiv \nu_r - \nu_e = k_\mu(t_e)v_e^\mu(t_e) - k_\mu(t_r)v_r^\mu(t_r) \tag{1}$$

Here $k^\mu(t_e)$ is the wave four-vector of the electromagnetic beam at the event of emission at time $t_e$. Same definition applies to the wave vector $k^\mu(t_r)$ at the event of reception at time $t_r$. Finally $v_e^\mu(t_e)$ and $v_r^\mu(t_r)$ are the four velocities of the emitter and receiver respectively, evaluated at the appropriate times.

As scalars are not affected by parallel transport, and as the wave vector is parallel transported along the beam, one can re-write (1) as:

$$\Delta\nu = k_\mu(t_r)\left[v_e^\mu(\{x_e^\alpha\} \underset{\text{parallel}}{\rightarrow} \{x_r^\alpha\}) - v_r^\mu(t_r)\right] \tag{2}$$

Here $v_e^\mu(\{x_e^\alpha\} \underset{\text{parallel}}{\rightarrow} \{x_r^\alpha\})$ stands for the emitter four-velocity, parallel-transported along the beam from the event of emission to that of reception. $\{x_e^\alpha\}$ are the four-coordinates of the

[1] Usually interferometers measure the phase of the beam relative to a local reference. However for all the signals we deal with here, both phase and frequency are time varying and bring the same physical information.

event of emission, with $x_e^0 = ct_e$ then, and $\{x_r^\alpha\}$ are the four-coordinates of the event of reception.

In the case of gravitational waves, one can safely assume that the metric deviation $h_{\mu,\nu}$ from the underlying flat space-time metric, is so small, that only linear terms matter. With this prescription, and for low velocities, the effect of parallel transport can be treated as a perturbation, and (2) becomes:

$$\begin{aligned}\Delta\nu(t_r) &= k_\mu(t_r)[v_e^\mu(t_r - L/c) - v_r^\mu(t_r)] \\ &\quad + \frac{1}{2}\nu_o \cos^2(\theta/2)\{h_+[t_r] - h_+[t_r - (L/c)(1-\cos(\theta))]\}\end{aligned} \tag{3}$$

Here $\nu_o$ is the unperturbed frequency of the beam. The gravitational wave is propagating along a direction at an angle $\theta$ from the direction of propagation of the light. The projection of the direction of propagation of the light, onto a plane normal to that of propagation of the gravitational wave, is taken as one of the axes to calculate the polarization of the gravitational wave. With this prescription $h_+[t]$ becomes the amplitude of the wave with "+" polarization.

The four-velocity vector of the emitter appearing in (3) is evaluated at the time $t_r - L/c$, with $L$ the ordinary space distance between the particles and requires no additional parallel transport.

If the underlying metric is static but not flat, as is the case for the solar system, the emitter four velocity must still be parallel transported in the statically curved space. This part of the parallel transport produces all static gravitational effects, like gravitational shift or Shapiro delay. Still there are no other first order effect depending on the GW, except for the one already appearing in (3).

Equation (3) has a very clear meaning: the gravitational wave signal superimposes to the classical Doppler effect in flat space time or within the static metric. Thus all disturbances that make the relative velocities of the particles vary in time, within the bandwidth of the measurement, will appear as noise.

Equation (3) acquires an even clearer form by taking its time derivative at the receiver. In flat space time it reads:

$$\begin{aligned}\frac{d\Delta\nu}{dt_r} &= \frac{\nu_o}{c}\frac{d\hat{k}}{dt_r}\cdot[\vec{v}_e(t_r - L/c) - \vec{v}_r(t_r)] + \frac{\nu_o}{c}\hat{k}\cdot[\vec{a}_e(t_r - L/c) - \vec{a}_r(t_r)] \\ &\quad + \frac{1}{2}\nu_o\cos^2(\theta/2)\left\{\frac{dh_+}{dt_r}[t_r - (L/c)(1-\cos(\theta))] - \frac{dh_+}{dt_r}[t_r]\right\}\end{aligned} \tag{4}$$

In (4) we have taken the limit of low velocities. We have also introduced the ordinary three-velocities, $\vec{v}_e$ and $\vec{v}_r$, and three-accelerations, $\vec{a}_e$ and $\vec{a}_r$, of the emitter and receiver respectively. Finally we have defined the unit three-vector $\hat{k}$ along the direction of propagation of the beam. This vector has a non-zero derivative if the line of sight joining the emitter and the receiver rotates in space. It is straightforward to derive that the rotation takes place if the delayed difference of velocity appearing in (4) has a non-zero component normal to the line of sight. Using this results, (4) becomes:

$$\begin{aligned}-\frac{c}{\nu_o}\frac{d\Delta\nu}{dt} &= \frac{[v_e^n(t_r - L/c) - v_r^n(t_r)]^2}{L} + [a_r^\ell(t_r) - a_e^\ell(t_r - L/c)] \\ &\quad + \frac{c}{2}\cos^2(\theta/2)\left\{\frac{dh_+}{dt}[t_r] - \frac{dh_+}{dt}[t_r - (L/c)(1-\cos(\theta))]\right\}\end{aligned} \tag{5}$$

In (5) "$n$" in the apex indicates a component normal to the line of sight, while "$\ell$" indicates a component along that line. Also we have dropped, for time derivatives, the distinction between $t_r$ and $t_e$ as it does not affect the result within the current approximation.

Thus in a reference frame moving with the receiver, and with one of the space directions taken along the line of sight, the Doppler link measures the relative, time delayed acceleration of the two particles along the line of sight itself. This acceleration also includes the centrifugal term originating from the rotation of the line of sight.

The gravitational signal appears then as an effective relative acceleration of the particles. At frequencies low enough such that the time delay $(L/c)[1-\cos(\theta)]$ becomes negligible, (5) simplifies even further to:

$$c\frac{\Delta\nu}{\nu_o} \simeq \frac{(v_e^n - v_r^n)^2}{L} + (a_e^\ell - a_r^\ell) - \frac{1}{2}\ddot{h}L(1-\cos(\theta)) \tag{6}$$

Where the wave amplitude is defined as $h[t] = h_+[t]\cos^2(\theta/2)$.

Ordinary acceleration is caused by force. Thus one may even consider the effect of the gravitational wave as caused by an equivalent, time delayed difference of force between the two particles. It is convenient to express forces per unit mass.

The equivalent "acceleration" along the line of sight becomes then:

$$\Delta a_g^\ell[t_r] = \frac{c}{2}\left\{\frac{dh}{dt}[t_r] - \frac{dh}{dt}[t_r - (L/c)(1-\cos(\theta))]\right\} \tag{7}$$

## 4 Noise in the Doppler Link

Usually the output signal of an interferometer is converted into an equivalent relative displacement of its end points. Thus the Doppler link, within a block-scheme representation, becomes an instrument that converts a differential force at the input into a relative displacement signal at the output.

The disturbances affecting the instrument can be analyzed within this conceptual scheme. First, obviously parasitic forces cause ordinary accelerations of particles. These forces are indistinguishable in practice from the equivalent gravitational force in (7). Thus they act as an input disturbance:

$$\Delta a_n^\ell[t_r] = a_r^\ell[t_r] - a_e^\ell[t_r - L/c] \tag{8}$$

In addition, all interferometers are noisy. Overall the noise resulting from the combination of the three interferometers entering in the measurement, can be described as an effective displacement noise $x_n[t_r]$. As is the case for the force contribution, $x_n[t_r]$ is in reality a combination of various, time delayed terms. However this has no consequence for the following discussion.

The overall displacement noise is contributed by various physical sources. Most noticeable is that, for a single arm, the noise would be orders of magnitude larger than what can be tolerated in LISA due to the large frequency noise. However, as already explained, the frequency noise is common mode within the arms, and thus can very effectively be suppressed with data post-processing. Thus within the single arm representation we are following here, we consider only the residual, unsuppressed part of the frequency noise as being contributing to $x_n[t_r]$.

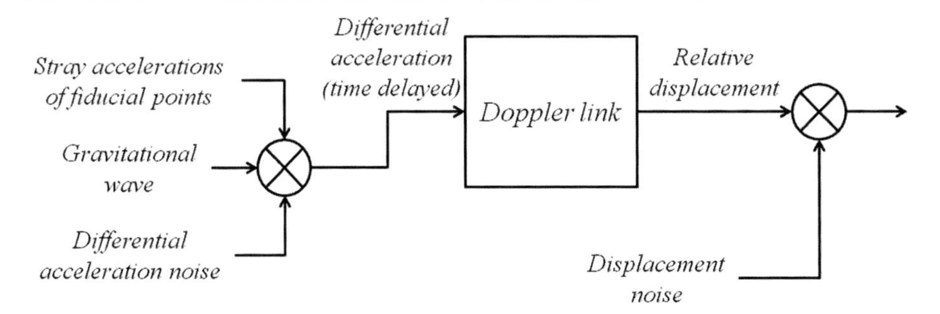


**Fig. 2** Block-scheme of the "Doppler link" model of one LISA arm. The Doppler link converts a differential, time delayed, acceleration signal, into an equivalent relative displacement. At the input, the gravitational wave signal is corrupted by differential acceleration noise acting on the TM, and by the leakage of relative acceleration of satellites, and of satellites relative to TM

Finally, if the various fiducial points of Fig. 1 are not perfectly aligned, the measurement is corrupted by the relative accelerations of all these points, due to the rotation of the SC or of the TM relative to the local inertial frame.

The rotation of the SC is by far the dominating effect. At tens of centimeters from the rotation axis, it can induce unwanted displacement noise with spectral density up to of $10\ \mathrm{nm}/\sqrt{\mathrm{Hz}}$, to be compared with LISA requirements in the range of $10\ \mathrm{pm}/\sqrt{\mathrm{Hz}}$. Thus the Doppler link ends measuring a combination of the relative acceleration of the distant TM, but also of the relative accelerations of the two satellites with respect to each other, and with respect to the TM hosted within them.

Overall the block scheme is represented in Fig. 2.

As in all input-output schemes, all sources of noise can be referred to a single port, that is, either to the input or to the output. We prefer here to discuss LISA performance "at input", that is, we describe the noise as an equivalent differential acceleration disturbance.

It is straightforward to calculate that the output displacement noise, with power spectral density $S_x(\omega)$, with $\omega$ the angular frequency, is equivalent to a differential acceleration noise with power spectral density $S_{\Delta a_x}(\omega) = \omega^4 S_x(\omega)$. We also define the spectral densities $S_{\Delta a_n}(\omega)$ and $S_{\Delta a_{\mathrm{SC}}}(\omega)$ of the differential acceleration noise in (8), and of the noise due to the stray accelerations of the fiducial points respectively. In this simplified model then, the performance of the time-delayed, differential accelerometer is described by a total noise spectral density given by:

$$S_{\Delta a_{\mathrm{tot}}}(\omega) = S_{\Delta a_n}(\omega) + S_{\Delta a_{\mathrm{SC}}}(\omega) + \omega^4 S_x(\omega) \tag{9}$$

Equation (9) shows the Doppler link shares the basic spectral structure of the noise with any other, delayed or not delayed, differential accelerometer. At low frequency the instrument is usually dominated by input noise, that is, by parasitic forces acting on the TM. At higher frequency the input equivalent of the output noise, that increases, in amplitude, as the square of the frequency or faster, dominates, giving origin to the characteristic ramp.

The possibility of formulating an overall performance figure of merit as an equivalent differential acceleration noise figure, allows for a direct comparison among many apparently unrelated space mission. Indeed this performance figure can be derived for Gravity Probe A, Cassini, Grace, Goce, Microscope, LISA pathfinder and LISA. For the details of such of an evaluation and references for these missions (see Danzmann et al. 2007; Vitale 2009).

## 5 Noise in LISA

The above scheme of analysis can be applied to the case of LISA. Indeed it is an established practice to express the instrument performance in term of single arm sensitivity. This is expressed more frequently as an equivalent TM to TM displacement noise, that is, within the conceptual scheme of Fig. 2, as an equivalent noise at the output. We prefer here instead to maintain the formulation at input, as an equivalent acceleration noise.

An estimate of the expected performance is reported in Fig. 3. The estimate is a miscellanea of experimental determination and analysis based calculation. Each contribution includes margin in various forms, either as an explicit one or by keeping a possible noise subtraction procedure in reserve. Examples will be given in the following.

### 5.1 Acceleration Noise

Acceleration noise is dominated by a few larger contributions. They are listed in the following in order of importance at the lowest frequency, 0.1 mHz, of LISA band. Performance estimations from different sources may differ in the details, there is however a general agreement on the overall result (Stebbins et al. 2004; Shulte 2007).

1. Brownian force noise due to residual gas pressure. The expected figure has been experimentally assessed very recently and is dominating above all other sources of Brownian noise (Cavalleri et al. 2009b).
2. Magnetic force. The TM is made out of a low magnetic susceptibility Au-Pt alloy. Nevertheless, the fluctuations of magnetic field and field gradient, acting on the magnetic moment induced by the dc-part of the field, exert a sizeable fluctuating force. In addition,

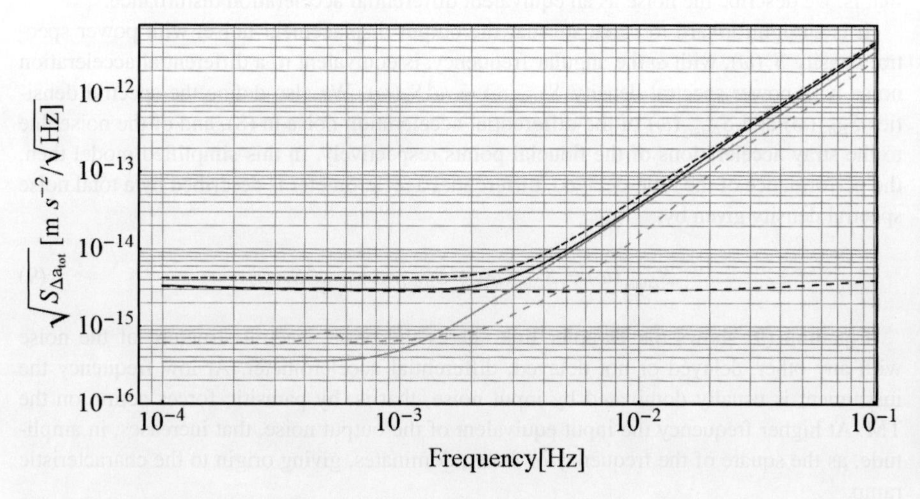


**Fig. 3** LISA single arm performance expressed as an equivalent acceleration noise spectral density within the measurement band. *Gray solid line* (the lowest at the low frequency end): total interferometer phase noise. *Gray dashed line*: contribution of the acceleration of fiducial points. *Black solid line*: differential acceleration of TM due to parasitic forces. *Black solid line*: quadratic sum of all the contributions. *Upper line with fine dashing*: performance requirements

as the TM is conducting, high frequency magnetic field of fluctuating amplitude, exerts a corresponding fluctuating force via the skin effect due to eddy currents. This figure could be reduced at the expense of more stringent requirements on magnetic cleanliness, and on electromagnetic interference suppression.

3. Noisy actuation force. Electric fields are used to apply force and torques to the TM along the electrostatically controlled degrees of freedom. The voltage used are noisy, and generate noisy forces and torques. Geometrical imperfection are expected to create a certain leakage of these forces along the sensitive direction. The overall contribution are calculated assuming conservative leakages of order of 0.5–1%, depending on the degree of freedom.
4. Force due random charge. The TM charges up because of cosmic rays. The net charge will be neutralized by the UV charge management system. However charging is a random process. The dc electric field within the electrode housing is known from experiment to be of order of 10–20 V/m, because of the parasitic potentials known as charge patches. The random charge within this field creates a random force. Parasitic potentials can be compensated, by applying voltages to the electrodes to better than 250 mV/m. However for the noise estimate, it is conservatively assumed that compensation can only be maintained at 1 V/m.
5. Fluctuating force due to SC jitter. Static force gradients, mostly gravitational and electrical, convert any relative jitter between the SC and the TM into a jittering force. A residual $4 \times 10^{-7}\ \mathrm{s}^{-2}$ force gradient per unit mass is conservatively estimated after gravitational balancing of the system. This couples to a residual jitter, estimated, from the end-to-end simulation of the SC controls, in the order of $\mathrm{nm}/\sqrt{\mathrm{Hz}}$. This jitter will be measured, with $10\ \mathrm{pm}/\sqrt{\mathrm{Hz}}$ accuracy, by the local interferometer, and force gradients can be measured by properly exciting the relative motion of SC and TM (see next section). Thus this noise could effectively be coherently subtracted from the data. However this possibility is currently kept in reserve as margin.
6. Thermal effects. Thermal gradients across the TM generate forces by a variety of phenomena. The leading ones are the radiometer effect within the residual gas, and the molecular impacts form the temperature enhanced residual out-gassing from the hot side of the electrode housing. A minor contribution is also expected form radiation pressure asymmetry. The contributions of these effects have been determined experimentally (Carbone et al. 2007).
7. Charge Patches. This is a notorious threat for the entire field of precise force measurements. Metal surfaces carry domains with different work-function values. Electron migrate to equalize the chemical potential, thus creating patches of charge on the surface. The phenomenon is suppressed with the gap size, because the field created by patches of size much smaller that the gap, decays at least with the cube of the gap width. This is the reason of the comparatively large gaps around the TM. In recent years, in the context of LISA development, it has been demonstrated that patches of many mm size can be created by non-homogeneity of surface coating and by contamination. These patches are responsible for the stray electrical gradients mentioned above. Time fluctuation of these voltages may create a fluctuating force. There are running studies (Carbone et al. 2005; Pollack et al. 2008) to measure the charge patch potential fluctuations and the related force. Though the studies are not completed yet, the current estimate for this contribution appears to meet the apportioned value with margin.

There are obviously many more candidate sources for fluctuating forces. Next to the mentioned ones follow, for example, the gravitational noise generated by thermo-elastic deformation of the SC, and the laser radiation pressure fluctuations caused by amplitude noise. The current estimate for these effect though, finds their contribution negligible.

Most of the disturbance mentioned above act at the surface of the TM. For these, over the last ten years, a quite powerful test method has been developed based on a torsion pendulum (Carbone et al. 2003, 2007). The TM is made part of the inertial member of the pendulum, and its motion in the horizontal plane imitates with good accuracy, a free fall condition. The mass can be surrounded by a faithful replica of the electrode housing. This effectively shields out external disturbances, but also reproduces all sources of noise, the majority, that sit within the electrode housing. Special pendulums have also been developed for the study of particular effects like charge patches (Pollack et al. 2008).

LISA torsion pendulums have now reached sensitivities of 20–30 fN/$\sqrt{\text{Hz}}$ at, or just below 1 mHz (Cavalleri et al. 2009a). However this extremely sensitive device cannot support a full size, 2 kg TM. Thus hollow TM are used that reproduce the surface of the real TM but cannot reproduce its bulk. Thus most of the magnetics, and gravitation cannot be tested with this instrument. Nevertheless the method has been extremely powerful in anchoring almost all surface effects mentioned above, to experimental values. The detail of many years of testing activity can be found in the cited literature.

### 5.2 Acceleration of Fiducial Points

The practical implementation of each LISA arm contains a series of non-idealities. As already explained, each of the three interferometers composing one arm, reads out the relative displacement of two points on two different extended bodies along some effective directions. Thus if these points do not lie exactly on the line joining the centers of mass of the two TM, and if the effective sensitive directions of the interferometers are not parallel to the same line, then the output of the three-interferometer chain will also be sensitive, to different extents, to:

- Rotation of the SC relative to the wave front of the laser beam.
- Rotation of each TM relative to its hosting SC.
- Displacement of the SC relative to the TM hosted into it.
- In addition, as a steerable mirror is also used to accommodate the breathing of the angles of the formation over the year, the unwanted translational motion of this mirror may also be detected.

These effects are minimized by making all the above relative motions as small as possible by the already mentioned control loops. Thus for instance, also the rotation of the SC relative to the beam wave-front is measured, and the SC is accordingly rotated via the micro-Newton thrusters to keep the error at a minimum. As always with disturbances, besides minimizing the input, also the "susceptibility" must be kept small. In all conversion from rotation to translation, some characteristic length enter. Thus the TM must be kept centered relative to the laser beam within 50 μm, while the phase center of the SC-to-SC interferometer detector must be kept aligned with the center of the TM to within 0.5 mm.

For many of these disturbances, a correction signal is available. For instance the rotation of the SC relative to the wave front is measured with nrad/$\sqrt{\text{Hz}}$ accuracy. Thus in principle the effect can be calibrated and subtracted. However again this possibility is kept in reserve and constitutes then an hidden margin.

In addition to the effects above, the system carrying the interferometers is not infinitely rigid. Thermo-mechanical expansion introduces an effective random wandering of the fiducial points. This effect is suppressed by the ultra-stable structure and thermal environment of LISA optics. The residual is included in the overall figure reported in Fig. 3.

### *5.2.1 Interferometer Noise*

Finally all three interferometers are noisy. The local interferometer, discussed a bit more in the next session, brings close similarities to the one being used on LISA Pathfinder (see next session). Actually using exactly the LISA Pathfinder interferometer is one of the open options for LISA, though for the moment is not the base line. Its performance, with a few $\mathrm{pm}/\sqrt{\mathrm{Hz}}$ have been already demonstrated on engineering models (Heinzel et al. 2006). Noise is contributed by a few larger source including frequency noise, electronic noise, thermo-optical effects and stray light.

The most noticeable part of LISA interferometer is the 5 million kilometer SC to SC laser transponder. There are a few peculiar requirement for an heterodyne interferometer $5 \times 10^9$ m wide.

As phase must be measured coherently over the full constellation, a common time reference needs to be distributed over it. This is achieved by synchronizing all local ultra-stable reference oscillators, using the laser beams as transmission channels. Specifically some purposely generated sidebands are used for clock distribution over the constellation.

The required clock stability exceeds that of an off-the-shelf ultra-stable oscillator. In the most critical frequency band, around 3 mHz, the clock requires a relative frequency stability better than $10^{-14}/\sqrt{\mathrm{Hz}}$, roughly one order of magnitude better than the state of the art space-qualified oscillators. The missing factor is gained by using the arm as a delay line. The difference between the instantaneous value of the oscillator phase, and the one coming back after a round trip over one arm is kept nominally at zero by closed loop control. This effectively suppresses fluctuations at frequencies lower that the round trip time.

A similar stabilization scheme, known as arm-locking, can be used to stabilize the frequency of the laser, after an initial stabilization obtained with a more standard cavity within each SC. Experiments and simulations show that the method is extremely powerful and suppress frequency noise up to 2–3 orders of magnitude at the lowest frequencies. However in the current configuration, the method is kept in reserve and noise performance is estimated without assuming its adoption.

As anticipated, equal arm interferometers are hard to achieve in space and on such large distances. Thus light collected in one SC, at a given time, from the arms converging into it, has not been generated at the same time in both arms. As laser frequency fluctuates in time, different beams are at frequencies that may differ by an amount many orders of magnitude larger that the differential modulation due to the gravitational signal. To cope with this problem, signals from different arms are combined in data post-processing on ground. Appropriate combinations of time-delayed data from different arms, preserve the differential gravitational wave signal but effectively suppress the frequency noise.

The method requires a good knowledge of the intervening delays, that is of the round trip time of each arm. The length of each arm is measured, to a $\approx$50 m accuracy, by continuously transmitting over the laser beam, a pseudo random code pulse, with some similarity with a GPS positioning algorithm. Simulation and experimental tests of the required electronic fidelity, show that the algorithm indeed achieves the required performance.

The final noise of the transponder is contributed by the limits of the above procedures, oscillator stabilization and frequency noise suppression. In addition the shot noise in the $\approx$1 W laser beam, widened and diluted by the 5 million kilometer transmission, is the largest single contribution to the overall noise. The sum of this noise, and of that of the two local interferometers, gives an overall performance within the required 12 $\mathrm{pm}/\sqrt{\mathrm{Hz}}$ at the highest corner of the LISA band. This performance, translated to input as an equivalent acceleration noise, is reported in Fig. 3.

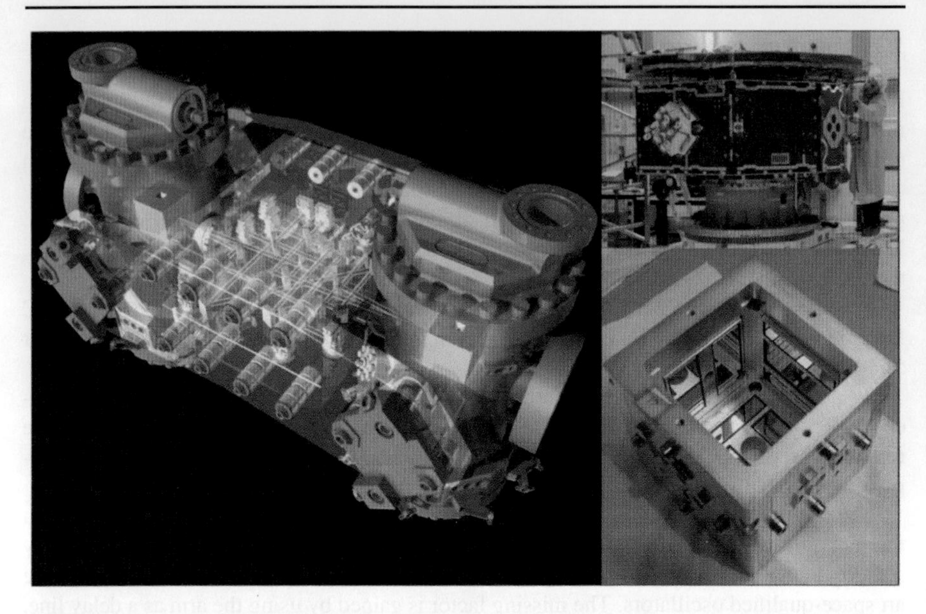

**Fig. 4** *Left*: View of the LTP. Shown in transparency are the TM and the interferometer. *Upper inset*: LPF SC. *Lower inset*: the electrode housing for LISA and LPF (Courtesy of ASTRIUM GmbH, ESA-ESTEC and INFN)

## 6 Proving Space-Time Metrology with LISA Pathfinder

Most of LISA metrology will be tested in flight on LISA Pathfinder (LPF) (Vitale et al. 2002; Anza et al. 2005). Indeed LPF consists of one entire LISA arm, except for the 5 million kilometer laser transponder, which is taken out in order to accommodate the full instrument within one SC.

The scope is to quantitatively validate all possible aspects of the model for the physical disturbances that affect the LISA arm described so far, with particular emphasis on those that cannot be tested in 1 g on ground. In addition this is achieved by using as much as possible the exact hardware to be used on LISA.

Thus LPF flies, in interplanetary orbit, the LISA technology Package (LTP) (Fig. 4): two TM, and two local interferometers measuring both the TM relative displacement and their displacement relative to the SC. TM, electrode-housing, TM launch-lock device, UV charge management system, and micro-Newton thrusters, are nominally equal to those to be used on LISA.

The local interferometers differ by a few details from those to be used on LISA. The need to achieve, in LISA, a coherent, easy to be implemented, overall optical design, as brought about some difference relative to LPF interferometers design. For example in LISA the local interferometer uses polarizing optics, while this is not the case in LPF. However LPF interferometer performance is the same as that required for LISA, and thus its design could directly be transferred to the latter.

The control logic of the SC in LPF is different from that of LISA. Indeed the LPF SC contains a full arm, and thus, along the sensitive direction, it can only chase one of the two TM. The other one needs then to be controlled electrostatically. As the stability of the electronics used is limited, the applied force is noisy. This is one of the limitation of the test. In addition, to stay within a reasonable budget, some of the environmental requirements, like for instance, magnetic field stability have been relaxed. All in all then the requirement for LPF is to demonstrate a differential acceleration noise of the arm of better than

$$S_{\Delta a_{tot}}^{1/2}(f) \leq 3 \times 10^{-14}\ (\mathrm{m\,s^2/\sqrt{Hz}})[1 + (f/3\ \mathrm{mHz})^2]$$

at frequencies between 1 and 30 mHz. The current estimate for the performance is currently below this figure and is reported in Fig. 1.

In reality the experimental plan for LPF is to perform a series of experiments aimed at quantifying all major contributions to the total differential acceleration noise budget. Some of the noise contributions can be estimated in real time. For instance, as already mentioned, the coupling to SC jitter, both via the force gradients that create a jittering force, and via the interferometer fiducial point jitter, can be estimated and subtracted from the data in the time domain. The jitter of the SC relative to the TM that drives the drag-free control, is indeed measured by one of the interferometer. This jitter, filtered by the proper transfer function, is then a real time estimate of the contribution of this term to the total acceleration budget. The transfer function, that is the force gradients and the interferometer pick-up coefficient of the SC jitter, will be measured by a separated calibration experiment.

This way a few major contributions can be identified and measured quantitatively. For the remaining ones, various parameters entering in their physical model will be measured either on ground or on orbit or on both. This will allow a quantitative prediction of their contribution.

By means of this detailed procedure of "noise projection" one should be able to compare the experimental noise with the physical model with a relative accuracy of a factor 2 on the power spectral density and then of ≈ 40% on its square root. Thus any unpredicted disturbance producing a noise larger than this limit should be detectable in the experiment. This 40% line is reported in Fig. 5, and gets very close to LISA requirements above 1 mHz.

It must be stressed that even if LISA could only achieve the performance represented by such a line, or if, even worse, that represented by LPF requirement, still the science return of the mission will be extremely significant. A large fraction of the galactic binary signals

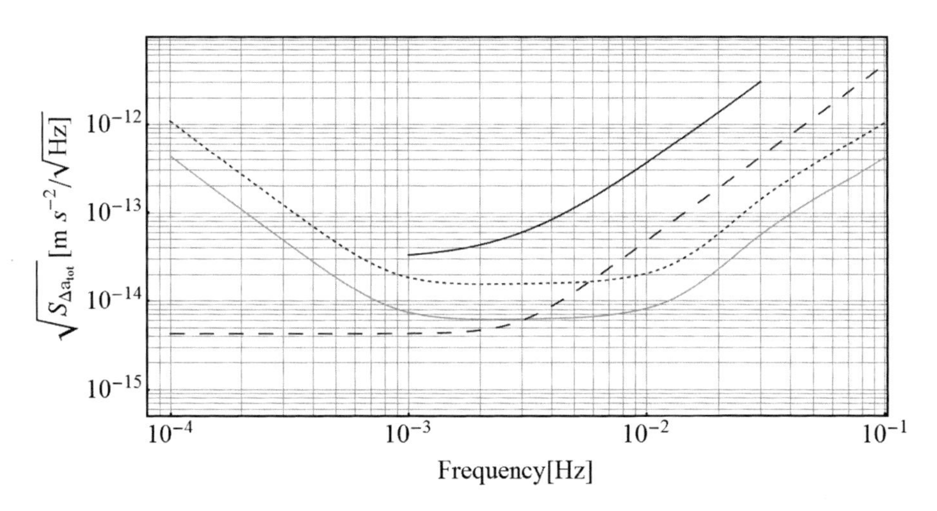


**Fig. 5** LPF performance. The *upper black solid line* represents LPF differential acceleration requirements. The *lowest coarsely dashed black line* represents LISA requirements. The *black, finely dashed line* is the current estimate for the LPF budget. The *gray solid line* is 40% of the noise budged, the estimate residual uncertainty on the difference between the measured noise and the physical model for it

and of those form super-massive black-hole would indeed still be detectable with significant signal to noise ratio.

**Acknowledgements** I warmly thank Karsten Danzmann, Oliver Jennrich and Bill Weber for reading the manuscript, and for the many useful discussions and suggestions.

## References

S. Anza et al., Class. Quantum Gravity **22**, S125 (2005)
P. Bender et al., LISA, Laser Interferometer Space Antenna for the detection and observation of gravitational waves. Pre-Phase A Report 2nd edition MPQ 233, July (1998)
L. Carbone et al., Phys. Rev. Lett. **91**, 151101 (2003)
L. Carbone et al., Class. Quantum Gravity **22**, S509 (2005)
L. Carbone et al., Phys. Rev. D **76**, 102003 (2007)
A. Cavalleri et al., Class. Quantum Gravity **26**, 094017 (2009a)
A. Cavalleri et al. (2009b, in preparation). Also in: Excess test-mass motion damping from residual gas pressure, and the resulting thermal noise S2-UTN-TN-3067, Iss. 2.0 (2009)
B. Chauvineau, T. Regimbau, J.Y. Vinet, S. Pireaux, Phys. Rev. D **72**, 122003 (2005)
K. Danzmann et al., LISA Pathfinder: Einstein's Geodesic Explorer. The Science Case for LISA Pathfinder., ESA-SCI(2007) 1, (2007)
G. Heinzel et al., J. Phys. Conf. Ser. **32**, 132–136 (2006)
O. Jennrich, Class. Quantum Gravity Topical Paper (2009, to appear)
S.E. Pollack et al., Phys. Rev. Lett. **101**, 071101 (2008)
T. Prince et al., LISA Science Case: Probing the Universe with Gravitational Waves, http://www.srl.caltech.edu/lisa/mission_documents.html. Accessed April 12 (2009)
H.R. Shulte, System Parameters and Error Budgets, LISA-ASD-BR-5002, Issue 1.4 (ESA LISA formulation study document) (2007)
R.T. Stebbins et al., Class. Quantum Gravity **21**, S653 (2004)
J.L. Synge, *Relativity: The General Theory* (North-Holland, Amsterdam, 1964)
S. Vitale, Space Res. Today (2009, in preparation)
S. Vitale et al., Nucl. Phys. (Proc. Suppl.) B **110**, 209 (2002)
J.M. Weisberg, J.H. Taylor, Astron. Soc. Pac. Conf. Ser. **328**, 25 (2005)

Space Sci Rev (2009) 148: 455–474
DOI 10.1007/s11214-009-9565-y

# The Microscope Mission and Its Uncertainty Analysis

**Pierre Touboul**

Received: 10 December 2008 / Accepted: 7 July 2009 / Published online: 29 July 2009
© Springer Science+Business Media B.V. 2009

**Abstract** The accurate test of the Universality of Free Fall may demonstrate a violation of Einstein Equivalence Principle (EP) as most attempts of Grand Unification theories seem to conduct. The MICROSCOPE space mission aims at an accuracy of $10^{-15}$ with a small drag free satellite and a payload based on electrostatic inertial sensors. The two test-masses made of Platinum and Titanium alloys are forced to follow accurately the same orbit. The sets of surrounding electrodes carried by gold coated silica parts allows the generation of electrical fields and electrostatic pressures on the masses. Common forces and torques are exploited to control the satellite drag compensation system and its fine inertial or rotating pointing. Difference in the force along the Earth gravity monopole is accurately measured and interpreted for the test. After a short presentation of the mission and the instrument, most of the relevant parameters to the experiment performance are detailed as well as the associated technologies to reach the expected levels of accuracy. Present error budgets confirm the test expected accuracy of better than $10^{-15}$.

**Keywords** Equivalence principle · Universality of free fall · MICROSCOPE space mission · Electrostatic inertial sensors · Space accelerometers

## 1 Introduction

The accurate test of the universality of free fall, as depicted by Galileo and considered by Newton, is today much than the verification of this well known property. The violation of the universality of free fall leads to the violation of the Equivalence Principle (EP), fundamental basis of the Einstein General Relativity. Einstein, himself considered this symmetry as enacted by the experimental results: "*The ratio of the masses of two bodies is defined in two ways which differ from each other fundamentally... The equality of these two masses, so*

P. Touboul (✉)
Physics and Instrumentation Department, ONERA, BP 72, 92322 Châtillon Cedex, France
e-mail: pierre.touboul@onera.fr
url: http://www.onera.fr

*differently defined, is a fact which is confirmed by experiments... The possibility of explaining the numerical equality of inertia and gravitation by the unity of their nature, gives to the general theory of relativity, according to my conviction, such a superiority over the conception of classical mechanics...*" (Einstein 1922). Furthermore, the Equivalence Principle lies at the foundation of other metric theories of gravity like Brans and Dicke one (Brans and Dickes 1961; Will 1985). Today, most attempts of Grand Unification like String theory and M-theory allow the violation of this principle (Damour 1996, Damour et al. 2002), introducing in particular scalar fields, while the experimental investigation of quantum gravity does appear directly very weakly accessible, see for instance the Plank length of $1.6 \times 10^{-35}$ m. The test of the Equivalence Principle is thus more then the test of general relativity but also the look for new experimental results as the necessary support of new theory. In addition, supersymmetry might be confirmed by the results which should be obtained in the near future with CERN LHC and new particles to be taken into consideration (Feldman et al. 2008).

The test of the universality of free fall with ultimate accuracy is then an important challenge regarding also the dark matter query. In this context, many efforts have been payed to perform important progress in the performance of ground tests (Schäfer 2003). Previous results have been obtained in the last decade with torsion balance and small laboratory bodies up to an accuracy of composition-dependant relative acceleration of $+/-4 \times 10^{-13}$ (Baessler et al. 1999). Moon laser ranging and the comparison of the Earth and Moon relative motions in the Sun field lead to the same order of performance (Williams et al. 1996). Concerning the laser ranging of passive satellite, no better performance can be expected in the near future (Iorio 2007).

Recent results have been obtained with continuously rotating balance in order to better reject the disturbances from the laboratory activities and the gravity gradient fluctuations: Eötvös parameter between Beryllium and Titanium has been evaluated to $(0.3 +/- 1.8) \times 10^{-13}$ (Schlamminger et al. 2008).

Other developments have been undertaken to increase the performance of such apparatus by magnetic superconducting levitation of the balance rotor in order to suppress the disturbances introduced by the 20 μm tungsten wire (Hammond et al. 2007). Reduction of the thermal environment has also been considered by implementing the apparatus in a helium Dewar (Newman 2001). New laser ranging telescope is also under implementation (Currie et al. 2008). In the same time, the MICROSCOPE space mission has been conceived and designed in order to reach a test accuracy level of $10^{-15}$. The definition of the mission takes advantage of existing space accelerometers with dedicated configurations and technologies for ultimate performances (Touboul et al. 2001). This space experiment exploits the specific soft environment that can be reached on board a drag free satellite with very weak residual accelerations and thermal fluctuations, and considers the possibility of a long duration experiment in a free fall laboratory when in orbit around the Earth.

## 2 The MICROSCOPE Space Experiment

The MICROSCOPE space mission has been selected since several years in the Cnes scientific program and the production of the satellite payload is now undertaken. The satellite is a rather small satellite of 200 kg that has been fully defined and dedicated to the MICROSCOPE fundamental physics experiment. The constraints of the available mass (35 kg), volume and power (40 W) for the dedicated scientific payload have been considered from the initial definition leading to a non cryogenic experiment, with a limited couple of tested materials, and a test performance that should be even better in

**Table 1** Proof mass properties

| Material | $N/\mu$ | $Z/\mu$ | $(N+Z)/\mu$ | $(N-Z)/\mu$ |
|---|---|---|---|---|
| PtRh10 | 0.59613 | 0.40357 | 0.999704 | 0.192555 |
| TA6V | 0.54197 | 0.46061 | 1.002588 | 0.081358 |

the future but with more complex mission, satellite and instrument (Sumner et al. 2007; Lafargue et al. 2002).

The concept of this space experiment is a basic free-fall test around the Earth, with the availability of long measurement duration, reduced test-mass disturbing accelerations, very precise instruments optimised for micro-gravity operation and the Earth gravity source signal modulated by the orientation of the rotating instrument leading to reach a precision of $10^{-15}$. In fact, two masses of different composition will be precisely positioned on the same orbit. In absence of Equivalence Principle violation, the two masses, submitted to the same Earth gravity field, will continue on the same common trajectory. The satellite which carries the instruments including the masses will be controlled to follow this common trajectory by acting the thrusters of its propulsion system, protecting the masses from Earth and Sun radiation pressures and from residual atmospheric drag. More precisely, the motion of the two masses will be accurately measured with respect to the same on board instrument reference frame and servo-controlled thanks to electrostatic actuators: the test-masses are so maintained relatively motionless with a better accuracy than $10^{-11}$ m. In that way, the stability of the test-mass/instrument configuration limits the fluctuations of the parasitic forces applied on the mass due for instance to the gravity field gradients or the electro-magnetic field. It benefits also to the operation linearity of the instrument capacitive sensing and electrostatic actuations, mainly depending on the configuration geometry. The common applied electrostatic acceleration will be nullified by acting the satellite thrusters in such a way that the common instrument reference frame is following the two masses in their in orbit motion. The difference of the applied electrostatic acceleration will be accurately observed in the direction of the Earth gravity monopole to be analysed as an eventual Equivalence Principle violation signal.

So, the two test-masses are the two proof-masses of two concentric ultra-sensitive inertial sensors composing a differential accelerometer, called SAGE (Space Accelerometer for Gravitational Experimentation). MICROSCOPE satellite can operate two SAGE instruments that will be identical except the mass materials. The two materials used for the test will be a Platinum Rhodium alloy, PtRh10 (90% Pt, 10% Rh), and a Titanium alloy, TA6V (90% Ti, 6% Al, 4% Va). The selection of these materials takes into account both instrument performance concerns (homogeneity, accuracy of machining, stability, electrical and magnetic properties...) and theoretical test aspects (see Table 1). A significant difference in subatomic particles may increase the likelihood of a detectable EP violation: Damour and Blaser mention parameters to be considered like baryon number over the atomic mass $(N+Z)/\mu$, the neutron excess $(N-Z)/\mu$, and nuclear electrostatic energy in $Z(Z-1)/(N+Z)^{1/3}$ (Damour and Blaser 1994).

The second SAGE instrument includes two masses made of the same Pt-Rh alloy. This instrument is only devoted to the in orbit verification of the systematic experiment errors. Pt-Rh alloy has been preferred for its high density, leading to a better rejection of the spurious surface effects. This second instrument exhibits so a better precision than the first one with test-masses of different composition: confidence in the obtained result is mandatory in such experiment, that is why we have not selected another pair of masses with different composition.

The MICROSCOPE satellite is scheduled to be launched in 2012. The selected quasi-circular heliosynchronous orbit has an altitude of 810 km. This altitude, $h$, is a compromise between the $1/(R+h)^2$ gravity signal, the residual atmospheric density that may disturb the satellite motion and the satellite thrusters and launcher possibilities. The heliosynchronism leads to a fixed satellite Sun side and so to an optimised rigid solar panel configuration: a maximum power is delivered for a minimum size and, more important, the fixed thermal external conditions of the satellite are very favourable for its thermo-elastic behaviour and its internal fluctuations of temperature.

The one year mission consists in different instrument calibration and measurement sequences. In orbit, the symmetry of the electrostatic actuations on the masses will be verified to a relative level of about $10^{-4}$ in order to reject common motion disturbances. The off-centring of the masses, limited to less than 20 μm by construction, will be also surveyed to limit the gravity gradient effects. Sequences with inertial and rotating pointing of the satellite are considered. In the first case, the Earth gravity field is modulated along the test-mass axes, at the orbital frequency, i.e. $f_{EPi} = 1.7 \times 10^{-4}$ Hz. In the second case, it is modulated at the sum of the orbital frequency and the satellite spin rate, i.e. $f_{EPs}$. Two spin rates and two 90° phases of the satellite pointing versus the Earth, defined at the ascendant node of the orbit, will be considered. So, the heterodyne detection of an eventual EP violating signal will be performed at these frequencies (EP frequencies). A minimum of 20 orbits integrating period is considered for the rejection of the stochastic disturbing signals.

Other space missions aiming at the test of the Equivalence Principle are under study and proposed to be selected in the future. Among them, the STEP mission, leaded by Stanford University is proposed to NASA and ESA in the frame of an international cooperation (Mester et al. 2001). This mission, much more ambitious than MICROSCOPE, envisages the implementation of four superconductive differential accelerometers, and so four pairs of test-masses, inside a 2 K Helium Dewar: superconducting loops insure the very steady passive magnetic levitation of the masses and SQUIDs devices are exploited in specific differential circuits to perform the measurement of the actual common and relative motions of the test masses which are not servo-controlled. The thermodynamic instrument noise is reduced by the operating temperature allowing an expected EP test accuracy of $10^{-18}$. This accuracy is very demanding with respect to the instrument and satellite production and operation, which is, as it is shown hereafter for the MICROSCOPE mission, already at the limit of the state of the art of many spacecraft technologies. Another mission, called Galileo Gallilei (Nobili et al. 2000), takes advantage, like in MICROSCOPE, of room temperature capacitive position sensing but considers the rotation of the test masses at rather high frequency, a few Hz, in order to better decouple their relative motion to the satellite disturbances in presence of mechanical spring between them, contrarily to MICROSCOPE. A laboratory model of the instrument is also developed to assess on ground the instrument concept and configuration.

Beside the scientific objectives of the MICROSCOPE mission, it remains, by its experiment relative simplicity and by its already demonstrated technologies with presently nine accelerometers in orbit (on board CHAMP, GRACE and GOCE satellites), a fine preparatory program for more ambitious space tests with a limited risk.

The major scientific instrument outputs are derived from the applied electrostatic forces on each test-mass which are accurately measured: $\overrightarrow{\Gamma}_{App,k}$, for the test-mass $k$ (centre $O_k$) can be expressed according to the acceleration of the mass, the Earth gravity field acceleration and the resultant of the parasitic forces induced by the environment (depicted in Sect. 5):

$$\overrightarrow{\Gamma}_{App,k} = \frac{\overrightarrow{F}el_k}{m_{Ik}} = \overrightarrow{\gamma}(O_k) - \frac{m_{Gk}}{m_{Ik}}\overrightarrow{g}(O_k) - \frac{\overrightarrow{F}pa_k}{m_{Ik}} \tag{1}$$

**Table 2** Main specifications of the satellite attitude control in the two operating modes, inertial and rotating pointing; in addition the drag compensation system limits the linear acceleration measured by the inertial sensors in common mode; both controls are performed by actuating the satellite micro-thrusters according to the outputs delivered by the 6-axes inertial sensors and the star trackers

| | Stochastic signal | At DC (inertial s/c – rotating s/c) | At EP frequency (inertial s/c – rotating s/c) |
|---|---|---|---|
| Pointing | 0.6 – 0.2 mrad/Hz$^{1/2}$ | 2.5 – 2.5 mrad a priori<br>1.0 – 1.0 mrad a posteriori | 10 – 10 µrad a priori<br>6 – 1.0 µrad a posteriori |
| Angular velocity Ω | $1 \times 10^{-4}$<br>$-1 \times 10^{-6}$ rads$^{-1}$ Hz$^{-1/2}$ | $1 \times 10^{-6}$<br>$-\sim 4 \times 10^{-3}$ rads$^{-1}$ | $7 \times 10^{-7}$<br>$-1 \times 10^{-9}$ rads$^{-1}$ |
| Angular acceleration $d\Omega/dt$ | $5 \times 10^{-9}$<br>$-5 \times 10^{-9}$ rads$^{-2}$ Hz$^{-1/2}$ | $3 \times 10^{-8}$ rads$^{-2}$<br>over 20 orbits | $1 \times 10^{-11}$<br>$-5 \times 10^{-12}$ rads$^{-2}$ |
| Linear acceleration | $3 \times 10^{-10}$<br>$-3 \times 10^{-10}$ ms$^{-2}$ Hz$^{-1/2}$ | $3 \times 10^{-8}$ ms$^{-2}$<br>over 20 orbits | $10^{-12}$ ms$^{-2}$ |

with $m_I$ and $m_G$ the inertial and gravitational mass: $\frac{m_{Gk}}{m_{Ik}} = 1 + \delta_k$. Let us recall that the Eötwös parameter for two test-masses $k$ and $l$ is defined by: $2\frac{|\delta_k - \delta_l|}{\delta_k + \delta_l}$.

So, the optimisation of the instrument consists in limiting all disturbances on the test-mass motion, especially about the EP test frequency and phase (or identical on both masses), predicting the exact (or the same) gravity field integrated on both masses, measuring the exact electrostatic force applied on it (amplitude, frequency and direction). The acceleration of each test-mass can be then expressed by considering the satellite linear and attitude motion, $(O_{sat}, M_{sat})$:

$$\overrightarrow{\Gamma_{App,k}} = \overrightarrow{\gamma}\,(O_{sat}) + H_{In,COR}(\overrightarrow{O_{sat}O_k}) - (1+\delta_k)\,\overrightarrow{g}\,(O_k) - \frac{\overrightarrow{F}\,pa_k}{m_{Ik}} \tag{2}$$

with the operator $H$ representing the sum of three dynamics effects to be controlled during the experiment: the inertia effect due to the satellite angular acceleration and velocity, both controlled by the satellite system (see Table 2), the Corriolis effect and the relative acceleration of the test-mass versus the instrument/satellite frame. The two letters are limited because of the stability of the silica instrument frame and the accuracy of the mass electrostatic servo-control to this frame with a stability of better than $10^{-16}$ ms$^{-2}$ at EP frequencies.

$$\begin{aligned} H_{In,COR}(\overrightarrow{O_{sat}O_k}) &= \left[\dot{\overrightarrow{\Omega}} \wedge \overrightarrow{O_{sat}O_k} + \overrightarrow{\Omega} \wedge (\overrightarrow{\Omega} \wedge \overrightarrow{O_{sat}O_k})\right] + 2[\Omega]\,\dot{\overrightarrow{O_{sat}O_k}} + \ddot{\overrightarrow{O_{sat}O_k}} \\ &= [In]\overrightarrow{O_{sat}O_k} + 2[\Omega]\,\dot{\overrightarrow{O_{sat}O_k}} + \ddot{\overrightarrow{O_{sat}O_k}} \end{aligned} \tag{3}$$

The satellite undergoes the Earth gravity, as well as external surface forces $\overrightarrow{F}\,ext$ like the atmospheric drag and the radiation pressures and also the thrust of its propulsion system $\overrightarrow{F}\,th$.

$$\underbrace{\left(M_I + \sum_k m_{Ik}\right)}_{M_{Isat}} \overrightarrow{\gamma}\,(O_{sat}) = \overrightarrow{F}\,ext + \overrightarrow{F}\,th + \underbrace{\left(M_G + \sum_k m_{Gk}\right)}_{M_{Gsat}} \overrightarrow{g}\,(O_{sat}) \tag{4}$$

Then, the expression of each inertial sensor output is:

$$\overrightarrow{\Gamma_{App,k}} = \frac{M_{Gsat}}{M_{Isat}}\overrightarrow{g}(O_{sat}) - (1+\delta_k)\overrightarrow{g}(O_k) + R_{In,COR}(\overrightarrow{O_{sat}O_k}) - \frac{\vec{F}pa_k}{m_{Ik}} + \frac{\vec{F}ext}{M_{Isat}} + \frac{\vec{F}th}{M_{Isat}}$$

$$= \overrightarrow{\Gamma}_{App,k/sat} - \frac{\vec{F}pa_k}{m_{Ik}} + \frac{\vec{F}ext_{/sat}}{M_{Isat}} + \frac{\vec{F}th_{/sat}}{M_{Isat}} \tag{5}$$

The detail of the instrument operation leads to introduce matrices of sensitivities $dK_{ij}$, alignments ($\theta_{ij} = -\theta_{ji}$) and couplings ($\eta_{ij} = \eta_{ji}$) between the axes, plus biases, $b_{0k}$, due to the read-out circuits, electrostatic disturbing forces corresponding to any differences between the measured and the actual ones and quadratic non linear terms:

$$\overrightarrow{\Gamma}_{mes,k} = \underbrace{\overrightarrow{b}_{0k}}_{bias} + \Big([1+\underbrace{dK_{1k}}_{scale}] + \underbrace{[\eta_k]}_{coupl.}\Big)\cdot\underbrace{[\theta_k]}_{align.}\cdot\left(\overrightarrow{\Gamma}_{App,k/sat} + \frac{\vec{F}ext_{/sat}}{M_{Isat}} + \frac{\vec{F}th_{/sat}}{M_{Isat}}\right)$$

$$+\Big([1+\underbrace{dK_{1k}}_{scale}] + \underbrace{[\eta_k]}_{coupl.}\Big)\cdot\left(-\frac{\vec{F}pa_{k/inst,k}}{m_{Ik}} - \frac{\vec{F}el,par_{k/inst,k}}{m_{Ik}}\right) + \underbrace{K_{2k}}_{quad}\,\Gamma\Sigma^2_{App,k/sat} \tag{6}$$

The stability of the introduced parameters of the instrument is of high importance for the accuracy of the experiment. In addition, common contributions and differences in the measured signals for both test-masses have to be separated. The first is used by the satellite computer for the compensation of the satellite drag. The second provides the EP test measure.

$$\begin{cases}\Gamma_{mes,c} = \frac{1}{2}\left(\Gamma_{mes,1} + \Gamma_{mes,2}\right)\\ \Gamma_{mes,d} = \frac{1}{2}\left(\Gamma_{mes,1} - \Gamma_{mes,2}\right)\end{cases} \tag{7}$$

with:

$$\overrightarrow{\Gamma}_{mes,d} = \underbrace{\overrightarrow{b}_{0d}}_{bias} + \Big([1+\underbrace{dK_{1c}}_{scale}] + \underbrace{[\eta_c]}_{coupl.}\Big)\,\overrightarrow{b}_{1d}$$

$$+\Big([1+\underbrace{dK_{1c}}_{scale}] + \underbrace{[\eta_c]}_{coupl.} + \underbrace{[d\theta_c]}_{align.}\Big)\cdot\overrightarrow{\Gamma}_{app,d/sat} + \Big(\underbrace{[dK_{1d}]}_{scale} + \underbrace{[\eta_d]}_{coupl.}\Big)\cdot\overrightarrow{b}_{1c}$$

$$+\Big(\underbrace{[dK_{1d}]}_{scale} + \underbrace{[\eta_d]}_{coupl.} + \underbrace{[\theta_d]}_{align.}\Big)\cdot\left(\overrightarrow{\Gamma}_{app,c/sat} + \frac{\vec{F}ext_{/sat}}{M_{Isat}} + \frac{\vec{F}th_{/sat}}{M_{Isat}}\right)$$

$$+\frac{1}{2}\underbrace{K_{2i}}_{quad}\,\Gamma^2_{App,i/sat} - \frac{1}{2}\underbrace{K_{2j}}_{quad}\,\Gamma^2_{App,j/sat} + O(dK, d\eta, d\theta)^2 \tag{8}$$

with index $c$ and $d$ specifying common and difference of the considered parameter defined as for the measured acceleration, and with:

$$\overrightarrow{b}_{1k} = \left(-\frac{\vec{F}pa_{k/inst,k}}{m_{Ik}} - \frac{\vec{F}el,par_{k/inst,k}}{m_{Ik}}\right) \tag{9}$$

considered in the instrument frame like $\overrightarrow{b}_{0k}$

$$\Gamma_{app,d/sat} = \frac{1}{2}\underbrace{(\delta_2 - \delta_1)}_{\delta} g + \frac{1}{2}\left(T_{/sat} - In_{/sat}\right).\Delta_{/sat} \tag{10}$$

where $T$ is the gravity gradient tensor (Earth and satellite source), $I_n$ is defined in (3) and $\Delta$ the distance between the two masses, all terms to be minimised; the signal proportional to $\delta$ is the one to be detected;

$$\begin{aligned}\Gamma_{App,c/sat} &= \left(\frac{M_{Gsat}}{M_{Isat}} - \frac{1}{2}(2+\delta_1+\delta_2)\right)\overrightarrow{g}(O_{sat})_{/sat} \\ &\quad - \frac{1}{2}[T]_{/sat}\left(\delta_1\overrightarrow{O_{sat}O_1} + \delta_2\overrightarrow{O_{sat}O_2}\right) - [T]_{/sat}\overrightarrow{O_{sat}O_M} \\ &\quad + \frac{1}{2}R_{In,COR/sat}\left(\overrightarrow{O_{sat}O_M}\right)\end{aligned} \tag{11}$$

with $M$ the middle of $O_1O_2$.

The term in $\overrightarrow{\Gamma}_{App,c/sat} + \frac{\overrightarrow{F}ext_{/sat}}{M_{Isat}} + \frac{\overrightarrow{F}th_{/sat}}{M_{Isat}}$ is fully rejected, only when the sensitivities, alignments and couplings of both sensors are matched. That is why the satellite drag-free system must minimise this term in presence of residual instrument asymmetries. *Fth/sat* is then managed by the drag-free servo-loop control to maintain, in the satellite control bandwidth and especially in the measurement one:

$$\rho\overrightarrow{\Gamma}_{mes,i} + (1-\rho)\overrightarrow{\Gamma}_{mes,j} = \overrightarrow{\Gamma}_{resDF} + \overrightarrow{C} \tag{12}$$

with $0 \le \rho \le 1$ and $\overrightarrow{\Gamma}_{resDF}$, $\overrightarrow{C}$ respectively the residue of the servo-control and the acceleration to be followed, according to the linear combination of the two inertial sensor outputs (see specification in Table 2).

## 3 Satellite Configuration

The MICROSCOPE satellite is derived from Cnes MYRIADE cubic platform with five rectangular aluminium honeycomb panels and one launcher interface. All equipments are fixed on these lateral and upper panels and the payload case is integrated at the centre on the lower structure. Two symmetric solar panels with half a meter square of AsGA cells each are located on the topside of the satellite maintained in the orbital plane: limited sizes, fixed panels and fixed solar orientations should guarantee a soft microvibration environment (see Fig. 1). In addition no moving masses like tank carburant sloshing or operating reaction wheels have been accepted during the EP experiment phase. A finite element model of the self gravity of the satellite has also been computed, with 31800 elements, leading to a DC gravity field on each test-mass less than 6.5 $\times 10^{-10}$ ms$^{-2}$ and fluctuations at the EP frequency in the test direction of less than $1.1 \times 10^{-16}$ ms$^{-2}$.

The two opposite sides of the satellite, $X_s$ (in the orbital plane), carries the propulsion system: twelve Cesium electrical microthrusters are presently considered as well as cold gas proportional thrusters. The first solution avoids 20 kg of tank mass but requires 50 W more electrical power, i.e. 50% more solar panel area. The resolution shall be better than 0.1 μN. The second solution requires on each side two symmetric sets of three pressurised Nitrogen

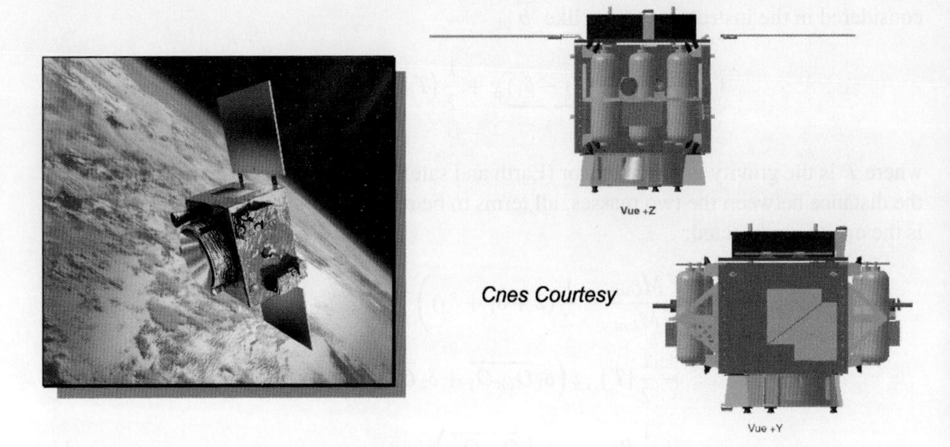


**Fig. 1** The two satellite configurations with Field Effect Electrical Propulsion thrusters (*left*) or cold gas proportional thrusters (*right*), 150 μN maximum range each; the two opposite $X_S$ panels carries the propulsion system; $X_S$, $Y_S$, axes are in the orbital plane, $Z_S$ is the spin axis. The Cnes innovative control system compensates for the contact forces on the satellite, including atmospheric drag and radiation pressure, so that the satellite actually follows the two masses on their natural free-fall trajectory. The control system uses as input the linear and angular acceleration measurements from the two masses of the instrument under test, as well as the star trackers measurements

tanks compatible with the 5 mm positioning accuracy of the satellite centre of mass and the 1% inertia accuracy: many disturbing sources have been analysed to ensure this solution, like self gravity gradients, temperature and thermal gradients, centrifugal acceleration, gas convection and consumption, piezo-servo valves motions... All effects are maximised by $10^{-16}$ ms$^{-2}$ (André et al. 2007).

The payload is composed of three units, the sensor unit (SU), the functional electronics unit (FEU) and the interface and control unit (ICU). The later manages the power supply of the instrument and the experiment processes, it interprets the telecommands and delivers the instrument data to the satellite bus. The two others require high thermal stability in order to meet high geometrical and electrical stability of the instrument arrangement, reduced radiometer and radiation pressure disturbances and electronics stability. They are mounted in a finally insulated two stage structure made in titanium with multi layer envelopes. The power consumption stability of the instrument is better than 5 mW for the electronics and one hundred times better for the sensor itself. So, a passive thermal control is compatible with the required stabilities, respectively:

- 1 mK for the tone fluctuations of the temperature, at the EP frequencies, $f_{EPi}$ or $f_{EPs}$, for the second stage on which are mounted the inertial sensor heads at the satellite centre; and stochastic fluctuations of 0.3 KHz$^{-1/2}$ from 0.1 mHz to 0.1 Hz;
- 5 mK/m at $f_{EPi}$ (2 mK/m at $f_{EPs}$) for the tone fluctuations of the thermal gradient of the same second stage; and 3 K/mHz$^{-1/2}$ from 0.1 mHz to 0.1 Hz in the axial direction;
- 10 mK at $f_{EPi}$ (3 mK at $f_{EPs}$) for the tone fluctuations of the first stage on which are mounted the functional electronics and 1 KHz$^{-1/2}$ from 0.1 mHz to 0.1 Hz for the stochastic fluctuations.

Expected tone fluctuations of temperature are even lower when the satellite is rotating and with an EP frequency at least four times larger.

The first stage is only conductively linked to its own external radiator, pointing anti-sun and protected from the Earth flux instabilities by a 54 cm diameter baffle. Because the star trackers and the inertial sensors measurements are jointly exploited on board, in the satellite attitude estimator, and also on ground, in the a posteriori data correction, the star trackers are fastened on the same anti-sun panel in order to obtain a fine alignment stability of its view cone axis with respect to the instrument spin axis.

## 4 Instrument Concept and Design

The MICROSCOPE instrument of the EP test experiment is mainly composed of four space inertial sensors operating at finely stabilised room temperature and associated by pairs constituting two differential accelerometers (Touboul et al. 2002). The sensor unit (SU) of each accelerometer comprises two quasi cylindrical and co-axial masses inside two concentric silica cores which are integrated in the same vacuum tight housing (see Fig. 2). At the top of the housing, a getter pumping device maintains during the entire mission the residual pressure level, $P_g$, of less than $10^{-5}$ Pa. All parts inside are backed out before integration to reduce their outgazing. At the bottom, a pair of three moving fingers clamps the two masses during the launch in order to sustain the vibrations. In orbit, the pyrotechnic valve of the blocking mechanism opens its pressurised reservoir and release the spring that move out the fingers.

Each test-mass is so surrounded by two gold coated silica cylinders with cuts away of the gold for defining six pairs of electrodes which are used for both, the capacitive sensing of the position and attitude of the mass, and the servo-loop control of its six degrees of freedom (see Fig. 3). The distance between the mass and the cylinders is 600 μm corresponding to a difficult compromise between on one hand, the capacitive sensing resolution and the capacity of control of the electrostatic actuators and, on the other end, the reduction of the electrical test-mass disturbances induced for instance by contact potential differences.

The four quadrant electrodes of the inner cylinder allows the control of the radial axes: two translations, $Y_i$ and $Z_i$, and two rotations about these axes. The cylindrical electrodes of the outer cylinder are symmetrically positioned around each end of the mass for the control

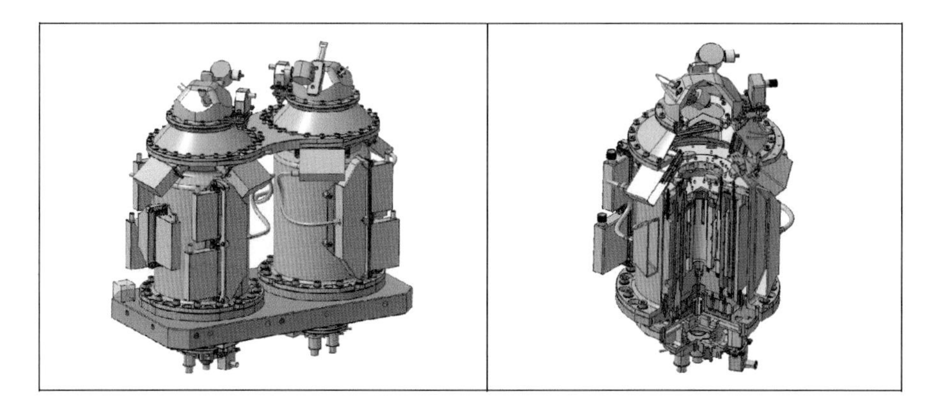

**Fig. 2** Two twin sensor units are integrated at the centre of the MICROSCOPE satellite; the two tight housing (*left*) includes at the top the vacuum getter system and at the bottom the blocking mechanism of the masses; each one includes two concentric quasi-cylindrical test-masses surrounded by their gold coated silica core constituting two inertial sensors (*right*)

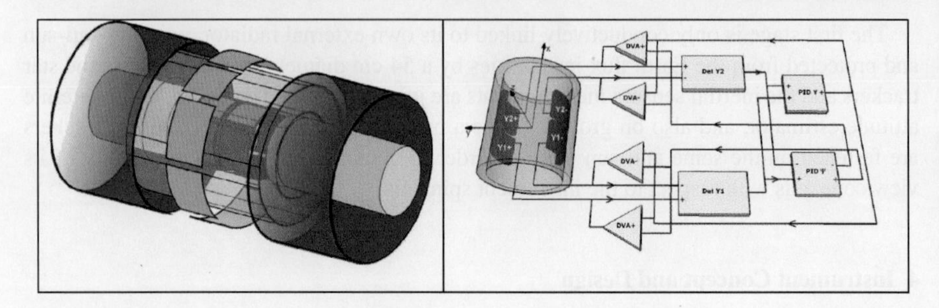


**Fig. 3** *Left*—Electrode configuration of one sensor unit: the mass *(in yellow)* includes an inner silica cylinder with four pairs of quadrant electrodes (*in blue* and *red*) devoted to the radial test-mass control; the outer gold coated cylinder supports two cylindrical electrodes (*in pink*) at the ends of the mass and four pairs of electrodes at the centre (*in green*) in regard to the flat areas for the axial control. *Right*—As an example, the two servo-loops associated to the two pairs of electrodes for the control of $Y_i$ and the rotation about $Z_i$

of the $X_i$ axis. The EP test is performed along this axis that is optimised to exhibit the best accuracy with a reduced electrostatic stiffness due to difference of electrical potentials between the mass and the electrode cylinders. The rotation of the mass about its axis of revolution, $X_i$, is measured by the last pair of electrodes: in fact, this pair is distributed all around the mass with eight quadrants in regard to four flat areas in order to break the symmetry of revolution of the mass. The width of these flat areas is a compromise between the accuracy of the capacitive sensing and the importance of the gravitational multi-poles of the mass.

The two silica electrode cylinders of the inner mass are finely grinded and represent the reference frame of the first inertial sensor. Cylindricity of the parts, better than 3 μm, and concentricity, better than 5 μm, are demanded. Great care is also paid to position and align them with respect to the two others for the outer mass.

When the inertial sensor is on, its test-mass is electrostatically levitated at the centre of its cage without any mechanical contact. Each capacitive sensor is maintained at null, i.e. the capacitances between the mass and the two electrodes are identical: a 100 kHz sine wave pumping signal, $V_d$, is applied to the mass and the collected signals on each electrode are compared. Six channels including digital controllers generate from the six capacitive sensing outputs the opposite electrical voltages $V_i$ which are applied on the opposite electrodes. Cold damping of all degrees of motion is provided in addition to a very accurate mass positioning (Grassia et al. 1999).

The electrical charge of the test-mass has to be controlled, in particular because of the high energy protons fluxes bombarding in orbit the satellite. That is why a thin gold wire of 5 μm diameter is used to fix the mass electrical potential to the electronics reference voltage: the stiffness and the damping factor of the wire have been carefully analysed and measured (Willemenot and Touboul 1999a, 1999b). Then, this wire is also used to apply on the mass the 100 kHz detection signal and a 5 V biasing voltage $V_p$: in presence of a perfect symmetric geometrical configuration and an anti-symmetric electrical one around the mass, the applied electrostatic force is proportional to $V_p$ and to $V_i$. By measuring $V_i$, one can deduce forces and torques applied to the mass, which is maintained motionless to the cage. So, the difference of acceleration and gravity applied to the test-mass is obtained and also the angular acceleration of the instrument.

The already mentioned FEU box includes the capacitive sensing of both of the masses, the reference voltage sources and the analogue electronics to generate the electrical voltages applied on the electrodes. The ICU comprises the digital electronics associated to the control

**Table 3** Local gravity gradient in the neighbourhood of the centres of the test-masses: this tensor can only be considered for very weak displacement about this centre

| Gravity gradient ($s^{-2}$) | | | |
|---|---|---|---|
| $T_{ij}$ | $X_i$ | $Y_i$ | $Z_i$ |
| $X$ | $-9.15\times10^{-7}$ | $-2.19\times10^{-8}$ | $1.72\times10^{-9}$ |
| $Y$ | $-2.19\times10^{-8}$ | $6.09\times10^{-7}$ | $-1.30\times10^{-9}$ |
| $Z$ | $1.72\times10^{-9}$ | $-1.30\times10^{-9}$ | $3.06\times10^{-7}$ |

laws of the servo-loops and the interfaces to the data bus of the satellite. It includes also the power converters.

The test-mass characteristics are major parameters for the success of the experiment. The geometry is machined with an accuracy of 1.5 µm by means and procedures developed in Physikalisch-Technische Bundesanstalt, Braunschweig, Germany. The material shall not include any non homogeneity larger than $5 \times 10^{-5}$ and the shape is selected to present a spherical matrix of inertia with a relative accuracy of $5.3 \times 10^{-6}$ for the inner mass and $2 \times 10^{-6}$ for the outer mass, out of diagonal component being less than $2.6 \times 10^{-6}$. In addition, the gravity multipoles are compatible with a weak sensitivity to the change of the nearest masses distribution due to thermoelastic behaviour of the surrounded parts. These effects have been computed with finite elements representation of the overall two accelerometers package with 128 286 elements for the instrument and 3840 mass points for the test-mass. The computed DC gravity field along $X_i$, $Y_i$, $Z_i$ is respectively ($-9.92 \times 10^{-9}$ ms$^{-2}$, $1.95 \times 10^{-08}$ ms$^{-2}$, $-1.39 \times 10^{-10}$ ms$^{-2}$). More important are the gravity gradients (see Table 3) and their fluctuations with temperature.

Sensitivities of the difference of the instrument gravity field, applied on each test-mass, have been evaluated, first to the relative mass position and rotation: sensitivities are less than $2.9 \times 10^{-7}$ ms$^{-2}$/m and $3.1 \times 10^{-10}$ ms$^{-2}$/rad leading respectively to differential perturbations at the EP frequencies of $1.8 \times 10^{-18}$ ms$^{-2}$ and $8.7 \times 10^{-20}$ ms$^{-2}$ when considering the expected temperature fluctuations. Displacement of one sensor with respect to the other has also been considered leading to a maximum of $5.7 \times 10^{-19}$ ms$^{-2}$. Dilatations of the parts introducing larger asymmetries have also been considered and leads in the worst case configuration to a maximum perturbation of $2.5 \times 10^{-16}$ ms$^{-2}$. Improvement in the selection of the material (from invar to titanium) should even lower this value.

The magnetic susceptibility of the selected material, $\chi_m$, is limited, $2.8 \times 10^{-4}$ for Platinum alloy and $1.4 \times 10^{-4}$ for titanium one, but requires nevertheless a sufficient shielding with respect to the Earth and satellite magnetic field fluctuations at EP frequencies. We have so evaluated, from the data of the Oersted space mission, the fluctuations of the Earth field that may be encountered along the orbit. We have in addition considered the worst case of the residual magnetic momentum of the MICROSCOPE satellite, i.e. 1 A m$^2$ with tone fluctuations at $f_{EP}$ of $10^{-3}$ A m$^2$ and stochastic variations of $4 \times 10^{-2}$ A m$^2$ Hz$^{1/2}$ at 0.3 m distance from the masses. We have computed the efficiency of the µmetal specific shield enclosing a large volume with both accelerometers in order not only to reduce the magnetic field but also its gradients. In complement, the sensor tight housing made in invar constitutes also a magnetic shield.

$$(\vec{\Gamma}_{m_{in}} - \vec{\Gamma}_{m_{out}}) \cdot \vec{X}_i = \frac{\chi_{m_{in}}}{2\mu_0 \cdot \rho_{m_{in}}} \vec{\nabla}\left[(\overrightarrow{B_{s/c}} + \overrightarrow{B_{Earth}})^2_{in}\right] \cdot \vec{X}_i - \frac{\chi_{m_{out}}}{2\mu_0 \cdot \rho_{m_{out}}} \vec{\nabla}\left[(\overrightarrow{B_{s/c}} + \overrightarrow{B_{Earth}})^2_{out}\right] \cdot \vec{X}_i \qquad (13)$$

In these conditions, the differential applied acceleration along the EP axis, $X_i$, is evaluated depending on the cross product of DC and EP frequency harmonics of both residual fields of the Earth and the satellite. In the worst case configuration, when the DC residual magnetic moment of the satellite is along $Y_i$, a maximum difference of acceleration of $2.08 \times 10^{-16}$ $\text{ms}^{-2}$ is obtained in inertial pointing of the satellite; and $1.19 \times 10^{-17}$ $\text{ms}^{-2}$ in case of rotating satellite when the DC residual magnetic moment is along $X_i$.

## 5 Instrument Performances

The performances of the electrostatic inertial sensors depend on: (i) the parasitic forces and damping applied on the test-masses; in absence of electrostatic control, the mass is not fully in free fall; (ii) the electronics noise and bias drifts to be considered through the transfer functions of the servo-loops; the mass is then not perfectly controlled motionless to the instrument frame and so to the other one; (iii) the performance of the pick-up electronics to extract, digitalise and store the measurement signal; the measured signal is not fully representative to the applied control acceleration with the right sensitivity.

Radiometer effects, involving residual low pressure gas and temperature gradients, as depicted first by Einstein (1924), and radiation pressures have been over-estimated according to the specified temperature gradients of the instrument gold coated silica cage surrounding the masses along the considered axis (see in Sect. 3, the thermal environment of the sensor; temperature to be considered here are inside the core, filtered out by the own instrument thermal inertia and insulation) by:

$$\Gamma_n = \frac{S}{m} \cdot \Delta T_{Si} \cdot \left( \frac{1}{2} \cdot \frac{P_g}{T} \oplus \frac{4\sigma}{3c} \cdot 4T^3 \right) \tag{14}$$

Tone radiation acceleration are evaluated in worst case to $1.2 \times 10^{-16}$ $\text{ms}^{-2}$ and radiometer to $7.9 \times 10^{-17}$ $\text{ms}^{-2}$ without considering any pressure re-equilibrium (Pollack et al. 2007). Outgassing of the parts has been also evaluated and leads to lower disturbances because of the selected materials and the backing of the parts.

Contact potential differences (CPD) between the mass and the silica coatings are also sources of DC force and fluctuations as well as asymmetries of the geometrical or electrical configuration around the mass. But fortunately, the mass is controlled motionless with respect to the silica cylinders, thus one only has to take into account the thermal and time variations at the EP test frequency and not in particular the DC values. This makes a large difference for the CPD which fluctuations are much smaller then their eventual DC patch. Variations of 15 $\mu$V/K and $(1 + 10^{-2}/f)$ $\mu\text{V/Hz}^{1/2}$ have been considered from in orbit results of the GRACE accelerometers (Touboul et al. 2004; Flury et al. 2008): larger levels should not be compatible with the exhibited $10^{-10}$ $\text{ms}^{-2}\,\text{Hz}^{-1/2}$ and bias stabilities.

Stiffness and damping of the thin gold wire glued on the mass and the silica support are also considered and remains a major limitation of the inertial sensor performance.

Capacitive position sensing fortunately exhibits very high performance, of a few $10^{-11}$ $\text{mHz}^{-1/2}$ depending on the electrical gain of the sensor and the geometrical configuration (Josselin et al. 1999). Details of the sensor characteristics are presented in Table 4 according to the mass and the considered axis. The back action of the sensor is also very important to be estimated. With 5 V detection signal on the mass, the induced electrostatic stiffness and force depend on the defects of symmetry of the electrostatic geometry and

**Table 4** Position capacitive sensing characteristics: gains, resolutions and thermal sensitivities (**a**); Power Spectral Density (PSD) of the sensor outputs in case of 40 V/pF gain (**b**)

| | Electronics detection gain (V/pF) | Physical detection gain (pF/µm or pF/mrad) | Global detection gain (V/m or V/rad) | Output noise (V/Hz$^{1/2}$) | Output noise (pm/Hz$^{1/2}$ or nrad/Hz$^{1/2}$) | Thermal Sensitivity (ppm/°C) |
|---|---|---|---|---|---|---|
| Outer mass | | | | | | |
| $X$ | 40 | $6.49 \times 10^{-3}$ | $2.60 \times 10^5$ | $1\times10^{-5}$ | 38.5 | 500 |
| $y/z$ | 5 | $6.06 \times 10^{-2}$ | $3.03 \times 10^5$ | $5\times10^{-6}$ | 16.5 | 500 |
| Tetha/psi | 5 | $9.54 \times 10^{-1}$ | $4.77 \times 10^3$ | $5\times10^{-6}$ | 1.05 | 500 |
| Phi | 80 | $1.24 \times 10^{-1}$ | $9.92 \times 10^3$ | $1.20\times10^{-5}$ | 1.21 | 500 |
| Inner mass | | | | | | |
| $X$ | 80 | $3.71 \times 10^{-3}$ | $2.97 \times 10^5$ | $1.20\times10^{-5}$ | 40.4 | 500 |
| $y/z$ | 16 | $1.37 \times 10^{-2}$ | $2.19 \times 10^5$ | $6.50\times10^{-6}$ | 29.7 | 500 |
| Tetha/psi | 16 | $1.06 \times 10^{-1}$ | $1.70 \times 10^3$ | $6.50\times10^{-6}$ | 3.83 | 500 |
| Phi | 80 | $1.76 \times 10^{-2}$ | $1.41 \times 10^3$ | $1.20\times10^{-5}$ | 8.52 | 500 |

(a)

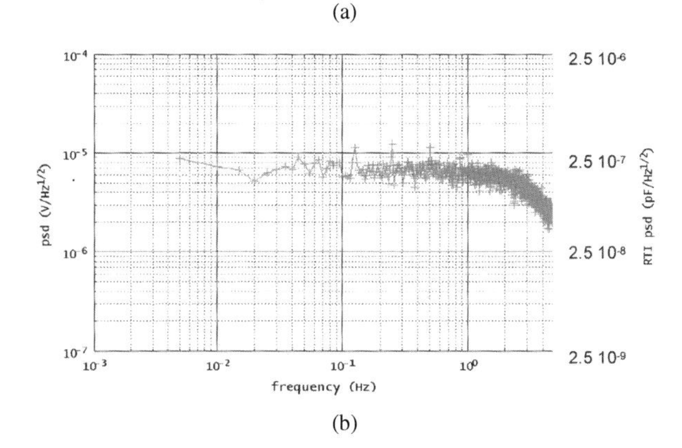


(b)

correspond, for the more sensitive outer mass and along the axial direction $X_i$, to a maximum noise of $8 \times 10^{-15}$ ms$^{-2}$ Hz$^{1/2}$ and a thermal drift of $7 \times 10^{-14}$ ms$^{-2}$/K with respect to the electronics temperature and $1 \times 10^{-14}$ ms$^{-2}$/K with respect to the sensor temperature.

For what concerns the electrostatic actuators, along the EP test axis, $X_i$, gains are respectively $3.25 \times 10^{-8}$ ms$^{-2}$/V and $1.83 \times 10^{-8}$ ms$^{-2}$/V for the outer and inner mass, the electrical noise of the electronics corresponding to $78.1 \times 10^{-15}$ N/Hz$^{1/2}$ and $43.9 \times 10^{-15}$ N/Hz$^{1/2}$.

The inertial sensor noise is so computed by quadratic sum of all stochastic contributions to the three above mentioned terms of the error budget. Figure 4 presents the obtained PSD of both sensors. At upper frequencies, the major contributor corresponds to the double derivation of the capacitive position sensing noise leading to $16 \cdot \pi^4 \cdot f^4 \cdot \mathrm{PSD}(x)$ law.

In the range from $10^{-3}$ to $10^{-2}$ Hz, read-out electronics noise of only 0.8 µV/Hz$^{1/2}$ have been obtained corresponding to $1 \times 10^{-13}$ ms$^{-2}$/Hz$^{1/2}$ and CPD fluctuations are major contributors.

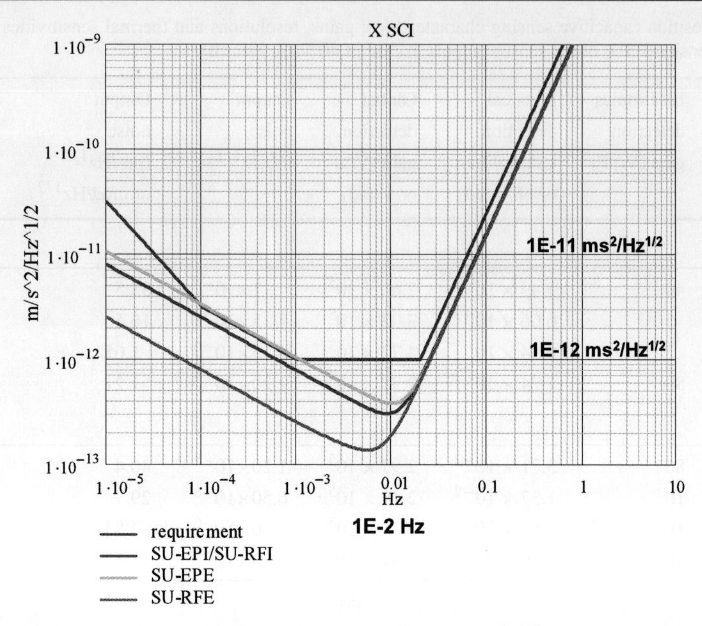


**Fig. 4** $(PSD)^{1/2}$ expressed in $\mathrm{ms}^{-2}\,\mathrm{Hz}^{-1/2}$ of all inertial sensors: Platinum external mass (*in purple*), Platinum internal mass (*in red*), Titanium external mass (*in green*), objectives to be reached (*in blue*)

In the lower frequencies domain, Niquist fluctuation/dissipation due to gold wires is preponderant corresponding to $1/f$ PSD law (Willemenot and Touboul 1999b).

$$\Gamma_n = \frac{1}{m}\sqrt{4k_B T \cdot \frac{S}{2\pi f \cdot Q}} \tag{15}$$

with $S$, gold wire stiffness, evaluated to $3.8 \times 10^{-8}$ N/m and quality factor $Q$ to one hundred at the EP test frequencies. At frequencies much lower then the orbital frequency, thermal fluctuations become more important and correspond to $1/f^2$ PSD.

## 6 Experiment Systematic Errors and in Orbit Calibration

In addition to the stochastic errors that can be reduced by the integration of the measured signal or the number of the experiment sessions, systematic errors have been considered: they are eventual sine wave signals at same frequency and phase as the Earth point gravity projection on the test-mass axial direction. They are mainly due to thermal sensitivities of the electronics or of the instrument core associated with sine wave fluctuations of the temperature at the EP test frequency, $f_{EP}$. Magnetic field fluctuations have been already considered above and self gravity of the satellite and the instrument too. Earth gravity gradients are also major sources of systematic errors.

The instrument core benefits of a temperature environment stability of 1 mK at $f_{EP}$. In addition, inside the tight housing, parts are insulated: the response time of the levitated mass temperature is larger than 60 hours and of the silica core larger than 1 hour and half. In these conditions, the two major contributors are the fluctuations of the gold wire stiffness due to the relative Yong Modulus sensitivity ($1/E \cdot dE/dT = 4 \times 10^{-4}$/K) and the variations of contact potential differences leading respectively to $1.6 \times 10^{-16}$ ms$^{-2}$ and $5 \times 10^{-17}$ ms$^{-2}$ disturbances. Tone fluctuations of the thermal gradient in the tight housing generate also systematic errors, respectively evaluated to $8 \times 10^{-17}$ ms$^{-2}$ for the radiometer effect and $1.2 \times 10^{-16}$ ms$^{-2}$ for the radiation pressure.

The 3 mK tone thermal variations of the electronics leads in particular to variations of the reference voltages applied on the mass, $V_p$ at DC and $V_d$ at 100 kHz. This corresponds, in presence of geometrical defects of symmetry of the mass or of the electrode parts, to $1.5 \times 10^{-16}$ ms$^{-2}$ tone error.

A direct sum of all these terms and others (not mentioned here but smaller) corresponds to an artefact signal of $6.2 \times 10^{-16}$ ms$^{-2}$ for the Titanium mass sensor and $2.9 \times 10^{-16}$ ms$^{-2}$ for the Platinum one with a full scale range of respectively $2.6 \times 10^{-7}$ ms$^{-2}$ and $2.3 \times 10^{-7}$ ms$^{-2}$.

Thanks to the drag-free loop of the satellite, the instrument output is maintained at a mean null value to better than $10^{-8}$ ms$^{-2}$. So, the evaluated thermal variations of the instrument sensitivity, limited to less than $10^{-8}$, do not introduce a major disturbance in addition.

Effects of the Earth gravity gradient, tensor $[T]$, requires beside the already mentioned characteristics of the masses, the centring of the two masses (see (10)). In fact, when considering only the Earth monopole, $[T]$ can be expressed in the instrument reference frame, and in the case of an inertial pointing satellite, by:

$$
\begin{aligned}
T_{11} &\approx \frac{1}{2}\frac{\mu}{R^3}\left[1 + 3\cos(4\pi f_{orb}t) + e\left(-\frac{3}{2}\cos(2\pi f_{orb}t - \psi) + 3\cos(2\pi f_{orb}t + \psi)\right.\right.\\
&\quad \left.\left. + \frac{21}{2}\cos(6\pi f_{orb}t + \psi)\right)\right]\\
T_{12} &= T_{21} \approx 0\\
T_{13} &= T_{31} \approx \frac{1}{2}\frac{\mu}{R^3}\left[-3\sin(4\pi f_{orb}t) + e\left(\frac{3}{2}\sin(2\pi f_{orb}t - \psi) - \frac{21}{2}\sin(6\pi f_{orb}t + \psi)\right)\right]\\
T_{22} &\approx -\frac{\mu}{R^3}\left[1 + 3e\cos(2\pi f_{orb}t + \psi)\right]\\
T_{23} &= T_{32} \approx 0\\
T_{33} &\approx \frac{1}{2}\frac{\mu}{R^3}\left[1 - 3\cos(4\pi f_{orb}t) + e\left(\frac{3}{2}\cos(2\pi f_{orb}t - \psi) + 3\cos(2\pi f_{orb}t + \psi)\right.\right.\\
&\quad \left.\left. - \frac{21}{2}\cos(6\pi f_{orb}t + \psi)\right)\right]
\end{aligned}
\tag{16}
$$

With $R$ the orbit means radius and $e$, the orbit eccentricity. The major terms of $[T]$ are at DC and $2f_{orb}$. Nevertheless, the eccentricity creates a term at $3f_{orb}$ and overall at $f_{orb}$, i.e. the EP experiment frequency. Eccentricity is constrained by the launcher to less than $5 \times 10^{-3}$ but not better. That is why the relative off centring of the two masses, $\Delta$, less than 20 μm after careful integration, is not sufficiently weak to make this term negligible in the differential measurement (see (8)). The MICROSCOPE approach is so to evaluate in orbit the off centring along $X_\iota$ and $Z_\iota$ in order to correct the measurement with a model of the Earth gravity field. To do so, the a posteriori knowledge of the satellite position and attitude must be sufficiently accurate as well as the timestamp of the measurements. Due to uncertainties in the changes of reference frame between the instrument measurement axes and the gravity gradient model corrections, the centring of the masses in the $Y_i$ is also evaluated. Off-centrings can be measured in orbit, by the observation of the signal, either induced by $[T]$ at 2 $f_{orb}$, or induced by $[In]$ (see (3)) when the satellite is voluntarily excited in attitude at a calibration frequency (Guiu et al. 2007).

By biasing the attitude control and the drag compensation servo-loops of the satellite, linear or angular sine wave accelerations can be applied to the satellite via its propulsion system. This will be done during the in orbit calibration phase of the instrument that helps to match the sensitivity matrices of the sensors reducing in (8) the terms in $d\Theta_c$ and in $dK_d$ and $\Theta_d$.

Performances better than the following specifications have been demonstrated by simulations: $d\Theta_c(Y_i \text{ or } Z_i) < 10^{-3}$ rad; $dK_d < 1.5 \times 10^{-4}$; $\Theta_d \cdot (Y_i \text{ or } Z_i) < 5 \times 10^{-5}$ rad; $\Delta_y < 4$ µm; $(1 + dKc) \cdot \Delta < 0.1$ µm ($X_i$ or $Z_i$).

The stability of these sensor characteristics is sufficient between two in orbit calibrations. The frequency of these calibration phases has been selected according to the long term temperature stability of the payload case depending on the orbital conditions. Corrections of the measurement files are performed on ground by taking into account the calibrated factors. Then the rejection of the common mode accelerations of the test-masses is improved and the differential accelerations reduced.

## 7 Mission Performance and Discussion

From (10) and (11), common mode and differential accelerations can be evaluated taking into account the satellite orbital and attitude motions and the instrument characteristics. First, the overall error budget of all six measurements of each sensor have been performed in such a way to verify the expected performance of the attitude and drag compensation system of the satellite, and also to take into account alignment and coupling between axes in the differential measurement. Then, the differential measured acceleration is evaluated at the EP frequency after the ground calibration: this measure includes or not the violating signal.

In the performed uncertainty analysis, worst case computations have been performed and stochastic errors are all summed quadratically while systematic errors are directly summed without considering statistic distributions of the amplitude and phase of the computed tone errors. More then two hundred terms have been evaluated.

For the stochastic errors, the first four error sources are summarised in Table 5 in the case of a rotating pointing satellite corresponding to $f_{EP}$ frequency of about 8 $\times 10^{-4}$ Hz. The major term is due to Niquist noise introduced by the damping of the 5 µm diameter, 2.5 cm length gold wires between the masses and the instrument cages. This term is quite identical to the quadratic sum of all terms that is 1.48 $\times 10^{-12}$ ms$^{-2}$ Hz$^{1/2}$. Increasing the frequency

**Table 5** Four major stochastic errors of the MICROSCOPE EP test experiment

| Terms in the equation of measure | PSD$^{1/2}$ at $f_{EP}$ | Effects, sources |
|---|---|---|
| Differential accelerometer noise | $1.40 \times 10^{-12}$ ms$^{-2}$ Hz$^{1/2}$ | Mass damping |
| Bias sensitivity to SU thermal gradient | $0.38 \times 10^{-12}$ ms$^{-2}$ Hz$^{1/2}$ | Radiation & radiometer force |
| Satellite AOCS & mass off centring | $0.24 \times 10^{-12}$ ms$^{-2}$ Hz$^{1/2}$ | Centrifugal accelerations Diff. |
| Scale factor sensitivity to FEU temp. | $0.13 \times 10^{-12}$ ms$^{-2}$ Hz$^{1/2}$ | Voltage ref. source instability |

**Table 6** Six major systematic errors of the MICROSCOPE Equivalence Principle test experiment

| Terms in the equation of measure | Tone error amplitude at $f_{EP}$ | Effects, sources |
|---|---|---|
| Projection of diff. accel. | $5.2 \times 10^{-16}$ ms$^{-2}$ | Instability of star tracker axis versus instrument axis |
| Scale matching & AOCS | $5.0 \times 10^{-16}$ ms$^{-2}$ | Drag free residual acceleration |
| Bias fluctuations, $b_1$ | $4.0 \times 10^{-16}$ ms$^{-2}$ | Magnetic parasitic force fluctuations |
| Bias sensitivity to SU temp. | $3.6 \times 10^{-16}$ ms$^{-2}$ | Gold wire stiffness thermal variations |
| Scale factor sensitivity to FEU temp. | $3.1 \times 10^{-16}$ ms$^{-2}$ | Thermal fluctuations of the electrostatic actuator gains |
| Bias sensitivity to FEU temp. | $2.8 \times 10^{-16}$ ms$^{-2}$ | Electrostatic configuration dissymmetry: thermal variations of the resultant pressure |

of the EP test should reduce this term but the spin of the satellite is limited by the strength of the thrusters and the satellite centring.

The second term corresponds to thermal gradient noise inside the accelerometer cores which generate fluctuations of the radiation and radiometer forces. The third one is the effect of the residual angular acceleration noise of the satellite that is not identical on the two masses because of the off centring. The last one corresponds to the reference voltage thermal fluctuations: this is the reference of the whole electrostatic configuration around the mass and it contributes to the scale factor of the performed acceleration measurement.

Table 6 presents the major systematic errors, at the EP frequency and computed in a worst case approach. In fact the direct sum of all these errors leads to 4.81 $\times 10^{-15}$ ms$^{-2}$. This is a majoring value of the actual disturbance that could be interpreted as a $5\sigma$ occurrence ($1\sigma$ value is then 0.96 $\times 10^{-15}$ ms$^{-2}$). All sine wave disturbing signals (more than fifty) should not be in the experiment at worst cases, so at maximum amplitudes, neither exactly at the same phase than the expected EP signal. For instance, the temperature of the sensor parts and the electronics components are submitted to fluctuations, with a contribution at EP frequency, but depending of the thermal flux propagations, not at the same phase. In addition, different EP test phases and frequencies are performed during the one year mission. By performing a quadratic sum of these systematic errors, a more reduced contribution is obtained, 1.17 $\times 10^{-15}$ ms$^{-2}$, this value seems to be coherent with the previous estimation of the $1\sigma$ error from the direct sum.

The first term corresponds to the instabilities of orientation of the sensor star trackers mounted on the satellite anti-sun face with respect to the instrument axes. The satellite structure is specifically rigid here but thermal gradients effects may introduce fluctuations of the orientations that will limit the rejection of the differential accelerations applied on both masses. The second term is introduced in the differential acceleration, $\Gamma d$, by the remaining defects (after in orbit calibration and ground correction) in the sensor axes scale factor matching and alignment, leading to a contribution of the residual common mode acceleration which is not fully nullified by the satellite drag compensation system. The third is induced by the fluctuations of the residual magnetic field distribution, around the two masses exhibiting different volumes and susceptibilities. The forth is due to the thermal sensitivity of the gold wire stiffness, the mass being controlled at null of the electrostatic configuration and not at null of the wire spring. The two last mentioned terms correspond to the thermal fluctuations of the reference voltage $V_p$ and $V_d$: in presence of a configuration asymmetry, these voltage fluctuations result in variations of electrostatic pressures on the test-masses

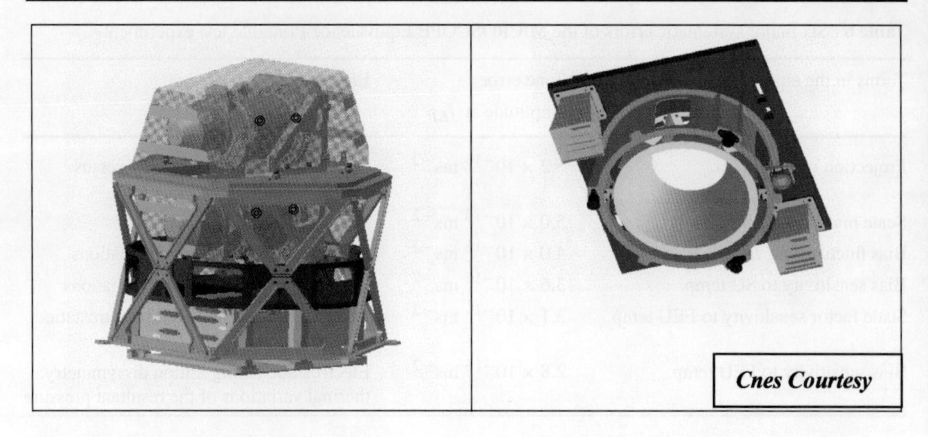


**Fig. 5** Payload case (*left*) and bottom satellite plate (*right*); the two differential accelerometer including each two inertial sensor (*in gold*) are included inside the magnetic shield; the purple structure separates the accelerometers to the electronics unit; the bottom plate support the radiator for the electronics (not drawn here) at the centre of a conic baffle to protect it from the Earth radiation; the two sensor star trackers are mounted on the same plate

and variations of the gain of the electrostatic actuators. It appears clearly that the thermal behaviour of the payload case with its satellite bottom face, supporting the radiator with its baffle and the two star sensor trackers, is a key element of the experiment success. That is why, dedicated tests will be performed in the near future to assess the first obtained experimental results and calibrate the numerical thermal simulations.

By considering the duration (20 orbits) of one experiment session, the stochastic error corresponds to $4.3 \times 10^{-15}$ ms$^{-2}$ rms value ($1\sigma$), to be considered in addition to the systematic resultant error evaluation. This results in a total of $4.4 \times 10^{-15}$ ms$^{-2}$ or $6.4 \times 10^{-15}$ ms$^{-2}$ depending on the selected evaluation for the systematic error (quadratic or direct sum). This level must be compared to the in orbit Earth gravity at 810 km of about 8 ms$^{-2}$: EP test accuracy should be then from $0.55 \times 10^{-15}$ to $0.80 \times 10^{-15}$.

By considering longer experiment sessions of 120 orbits, as it is presently the objective of the satellite operation, the total evaluated error is $2.1 \times 10^{-15}$ ms$^{-2}$ or $5.1 \times 10^{-15}$ ms$^{-2}$ corresponding to EP test accuracy of $0.26 \times 10^{-15}$ and $0.64 \times 10^{-15}$. During the one year mission and in addition to the commissioning phase of the satellite and the instrument, at least fifty measurement sessions are presently foreseen in the in orbit experiment scenario. This is to the benefit of the rejection of the stochastic errors and to the characterisation of the instrument sensitivities and the systematic errors, in particular through the differential accelerometer including two test-masses made of the same Platinum alloy.

This uncertainty analysis shows that the present definition of the mission and the instrument are in agreement with the objectives of the test of the Equivalence Principle with $10^{-15}$ accuracy or better. Numerical simulations of the satellite operation agree with the specification of the attitude control and of the drag compensation. Electronics of the instrument have been tested and exhibits noise, bandwidth, gains and linearities in accordance. Prototype models of the instrument have been integrated with silica masses (small density) in order to be accurately tested in laboratory under one $g$. Operation and dynamics behaviour of the electrostatic servo-loops have been already verified, effort is now paid in a more detailed experimental analysis of the performance. In addition, free fall tests of the instrument are prepared with the ZARM drop tower facility in Bremen. This facility gives the opportunity to observe on board a capsule and during its 9 s vertical parabolic motion, the actual operation of the four inertial sensors, in their flight configuration and in microgravity. Flight models of the accelerometers should be ready for the integration in the MICROSCOPE satellite in 2010 for a mission in the following years.

**Acknowledgements** This work was supported primarily by Onera, the French Aerospace Lab in Châtillon, France. The author would like to express his gratitude to the MICROSCOPE mission staff of Cnes in Toulouse, the Observatoire de la Côte d'Azur in Grasse and Onera in Châtillon, for all exchanges about the mission, the satellite and the payload. He is particularly grateful to G. Métris, A. Robert and R. Chhun for many discussions. The research described in this paper was carried out in Onera Châtillon, the development of the instrument is performed under contract with Cnes. MICROSCOPE is a Cnes mission with ESA cooperation.

## References


Y. André et al., Impact de la propulsion gaz froid sur la mission MICROSCOPE. *Cnes Report* MIC-NT-S-0-962-CNS (2007)

S. Baessler et al., Improved test of the equivalence principle for gravitational self- energy. Phys. Rev. Lett. **83**, 18 (1999)

C. Brans, R.H. Dickes, Mach's principle and a relativistic theory of gravitation. Phys. Rev. **124**, 925 (1961)

D.G. Currie et al., A lunar laser ranging RetroReflector array for the 21st century, in *NLSI Lunar Science Conference* (2008)

T. Damour, Testing the equivalence principle: Why and how? Class. Quantum Grav. **13**, A33–A41 (1996)

T. Damour, J.P. Blaser, Optimizing the choice of materials in equivalence principle experiments, in *Particle Astrophysics, Atomic Physics and Gravitation*, ed. by J. Tran Than Van, G. Fontaine, E. Hinds (Frontières, Gif-sur-Yvette, 1994), pp. 433–440

T. Damour, F. Piazza, G. Veneziano, Violation of the equivalence principle in a dilaton-runaway scenario. Phys. Rev. D **66**, 046007 (2002)

A. Einstein, *The Meaning of Relativity* (Princeton University Press, Princeton, 1922) (1988)

A. Einstein, Theory of radiometer energy source. Z. Phys. **27**, 1–6 (1924)

D. Feldman, Z. Liu, P. Nath, Sparticles at the LHC. JHEP **04**, 054 (2008)

J. Flury, S. Bettadpur, B.D. Tapley, Precise accelerometry onboard the GRACE gravity field satellite mission. Adv. Space Res. **42**, 1414–1423 (2008)

F. Grassia et al., Quantum theory of fluctuations in a cold damped accelerometer. Eur. Phys. J. D **8**, 101–110 (1999)

E. Guiu et al., Calibration of MICROSCOPE. Adv. Space Res. **39**, 315–323 (2007)

G.D. Hammond et al., New constraints on short-range forces coupling mass to intrinsic spin. Phys. Rev. Lett. **98**, 081101 (2007)

L. Iorio, LARES/WEBER-SAT and the equivalence principle. Europhys. Lett. **80**, 40007 (2007)

V. Josselin, P. Touboul, R. Kielbasa, Capacitive detection scheme for space accelerometers applications. Sens. Actuators **78**, 92–98 (1999)

L. Lafargue, M. Rodrigues, P. Touboul, Towards low temperature electrostatic accelerometry. Rev. Sci. Instrum. **73**, 1 (2002)

J. Mester, R. Torii, P. Worden, N. Lockerbie, S. Vitale, C.W.F. Everitt, The STEP mission: principles and baseline design. Class. Quantum Grav. **18**, 2475–2486 (2001). See also http://einstein.stanford.edu/

R. Newman, Prospects for terrestrial equivalence principle tests with a cryogenic torsion pendulum. Class. Quantum Grav. **18**, 2407–2415 (2001)

A. Nobili et al., The GG project: Testing the Equivalence Principle in space and on Earth. Adv. Space Res. **25**, 1231–1235 (2000)

S.E. Pollack, S. Schlamminger, J.H. Gundlach, Outgassing, temperature gradients and the radiometer effect in LISA: a torsion pendulum investigation. arXiv:gr-qc/0702051v2 (2007)

G. Schäfer, Where do we stand in testing general relativity? Adv. Space Res. **32**(7), 1203–1208 (2003)

S. Schlamminger et al., Test of the equivalence principle using a rotating torsion balance. Phys. Rev. Lett. **100**, 04110 (2008)

T.J. Sumner et al., STEP (satellite test of the equivalence principle). Adv. Space Res. **39**, 254–258 (2007)

P. Touboul et al., MICROSCOPE, testing the equivalence principle in space. C. R. Acad. Sci. Paris, Sér. IV **2**, 1271–1286 (2001)

P. Touboul et al., The MICROSCOPE mission. Acta Astronaut. **50**(7), 433–443 (2002)

P. Touboul et al., In orbit nano-g measurements, lessons for future space missions. Aerospace Sci. Technol. **8**, 431–44 (2004)

C.M. Will, *Theory and Experiment in Gravitational Physics* (Cambridge University Press, Cambridge, 1985)
E. Willemenot, P. Touboul, On-ground investigations of space accelerometers noise with an electrostatic torsion pendulum. Rev. Sci. Instrum. **71**(1), 302–309 (1999a)
E. Willemenot, P. Touboul, Electrostatically suspended torsion pendulum. Rev. Sci. Instrum. **71**(1), 310–314 (1999b)
J.G. Williams, X.X. Newhall, J.O. Dickey, Relativity parameters determined from lunar laser ranging. Phys. Rev. D **53**, 6730 (1996)

Space Sci Rev (2009) 148: 475–487
DOI 10.1007/s11214-009-9558-x

# The *STEP* and *GAUGE* Missions

**Timothy J. Sumner**

Received: 19 May 2009 / Accepted: 19 June 2009 / Published online: 25 July 2009
© Springer Science+Business Media B.V. 2009

**Abstract** *STEP* is one of a number of missions now being developed to take advantage of the quiet space environment to carry out very sensitive gravitational experiments. Using pairs of concentric free-falling proof-masses, *STEP* will be able to test the Equivalence Principle (EP) to a sensitivity at least five orders of magnitude better than currently achievable on ground. The EP is a founding principle of general relativity and *STEP* is the most sensitive experiment of this type planned so far, aiming at 1 part in $10^{18}$. Recently the *GAUGE* mission was proposed to the European Space Agency and this contained a variant of the *STEP* experiment as its central payload element with a number of others probing various aspects of the quantum gravity interface. Both *STEP* and *GAUGE* payload concepts will be presented, together with their performance parameters.

**Keywords** STEP · GAUGE · Equivalence principle · Quantum gravity · Space missions

## 1 Introduction

In 1970 Chapman and Hanson made the first proposal to test the EP in space, by measuring the relative motion between two test bodies of different materials. One was mounted inside the other on springs and the combination rotated to bring the signal into a region of low sensor noise. This room temperature experiment intended a precision of one part in $10^{15}$. Following a collaboration with Chapman and Hanson, P.W. Worden and C.W.F. Everitt developed a somewhat different concept (Worden and Everitt 1972), which proposed independent masses and cryogenic technology to achieve sensitivities better than one part in $10^{17}$. Concept development continued over the next several years and a ground-based prototype was built by P.W. Worden. The name *STEP* (Satellite Test of the Equivalence Principle) was given to this project in 1989 when it was proposed to ESA as a NASA/ESA collaborative mission, and from 1990 to 1996, various concepts for *STEP*

On behalf of the *STEP* and *GAUGE* collaborations.

T.J. Sumner (✉)
Imperial College London, London, UK
e-mail: t.sumner@imperial.ac.uk

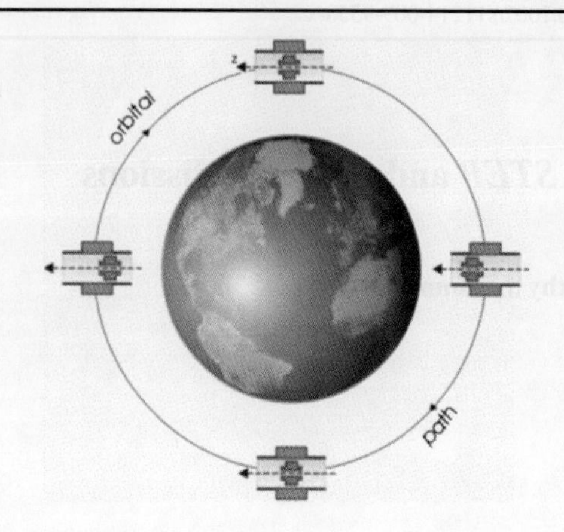


**Fig. 1** Cartoon of the *STEP* free-fall measurement concept in a near-Earth orbit. The illustration shows how a relative motion between two coaxial proof-masses would happen as a result of an equivalence principle violation causing the inner proof-mass to suffer a larger acceleration towards the Earth

were studied at Assessment and Phase A level for an ESA medium-sized project, M2 or M3. Since 1996 the concept has remained stable as a cryogenic instrument with four pairs of coaxial proof-mass cylinders forming independent differential accelerometers (DA) in undisturbed gravitational free-fall inside a drag-free satellite in low-Earth orbit as illustrated in Fig. 1. Lockerbie et al. (2001), Mester et al. (2001, 2004), Sumner et al. (2007), Kolodziejczak and Mester (2007) provide reviews of the *STEP* mission. A sensitivity of one part in $10^{18}$, a 5–6 orders of magnitude improvement over current knowledge, will be achieved.

The EP states the 'equivalence' between gravitational mass, $m_g$, and inertial mass, $m_i$, as they appear in $F = GMm_g/r^2$ and $F = m_i a$. This leads to bodies of different composition and/or mass falling with the same acceleration ("Universality of Free Fall"). This postulate ($m_g = m_i$) cannot be proven, it can only be tested to higher and higher precision. There is no a priori reason why it should be strictly valid. Today, the best experiments in the world have reached a level of a few parts in $10^{13}$, using two very different methods: a torsion balance in a ground-based laboratory (Baessler et al. 1999) and lunar-laser ranging (Williams et al. 1996). A non-null result would constitute the discovery of a new fundamental interaction of nature. A wide range of experimental tests trying to exploit the quiet environment of space are planned or ongoing (Sumner 2004).

Einstein was convinced that the EP is strictly valid and made it the foundation of General Relativity (GR). Although, up to now, GR has withstood all tests, there are reasons to believe that it may ultimately need some revision. The EP implies that gravity can be treated like any other force. However, the other known forces are quantized whereas gravity manifests itself as a distortion of space-time which cannot be quantized. Indeed, approaches to unified theories—'supersymmetry' and string theory—strongly suggest that there should be, associated with gravity, new very weak, long-range fields or 'moduli' leading directly to EP violations (Damour and Polyakov 1994; Damour et al. 2002a, 2002b). Violations are also expected in scenarios allowing time-variations of the fine structure constant $\alpha$ (Sandvik et al. 2002; Barrow and Li 2008; Barrow and Shaw 2008), anomalous couplings to the weak interaction (Fischbach et al. 1995), and from space–time fluctuations at the Planck scale (Göklü and Lämmerzahl 2008). To detect any of these violations would provide valuable new insight into possible modifications to General Relativity (Will 2006).

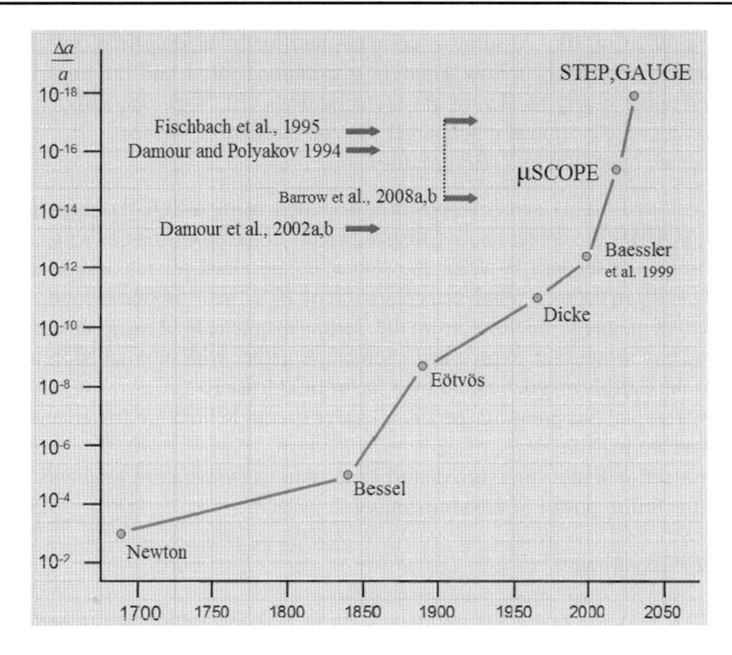


**Fig. 2** Historical progress in equivalence principle measurements. The potential of STEP/GAUGE to make significant and rapid progress is clear. Predicted theoretical levels of violation discussed in the text are indicated in the figure. The sensitivity of the MICROSCOPE ($\mu$SCOPE) mission being built by CNES/ESA (Hudson et al. 2007) is also shown

## 2 The *STEP* Mission

### 2.1 Introduction

The theoretical studies by Damour and Polyakov (1994) indicate that string dilaton effects could produce violations in the range parts in $10^{13}$ to parts in $10^{21}$. The new interactions would couple to a "charge" which is some linear combination of the atomic proton number, $Z$, and neutron number, $N$. Therefore the experiment must be designed with enough different materials to allow the coupling to both $Z$ and $N$ to be independently determined, while simultaneously providing some redundancy as a check against disturbances. To this end our baseline experiment has four DAs each with a pair of proof-masses in the form of concentric, hollow cylinders. Superconductivity provides frictionless magnetic bearings, ultra-stable magnetic shields, high accuracy SQUID position detectors, very low residual gas pressure, and small thermal gradients. A 230 litre superfluid helium dewar enables a lifetime of 8 months. The boil-off helium is used to feed 16 proportional gas thrusters which, controlled by the accelerometers in a feed-back loop, provide compensation for atmospheric-drag and solar radiation pressure. A drag-free performance of $10^{-13}$ ms$^{-2}$ averaged over the measurement bandwidth of $10^{-5}$ Hz around the signal frequency is required to do a 1 part in $10^{18}$ EP test. Data collection will last 6 months ($\approx$2800 orbits), although 1 part in $10^{18}$ requires only 20 orbits. Repeating the test many times, while systematically varying experimental parameters will strengthen the reliability of the final result.

The use of space offers two major advantages over any laboratory on the ground for fundamental physics measurements (Sumner 2008). Firstly the level of micro-seismicity is reduced by a factor of $10^3$. Air-drag is compensated for by a 'drag-free' satellite so that the proof-masses essentially follow a purely gravitational orbit and the experiments can be performed under nearly ideal zero-gravity conditions ($10^{-13}$ g). Secondly, in low-Earth orbit, the driving acceleration is much larger than on the Earth's surface.

The cryogenic operation of *STEP* provides:

- a very stable and highly sensitive position detection using a SQUID magnetometer. In principle, SQUIDs can detect relative displacements of the proof-masses with a sensitivity of $10^{-15}$ m in one second. In *STEP* the sensor is optimized for maximum acceleration sensitivity and can detect a relative acceleration of $10^{-15}$ ms$^{-2}$ in one second.
- almost perfect magnetic shielding from the Earth's magnetic field using superconductors. A thin lead bag around the experiment chamber, together with internal Niobium shields, attenuates the Earth's magnetic field by a factor of at least $10^{10}$.
- a very low residual gas pressure. At 2 K all gases except helium are frozen and pressures $< 10^{-10}$ torr are feasible.
- greatly reduced radiation pressure disturbances due to temperature gradients which decrease as the fourth power of the temperature.

### 2.2 The *STEP* Experiment

The four *STEP* DAs are operated simultaneously to maximize the quantity and quality of data returned. They work at a nominal temperature of approximately 2 K. Each DA (see Fig. 3) contains two cylindrical proof-masses that are constrained from moving radially by superconducting magnetic bearings. Motion along the cylinder axis is measured precisely by SQUID magnetometers, and a capacitance pick-off measures motion in all degrees of freedom at lower resolution. The magnetic reaction force from the SQUID sensing coils provides a weak restoring spring in the axial direction. The SQUID outputs are arranged to measure differential mode acceleration and common mode acceleration independently (Fig. 4). On the left of the figure is shown a single proof-mass with superconducting magnetic sensing coils at either end. As the superconducting proof-mass moves along its bearing the inductances of the coils change (one increases and the other decreases) and the currents through coils will then change differentially to maintain a constant magnetic flux through them. A current adjustment then flows through the inductance coupling into the SQUID. This provides a very sensitive measurement of the single proof-mass motion. On the right of the figure is shown how the circuits from the pair of nested proof-mass within each DA are combined. This uses two separate SQUIDs to simultaneously measure the differential and common-mode motions of the proof-masses. The coils themselves are deposited using thin-film technology onto the sensor coil formers shown in Fig. 3. Segmented capacitance electrodes are deposited onto the electrode surfaces surrounding the proof-masses and these allow the mass positions to be manipulated by applied voltages. The free charge on the masses can be estimated from the response of the masses to specific voltages used for this purpose, and the level of charge can be controlled by an ultraviolet discharging mechanism (Buchman et al. 1995). This will need to be done periodically throughout the mission due to the charging action of trapped radiation belt particles and cosmic-rays (Jafry et al. 1996). The masses will be caged during launch by a hydraulic actuator using pressurized helium as the working fluid.

The baseline choice of three materials, Pt–Ir, Nb, and Be, allows a cyclic closure condition; one pair, Pt–Ir/Be, is duplicated. This choice conservatively presumes that sensitive comparisons of two materials will be possible only within a single DA and the pairs of materials where chosen to maximize, as far as possible, the 'spread' of nuclear properties

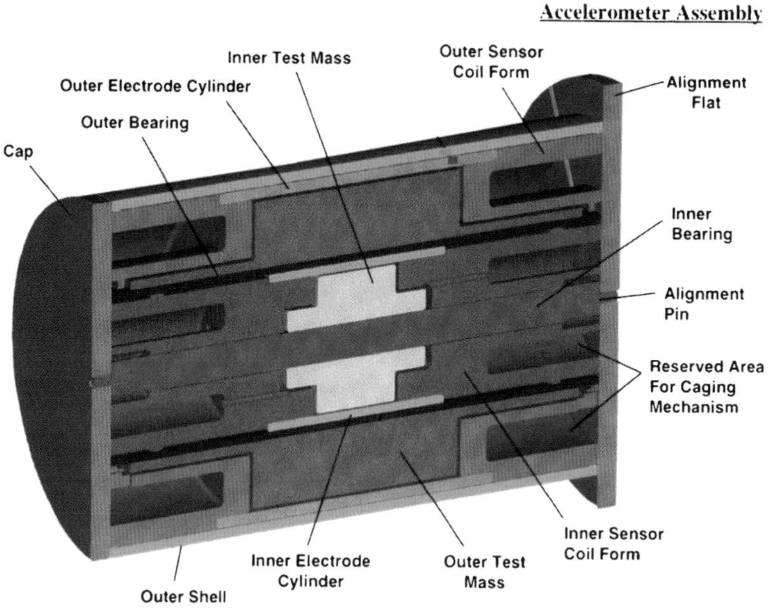

Accelerometer Assembly

Inner Test Mass
Outer Sensor Coil Form
Outer Electrode Cylinder
Alignment Flat
Outer Bearing
Cap
Inner Bearing
Alignment Pin
Reserved Area For Caging Mechanism
Inner Sensor Coil Form
Inner Electrode Cylinder
Outer Test Mass
Outer Shell

**Fig. 3** A cutaway representation of a differential accelerometer

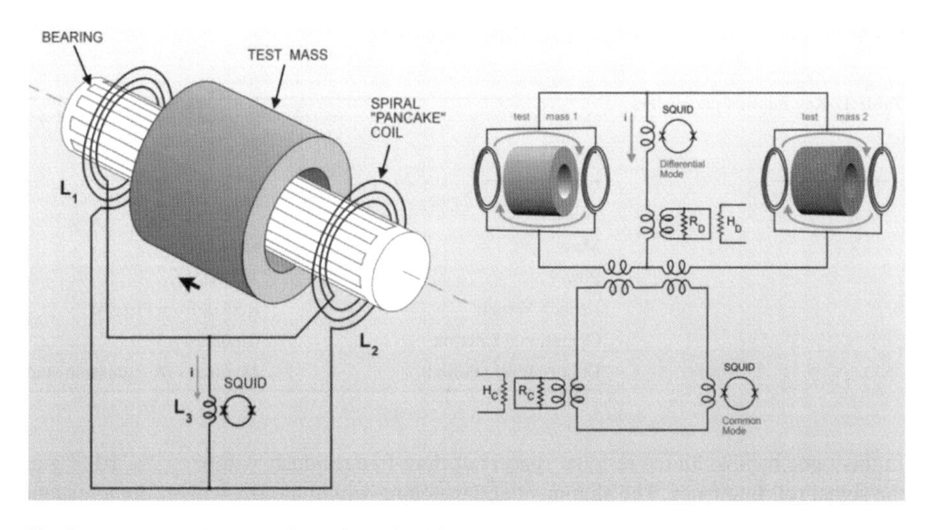


**Fig. 4** Arrangement for magnetic sensing using SQUIDs

(Blaser 2001; Blaser and Damour 2003). The proof-mass shapes (see Fig. 5) have been designed to minimize the coupling of their higher order gravitational multipole moments to the spacecraft (Lockerbie et al. 1993). This approach greatly reduces spurious gravitational effects due to motion and deformation of the spacecraft and particularly to systematic tidal movement of bubbles within the helium dewar (Lockerbie et al. 1994). The proof-masses are coated with a thin film of superconductor. The SQUID displacement sensors have a readout resolution in differential mode of $\sim 4 \times 10^{-19}$ g in an averaging time of 20 orbits.

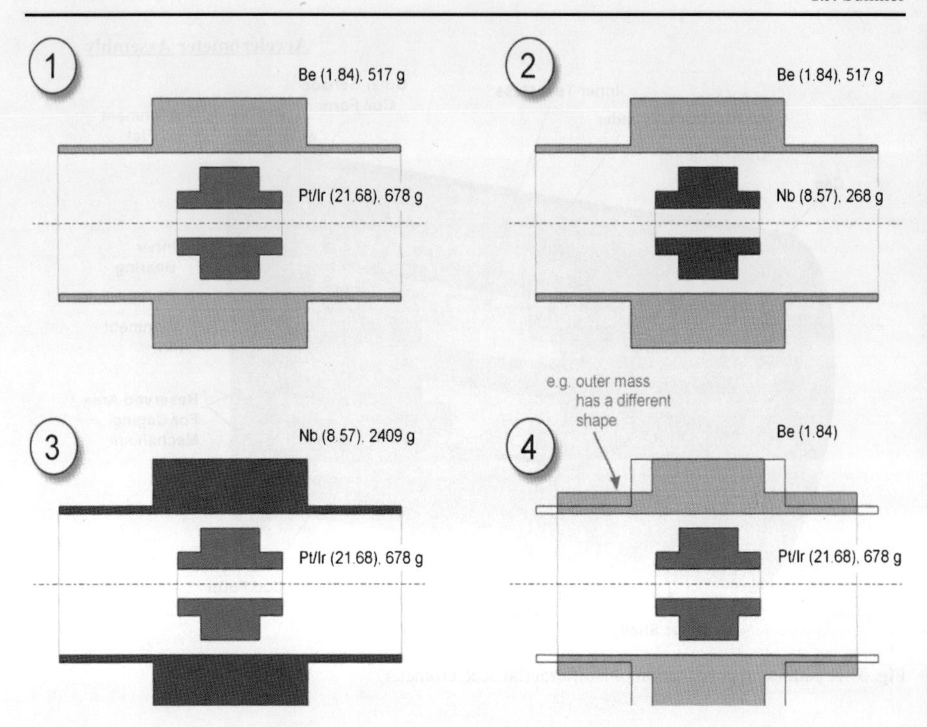


**Fig. 5** Proof-mass details for the four differential accelerometers

**Table 1** Key mission parameters

| | |
|---|---|
| Orbit type | Sun synchronous ($I = 97°$) |
| Altitude | 550 km |
| Eccentricity | <2% |
| Pointing | 3-axis stabilized to 2 arc sec |
| Mass | 819 kg |
| Power | 301 W |
| Launch Vehicle | Rockot from Plesetsk |
| Operational Lifetime | 6 months |
| Data Analysis Period | 18 months from mission start |

In this same bandwidth the residual spacecraft drag-free residuals will be $\sim 2 \times 10^{-15}$ g at the signal roll frequency. The alignment and matching within the DA will ensure a readout common mode rejection better than one part in $10^4$. All non-EP differential accelerations will be kept below $2 \times 10^{-19}$ g (Worden et al. 2001). The SQUIDs can also measure the common mode acceleration to about $1 \times 10^{-18}$ g over the same time span.

## 2.3 Spacecraft and Operations

The four DAs are held within a quartz block inserted in the helium dewar. They are arranged in two orthogonal pairs with orientation as shown in Fig. 6.

The accelerometers are contained in a superfluid helium dewar, part of a "drag-free" satellite (Fig. 7), which completely surrounds the entire instrument, protecting it from dis-

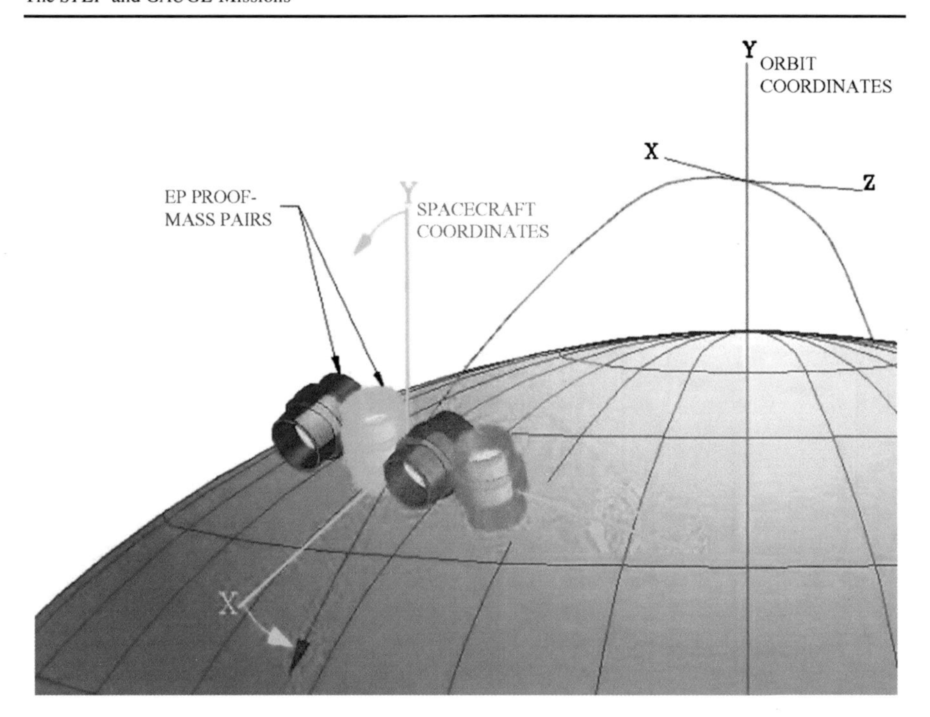


**Fig. 6** Arrangement of the four differential accelerometers. Payload and orbit configuration

turbances such as air drag, magnetic field, and solar pressure. The dewar incorporates a superconducting magnetic shield and ultrahigh vacuum chamber. It also includes aerogel packing to further mitigate any spurious effects due to tidal motion of bubbles developing within the helium (Dolesi et al. 1999, 2000). The satellite will precisely follow the proof-masses by using linear cold gas thrusters that cancel the drag, using boil-off from the helium dewar (Jafry 1992). The mission will also include a radiation particle sensor to improve the estimate of the proof-mass charging rates, an external magnetic field sensor, electronics to control the payload, communications equipment, a power supply, an on-board computer, and other support equipment.

The overall design attempts to avoid disturbances that could imitate an EP signal. The inherent stability and low losses in superconductors reduce the intrinsic sensor noise. Superconducting shielding suppresses external magnetic and electrical disturbances. Likewise the very low temperature ($\sim$2 K), temperature stability ($<$0.5 mK per orbit), uniformity, and pressure reduce most thermal disturbances to insignificant levels. The proof-mass shapes minimize disturbing gravity forces from the satellite, and the masses will be centered within $10^{-10}$ m to reduce the effect of the Earth's gravity gradient. A charge control system counteracts disturbances from particle radiation causing electrical charging (Buchman et al. 1995).

Active drag compensation reduces satellite acceleration by seven orders of magnitude across the measurement bandwidth (Wiegand et al. 2001). Helium boil-off gas is used as the propellant, and is vented through continuously variable thrusters.

Temperature control of the dewar can be assisted by a "null-dump" of excess helium for cooling without changing the net thrust. A sun-synchronous orbit (see Fig. 7) minimizes thermal changes to the satellite. The satellite requires no deployments and has no moving parts. The solar array acts as a sun shield for the dewar.

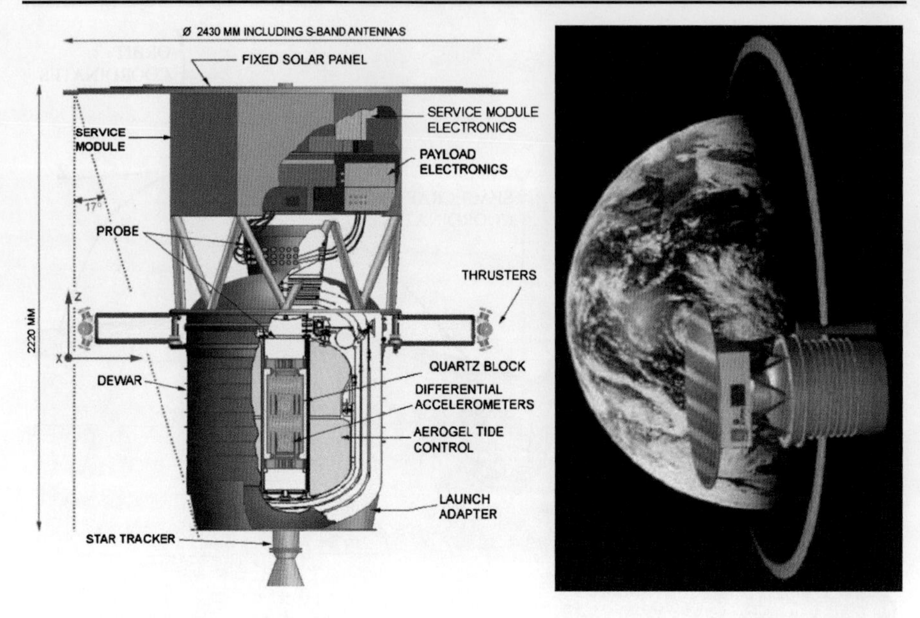


**Fig. 7** The *STEP* payload and orbit configuration

**Fig. 8** The simulated *STEP* drag-free performance. A $10^{-18}$ violation signal is shown for comparison. Note the disturbance level at the orbital frequency and how the use of a spacecraft roll moves the EP signal into a lower noise frequency band

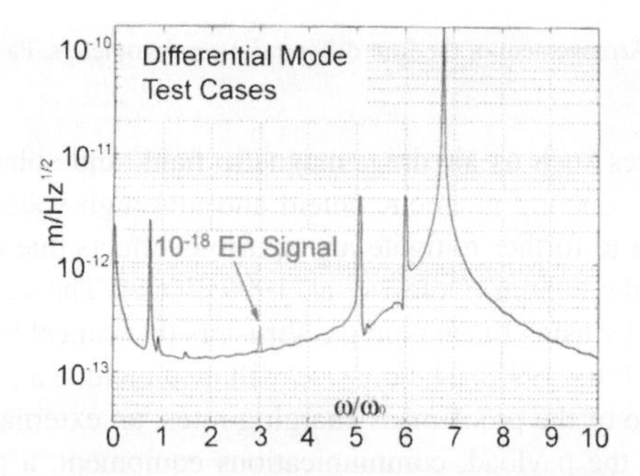


To distinguish an EP violation the satellite is rotated about the spacecraft $Z$-axis, normal to a plane parallel to the sensitive directions (or cylinder axis) of the accelerometers (Fig. 6). Any EP signal would appear at the difference in frequency between the orbit frequency and the rotation frequency. The rotation frequency will be changed between measurements to move the signal frequency away from fixed frequency disturbances. The orthogonal accelerometer pairs will have a 90 degree phase shift in their EP signals.

Validation of a measured result will be done by repeating the 20-orbit experiments many times over a six-month mission timeline. Some experiments would vary the experimental conditions, to identify systematic disturbances; for example, the electric charge on the proof-mass or the center of mass displacement can be varied, allowing direct measurement of electrical and gravitational forces. Towards the end of the mission the experimental parameters would be pushed to extremes to look for exaggerated systematics. In this way the experimental result should be very robust with a combination of an experiment design with a closure check, repetition and extensive exploration of systematics.

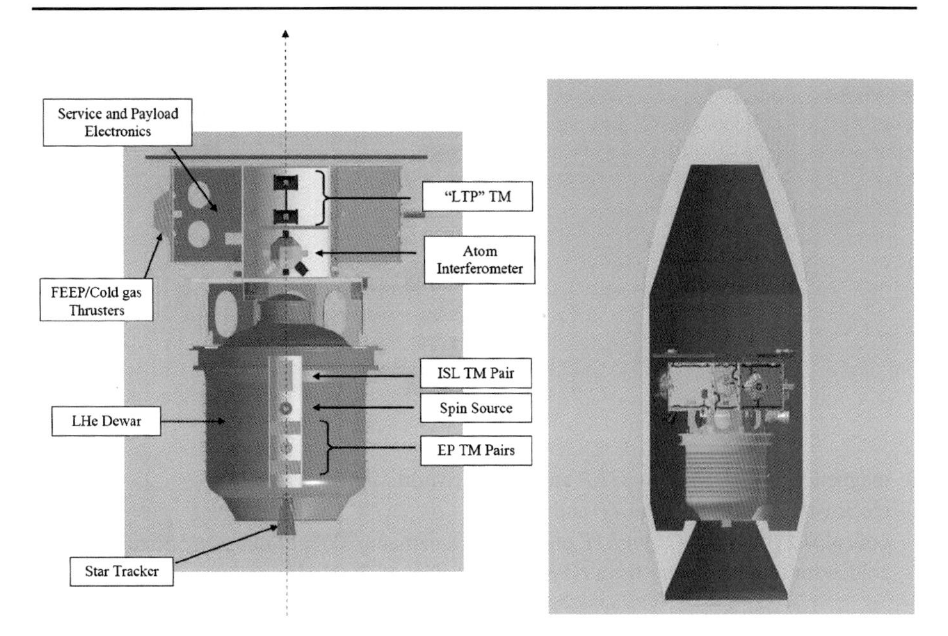


**Fig. 9** The *GAUGE* payload element identification and launch configurations

## 3 The *GAUGE* Mission Proposal

### 3.1 Introduction

*GAUGE* (Gr And Unification and Gravity Explorer) was submitted as a proposal to ESA for a medium-sized mission in fundamental physics (Amelino-Camelia et al. 2008). The concept was for a drag-free spacecraft platform onto which would be attached a number of modular experiments. The possible complement of experiments was designed to address a number of key issues at the interface between gravity and unification with the other forces of nature. Included are:

- A test of string-dilaton theories using a high precision macroscopic equivalence principle experiment.
- A test of the effect of quantum space–time fluctuations in a microscopic equivalence principle experiment.
- An inverse-square law test at intermediate ranges.
- An axion-like mass-spin coupling search.
- Measurement of quantum decoherence from Planck-scale space–time fluctuations.

### 3.2 Technological Concept

The list of proposed experiments was extensive and was selected to benefit from commonalities and heritages. For example the assumed drag-free platform was a replica of that of *LISAPathfinder* (*LISAPF*) including its gravitational reference sensor, the *LTP* (Vitale et al. 2002; Dolesi et al. 2003; Anza et al. 2005; McNamara 2006). Each experiment uses some variant of precision proof-mass displacement sensing with prior heritage:

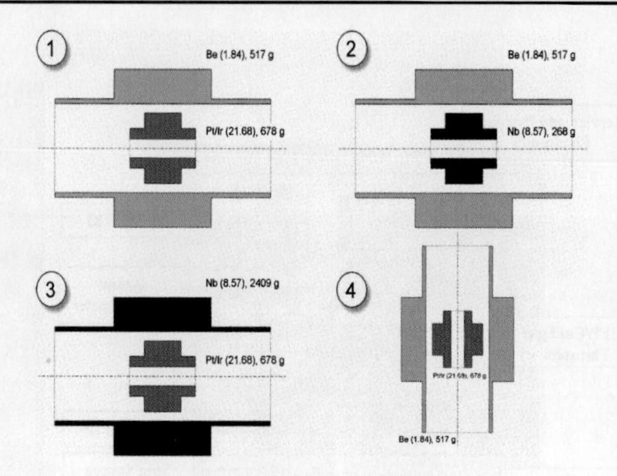


**Fig. 10** The modified *GAUGE* differential accelerometer selection

- magnetic (SQUIDS) from *STEP* and *GP-B* (Everitt 1988),
- electrostatic sensing from *LTP* and *STEP*,
- optical interferometry from *LTP* and *LISA* (Danzmann 2003; Sumner and Shaul 2004),
- cold atom interferometry from *HYPER* (Jentsch et al. 2004).

More specifically, the macroscopic equivalence principle test is closely derived from *STEP* (and *GP-B*), the microscopic equivalence principle and quantum decoherence tests are derived from *HYPER*, and the inverse-square law and spin-coupling tests derive partly from an early version of *STEP* (Speake et al. 1993a, 1993b) (and *SSPIN*, a specific spin-coupling proposal to ESA for M2) and partly from recent laboratory developments (Hammond et al. 2007; Speake et al. 1999). Variants of the inverse-square law and spin-coupling experiments using atom interferometry would be especially interesting with atom laser technology (Guerin et al. 2006).

The VEGA rocket would provide a launch into a low-Earth Sun-synchronous orbit.

## 3.3 Scientific Objectives

### *3.3.1 Macroscopic Equivalence Principle*

*GAUGE* will achieve a sensitivity of one part in $10^{18}$ using three proof-mass pairs working as differential accelerometers. Materials (Be:Pt/Ir, Be:Nb, Nb:Pt/Ir) are chosen to maximize sensitivity within a string-dilaton scenario (Damour and Polyakov 1994) and to provide a cyclic closure cross-check. The sensitivity grasp is also within the range required by theories incorporating fine structure time variation (see Fig. 2) and, with *STEP*, represents the most aggressive on offer (Sumner 2004).

### *3.3.2 Microscopic Equivalence Principle*

Time domain atom interferometry using cohabiting atomic species (Cs/Rb) offers an equivalence principle test with rather more modest sensitivity of (Dimopoulos et al. 2007), but which nonetheless could test theoretical scenarios in which quantum fluctuations induce violations for microscopic particles in this range (Göklü and Lämmerzahl 2008), but would be much reduced in macroscopic proof-mass.

### *3.3.3 Inverse Square Law*

At intermediate ranges (10–40 cm) the macroscopic *LTP* proof-masses can be used in conjunction with the macroscopic equivalence principle masses and one additional differential accelerometer aligned along the spacecraft axis. This enables an intermediate range test two orders of magnitude more sensitive than existing limits. In the very short-range regime atom interferometry using a atom laser source looks as though it could well be more competitive than the already very interesting ground-based limits (Kapner et al. 2007; Adelberger et al. 2007; Dimopoulos and Geraci 2003).

### *3.3.4 Spin-Coupling*

At intermediate ranges (10–40 cm) a specialised spin-source within the cryogenic macroscopic equivalence principle enclosure can be used in conjunction with the macroscopic equivalence principle masses to look for coupling between quantum mechanical spin and matter (Moody and Wilcek 1984). This would achieve an improvement of a few orders of magnitude over existing mass-spin coupling limits (Hammond et al. 2007). Again in the very short-range regime atom interferometry using a atom laser source looks like it could be much more sensitive even though it would still probably struggle to reach axion search sensitivity.

### *3.3.5 Quantum Decoherence*

Understanding fluctuations in the fabric of space time may lead the way to quantum-gravity. If they exist their effect may well be first seen as a limitation in matter wave interferometry (Wang et al. 2006). Demonstrating this requires proper understanding and control of all other limiting disturbances and what could be better than doing this in the quiet environment of space.

## 4 Summary

The scientific case for testing the universality of free-fall to much better sensitivity is overwhelming. Such a measurement could provide vital clues about quantum gravity which is proving rather difficult to progress on a purely theoretical front. Taking such measurements into space seems an obvious step for the future. However, the cost of doing this is significant and this poses a difficult catch-22 situation as it is clearly always going to be impossible to fully test the instrument on the ground prior to launch to verify its sensitivity, otherwise the experiment itself could be done on the ground. This presents a nightmare scenario for space agencies which assume they can validate every aspect of a mission prior to launch. In addition the type of payload required is unlike any conventional mission as the satellite control system is an essential part of the experiment. This also moves such missions outside of the comfort zone of the agencies. *STEP* has now been studied more carefully than any other mission of its type and this study has been backed up by laboratory verifications of many aspects and has now seen many of its intended technologies verified in space by the successful *GP-B* mission. In addition there have been a handful of geodesy missions which have successfully flown accelerometers. *MICROSCOPE* and *LISAPF* will be the first of the next generation of more advanced accelerometers specifically designed for physics investigations. After that LISA will be a major milestone for the community and promises to bring fundamental physics into its own as a space-based activity. Maybe by then *STEP* and/or *GAUGE* will have succeeded in becoming approved missions for NASA and/or ESA. Their 'conventional' use of macroscopic proof-masses offers a more secure and better studied route to equivalence principle measurements at the moment, but in the longer term still it is highly probable that the use of atom interferometry will join the suite of techniques which could be used in space, and, moreover may become scientifically complementary for some types of measurement in which the behaviour of microscopic and macroscopic test bodies could be different.

## References

E.G. Adelberger et al., Phys. Rev. Lett. **98**, 131104 (2007)
G. Amelino-Camelia et al., Exp. Astron. **23**, 549 (2008)
S. Anza et al., Class. Quant. Grav. **22**, S125 (2005)
S. Baessler, B.R. Heckel, E.G. Adelberger, J.H. Gundlach, U. Schmidt, H.E. Swanson, Phys. Rev. Lett. **83**, 3585 (1999)
J.D. Barrow, B. Li, Phys. Rev. D **78**, 083536 (2008)
J.D. Barrow, D.J. Shaw, Phys. Rev. D **78**, 067304 (2008)
J.P. Blaser, Class. Quant. Grav. **18**, 2515 (2001)
J.P. Blaser, T. Damour, Adv. Space. Res. **32**, 1335 (2003)
S. Buchman, T. Quinn, G.M. Keiser, D. Gill, T.J. Sumner, Rev. Sci. Instrum. **66**, 120 (1995)
T. Damour, A.M. Polyakov, Nucl. Phys. B **423**, 532 (1994)
T. Damour, F. Piazza, G. Veneziano, Phys. Rev. D **66**, 046007 (2002a)
T. Damour, F. Piazza, G. Veneziano, Phys. Rev. L **89**, 081601 (2002b)
K. Danzmann (LISA Science Team), Adv. Space Res. **32**, 1233 (2003)
S. Dimopoulos, A.A. Geraci, Phys. Rev. D **68**, 124021 (2003)
S. Dimopoulos et al., Phys. Rev. Lett. **98**, 111102 (2007)
R. Dolesi, M. Bonaldi, S. Vitale, Cryogenics **39**, 691 (1999)
R. Dolesi, R. Rossi, R. Torii, S. Vitale, Adv. Space Phys. **25**, 1215 (2000)
R. Dolesi et al., Class. Quant. Grav. **20**, S99 (2003)
C.W.F. Everitt, The Stanford Relativity Gyroscope Experiment (A): History and Overview, in *Near Zero: New Frontiers of Physics*, ed. by J.D. Fairbank, B.S. Deaver Jr., C.W.F. Everitt, P.F. Michelson (Freeman, New York, 1988), pp. 587–639
E. Fischbach, D.E. Krause, C. Talmadge, D. Tadić, Phys. Rev. D **52**, 5417 (1995)
E. Göklü, C. Lämmerzahl, Class. Quant. Grav. **25**, 105012 (2008)
W. Guerin et al., Phys. Rev. Lett. **97**, 200402 (2006)
G.D. Hammond et al., Phys. Rev. Lett. **98**, 081101 (2007)
D. Hudson, R. Chhun, P. Touboul, Adv. Space Res. **39**, 307 (2007)
Y.R. Jafry, Ph.D. Thesis, Stanford University, CA (1992)
Y. Jafry, T.J. Sumner, S. Buchman, Class. Quant. Grav. **13**, A97 (1996)
C. Jentsch et al., Gen. Rel. Grav. **36**, 2197 (2004)
D.J. Kapner et al., Phys. Rev. Lett. **98**, 021101 (2007)
J.J. Kolodziejczak, J. Mester, Int. J. Mod. Phys. D **16**, 2215 (2007)
N.A. Lockerbie, A.V. Veryaskin, X. Xu, Class. Quant. Grav. **10**, 2419 (1993)
N.A. Lockerbie, X. Xu, A.V. Veryaskin, M.A. Hosey, Class. Quant. Grav. **11**, 1575 (1994)
N. Lockerbie, J. Mester, R. Torii, S. Vitale, P.W. Worden, in *Gyros, Clocks, Interferometers: Testing Relativistic Gravity in Space*. Lecture Notes in Physics, vol. 562 (Springer, Berlin, 2001), p. 213
P.W. McNamara, AIP Conf. Ser. **873**, 49 (2006)
J. Mester, R. Torii, P. Worden, N. Lockerbie, S. Vitale, C.W.F. Everitt, Class. Quant. Grav. **18**, 2475 (2001)
J. Mester et al., Nucl. Phys. B Proc. Suppl. **134**, 147 (2004)
J.E. Moody, F. Wilcek, Phys. Rev. D **30**, 130 (1984)
H.B. Sandvik, J.D. Barrow, J. Magueijo, Phys. Rev Lett. **88**, 031302 (2002)
C.C. Speake et al., STEP spin-coupling experiment, in *Perspectives on Neutrinos, Atomic Physics and Gravitation, Proc. XXVIIIth Rencontre de Moriond*, ed. by J. Tran Thanh Van, T. Damour, E. Hinds, J. Wilkerson (1993a), pp. 445–451
C.C. Speake et al., Proc. STEP Symposium, PISA. **ESA WPP-115**, 133–147 (1993b)
C.C. Speake et al., J. Meas. Sci. Tech. **10**, 508 (1999)

T.J. Sumner, Gen. Rel. Grav. **36**, 2331 (2004)
T.J. Sumner, D.N.A. Shaul, Mod. Phys. Lett. A **19**, 785 (2004)
T.J. Sumner et al., Adv. Space Res. **39**, 254 (2007)
T.J. Sumner, in *Proc. 4th Meeting on CPT and Lorentz Symmetry* (Bloomington, 2007) (World Scientific, 2008)
S. Vitale et al., Nucl. Phys. B Proc. Suppl. **110**, 209–216 (2002)
C. Wang, R. Bingham, T. Mendonca, Class. Quant. Grav. **23**, L59 (2006)
S. Wiegand, S. Theil, J. Mester, in *Proc. 2nd Pan-Pacific Basin Workshop on Microgravity Sciences* (Pasadena, CA, 2001)
C. Will, Living Rev. Relativ. **9**, 3 (2006). http://www.livingreviews.org/lrr-2006-3
J.G. Williams, X.X. Newhall, J.C. Dickey, Phys. Rev. D **53**, 6730 (1996)
P.W. Worden, C.W.F. Everitt, in *Proc. Int. School of Physics*, "Enrico Fermi", Course LVI (1972), p. 382
P. Worden, J. Mester, R. Torii, Class. Quant. Grav. **18**, 2543 (2001)

Space Sci Rev (2009) 148: 489–499
DOI 10.1007/s11214-009-9568-8

# Satellite Test of the Equivalence Principle Uncertainty Analysis

**Paul Worden · John Mester**

Received: 2 April 2009 / Accepted: 9 July 2009 / Published online: 29 July 2009
© Springer Science+Business Media B.V. 2009

**Abstract** STEP, the Satellite Test of the Equivalence Principle, is intended to test the apparent equivalence of gravitational and inertial mass to 1 part in $10^{18}$ (Worden et al. in Adv. Space Res. 25(6):1205–1208, 2000). This will be an increase of more than five orders of magnitude over ground-based experiments and lunar laser ranging observations (Su et al. in Phys. Rev. D 50:3614–3636, 1994; Williams et al. in Phys. Rev. D 53:6730–6739, 1996; Schlamminger et al. in Phys. Rev. Lett. 100:041101, 2008). It is essential to have a comprehensive and consistent model of the possible error sources in an experiment of this nature to be able to understand and set requirements, and to evaluate design trade-offs. In the following pages we describe existing software for such an error model and the application of this software to the STEP experiment. In particular we address several issues, including charge and patch effect forces, where our understanding has improved since the launch of GP-B owing to the availability of GP-B data and preliminary analysis results (Everitt et al. in Space Sci. Rev., 2009, this issue; Silbergleit et al. in Space Sci. Rev., 2009, this issue; Keiser et al. in Space Sci. Rev., 2009, this issue; Heifetz et al. in Space Sci. Rev., 2009, this issue; Muhlfelder et al. in Space Sci. Rev., 2009, this issue).

**Keywords** STEP · Equivalence principle · Error analysis · GP-B · Electrostatic patch effect

## 1 Motivation

The success of GP-B has provided a wealth of information regarding conditions in orbit, drag free performance, and instrument performance, much of which can be directly applied in the STEP error analysis program. This data can replace what were previously only assumptions or extrapolations, and gives significant additional credibility to the final performance estimate. Specifically, the GP-B SQUID measurement systems performed as well as, or better than required; the performance of the drag free and charge control systems

P. Worden (✉) · J. Mester
Hansen Experimental Physics Laboratory (HEPL), Stanford, CA 94305-4085, USA
e-mail: worden@relgyro.stanford.edu

were within requirements; magnetic field and manufacturing requirements were met, and the pressure requirement was met with very large margin. Electrostatic patch effect was an unexpected issue in GP-B which was previously anticipated in STEP.

STEP's requirements on magnetic field, manufacturing accuracy, and spacecraft environment are less stringent than GP-B; For SQUID noise, pressure, and electric charge, STEP's requirements are very similar. GP-B's drag free and attitude requirements are not directly comparable to STEP but requirements on the thrusters are very similar. Electrostatic patch effect has been considered in the STEP error analysis since the beginning, and it has a very different and moderate effect compared to GP-B. These and other comparisons with GP-B are elaborated below.

The most common approach to error modelling when designing physics experiments is to analyze a highly simplified and idealized version of the experiment, allowing the designer to focus on a manageable number of parameters. In STEP we found that this often led to inconsistent requirements because of the implicit assumption of independent subsystems. Subsystems which are *prima facie* independent often depend on a common input, or interact through an unanticipated feedback. It is particularly important for complex, very sensitive missions like GP-B and STEP that such disturbances (which would ordinarily be considered small) are properly anticipated. Although STEP is much simpler than GP-B in many regards, we must still regard STEP as a single integrated system for design purposes, and in order to find an optimum set of requirements it is necessary to have a correspondingly integrated error model. We have actively developed and used such an error model for many years. We are presently in the process of incorporating the results from GP-B and evaluating their impact on the STEP error estimate.

The STEP Error Model is intended to model the experiment and apparatus as a self-consistent whole. The ability to easily include new tradeoffs and revised assumptions is needed in order to determine their effect in a timely and reliable manner. Likewise explicit relationships and traceability, in the sense of being able to determine how changes propagate throughout the experiment, are required. These requirements suggest that ideal error model software should not be structured as a conventional simulation, in which the time evolution of the system is calculated for one or more cases. Instead we estimate the experiment response in the frequency domain, as a collection of transfer functions whose effects can be summed to find the total response. The transfer functions describe the known or estimated response of the apparatus to random and systematic errors as well as experimental input. Implemented in a spreadsheet, with common input parameters for all functions, this approach provides a very efficient design and evaluation procedure, giving as output the expected uncertainty in the experiment as a spectral density. Because it must explicitly incorporate any known or proposed tradeoffs it can be used to set experiment requirements.

Our program systematically evaluates the known errors starting from a relatively small and consistent set of assumptions. Many experimental parameters which are often considered independent for purposes of a quick analysis, are actually dependent on other parameters or requirements. The program automatically calculates these: for example signal frequency from orbit height and satellite spin rate, or less directly the minimum possible common mode rejection ratio from the setup currents and coil dimensions in the SQUID position sensor. This prevents much of the inconsistency that can result from independent assumptions and separate analyses of disconnected experiment systems. New error sources, assumptions, or design concepts can be straightforwardly included.

## 2 Program Design

Our current understanding of the STEP experiment is implemented in a single program comprising models of each subsystem, with a unified database and assumptions. The single set of inputs and assumptions ensures consistent treatment of the systems and error sources in our models. Maintaining consistency between different subsystem analyses is essential if the program is to be both useful and reliable, because very often, outputs from different analyses feed into each other in unexpected ways. The utility of this tool is not simply in quickly calculating an estimate of the error, but in explicitly and forcibly modelling the important interconnections and feedbacks between different subsystems, including some which appear at first to be unrelated.

Our error analysis program is rooted in an actively maintained text document which lists and describes its components. The *STEP Error Analysis* describes analytic models of specific disturbances to the masses in the general categories of thermal noise, gas pressure forces, electrical forces, magnetic forces, gravitational forces, radiation pressure, and vibration. The models are simplified mathematical descriptions of the quantities being estimated, which can be elaborated if more detail is required. The models normally include disturbances to each subsystem, for example process and measurement noise, variations in superconducting penetration depth, and effects of thermal and mechanical stability. The next level of analysis, typically done in a higher-level language, uses results from the STEP Error Analysis and other documents to produce response formulas and approximations suitable to be used as components of the final product. These solutions are pasted from the text document into the final error analysis program.

The *STEP Error Analysis Program* in effect calculates the spectral response of the entire system. This approach is more efficient in determining the desired results (the noise spectral density at signal frequency, i.e. the experiment's performance) than simulating the entire time development, and is no more subject to the usual mistakes in assumptions and interpretation than a full simulation would be. The Error Analysis Program is implemented in a spreadsheet with each cell corresponding to a unique input variable (for example a test mass dimension) or to an intermediate result calculated from prior inputs, (e.g. the position sensitivity of a SQUID sensor from currents and inductance values), or to an output containing a desired result (the experiment sensitivity, for one). The analysis is to an extent self documenting since the spreadsheet structure makes interactions and dependencies explicit and traceable.

As inputs we have considered a number of specific disturbances which include

- electric potentials in the housing around the masses (including the patch effect),
- the radiometer effect,
- other gas pressure effects including streaming and viscous coupling,
- losses from eddy current damping,
- gravitational coupling from helium tide,
- drag free residual vibration coupled to the differential mode,
- SQUID noise,
- charge sensor noise,
- penetration depth changes in superconductors,
- and momentum transfer from penetrating particle radiation.

Environmental parameters include

- the gravitational field of the earth including first-order gradients,
- the magnetic field of the earth,

- the radiation environment as a function of orbit height,
- and atmospheric drag and its variation.

Instrument parameters include

- details of the instrument and spacecraft geometry,
- temperature and temperature gradients,
- emissivity of test masses,
- superconducting penetration depth variation,
- spacecraft rotation rate stability,
- drag-free response to environmental noise
- attitude control laws

The output is not limited to the experiment error. As a side effect of this style of calculation, many useful intermediate parameters (masses of test objects, SQUID sensitivity estimates, expected drag-free performance) result.

## 3 Description of Representative Analysis

Figure 1 illustrates a typical output, which we will use to describe the application of our error analysis program. In this particular trial the dominant noise sources are SQUID noise, thermal (Nyquist) noise, and residual drift in the (temperature regulated) SQUID carriers. Single and multiple trials with this program can be used to set requirements on the modelled spacecraft systems' performance, by quantitatively studying their interconnectedness and performing tradeoffs on the model assumptions. These in turn lead to more meaningful error budgets (with alternatives) and more realistic performance estimates for the experiment.

### 3.1 Connectivity

We will illustrate the unity of the STEP error analysis with an example. One cell in the spreadsheet calculates "SQUID noise" which is the disturbance to measured acceleration caused by noise in the SQUID. The SQUID noise spectral density is calculated according to a simplified model, $E_{sq} = E\sqrt{1 + (\omega_f/\omega)^2}$. With appropriate choices for the $1/f$ cutoff frequency $\omega_f$ and high frequency limit $E$, this closely matches the measured GP-B SQUID noise spectrum (Fig. 2). This noise energy must be converted to an acceleration in the bandwidth of the Equivalence Principle measurement. As a first step, the noise flux in the SQUID from this source is $\sqrt{2L_f E_{sq}}$ where $L_f$ is the effective inductance coupled to the SQUID.

Flux in the SQUID corresponds to acceleration in the Equivalence Principle measurement. The model is basically a mass on a spring, magnetically coupled to an inductance connected to a SQUID (the actual model (Fig. 3) has two coupled masses). A displacement (measured by the change in flux $\phi$ in the SQUID) corresponds to an acceleration $A$, and the proportionality can be calculated from the inductance and other parameters. Multiplying the noise flux by the acceleration per unit flux in the SQUID, $dA/d\phi$, and averaging over the observation time $T_{obs}$, gives the limit to acceleration measurement from SQUID noise, $(dA/d\phi)\sqrt{2L_f E_{sq}/T_{obs}}$. $dA/d\phi$ for the differential mode of the masses is expressed in terms of circuit parameters, for example inductances. Normally an error analysis would stop here and make numerical assumptions about these quantities. In the STEP analysis $(dA/d\phi)$ is a precomputed expression involving the inductances, currents and stiffnesses originating in the position sensor circuit. These are in turn calculated from other inputs such as sense coil and test mass dimensions. At any one level the formulas are usually very simple, yet

| DFC Reference accelerometer | Systematic component at signal frequency | | |
|---|---|---|---|
| Disturbance | m/sec^2 | | Comment |
| SQUID noise | 1.72E-18 | | acceleration equivalent to intrinsic noise |
| SQUID temp. drift | 7.26E-19 | | regulation of SQUID carriers |
| Thermal expansion | 8.47E-22 | | gradient along DAC structure |
| Differential Thermal expansion | 3.35E-23 | | Radial gradient in DAC structure |
| Nyquist Noise | 1.69E-18 | | RMS acceleration equivalent |
| Gas Streaming | 5.63E-19 | | decaying Gas flow, outgassing |
| Radiometer Effect | 2.59E-21 | | gradient along DAC structure |
| Thermal radiation on mass | 4.73E-25 | | Radiation pressure, gradient |
| Var. Discharge uv light | 5.01E-19 | | unstable source, opposite angles on masses |
| Earth field leakage to SQUID | 8.65E-19 | | estimate for signal frequency component |
| Earth Field force | 7.49E-22 | | estimate for signal frequency component |
| Penetration depth change | 5.19E-23 | | longitudinal gradient |
| Electric Charge | 1.08E-19 | | Assumptions about rate |
| Electric Potential | 3.95E-19 | | variations in measurement voltage |
| Sense voltage offset | 3.39E-20 | | bias offset |
| Drag free residual in diff. Mode | 3.12E-21 | | estimated from squid noise |
| Viscous coupling | 9.47E-25 | | gas drag + damping |
| Cosmic ray momentum | 4.79E-21 | | mostly directed downward |
| Proton radiation momentum | 4.65E-19 | | unidirectional, downward |
| dynamic CM offset | 2.98E-19 | | vibration about setpoint, converted |
| static CM offset limit | 1.28E-21 | | A/D saturation by 2nd harmonic gg |
| Trapped flux drift acceleration | 1.06E-22 | | actual force from Internal field stability |
| Trapped flux changes in squid | 5.39E-20 | | apparent motion from internal field stability |
| S/C gradient + CM offset | 3.82E-35 | | gravity gradient coupling to DFC residual of S/C |
| rotation stability | 8.88E-23 | | centrifugal force variation + offset from axis |
| Eccentricity subharmonic. | 9.40E-20 | | real part at signal frequency |
| Helium Tide | 7.00E-20 | | placeholder |
| position sensor gap, mm | 1.00 | 400000 | Orbit height |
| common mode period | 2116 STABILITY MARGIN WARNING | 0.0086 | Sensor current, A |
| differential mode period | 1404 | 7.4E-12 | CM distance, m |
| S/C rotation per orbit | -2.70E+00 | | |
| **Total error** | 7.53E-18 | **RMS error** 2.85E-18 | m/sec^2 |

**Fig. 1** Representative output of the STEP error analysis

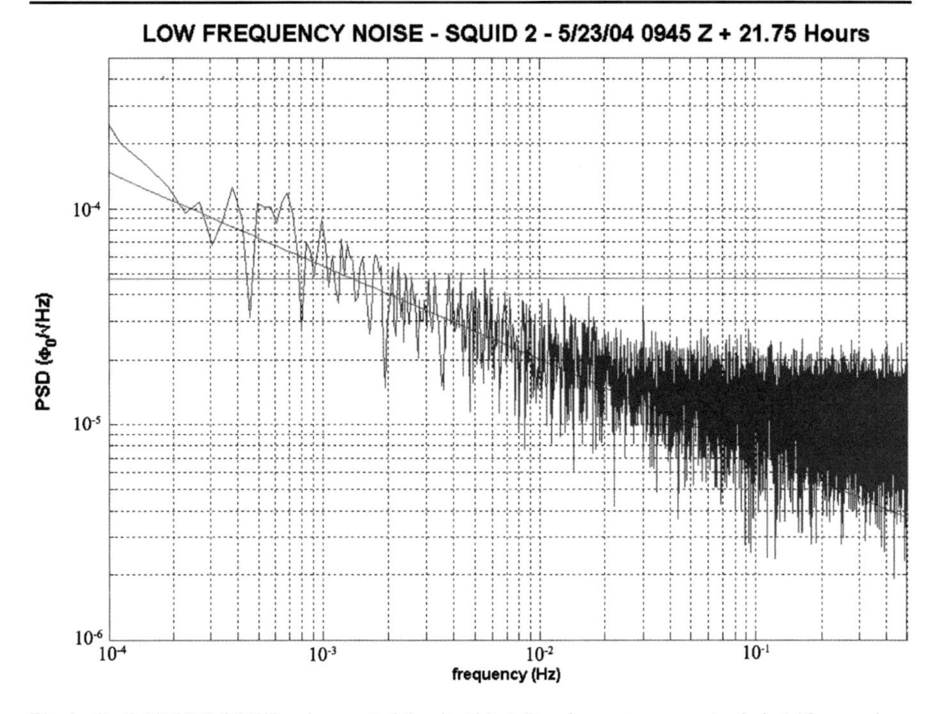


**Fig. 2** On-Orbit GP-B SQUID noise spectral density (*blue*). Requirement—*green*; Analytic $1/f$ approximation—*red*

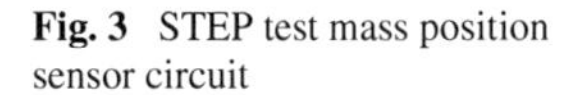

**Fig. 3** STEP test mass position sensor circuit

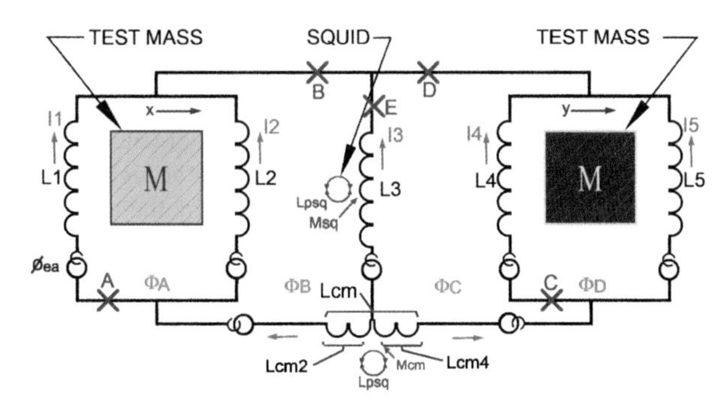


together they capture many more details than could be practically included in a single expression.

The top level result of the SQUID noise spectral density shows that, in this example, it directly contributes about $1.72 \times 10^{-18}$ m/s$^2$ to the overall error budget (Fig. 1). But the effect of the SQUID noise spectral density is not confined to the cell labelled “SQUID noise”. It is also an input to the differential and common mode position sensitivity of the accelerometer. The differential mode sensitivity is an output which is used nowhere in the final result, but which is commonly requested: in this example it is about $4 \times 10^{-19}$ g (not shown), for setup values that give a natural frequency for the mass of $\sim 10^{-3}$ Hz and a displacement sensitivity of $\sim 10^{-13}$ m.

The common mode sensitivity (in this example, about $1.2 \times 10^{-10}$ m/Hz$^{0.5}$, not shown) determines the ultimate performance limit for the drag free system and the charge measurement system. These each affect the experiment sensitivity through imperfect common mode rejection, and several other effects which are not included in the "SQUID noise", resulting in significant contributions in the cells "Electric Potential" (a disturbance from sense voltage variations) and "Dynamic CM offset" (an intermode coupling effect related to changes in the center of mass displacement). In all, eleven of the 27 disturbances in the table depend in some degree on the SQUID noise spectral density. All of the effects that result from a change in the SQUID noise spectral density, including many which would have been overlooked in a conventional treatment, are included in our final summation.

### 3.2 Tradeoffs

As a simple example of a trade study, consider the impact of orbit altitude on the performance of a proposed active charge control system. This type of charge control system estimates the charge on a freely suspended mass by applying a small electric field and measuring the mass's acceleration. Then it corrects the charge toward zero by an ultraviolet discharge method. At an orbit height of 400 km, the test mass charging rate is reduced relative to that in higher orbits, and much reduced from what it would be for orbits $\gg$700 km. Naively one expects that the charge on the test masses could be better controlled in lower orbits because the charging rate is smaller. Unexpectedly the error analysis shows this is not the case. We can trace the reasons.

Variations in atmospheric drag increase very rapidly at lower altitude as well as being very dependent on solar cycle and other factors. At the lower altitude, atmospheric drag is greater, and the variation in the drag (in our model) causes more disturbance to the drag free control system than in higher orbits. The residual acceleration limits the charge measurement: spacecraft acceleration cannot be distinguished from test mass acceleration caused by a small charge. At 400 km the smallest measurable charge is much greater than at higher altitude, because the residual acceleration is greater due to the rapid increase in drag variation.

The charge cannot be controlled to a smaller value than can be measured, so the charge control performance is worse at 400 km than at higher altitudes. But the drag free system is also affected (in somewhat lesser degree) by the test mass charge, which causes a force between the test mass and its housing. In this case, independent estimates of charge control and drag free performance could both be highly unreliable, especially when the ultimate performance is required of both. Accurate modelling of each subsystem, and the disturbances on them, is required, but accurately modelling the interaction between the subsystems is essential. In this case, tracing the connections between program components usefully points out a potential weakness in our analysis.

With the explicit modelling in our analysis we can quickly trace which subsystems are most important for controlling a disturbance, and find which inputs they have in common and which they do not. Common inputs (in this example the orbit height) can be optimized quickly by a little thought backed up by a systematic search or trial and error. Other inputs (here, control parameters and bandwidths for the charge and drag free systems) can be changed to optimize performance or increase margin. The conclusion of this example is discussed below.

### 3.3 Error Budgets

Assigning error budgets with this system is straightforward. Where an input is known, say the Nyquist thermal fluctuation noise, that input is included in the analysis. The cell

"Nyquist noise" in Fig. 1 contains the acceleration of the modelled oscillator in response to the thermal fluctuation force:

$$dA^2 = 2\omega_d K_b T/(QMT_{obs})$$

where $\omega_d$ is the differential mode frequency, $K_b$ the Boltzmann constant, $T$ the temperature, $M$ the effective mass, and $T_{obs}$ the observation time. $\omega_d$ depends on the same input parameters that went into the acceleration sensitivity calculation, as well as the mass. Changing the temperature and observation time also changes the acceleration sensitivity, so the only really independent factor is the quality factor $Q$. The other parameters are automatically traded. $Q$ depends on some fundamental inputs: gas damping and electrical (and any other) losses. These inputs can be traded against each other until a stable, achievable configuration is established, then if desired they can be fixed as requirements. Note that in a simpler analysis we could have set a requirement on $Q$ directly, based on engineering judgement, at the risk of not being able to meet it because of excess damping.

Note that this error budgeting is a dynamic process: it can be quickly repeated at any time, if the original goals for any input are found to be unrealistic.

Margin is necessary to avoid unstable requirements. Requirements on fundamental inputs (e.g. SQUID noise spectral density or gas pressure) are considered lower-level requirements and should be set at or slightly above the best achievable level. The STEP requirement on gas pressure ($10^{-11}$ Pa) is an example. Margin on higher-level requirements such as $Q$ is often a matter of engineering judgment.

Higher level requirements should be set for inputs which are affected by future tradeoffs to be performed, for which the models are incomplete, and which directly affect the experimental outcome. For example, STEP's requirement on the common mode rejection ratio CMRR of the differential accelerometers ($<10^{-4}$) is a single number which enters into the sensitivity in many places, and is determined by tradeoffs restricted to the instrument. But the quality factor $Q$ of the differential accelerometers ($>10^6$, an achievable value) is a purely descriptive quantity which does not enter into further calculations (although its inputs do) so no requirement has been set for it.

### 3.4 Description of Error Terms with Specific Conclusions from GP-B

In the example in Fig. 1, the result is dominated by the acceleration equivalent of the SQUID noise, i.e. the sensor. The SQUID noise is partly determined by the SQUID itself, which is based very closely on the actual GP-B SQUIDs, as described above and shown in Fig. 2. The setup of the sensor is the largest determinant of performance. Large setup currents constrain the test mass tightly between the sensing coils, and restrict the mass motion more than they improve the position sensitivity; hence the acceleration sensitivity is reduced. Very small currents improve the acceleration sensitivity but reduce the stability of the oscillator. There is an optimum current. This example used slightly more than the optimum current, for margin.

The "SQUID temperature drift" is also based on the actual performance of the GP-B thermal regulator for the SQUID. A regulator is necessary because the SQUID characteristics are somewhat temperature dependent. The SQUID temperature drift noise also depends strongly on the setup parameters for the sensor circuit, especially the setup currents.

The second largest contribution to the total experiment error is the Nyquist noise. Although the response to thermal fluctuations depends indirectly on the SQUID setup, the limit to the experiment itself from this cause is determined by the damping time. The damping time is determined by the sum of gas damping and electrostatic damping. GP-B achieved a pressure of $\sim 1.3 \times 10^{-12}$ Pa, much less than the STEP requirement of $10^{-11}$. This pressure gives an estimated gas damping time of about 7 million years, well in excess of STEP's original requirement. For this example we assumed damping from all other sources, including induced currents from charge and electromagnetic patch effect, is about $10^5$ times larger, corresponding to a damping time of the order of 70 years. The actual damping times for GP-B's gyroscopes (mostly from induced currents caused by the patch effect) ranged from 7000 to 25 000 years. The disturbance scales as the square root of the damping time, so in the worst (7000-year) case the patch effect damping would be about 10% of what we have already assumed, or about $10^{-19}$ m/s$^2$. The GP-B results suggest that we have sufficient margin on damping.

The third largest contribution in Fig. 1 is the leakage from the Earth's magnetic field through gaps in the superconducting shielding to the instrument. Because this depends on both the position in orbit and attitude of the spacecraft, it may have a component at STEP's signal frequency, and has previously been considered potentially important. We have not yet revised our model with GP-B results, because no measureable field penetration has been found in the GP-B data.

Electric charge *per se* was not a problem for GP-B. Because STEP's requirements on total charge and time variation of charge are similar, it is not expected to be a factor for STEP: If anything GP-B simplified the situation by showing that active charge control is not required. The charge on GP-B's rotors was controllable to less than 5 millivolts and changed about 0.1 mV per day (Fei 2007; Buchman 2007). This corresponds to a charge of about $2.7 \times 10^{-12}$ C on the rotor changing about $3 \times 10^{-17}$ C/s. STEP's conservatively set requirement is less than $8 \times 10^{-13}$ C, changing less than $1 \times 10^{-14}$ C at signal frequency ($\sim 1.4 \times 10^{-16}$ C/s; somewhat more can be tolerated on the outer test mass). The charge and rate requirements are closely related and can be traded in future revisions. In the particular example of Fig. 1, electric charge is a minor contributor to the overall error budget.

The remaining disturbances contribute much less to the STEP error budget. Gas streaming did not measureably affect GP-B, and the variation in UV light pressure (for charge control) is easily eliminated by turning it off during Equivalence Principle measurements. GP-B showed active charge control will not be needed during STEP measurements. There is no GP-B information about momentum transfer from proton radiation. The Dynamic Center of Mass offset converts vibration of the masses about equilibrium into a disturbance at the signal frequency by a mixing process. Relative radial motion of the test masses, at harmonics of the signal frequency, converts to axial acceleration in the earth's gravity gradient, at the signal frequency. The amplitude of this disturbance depends directly on the performance of the drag-free control system and the electrostatic mass positioning system. GP-B's drag free performance met requirements which are not directly comparable to the more stringent STEP requirements, but which suggest that an equivalent care in design will result in a successful mission with little disturbance from this source.

GP-B suffered some unexpected disturbances from the electrostatic patch effect, which are not expected to directly impact STEP's performance. These included added damping (which caused decay of the gyro polhode motion) and extra torques (which caused significant spin axis motion when roll frequency came into resonance with the polhoding). Similarly, the electrostatic patch effect affects STEP through added damping (and the resulting thermal noise) and additional restoring forces. The added damping is trivially included in our analysis as part of the Nyquist Noise (see above), and based on the observed GP-B damping, it is not very important. The electrical patches in GP-B evidently did not change over periods of weeks, and in STEP are not expected to change over the shorter duration of a measurement (1–5 days). The variability of GP-B's patches will be studied in future analysis. The effect on STEP is similar to that of a changing charge and can be included in that analysis.

The static patch effect will not directly disturb the STEP measurement, because the Equivalence Principle signal is periodic. There is, however, an indirect patch effect disturbance arising from the gradient of the patch effect forces. It has no effect on STEP's measurement provided the setup can be made correctly and the patch effect is constant. This disturbance is not to the Equivalence Principle measurement, but to the setup procedure.

STEP requires that the test masses have very long periods and careful setup is needed to achieve them. This setup could be perturbed (or impossible) if the patch effect forces are too large or irregular. An allowance for unmodelled position-dependent forces has been in the analysis from the beginning, because of prior experimental studies of much larger, equally irregular magnetic forces (Worden 1987). Patch effect forces can be trivially included in this allowance. These added stiffnesses affect almost all the listed disturbances since they involve the basic properties of the measurement system. Therefore, they do not appear explicitly in the output of our analysis.

The setup procedure tries to achieve certain values of period, mass position, and SQUID sensitivity which allow Equivalence Principle measurements to be made with the desired sensitivity. No disturbance to the sensitivity is expected if these setup values can be met. Preliminary work suggests the patch effect disturbance to setup is manageable without modification to the baseline instrument or procedures in all but the very worst cases. The STEP masses will be gold coated which is known to produce smaller patch effect than the bare niobium metal of the GP-B rotors.

Additional study of the need for mitigation of the patch effect is in progress. The disturbance has the form of an extra addition to the spring constant, with added nonlinearity for motions much larger than we anticipate in orbit. Analytic study shows that this disturbance should be insignificant if the patches are small relative to the gap between the mass and its housing. In this case the extra restoring force is approximately $\varepsilon_o l_p^3 V^2/(4\pi r^4 A^{1/2})$, where $l_p$ is the size of a patch, $r$ the gap, and $A$ the surface area; A STEP test mass with GP-B patches 10 microns across of 0.3 V would have a period from this cause of about 28 000 seconds, much longer than the requirement of 1000 seconds, and unlikely to cause much perturbation. The case for small gap and large patches is less favorable, $2\varepsilon_o A^{1/2} V^2/(l_p r)$, giving 1800 seconds period for the disturbing restoring force. This has an uncomfortably small margin and further study is indicated.

Helium tide in the STEP error analysis is a placeholder. There is no credible calculation of the amplitude of motion of a free surface under STEP conditions, so a requirement is given. However, in STEP the helium surface will be anything but free; it will be contained in an aerogel. We have conducted experimental studies demonstrating the ability of aerogel to restrain superfluid helium completely in 1 g conditions, and documented its ability to withstand launch accelerations when fully loaded with helium (Wang et al. 2001). GP-B demonstrated the ability to fly drag-free with precision even though it contained ten times more helium than STEP with a free surface.

The Radiometer Effect originates in gas molecules emitted from hot surfaces with more momentum than from cold surfaces. In the molecular flow regime it can be modelled as $P(dT/dx)/(2T\rho)$ where $P$ is the pressure, $T$ the temperature, and $\rho$ the density of the test mass. Although at first glance this increases at low temperature, there is actually a gain: very minute temperature differences and pressure are relatively easy at cryogenic temperatures. This is ordinarily a very small effect in STEP, and the GP-B results confirm our assumptions about temperature and pressure.

## 4 Conclusions

We have incorporated experimental data and current results from GP-B into our analysis of the STEP mission and found that in the cases where the GP-B analysis is applicable and complete, STEP's requirements for a $10^{-18}$ measurement of Equivalence are met, in some cases with large margin. For the electrostatic patch effect, the disturbance to STEP differs qualitatively from the disturbance to GP-B. The (presently) incomplete GP-B results, together with our analysis, suggest that patch effect has no significant impact to STEP. Work on application of GP-B's patch effect results to STEP will continue until we reach a definite conclusion.

## References

S. Buchman, in *Workshop on Charging Issues in Experimental Gravity, Massachusetts Institute of Technology*, 26–27 July 2007. http://www.ligo.caltech.edu/docs/G/G070568-00/
C.W.F. Everitt et al., Space Sci. Rev. (2009), this issue
E. Fei, Stanford University Physics Department Senior Thesis, 2007
M. Heifetz et al., Space Sci. Rev. (2009), this issue
M. Keiser et al., Space Sci. Rev. (2009), this issue
B. Muhlfelder et al., Space Sci. Rev. (2009), this issue
S. Schlamminger, K.-Y. Choi, T.A. Wagner, J.H. Gundlach, E.G. Adelberger, Phys. Rev. Lett. **100**, 041101 (2008)
A. Silbergleit et al., Space Sci. Rev. (2009), this issue
Y. Su, B.R. Heckel, E.G. Adelberger, J.H. Gundlach, M. Harris, G.L. Smith, H.E. Swanson, Phys. Rev. D **50**, 3614–3636 (1994)
S. Wang, R. Torii, S. Vitale, Class. Quantum Gravity **18**, 2551–2559 (2001)
J.G. Williams, X.X. Newhall, J.O. Dickey, Phys. Rev. D **53**, 6730–6739 (1996)
P.W. Worden, in *Proc. International Symposium on Experimental Gravitational Physics*, Guangzhou, China, ed. by P. Michelson (World Scientific, Singapore, 1987)
P. Worden, R. Torii, J.C. Mester, C.W.F. Everitt, Adv. Space Res. **25**(6), 1205–1208 (2000)

Space Sci Rev (2009) 148: 501–522
DOI 10.1007/s11214-009-9559-9

# What Determines the Nature of Gravity? A Phenomenological Approach

**Claus Lämmerzahl**

Received: 27 March 2009 / Accepted: 24 June 2009 / Published online: 16 July 2009
© Springer Science+Business Media B.V. 2009

**Abstract** The gravitational field can only be explored through the motion of test objects. To achieve this one first has to set up the correct equations of motion. Initially these equations are based on Newton's laws. Corresponding experiments that support Newton's laws are described. Furthermore, the basic characteristics of the motion of test objects in gravitational fields are described. This leads to the notion of Einstein's Equivalence Principle which has as consequence a metric theory of gravity. One particular metric theory is General Relativity based on Einstein's field equations with its particular predictions for effects like periastron advance, light deflection, etc. An overview over the experimental confirmation of General Relativity, in particular those presented at this workshop, is given. This workshop summary ends with open problems. We also describe some of the strategies for the experimental search for a quantum gravity theory.

**Keywords** General relativity · Special relativity · Newton's axioms · Experimental relativity · Equivalence principle · Solar system tests · Quantum gravity phenomenology

## 1 Introduction

In this article we present a general frame of how to define and explore the nature of gravity as well as the mathematical formalism and the equations that represent gravitational phenomena. In principle there are two ways to state physical equations: The first way—a top–down scheme—is to postulate the equations. For General Relativity (GR), for example, one may postulate a Lorentzian space–time manifold, the Einstein field equations and the geodesic equations for point-like masses and light, as well as the observables of the theory. All observable consequences will follow from these statements. A second way—corresponding to a bottom–up approach—is to base the physical laws on a few basic observations and to build up the theory in a constructive manner. Here we like to proceed along the second way as far as possible.

C. Lämmerzahl (✉)
ZARM, University of Bremen, Am Fallturm, 28359 Bremen, Germany
e-mail: laemmerzahl@zarm.uni-bremen.de

The advantages of the second way are: (i) a physical understanding of mathematical schemes, (ii) each mathematical structure is directly related to an observable phenomenon and, thus, is immediately physically interpretable, and (iii) if generalizations are necessary (like those expected from quantum gravity) natural generalizations are offered through this way.

In what follows we can only give a very rough, short and incomplete description of the scheme. However, we hope to show all the ingredients and what in principle has to be taken into account in order to reach a certain stage of complete mathematical description of the nature of gravity. At many instances we refer to other contributions to this workshop which expand several short remarks given here.

## 2 How to Explore Gravity

The gravitational field and its properties can *only* be explored through the observation of the dynamics of test objects: point particles, light rays or quantum fields (Ehlers 2006). Their dynamics is governed by equations of motion. The simplest such equations are Newton's laws. Much more complicated laws are feasible: one may consider dynamical equations with higher order time derivative (Lämmerzahl and Rademaker 2009), for example.

After we setting the main structure of equations of motion for test objects we ask for characteristic features of the interaction of these objects with the gravitational field. These are governed by the Einstein Equivalence Principle. From that we conclude that gravity is a metric theory. Each of these theories shows the typical effects like perihelion shift, light deflection, time delay, Lense–Thirring and Schiff effects, etc. Only for certain values of these effects gravity is described by the Einstein field equations. These field equations are then extrapolated to the strong field regime and can be confronted with observations of binary systems and black holes.

We always consider fundamental equations only. Effective equations for the motion of test objects with largely different features may come out from complicated calculations that take, for example, radiation reaction into account.

## 3 The Structure of Dynamics

The notion of an inertial system, of the inertial law and the law of reciprocal actions (*actio* = *reactio*) is assumed in all equations of motion, either non-relativistic or relativistic. Any test or exploration of the gravitational interaction has to account for the structure of the equations of motion which are used to measure gravitational effects.

In many instances we have equations of motion which are more general than those related to Newton's axioms. Examples are the equations taking into account radiation reaction, or dynamics with memory. However, these are effective equations of motion. In our 'bottom-up' approach we are interested in the fundamental equations of motion only.

### 3.1 Existence of Inertial Frames

A condition for the existence of inertial frames is content of the first of Newton's laws. In an intuitive sense an inertial frame is a local reference frame where all freely falling particles move uniformly along straight lines. Here we already have to put in an intuitive understanding of "free" motion. Though it is not clear of how to characterize a force-free motion (non-gravitational forces) uniquely in an experimental way, Finsler space–time (Lämmerzahl and Perlick 2009) provides a model for the non-existence of inertial frames.[1]

An indefinite Finslerian geometry is given by the line element

$$ds^2 = F(x, dx) \quad \text{with} \quad F(x, \lambda dx) = \lambda^2 F(x, dx), \tag{1}$$

for all $\lambda \in \mathbb{R}$ and where $F$ is a function homogeneous of degree two. Then

$$ds^2 = g_{\mu\nu}(x, dx) dx^\mu dx^\nu \quad \text{with} \quad g_{\mu\nu}(x, y) = \frac{\partial^2 F(x, y)}{\partial y^\mu \partial y^\nu}, \tag{2}$$

where $g_{\mu\nu}(x, dx)$ is a Finslerian metric which, however, depends on the vector it is acting on. The motion of light rays and point particles is given by the action principle $0 = \delta \int ds^2$ and leads to the equation of motion

$$\frac{d^2 x^\mu}{ds^2} = \left\{ {\mu \atop \rho\sigma} \right\}(x, \dot{x}) \frac{dx^\rho}{ds} \frac{dx^\sigma}{ds}. \tag{3}$$

Since the Finsler–Christoffel connection

$$\left\{ {\mu \atop \rho\sigma} \right\}(x, \dot{x}) = \frac{1}{2} g^{\mu\nu}(x, \dot{x}) \left( \partial_\rho g_{\sigma\nu}(x, \dot{x}) + \partial_\sigma g_{\rho\nu}(x, \dot{x}) - \partial_\nu g_{\rho\sigma}(x, \dot{x}) \right) \tag{4}$$

(here $g^{\mu\nu}(x, \dot{x})$ is the inverse of $g_{\mu\nu}(x, \dot{x})$ defined through $g^{\mu\nu}(x, \dot{x}) g_{\nu\rho}(x, \dot{x}) = \delta^\mu_\rho$) depends on the velocity $\dot{x}^\mu$, it cannot be transformed away. As a consequence, there is no frame in which all particles move uniformly along straight lines. We always have accelerated particles. This provides a model for that gravity cannot be transformed away and, thus, for the non-existence of inertial systems. This is true for all equations of motion with a non-linear connection of the form (3). Another consequence of a Finslerian metric is the anisotropy of, e.g., light propagation violating Lorentz invariance (see below and Lämmerzahl et al. 2009).

Free fall experiments and orbits of planets and satellites yield that the order of magnitude of any hypothetical Finslerian deviation from ordinary Riemannian space–times should be smaller than $10^{-9}$ m/s$^2$ (Lämmerzahl and Perlick 2009).

### 3.2 The Inertial Law

The inertial law

$$\dot{\boldsymbol{p}} = m\ddot{\boldsymbol{x}} = \boldsymbol{F} \tag{5}$$

is characterized by its order of differentiation and the linear relation between force and acceleration. We highlight both properties. Any change in these characteristic features of the inertial law dramatically influences the interpretation of, e.g., orbits of satellites, planets or stars.

[1] We leave out Berwald–Finsler space–times where the space–time metric depends on the connecting vector while the equation of motion still is the ordinary Riemannian geodesics.

### 3.2.1 Order of Equations of Motion

Newtons second law (5) with $\boldsymbol{p} = m_{\rm i}\boldsymbol{v}$ where $m_{\rm i}$ is the inertial mass implies an equation of motion of second order for the position. The relation between momentum and velocity has the structure of a constitutive law and, thus, may be generalized to $\boldsymbol{p} = \boldsymbol{p}(m, \boldsymbol{v}, \dot{\boldsymbol{v}}, \ddot{\boldsymbol{v}}, \ldots)$. This implies higher-order equations of motion. Higher-order equations of motion may also come from metrical fluctuations with a certain time correlation. Here we are concerned only with fundamental equations: In the contrary effective equations of motion in general contain higher order derivatives from radiation reaction.

The most simple model for a higher order dynamics is based on a second order Lagrangian $L = L(t, \boldsymbol{x}, \dot{\boldsymbol{x}}, \ddot{\boldsymbol{x}})$. This gives an equation of motion of fourth order. With a small additional term of appropriate sign (see Lämmerzahl and Rademaker 2009 for more details) the solution of this fourth order equation gives the standard solution of the usual second order equation together with a kind of *zitterbewegung*. Therefore, the usual second order equations of motion seem to be rather robust against small higher-order additions. In order to be consistent we introduce interactions with external fields through a gauge principle. Such higher order gauge principles result in novel gauge fields.

As we expect from this approach only a small *zitterbewegung*, an experimental detection of such a phenomenon is rather difficult. One potentially feasible idea is to look for fundamental noise in electronic devices with characteristics differing from the standard Nyquist or $1/f$ noise. Corresponding proposals will be worked out.

### 3.2.2 Linearity

Newton's inertial law (5) is a definition of the force $\boldsymbol{F}$. Measuring the path of a test object and knowing its characteristic parameters determines the force.

Theories modifying this relation by introducing a function $f(\boldsymbol{a})$ on the left hand side, $m\boldsymbol{f}(\boldsymbol{a}) = \boldsymbol{F}(\boldsymbol{x})$, as MOND does, (Milgrom 2002), are equivalent to a theory of modified gravity provided the function $\boldsymbol{f}$ possesses an inverse. Then $m\boldsymbol{a} = m\boldsymbol{f}^{-1}(\boldsymbol{F}(\boldsymbol{x})/m)$, and this is a mere redefinition of the force equivalent to a modification of the gravitational influence. This modified Newtonian dynamics or modified gravity is rather successful in modeling galactic rotations curves. The function $\boldsymbol{f}(\boldsymbol{a})$ is mainly determined by a characteristic acceleration scale $a_0$ of the order $10^{-10}$ m s$^{-2}$.

Though the inertial law defines the force, there is one aspect which may be subject to experimental proof: If the force acting on a body is given by a gravitating mass, $F = m\nabla U$ with $U = G \int \rho(x')/|x - x'| dV'$ ($G$ is Newton's graviational constant and $\rho$ the mass density), then one may ask the question whether the acceleration decreases linearly with decreasing gravitating mass which can be measured through its weight. If the gravitating mass $M$ is spherically symmetric, $U = GM/r$, then the question is whether $\ddot{x} \to \alpha\ddot{x}$ for $M \to \alpha M$, in particular in the case of small $M$. This is an operationally well defined question which is worth to be explored experimentally.

A recent laboratory experiment performed tests of the linearity between force and acceleration in the extremely weak force regime, (Gundlach et al. 2007). No deviation from Newton's inertial law has been found for accelerations down to $5 \cdot 10^{-14}$ m s$^{-2}$. This experiment, however, does not test MOND. Within MOND it is required that the full acceleration has to be smaller than approx $10^{-10}$ m s$^{-2}$ while in the above experiment only two components of the acceleration were small while the acceleration due to the Earth attraction was still present. Therefore such tests of MOND have to be performed in space (for the constraints of tests on Earth, see Ignatiev 2007). An earlier test (Abramovici and Vager 1986) went down to accelerations of $3 \cdot 10^{-11}$ m s$^{-2}$. In both cases the applied force was non–gravitational. It might be speculated whether the MOND ansatz applies to all forces or to the gravitational force only.

It has also been speculated whether the MOND ansatz can describe the Pioneer anomaly (Milgrom 2002; Anderson et al. 2002) but this has not been convincingly confirmed. In any case, it is a very remarkable coincidence that the Pioneer acceleration, the MOND characteristic acceleration $a_0$ as well as the cosmological acceleration are all of the same order of magnitude, $a_{\mathrm{Pioneer}} \approx a_0 \approx cH$, where $H$ is the Hubble constant.

### 3.3 Law of Reciprocal Action

A key model for the violation of the law of reciprocal action is a difference in active and passive gravitational masses. The notion of active and passive masses and their possible non-equality has first been introduced and discussed by (Bondi 1957). The *active mass* $m_{\mathrm{a}}$ is the source of the gravitational field (here we restrict to the Newtonian case with the gravitational potential $U$) $\Delta U = 4\pi m_{\mathrm{a}} \delta(\boldsymbol{x})$, whereas the *passive mass* $m_{\mathrm{p}}$ reacts to it

$$m_{\mathrm{i}} \ddot{\boldsymbol{x}} = m_{\mathrm{p}} \nabla U(\boldsymbol{x}). \tag{6}$$

Here, $m_{\mathrm{i}}$ is the inertial mass and $\boldsymbol{x}$ the position of the particle. The equations of motion for a gravitationally bound two–body system then are

$$m_{1\mathrm{i}} \ddot{\boldsymbol{x}}_1 = G m_{1\mathrm{p}} m_{2\mathrm{a}} \frac{\boldsymbol{x}_2 - \boldsymbol{x}_1}{|\boldsymbol{x}_2 - \boldsymbol{x}_1|^3}, \qquad m_{2\mathrm{i}} \ddot{\boldsymbol{x}}_2 = G m_{2\mathrm{p}} m_{1\mathrm{a}} \frac{\boldsymbol{x}_1 - \boldsymbol{x}_2}{|\boldsymbol{x}_1 - \boldsymbol{x}_2|^3}, \tag{7}$$

where $1, 2$ refer to the two particles and $G$ is the gravitational constant. For the equation of motion of the center of mass $\boldsymbol{X}$, we find

$$\ddot{\boldsymbol{X}} = G \frac{m_{1\mathrm{p}} m_{2\mathrm{p}}}{M_{\mathrm{i}}} C_{21} \frac{\boldsymbol{x}}{|\boldsymbol{x}|^3} \quad \text{with} \quad C_{21} = \frac{m_{2\mathrm{a}}}{m_{2\mathrm{p}}} - \frac{m_{1\mathrm{a}}}{m_{1\mathrm{p}}} \tag{8}$$

where $M_{\mathrm{i}} = m_{1\mathrm{i}} + m_{2\mathrm{i}}$ and $\boldsymbol{x}$ is the relative coordinate. Thus, if $C_{21} \neq 0$ then active and passive masses are different and the center of mass shows a self-acceleration along the direction of $\boldsymbol{x}$. This is a violation of Newton's *actio* equals *reactio*. A limit has been derived by Lunar Laser Ranging (LLR): no self-acceleration of the moon has been observed yielding a limit of $|\bar{C}_{\mathrm{Al-Fe}}| \leq 7 \cdot 10^{-13}$ (Bartlett and van Buren 1986).

The dynamics of the relative coordinate

$$\ddot{\boldsymbol{x}} = -G \frac{m_{1\mathrm{p}} m_{2\mathrm{p}}}{m_{1\mathrm{i}} m_{2\mathrm{i}}} \left( m_1 \frac{m_{1\mathrm{a}}}{m_{1\mathrm{p}}} + m_2 \frac{m_{2\mathrm{a}}}{m_{2\mathrm{p}}} \right) \frac{\boldsymbol{x}}{|\boldsymbol{x}|^3} \tag{9}$$

has been probed in the laboratory by (Kreuzer 1968) with the result $|C_{21}| \leq 5 \cdot 10^{-5}$.

Similar considerations have been made for active and passive charges or for magnetic moments (Lämmerzahl et al. 2007a).

## 4 The Structure of Gravity

After having set up the fundamental equations of motion for test objects one can start to explore the structure of interactions. The gravitational interaction is first characterized by a number of *universality principles* put together in the Einstein Equivalence Principle (EEP).

It consists of (i) the Universality of Free Fall (UFF), (ii) the Universality of the Gravitational Redshift (UGR), and (iii) Local Lorentz Invariance (LLI), see Will (1993). These principles further constrain the structure of the equations of motion of test objects: The EEP leaves only freedom for a symmetric second rank tensor field to couple to the equation of motion of test objects. As a consequence one arrives at a *metric theory of gravity*.

Any metric theory of gravity shows the standard Solar system effects like perihelion shift, light deflection, gravitational time delay, and the Lense–Thirring and Schiff effect. Only for GR given by the Einstein field equations these effects attain certain values. There is no constructive way to derive these Einstein equations. However, the Parametrized Post Newtonian formalism (PPN), Will (1993), with approximate 10 undetermined parameters (the number of parameters depends on the chosen version) provides a very powerful method of parameterizing deviations from GR. For GR these parameters attain certain values. The PPN formalism gives a theoretical frame within which by means of a finite number of observations and experiments it is possible to single out GR from other theories of gravity.

In the following two sections we describe the experiments and observations first leading to a metric theory of gravity and second singling out GR from all other metric theories (see also the contribution of C. Will).

## 5 The Foundations of Metric Gravity

### 5.1 Universality of Free Fall

The UFF states that all neutral point–like particles move in a gravitational field in the same way: The path of these bodies is independent of the composition of the body. The corresponding tests are described in terms of the acceleration of these particles in the reference frame of the gravitating body: the Eötvös factor compares the normalized accelerations of two bodies $\eta = \frac{a_2 - a_1}{\frac{1}{2}(a_2 + a_1)}$ in the same gravitational field. In the frame of Newtons theory this can be expressed as $\eta = \frac{\mu_2 - \mu_1}{\frac{1}{2}(\mu_2 + \mu_1)}$, where $\mu = m_{\mathrm{g}}/m_{\mathrm{i}}$ is the ratio of the (passive) gravitational and inertial mass.

There are two principal schemes to perform tests of UFF. The first scheme uses the free fall of bodies. In this case the full gravitational attraction towards the Earth can be exploited. However, these experiments suffer from the fact that the time-of-flight is limited to roughly 1 s and that a repetition needs new adjustment. The other scheme uses a restricted motion confined to one dimension only, namely a pendulum or a torsion balance. The big advantage is the periodicity of the motion which by far outweights the disadvantage that only a fraction of the gravitational attraction is used. In fact, the best test today of the UFF uses a torsion pendulum and confirms it to the order of $2 \cdot 10^{-13}$. Altogether we then have the amazing equality $m_{\mathrm{i}} = m_{\mathrm{g}} = m_{\mathrm{a}} = m_{\mathrm{p}}$. New proposed tests in space, the approved mission MICROSCOPE, (Touboul 2001 and the contribution of P. Touboul), and the proposal STEP, (Lockerbie et al. 2001), will combine the advantages of free fall and periodicity (see also the contributions by J. Mester and T. Sumner).

There are hints from quantum gravity inspired scenarios that the UFF might be violated below the $10^{-13}$ level, Damour et al. (2002a, 2002b) and the contribution of T. Damour. Also from cosmology with a dynamical vacuum energy (quintessence) one can derive a violation of UFF at the $10^{-14}$ level (Wetterich 2003). The validity of the UFF has also been used for setting bounds on the time variability of various constants such as the fine structure constant and the electron-to-proton mass ratio (Dent 2006).

According to GR, spinning particles couple to the space–time curvature (Hehl 1971; Audretsch 1981) and, thus, violate the UFF. However, the effect is far beyond any experimental detectability. Therefore testing the UFF for spinning matter amounts to a search for an anomalous coupling of spin to gravity. Motivations for anomalous spin couplings came from the search for the axion, a candidate for the dark matter in the universe which also can resolve the strong PC puzzle in Quantum Chromodynamics (Moody and Wilczek 1984). In these models spin may couple to the gradient of the gravitational potential or to gravitational fields generated by the spin of the gravitating body. The first case can easily be tested by weighting polarized bodies what showed that for polarized matter the UFF is valid up to the order of $10^{-8}$ (Hsieh et al. 1989).

Also charged particles do couple to the space–time curvature (DeWitt and Brehme 1960) but this effect is again too small to be detectable. Further, it is possible to introduce a charge-dependent violation of the UFF by assuming a charge-dependent anomalous inertial and/or gravitational mass. It is also possible to choose the model such that for a neutral atom UFF is fulfilled exactly while it is violated for isolated charges (Dittus et al. 2004). It has been suggested to carry out a corresponding experiment in space (Dittus et al. 2004).

## 5.2 Universality of Gravitational Redshift

A test of the universal influence of the gravitational field on clocks based on different physical principles requires clock comparison during their common transport through different gravitational potentials. There is a large variety of clocks which can be compared: (i) light clocks (optical resonators), (ii) various atomic clocks, (iii) various molecular clocks, (iv) gravitational clocks based on the revolution of planets or binary systems, (v) the rotation of the Earth, (vi) pulsar clocks based on the spin of stars, and (vii) clocks based on particle decay.

On a phenomenological level the comparison of two collocated clocks is given by

$$\frac{\nu_{\text{clock}\,1}(x_1)}{\nu_{\text{clock}\,2}(x_1)} = \left(1 - (\alpha_{\text{clock}\,2} - \alpha_{\text{clock}\,1})\,\frac{U(x_1) - U(x_0)}{c^2}\right)\frac{\nu_{\text{clock}\,1}(x_0)}{\nu_{\text{clock}\,2}(x_0)} \tag{10}$$

where $\alpha_{\text{clock}\,i}$ are clock-dependent parameters. If this frequency ratio does not depend on the gravitational potential then the gravitational redshift is universal. This is a null-test of $\alpha_{\text{clock}\,2} - \alpha_{\text{clock}\,1}$. It is obviously preferable to use large differences in the gravitational potential which clearly shows the need for space experiments. In experiments today the variation of the gravitational field is induced by the motion of the Earth around the Sun.

The best test up to date has been performed by comparing the frequency ratio of the 282 nm $^{199}\text{Hg}^+$ optical clock transition to the ground state hyperfine splitting in $^{133}\text{Cs}$ over 6 years. The result is $|\alpha_{\text{Hg}} - \alpha_{\text{Cs}}| \leq 5 \cdot 10^{-6}$ (Ashby et al. 2007; Fortier et al. 2007). Other tests compare Cs clocks with the hydrogen maser, Cs or electronic transitions in $\text{I}_2$ with optical resonators. We are looking forward to ultrastable clocks on the ISS and on satellites in Earth orbit or even in deep space as proposed by SPACETIME (Maleki 2001), OPTIS (Lämmerzahl et al. 2004) and SAGAS (Wolf et al. 2008), which should considerably improve the scientific results (see also the contribution by S. Reynaud).

So far there are no tests using ‘anti clocks’, that is, clocks made of anti-matter. However, since the production of anti-hydrogen is a working technique today, there are attempts to perform high-precision spectroscopy of anti-hydrogen. These measurements first should test special relativistic CPT invariance but, as a long-term goal, they could also be used to test the Universality of the Gravitational Redshift for a clock based on anti-hydrogen.

In many scenarios it is assumed that constants vary with (cosmological) time. Since different atomic or molecular states depend differently on these constants the question of th constancy of constants is related to the UGR, cf. the contributions by J.-P. Uzan, N. Kolachevsky, P. Petitjean, and E. Fischbach.

## 5.3 Local Lorentz Invariance

Lorentz invariance, the symmetry of Special Relativity (SR) which also holds locally in GR, is based on the constancy of the speed of light and the relativity principle. For a recent review, see Amelino-Camelia et al. (2005).

### *5.3.1 The Constancy of the Speed of Light*

The constancy of the speed of light has many aspects:

1. The speed of light does not depend on the velocity of the source. Using the model $c' = c + \kappa v$, where $v$ is the velocity of the source and $\kappa$ some parameter one gets from astrophysical observations $\kappa \leq 10^{-11}$ (Brecher 1977).
2. The speed of light does not depend on the frequency and polarization. The best results come from astrophysics. From radiation at frequencies $7.1 \cdot 10^{18}$ Hz and $4.8 \cdot 10^{19}$ Hz of Gamma Ray Burst GRB930229 one obtains $\Delta c/c \leq 6.3 \cdot 10^{-21}$ (Schaefer 1999). Analysis of the polarization of light from distant galaxies yielded an estimate $\Delta c/c \leq 10^{-32}$ (Kostelecky and Mewes 2002).
3. The speed of light is universal. This means that the velocity of all other massless particles as well as the limiting maximum velocity of all massive particles coincides with $c$. The maximum speed of electrons, neutrinos and muons has been shown in various laboratory experiments to coincide with the velocity of light at a level $(c_{\text{particle}} - c)/c \leq 10^{-6}$ (Brown et al. 1973; Guiragossian et al. 1975; Alspector et al. 1976; Kalbfleisch et al. 1979). Astrophysical observations of radiation from the supernova SN1987A yield for the comparison of photons and neutrinos an estimate which is two orders of magnitude better (Stodolsky 1988; Longo 1987).
4. The speed of light does not depend on the velocity of the laboratory. This can be tested in Kennedy–Thorndike experiments which is a clock–clock comparison experiment where the laboratory moves with varying speed (e.g. a laboratory on the surface of the Earth moves with a velocity consisting of the rotation around its own axis and its revolution around the Sun). The two clocks can be either two light clocks (different resonators or a Michelson-type interferometer with different arm lengths) or a light clock and an atomic clock. The best comparison yields $\Delta c/c \leq 10^{-16}$ (Müller et al. 2007).
5. The speed of light depends not on the direction of propagation. This has been confirmed by modern Michelson–Morley experiments using optical resonators to a relative accuracy of $\Delta c/c \leq 10^{-16}$ (Müller et al. 2007).
6. A bit more involved is the combination of a finite velocity of signal propagation with quantum systems and quantum measurements involving entanglement ("spooky action at a distance"). Though quantum systems may be entangled over long distances and a measurement of one part of the system has some influence on the properties of the other part of the quantum system it is not possible to communicate with velocities larger than the velocity of light.

This altogether means that the velocity of light is a universal structure and, thus, can be interpreted as part of a space–time geometry.

### 5.3.2 *The Relativity Principle*

The relativity principle states that the outcome of all experiments when performed identically within a laboratory without reference to the external word, is independent of the orientation and the velocity of the laboratory. For the photon sector this can be tested with Michelson–Morley and Kennedy–Thorndike type experiments already discussed above. Regarding the matter sector the corresponding tests are Hughes–Drever type experiments. In general, these are nuclear or electronic spectroscopy experiments. Such effects can be modeled by an anomalous inertial mass tensor (Haugan 1979) of the corresponding particle. For nuclei one then gets estimates of the order $\delta m/m \leq 10^{-30}$ (Chupp et al. 1989). Also an anomalous coupling of the spin to some given cosmological vector or tensor fields destroys the Lorentz invariance. All anomalous spin couplings are absent to the order of $10^{-31}$ GeV, see Walsworth (2006) for a review. Also higher-order derivatives in the Dirac and Maxwell equations in general lead to anisotropy effects (Lorek and Lämmerzahl 2008).

A further aspect of anisotropy is that there might be some anisotropies in the Coulomb or Newtonian potential (Kostelecky and Mewes 2002; Kostelecky 2004). Anisotropies in the Coulomb potential may affect the length of, e.g., optical cavities which may influence the frequency of light in the cavity. However, it has been shown that the influence of the anisotropies of the Coulomb potential are smaller than the corresponding anisotropies in the velocity of light (Müller et al. 2003). Anisotropies in the Newtonian potential of the Earth has recently been looked for by means of atomic interferometry; these measurements constrain the anisotropies to the $10^{-8}$ level (Müller et al. 2008).

Future spectroscopy of anti-hydrogen may yield further information about the validity of the PCT symmetry.

### 5.4 The Consequence

The consequence of the validity of the EEP is that gravity is described by a Riemannian metric $g_{\mu\nu}$, a symmetric second rank tensor defined on a differentiable manifold being the collection of all possible physical events. The purpose of this metric is twofold: First, it governs the rate of clocks, that is,

$$s = \int ds, \quad ds = \sqrt{g_{\mu\nu} dx^\mu dx^\nu} \tag{11}$$

is the time shown by clocks where the integration is along the worldline of these clocks. Second, the metric gives the equation of motion for massive point particles as well as for light rays,

$$0 = \frac{d^2 x^\mu}{ds^2} + \left\{ {\mu \atop \rho\sigma} \right\} \frac{dx^\rho}{ds} \frac{dx^\sigma}{ds} \tag{12}$$

where $\left\{ {\mu \atop \rho\sigma} \right\} = \frac{1}{2} g^{\mu\nu} (\partial_\rho g_{\nu\sigma} + \partial_\sigma g_{\nu\rho} - \partial_\nu g_{\rho\sigma})$ is the Christoffel symbol. Here $x = x(s)$ is the worldline of the particle parametrized by its proper time. It can be shown that the metric also describes the propagation of, e.g, the spin vector, $D_v S = 0$, where $S$ is a particle spin.

## 6 Motivating Einstein's Field Equations

There is no derivation of Einstein's field equations from a few key observations. However, a PPN formalism (Will 1993), makes it possible to parametrize in terms of ten or more parameters deviations of the metric from a metric following from Einstein's field equations

$$R_{\mu\nu} - \frac{1}{2} R g_{\mu\nu} = \kappa T_{\mu\nu} \tag{13}$$

where $R_{\mu\nu}$ and $R$ are the Ricci tensor and scalar, respectively, $T$ is the energy-momentum tensor of the matter creating the gravitational field, and $\kappa$ the relativistic coupling constant. In the case of the validity of Einstein's field equations these parameters take specific values. As a consequence the precise measurement of the effects described in the next section will also give a justification of the validity of Einstein's field equations.

While all this is going in the weak–field and low-velocity regime, one extrapolates the field equation (13) to the strong field and large velocity regime. This extrapolation then can be examined by observations of binary systems and effects near black holes where higher order terms are needed for their correct description (Blanchet 2006).

One largely discussed generic deviation from GR is a modification of the Newtonian $1/r$ potential. Such deviations described by $V(r) = \frac{M}{r}(1 + \alpha e^{r/\lambda})$ are parametrized by the strength $\alpha$ and range $\lambda$. Various experiments yield estimates for $\alpha$ for a given range $\lambda$. The high precision of LLR and ephemerides give very tight restrictions on $\alpha$ for interplanetary ranges. Higher-dimensional models predict deviations from the $1/r$ potential at short distances which motivated big experimental efforts in that direction, see e.g. the contribution by R. Newman.

## 7 Proving Consequences of General Relativity

Gravity can be explored only through its action on test particles (or test fields). Accordingly the gravitational interaction has been studied through the motion of stars, planets, satellites and of light. There are only very few experiments which demonstrate the effects of gravity on quantum fields.

There are two classes of tests: Weak gravity effects, mostly observed within the Solar system, and strong gravity effects present in binary systems and near black holes.

### 7.1 Solar System Effects

For the calculation of the effects to be described one needs a solution of Einstein's field equations or an approximate solution in the frame of the PPN formalism.

#### *7.1.1 The Gravitational Redshift*

In a stationary gravitational field the gravitational redshift between two positions with radial coordinates $r_1$ and $r_2$ is given by

$$\frac{\nu_2}{\nu_1} = \sqrt{\frac{g_{tt}(r_1)}{g_{tt}(r_2)}} \approx 1 - \frac{GM}{c^2}\left(\frac{1}{r_1} - \frac{1}{r_2}\right), \tag{14}$$

where $r_1$ and $r_2$ are the radial positions of the two observers. The right hand side of the equation comes out if we assume the validity of Einstein theory of gravity. This effect has best been observed in a space experiment where the time of a hydrogen maser in a rocket has been compared with the time of an identical hydrogen maser on ground yielding a confirmation of GR at the level of 1 part in $10^4$ (Vessot et al. 1980).

### 7.1.2 Light Deflection

The deflection of light was the first prediction of Einstein's GR; it has been confirmed by observation four years after the theory has been completed. In the frame of the PPN formalism we obtain

$$\Delta\varphi = \frac{1}{2}(1+\gamma)\frac{M}{b}, \tag{15}$$

where $M$ is the mass of the Sun and $b$ the impact parameter. Today's observations use Very Long Baseline Interferometry (VLBI), and this has led to $|\gamma - 1| \leq 10^{-4}$ (Shapiro et al. 2004).

### 7.1.3 Perihelion/Periastron Shift

Within the PPN formalism we obtain the perihelion shift

$$\delta\varphi = \frac{1}{3}(2+2\gamma-\beta)\frac{6\pi M}{a^2(1-e^2)}, \tag{16}$$

where $a$ is the semimajor axis and $e$ the eccentricity of the orbit. Today this post-Newtonian perihelion shift has been determined as 42″98 per century with an error of the order $10^{-4}$ (Pitjeva 2005). Recently a huge periastron shift of a candidate binary black hole in the quasar OJ287 has been observed where one black hole is small compared to the other (Valtonen et al. 2008). The observed perihelion shift is approximately 39° per revolution, which takes 12 years.

### 7.1.4 Gravitational Time Delay

In the vicinity of masses, electromagnetic signals move slower than in empty space, when compared in a coordinate system attached to spatial infinity. This is the gravitational time delay. There are two ways to confirm this effect: (i) direct observation, that is, by comparing the time of flight of light signals in two situations for fixed sender and receiver, and (ii) by observing the change in the frequency induced by this gravitational time delay.

*Direct measurement* The gravitational time delay for signals which pass through the vicinity of a body of mass $M$ is given by

$$\delta t = 2(1+\gamma)\frac{GM}{c^3}\ln\frac{4x_{\mathrm{Sat}}x_{\mathrm{Earth}}}{b^2}, \tag{17}$$

where $x_{\mathrm{Sat}}$ and $x_{\mathrm{Earth}}$ are the distances of the satellite and the Earth, respectively, from the gravitating mass. If the gravitating body is the Sun and if we take $b$ to be the radius of the Sun then the effect would be of the order $10^{-4}$ s which is clearly measurable. This has been measured using Mars ranging data of the Viking Mars mission giving $|\gamma - 1| \leq 10^{-4}$ (Reasenberg et al. 1979).

*Measurement of frequency change* Though the time delay is comparatively small, the induced modification of the received frequency can indeed be measured with higher precision. The reason is that clocks are very precise and, thus, can resolve frequencies also very precisely.

The corresponding change in the frequency is

$$y(t) = \frac{\nu - \nu_0}{\nu_0} = 2(1+\gamma)\frac{GM_\odot}{c^3}\frac{1}{b(t)}\frac{db(t)}{dt}, \tag{18}$$

where $\nu_0$ is the emitted frequency. It is the time dependence of the impact parameter which is responsible for the effect. This effect has been measured by the Cassini mission. One important issue in the actual measurement was that three different wavelengths for the signals have been used. This made it possible to eliminate dispersion effects near the Sun and to verify with this time delay GR with an accuracy of $|\gamma - 1| \leq 2.5 \cdot 10^{-5}$ (Bertotti et al. 2003).

### 7.1.5 *Lense–Thirring Effect*

For the Einstein field equation as well as within the PPN formalism a rotating gravitating body gives metric components $J_i\, dt\, dx^i$, where $J$ is the angular momentum of the rotating body. On the level of the equations of motion this results in a Lorentz type gravitational force acting on bodies called gravitomagnetism, see also the contribution of G. Schäfer. The influence of this field on the trajectory of satellites results in a motion of the nodes, which has been measured by observing the LAGEOS satellites via laser ranging. Together with new data of the Earth's gravitational field obtained from the CHAMP and GRACE satellites the confirmation recently reached the 10% level (Ciufolini 2004, see also the contribution of I. Ciufolini for the LAGEOS results and the contribution of L. Stella for the Lense–Thirring effect in astrophysics). In the meantime the LARES mission has been approved. This is another satellite of the same tye as LAGEOS which orbit will have a different inclination than LAGEOS. This makes is possible to eliminate multipole moments of the Earth from the joint LAGEOS and LARES data. The launch is scheduled for early 2011.

This gravitomagnetic field also influences the proper time and, thus, the rate of clocks. It can be shown that the difference of the proper time of two counterpropagating clocks is $s_+ - s_- = 4\pi J/M$. It should be remarked that this quantity does not depend on $G$ and $r$. This effect for clocks in satellites orbiting the Earth can be as large as $10^{-7}$ s per revolution (Mashhoon et al. 2001).

### 7.1.6 *Schiff Effect*

The gravitational field of a rotating gravitating body also influences the rotation of gyroscopes. This effect is right now under consideration by the data analysis group of the GP-B mission flown in 2004. Data analysis is expected to be completed early 2010. Though the mission met all requirements and, thus, was a big technological success it turned out after the mission that contrary to all expectations and requirements the gyroscopes lost more energy than calculated. This requires the determination of further constants characterizing this spinning down effect which effects the overall accuracy of the measurement of the Schiff effect which was expected to be of the order of 0.5%. Nevertheless, recent reports of the GP-B data analysis group indicate that finally the error may go down to 1% (see the contribution of F. Everitt and the GP-B team). For updates of the data analysis one may contact GP-B's website.[2]

It should be noted that though both effects within GR are related to the gravitomagnetic field of a rotating gravitational source, the Lense–Thirring effect and the Schiff effect are

[2] See http://einstein.stanford.edu/.

conceptually different and measure different quantities and, thus, should be regarded as *independent tests* of GR. In a generalized theory of gravity spinning objects may couple to different gravitational fields (like torsion) than the trajectory of orbiting satellites. Furthermore, the Lense–Thirring effect is a global effect related to the whole orbit while the Schiff effect observes the Fermi-propagation, a characterization of a torque-free dynamics, of the spin of the gyroscope.

#### *7.1.7 The Strong Equivalence Principle*

The gravitational field of a body contains energy which adds to the rest mass of the gravitating body. The strong equivalence principle now states that EEP is valid also for self gravitating systems, that is, that UFF is valid for the gravitational energy, too. This has been confirmed by LLR with an accuracy of $10^{-3}$ (Will 1993), where the validity of UFF had to be assumed. The latter has been tested separately for artificial bodies of a composition similar to that of the Earth and the Moon yielding a confirmation with an accuracy of $1.4 \cdot 10^{-13}$ (Baeßler et al. 1999).

### 7.2 Strong Gravity Effects

While most of the observations and tests of gravity are being performed in weak fields: Solar system tests, galaxies, galaxy clusters, recently it became possible to observe phenomena in strong gravitational fields: in binary systems and in the vicinity of black holes.

The observation of stars in the vicinity of black holes (Schoedel et al. 2007) may in one or two decades give new improved measurements of the perihelion shift or of the Lense–Thirring effect. Binary systems present an even better laboratory for observing strong field effects. See, e.g., the binary black hole candidate observed by (Valtonen et al. 2008).

The inspiral of binary systems which has been observed with very high precision can be completely explained by the loss of energy through the radiation of gravitational waves as calculated within GR (Blanchet 2006). The various data from such systems can be used to constrain hypothetical deviations from GR. As an example, it can be used for a test of the strong equivalence principle (Damour and Schäfer 1991) and of preferred frame effects and conservation laws Bell and Damour (1996) in the strong field regime.

Recently, double pulsars have been detected and studied. These binary systems offer the new possibility to analyze spin effects and, thus, open up a new domain of exploration of gravity in the strong field regime (Kramer et al. 2006a, 2006b). Accordingly, the dynamics of spinning binary objects has been intensively analyzed recently (Faye et al. 2006; Blanchet et al. 2006; Steinhoff et al. 2008).

A consequence of strong gravity is the emission of gravitational waves. At present ground experiments are reaching their projected sensitivity and collect data. The space mission LISA is sensitive to a lower frequency range more adapted to the long inspiral period of binary systems and is a cornerstone mission of ESA/NASA. LISA is presently prepared through the technology–testing LISA–Pathfinder mission (see the contribution of S. Vitale).

## 8 Open Problems—Unexplained Observations

There are several observations which have not yet found a convincing explanation. In most cases there is no doubt concerning the data. The main problem is the interpretation of the observations and measurements.

### 8.1 Dark Matter

Dark matter is needed to describe the motion of galaxy clusters, as has been first speculated by F. Zwicky (1933), and for stars in galaxies, and has been also confirmed with gravitational lensing, see e.g. Sumner (2002). Also structure formation needs this dark matter. However, until now there is no single observational hint at particles which could make up this dark matter. As a consequence, there are attempts to describe the same effects by a modification of the gravitational field equations, e.g., of Yukawa form (Sanders 1984), or nonlocal gravity (Hehl and Mashhoon 2008), or by a modification of the dynamics of particles, like the MOND ansatz (Milgrom 2002; Sanders and McGough 2002), recently formulated in a relativistic frame (Bekenstein 2004). Due to the lack of direct detection of Dark Matter particles, all those attempts are on the same footing. There are suggestions that at least a considerable part of the observations which usually are "explained" by dark matter can be related to a stronger gravitational field which come out while taking the full Einstein equations into account (Cooperstock and Tieu 2005; Balasin and Grumiller 2006).

### 8.2 Dark Energy

Observations of type Ia supernovae, (Riess et al. 1998; Perlmutter et al. 1999), WMAP measurements of the cosmic microwave background (Spergel et al. 2007), the galaxy power spectrum and the Lyman–alpha forest data lines (van de Bruck and Priester 1998; Overduin and Priester 2001; Tegmark et al. 2004), indicate an accelerating expansion of the universe and that 75% of the total energy density consist of a dark energy component with negative pressure (Peebles and Ratra 2003).

Buchert and Ehlers (1997) have shown first in a Newtonian framework that within a spatial averaging of matter and the gravitational field, rotation and shear of matter can influence the properties of the averaged gravitational field which are described in effective Friedman equations. This also holds in the relativistic case (Buchert 2008). Therefore it is an open question whether dark energy is just a result of a correct averaging procedure. An influence of the averaging has been found in existing data (Li and Schwarz 2007; Li et al. 2008). These topics are illuminated in more detail in the contributions by Zakharov, Lasenby, Caldwell, and Goobar.

### 8.3 Pioneer Anomaly

The Pioneer anomaly, an unexplained anomalous acceleration of the Pioneer 10 and 11 spacecraft of $a_{\text{Pioneer}} = (8.74 \pm 1.33) \cdot 10^{-10}\ \text{m/s}^2$ toward the Sun, is discussed in Anderson et al. (2002) and the contribution of S. Turyshev. This acceleration seemed to turn on after the last flyby at Jupiter and Saturn and stayed constant within a 3% range. Until now no convincing explanation has been found. An anisotropy of the thermal radiation might explain the acceleration. However, while the power provided by the plutonium decay decreases exponentially, the acceleration stays constant. Nevertheless, further work on a good thermal modeling of the spacecraft is going on at ZARM (Rievers et al. 2008). Moreover, an analysis of the early tracking data is on the way. Improvements of ephemerides also helps to rule out various suggested explanations and theories (Standish 2008).

### 8.4 Flyby Anomaly

It has been observed at various occasions that satellites having been subjected to an Earth swing-by possess a significant unexplained velocity increase by a few mm/s. This unexpected and still unexplained velocity increase is called the *flyby anomaly*. For a summary of recent investigations of this phenomenon, see Lämmerzahl et al. (2007b). Anderson et al. (2008) have proposed the heuristic formula

$$\Delta v = v \frac{\omega R}{c^2} (\cos \delta_{\text{in}} - \cos \delta_{\text{out}}) \tag{19}$$

which describes all flybys. Here $R$ and $\omega$ are the radius and the angular velocity, respectively, of the Earth, and $\delta_{\text{in}}$ and $\delta_{\text{out}}$ are the inclinations of the incoming and outgoing trajectory. However, the recent observation of a Rosetta flyby could not verify this empirical formula.[3]

Until now no explanation has been found but, currently, it is expected that it is a mismodeling of either (i) the thermal influence of the Earth and the Sun's radiation on the satellite, (ii) of reference systems (this is supported by the fact that all the flybys can be modeled by (19), which contains geometrical terms only), (iii) of the flyby since this takes place at an accelerated body, or (vi) of the satellite's body being described by a point mass. There was an ISSI workshop on this topic in March 2009 (cf. footnote 3).

### 8.5 Increase of Astronomical Unit

From the analysis of radiometric measurements of distances between the Earth and the major planets including observations from Martian orbiters and landers from 1961 to 2003 a secular increase of the Astronomical Unit of approximately 10 m per century has been reported (Krasinsky and Brumberg 2004) (see also the article Standish 2005 and the discussion therein). This increase cannot be explained by a time-dependent gravitational constant $G$ because the $\dot{G}/G$ that would be needed is larger than the restrictions obtained from LLR. Such an increase might be mimicked, e.g., by a long-term increase of the density of the Sun plasma.

### 8.6 Quadrupole and Octupole Anomaly

Recently an anomalous behavior of the low-$l$ contributions to the cosmic microwave background has been reported. It has been shown that (i) there exists an alignment between the quadrupole and octupole with $>99.87\%$ C.L. (de Oliveira-Costa et al. 2005), and (ii) that the quadrupole and octupole are aligned to Solar system ecliptic to $>99\%$ C.L. (Schwarz et al. 2004). No correlation with the galactic plane has been found.

The reason for this is totally unclear. One may speculate that an unknown gravitational field within the Solar system slightly redirects the incoming cosmic microwave radiation (in the similar way as a motion with a certain velocity with respect to the rest frame of the cosmological background redirects the cosmic background radiation and leads to modifications of the dipole and quadrupole parts). Such a redirection should be more pronounced for low-$l$ components of the radiation. It should be possible to calculate the gravitational field needed for such a redirection and then to compare that with the observational data of the Solar system and the other observed anomalies.

[3]Team meeting "Investigation of the flyby anomaly", ISSI, Bern, March 2–6, 2009; http://www.issibern.ch/teams/investflyby/.

## 9 The Search for Signals of Quantum Gravity

There are many experiments proving that matter has to be quantized and, in fact, all experiments in the quantum domain are in full agreement with quantum theory with all its somehow strange postulates and consequences. Consistency of the theory also requires that the fields to which quantized matter field couple has to be quantized, too. Therefore, also the gravitational interaction has to be quantized. In particular, there is no meaning of the Einstein equation if the right hand side consists of quantized matter while the left hand side is purely classical. Also the semiclassical Einstein equation with an expectation value on the right hand side has been shown to lead to unwanted effects like faster than light propagation. However, though gravity is an interaction between particles it also deforms the underlying geometry. This double-role of gravity seems to prevent all quantization schemes from being successful in the gravitational domain.

The incompatibility of quantum mechanics and GR also shows up in the role of time which plays a different role in quantum mechanics and in GR. Furthermore, it is expected that a quantum theory of gravity would solve the problem of the singularities appearing within GR. As a last issue, it is the wish that such a new theory also would lead to a true unification of all interactions and, thus, to a better understanding of the physical world.

Any theory is characterized by their own set of constants. It is believed that the Planck energy $E_{\mathrm{Pl}} \approx 10^{28}$ eV sets the scale of quantum gravity effects. As a consequence, all expected effects scale with this energy or the corresponding Planck length, Planck time, etc. In string theory other scales influence the modifications (as is explained in the contributions by T. Damour and by B. Schutz). The implications of deviations from the standard model of cosmology is the subject of the article by S. Sarkar.

### 9.1 Theoretical Approaches

The low energy limit of string theory, a quasiclassical limit of loop quantum gravity as well as results from noncommutative geometry suggest that many of the standard laws of physics will suffer modifications. At a basic level these modifications show up in the equations of the standard model and in Einstein's field equations. These modifications then result in

- violation of Lorentz invariance
  - different limiting velocities of different particles
  - modified dispersion relation leading to birefringence in vacuum
  - modified dispersion relation leading to frequency-dependent velocity of light in vacuum
  - orientation and velocity dependence of effects
- time– and position-dependence of constants (varying $\alpha$, $G$, etc.)
- modified Newton potential at short and large distances.

In recent years there have been increased activities to search for these possible effects. However, until now nothing has been found.

### 9.2 Experimental Approaches

The experimental search for signals of a new theory requires to measure effects which have never been measured before. A strategy to find new things is (i) to explore new parameter regions in extreme situations, (ii) to use more precise devices, (iii) to use high-precision methods for new tests, or (iv) to test or measure "exotic" things.

### 9.2.1 *Extreme Situations*

Rather often some kind of "new physics" has been discovered when exploring new situations. We discuss various situations of this kind.

*Extremely high energy* One possibility to explore new physics is to probe the physical processes at very high energies. One example is the LHC where in future energies of the order of $10^{13}$ eV should be achievable. It is the hope to find signals of the Higgs particle and of supersymmetry. However, this energy range is still far away from the quantum gravity scale. The best what one can do is to observe high energy cosmic rays which have energies of up to $10^{21}$ eV. In fact, it has been speculated that the observations of high energy cosmic rays which according to standard theories are forbidden owing to the Greisen–Zatsepin–Kuzmin–cutoff (Greisen 1966; Zatsepin and Kuzmin 1966) could indicate a modified dispersion relation.

*Extremely low energy* The other extreme, very low temperatures, might also be a tool to investigate possible signals of quantum gravity. One may speculate that the influence of possible space–time fluctuations on the dynamics of quantum systems is more pronounced for very low temperatures. One may even speculate that such space–time fluctuations may give rise to a temperature threshold above the absolute zero.

Very low temperatures may be achievable in Bose–Einstein condensates (BEC) during long time of free evolution. Recently, at the Bremen drop tower, within the DLR-funded QUANTUS project a freely falling BEC has been created and its free expansion of longer than 1 s has been observed. These BECs then may be used for novel investigations, including a search for deviations from standard predictions. The contributions of W. Schleich and M. Kasevich explore the importance and use of ultracold atoms.

*Large distances* The unexplained phenomena, dark matter, dark energy, and the Pioneer anomaly are related to large distances. This could indicate that the laws of gravity have to be modified at large distances. Recently, some suggestions have been made:

- It has been discussed whether a Yukawa modification of the Newtonian potential may account for galactic rotations curves (Sanders 1984).
- In the context of higher dimensional braneworld theories deviations from Newtons potential also occurs (Dvali et al. 2000). At large distances the potential behaves like $1/r^2$, as one would expect from the Poisson equation in 5 dimensions. A comparison with cosmological and astrophysical observations has been reviewed by (Lue 2005).
- From considering a running coupling constant it has been suggested that the spatial parts of the space–time metric posses a part which grows linearly with distance (Jaekel and Reynaud 2005). This approach is in agreement with present solar system tests and also describes the Pioneer anomaly (Jaekel and Reynaud 2007).

*Weak accelerations* An acceleration $a$, being of physical dimension $\mathrm{m\,s^{-2}}$ can be related to a length scale $l_0 = c^2/a$. Now, the largest length scale in our universe is the Hubble length $L_\mathrm{H} = c/H$, where $H$ is the Hubble constant. The corresponding acceleration is $cH$ whose order of magnitude remarkably coincides with the Pioneer acceleration and the MOND acceleration scale. As a consequence, it seems mandatory to perform experiments to explore the physics for such small accelerations, as discussed on page 554.

*Strong accelerations* Analogously, since the smallest length scale is the Planck length $l_{\rm Pl}$, the corresponding acceleration is $a = 2 \cdot 10^{51}\ {\rm m\,s^{-2}}$ which, however, is far out of experimental reach. For smaller acceleration which might be reached by electrons in the fields of strong lasers one might be able to detect Unruh radiation or to probe the physics near black holes (Schäfer and Sauerbrey 1998; Schützhold et al. 2006, respectively).

*Strong gravitational fields* See discussion in Sect. 7.2.

### *9.2.2 Better Accuracy and Sensitivity*

It is clear that for a search of tiny effects a better accuracy always is a good strategy. In fact, it is amazing how the accuracy for testing Lorentz invariance, for example, has increased over the last years. It took more than 20 years to improve the results of the experiment by (Brillet and Hall 1979) and within a few years the accuracy improved by two orders of magnitude better and is still improving further.

Similar developments can be observed in other areas of quantum optics. There are improvements in the performance of atomic interferometry which are expected to give an improvement of measuring the fine structure constant $\alpha$ by one or two orders of magnitude. New optical clocks will show an improvement in accuracy by three orders of magnitude. A further improvement of experiments make use of atoms from ultracold BECs, which can be created in free fall. A particular effort in this direction is made by the Center of Excellence QUEST where new quantum–optical devices are being developed for novel space–time research.[4]

### *9.2.3 New Tests "Misusing" High Precision Devices*

It might also be of interest to identify devices which have, at least in principle, the sensitivity to find quantum gravity effects. One example for that are gravitational wave interferometers, (Amelino-Camelia and Lämmerzahl 2004). Todays already running gravitational wave interferometers have a strain sensitivity of $10^{-21}$. With the advanced LIGO the sensitivity will become $10^{-24}$. Thus, for a continuous gravitational wave with a frequency in the maximum sensitivity range between 10 and 1000 Hz a continuous observation over one year would reach a sensitivity of a bit less than $10^{-28}$. This is the sensitivity needed for observing Planck scale effects ($10^{28}$ eV) by optical laboratory devices (which have an energy scale of $\sim 1$ eV). This sensitivity is just the sensitivity needed to detect Planck-scale modifications in the dispersion relation for photons (Amelino-Camelia and Lämmerzahl 2004).

Another example for that is to misuse stable devices in order to search for a fundamental noise. Such fundamental noise scenarios with a power spectral density for the strain $\Delta L/L$, the relative length uncertainty, of the form

$$S(\nu) = \frac{L}{c}\left(\frac{L_{\rm Pl}}{L}\right)^{\alpha}\left(\frac{\nu}{c/L}\right)^{\gamma}, \tag{20}$$

have first been discussed by Amelino-Camelia (2000) in relation to gravitational wave interferometers like GEO600. Here $L_{\rm Pl}$ is the Planck length, $L$ a characteristic length of the device, $\nu$ the frequency of the radiation involved (laser frequency in the gravitational wave interferometer or in an optical cavity), and $\alpha$ and $\gamma$ are arbitrary exponents related to the

[4] See http://www.questhannover.de.

noise scenario (Amelino-Camelia 2000; Ng 2003). Perhaps more suited for the search for such fundamental noise are ultrastable cavities. A first experimental search using such devices has been carried out by Schiller et al. (2004). Noise scenarios of this kind may also influence the dynamics of massive particles leading to an apparent violation of the UFF measurable by atomic interferometry (Göklü and Lämmerzahl 2008).

A particular noise with the exponents $\alpha = \frac{1}{2}$ and $\gamma = 0$ is considered as holographic noise related to the information stored on the surface of black holes. This case has been discussed in Hogan (2008a, 2008b) with respect to its detectability in GEO600. The same scenario may give a violation of the UFF of up to an order of $10^{-10}$ (Göklü and Lämmerzahl 2008).

## 10 The Need for Improved Tests

One may think that improved tests of the foundations and the predictions of SR and GR are just the wish of some esoteric very specialized physicists. However, there are many aspects and reasons for trying to improve experiments:

Physical reason. Fundamental theories always have to be tested as much and as far as possible.

Practical reason. Metrology is the definition, preparation and dissemination of physical units like the second, the meter, the kilogram, etc. with the highest possible precision. The definition of units in most cases depends on fundamental symmetries. The definition of the meter, for example, depends on the constancy of the speed of light. The definition of the international atomic time (TAI) depends on the special relativistic time dilation as well as on the gravitational redshift. Therefore each high precision test also contributes to metrology. This is also the reason why so many fundamental tests are carried through at the national bureaus of standard like BIPM, NIST or PTB. It is also well known that the Global Positioning Systems relies on SR and GR.

Theoretical reason. Since Quantum Gravity should show effectively some measurable deviations from standard physics, each high precision test of SR and GR can in principle, be interpreted as a search for Quantum Gravity.

**Acknowledgements** I thank H. Dittus and V. Perlick for discussions. Financial support from the German aerospace Center DLR and the German Research Foundation DFG is acknowledged.

## References

A. Abramovici, Z. Vager, Phys. Rev. D **34**, 3240 (1986)

J. Alspector, G. Kalbfleisch, N. Baggett, E. Fowler, B. Barish, A. Bodek, D. Buchholz, F. Sciulli, E. Siskind, L. Stutte, H. Fisk, G. Krafczyk, D. Nease, O. Fackler, Phys. Rev. Lett. **36**, 837 (1976)

G. Amelino-Camelia, Phys. Rev. D **62**, 0240151 (2000)

G. Amelino-Camelia, C. Lämmerzahl, Class. Quantum Grav. **21**, 899 (2004)

G. Amelino-Camelia, C. Lämmerzahl, A. Macias, H. Müller, In *Gravitation and Cosmology*, ed. by A. Macias, C. Lämmerzahl, D. Nunez. AIP Conference Proceedings, vol. 758 (Melville, New York, 2005), p. 30

J. Anderson, P. Laing, E. Lau, A. Liu, M. Nieto, S. Turyshev, Phys. Rev. D **65**, 082004 (2002)

J. Anderson, J. Campbell, J. Ekelund, J. Ellis, J. Jordan, Phys. Rev. Lett. **100**, 091102 (2008)

N. Ashby, T. Heavner, T. Parker, A. Radnaev, Y. Dudin, Phys. Rev. Lett. **98**, 070802 (2007)

J. Audretsch, J. Phys. A **14**, 411 (1981)

S. Baeßler, B. Heckel, E. Adelberger, J. Gundlach, U. Schmidt, H. Swanson, Phys. Rev. Lett. **83**, 3585 (1999)

H. Balasin, D. Grumiller, Significant reduction of galactic dark matter by general relativity, 2006. arXiv:astro-ph/0602519

D. Bartlett, D. vanBuren, Phys. Rev. Lett. **57**, 21 (1986)
J. Bekenstein, Phys. Rev. D **70**, 083509 (2004)
J. Bell, T. Damour, Class. Quantum Grav. **13**, 3121 (1996)
B. Bertotti, L. Iess, P. Tortora, Nature **425**, 374 (2003)
L. Blanchet, Living Rev. Relativ. **9**(4) (2006). http://www.livingreviews.org/lrr-2006-4
L. Blanchet, A. Buonanno, G. Faye, Higher-order spin effects in the dynamics of compact binaries II. Radiation field, 2006. arXiv:gr-qc/0605140
H. Bondi, Rev. Mod. Phys. **29**, 423 (1957)
K. Brecher, Phys. Rev. Lett. **39**, 1051 (1977)
A. Brillet, J. Hall, Phys. Rev. Lett. **42**, 549 (1979)
B. Brown, G. Masek, T. Maung, E. Miller, H. Ruderman, W. Vernon, Phys. Rev. **30**, 763 (1973)
T. Buchert, Gen. Relativ. Grav. **40**, 467 (2008)
T. Buchert, J. Ehlers, Astron. Astrophys. **320**, 1 (1997)
T. Chupp, R. Hoara, R. Loveman, E. Oteiza, J. Richardson, M. Wagshul, Phys. Rev. Lett. **63**, 1541 (1989)
I. Ciufolini, Gen. Relativ. Grav. **36**, 2257 (2004)
F. Cooperstock, S. Tieu, General relativity resolves galactic rotation without exotic dark matter, 2005. arXiv:astro-ph/0507619
T. Damour, G. Schäfer, Phys. Rev. Lett. **66**, 2549 (1991)
T. Damour, F. Piazza, G. Veneziano, Phys. Rev. Lett. **89**, 081601 (2002a)
T. Damour, F. Piazza, G. Veneziano, Phys. Rev. D **66**, 046007 (2002b)
A. de Oliveira-Costa, M. Tegmark, M. Devlin, L. Page, A. Miller, C. Netterfield, Y. Xu, Phys. Rev. D **71**, 043004 (2005)
T. Dent, *Varying "constants" in astrophysics and cosmology and ... in Proceedings of SUSY06* (2006, to appear). arXiv:hep-ph/0610376
B. DeWitt, R. Brehme, Ann. Phys. (NY) **9**, 220 (1960)
H. Dittus, C. Lämmerzahl, H. Selig, Gen. Relativ. Grav. **36**, 571 (2004)
D. Dvali, G. Gabadadze, M. Porrati, Phys. Lett. B **485**, 208 (2000)
J. Ehlers, Gen. Relativ. Grav. **38**, 1059 (2006)
G. Faye, L. Blanchet, A. Buonanno, Phys. Rev. D **74**, 104033 (2006)
T.M. Fortier, N. Ashby, J. Bergquist, M. Delaney, S. Diddams, T. Heavner, L. Hollberg, W. Itano, S. Jefferts, K. Kim, F. Levi, L. Lorini, W. Oskay, T. Parker, J. Shirley, J. Stalnaker, Phys. Rev. Lett. **98**, 070801 (2007)
E. Göklü, C. Lämmerzahl, Class. Quantum Grav. **25**, 105012 (2008)
Z. Guiragossian, G. Rothbart, M. Yearian, R. Gearhart, J. Murray, Phys. Rev. Lett. **34**, 335 (1975)
J. Gundlach, S. Schlamminger, C. Spitzer, K.Y. Choi, B. Woodahl, J. Coy, E. Fischbach, Phys. Rev. Lett. **98**, 150801 (2007)
K. Greisen, End of the cosmic ray spectrum? Phys. Rev. Lett. **16**, 748 (1966)
M. Haugan, Ann. Phys. **118**, 156 (1979)
F. Hehl, Phys. Lett. A **36**, 225 (1971)
F. Hehl, B. Mashhoon, (2008). arXiv:0812.1059v3[gr-qc]
C. Hogan, Phys. Rev. D **78**, 087501 (2008a)
C. Hogan, Phys. Rev. D **77**, 104031 (2008b)
C.H. Hsieh, P.Y. Jen, K.L. Ko, K.Y. Li, W.T. Ni, S.S. Pan, Y.H. Shih, R.J. Tyan, Mod. Phys. Lett. **4**, 1597 (1989)
A. Ignatiev, Phys. Rev. Lett. **98**, 101101 (2007)
M.T. Jaekel, S. Reynaud, Mod. Phys. Lett. A **20**, 1047 (2005)
M.T. Jaekel, S. Reynaud, In *Lasers, Clocks and Drag-Free*, ed. by H. Dittus, C. Lämmerzahl, S. Turyshev (Springer, Berlin, 2007), p. 193
G. Kalbfleisch, N. Baggett, E. Fowler, J. Alspector, Phys. Rev. Lett. **43**, 1361 (1979)
V. Kostelecky, Phys. Rev. **69**, 105009 (2004)
A. Kostelecky, M. Mewes, Phys. Rev. D **66**, 056005 (2002)
M. Kramer, I. Stairs, R. Manchester, M. MacLaughlin, A. Lyre, R. Ferdman, M. Burgag, D. Lorimer, A. Possenti, N. D'Amico, J. Sarkission, B. Joshi, P. Freire, F. Camilo, Ann. Phys. (Leipzig) **15**, 34 (2006a)
M. Kramer, I. Stairs, R. Manchester, M. MacLaughlin, A. Lyre, R. Ferdman, M. Burgag, D. Lorimer, A. Possenti, N. D'Amico, J. Sarkission, G. Hobbs, J. Reynolds, P. Freire, F. Camilo, Science **314**, 97 (2006b)
G. Krasinsky, V. Brumberg, Celest. Mech. Dyn. Astron. **90**, 267 (2004)
L. Kreuzer, Phys. Rev. **169**, 1007 (1968)
C. Lämmerzahl, V. Perlick, Confronting Finsler gravity with experiment, 2009. Preprint Univ. Bremen
C. Lämmerzahl, P. Rademaker, Gravity, equivalence principle and clocks, 2009. arXiv:0904.4779 [gr-gc]. Preprint, University of Bremen

C. Lämmerzahl, C. Ciufolini, H. Dittus, L. Iorio, H. Müller, A. Peters, E. Samain, S. Scheithauer, S. Schiller, Gen. Relativ. Grav. **36**, 2373 (2004)
C. Lämmerzahl, A. Macias, H. Müller, Phys. Rev. A **75**, 052104 (2007a)
C. Lämmerzahl, O. Preuss, H. Dittus, in *Lasers, Clocks, and Drag-Free Exploration of Relativistic Gravity in Space*, ed. by H. Dittus, C. Lämmerzahl, S. Turyshev (Springer, Berlin, 2007b), p. 75
C. Lämmerzahl, D. Lorek, H. Dittus, Gen. Relativ. Grav. **41** (2009)
N. Li, D. Schwarz (2007). arXiv:gr-gc/0702043v3 [gr-gc]
N. Li, M. Seikel, D.J. Schwarz, Is dark energy an effect of averaging? 2008. arXiv.org:0801.3420
N. Lockerbie, J. Mester, R. Torii, S. Vitale, P. Worden, In *Gyros, Clocks, and Interferometers: Testing Relativistic Gravity in Space*, ed. by C. Lämmerzahl, C. Everitt, F. Hehl (Springer, Berlin, 2001), p. 213
M. Longo, Phys. Rev. D **36**, 3276 (1987)
D. Lorek, C. Lämmerzahl, In *Proceedings of the 11th Marcel Grossmann Meeting*, ed. by R. Jantzen, H. Kleinert, R. Ruffini (World Scientific, Singapore, 2008), p. 2618
A. Lue, Phys. Rep. **423**, 1 (2005)
L. Maleki, SPACETIME–a midex proposal, 2001. JPL
B. Mashhoon, F. Gronwald, H. Lichtenegger, In *Gyroscopes, Clock, Interferometers, ...: Testing Relativistic Gravity in Space*, ed. by C. Lämmerzahl, C. Everitt, F. Hehl. LNP, vol. 562 (Springer, Berlin, 2001), p. 83
M. Milgrom, New Astron. Rev. **46**, 741 (2002)
J. Moody, F. Wilczek, Phys. Rev. D **30**, 130 (1984)
H. Müller, C. Braxmaier, S. Herrmann, A. Peters, C. Lämmerzahl, Phys. Rev. D **67**, 056006 (2003)
H. Müller, P. Stanwix, M. Tobar, E. Ivanov, P. Wolf, S. Herrmann, A. Senger, E. Kovalchik, A. Peters, Phys. Rev. Lett. **99**, 050401 (2007)
H. Müller, S.W. Chiow, S. Herrmann, S. Chu, K.Y. Chung, Phys. Rev. Lett. **100**, 031101 (2008)
Y. Ng, Mod. Phys. Lett. A **18**, 1073 (2003)
J. Overduin, W. Priester, Naturwiss **88**, 229 (2001)
P. Peebles, B. Ratra, Rev. Mod. Phys. **75**, 559 (2003)
S. Perlmutter, G. Aldering, G. Goldhaber et al., Astrophys. J. **517**, 565 (1999)
E. Pitjeva, Astron. Lett. **31**, 340 (2005)
R. Reasenberg, I. Shapiro, P. MacNeil, R. Goldstein, J. Breidenthal, J. Brenkle, D. Cain, T. Kaufman, T. Komarek, A. Zygielbaum, Astrophys. J. Lett. **234**, 219 (1979)
A. Riess, A. Filippenko, P. Challis, et al., Astron. J. **116**, 1009 (1998)
B. Rievers, C. Lämmerzahl, M. List, S. Bremer (2008). Preprint, Univercity of Bremen
R. Sanders, Astron. Astrophys. **136**, 21 (1984)
R. Sanders, S. McGough, Ann. Rev. Astron. Astrophys. **40**, 263 (2002)
B. Schaefer, Phys. Rev. Lett. **82**, 4964 (1999)
G. Schäfer, R. Sauerbrey, Probing black-hole physics in the laboratory using high intensity femtosecond lasers, 1998. arXiv:astro-ph/9805106
S. Schiller, C. Lämmerzahl, H. Müller, C. Braxmaier, S. Herrmann, A. Peters, Phys. Rev. D **69**, 027504 (2004)
R. Schoedel, A. Eckart, T. Alexander, D. Merritt, R. Genzel, A. Sternberg, L. Meyer, F. Kul, J. Moultaka, T. Ott, C. Straubmeier, Astron. Astroph. **469**, 125 (2007)
R. Schützhold, G. Schaller, D. Habs, Phys. Rev. Lett. **97**, 121302 (2006)
D. Schwarz, G. Starkman, D. Huterer, C. Copi, Phys. Rev. Lett. **93**, 221301 (2004)
S. Shapiro, J. Davis, D. Lebach, J. Gregory, Phys. Rev. Lett. **92**, 121101 (2004)
D.N. Spergel, R. Bean, O. Dore, M.R. Nolta, C.L. Bennett, J. Dunkley, G. Hinshaw, N. Jarosik, E. Komatsu, L. Page, H.V. Peiris, L. Verde, M. Halpern, R.S. Hill, A. Kogut, M. Limon, S.S. Meyer, N. Odegard, G.S. Tucker, J.L. Weiland, E. Wollack, E.L. Wright, Astroph. J. **170**, 377 (2007)
E. Standish, in *Transits of Venus: New Views of the Solar System and Galaxy, Proceedings IAU Colloquium No. 196*, ed. by D. Kurtz (Cambridge University Press, Cambridge, 2005), p. 163
E. Standish, in *Gravitation and Cosmology*, ed. by A. Macias, C. Lämmerzahl, A. Camacho. AIP Conference Proceedings, vol. 977 (Melville, New York, 2008), p. 254
J. Steinhoff, G. Schäfer, S. Hergt, Phys. Rev. D **77**, 104018 (2008)
L. Stodolsky, Phys. Lett. B **201**, 353 (1988)
T. Sumner, Living Rev. Relativ. **5**, 2002–420112005 (2002)
M. Tegmark, et al., Phys. Rev. D **69**, 103501 (2004)
P. Touboul, Comptes Rendus de l'Acad. Sci. Série IV: Phys. Astrophys. **2**, 1271 (2001)
M.J. Valtonen, H.J. Lehto, K. Nilsson, J. Heidt, L. Takalo, A. Sillanpää, C. Villforth, M. Kidger, G. Poyner, T. Pursimo, S. Zola, J.H. Wu, X. Zhou, K. Sadakane, M. Drozdz, D. Koziel, D. Marchev, W. Ogloza, C. Porowski, M. Siwak, G. Stachowski, M. Winiarski, V.P. Hentunen, M. Nissinen, A. Liakos, S. Dogru, Nature **452**, 851 (2008)

C. van de Bruck, W. Priester, in *Dark Matter in Astrophysics and Particle Physics 1998: Proceedings of the Second International Conference on Dark Matter in Astrophysics and Particle*, ed. by H. Klapdor-Kleingrothaus (Inst. of Physics, London, 1998)

R. Vessot, M. Levine, E. Mattison, E. Blomberg, T. Hoffmann, G. Nystrom, B. Farrel, R. Decher, P. Eby, C. Baughter, J. Watts, D. Teuber, F. Wills, Phys. Rev. Lett. **45**, 2081 (1980)

R. Walsworth, in *Special Relativity*, ed. by J. Ehlers, C. Lämmerzahl (Springer, Berlin, 2006), p. 493

C. Wetterich, Phys. Lett. B **561**, 10 (2003)

C. Will, *Theory and Experiment in Gravitational Physics (revised edition)* (Cambridge University Press, Cambridge, 1993)

P. Wolf, C.J. Borde, A. Clairon, L. Duchayne, A. Landragin, P. Lemonde, G. Santarelli, W. Ertmer, E. Rasel, F.S. Cataliotti, M. Inguscio, G.M. Tino, P. Gill, H. Klein, S. Reynaud, C. Salomon, E. Peik, O. Bertolami, P. Gil, J. Paramos, C. Jentsch, U. Johann, A. Rathke, P. Bouyer, L. Cacciapuoti, D. Izzo, P. de Natale, B. Christophe, P. Touboul, S.G. Turyshev, J.D. Anderson, M.E. Tobar, F. Schmidt-Kaler, J. Vigue, A. Madej, L. Marmet, M.C. Angonin, P. Delva, P. Tourrenc, G. Metris, H. Muller, R. Walsworth, Z.H. Lu, L. Wang, K. Bongs, A. Toncelli, M. Tonelli, H. Dittus, C. Lämmerzahl, G. Galzerano, P. Laporta, J. Laskar, A. Fienga, F. Roques, K. Sengstock, Exp. Astron. **23** (2008)

G.T. Zatsepin, V.A. Kuzmin, Upper limit of the spectrum of cosmic rays. JETP Lett. **4**, 78 (1966)

F. Zwicky, Helv. Phys. Acta **6**, 110 (1933)

# Space Science Series of ISSI

1. R. von Steiger, R. Lallement and M.A. Lee (eds.): *The Heliosphere in the Local Interstellar Medium*. 1996 ISBN 0-7923-4320-4
2. B. Hultqvist and M. Øieroset (eds.): *Transport Across the Boundaries of the Magnetosphere*. 1997 ISBN 0-7923-4788-9
3. L.A. Fisk, J.R. Jokipii, G.M. Simnett, R. von Steiger and K.-P. Wenzel (eds.): *Cosmic Rays in the Heliosphere*. 1998 ISBN 0-7923-5069-3
4. N. Prantzos, M. Tosi and R. von Steiger (eds.): *Primordial Nuclei and Their Galactic Evolution*. 1998 ISBN 0-7923-5114-2
5. C. Fröhlich, M.C.E. Huber, S.K. Solanki and R. von Steiger (eds.): *Solar Composition and its Evolution – From Core to Corona*. 1998 ISBN 0-7923-5496-6
6. B. Hultqvist, M. Øieroset, Goetz Paschmann and R. Treumann (eds.): *Magnetospheric Plasma Sources and Losses*. 1999 ISBN 0-7923-5846-5
7. A. Balogh, J.T. Gosling, J.R. Jokipii, R. Kallenbach and H. Kunow (eds.): *Co-rotating Interaction Regions*. 1999 ISBN 0-7923-6080-X
8. K. Altwegg, P. Ehrenfreund, J. Geiss and W. Huebner (eds.): *Composition and Origin of Cometary Materials*. 1999 ISBN 0-7923-6154-7
9. W. Benz, R. Kallenbach and G.W. Lugmair (eds.): *From Dust to Terrestrial Planets*. 2000 ISBN 0-7923-6467-8
10. J.W. Bieber, E. Eroshenko, P. Evenson, E.O. Flückiger and R. Kallenbach (eds.): *Cosmic Rays and Earth*. 2000 ISBN 0-7923-6712-X
11. E. Friis-Christensen, C. Fröhlich, J.D. Haigh, M. Schüssler and R. von Steiger (eds.): *Solar Variability and Climate*. 2000 ISBN 0-7923-6741-3
12. R. Kallenbach, J. Geiss and W.K. Hartmann (eds.): *Chronology and Evolution of Mars*. 2001 ISBN 0-7923-7051-1
13. R. Diehl, E. Parizot, R. Kallenbach and R. von Steiger (eds.): *The Astrophysics of Galactic Cosmic Rays*. 2001 ISBN 0-7923-7051-1
14. Ph. Jetzer, K. Pretzl and R. von Steiger (eds.): *Matter in the Universe*. 2001 ISBN 1-4020-0666-7
15. G. Paschmann, S. Haaland and R. Treumann (eds.): *Auroral Plasma Physics*. 2002 ISBN 1-4020-0963-1
16. R. Kallenbach, T. Encrenaz, J. Geiss, K. Mauersberger, T.C. Owen and F. Robert (eds.): *Solar System History from Isotopic Signatures of Volatile Elements*. 2003 ISBN 1-4020-1177-6
17. G. Beutler, M.R. Drinkwater, R. Rummel and R. von Steiger (eds.): *Earth Gravity Field from Space – from Sensors to Earth Sciences*. 2003 ISBN 1-4020-1408-2
18. D. Winterhalter, M. Acuña and A. Zakharov (eds.): *"Mars" Magnetism and its Interaction with the Solar Wind*. 2004 ISBN 1-4020-2048-1
19. T. Encrenaz, R. Kallenbach, T.C. Owen and C. Sotin: *The Outer Planets and their Moons* ISBN 1-4020-3362-1
20. G. Paschmann, S.J. Schwartz, C.P. Escoubet and S. Haaland (eds.): *Outer Magnetospheric Boundaries: Cluster Results* ISBN 1-4020-3488-1
21. H. Kunow, N.U. Crooker, J.A. Linker, R. Schwenn and R. von Steiger (eds.): *Coronal Mass Ejections* ISBN 978-0-387-45086-5

22. D.N. Baker, B. Klecker, S.J. Schwartz, R. Schwenn and R. von Steiger (eds.): *Solar Dynamics and its Effects on the Heliosphere and Earth* ISBN 978-0-387-69531-0
23. Y. Calisesi, R.-M. Bonnet, L. Gray, J. Langen and M. Lockwood (eds.): *Solar Variability and Planetary Climates* ISBN 978-0-387-48339-9
24. K.E. Fishbaugh, P. Lognonné, F. Raulin, D.J. Des Marais and O. Korablev (eds.): *Geology and Habitability of Terrestrial Planets* ISBN 978-0-387-74287-8
25. O. Botta, J.L. Bada, J. Gomez-Elvira, E. Javaux, F. Selsis and R. Summons (eds.): *Strategies of Life Detection* ISBN 978-0-387-77515-9
26. A. Balogh, L. Ksanfomality and R. von Steiger (eds.): *Mercury* ISBN 978-0-387-77538-8
27. R. von Steiger, G. Gloeckler and G.M. Mason (eds.): *The Composition of Matter* ISBN 978-0-387-74183-3
28. H. Balsiger, K. Altwegg, W. Huebner, T.C. Owen and R. Schulz (eds.): *Origin and Early Evolution of Comet Nuclei, Workshop honouring Johannes Geiss on the occasion of his 80th birthday* ISBN 978-0-387-85454-0
29. A.F. Nagy, A. Balogh, T.E. Cravens, M. Mendillo and I. Mueller-Wodarg (eds.): *Comparative Aeronomy* ISBN 978-0-387-87824-9
30. F. Leblanc, K.L. Aplin, Y. Yair, R.G. Harrison, J.P. Lebreton and M. Blanc (eds.): *Planetary Atmospheric Electricity* ISBN 987-0-387-87663-4
31. J.L. Linsky, V. Izmodenov, E. Möbius and R. von Steiger (eds.): *From the Outer Heliosphere to the Local Bubble: Comparison of New Observations with Theory* ISBN 978-1-4419-0246-7
32. M.J. Thompson, A. Balogh, J.L. Culhane, Å. Nordlund, S.K. Solanki and J.-P. Zahn (eds.): *The Origin and Dynamics of Solar Magnetism* ISBN 978-1-4419-0238-2
34. C.W.F. Everitt, M.C.E. Huber, R. Kallenbach, G. Schäfer, B.F. Schutz and R.A. Treumann (eds.): *Probing The Nature of Gravity: Confronting Theory and Experiments in Space* ISBN 978-1-4419-1361-6

---

Springer – Dordrecht / Boston / London